Biochemistry, Molecular Biology, and Genetics

SIXTH EDITION

Biochemistry, Molecular Biology, and Genetics

SIXTH EDITION

Michael A. Lieberman, PhD
Distinguished Teaching Professor
Department of Molecular Genetics, Biochemistry, and Microbiology
University of Cincinnati College of Medicine
Cincinnati, Ohio

Rick Ricer, MD
Professor Emeritus
Department of Family Medicine
University of Cincinnati College of Medicine
Cincinnati, Ohio

Philadelphia • Baltimore • New York • London
Buenos Aires • Hong Kong • Sydney • Tokyo

Publisher: Michael Tully
Acquisitions Editor: Susan Rhyner
Product Manager: Stacey Sebring
Marketing Manager: Joy Fisher-Williams
Vendor Manager: Bridgett Dougherty
Designer: Holly Reid McLaughlin
Manufacturing Coordinator: Margie Orzech
Compositor: S4 Carlisle

6th Edition

351 West Camden Street
Baltimore, MD 21201

Two Commerce Square
2001 Market Street
Philadelphia, PA 19103

Printed in China

Library of Congress Cataloging-in-Publication Data

Lieberman, Michael, 1950-
Biochemistry, molecular biology, and genetics. — 6th ed. / Michael A. Lieberman.
p. ; cm. — (Board review series)
Includes index.
Rev. ed. of: Biochemistry, molecular biology, and genetics / Todd A. Swanson, Sandra I. Kim, Marc J. Glucksman. 5th ed. c2010.
ISBN 978-1-4511-7536-3
I. Swanson, Todd A. Biochemistry, molecular biology, and genetics. II. Title. III. Series: Board review series.
[DNLM: 1. Biochemical Phenomena—Examination Questions. 2. Biochemical Phenomena—Outlines. 3. Genetic Processes—Examination Questions. 4. Genetic Processes—Outlines. QU 18.2]
QP518.3
572.8076—dc23

2013007054

DISCLAIMER

Care has been taken to confirm the accuracy of the information present and to describe generally accepted practices. However, the authors, editors, and publisher are not responsible for errors or omissions or for any consequences from application of the information in this book and make no warranty, expressed or implied, with respect to the currency, completeness, or accuracy of the contents of the publication. Application of this information in a particular situation remains the professional responsibility of the practitioner; the clinical treatments described and recommended may not be considered absolute and universal recommendations.

The authors, editors, and publisher have exerted every effort to ensure that drug selection and dosage set forth in this text are in accordance with the current recommendations and practice at the time of publication. However, in view of ongoing research, changes in government regulations, and the constant flow of information relating to drug therapy and drug reactions, the reader is urged to check the package insert for each drug for any change in indications and dosage and for added warnings and precautions. This is particularly important when the recommended agent is a new or infrequently employed drug.

Some drugs and medical devices presented in this publication have Food and Drug Administration (FDA) clearance for limited use in restricted research settings. It is the responsibility of the health care provider to ascertain the FDA status of each drug or device planned for use in their clinical practice.

To purchase additional copies of this book, call our customer service department at **(800) 638-3030** or fax orders to **(301) 223-2320**. International customers should call **(301) 223-2300**.

Visit Lippincott Williams & Wilkins on the Internet: http://www.lww.com. Lippincott Williams & Wilkins customer service representatives are available from 8:30 am to 6:00 pm, EST.

9 8 7 6 5 4 3 2 1

RRS1306

Preface and Acknowledgements

This revision of *BRS Biochemistry, Molecular Biology, and Genetics* is intended to help students prepare for the United States Medical Licensing Examination (USMLE) Step 1, as well as other board examinations for students in health-related professions. The basic material of biochemistry is presented in an integrative fashion on the basis of the conviction that details are easier to remember if they are presented within the context of the physiologic functioning of the human body. It presents the essentials of biochemistry in the form of condensed descriptions and simple illustrations. Test questions at the end of the chapter emphasize important information and lead to a better understanding of the material. A comprehensive examination at the end of the book serves as a self-evaluation to help the student uncover areas of strength and weakness.

We hope that this edition will aid students not only with the immediate task of passing a set of examinations, but also with the more long-term objective of fitting the subject of biochemistry into the framework of basic and clinical sciences, so essential to understanding their future patients' problems.

In a book of this nature it is possible that certain questions will have mixed interpretations. Any errors in the book are the sole responsibility of the authors, and we would like to be informed of such errors, or alternative explanations. Through this feedback future print ings of the book will reflect the correction of these errors.

The authors would like to thank Dr. Anil Menon for his careful review of Chapter 10 (Human Genetics), and Stacey Sebring, our managing editor, for her patience with us as we worked on this revision of BRS Biochemistry, Molecular Biology, and Genetics.

How to Use this Book

Anyone who has been teaching for a number of years knows that students, particularly those in medical school or in other programs within the health sciences, do not have an infinite amount of time to study or to review any given course. Therefore, this book was designed to make it easier for you to review biochemistry only at the depth you require, depending on the purpose for your review and the amount of time you have available.

Each chapter begins with an overview in a shaded box. This overview serves as a summation of the topics that will be covered in the chapter. In addition, these overviews help you review essential information quickly and reinforce key concepts.

Clinical Correlates in each chapter provide additional clinical insight and relate basic biochemistry to actual medical practice. They are designed to challenge you and encourage assimilation of information.

After you finish a chapter, try the questions and compare your answers to those in the explanations. As biochemistry is being integrated with other disciplines on NBME exams, a number of clinical questions require knowledge that would have been learned outside of a biochemistry class, and has not been reviewed in this text. If you have difficulty with the questions, review the chapter again and also look up relevant material from other courses in your curriculum for those questions which integrate biochemistry with another discipline. In addition to the questions in the print book, there are bonus questions available on the Point for further self-assessment and exam practice.

By following the process outlined above, you can save time by reviewing only the topics you need to review and by concentrating only on the details you have forgotten.

Michael A. Lieberman, PhD
Rick Ricer, MD

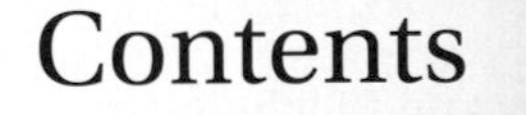

Contents

chapter **1**

Fuel Metabolism and Nutrition: Basic Principles

The main clinical uses of understanding the material in this chapter will be for nutritional counseling (e.g., patients trying to lose weight using "fad diets," patients on a diabetic diet, patients with nutritional misinformation, patients with anorexia, patients with chronic diseases, patients with malabsorption problems) and ordering appropriate diets for hospitalized patients (e.g., frail elderly, those with end-stage organ disease, or those on intravenous nutrition or tube feeding). Understanding the basic fuel metabolism is critical to understanding normal human functioning, and recognizing the abnormalities in basic fuel metabolism will allow for the diagnosis and treatment of a wide variety of disorders.

OVERVIEW

- The major fuels of the body—carbohydrates, fats, and proteins—are obtained from the diet and stored in the body's fuel depots.
- In the fed state (after a meal), ingested fuel is used to meet the immediate energy needs of the body and excess fuel is stored as either glycogen or triacylglycerol.
- During fasting (e.g., between meals or overnight), stored fuels are used to derive the energy needed to survive until the next meal.
- In prolonged fasting (starvation), changes occur in the use of fuel stores that permit survival for extended periods.
- The level of insulin in the blood increases in the fed state and promotes the storage of fuel, whereas the level of glucagon increases in the fasting state and promotes the release of stored fuel.

I. METABOLIC FUELS AND DIETARY COMPONENTS

- Carbohydrates, fats, and proteins serve as the major fuels of the body and are obtained from the diet. After digestion and absorption, these fuels can be oxidized for energy.
- The fuel consumed in excess of the body's immediate energy needs is stored, mainly as fat, but also as glycogen (a carbohydrate storage molecule). To some extent, body protein can also be used as fuel.
- The daily energy expenditure (DEE) of an individual includes the energy required for the basal metabolic rate (BMR) and the energy required for physical activity.

- In addition to providing energy, the diet also produces precursors for the synthesis of structural components of the body and supplies essential compounds that the body cannot synthesize (e.g., the essential fatty acids and amino acids, and the vitamins and minerals that often serve as cofactors for enzymes).

A. Fuels

When **fuels** are metabolized in the body, **heat** is generated and adenosine triphosphate **(ATP)** is synthesized.

1. **Energy is produced by oxidizing fuels to CO_2 and H_2O.**
   a. **Carbohydrates** produce about **4 kcal/g**.
   b. **Proteins** produce about **4 kcal/g**.
   c. **Fats** produce more than twice as much energy **(9 kcal/g)**.
   d. **Alcohol**, present in many diets, produces about **7 kcal/g**.
2. Physicians and nutritionists often use the term **"calorie"** in place of **kilocalorie**.
3. The **heat** generated by fuel oxidation is used to maintain **body temperature.**
4. **ATP** generated by fuel metabolism is used for biochemical reactions, muscle contraction, and other energy-requiring processes.

B. Composition of body fuel stores (Table 1.1)

1. **Triacylglycerol (triglyceride)**
   a. **Adipose triacylglycerol** is the major fuel store of the body.
   b. Adipose tissue stores fuel very efficiently. It has more stored calories per gram and less water (15%) than do other fuel stores. (Muscle tissue is about 80% water.)
2. **Glycogen** stores, although small, are extremely important.
   a. **Liver glycogen** is used to maintain blood glucose levels during the early stages of fasting.
   b. **Muscle glycogen** is oxidized for muscle contraction. It does not contribute to the maintenance of blood glucose levels under any conditions.
3. **Protein** does not serve solely as a source of fuel and can be degraded only to a limited extent.
   a. Approximately one-third of the total body protein can be degraded.
   b. If too much protein is oxidized for energy, body functions can be severely compromised.

C. DEE is the amount of energy required each day

1. **BMR** is the energy used by a person who has fasted for at least 12 hours and is awake but at rest. A rough estimate for calculating the BMR is

$$\text{BMR} = 24 \text{ kcal/kg body weight per day.}$$

2. **Diet-induced thermogenesis** (DIT) is the elevation in metabolic rate that occurs during digestion and absorption of foods. It is often ignored in calculations because its value is usually unknown and probably small ($<$10% of the total energy).
3. **Physical activity**
   a. The number of calories that physical activity adds to the DEE varies considerably. A person can expend about 5 calories (kcal) each minute while walking but 20 calories while running.
   b. The daily energy requirement for an extremely sedentary person is about 30% of the BMR. For a more active person, it may be 50% or more of the BMR.

table 1.1 Fuel Composition of the Average 70-kg Man after an Overnight Fast

Fuel	Amount (kg)	Percent of Total Stored Calories
Glycogen		
Muscle	0.15	0.4
Liver	0.08	0.2
Protein	6.0	14.4
Triglyceride	15	85

CLINICAL CORRELATES The thyroid gland produces thyroid hormone, which has profound effects on a person's BMR. One of the most common forms of **hyperthyroidism** is **Graves disease**. In this disease, the body produces antibodies that stimulate the thyroid gland to produce **excess thyroid hormone.** The disease is characterized by an **elevated BMR**, an enlarged thyroid (goiter), protruding eyes, nervousness, tremors, palpitations, excessive perspiration, and weight loss. **Hypothyroidism** results from a **deficiency of thyroid hormone.** The **BMR is decreased**, and mucopolysaccharides accumulate on the vocal cords and in subcutaneous tissue. The common symptoms are lethargy, dry skin, a husky voice, decreased memory, and weight gain.

D. Body Mass Index (BMI) is utilized to determine a healthy body weight

1. The BMI is defined as the value obtained when the weight (in kilogram) is divided by the height (in meters) squared:

$$\text{BMI} = \text{kg/m}^2$$

2. Table 1.2 indicates the interpretation of BMI values.

CLINICAL CORRELATES There are a number of disorders related to abnormal BMI values, some of which are lifestyle-induced. **Obesity** is associated with problems such as **hypertension, cardiovascular disease, and type 2 diabetes mellitus (DM)**. The treatment involves altering the lifestyle, particularly by decreasing food intake and increasing exercise. **Type 2 diabetes** is the result of reduced cellular responsiveness to insulin. Insulin production, initially, is normal or even increased when compared with normal. **Anorexia nervosa** is characterized by **self-induced weight loss**. Those frequently affected include women who, in spite of an emaciated appearance, often claim to be "fat." It is partially a behavioral problem; those afflicted are obsessed with losing weight. People with **bulimia** suffer from binges of **overeating**, followed by **self-induced vomiting** to avoid gaining weight.

E. Other dietary requirements and recommendations for normal adults

1. **Lipids**
   a. **Fat** should constitute between 20% and 35% of the total calories, with saturated fatty acids accounting for 10% or less of that total.
   b. **Cholesterol** intake should be no more than 300 mg/day for healthy individuals and <200 mg/day in those with established atherosclerosis.
   c. **Essential fatty acids (linoleic and α-linolenic acids)** are the precursors of the polyunsaturated fatty acids required for the **synthesis of** prostaglandins and other **eicosanoids**, such as

table 1.2 Interpretation of BMI Values

Classification	BMI (kg/m²)
Underweight	<18.50
Normal range	18.50–24.99
Overweight	>25.00
Preobese	25.00–29.99
Obese	≥30.00
Obese, class I	30.00–34.99
Obese, class II	35.00–39.99
Obese, class III (morbidly obese)	≥40.00

arachidonic acid and eicosapentaenoic acid (EPA). These essential fatty acids can be found in high levels in fish oils.

2. **Protein**

The recommended protein intake is **0.8 g/kg body weight** per day. Protein can be of high or low quality. High-quality protein contains many of the essential amino acids, and is usually obtained from dry beans and meat, chicken, or fish products. Low-quality protein is found in many vegetables. It lacks some of the essential amino acids required for the human diet.

CLINICAL CORRELATES A number of diet plans call for **high-protein diets**. When high-protein diets are very low in calories and the protein is of **low biologic value, or quality** (i.e., lacking in essential amino acids), **negative nitrogen balance** results. Body protein is degraded as amino acids are converted to glucose. A decrease in heart muscle can lead to death. Even if the protein is of high quality, ammonia and urea levels rise, putting increased stress on the kidneys. Vitamin deficiencies may occur due to a lack of intake of fruits and vegetables.

a. **Essential amino acids**
   (1) **Nine** amino acids cannot be synthesized in the body and, therefore, must be present in the diet in order for protein synthesis to occur. These essential amino acids are **histidine, isoleucine, leucine, lysine, methionine, phenylalanine, threonine, tryptophan,** and **valine**.
   (2) Only a small amount of **histidine** is required in the diet; however, **larger amounts** are required for **growth** (e.g., for children, pregnant women, people recovering from injuries).
   (3) Because **arginine** can be synthesized only in limited amounts, it is required in the diet for **growth**.

b. **Nitrogen balance**
   (1) **Dietary protein**, which contains about 16% nitrogen, is the body's primary source of nitrogen.
   (2) **Proteins** are constantly being synthesized and degraded in the body.
   (3) As amino acids are oxidized, the nitrogen is converted to **urea** and excreted by the kidneys. Other nitrogen-containing compounds produced from amino acids are also excreted in the urine (**uric acid, creatinine**, and NH_4^+).
   (4) **Nitrogen balance** (the normal state in the adult) occurs when **degradation** of body protein **equals synthesis** of new protein. The amount of nitrogen excreted in the urine each day equals the amount of nitrogen ingested daily.
   (5) **A negative nitrogen balance** occurs when **degradation** of body protein **exceeds synthesis** of new protein. More nitrogen is excreted than ingested. It results from an inadequate amount of protein in the diet or from the absence of one or more essential amino acids.
   (6) **A positive nitrogen balance** occurs when **degradation** of body protein is **less than synthesis** of new protein. Less nitrogen is excreted than ingested. It occurs during growth and synthesis of new tissue.

3. **Carbohydrates**
   a. There is no requirement for carbohydrates in the diet, as the body can synthesize all required carbohydrates from amino acid carbons.
   b. A healthy diet should consist of 45% to 65% of the total calories in the diet as carbohydrates.

4. **Vitamins and minerals**
   a. **Vitamins and minerals** are required in the diet. Many serve as **cofactors for enzymes.**
   b. **Minerals** required in large amounts include **calcium and phosphate**, which serve as structural components of bone. Minerals required in trace amounts include **iron**, which is a component of heme.

II. THE FED OR ABSORPTIVE STATE (FIG. 1.1)

- Dietary carbohydrates are cleaved during digestion, forming monosaccharides (mainly glucose), which enter the blood. Glucose is oxidized by various tissues for energy or is stored as glycogen in the liver and in muscle. In the liver, glucose is also converted to triacylglycerols, which are packaged in very low density lipoproteins (VLDL), and released into the blood. The fatty acids of the VLDL are stored in adipose tissue.
- Dietary fats (triacylglycerols) are digested to fatty acids and 2-monoacylglycerols (2-monoglycerides). These digestive products are resynthesized to triacylglycerols by intestinal epithelial cells, packaged in chylomicrons, and secreted via the lymph into the blood. The fatty acids of chylomicrons are stored in adipose triacylglycerols. Dietary cholesterol is absorbed by the intestinal epithelial cells and then follows the same fate as the dietary triacylglycerols.
- Dietary proteins are digested to amino acids and absorbed into the blood. The amino acids are used by various tissues to synthesize proteins and to produce nitrogen-containing compounds (e.g., purines, heme, creatine, epinephrine), or they are oxidized to produce energy.

A. Digestion and absorption

1. Carbohydrates

a. Starch, the storage form of carbohydrate in plants, is the major dietary carbohydrate.

(1) Salivary α-amylase (in the mouth) and **pancreatic α-amylase** (in the intestine) cleave starch to disaccharides and oligosaccharides.

(2) Enzymes with **maltase** and **isomaltase** activity are found in complexes located on the surface of the brush border of intestinal epithelial cells. They complete the conversion of starch to glucose.

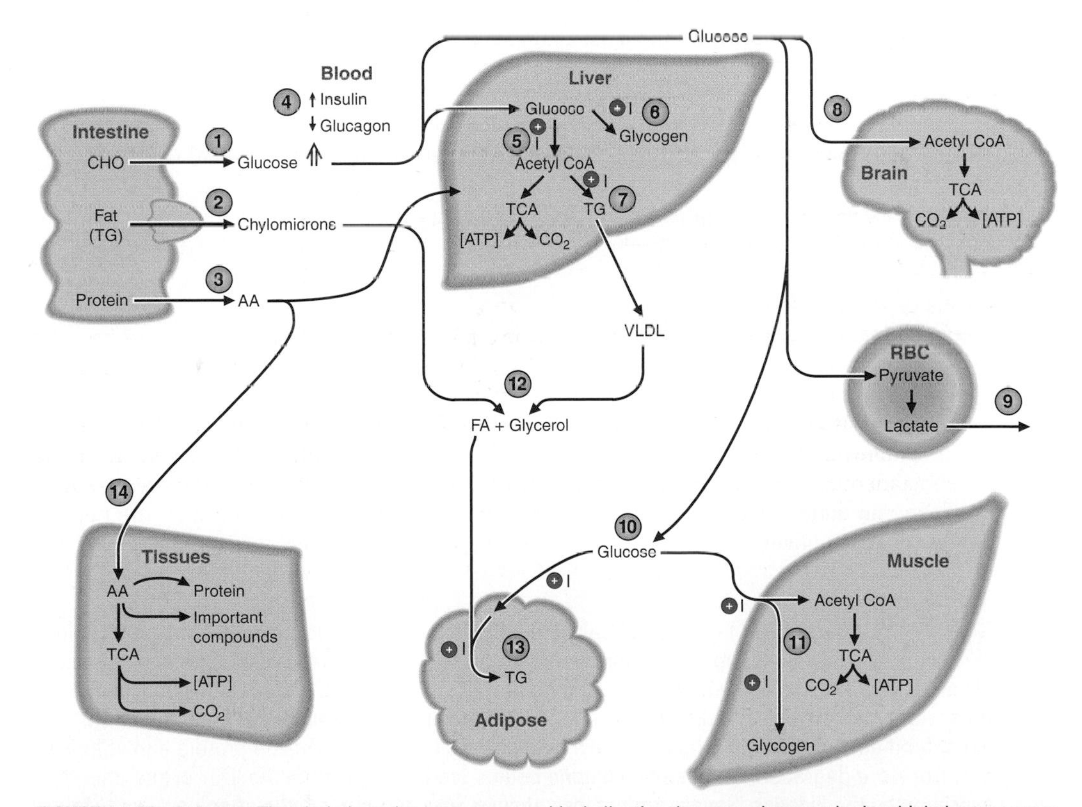

FIGURE 1.1. The fed state. The circled numbers serve as a guide, indicating the approximate order in which the processes begin to occur. AA, amino acid; FA, fatty acid; I, insulin; RBC, red blood cells; TG, triacylglycerols; VLDL, very low density lipoprotein; ⊕, stimulated by.

b. **Sucrose and lactose** (ingested disaccharides) are cleaved by enzymes that are part of the complexes on the surface of intestinal epithelial cells.

(1) **Sucrase** converts sucrose to fructose and **glucose.**

(2) **Lactase** converts lactose to glucose and **galactose**.

c. **Monosaccharides** (mainly **glucose** and some **fructose** and **galactose**) are absorbed by the intestinal epithelial cells and pass into the blood.

2. **Fats**

a. **Triacylglycerol** is the primary dietary fat. It is obtained from the fat stores of the plants and animals that serve as food.

b. The triacylglycerols are **emulsified** in the intestine by **bile salts** and **digested** by **pancreatic lipase** to **2-monoacylglycerols** and **fatty acids**, which are packaged into **micelles** (solubilized by bile salts) and absorbed into intestinal epithelial cells, where they are reconverted to **triacylglycerols**.

c. After digestion and resynthesis, the triacylglycerols are packaged in **chylomicrons**, which first enter the lymph and then the blood.

3. **Proteins**

a. Proteins are **digested** first by **pepsin** in the stomach and then by a series of enzymes in the intestine.

(1) The pancreas produces **trypsin, chymotrypsin, elastase,** and **carboxypeptidases,** which act in the lumen of the intestine.

(2) **Aminopeptidases, dipeptidases, and tripeptidases** are associated with the intestinal epithelial cells.

b. Proteins are ultimately **degraded** to a mixture of **amino acids**, which then enter intestinal epithelial cells, where some amino acids are metabolized. The remainder pass into the blood.

CLINICAL CORRELATES **Cystic fibrosis** is the most common lethal **genetic disease** among the white population of the United States. Proteins of **chloride ion channels** are **defective**, and both endocrine and exocrine **gland** functions are affected. Pulmonary disease and pancreatic insufficiency frequently occur. Food, particularly fats and proteins, are only partially digested, and **nutritional deficiencies** result. **Nontropical sprue** (adult celiac disease) results from a **reaction to gluten**, a protein found in grains. The intestinal epithelial cells are damaged, and **malabsorption** results. Common symptoms are steatorrhea, diarrhea, and weight loss.

B. Digestive products in the blood

1. **Hormone levels change** when the products of digestion enter the blood.

a. **Insulin levels rise** principally as a result of **increased blood glucose levels** and, to a lesser extent, increased blood levels of amino acids.

b. **Glucagon levels fall** in response to glucose but rise in response to amino acids. Overall, after a mixed meal (containing carbohydrate, fat, and protein), glucagon levels remain fairly constant or are reduced slightly in the blood.

2. **Glucose** and **amino acids** leave the intestinal epithelial cells and travel through the hepatic portal vein to the **liver**.

CLINICAL CORRELATES **Type 1 DM** leads to difficulty in maintaining appropriate blood glucose levels. In untreated type 1 DM, insulin levels are low or nonexistent because of destruction of β cells of the pancreas, usually by an autoimmune process. Before insulin became widely available, individuals with type 1 DM behaved metabolically as if they were in a constant state of starvation. Ingestion of food did not result in a rise in insulin, so fuel was not stored. Muscle protein and adipose triacylglycerol were degraded. Glucose and ketone bodies were produced by the liver in amounts that led to excretion by the kidneys. Severe weight loss ensued, and death occurred at an early age. After insulin became available, these metabolic derangements have been controlled to some extent.

C. The fate of glucose in the fed (absorptive) state

1. **The fate of glucose in the liver:** Liver cells either oxidize glucose or convert it to glycogen and triacylglycerols.
   a. **Glucose** is **oxidized** to CO_2 and H_2O to meet the immediate energy needs of the liver.
   b. **Excess glucose** is **stored** in the liver as **glycogen**, which is used during periods of fasting to maintain blood glucose levels.
   c. **Excess glucose** can be **converted to fatty acids** and a **glycerol** moiety, which combine to form **triacylglycerols**, which are released from the liver into the blood as **VLDL**.
2. **The fate of glucose in other tissues**
   a. The **brain**, which depends on glucose for its energy, **oxidizes glucose to CO_2 and H_2O**, producing ATP.
   b. **Red blood cells**, lacking mitochondria, oxidize glucose to **pyruvate** and **lactate**, which are released into the blood.
   c. **Muscle cells** take up glucose by a **transport** process that is **stimulated by insulin**. They **oxidize glucose** to CO_2 and H_2O to generate ATP for contraction, and they also **store** glucose as **glycogen** for use during contraction.
   d. **Adipose cells** take up glucose by a **transport** process that is **stimulated by insulin.** These cells oxidize glucose to produce energy and convert it to the glycerol moiety used to produce triacylglycerol stores.

D. The fate of lipoproteins in the fed state

1. The triacylglycerols of **chylomicrons** (produced from dietary fat) and **VLDL** (produced from glucose by the liver) are **digested** in capillaries by **lipoprotein lipase** to form fatty acids and glycerol.
2. The **fatty acids** are taken up by **adipose tissue**, converted to **triacylglycerols**, and stored.

E. The fate of amino acids in the fed state

Amino acids from dietary proteins enter the cells and are

1. used for **protein synthesis** (which occurs on ribosomes and requires mRNA). Proteins are constantly being synthesized and degraded.
2. used to make **nitrogenous compounds** such as heme, creatine phosphate, epinephrine, and the bases of DNA and RNA.
3. oxidized to generate **ATP**.

III. FASTING (FIG. 1.2)

- As blood glucose levels decrease after a meal, insulin levels decrease and glucagon levels increase, stimulating the release of stored fuels into the blood.
- The liver supplies glucose and ketone bodies to the blood. The liver maintains blood glucose levels by glycogenolysis and gluconeogenesis and synthesizes ketone bodies from fatty acids supplied by adipose tissue. **Hypoglycemia** refers to low blood glucose levels (normal blood glucose levels are 80 to 100 mg/dL); **hyperglycemia** refers to elevated blood glucose levels when compared with normal.
- Adipose tissue releases fatty acids and glycerol from its triacylglycerol stores. The fatty acids are oxidized to CO_2 and H_2O by tissues. In the liver, they are converted to ketone bodies. The glycerol is used for gluconeogenesis. **Hyperlipidemia** refers to elevated blood lipid levels (normal is $\leq$150 mg/dL for triglycerides).
- Muscle releases amino acids. The carbons are used by the liver for gluconeogenesis, and the nitrogen is converted to urea.

A. The liver during fasting

The liver produces **glucose** and **ketone bodies**, which are released into the blood and serve as sources of energy for other tissues.

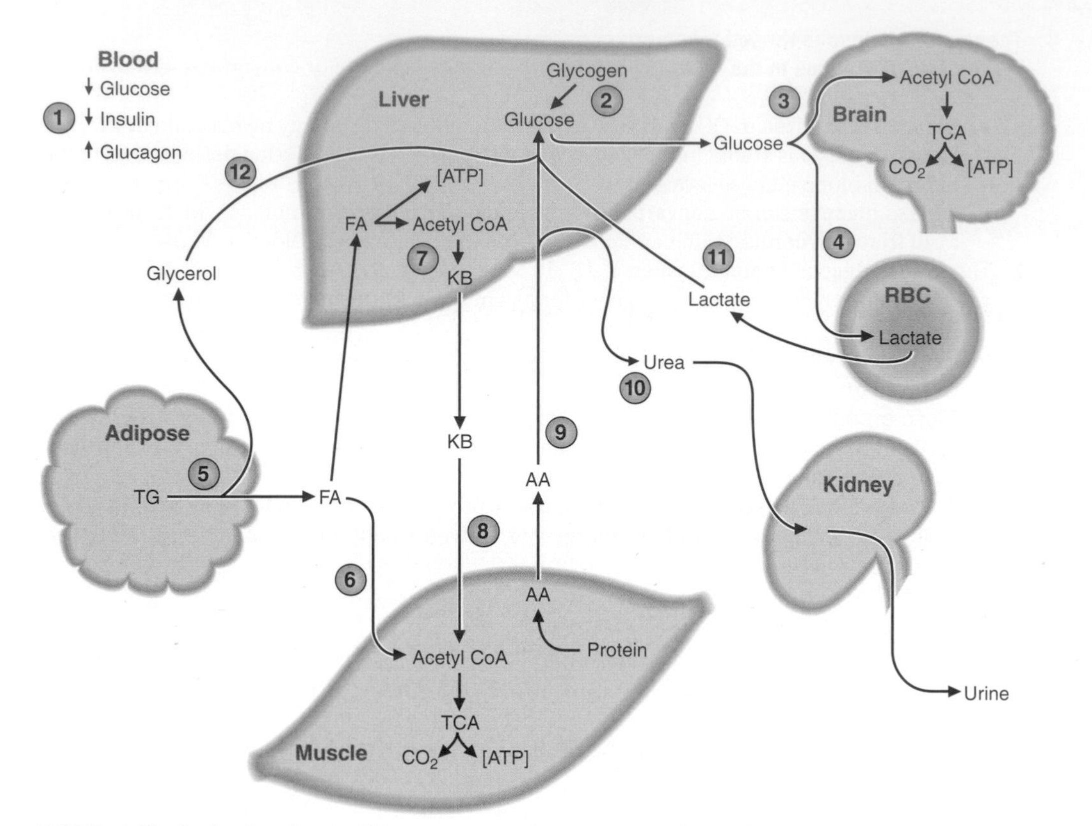

FIGURE 1.2. **The fasting (basal) state.** This state occurs after an overnight (12-hour) fast. The circled numbers serve as a guide, indicating the approximate order in which the processes begin to occur. KB, ketone bodies; AA, amino acid; FA, fatty acid; I, insulin; RBC, red blood cells; TG, triacylglycerols; VLDL, very low density lipoprotein; ⊕, stimulated by.

1. **Production of glucose by the liver:** The liver has the major responsibility for **maintaining blood glucose levels**. Glucose is required particularly by tissues such as the brain and red blood cells. The brain oxidizes glucose to CO_2 and H_2O, whereas red blood cells oxidize glucose to pyruvate and lactate.
   a. **Glycogenolysis:** About 2 to 3 hours after a meal, the liver begins to break down its glycogen stores by the process of glycogenolysis, and glucose is released into the blood. The glucose is then taken up by tissues and oxidized.
   b. **Gluconeogenesis**
      (1) After about 4 to 6 hours of fasting, the **liver** begins the process of gluconeogenesis. Within 30 hours, liver glycogen stores are depleted, leaving gluconeogenesis as the major process responsible for maintaining blood glucose levels.
      (2) **Carbon sources** for gluconeogenesis are as follows:
         (a) **Lactate** produced by tissues like red blood cells or exercising muscle
         (b) **Glycerol** from breakdown of triacylglycerols in adipose tissue
         (c) **Amino acids**, particularly alanine, from muscle protein
         (d) **Propionate** from oxidation of odd-chain fatty acids (minor source)

CLINICAL CORRELATES

Intravenous feeding. Solutions containing **5 g/dL glucose** are frequently infused into the veins of hospitalized patients. These solutions should be administered only for brief periods, because they lack the essential fatty and amino acids and because a high enough volume cannot be given each day to provide an adequate number of calories. More nutritionally complete solutions are available for long-term parenteral administration.

2. **Production of ketone bodies by the liver**
   a. As glucagon levels rise, adipose tissue breaks down its **triacylglycerol stores** into fatty acids and glycerol, which are released into the blood.
   b. Through the process of **β-oxidation**, the liver converts the fatty acids to acetyl CoA.
   c. **Acetyl CoA** is used by the liver for the synthesis of the ketone bodies, **acetoacetate and β-hydroxybutyrate**. The liver cannot oxidize ketone bodies, and hence releases them into the blood.

B. Adipose tissue during fasting

1. As glucagon levels rise, adipose **triacylglycerol stores** are **mobilized**. The triacylglycerol is degraded to three free fatty acids and glycerol, which enter the circulation. The liver converts the fatty acids to ketone bodies and the glycerol to glucose.
2. Tissues such as muscle oxidize the fatty acids to CO_2 and H_2O.

C. Muscle during fasting

1. **Degradation of muscle protein**
   a. During fasting, muscle protein is degraded, producing amino acids, which are partially metabolized by muscle and released into the blood, mainly as **alanine** and **glutamine**.
   b. Tissues, such as **gut** and **kidney**, metabolize the glutamine.
   c. The products (mainly **alanine** and **glutamine**) travel to the **liver**, where the carbons are converted to glucose or ketone bodies and the nitrogen is converted to urea.
2. **Oxidation of fatty acids and ketone bodies**
   a. During **fasting**, muscle oxidizes fatty acids released from adipose tissue and ketone bodies produced by the liver.
   b. During **exercise**, muscle can also use its own glycogen stores as well as glucose, fatty acids, and ketone bodies from the blood.

IV. PROLONGED FASTING (STARVATION)

- In starvation (prolonged fasting), muscle decreases its use of ketone bodies. As a result, ketone body levels rise in the blood, and the brain uses them for energy. Consequently, the brain needs less glucose, and gluconeogenesis slows down, sparing muscle protein. This occurs after approximately 3 to 4 days of starvation.
- These changes in the fuel utilization patterns of various tissues enable us to survive for extended periods of time without food.

A. Metabolic changes in starvation (Fig. 1.3)

When the body enters the **starved state**, after **3 to 5 days of fasting**, changes occur in the use of fuel stores.

1. Muscle decreases its use of ketone bodies and oxidizes fatty acids as its primary energy source.
2. Because of the decreased use by muscle, **blood ketone body levels rise**.
3. The **brain** then takes up and **oxidizes the ketone bodies** to derive energy. Consequently, the brain decreases its use of glucose, although glucose is still a major fuel for the brain.
4. Liver **gluconeogenesis decreases**.
5. **Muscle protein is spared** (i.e., less muscle protein is degraded to provide amino acids for gluconeogenesis).
6. Because of decreased conversion of amino acids to glucose, **less urea is produced** from amino acid nitrogen in starvation than after an overnight fast.

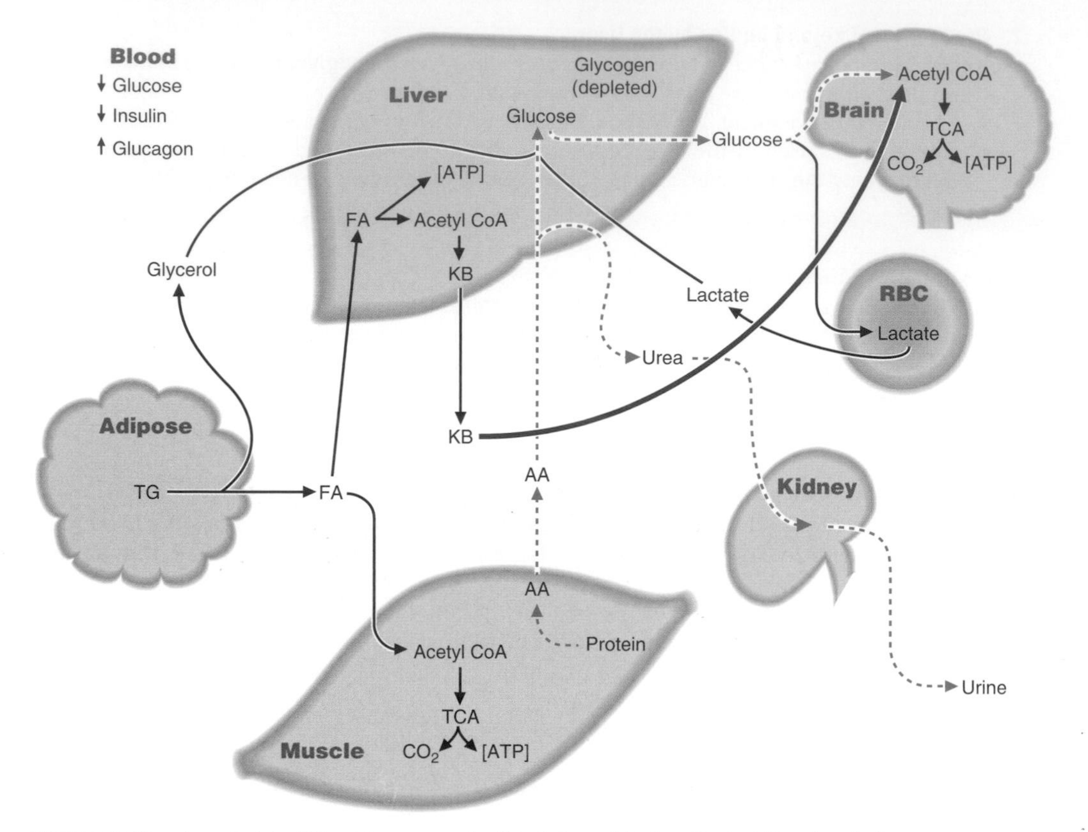

FIGURE 1.3. **The starved state**. This state occurs after 3 to 5 days of fasting. Dashed blue lines indicate processes that have decreased, and the red solid line indicates a process that has increased relative to the fasting state. KB, ketone bodies; AA, amino acid; FA, fatty acid; I, insulin; RBC, red blood cells; TG, triacylglycerols; VLDL, very low density lipoprotein; ⊕, stimulated by.

CLINICAL CORRELATES Diseases of malnutrition and starvation include **kwashiorkor** and **marasmus**. Kwashiorkor commonly occurs in children in third-world countries, where the **diet**, which is **adequate in calories**, is **low in protein**. A deficiency of dietary protein causes a decrease in protein synthesis (which can be observed through the measurement of serum albumin levels), which eventually affects the regeneration of intestinal epithelial cells, and thus, the problem is further compounded by **malabsorption**. Hepatomegaly and a **distended abdomen** are often observed. The lack of albumin in the blood leads to osmotic pressure differences between the blood and interstitial spaces, leading to water accumulation in the interstitial spaces, and the appearance of bloating. Marasmus results from a **diet deficient** in both protein and calories. Persistent starvation ultimately results in death.

B. Fat: the primary fuel

The body uses its fat stores as its primary source of energy during starvation, conserving functional protein.

1. Overall, fats are quantitatively the most important fuel in the body.
2. The length of time that a person can survive without food depends mainly on the amount of fat stored in the adipose tissue.

Review Test

Questions 1 to 10 examine your basic knowledge of fuel metabolism and are not in the standard clinical vignette format.

Questions 11 to 35 are clinically relevant, USMLE-style questions.

Basic Knowledge Questions

Questions 1 to 4

Match each of the characteristics below with the source of stored energy that it best describes. An answer (choices A through D) may be used once, more than once, or not at all.

1. The largest amount of stored energy in the body
2. The energy source reserved for strenuous muscular activity
3. The primary source of carbon for maintaining blood glucose levels during an overnight fast
4. The major precursor of urea in the urine

A. Protein
B. Triacylglycerol
C. Liver glycogen
D. Muscle glycogen

5. A 32-year-old male is on a weight-maintenance diet, so he does not want to lose or gain any weight. Which amino acid must be present in the diet so the patient does not go into a negative nitrogen balance?

(A) Alanine
(B) Arginine
(C) Glycine
(D) Threonine
(E) Serine

Questions 6 to 10

Match each of the characteristics below with the tissue it best describes. An answer (choices A through D) may be used once, more than once, or not at all.

6. After a fast of a few days, ketone bodies become an important fuel
7. Ketone bodies are used as a fuel after an overnight fast
8. Fatty acids are not a significant fuel source at any time
9. During starvation, this tissue uses amino acids to maintain blood glucose levels
10. This tissue converts lactate from muscle to a fuel for other tissues

A. Liver
B. Brain
C. Skeletal muscle
D. Red blood cells

Board-style Questions

Questions 11 to 15 are based on the following patient:

A young woman (5' 3" tall, 1.6 m) who has a sedentary job and does not exercise consulted a physician about her weight, which was 110 lb (50 kg). A dietary history indicates that she eats approximately 100 g of carbohydrate, 20 g of protein, and 40 g of fat daily.

11. What is this woman's BMI?

(A) 16.5
(B) 17.5
(C) 18.5
(D) 19.5
(E) 20.5

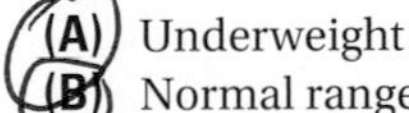

12. According to the woman's BMI, into what classification does her weight and height place her?

(A) Underweight
(B) Normal range
(C) Overweight (preobese)
(D) Class I obese range
(E) Class II obese range

13. How many calories (kcal) does this woman consume each day?

(A) 1,440
(B) 1,340
(C) 940
(D) 840
(E) 640

14. What is the woman's approximate DEE in calories (kilocalories) per day at this weight?

(A) 1,200
(B) 1,560
(C) 1,800
(D) 2,640
(E) 3,432

15. On the basis of the woman's current weight, diet, and sedentary lifestyle, which one of the following does the physician correctly recommend that she should undertake?

(A) Increase her exercise level
(B) Decrease her protein intake
(C) Increase her caloric intake
(D) Decrease her fat intake to <30% of her total calories
(E) Decrease her caloric intake

16. Consider a normal 25-year-old man, about 70 kg in weight, who has been shipwrecked on a desert island, with no food available, but plenty of freshwater. Which of the following fuel stores is least likely to provide significant calories to the man?

(A) Adipose triacylglycerol Most
(B) Liver glycogen
(C) Muscle glycogen
(D) Muscle protein
(E) Adipose triacylglycerol and liver glycogen

17. The shipwrecked man described in the previous question will have most of his fuel stored as triacylglycerol instead of protein in muscle due to triacylglycerol stores containing which of the following as compared to protein stores?

(A) More calories and more water
(B) Less calories and less water
(C) Less calories and more water
(D) More calories and less water
(E) Equal calories and less water

18. A vegan has been eating low-quality vegetable protein for many years, and is now exhibiting a negative nitrogen balance. This may be occurring due to a lack of which one of the following in his/her diet?

(A) Linoleic acid
(B) Starch
(C) Serine
(D) Lysine
(E) Linolenic acid

19. A medical student has been studying for exams, and neglects to eat anything for 12 hours. At this point, the student opens a large bag of pretzels and eats every one of them in a short period. Which one of the following effects will this meal have on the student's metabolic state?

(A) Liver glycogen stores will be replenished.
(B) The rate of gluconeogenesis will be increased.
(C) The rate at which fatty acids are converted to adipose triacylglycerols will be reduced.
(D) Blood glucagon levels will increase.
(E) Glucose will be oxidized to lactate by the brain and to CO_2 and H_2O by the red blood cells.

20. After a stressful week of exams, a medical student sleeps for 15 hours, then rests in bed for an hour before getting up for the day. Under these conditions, which one of the following statements concerning the student's metabolic state would be correct?

(A) Liver glycogen stores are completely depleted.
(B) Liver gluconeogenesis has not yet been activated.
(C) Muscle glycogen stores are contributing to the maintenance of blood glucose levels.
(D) Fatty acids are being released from adipose triacylglycerol stores.
(E) The liver is producing and oxidizing ketone bodies to CO_2 and H_2O.

21. A physician working in a refugee camp in Africa notices a fair number of children with emaciated arms and legs, yet a large protruding stomach and abdomen. An analysis of the children's blood would show significantly reduced levels of which one of the following as compared with those in a healthy child?

(A) Glucose
(B) Ketone bodies

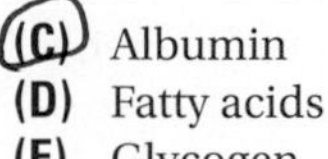

(C) Albumin
(D) Fatty acids
(E) Glycogen

Questions 22 to 25 are based on the following patient:

A 50-year-old male with a "pot belly" and a strong family history of heart attacks is going to his physician for advice on how to lose weight. He weighs 220 lb (100 kg) and is about 6' tall (1.85 m). His lifestyle can be best described as sedentary.

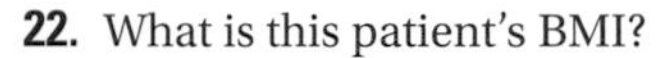

22. What is this patient's BMI?

(A) 24
(B) 29
(C) 31
(D) 36
(E) 40

23. Into which of the following categories does his BMI place him?

(A) Underweight
(B) Healthy
(C) Overweight (preobese)
(D) Obese (class I)
(E) Obese (class II)

24. How many kilocalories per day would the patient need to maintain this weight?

(A) 2,400
(B) 2,620
(C) 3,120
(D) 3,620
(E) 3,950

25. For which of the following disease processes is this patient at higher risk?

(A) Diabetes mellitus, type 1
(B) Insulin resistance syndrome
(C) Gaucher disease
(D) Low blood pressure
(E) Sickle cell disease

26. Which of the following metabolic patterns would be observed in a person after 1 week of starvation? Choose the one best answer.

	Brain use of fuels	Liver glycogen content (% of normal)	Nitrogen balance	Gluconeogenesis
A	Glucose only	<5	Positive	Inhibited
B	Glucose and ketone bodies	<5	Negative	Activated
C	Ketone bodies only	<5	Negative	Inhibited
D	Fatty acids and ketone bodies	50	Positive	Activated
E	Glucose and fatty acids	50	In balance	Inhibited

27. When compared with an individual's state after an overnight fast, a person who fasts for 1 week will have which one of the following patterns expressed?

	Blood glucose level	Amount of muscle protein	Amount of adipose triacylglycerol	Level of blood ketone bodies
A	Higher	Greater	Greater	Lower
B	Higher	Lower	Lower	Lower
C	Lower	Greater	Greater	Greater
D	Lower	Lower	Lower	Greater
E	The same	The same	The same	The same

28. Which one of the following is a common metabolic feature of patients with anorexia nervosa, untreated type 1 DM, hyperthyroidism, and nontropical sprue?

(A) A high BMR
(B) Elevated insulin levels in the blood
(C) Loss of weight
(D) Malabsorption of nutrients
(E) Low levels of ketone bodies in the blood

29. An 18-year-old person with type 1 diabetes has not injected her insulin for 2 days. Her blood glucose is currently 600 mg/dL (normal values are 80 to 100 mg/dL). Which one of the following cells of her body can still utilize the blood glucose as an energy source?

(A) Brain cells
(B) Muscle cells
(C) Adipose cells

30. A patient is brought to the emergency room after being found by search and rescue teams. He was mountain climbing, got caught in a sudden snowstorm, and had to survive in a cave. He had no food for 6 days. In adapting to these conditions, which metabolic process has **increased** rather than decreased?

(A) The brain's use of glucose
(B) Muscle's use of ketone bodies
(C) The red blood cells' use of glucose
(D) The brain's use of ketone bodies
(E) The red blood cells' use of ketone bodies
(F) Muscle's use of glucose

Questions 31 to 35 are based on the following case:

A 27-year-old male got lost while hiking in Yosemite National Park. He was found 8 days later. He had nothing to eat and only water to drink before being rescued.

31. Which one of the following would be his primary source of carbons for maintaining blood glucose levels when he was found?

(A) Liver glycogen
(B) Muscle glycogen
(C) Fatty acids
(D) Triacylglycerol
(E) Ketone bodies

32. Which cell can only use glucose for energy needs?

(A) Brain
(B) Red blood cells
(C) Hepatocyte
(D) Heart
(E) Muscle

33. Which one of the following is an essential nutrient that he has not received over the last 8 days?

(A) Lactic acid
(B) Oleic acid
(C) Steric acid
(D) EPA
(E) Palmitic acid

34. The man's brain would attempt to decrease consumption of glucose and increase consumption of ketones in order to protect the breakdown (catabolism) of which one of the following?

(A) Muscle glycogen
(B) Liver glycogen
(C) Muscle protein
(D) Red blood cells (to provide heme)
(E) Adipose triacylglycerol

35. Which one of the following lab tests should be run on the patient to determine whether he is suffering from overall protein malnutrition?

(A) Albumin
(B) Blood urea nitrogen (BUN)
(C) Creatinine
(D) Ferritin
(E) Creatine phosphokinase (CPK)

Answers and Explanations

1. **The answer is B.** Adipose triacylglycerols contain the largest amount of stored energy in humans, followed by protein (even though loss of too much protein will lead to death), muscle glycogen, and liver glycogen (see Table 1.1).
2. **The answer is D.** Muscle glycogen is used for energy during exercise. The glycogen is degraded to a form of glucose that can enter metabolic pathways for energy generation. Because exercise is strenuous, muscle requires large amounts of energy, and this can be generated at the fastest rate by converting muscle glycogen to pathway precursors within the muscle. Liver glycogen will produce glucose that enters the circulation. Once in the circulation, the muscle can take up that glucose and use it to generate energy; however, the rate of energy generation from liver-derived glucose is much slower than that from muscle-derived glucose.
3. **The answer is C.** Liver glycogenolysis is the major process for maintaining blood glucose levels after an overnight fast. The muscle cannot export glucose to contribute to the maintenance of blood glucose levels, and fatty acid carbons cannot be utilized for the net synthesis of glucose.
4. **The answer is A.** The nitrogen in amino acids derived from protein is converted to urea and excreted in the urine. Uric acid, another excretion product that contains nitrogen, is derived from purine bases (found in nucleic acids), not from protein.
5. **The answer is D.** The lack of one essential amino acid will lead to a negative nitrogen balance due to increased protein degradation to supply that amino acid for the ongoing protein synthesis. Of the amino acids listed, only threonine is an essential amino acid (alanine can be synthesized from pyruvate [which can be derived from glucose], arginine is produced in the urea cycle using aspartic acid and the amino acid ornithine, glycine is derived from serine, and serine is derived from 3-phosphoglycerate, which can be produced from glucose).
6. **The answer is B.** The brain begins to use ketone bodies when levels start to rise after 3 to 5 days of fasting. Normally, the brain will use only glucose as a fuel (most fatty acids cannot cross the blood–brain barrier to be metabolized by the brain), but when ketone bodies are elevated in the blood, they can enter the brain and be used for energy.
7. **The answer is C.** Skeletal muscle oxidizes ketone bodies, which are synthesized in the liver from fatty acids derived from adipose tissue. As the fast continues, the muscle will switch to oxidizing fatty acids, which allows ketone body levels to rise such that the brain will begin using them as an energy source.
8. **The answer is D.** Oxidation of fatty acids occurs in mitochondria. Red blood cells lack mitochondria and therefore cannot use fatty acids. The brain will not transport most fatty acids across the blood–brain barrier (the essential fatty acids are a notable exception). Therefore, the brain cannot use fatty acids as an energy source. The brain does, however, synthesize its own fatty acids, and will oxidize those fatty acids when appropriate. Red blood cells can never use fatty acids as an energy source due to their lack of mitochondria.
9. **The answer is A.** The liver converts amino acids to blood glucose by gluconeogenesis. The other substrates for gluconeogenesis are lactate from the metabolism of glucose within the red blood cells and glycerol from the breakdown of triacylglycerol to free fatty acids and glycerol. Neither the brain, nor the skeletal muscle, nor the red blood cell can export glucose into the circulation.
10. **The answer is A.** Exercising muscle produces lactate, which the liver can convert to glucose by gluconeogenesis. Blood glucose is oxidized by red blood cells and other tissues. Only the liver and kidney (to a small extent) can release free glucose into the circulation for use by other tissues.
11. **The answer is D.** The BMI is calculated by dividing the weight of the individual (in kilograms) by the square of the height of the individual (in meters). For this woman, BMI $= 50/1.6^2 = 19.5$.

12. The answer is B. According to Table 1.2, a BMI of 19.5 places the woman at the lower end of the normal range. Underweight is indicated by a BMI of <18.5; preobesity occurs above a BMI of 25, but <30. Class I obesity is indicated by a BMI between 30 and 35, and class II obesity by a BMI between 35 and 40.

13. The answer is D. The woman consumes 400 calories (kcal) of carbohydrate (100 g × 4 kcal/g), 80 calories of protein (20 × 4), and 360 calories of fat (40 × 9) for a total of 840 calories daily.

14. The answer is B. This woman's DEE is 1,560 calories (kcal). DEE equals BMR plus physical activity. Her weight is 110 lb/2.2 = 50 kg. Her BMR (about 24 kcal/kg) is 50 kg × 24 = 1,200 kcal/day. She is sedentary and needs only 360 additional kcal (30% of her BMR) to support her physical activity. Therefore, she needs 1,200 + 360 = 1,560 kcal each day.

15. The answer is C. Because her caloric intake (840 kcal/day) is less than her expenditure (1,560 kcal/day), the woman is losing weight. She needs to increase her caloric intake. Exercise would cause her to lose more weight. She is probably in negative nitrogen balance because her protein intake is low (0.8 g/kg/day is recommended). Although her fat intake is 43% of her total calories and recommended levels are <30%, she should increase her total calories by increasing her carbohydrate and protein intake rather than decreasing her fat intake.

16. The answer is B. As indicated in Table 1.1, in the average (70 kg) man, adipose tissue contains 15 kg of fat or 135,000 calories (kcal). Liver glycogen contains about 0.08 kg of carbohydrate (320 calories), and muscle glycogen contains about 0.15 kg of carbohydrate (600 calories). In addition, about 6 kg of muscle protein (24,000 calories) can be used as fuel. Therefore, liver glycogen contains the fewest available calories.

17. The answer is D. Adipose tissue contains more calories (kilocalories) and less water than does muscle protein. Triacylglycerol stored in adipose tissue contains 9 kcal/g, and adipose tissue has about 15% water. Muscle protein contains 4 kcal/g and has about 80% water.

18. The answer is D. A negative nitrogen balance will result from a diet deficient in one essential amino acid, or in a very diseased state. Linoleic and linolenic acids are the essential fatty acids in the diet, and a lack of these fatty acids will not affect nitrogen balance. Starch is a glucose polymer, and the lack of starch will not affect nitrogen balance. Lysine is an essential amino acid, whereas serine can be synthesized from a derivative of glucose. Lack of lysine in the diet will lead to a negative nitrogen balance as existing protein is degraded to provide lysine for new protein synthesis.

19. The answer is A. After a meal of carbohydrates (the major ingredient of pretzels), glycogen is stored in the liver and in muscle, and triacylglycerols are stored in adipose tissue. Owing to the rise in glucose level in the blood (from the carbohydrates in the pretzels), insulin is released from the pancreas and the level of glucagon in the blood decreases. Since blood glucose levels have increased, there is no longer a need for the liver to synthesize glucose, and gluconeogenesis decreases. The change in insulin-to-glucagon ratio also inhibits the breakdown of triacylglycerols and favors their synthesis. The brain oxidizes glucose to CO_2 and H_2O, whereas the red blood cells produce lactate from glucose, since red blood cells cannot carry out aerobic metabolism.

20. The answer is D. During fasting, fatty acids are released from adipose tissue and oxidized by other cells. Liver glycogen is not depleted until about 30 hours of fasting. After an overnight fast, both glycogenolysis and gluconeogenesis by the liver help maintain blood glucose levels. Muscle glycogen stores are not used to maintain blood glucose levels. The liver produces ketone bodies but does not oxidize them, but under the conditions described in this question, ketone body formation would be minimal.

21. The answer is C. The children are exhibiting the effects of kwashiorkor, a disorder resulting from adequate calorie intake but insufficient calories from protein. This results in the liver producing less serum albumin (due to the lack of essential amino acids), which affects the osmotic balance of the blood and the fluid in the interstitial spaces. Owing to the reduction in osmotic pressure of the blood, water leaves the blood and enters the interstitial spaces, producing edema in the children (which leads to the expanded abdomen). The children are degrading muscle protein to allow the synthesis of new protein (due to a lack of essential amino acids), and this leads to the wasting of the arms and legs of children with this disorder. The children will exhibit normal or slightly elevated levels of ketone bodies and fatty acids in the blood, as the diet is calorie sufficient. Glycogen levels may only be slightly reduced (since the diet is calorie sufficient), but glycogen is not found in the blood. Glucose levels will be only slightly reduced, as gluconeogenesis will keep glucose levels near normal.

22. **The answer is B.** The BMI is equal to kg/m^2, which in this case is equal to $100/1.85^2$, which is about 29.

23. **The answer is C.** The patient is in the overweight (preobese) category with a BMI of 29. As indicated in Table 1.2, a BMI of <18.5 is the underweight category, a BMI between 18.5 and 24.9 is the healthy range, a BMI between 25 and 30 is the overweight (preobese) category, and any BMI of 30 or above is considered the obese range. Class I obese is between 30 and 35, whereas class II obese is between 35 and 40. Class III obesity, or morbidly obese, is the classification for individuals with a BMI of 40 or higher.

24. **The answer is C.** The DEE is equal to the BMR plus physical activity factor. For the patient in question, the BMR = 24 kcal/kg/day × 100 kg, or 2,400 kcal/day. Since the patient is sedentary, the activity level is 30% that of the BMR, or 720 kcal/day. The overall daily needs are therefore 2,400 + 720 kcal/day, or 3,120 kcal/day. If the patient consumes $<3,000$ kcal/day, or increases his physical activity level, then weight loss would result.

25. **The answer is B.** The patient's weight, age, and activity put him at higher risk for insulin resistance syndrome. The entire syndrome includes hypertension, diabetes mellitus (type 2), decreased high-density lipoprotein levels, increased triglyceride levels, increased urate, increased levels of plasminogen activator inhibitor 1, nonalcoholic fatty liver, central obesity, and polycystic ovary syndrome (PCOS) (in females). Insulin resistance syndrome leads to early atherosclerosis throughout the entire body. The patient is not at increased risk for diabetes mellitus, type 1, as that is the result of an autoimmune condition that destroys the β cells of the pancreas such that insulin can no longer be produced. The lifestyle exhibited by the patient has not been linked to autoimmune disorders. Gaucher disease is a disorder of the enzyme β-glucocerebrosidase, and is an autosomal recessive disorder. Since this disease is an inherited disorder, the patient's lifestyle does not increase his risk of having this disease. The patient's increasing weight might lead to increased blood pressure, but not to reduced blood pressure. Sickle cell disease is another autosomal recessive disorder leading to an altered β-globin gene product, and like Gaucher disease, it is an inherited disorder that is not altered by the patient's lifestyle.

26. **The answer is B.** After 3 to 5 days of starvation, the brain begins to use ketone bodies, in addition to glucose, as a fuel source. Glycogen stores in the liver are depleted ($<5\%$ of normal) during the first 30 hours of fasting. Inadequate protein in the diet results in a negative nitrogen balance. Blood glucose levels are being maintained by gluconeogenesis, using lactate (from red blood cells), glycerol (from triacylglycerol), and amino acids (from the degradation of muscle proteins) as carbon sources.

27. **The answer is D.** If a person who has fasted overnight continues to fast for 1 week, muscle protein will continue to decrease because it is being converted to blood glucose. However, it will not decrease at as rapid a rate as with a shorter fast, because the brain is using ketone bodies and, therefore, less glucose. The individual's blood glucose levels will decrease about 40%, because initially glycogenolysis and then gluconeogenesis by the liver help to maintain blood glucose levels, but oxidation of ketone bodies by the brain will reduce the brain's overall dependence on glucose. Adipose tissue will decrease as triacylglycerol is mobilized. Fatty acids from adipose tissue will be converted to ketone bodies in the liver. Blood ketone body levels will rise, and the brain will use ketone bodies as an alternative energy source, to reduce its dependency on glucose (during starvation, about 40% of the brain's energy needs can be met by oxidizing ketone bodies, whereas the other 60% still requires glucose oxidation).

28. **The answer is C.** All of these patients will lose weight—the anorexic patients because of insufficient calories in the diet, the patients with type 1 DM because of low insulin levels that result in the excretion of glucose and ketone bodies in the urine, those with hyperthyroidism because of an increased BMR, and those with nontropical sprue because of decreased

absorption of food from the gut. The untreated diabetic patients will have high ketone levels because of low insulin. Ketone levels may be elevated in anorexia and also in sprue, due to a reduction in levels of gluconeogenic precursors. An increased BMR would be observed only in hyperthyroidism. Nutrient malabsorption would occur only in nontropical sprue and anorexia.

29. The answer is A. Muscle and adipose cells require insulin to stimulate the transport of glucose into the cell, whereas the glucose transporters for the blood–brain barrier are always present, and are not responsive to insulin. Thus, the brain can always utilize the glucose in circulation, whereas muscle and adipose tissue are dependent on insulin for glucose transport into the tissue.

30. The answer is D. In the starvation state, muscle decreases the use of ketone bodies, causing an elevation of ketone bodies in the bloodstream. The brain uses the ketone bodies for energy and uses less glucose, which decreases the need for gluconeogenesis, thus sparing muscle protein degradation to provide the precursors for gluconeogenesis. Red blood cells cannot use ketone bodies and must utilize glucose. Therefore, the use of glucose by red blood cells would be unchanged under these conditions.

31. The answer is D. The glycerol component of triacylglycerol would be the major contributor of carbons for gluconeogenesis among the answer choices provided. Substrates for hepatic gluconeogenesis are lactate (from red blood cells), amino acids (from muscle), and glycerol (from adipose tissue). Fatty acids would be used for energy, but the carbons of fatty acids cannot be used for the net synthesis of glucose. Hepatic glycogen stores are exhausted about 30 hours after the initiation of the fast, and muscle glycogen stores contribute only to muscle energy needs and not to the maintenance of blood glucose levels.

32. The answer is B. Red blood cells lack mitochondria, so they can use only glucose for fuel (fatty acids and ketone bodies require mitochondrial proteins for their oxidative pathways). The brain can also use ketone bodies, along with glucose. The liver can use glucose, fatty acids, and amino acids as energy sources. The heart can use glucose, fatty acids, amino acids, and lactic acid as potential energy sources.

33. The answer is D. Eicosapentaenoic acid (EPA, a 20-carbon fatty acid containing five double bonds) can be derived from an essential fatty acid found in fish oils (linolenic acid), and is a precursor of eicosanoids (prostaglandins, leukotrienes, and thromboxanes). EPA is also ingested from fish oils. Lactic acid is produced from muscle and red blood cells, and is not an essential nutrient. Palmitic acid (a fatty acid containing 16 carbons, with no double bonds), oleic acid (a fatty acid containing 18 carbons, with one double bond), and stearic acid (a fatty acid containing 18 carbons, with no double bonds) can all be synthesized by the mammalian liver through the normal pathway of fatty acid synthesis.

34. The answer is C. In an attempt to save muscle tissue (amino acids used for gluconeogenesis), the brain in starvation mode will utilize ketone bodies for a portion of its energy needs. Liver glycogen stores would be depleted under the conditions described. Heme is not used for energy production, and produces bilirubin when degraded, which cannot be used to generate energy or ketone bodies. Muscle glycogen cannot contribute to blood glucose levels, as muscle tissue lacks the enzyme that allows free glucose to be produced within the muscle.

35. The answer is A. Albumin, though nonspecific, is considered the standard for assessing overall protein malnutrition. Albumin is made by the liver and is found in the blood. It acts as a nonspecific carrier of fatty acids and other hydrophobic molecules. When amino acid levels become limiting, the liver reduces its levels of protein synthesis, and a reduction in albumin levels in the circulation is an indication of liver dysfunction. Ferritin is an iron storage protein within tissues, and its circulating levels are low at all times. Creatinine is a degradation product of creatine phosphate (an energy storage molecule in muscle), and its presence in the circulation reflects the rate of creatinine clearance by the kidney. High levels of creatinine indicate a renal insufficiency. Creatine phosphokinase is a muscle enzyme that is released into circulation only when there is damage to the muscle. Blood urea nitrogen indicates the rate of amino acid metabolism to generate urea, but does not indicate protein malnutrition.

chapter 2

Basic Aspects of Biochemistry: Organic Chemistry, Acid–Base Chemistry, Amino Acids, Protein Structure and Function, and Enzyme Kinetics

The main clinical uses of this chapter are in understanding the basics behind acidoses, alkaloses, treatments in renal failure, actions of pharmaceuticals, dose and frequency of certain medications, and the hemoglobinopathies (protein structure and function).

OVERVIEW

- Acids dissociate, releasing protons and producing their conjugate bases.
- Bases accept protons, producing their conjugate acids.
- Buffers consist of acid–base conjugate pairs that can donate and accept protons, thereby maintaining the pH of a solution.
- Proteins, which are composed of amino acids, serve in many roles in the body (e.g., as enzymes, structural components, hormones, and antibodies).
- Interactions between amino acid residues produce the three-dimensional conformation of a protein, starting with the primary structure, leading to secondary and tertiary structures, and for multisubunit proteins, a quaternary structure.
- Enzymes are proteins that catalyze biochemical reactions.
- Enzymes accelerate reactions by reducing the Gibbs free energy of activation.
- Enzyme-catalyzed reactions can be described by the Michaelis–Menten equation, in which the K_m is the substrate concentration at which the rate of formation of the product of the reaction (the velocity) is equal to one-half of the maximal velocity (V_{max}).
- Reversible enzyme inhibitors can be classified as either competitive or noncompetitive, and can be distinguished via a Lineweaver–Burk plot.

I. A BRIEF REVIEW OF ORGANIC CHEMISTRY

- Biochemical reactions involve the functional groups of molecules.

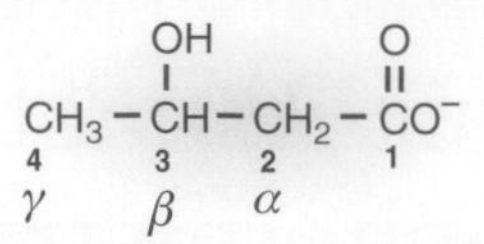

FIGURE 2.1. Identification of carbon atoms in an organic compound. Carbons are numbered starting from the most oxidized carbon-containing group, or they are assigned Greek letters, with the carbon next to the most oxidized group designated as the α-carbon. This compound is 3-hydroxybutyrate or β-hydroxybutyrate. It is a ketone body.

A. Identification of carbon atoms

As indicated in Figure 2.1, carbon atoms are either **numbered** or given **greek letters**.

B. Functional groups in biochemistry

Types of groups: Alcohols, aldehydes, ketones, carboxyl groups, anhydrides, sulfhydryl groups, amines, esters, and amides are all important components of biochemical compounds (Fig. 2.2).

C. Biochemical reactions

1. Reactions are classified according to the functional groups that react (e.g., **esterifications, hydroxylations, carboxylations, and decarboxylations**).
2. **Oxidations** of sulfhydryl groups to disulfides, of alcohols to aldehydes and ketones, and of aldehydes to carboxylic acids frequently occur.
   a. Many of these oxidations are reversed by **reductions**.
   b. In **oxidation** reactions, electrons are lost. In **reduction** reactions, electrons are gained.
   c. As **foods are oxidized**, electrons are released and passed through the electron transport chain. Adenosine triphosphate (**ATP**) is generated, and it supplies the energy to drive various functions of the body.

II. ACIDS, BASES, AND BUFFERS

- Many biochemical compounds, ranging from small molecules to large polymers, are capable of releasing or accepting protons at physiologic pH, and as a consequence, may carry a charge.
- Most biochemical reactions occur in aqueous solutions.

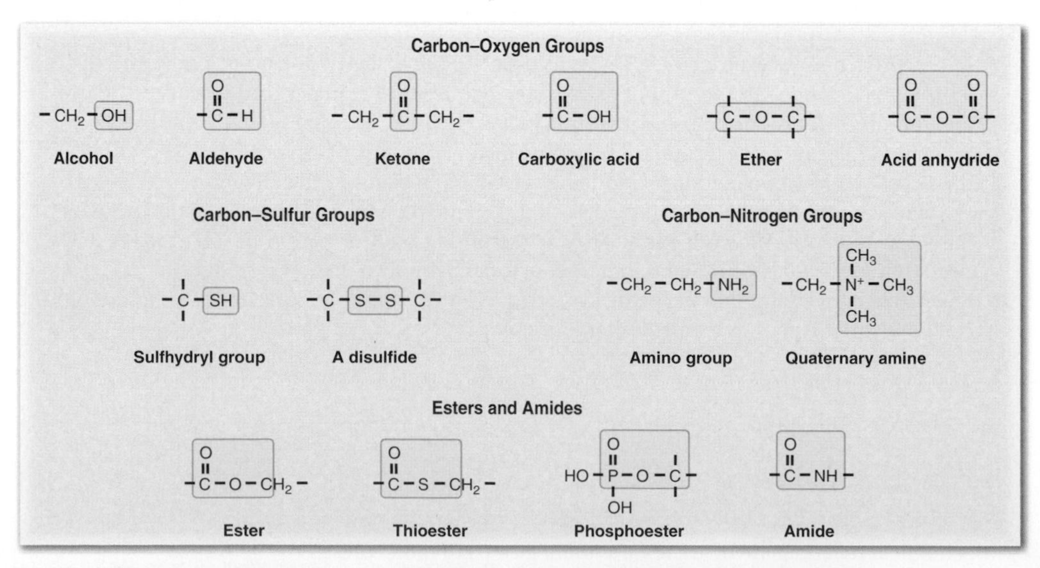

FIGURE 2.2. Major types of functional groups found in biochemical compounds of the human body.

- The pH of a solution is the negative $\log_{10}$ of its hydrogen ion concentration $[H^+]$.
- Acids are proton donors, and bases are proton acceptors.
- The Henderson–Hasselbalch equation describes the relationship between pH, pK (the negative log of the dissociation constant), and the concentrations of an acid and its conjugate base.
- Buffers consist of solutions of acid–base conjugate pairs that resist changes in pH when H^+ or OH^- are added.
- Acids that are ingested or produced by the body are buffered by bicarbonate and by proteins, particularly hemoglobin. These buffers help to maintain the pH in the body within the range compatible with life.

A. Water

1. Water is the **solvent of life.** It dissociates:

$$H_2O \rightleftharpoons H^+ + OH^-$$

with an equilibrium constant:

$$K = \frac{[H^+][OH^-]}{[H_2O]}$$

2. Because the extent of dissociation is not appreciable, H_2O remains constant at 55.5 M, and the ion product of H_2O is:

$$K_w = [H^+][OH^-] = 1 \times 10^{-14}$$

3. The **pH** of a solution is the negative $\log_{10}$ of its hydrogen ion concentration $[H^+]$:

$$pH = -\log_{10} [H^+]$$

For pure water,

$$[H^+] = [OH^-] = 1 \times 10^{-7}$$

Therefore, **the pH of pure water is 7.**

B. Acids and bases

Acids are compounds that donate protons, and bases are compounds that accept protons.

1. **Acids dissociate.**
   a. **Strong acids,** such as hydrochloric acid (HCl), dissociate completely.
   b. **Weak acids,** such as **acetic acid, dissociate only to a limited extent:**

$$HA \rightleftharpoons H^+ + A^-$$

   where HA is the acid and A^- is its conjugate base.

   c. The **dissociation constant** for a weak acid is:

$$K = \frac{[H^+][A^-]}{[HA]}$$

2. The **Henderson–Hasselbalch equation** was derived from the equation for the dissociation constant:

$$pH = pK + \log_{10} \frac{[A^-]}{[HA]}$$

where pK is the negative $\log_{10}$ of K, the dissociation constant.

3. The **major acids** produced by the body include **phosphoric acid, sulfuric acid, lactic acid,** and the ketone bodies, **acetoacetic acid** and **β-hydroxybutyric acid.** CO_2 is also produced, which combines with H_2O to form **carbonic acid** in a reaction catalyzed by carbonic anhydrase:

$$CO_2 + H_2O \overset{\text{Carbonic anhydrase}}{\rightleftharpoons} H_2CO_3 \rightleftharpoons H^+ + HCO_3^-$$

C. Buffers

1. Buffers consist of **solutions of acid–base conjugate pairs**, such as acetic acid and acetate.
   a. Near its p*K*, a buffer maintains the pH of a solution, resisting changes due to addition of acids or bases. For a weak acid, the p*K* is often designated as pK_a.
   b. At the pK_a, $[A^-]$ and [HA] are equal, and the buffer has its maximal capacity.
2. **Buffering mechanisms in the body**
   a. The **normal pH range** of arterial blood is **7.37 to 7.43**.
   b. The major buffers of blood are **bicarbonate** (HCO_3^-/H_2CO_3) and **hemoglobin** (Hb/HHb).
   c. These buffers act in conjunction with mechanisms in the kidneys for excreting protons and mechanisms in the lungs for exhaling CO_2 to maintain the pH within the normal range.

CLINICAL CORRELATES

Acid–base disturbances occur under a variety of conditions. Hypoventilation causes retention of CO_2 by the lungs, which can lead to a **respiratory acidosis**. Hyperventilation can cause a **respiratory alkalosis**. **Metabolic acidosis** can result from accumulation of metabolic acids (lactic acid or the ketone bodies, β-hydroxybutyric acid and acetoacetic acid), or ingestion of acids or compounds that are metabolized to acids (e.g., methanol, ethylene glycol). **Metabolic alkalosis** is due to increased HCO_3^-, which is accompanied by an increased pH. Acid–base disturbances lead to compensatory responses that attempt to restore the normal pH. For example, a metabolic acidosis causes hyperventilation and the release of CO_2, which tends to lower the pH. During metabolic acidosis, the kidneys excrete NH_4^+, which contains H^+ buffered by ammonia:

$$H^+ + NH_3 \rightleftharpoons NH_4^+$$

Failure of the gastroesophageal sphincter can lead to gastric reflux disease, in which the acid (HCl) contents of the stomach travel up the esophagus. The consequences of this disorder (esophageal damage due to acid refluxing up into the esophagus) can be treated, in part, by use of drugs that inhibit the gastric proton-translocating H^+/K^+ ATPase of the parietal cells, which pumps protons into the stomach lumen (in exchange for K^+ outside the cell) against a concentration gradient using the energy of ATP. The use of these drugs increases the pH of the stomach contents, which lessens esophageal damage and allows the tissue to heal.

III. AMINO ACIDS AND PEPTIDE BONDS

- An amino acid usually contains a carboxyl group, an amino group, and a side chain, all bonded to the α-carbon atom.
- Amino acids are usually of the L-configuration.
- At physiologic pH, amino acids carry a positive charge on their amino groups and a negative charge on their carboxyl groups.
- The side chains of the amino acids contain different chemical groups. Some side chains carry a charge.
- Peptide bonds link adjacent amino acid residues in a protein chain.

A. Amino acids

a. There are 20 amino acids commonly found in proteins. Figure 2.3 indicates the structures of the amino acids, their three-letter abbreviations, and their single-letter code. These 20 amino acids are used for the synthesis of proteins by the mRNA-directed process that occurs on ribosomes (see Chapter 3).
b. Other amino acids exist for which there is no genetic code, for example, in the urea cycle or in proteins where they are generated by posttranslational modifications (such as hydroxyproline in collagen).
c. Selenocysteine is unique in that a serine residue is converted to selenocysteine while attached to a transfer RNA. Selenium is a necessary metal ion for certain enzymes, such as glutathione peroxidase.

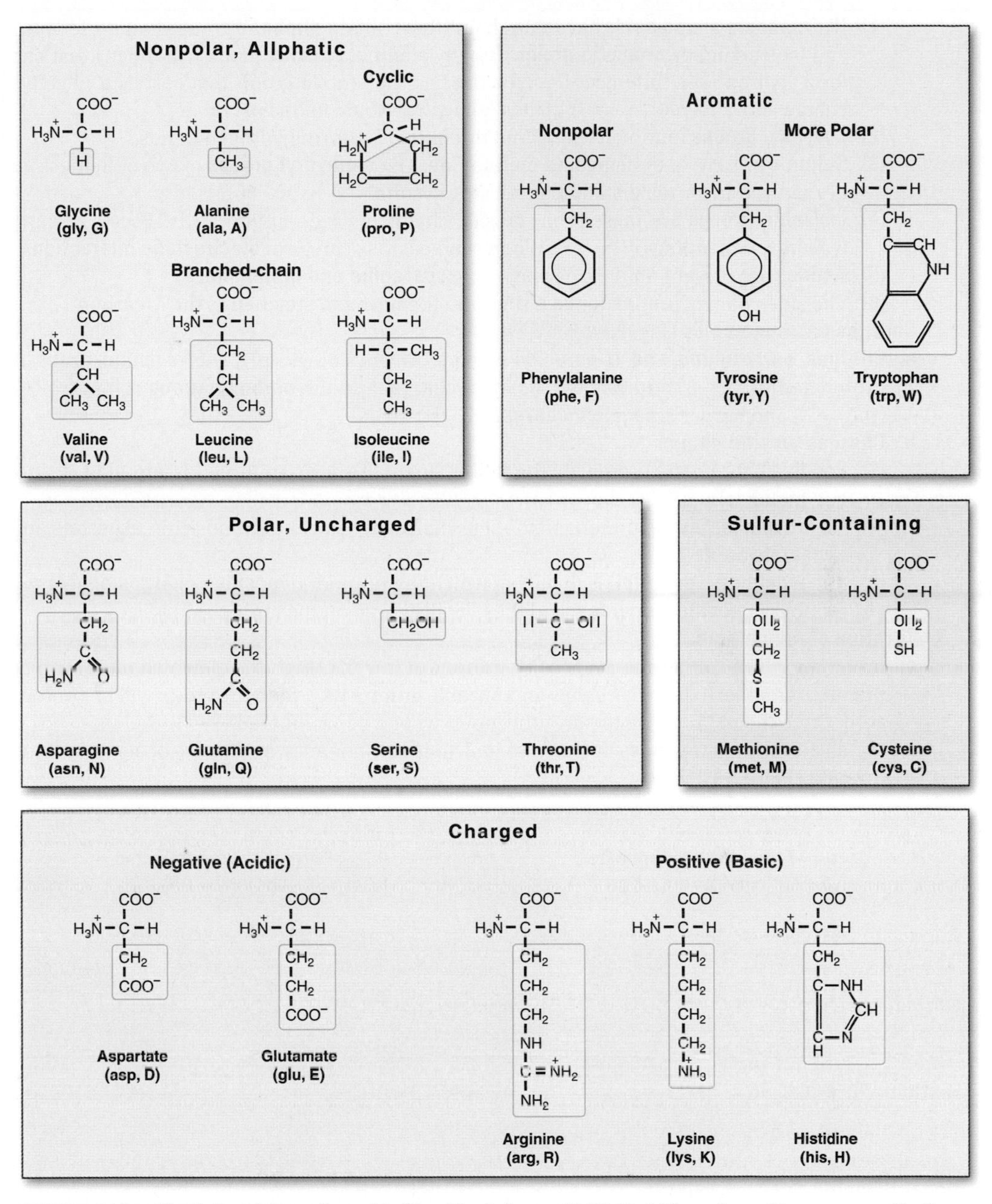

FIGURE 2.3. **The side chains of the amino acids.** The side chains are highlighted. The amino acids are grouped by the polarity and structural features of their side chains. These groupings are not absolute, however. Tyrosine and tryptophan, often listed with the nonpolar amino acids, are more polar than other aromatic amino acids because of their phenolic and indole rings, respectively. The single- and three-letter codes are also indicated for each amino acid.

1. **Structures of the amino acids** (see Fig 2.3)
   a. Most amino acids contain a **carboxyl group**, an **amino group**, and a **side chain** (R group), all attached to the α-carbon. Exceptions are:
      (1) **Glycine**, which does not have a side chain. Its α-carbon contains two hydrogens.
      (2) **Proline**, in which the nitrogen is part of a ring.
   b. All of the 20 amino acids except glycine are of the **L-configuration**, as for all but one amino acid the α-carbon is an asymmetric carbon. Because glycine does not contain an asymmetric carbon atom, it is not optically active and, thus, it is neither D nor L.
   c. The **classification** of amino acids is based on their side chains.

(1) **Hydrophobic amino acids** have side chains that contain **aliphatic** groups (valine, leucine, and isoleucine) or **aromatic groups** (phenylalanine, tyrosine, and tryptophan) that can form hydrophobic interactions. **Tyrosine** has a phenolic group that carries a negative charge above its pK_a (~ 10.5), so it is not hydrophobic in this pH range.

(2) **Hydroxyl groups** found on serine and threonine can form hydrogen bonds.

(3) **Sulfur** is present in cysteine and methionine. The **sulfhydryl groups** of two cysteines can be oxidized to form a **disulfide**, producing cystine.

(4) **Ionizable groups** are present on the side chains of seven amino acids. They can **carry a charge**, depending on the pH. When charged, they can form **electrostatic interactions**.

(5) **Amides** are present on the side chains of **asparagine and glutamine**.

(6) The side chain of **proline forms a ring** with the nitrogen attached to the α-carbon.

2. Charges on amino acids (Fig. 2.4)

a. Charges on α-amino and α-carboxyl groups: At physiologic pH, the **α-amino group** is protonated (pK_a ~ 9) and carries a **positive charge**, and the **carboxyl group** is dissociated (pK_a ~ 2) and carries a **negative charge**.

b. Charges on side chains

(1) **Positive charges** are present on the side chains of the basic amino acids, **arginine, lysine,** and **histidine** at pH 7.

(2) **Negative charges** are present on the side chains of the acidic amino acids, **aspartate and glutamate** at pH 7.

(3) The **isoelectric point (pI)** is the pH at which the number of **positive charges equals** the number of **negative charges**, and the overall charge on the amino acid is zero.

3. Titration of amino acids

a. Ionizable groups on amino acids carry protons at low pH (high $[H^+]$), which dissociate as the pH increases. If the pH is below an ionizable group's pK_a, then the group will be protonated. Once the pH is above the pK_a, the group will be deprotonated.

b. For an amino acid that does not have an ionizable side chain, two pK_a's are observed during titration (Fig. 2.5A).

	Form that predominates below the pK_a	pK_a	Form that predominates above the pK_a
Aspartate	$-CH_2-COOH$	3.9	$-CH_2-COO^-$ + H^+
Glutamate	$-CH_2-CH_2-COOH$	4.1	$-CH_2-CH_2-COO^-$ + H^+
Histidine	$-CH_2$–imidazole (HN^+, NH)	6.0	$-CH_2$–imidazole (N, NH) + H^+
Cysteine	$-CH_2SH$	8.4	$-CH_2S^-$ + H^+
Tyrosine	$-CH_2-C_6H_4-OH$	10.5	$-CH_2-C_6H_4-O^-$ + H^+
Lysine	$-CH_2-CH_2-CH_2-CH_2-\overset{+}{N}H_3$	10.5	$-CH_2-CH_2-CH_2-CH_2-NH_2$ + H^+
Arginine	$-CH_2-CH_2-CH_2-NH-C(=\overset{+}{N}H_2)NH_2$	12.5	$-CH_2-CH_2-CH_2-NH-C(=NH)NH_2$ + H^+

FIGURE 2.4. **Dissociation of the side chains of the amino acids.** As the pH increases, the charge on the side chain goes from zero to negative or from positive to zero. The pK_a is the pH at which one-half of the molecules of an amino acid in solution have side chains that are charged. The other half are uncharged.

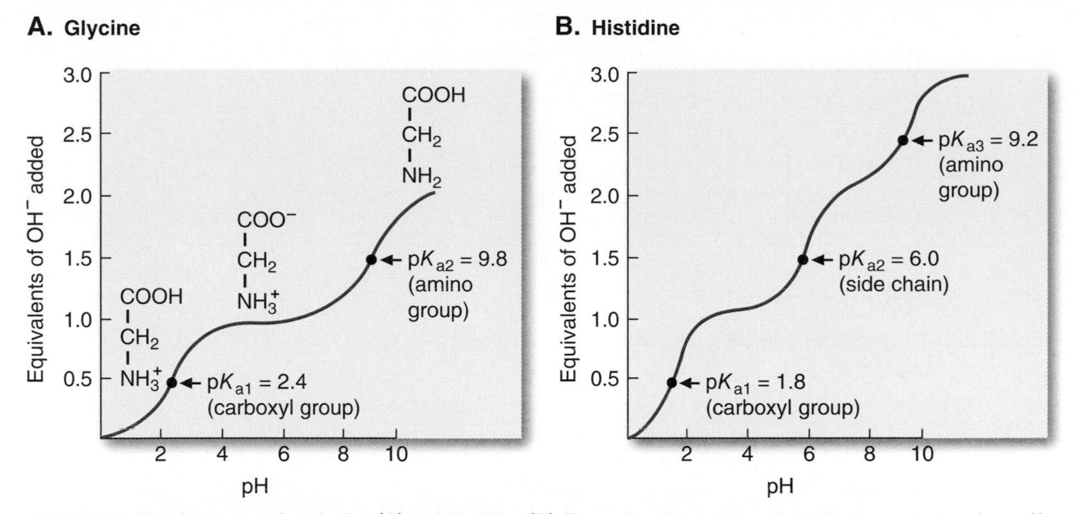

FIGURE 2.5. **Titration curves for glycine (A) and histidine (B).** The molecular species of glycine present at various pHs are indicated by the molecules above the curve. For histidine, pK_{a2} is the dissociation constant of the imidazole (side chain) group.

(1) The first (**pK_{a1}**) corresponds to the α-**carboxyl group** (pK_{a1}~2). As the proton dissociates, the carboxyl group goes from a zero to a minus charge.

(2) The second (**pK_{a2}**) corresponds to the α-**amino group** (pK_{a2} ~ 9). As the proton dissociates, the amino group goes from a positive to a zero charge.

c. For an amino acid with an ionizable side chain, **three pK_a's** are observed during titration (see Fig. 2.5B).

(1) The α-carboxyl and α-amino groups have pK_a's of about 2 and 9, respectively.

(2) The **third pK_a** varies with the amino acid and depends on the pK_a of the side chain (see Fig. 2.4).

B. Peptide bonds

Peptide bonds covalently join the α-carboxyl group of each amino acid to the α-amino group of the next amino acid in the protein chain (Fig. 2.6).

1. Characteristics

a. The **atoms** involved in the peptide bond form a **rigid, planar unit.**

b. Because of its **partial double-bond character**, the peptide bond has **no freedom of rotation.**

c. However, the bonds involving the **α-carbon** can **rotate freely** although there are only a limited number of angles that these bonds can form within a protein.

2. Peptide bonds are extremely stable. Cleavage generally involves the action of proteolytic enzymes.

IV. PROTEIN STRUCTURE

- The primary structure of a protein consists of the amino acid sequence along the chain.
- The secondary structure involves α-helices, β-sheets, and other types of folding patterns that occur due to a regular repeating pattern of hydrogen bond formation.
- The tertiary structure (the three-dimensional conformation of a protein) involves electrostatic and hydrophobic interactions, van der Waals interactions, and hydrogen and disulfide bonds.
- Quaternary structure refers to the interaction of one or more subunits to form a functional protein, using the same forces that stabilize the tertiary structure.
- Proteins serve in many roles (e.g., as enzymes, hormones, receptors, antibodies, structural components, transporters of other compounds, and contractile elements in muscle).

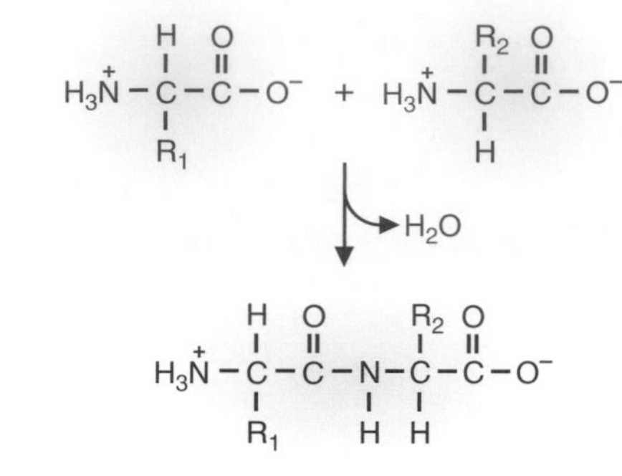

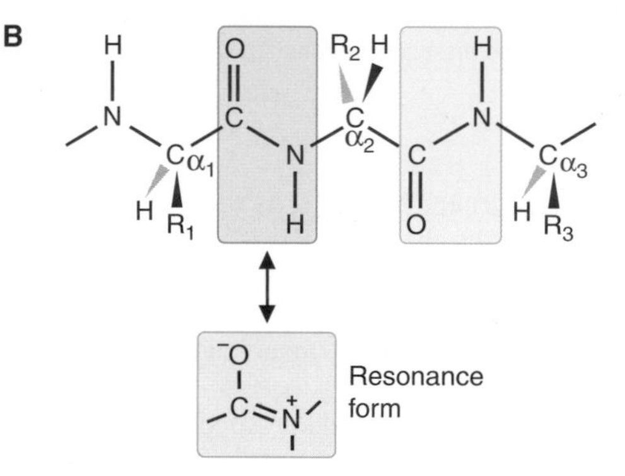

FIGURE 2.6. **The peptide bond. A.** Amino acids in a polypeptide chain are joined through peptide bonds between the carboxyl group of one amino acid and the amino group of the next amino acid in the sequence. **B.** Because of the resonance nature of the peptide bond, the C and N of the peptide bonds form a series of rigid planes. Rotation within allowed torsion angles can occur around the bonds attached to the α-carbon.

A. General aspects of protein structure (Fig. 2.7)

The **linear sequence** of amino acid residues in a polypeptide chain determines the three-dimensional configuration of a protein, and the **structure** of a protein determines its function.

1. The **primary structure** is the sequence of amino acids along the polypeptide chain.
   - **a.** By convention, the **sequence** is written from left to right, starting with the **N-terminal** amino acid.
   - **b.** Because there are no dissociable protons in peptide bonds, the **charges** on a polypeptide chain are due only to the N-terminal amino group, the C-terminal carboxyl group, and the side chains on amino acid residues (see Fig. 2.4).
   - **c.** A protein will **migrate** in an **electric field**, depending on the sum of its charges at a given pH (the net charge).
     - **(1) Positively charged** proteins are cations and migrate toward the **cathode** (–).
     - **(2) Negatively charged** proteins are anions and migrate toward the **anode** (+).
     - **(3)** At the **isoelectric pH** (the pI), the net charge is zero, and the protein does not migrate.
2. The **secondary structure** includes various types of local conformations in which the atoms of the side chains are not involved. Secondary structures are formed by a regular repeating pattern of hydrogen bond formation between backbone atoms.
   - **a.** An **α-helix** is generated when each carbonyl of a peptide bond forms a **hydrogen bond** with the–NH of a peptide bond four amino acid residues further along the chain (Fig. 2.8).
     - **(1)** The side chains of the amino acid residues extend outward from the central axis of the rodlike structure.
     - **(2)** The α-helix is disrupted by proline residues, in which the ring imposes geometric constraints, and by regions in which numerous amino acid residues have charged groups or large, bulky side chains.

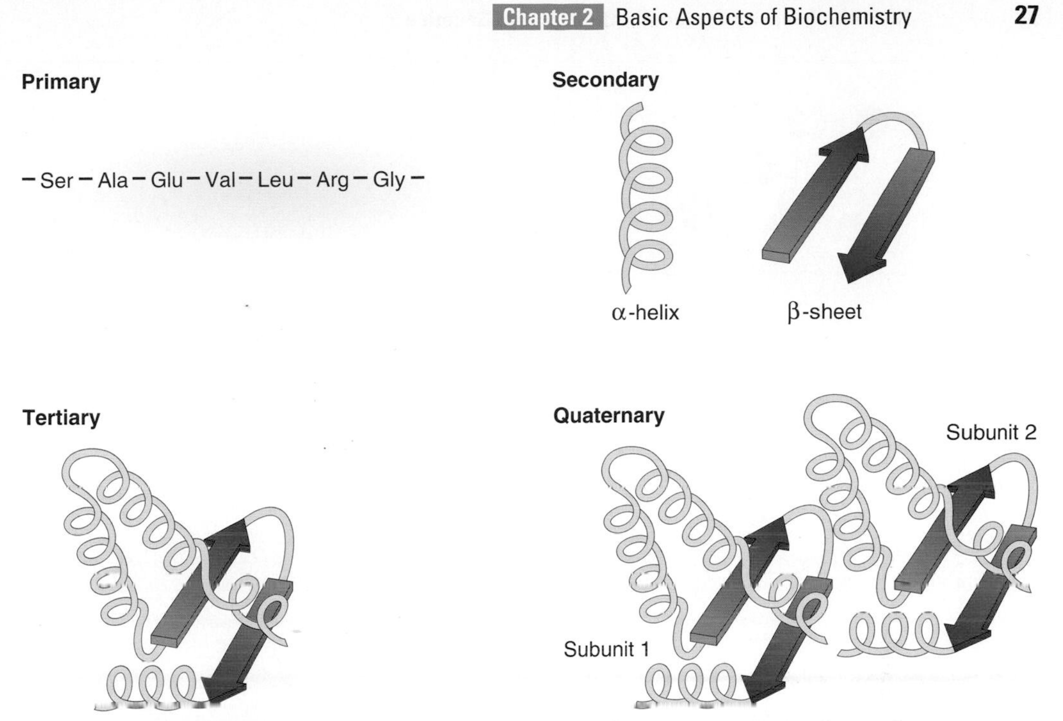

FIGURE 2.7. Schematic diagram of the primary, secondary, tertiary, and quaternary structures of a protein.

b. **β-Sheets** are formed by **hydrogen bonds** between two extended polypeptide chains or between two regions of a single chain that folds back on itself (Fig. 2.9).

(1) These **interactions** are between the **carbonyl** of one peptide bond and the **–NH** of another.

(2) The chains may run in the same direction (parallel) or in opposite directions (antiparallel).

c. **Supersecondary structures**

(1) Certain folding patterns involving α-helices and β-sheets are frequently found, and include the **helix-turn-helix**, the **leucine zipper**, and the **zinc finger**.

(2) Other types of **helices** or **loops** and **turns** can occur that differ from one protein to another (random coils).

3. The **tertiary structure** of a protein refers to its overall **three-dimensional conformation**. It is produced by interactions between amino acid residues that may be located at a considerable distance from each other in the primary sequence of the polypeptide chain (Fig. 2.10).

a. **Hydrophobic amino acid residues** tend to collect in the **interior** of globular proteins, where they exclude water, whereas **hydrophilic residues** are usually found on the **surface**, where they interact with water.

b. The types of interactions between amino acid residues that produce the **three-dimensional shape** of a protein include **hydrophobic** interactions, **electrostatic** interactions, and **hydrogen bonds**, all of which are **noncovalent**. **Covalent disulfide bonds** also occur.

4. **The quaternary structure** refers to the spatial arrangement of **subunits** in a protein that consists of more than one polypeptide chain (see Fig. 2.10). The subunits are joined together by the same types of **noncovalent interactions** that join various segments of a single chain to form its tertiary structure, as well as disulfide bonds.

5. **Denaturation and renaturation**

a. Proteins can be **denatured** by agents such as **heat** and **urea** that cause **unfolding** of polypeptide chains without causing hydrolysis of peptide bonds.

b. The denaturing agents destroy secondary and tertiary structures, without affecting the primary structure.

c. If a denatured protein returns to its native state after the denaturing agent is removed, the process is called **renaturation**.

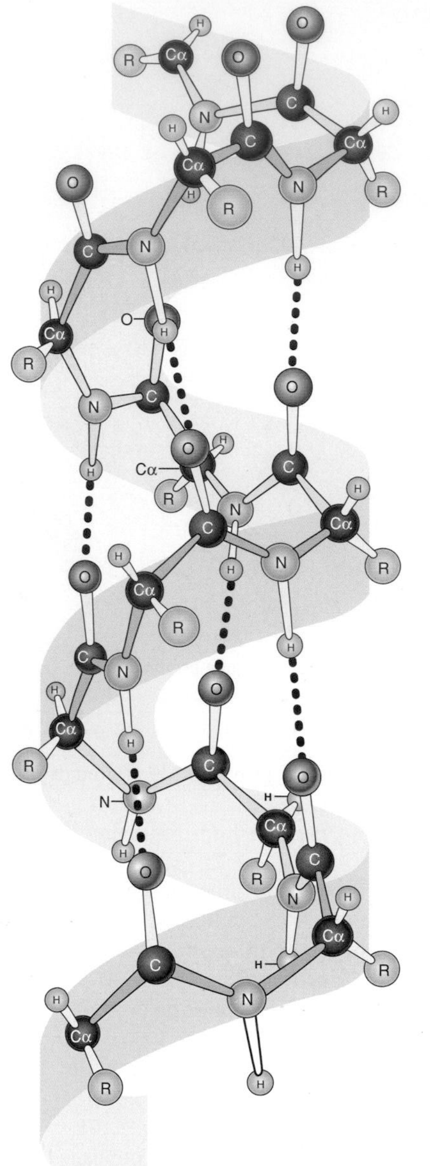

FIGURE 2.8. The α-helix. Each oxygen atom of a carbonyl group of a peptide bond forms a hydrogen bond with the hydrogen atom attached to a nitrogen atom in a peptide bond four amino acids further along the chain. The result is a highly compact and rigid structure.

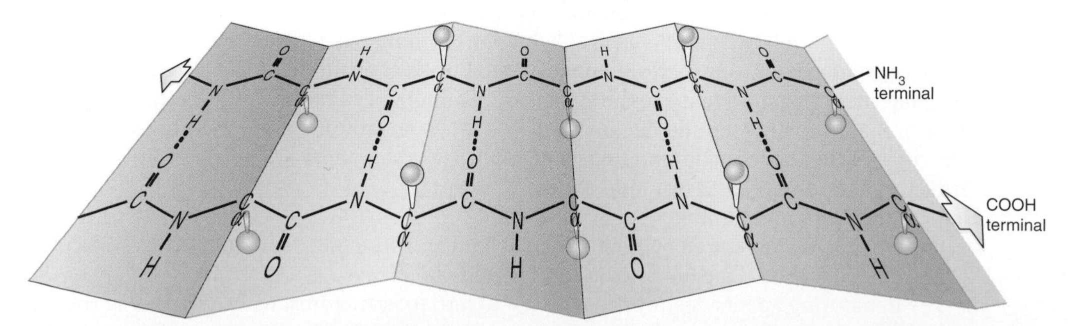

FIGURE 2.9. The structure of an antiparallel β-sheet. In this case, the chains are oriented in opposite directions. The large arrows show the direction of the carboxy terminal. The amino acid side chains (R) in one strand are trans to each other, and alternate above and below the plane of the sheet.

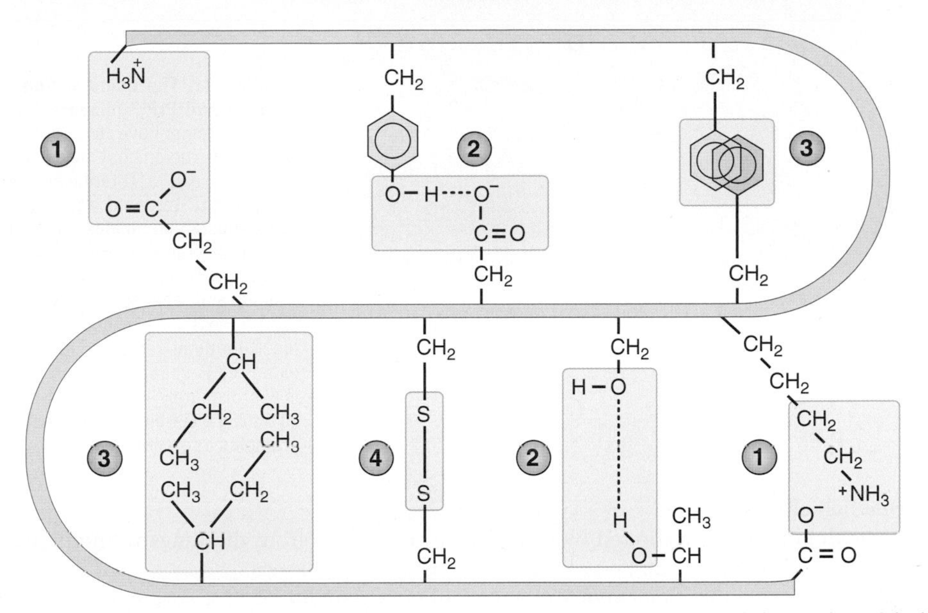

FIGURE 2.10. **Interactions between amino acid residues in a polypeptide chain.** ❶ Electrostatic interactions; ❷ hydrogen bonds; ❸ hydrophobic interactions; ❹ a disulfide bond.

6. **Posttranslational modifications** of proteins occur after the protein has been synthesized on the ribosome. Phosphorylation, glycosylation, ADP ribosylation, methylation, hydroxylation, and acetylation affect the charge and the interactions between amino acid residues, **altering the three-dimensional configuration** and, thus, the function of the protein.

CLINICAL CORRELATES

Accumulated misfolded proteins occur in a variety of diseases. This occurs from the precipitation of proteins within tissues, thereby interfering with tissue function. **Prion diseases** (such as mad cow disease, scrapie, Kuru, and Creutzfeldt–Jakob disease, all examples of **spongiform encephalopathies**) result from an alternative tertiary structure for a neuronal protein, which will polymerize into an insoluble structure that cannot be degraded (Fig. 2.11). These diseases can result from mutations (familial Creutzfeldt–Jakob disease) or infection with an altered prion product (mad cow disease and Kuru), which usually occurs through ingestion of the altered protein. **Amyloidosis** is a term used to describe a variety of disorders, the common feature of which is that **amyloid proteins** are produced, which accumulate and precipitate in various organs owing to alterations in secondary and tertiary structures. Examples include AL amyloidosis, consisting of immunoglobulin light chains that are overproduced and precipitate, and Aβ, a β-amyloid protein found in brain lesions of Alzheimer disease.

B. Hemoglobin (Fig. 2.12)

1. Structure of hemoglobin

Adult hemoglobin (HbA) consists of **four polypeptide chains** (two α chains and two β chains), each containing a molecule of **heme**.

a. The **α chains and β chains** of HbA are **similar** in three-dimensional configuration to each other and to the single chain of muscle myoglobin although their amino acid sequences differ.

b. **Eight regions of α-helix** occur in each chain, labeled A through H.

c. **Heme** fits into a crevice in each globin chain and interacts with two histidine residues.

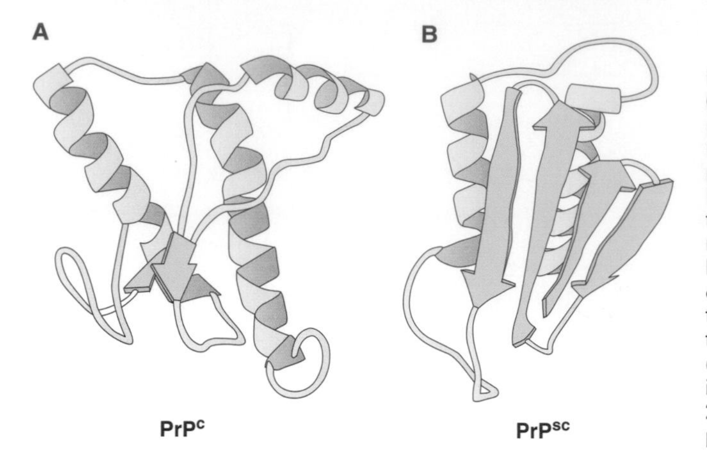

FIGURE 2.11. **The conformation of PrPc (normal) and PrPsc (disease form).** The prion proteins have two domains: an N-terminal region that binds four Cu^{+2} per chain and a C-terminal region. In PrPc, the C-terminal regions contain three substantial helices and two three-residue β-strands joined by two to three hydrogen bonds (approximately 40% α-helix and almost no β-sheet structure). It exists as a monomer. In PrPsc, the C-terminal region is folded into an extensive β-sheet. The overall structure is approximately 40% to 50% β-sheet and 20% to 30% α-helices. This conformation promotes aggregation.

2. Function of hemoglobin

Hemoglobin, found in the red blood cells, carries oxygen from the lungs to the tissues, and returns carbon dioxide and protons from the tissues to the lungs.

a. The **oxygen saturation curve** for hemoglobin is **sigmoidal** (Fig. 2.13).

(1) **Each heme** binds **one O_2** molecule, for a total of four O_2 molecules per HbA molecule. HbA changes from the taut or tense **(T) form** to the relaxed **(R) form** when oxygen binds.

(2) Binding of O_2 to one heme group in hemoglobin increases the affinity for O_2 of its other heme groups. This effect produces the sigmoidal oxygen saturation curve, and is known as positive cooperativity.

b. The binding of **protons** to HbA stimulates the release of O_2, a manifestation of **the Bohr effect** (see Fig. 2.13).

(1) Thus, O_2 is readily released in the tissues where $[H^+]$ is high due to the production of CO_2 by metabolic processes.

(2) These reactions are reversed in the lungs. O_2 binds to HbA, and CO_2 is exhaled (Fig. 2.14).

c. Covalent binding of **CO_2** to HbA in the tissues also causes the release of O_2.

d. Binding of **2,3-bisphosphoglycerate (BPG)**, a side product of glycolysis in red blood cells, decreases the affinity of HbA for O_2. Consequently, O_2 is more readily released in tissues when BPG is bound to HbA (see Fig. 2.13).

e. **Fetal hemoglobin (HbF)**, composed of two α subunits and and two γ subunits, has a lower affinity for BPG than does HbA, and, therefore, HbF has a higher affinity for O_2 than does HbA.

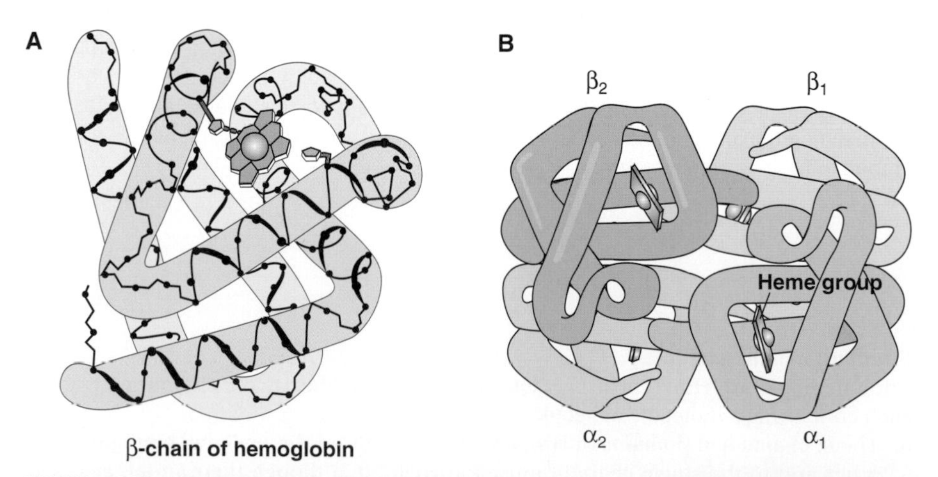

FIGURE 2.12. **The structure of the β chain of hemoglobin (panel A) and hemoglobin (panel B).** Cylindrical regions contain α-helices. The planar structure near the top center of the polypeptide chain is heme. (From: Ferscht A. *Structure and Mechanism in Protein Science.* New York, NY: W.H. Freeman and Company; 1999, with permission.)

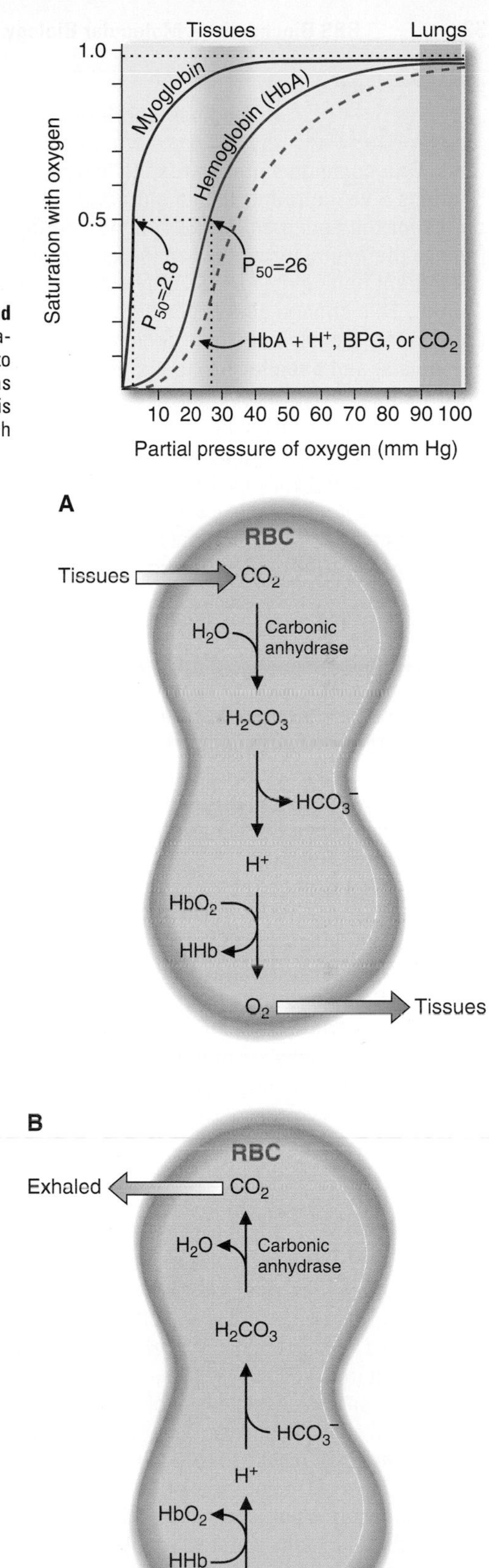

FIGURE 2.13. **Oxygen saturation curves for myoglobin and adult hemoglobin (HbA).** Myoglobin has a hyperbolic saturation curve. HbA has a sigmoidal curve. The HbA curve shifts to the right at lower pH, with higher concentrations of BPG, or as CO_2 binds to HbA in the tissues. Under these conditions, O_2 is released more readily. P_{50} is the partial pressure of O_2 at which half-saturation with O_2 occurs.

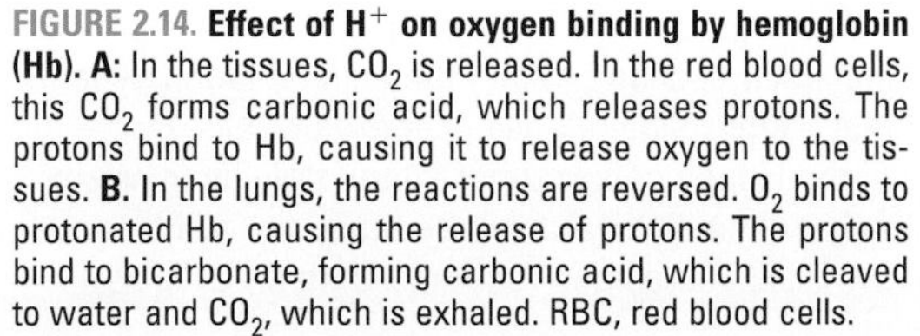

FIGURE 2.14. **Effect of H^+ on oxygen binding by hemoglobin (Hb). A:** In the tissues, CO_2 is released. In the red blood cells, this CO_2 forms carbonic acid, which releases protons. The protons bind to Hb, causing it to release oxygen to the tissues. **B.** In the lungs, the reactions are reversed. O_2 binds to protonated Hb, causing the release of protons. The protons bind to bicarbonate, forming carbonic acid, which is cleaved to water and CO_2, which is exhaled. RBC, red blood cells.

CLINICAL CORRELATES

Hemoglobinopathies result from **mutations** that produce alterations in the structure of hemoglobin. There are many described hemoglobinopathies. One common mutation results in **sickle cell anemia**, in which the β chain of hemoglobin contains a valine rather than a glutamate at position 6 (designated as E6V, using the single-letter codes for the amino acids. E6V means that the glutamate [E] at position 6 in the amino acid chain [where the amino terminal is amino acid number 1] has been replaced by a valine [V]). Thus, in the mutant hemoglobin (HbS), a hydrophobic amino acid replaces an amino acid with a negative charge. This change allows deoxygenated molecules of HbS to polymerize. Red blood cells that contain large complexes of HbS molecules can assume a sickle shape. These cells undergo **hemolysis**, and **anemia** results. Painful **vaso-occlusive crises** also occur, and **end-organ damage** may result.

In addition to hemoglobinopathies, which will interfere with oxygen delivery to the tissues, **carbon monoxide poisoning** will also do the same. Carbon monoxide poisoning results from the tight binding of CO to the iron in heme, preventing both oxygen delivery to the tissues and electron transfer through cytochrome oxidase (which requires oxygen) in the electron transfer chain. The affinity of CO for hemoglobin is 230 times greater than that of oxygen for hemoglobin. CO is a combustion product of fuels (such as cigarette smoke and internal combustion engines).

C. Collagen refers to a group of very similar structural proteins that are found, for example, in the extracellular matrix, the vitreous humor of the eye, and in bone and cartilage.

1. Structure of collagen

- **a.** Collagen consists of **three chains** that wind around each other to form a **triple helix**.
- **b.** Collagen contains approximately 1,000 amino acids, one-third of which are **glycine**. The sequence Gly-X-Y frequently occurs, in which X is often **proline** and Y is **hydroxyproline** or **hydroxylysine**.

2. Synthesis of collagen

- **a.** The polypeptide chains of **preprocollagen** are synthesized on the **rough endoplasmic reticulum**, and the signal (pre) sequence is cleaved.
- **b.** **Proline** and **lysine** residues are **hydroxylated** by a reaction that requires O_2 and **vitamin C**.
- **c.** **Galactose** and **glucose** are added to hydroxylysine residues.
- **d.** The **triple helix** forms, procollagen is secreted from the cell, and cleaved to form collagen.
- **e.** **Cross-links** are produced. The side chains of lysine and hydroxylysine residues are oxidized to form aldehydes, which can undergo aldol condensation or form Schiff bases with the amino groups of lysine residues.

CLINICAL CORRELATES

Problems associated with connective tissue and structural proteins are present in a number of diseases. Mutations in the synthesis and processing of **collagen** occur in **Ehlers–Danlos syndrome**, which is characterized by abnormalities of skin, ligaments, and internal organs. The skin is fragile and often stretches easily. Joint laxity occurs. Abnormalities in type I **collagen** genes have been found in **osteogenesis imperfecta**. In this condition, bones are fragile and are readily fractured. In **scurvy**, which is due to **vitamin C deficiency**, hydroxylation of proline residues is decreased, and an unstable form of **collagen** is produced. Bones, teeth, blood vessels, and other structures rich in collagen develop abnormally. Bleeding gums and poor wound healing are often observed. **Marfan syndrome** results from a defect in the protein **fibrillin**. Clinical manifestations are variable but often include a tall stature with arachnodactyly (long, thin fingers and toes), mitral valve prolapse, and lens dislocation. **Alport syndrome**, which leads to kidney failure, is due to mutations in collagen, type IV genes, which forms a meshlike collagen network that supports the kidney glomerular cells. In the absence of this collagen, the basement membranes of the kidney cells cannot adequately filter waste products from the blood. **Hereditary spherocytosis** leads to hemolytic anemia because the red blood cells have a sphere shape instead of being a biconcave disk. The spleen recognizes these cells as foreign and removes them from the circulation, leading to anemia. This disorder can be caused by mutations in a variety of red blood cell membrane proteins, such as **spectrin** and **ankyrin**. Splenomegaly is observed in hereditary spherocytosis, and the treatment often consists of removal of the spleen. **Familial hypertrophic cardiomyopathy (FHC)** that leads to a thickening of the heart muscle, which may lead to sudden death under exercising conditions, is due to a mutation in either of various muscle sarcomeric proteins, the most frequent of which is the β-myosin heavy chain. These sarcomeric proteins provide a structural role within the muscle, as well as participate in the contraction of the muscle.

D. Insulin

1. **Structure of insulin** (Fig. 2.15)
   a. Insulin is a polypeptide hormone that is produced by the β **cells** of the **pancreas**. It has 51 amino acids in **two polypeptide chains**, which are linked by two **disulfide bridges**.
2. **Synthesis of insulin**
   a. **Preproinsulin** is synthesized on the rough endoplasmic reticulum and the pre (signal) sequence is removed to form proinsulin.
   b. In secretory granules, **proinsulin** is cleaved, and the **C-peptide** is released. The remainder of the molecule forms the active hormone.
3. **Function of insulin**
   a. Insulin is released from the pancreas when blood glucose levels are elevated, such as eating a meal containing carbohydrates.
   b. Insulin promotes the **transport of glucose** into muscle and fat cells.
   c. Insulin promotes the **storage of energy**; glycogen synthesis is stimulated in the liver and muscle, and triacylglycerol synthesis is stimulated in the liver.

CLINICAL CORRELATES

Diabetes mellitus, which is due to a **deficiency of insulin** (Type 1) or to a decreased secretion of insulin or **resistance** of tissues to insulin action (Type 2), results in **hyperglycemia**. Autoimmunity plays a role in the etiology of Type 1 diabetes. In this condition, the plasma usually contains antibodies to islet cells of the pancreas, including those that produce insulin.

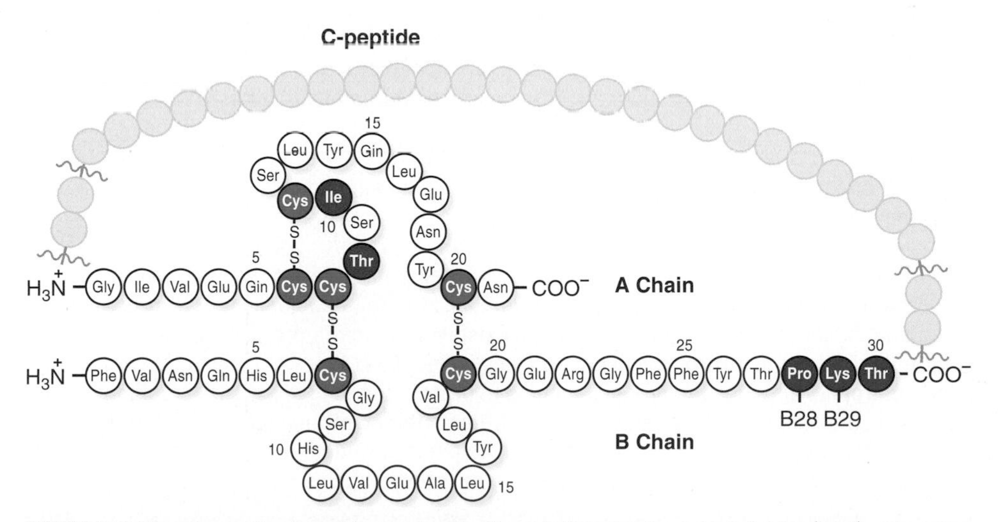

FIGURE 2.15. **The primary structure of human insulin.** The substituted amino acids in bovine (beef) and porcine (pork) insulin are shown in red. The cysteine residues, which form the disulfide bonds (in blue) are invariant. Proinsulin is converted to insulin by proteolytic cleavage of certain peptide bonds (squiggly lines in the figure). The cleavage removes a few amino acids and the 31 amino acid C-peptide that connects the A and B chains. The active insulin molecule thus has two nonidentical chains.

V. ENZYMES

- A major role of proteins is to serve as enzymes, the catalysts of biochemical reactions. They do so by reducing the Gibbs free energy of activation, $\Delta G^{\ddagger}$, making it easier for the reaction to reach its transition state.
- The active sites of enzymes are the regions where substrates bind and are converted to products, which are released.
- The rate (v) of many enzyme-catalyzed reactions can be described by the Michaelis–Menten equation. For enzymes that exhibit Michaelis–Menten kinetics, plots of velocity-versus-substrate concentration are hyperbolic.
- The Michaelis–Menten equation can be rearranged to give the Lineweaver–Burk equation.
- Competitive inhibitors compete with the substrate for binding at the active site of the enzyme.
- Noncompetitive inhibitors bind to the enzyme or the enzyme–substrate complex at a site different from the active site.
- Allosteric enzymes bind activators or inhibitors at sites other than the active site. Plots of the velocity-versus-substrate concentration for allosteric enzymes produce curves that are sigmoidal.

A. General properties of enzymes

1. The reactions of the cell would not occur rapidly enough to sustain life if enzyme catalysts were not present.
2. **Substrates** bind at the **active sites** of enzymes, where they are converted to products and released.
3. Enzymes are usually **highly specific** for their substrates and products.
   a. Many enzymes recognize only a single compound as a substrate.
   b. Some enzymes, such as those involved in digestion, are less specific.
4. Many enzymes require **cofactors** that frequently are **metal ions** or derivatives of **vitamins**.
5. Enzymes **decrease** the **energy of activation** for a reaction. They do not affect the equilibrium concentrations of the substrates and products, nor do they change the overall **Gibbs free energy change** for the reaction.

CLINICAL CORRELATES Thousands of diseases related to **deficient or defective enzymes** occur. Many of them are **rare**. For example, in **phenylketonuria** (which has an incidence of 1 in 10,000 births in whites and Asians), the enzyme **phenylalanine hydroxylase**, which converts phenylalanine to tyrosine, is deficient. Phenylalanine accumulates, and tyrosine becomes an essential amino acid that is required in the diet. **Mental retardation** is a result of this metabolic derangement, in part due to the brain lacking various essential amino acids owing to the elevated levels of phenylalanine in the circulation. A more **common problem** is **lactase deficiency**, which occurs in more than 80% of Native-, African-, and Asian Americans. Lactose is not digested at a normal rate and accumulates in the gut where it is metabolized by bacteria. **Bloating, abdominal cramps**, and **watery diarrhea** result. **Emphysema** may result from an inherited deficiency of **α1-antitrypsin**, an enzyme that inhibits elastase action in the lungs. Elastase is a serine protease found in neutrophils that utilize the enzyme to destroy inhaled organisms in the air. At times, the elastase may escape from the neutrophil, and then the protease begins to destroy the lung cells. The circulating protein α1-antitrypsin blocks the action of elastase, and protects the lung from damage. Cigarette smoke contains oxidizing agents that will destroy a key methionine residue in α1-antitrypsin, and destroys α1-antitrypsin activity. Smoking for an extended period of time may lead to emphysema.

B. Dependence of velocity on [E], [S], temperature, and pH

1. The **velocity** of a reaction, v, **increases with the enzyme concentration**, **[E]**, if the substrate concentration, **[S]**, is constant.

2. If [E] is constant, v increases with [S] until the **maximum velocity,** V_{max}, is attained. At V_{max}, all the **active sites** of the enzyme are **saturated** with substrate.
3. The **velocity** of a reaction **increases with temperature** until a maximum is reached, after which the velocity decreases owing to the denaturation of the enzyme.
4. Each enzyme-catalyzed reaction has an **optimal pH** at which appropriate charges are present on both the enzyme and the substrate, and the velocity is at a maximum. Changes in the pH can alter these charges so that the reaction proceeds at a slower rate. If the pH is too high or too low, the enzyme can also undergo **denaturation.**

C. The Michaelis–Menten equation

1. If, during a reaction, an **enzyme–substrate complex** is formed that **dissociates** (to re-form the free enzyme and the substrate) or **reacts** (to release the product and regenerate the free enzyme):

$$E + S \underset{k_2}{\overset{k_1}{\rightleftharpoons}} ES \overset{k_3}{\rightarrow} E + P$$

where E is the enzyme, S is the substrate, ES is the enzyme–substrate complex, P is the product, and k_1, k_2, and k_3 are rate constants.

2. From this concept, the **Michaelis–Menten** equation was derived:

$$v = \frac{V_{max}[S]}{K_m + [S]}$$

where $K_m = (k_2 + k_3)/k_1$ and V_{max} is the maximum velocity.

3. The rate of formation of products (the velocity of the reaction) is related to the **concentration of the enzyme–substrate complex:**

$$v = k_3[ES]$$

V_{max} is reached when all of the enzyme is in the enzyme–substrate complex.

4. K_m is the substrate concentration at which $v = ½\ V_{max}$.

When $[S] = K_m$, substitution of K_m for [S] in the Michaelis-Menten equation yields $v = ½\ V_{max}$.

5. When the velocity is plotted versus [S], a **hyperbolic curve** is produced (Fig. 2.16A).

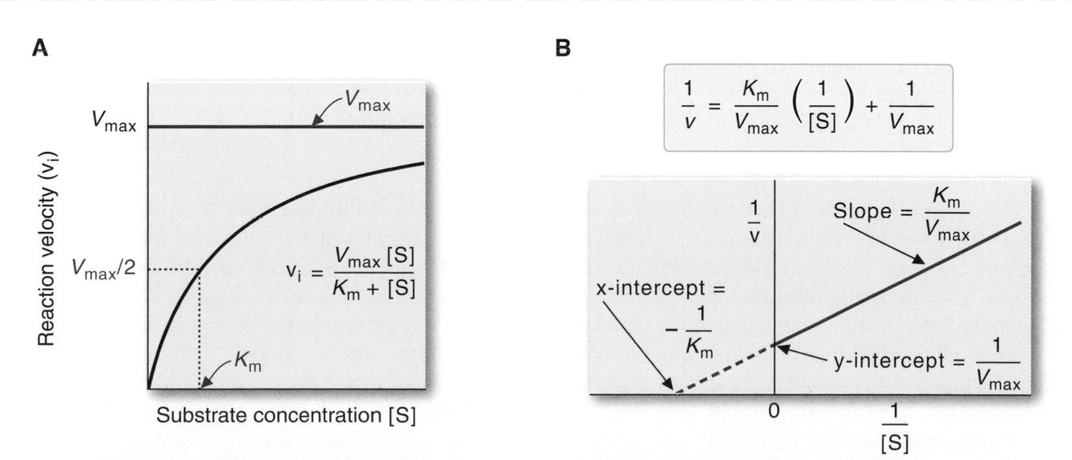

FIGURE 2.16. The velocity of an enzyme-catalyzed reaction that exhibits Michaelis–Menten kinetics. A. Velocity (v) versus substrate concentration ([S]). **B.** Lineweaver–Burk plot. Note the points on each plot from which V_{max} and K_m can be determined. V_{max}, maximum velocity; K_m, the substrate concentration at ½ V_{max}.

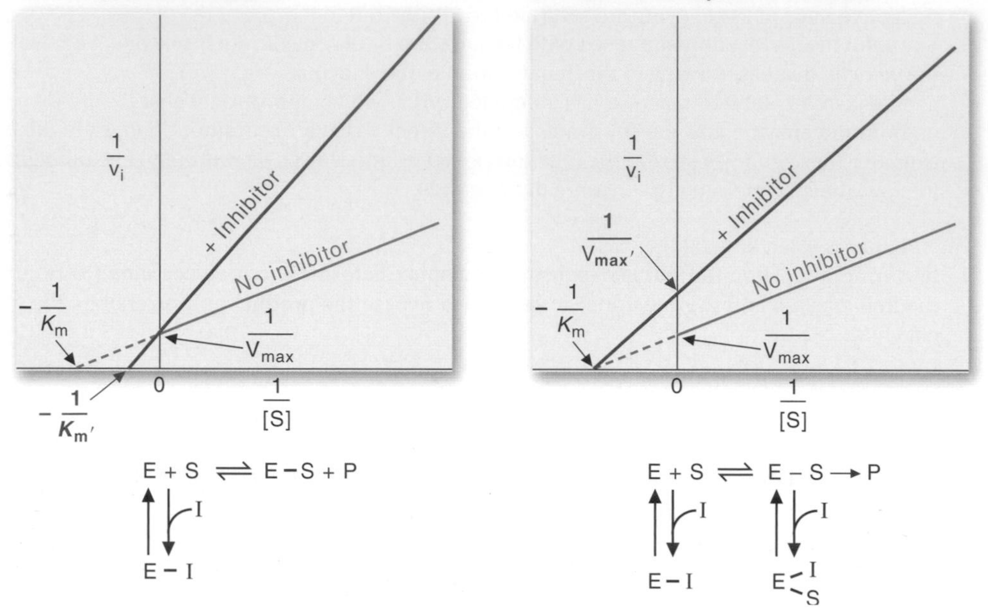

FIGURE 2.17. **Effect of inhibitors on Lineweaver–Burk plots. A.** Competitive inhibition. Note the intersection of lines on the *y*-axis, indicating the same V_{max}. **B.** Pure noncompetitive inhibition (in which the inhibitor binds to E and ES with the same affinity). Note the intersection of the lines on the *x*-axis, indicating the same K_m. If the affinities of the inhibitor differ in binding to the E or ES forms, the lines will not intersect on the *x*-axis and the apparent K_m ($K_{m'}$) will differ from K_m. $V_{max'}$, the apparent V_{max}.

D. The Lineweaver–Burk equation (see Fig. 2.16B)

Because of the difficulty in determining V_{max} from a hyperbolic curve, the Michaelis-Menten equation was transformed into an equation for a straight line by Lineweaver and Burk.

E. Inhibitors of enzymes decrease the rate of enzymatic reactions.

1. **Competitive inhibitors** compete with the substrate for the active site of the enzyme and form an enzyme-substrate complex, EI (see Fig. 2.17A).
   a. Competitive inhibition is reversed by increasing [S].
   b. V_{max} remains the same, but the **apparent K_m (K'_m) is increased.**
   c. For Lineweaver-Burk plots, **lines** for the inhibited reaction **intersect** on the ***y*-axis** with those for the uninhibited reaction.
2. **Noncompetitive inhibitors** bind to the enzyme or the enzyme-substrate complex at a site different from the active site, decreasing the activity of the enzyme (see Fig. 2.17B). Thus, V_{max} **is decreased.**
3. **Irreversible inhibitors** bind tightly to the enzyme and **inactivate** it.

CLINICAL CORRELATES **Drugs** are frequently used therapeutically to **inhibit enzymes**; for example, 5-fluorouracil (5-FU) is used to inhibit the enzyme thymidylate synthetase. This enzyme converts dUMP to dTMP, which ultimately provides the thymine for DNA synthesis. **5-FU** is used as a **chemotherapeutic agent** to inhibit the proliferation of cancer cells. **Aspirin** irreversibly inhibits the enzyme cyclooxygenase, and interferes with the generation of prostaglandins and thromboxanes, the secondary mediators of pain signaling. **Allopurinol** is a suicide inhibitor of xanthine oxidase, and is used to treat **gout**. Xanthine oxidase will recognize allopurinol as a substrate and oxidize it to create oxypurinol, which binds so tightly to the active site that it is not released, and then inhibits further reactions by the enzyme. A drug used to treat **alcoholism** irreversibly inhibits the enzyme aldehyde dehydrogenase, leading to an accumulation of acetaldehyde whenever alcohol is ingested. The acetaldehyde leads to "hangover" symptoms, and may deter the patient from drinking alcohol.

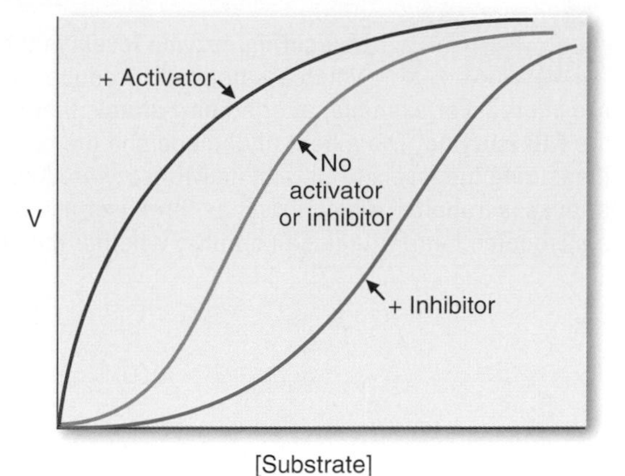

FIGURE 2.18. Effect of activators and inhibitors on an allosteric enzyme.

F. Allosteric enzymes

1. Allosteric enzymes **bind activators or inhibitors** at sites other than the active site (Fig. 2.18).
2. **Sigmoidal curves** are generated by plots of v versus [S].
   a. An allosteric enzyme has two or more subunits each with **substrate-binding sites** that exhibit **cooperativity**. Binding of a substrate molecule at one site facilitates binding of other substrate molecules at other sites.
      (1) **Allosteric activators** cause the enzyme to bind substrate more readily (shift the kinetic curve to the left, thereby decreasing the apparent K_m).
      (2) **Allosteric inhibitors** cause the enzyme to bind substrate less readily (shift the curve to the right, increasing the apparent K_m).
   b. Similar effects occur during O_2 binding to **hemoglobin** (see Fig. 2.13).

G. Regulation of enzyme activity by covalent modification

Enzyme activity may increase or decrease after the covalent addition of a chemical group.

1. Phosphorylation affects many enzymes.
   a. Pyruvate dehydrogenase and glycogen synthase are inhibited by phosphorylation.
   b. Glycogen phosphorylase is activated by phosphorylation.
2. Phosphatases that remove the phosphate groups alter the activities of these enzymes.
3. Phosphorylation introduces negative charges to the protein, which may alter the secondary and tertiary structures.

H. Regulation by protein-protein interactions

Proteins can bind to enzymes, altering their activity. For example, regulatory subunits inhibit the activity of protein kinase A. When these regulatory subunits bind cyclic AMP (cAMP) and are released from the enzyme, the catalytic subunits become active.

I. Isoenzymes

1. Isoenzymes (or isozymes) are enzymes that catalyze the same reaction but differ in their amino acid sequence and, therefore, in many of their properties.
2. Tissues contain **characteristic isozymes** or mixtures of isozymes. Enzymes such as lactate dehydrogenase and creatine kinase (CK) differ from one tissue to another.
   a. **Lactate dehydrogenase** contains four subunits. Each subunit may be either of the heart (H) or the muscle (M) type. Five isozymes exist (HHHH, HHHM, HHMM, HMMM, and MMMM).
   b. **CK** contains two subunits. Each subunit may be either of the muscle (M) or the brain (B) type. Three isozymes exist (MM, MB, and BB). The MB fraction is most prevalent in heart muscle.

CLINICAL CORRELATES **Measuring enzyme levels in the blood can monitor tissue damage.** Enzymes, which are normally produced in cells, are released into the blood when cells are injured. For example, after a **heart attack**, there is an increase in blood levels of **CK**, particularly the **MB isozyme**. The extent of damage and the rate of recovery can be estimated by periodically measuring the levels of CK and its MB isozyme. Another isozyme that is used diagnostically for heart attacks is troponin, which exists as three isozymes, I, C, and T. Measurement of the I or T forms after a suspected heart attack can confirm whether the heart muscle cells were damaged or not.

Review Test

Questions 1 to 10 examine your basic knowledge of structures and reaction types and are not in the standard clinical vignette format.

Questions 11 to 35 are clinically relevant, USMLE-style questions.

Basic Knowledge Questions

1. The bonds labeled A and B in the compound shown are best described as which one of the following?

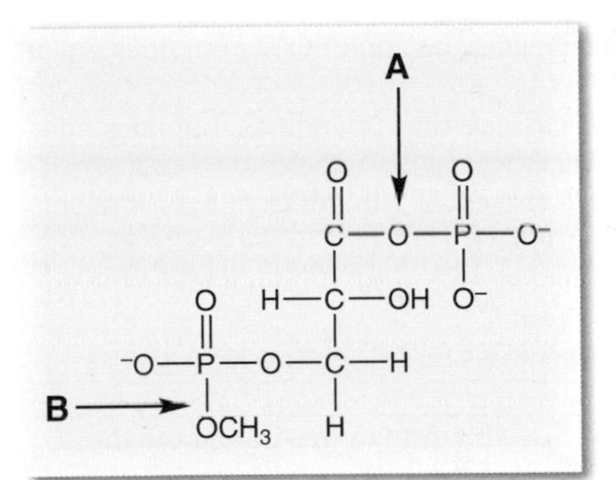

	Compound A	Compound B
A	Anhydride	Ester
B	Ester	Anhydride
C	Ether	Ester
D	Ester	Ether
E	Phosphodiester	Anhydride
F	Anhydride	Phosphodiester

2. Both β-hydroxybutyrate and acetoacetate are ketone bodies, which are produced by the liver under conditions of extended fasting. The two ketone bodies are easily interconverted depending upon the conditions present within the mitochondria, where they are synthesized. The conversion of β-hydroxybutyrate to acetoacetate occurs by what type of reaction?

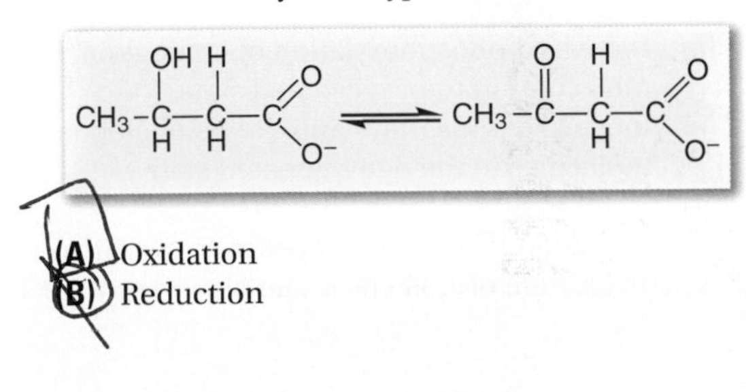

(A) Oxidation
(B) Reduction
(C) Dehydration
(D) Dehydroxylation
(E) Decarboxylation

3. When the pH of a solution of a weak acid, HA, is equal to the pK_a, the ratio of the concentrations of the salt and the acid ([A^-]/[HA]) is which one of the following?

(A) 0
(B) 1
(C) 2
(D) 3
(E) 4

4. Human serum albumin, the most abundant blood protein, has multiple roles, including acting as a buffer to help maintain blood pH. Albumin can act as a buffer because of which one of the following?

(A) The protein contains a large number of amino acids.
(B) The protein contains many amino acid residues with different pK_a values.
(C) The amino and carboxyl ends of albumin can donate and accept protons in the range of physiologic pH.
(D) Albumin contains peptide bonds that readily hydrolyze, consuming hydrogen and hydroxyl ions.
(E) Albumin contains a large number of hydrogen bonds in α-helices, which can accept and donate protons.

Questions 5 and 6 are based on the hexapeptide with the following sequence: D-A-S-E-V-R

5. The C-terminal amino acid of the hexapeptide is which one of the following?

(A) Ala
(B) Asn
(C) Asp
(D) Arg
(E) Glu

6. At physiologic pH (7.4), this hexapeptide will contain a net charge of which one of the following?

(A) -2
(B) -1

(C) 0
(D) +1
(E) +2

7. Which one of the following types of bonds is covalent?

(A) Hydrophobic
(B) Hydrogen
(C) Disulfide
(D) Electrostatic
(E) Van der Waals

8. One method to separate proteins is by charge, through a suitable gel-like substance. If normal HbA and sickle cell hemoglobin (HbS) were placed on such a gel, which molecule would migrate more rapidly to the positive pole of the gel?

(A) HbA
(B) HbS
(C) Both would have the same charge, so there is no difference in migration.

9. The active site of an enzyme will bind to which set of molecules indicated below?

	Substrate of the reaction	Allosteric inhibitors	Competitive inhibitors	Noncompetitive inhibitors
A	Yes	Yes	Yes	Yes
B	Yes	No	Yes	Yes
C	Yes	No	Yes	No
D	No	No	No	No
E	No	Yes	No	Yes

10. An enzyme catalyzing the reaction

$$E + A \rightleftharpoons EA \rightarrow E + P$$

was mixed with 4 mM substrate (compound A). The initial rate of product formation was 25% of V_{max}. The K_m for the enzyme is which one of the following?

(A) 2 mM
(B) 4 mM
(C) 9 mM
(D) 12 mM
(E) 25 mM

Board-style Questions

11. The liver enzyme glucokinase catalyzes the phosphorylation of glucose to glucose 6-phosphate. The value of K_m for glucose is about 7 mM. Blood glucose is 5 mM under fasting conditions, and can rise in the liver to 20 mM after a high-carbohydrate meal. Therefore, if a person who is fasting eats a high-carbohydrate meal, the velocity of the glucokinase reaction will change in which one of the following ways?

(A) Remain at $<50\% V_{max}$
(B) Remain above 80% V_{max}
(C) Increase from $<50\% V_{max}$ to $>50\% V_{max}$
(D) Decrease from $>50\% V_{max}$ to $<50\% V_{max}$
(E) Remain at V_{max}

12. Malonate is a competitive inhibitor of succinate dehydrogenase, a key enzyme in the Krebs tricarboxylic acid cycle. The presence of malonate will affect the kinetic parameters of succinate dehydrogenase in which one of the following ways?

(A) Increases the apparent K_m but does not affect V_{max}.
(B) Decreases the apparent K_m but does not affect V_{max}.
(C) Decreases V_{max} but does not affect the apparent K_m.
(D) Increases V_{max} but does not affect the apparent K_m.
(E) Decreases both V_{max} and K_m.

13. A young black man was brought to the emergency room (ER) owing to severe pain throughout his body. He had been exercising vigorously when the pain started. He has had such episodes about twice a year for the past 10 years. An analysis of the blood shows a reduced blood cell count (anemia), and odd-looking red blood cells that were no longer concave and looked like an elongated sausage. An underlying cause in the change of shape of these cells is which one of the following?

(A) Increased ionic interactions between hemoglobin molecules in the oxygenated state.
(B) Increased ionic interactions between hemoglobin molecules in the deoxygenated state.
(C) Increased hydrophobic interactions between hemoglobin molecules in the oxygenated state.
(D) Increased hydrophobic interactions between hemoglobin molecules in the deoxygenated state.
(E) Increased phosphorylation of hemoglobin molecules in the oxygenated state.
(F) Increased phosphorylation of hemoglobin molecules in the deoxygenated state.

14. An individual is visiting Mexico City, which is at an altitude of 7,350 feet. The person is having trouble breathing due to difficulty in getting sufficient oxygen to the tissues. Which one of the following treatments might the person try to get hemoglobin to release oxygen more readily?

(A) Take a drug that initiates a metabolic alkalosis.
(B) Take a drug that increases the production of BPG.
(C) Hyperventilate, which will lead to decreased levels of carbon dioxide in the blood.
(D) Take a drug that induces the synthesis of the γ subunits of hemoglobin.
(E) Take a drug that induces the synthesis of the β subunits of hemoglobin.

15. An environmentalist attempted to live in a desolate forest for 6 months, but had to cut his experiment short when he began to suffer from bleeding gums, some teeth falling out, and red spots on the thighs and legs. This individual is suffering from an inability to properly synthesize which one of the following proteins?

(A) Myoglobin
(B) Hemoglobin
(C) Collagen
(D) Insulin
(E) Fibrillin

16. A patient seen in the ER has ingested antifreeze in a suicide attempt. Other than bicarbonate, which one of the following is the major buffer of acids to help maintain the pH in the blood within the range compatible with life?

(A) Hemoglobin
(B) Acetoacetate
(C) β-Hydroxybutyrate
(D) Phosphate
(E) Collagen

17. Which one of the following is the amino acid in hemoglobin that accepts H^+ and allows hemoglobin to act as a buffer to acids?

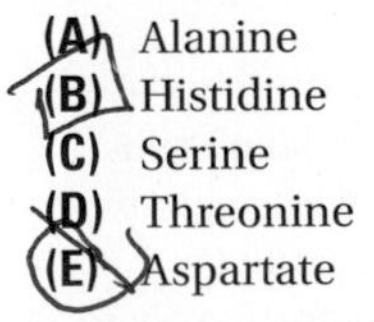

(A) Alanine
(B) Histidine
(C) Serine
(D) Threonine
(E) Aspartate

18. A patient is going skiing high in the Rockies, and is given acetazolamide to protect against altitude sickness. Unfortunately, the patient is also a Type 1 diabetic. He is admitted to the hospital in a worsening ketoacidosis. In which of the following cells has acetazolamide inhibited a reaction that has led to the severity of the metabolic acidosis?

(A) White blood cells
(B) Red blood cells
(C) Lens of the eye
(D) Hepatocyte
(E) Muscle

19. A 23-year-old female patient presents to the ER with a feeling of being unable to catch her breath, light-headedness, and "tingling" of her fingers, toes, and around her mouth. This happens whenever she drives through a tunnel, and that is what set off this episode. Which of the following arterial blood pHs would be most consistent with her diagnosis?

(A) 8.10
(B) 7.55
(C) 7.15
(D) 6.40
(E) 6.10

20. A new antibiotic has been developed that shows a strong affinity for attacking amino acids with a specific orientation in space. In order for it to work well in humans, the antibiotic must be effective against amino acids in which one of the following configurations?

(A) R-configuration
(B) L-configuration
(C) Aromatic ring configuration
(D) Polypeptide chain configuration
(E) D-configuration

21. A 40-year-old tobacco farmer is seen in the ER with bradycardia, profuse sweating, vomiting, increased salivation, and blurred vision. He was spraying his field with malathion when the hose ruptured and he was covered with the malathion. Which of the following types of inhibition of enzymes does this poisoning represent?

(A) Competitive
(B) Noncompetitive
(C) Irreversible
(D) Reversible
(E) The drug does not work by inhibiting enzyme activity.

22. A baby did well with no discernable problems until 3 months of age when he began having cyanotic spells and later had a myocardial infarction. An autopsy revealed a congenital defect, where the left main coronary artery arose from the pulmonary artery instead of the aorta. Of the following, which is the most likely reason that he showed no symptoms until 3 months of age?

(A) He had abnormal α chains of hemoglobin.
(B) He had abnormal β chains of hemoglobin.
(C) He had abnormal γ chains of hemoglobin.
(D) HbF has a lower affinity for BPG.
(E) HbF has a higher affinity for BPG.

23. A 28-year-old female presents with fluctuating fatigue, drooping of her eyelids, difficulty swallowing, and slurred speech. The patient is given a drug that affects an enzyme's activity, and kinetic analysis of the enzyme-catalyzed reaction, in the presence and absence of the drug, is shown below. The effect of this medication can best be described by which set of terms below?

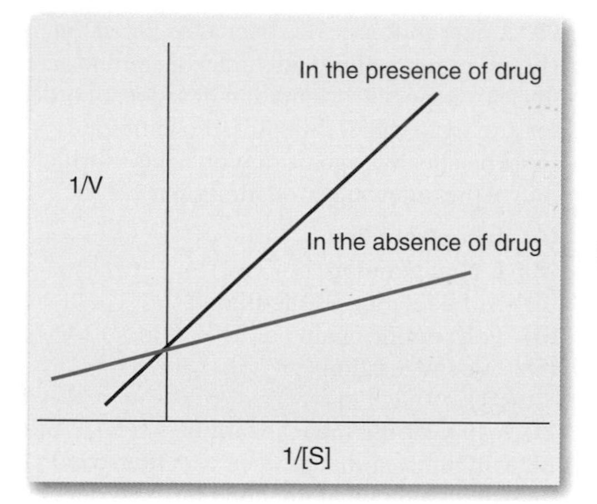

	Type of inhibition	Effect on K_m (as compared to no drug)	Effect on V_{max} (as compared to no drug)
A	Competitive	Increased	No change
B	Competitive	Decreased	Decreased
C	Noncompetitive	Increased	No change
D	Noncompetitive	Decreased	Increased
E	Irreversible	Increased	No change
F	Irreversible	Decreased	Decreased

24. A patient who wanted to go skiing in the Rockies took a medication to combat altitude sickness. A graph of this medication's mechanism of inhibition is shown below. What type of inhibitor is this medication?

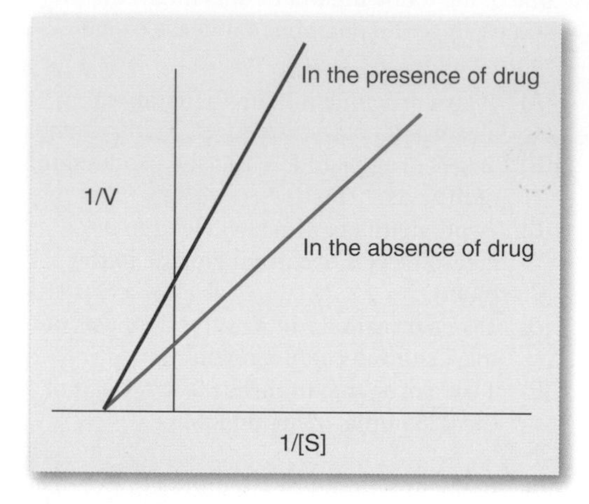

(A) Competitive
(B) Noncompetitive
(C) Irreversible
(D) Allosteric
(E) The drug is not an inhibitor, rather it is an activator.

25. A person with Type 1 diabetes ran out of her prescription insulin and has not been able to inject insulin for the past 3 days. An overproduction of which of the following could cause a metabolic acidosis?

(A) Hemoglobin
(B) Ketone bodies
(C) HCl
(D) Bicarbonate

26. The patient described in the previous question is hyperventilating to compensate for her metabolic acidosis. Which of the following reactions explains this partially compensating respiratory alkalosis?

(A) $H + NH_3 \rightleftharpoons NH_4^+$
(B) $CH_3CHOHCH_2COOH \rightleftharpoons CH_3CHOHCH_2COO^- + H^+$
(C) $CO_2 + H_2O \rightleftharpoons H_2CO_3 \rightleftharpoons H^+ + HCO_3^-$
(D) $H_2O \rightleftharpoons H^+ + HO^-$

27. A pathologist, while doing an autopsy of a patient who died from Creuzfeldt–Jakob syndrome, accidentally cut himself while examining the brain. The pathologist became very concerned for his well-being, due primarily to the possibility of which one of the following materials entering his circulation?

(A) A virus
(B) A protein

(C) A lipid
(D) A bacteria
(E) A polynucleotide

28. Unbeknownst to its owners, a cow recently sacrificed for meat production had mad cow disease. The precipitating event in the cow's brain that led to this disease is which one of the following?

(A) Altered gene expression
(B) Infection of the brain with a virus
(C) Proteolytic cleavage of an existing brain protein
(D) An altered secondary and tertiary structures for an existing brain protein
(E) Loss of the nuclear membrane

29. A patient is taking omeprazole for gastric reflux disease. Omeprazole contains a free sulfydryl group that is critical for its mechanism of action. This drug will most likely act in which one of the following ways?

(A) Reduce an existing sulfhydryl on the intestinal proton pump.
(B) Form a disulfide bond with a methionine on the gastric proton pump.
(C) Form a disulfide bond with a cysteine on the gastric proton pump.
(D) Reduce an existing sulfhydryl group on the gastric proton pump.
(E) Form a disulfide bond with a cysteine on the intestinal proton pump.

30. A family has been using an additional propane heater in their enclosed apartment during the winter months. One morning, a family member is difficult to awake, and when awake, complains of a splitting headache and being very tired. His mucous membranes are also a cherry red color. These symptoms are the result of which one of the following?

(A) Increased oxygen delivery to the tissues
(B) Decreased oxygen delivery to the tissues
(C) Increased blood flow to the brain
(D) Decreased blood flow to the brain
(E) Decreased oxygen affinity to hemoglobin

31. A 53-year-old man, who has been smoking for the past 35 years at a two-pack-a-day rate, visits his physician for a cough that will not go away, and for difficulty in breathing. A chest X-ray rules out cancer, but does display an increased anterior–posterior (AP) diameter, flattened diaphragms, and "air trapping." The patient is told that his condition will not improve, and that he needs to stop smoking to stop the progression of the disease. At the molecular level, this disease is due to which one of the following?

(A) Enhanced trypsin activity in the lung
(B) Decreased trypsin activity in the lung
(C) Enhanced α1-antitrypsin activity in the lung
(D) Decreased α1-antitrypsin activity in the lung
(E) Enhanced reduction of sulfhydryl groups in the lung
(F) Decreased reduction of sulfhydryl groups in the lung

32. A 16-year-old male high school student was playing basketball for his school when he collapsed on court and could not be resuscitated. An autopsy demonstrated increased thickness of the intraventricular septum and left ventricular wall. These findings could be explained by a mutation in which one of the following proteins?

(A) Spectrin
(B) α1-Antitrypsin
(C) Collagen
(D) Fibrillin
(E) β-Myosin heavy chain

33. A 28-year-old man presents to the ER with a large amount of blood and protein in his urine. He has had a sensorineural hearing loss since his teen years and has misshaped lenses (anterior lenticonus). The physician is suspicious of a genetic disorder that may lead to eventual kidney failure. If this is the case, the patient most likely has a mutation in which one of the following proteins?

(A) Spectrin
(B) α1-Antitrypsin
(C) Collagen
(D) Fibrillin
(E) β-Myosin heavy chain

34. A 45-year-old woman has been admitted to a substance abuse center for her alcoholism. As a first attempt to curb the patient's drinking, she is given a drug that will lead to an elevation of which one of the following metabolites if she drinks alcohol?

(A) Acetic acid
(B) Acetaldehyde
(C) Ethanol

(D) Carbon dioxide
(E) Carbon monoxide

35. A 23-year-old male presents to the ER with a fracture of his humerus, sustained in what appeared to be a minor fall. He has a history of multiple fractures after a seemingly minor trauma. He also has "sky blue" sclera and an aortic regurg murmur. His underlying problem is most likely due to a mutation in which one of the following proteins?

(A) Fibrillin
(B) Type 1 collagen
(C) Type IV collagen
(D) α1-Antitrypsin
(E) β-Myosin heavy chain

Answers and Explanations

1. **The answer is A.** Bond A is an anhydride bond, which is formed when a carboxylic acid and a phosphoric acid react, releasing H_2O. Bond B is a phosphate ester, formed when phosphoric acid reacts with an alcohol (methanol in this case), releasing water. An ether linkage is not found in this structure (a-C–O–C–linkage), nor is a phosphodiester (when a phosphate group contains two ester linkages, as in the structures of the nucleic acids).

2. **The answer is A.** An alcohol is oxidized to a ketone when β-hydroxybutyrate is converted to acetoacetate (look at the change in the β-carbon, carbon 3). These compounds are classically described as ketone bodies, although technically only acetoacetate contains a ketone.

3. **The answer is B.** The p*K*a is the pH at which the functional group is 50% dissociated, which in this case is the pH at which $[A^-] = [HA]$. The Henderson–Hasselbalch equation, $pH = pK_a + \log_{10} [A^-]/[HA]$, gives the relationship between these parameters. If $pH = pK_a$, $\log_{10} [A^-]/[HA] = 0$, and $[A^-]/[HA] = 1$.

4. **The answer is B.** The side chains of the amino acid residues in proteins contain functional groups with different pK_a's. Therefore, they can donate and accept protons at various pH values and act as buffers over a broad pH spectrum. There is only one N-terminal amino group ($pK_a \sim 9$) and one C-terminal carboxyl group ($pK_a \sim 3$) per polypeptide chain. At physiologic pH, these groups would not be accepting or donating protons, as the amino terminal group would always be protonated and the carboxy terminal carboxylic acid would always be deprotonated. Peptide bonds are not readily hydrolyzed, and such hydrolysis would not provide buffering action. Hydrogen bonds have no buffering capacity, as the hydrogen in these bonds is not donated or accepted once the bond is formed.

5. **The answer is D.** By convention, peptides are written with the N-terminal amino acid on the left and the C-terminal amino acid on the right. Therefore, this peptide contains arginine (single-letter code R, three-letter code arg) at its C-terminus. The sequence of this peptide is aspartic acid (D, asp), alanine (A, ala), serine (S, ser), valine (V, val), and arginine.

6. **The answer is B.** The N-terminal aspartate contains a positive charge on its N-terminal amino group and a negative charge on the carboxyl group of its side chain. The side chains of alanine and serine have no charge at physiologic pH. Glutamate contains a negative charge on the carboxyl group of its side chain. The valine side chain is hydrophobic, and has no charge. The C-terminal arginine contains a negative charge on its C-terminal carboxyl group and a positive charge on its side chain. Thus, the overall charges are +2 and -3, which gives a net charge of -1. The amino acids within the interior of this hexapeptide (alanine, serine, glutamate, and valine) have their amino and carboxy ends involved in peptide bond formation, so there are no charges associated with those groups of the internal amino acids.

7. **The answer is C.** Disulfide bonds are an example of covalent bonds. Hydrophobic interactions occur between hydrophobic groups as they come together in space to reduce their interactions with water, and to allow water to maximize its entropy. Hydrogen bonds are the sharing of a hydrogen atom between two electronegative atoms. While the hydrogen is covalently bound to one of those atoms, it is also attracted to the other electronegative group (which creates the hydrogen bond) via partial charge interactions, in a noncovalent manner. Electrostatic interactions are the attraction of fully charged groups between each other (one negatively charged, such as a carboxylic acid, and one positively charged, such as a primary amine), due to the opposite charges attracting each other. Van der Waals interactions are nonspecific interactions between two atoms as they approach each other up to a certain distance; once they get too close, repulsion will occur between the two atoms.

8. **The answer is A.** The mutation in sickle cell is E6V of the β-globin chain. The sickle hemoglobin molecules have a valine in place of a glutamate in the two β chains within the tetramer. All other amino acids are the same, and so in comparison to normal hemoglobin, the sickle variant has two fewer negative charges. This means that the normal form of hemoglobin (HbA) will migrate more rapidly toward the positive pole of a gel because it contains more negative charges than does HbS.

9. **The answer is C.** The active site is formed when the enzyme folds into its three-dimensional configuration, and may involve amino acid residues that are far apart in the primary sequence. Substrate molecules bind at the active site, as will competitive inhibitors (since the inhibitor reduces enzyme activity by competing with substrate for binding at the active site). Allosteric inhibitors bind at a site other than the active site, as do noncompetitive inhibitors (which reduce the V_{max} without affecting the K_m).

10. **The answer is D.** The Michaelis–Menten equation is $v = (V_{max} \times [S])/(K_m + [S])$. In this case, the velocity (v) is ¼ V_{max}, and [S] = 4 mM. Thus, the Michaelis–Menten equation becomes ¼ $V_{max} = (V_{max} \times 4)/(K_m + 4)$. When one solves that equation for K_m, K_m = 12 mM.

11. **The answer is C.** This problem is best solved using the Michaelis–Menton equation and comparing the velocity (as a function of maximal velocity) under fasting and nonfasting conditions. During fasting, [S] = 5 mM, and the K_m is 7 mM; so $v = (5 \times V_{max})/(7 + 5) = 42\%\ V_{max}$. In the fed state, [S] = 20 mM, and the K_m is 7 mM; so $v = (20 \times V_{max})/(7 + 20) = 74\%\ V_{max}$. Glucokinase is more active in the fed than in the fasting state, and the velocity will increase from <50% V_{max} to >50% V_{max}.

12. **The answer is A.** A competitive inhibitor competes with the substrate for binding to the active site of the enzyme, in effect increasing the apparent K_m (in the presence of inhibitor, it will require a higher concentration of substrate to reach ½ V_{max}, as the substrate is competing with the inhibitor for binding to the active site). As the substrate concentration is increased, the substrate, by competing with the inhibitor, can overcome its inhibitory effects, and eventually the normal V_{max} is reached. A noncompetitive inhibitor will decrease the V_{max} without affecting the binding of substrate to the active site, so the K_m is not altered under those conditions. An activator of an allosteric enzyme will decrease the apparent K_m without affecting V_{max} (less substrate is required to reach the maximal velocity).

13. **The answer is D.** The man has sickle cell disease, and his hemoglobin consists of mutated β chains, along with normal α chains. The glutamate at position 6 in the β chains of HbA is replaced by valine in HbS. Valine contains a hydrophobic side chain, whereas glutamate contains an acidic side chain. Under low oxygenation conditions (such as vigorous exercise), the HbS molecules will polymerize owing to hydrophobic interactions between the valine on the β chain and a hydrophobic patch on another HbS molecule. Under well-oxygenated conditions, the valine in the β chain is not exposed on the surface of the molecule, and it cannot form an interaction with the hydrophobic patch on another hemoglobin molecule. Once the HbS polymerizes, it forms a rigid rod within the red blood cells, which deforms the cell and gives it the "sickle" appearance. Once sickled, the red blood cells cannot easily deform and pass through narrow capillaries, leading to loss of oxygen to certain areas of the body, which is what leads to the pain experienced by the patient. The sickling is not due to increased or decreased ionic interactions between HbS molecules, or to phosphorylation of the HbS monomers.

14. **The answer is B.** In order for hemoglobin to release oxygen more readily, the deoxygenated state of hemoglobin needs to be stabilized. This can occur by decreasing the pH (the Bohr effect), increasing the CO_2 concentration, or increasing the concentration of BPG. Fetal hemoglobin (HbF = $\alpha_2\gamma_2$) has a greater affinity for O_2 than does HbA ($\alpha_2\beta_2$), so inducing the synthesis of the γ genes would have the opposite of the intended effect. Inducing the concentration of the β chains would not decrease oxygen binding to hemoglobin (in fact, if there is not a concurrent increase in α-gene synthesis, this may be quite detrimental to the individual, as an imbalance in the synthesis of the hemoglobin chains leads to a disorder known as thalassemia, and overall oxygen transport to the tissues would be decreased). Increased BPG would cause O_2 to be more

readily released. A metabolic alkalosis would raise the pH of the blood, which would stabilize the oxygenated form of hemoglobin. Reducing carbon dioxide levels in the blood through hyperventilation will also stabilize the oxygenated form of hemoglobin, and make it more difficult to deliver oxygen to the tissues.

15. **The answer is C.** The environmentalist is suffering from scurvy, a deficiency of vitamin C. The hydroxylation of proline and lysine residues in collagen requires vitamin C and oxygen. In the absence of vitamin C, the collagen formed cannot be appropriately stabilized (owing, in part, to reduced hydrogen bonding between subunits due to the lack of hydroxyproline) and is easily degraded, leading to the bleeding gums and loss of teeth. Globin synthesis might be indirectly affected because absorption of iron from the intestine is stimulated by vitamin C, but globin is not modified through a hydroxylation reaction. Iron is involved in heme synthesis, which regulates globin synthesis. Insulin and fibrillin synthesis are not dependent on vitamin C (lack of insulin will lead to diabetes, and mutations in fibrillin lead to Marfan syndrome).

16. **The answer is A.** Antifreeze contains ethylene glycol ($HO\text{-}CH_2\text{-}CH_2\text{-}OH$), and ingestion of ethylene glycol (which has a sweet taste) will lead to a metabolic acidosis due to the metabolism of ethylene glycol to glycolic and oxalic acids. As the acids form, protons are released, and bicarbonate and hemoglobin are the major buffers in the blood that will bind these protons to blunt the drop in blood pH. Acetoacetate and β-hydroxybutyrate are ketone bodies produced by the liver, and since they are both acids, their accumulation is often a cause of metabolic acidosis. Increasing their synthesis under acidotic conditions would only exacerbate the acidosis. Phosphate is an intracellular buffer, but its role is not as significant as that of either hemoglobin or bicarbonate.

17. **The answer is B.** The side chain of histidine has a p*Ka* of 6.0, which, of all amino acid side chains, is the one closest to physiologic pH. The local environment of the protein can raise this p*Ka* value closer to 7 such that the histidine side chains within hemoglobin will be the major groups that accept and donate protons when hemoglobin acts as a buffer. The alanine side chain (a methyl group) cannot accept or donate protons. The p*Ka* for the side chains of serine or threonine are above 10.0, so at physiologic pH these side chains are always protonated, and cannot act as a binding site for excess protons generated during an acidotic event. The p*Ka* for the side chain of aspartate is about 4.0, so at physiologic pH that group is always deprotonated, and will not accept protons generated during an acidotic event.

18. **The answer is B.** Acetazolamide is a carbonic anhydrase inhibitor, which is found primarily in red blood cells. The red blood cells contain carbonic anhydrase that catalyzes the reaction that forms carbonic acid from CO_2 and H_2O. Under high-altitude conditions, the inhibition of carbonic anhydrase will lead to a decrease in blood pH, which stabilizes the deoxygenated form of hemoglobin. This is due to an increased loss of bicarbonate in the urine by the inhibition of carbonic anhydrase within the kidney. The change in pH increases oxygen delivery to the tissues, and can overcome, in part, the symptoms of altitude sickness. However, in the case of the person with Type I diabetes who begins to produce ketone bodies, the body's main compensatory mechanism to overcome the acidosis is blocked. As ketone bodies are formed and protons generated, the H^+ will react with bicarbonate to form carbonic acid. Carbonic anhydrase, which catalyzes a reversible reaction, will then convert the carbonic acid to CO_2 and H_2O, with the CO_2 being exhaled. These reactions soak up excess protons, and help to buffer against the acidosis. If, however, carbonic anhydrase has been inhibited by acetazolamide, then the bicarbonate cannot buffer the blood pH and the acidosis could become more severe. White blood cells, muscle cells, liver cells, and the lens of the eye do not contribute to the buffering of the blood, and inhibition of carbonic anhydrase in those cells would not affect the ability to overcome an acidosis.

19. **The answer is B.** The patient is having a panic attack (due to driving in tunnels) and is hyperventilating, causing an acute respiratory alkalosis. The loss of CO_2 pushes the carbonic anhydrase reaction in the direction of CO_2 production, which reduces the proton concentration (and thereby raising the pH). The patient would lose consciousness with a more severe attack. A respiratory alkalosis is usually mild as compared to a metabolic alkalosis. For this reason,

the pH increase is smaller (7.55 is more likely than a pH of 8.10, which could occur via a metabolic alkalosis). The other choices given, 7.15, 6.40, and 6.10, are all lower than physiologic pH (7.4), and would be considered acidosis, instead of alkalosis. An example of a metabolic alkalosis is hypokalemia, a reduction in normal potassium values. Owing to low serum potassium levels, potassium leaves the cells and is replaced by protons from the circulation. The loss of protons from the blood leads to the alkalosis.

20. The answer is E. Amino acids in humans are in the L-configuration (except glycine which is neither L nor D), whereas bacterial amino acids can be in either the L- or D-configuration. An antibiotic would need to be effective against bacterial proteins and not human proteins, so developing an antibiotic that recognizes proteins or polypeptides that contain D amino acids would only be effective against bacterial products. All amino acids are in polypeptide chains, and phenylalanine, tyrosine, and tryptophan are amino acids that contain aromatic rings, and are present in both bacteria and humans. The R and S nomenclature is not commonly used in biochemistry to describe the configuration of amino acids.

21. The answer is C. Malathion is an organophosphate that inhibits the action of acetylcholinesterase in an irreversible manner. It is one of the most common causes of poisoning worldwide. Malathion forms an irreversible covalent bond between the inhibitor and the active site serine side chain of the enzyme. Without acetylcholinesterase, acetylcholine accumulates in the neuromuscular junction and causes the symptoms described in the case. Both competitive and noncompetitive inhibition are reversible forms of inhibition, and their mechanism of action does not apply to malathion.

22. The answer is D. Fetal hemoglobin (HbF) is composed of two α subunits and two γ subunits. It has a lower affinity for BPG and therefore a higher affinity for oxygen. The congenital defect in the child has led to a reduced oxygenation of the blood supplying the heart (the end organ of the coronary arteries). The coronary arteries normally receive oxygenated blood from the aorta, but in this case, the left main coronary artery is supplying the left ventricle with deoxygenated blood from the pulmonary artery. However, as the child matures and begins producing HbA instead of HbF, there is insufficient oxygen being delivered to the aorta, leading to a myocardial infarction due to the lack of oxygen reaching the heart. As the HbF was replaced with HbA (two α subunits and two β subunits), the lower affinity for oxygen became manifest since the pulmonary artery blood is lower in oxygen than the left ventricle, leading to a myocardial infarction from the lack of O_2 delivered to the myocardium. If the child had been born with abnormal γ chains, the deficiency would have been manifest at birth, and not first appeared at 3 months of age.

23. The answer is A. The patient has myasthenia gravis, and the treatment is pyridostigmine, a competitive, reversible inhibitor of acetylcholinesterase. Myasthenia gravis is caused by autoantibodies to the acetylcholine receptor, reducing the effectiveness of acetylcholine at the neuromuscular junction. By reversibly inhibiting acetylcholinesterase, the effective levels of acetylcholine are increased, thereby providing sufficient acetylcholine to bind to the few functional receptors that remain. The graph is classic for a competitive inhibitor. Competitive inhibitors display an increased apparent K_m, and a constant V_{max}.

24. The answer is B. In noncompetitive inhibition, the K_m is not altered, whereas the V_{max} is decreased. The drug used to treat altitude sickness is acetazolamide, which is a noncompetitive inhibitor of carbonic anhydrase. The graph is one of a pure noncompetitive inhibitor.

25. The answer is B. Ketone bodies are weak acids. In diabetic ketoacidosis, the liver produces ketone bodies, which will reduce the brain's dependency on glucose as its sole energy source. This is due to the lack of insulin, and the liver switching to starvation mode owing to the constant signaling by glucagon. Hemoglobin in the red blood cells and bicarbonate, both in the red blood cells and the plasma, are two of the body's major buffers, and their overproduction would not lead to an acidosis. HCl overproduction within the stomach might lead to duodenal ulcers or gastroesophageal reflux, but not to an overall metabolic acidosis, as the protons do not find their way into the circulation. A loss of chloride, if severe enough, could produce a metabolic alkalosis, but not an acidosis.

26. The answer is C. The patient is "blowing off CO_2" to reduce acid. The equation described in answer C is the conversion of carbon dioxide into a soluble form, then into bicarbonate. During an acidosis, the high levels of protons push the reaction described in answer C to the left, to the formation of water and carbon dioxide. As the carbon dioxide is exhaled, and the concentration of carbon dioxide decreases, more carbon dioxide is formed, thereby reducing the pool of free protons and raising the pH. The protonation of ammonia to form ammonium ion takes place in the kidney, and not the lungs. Its primary purpose is to alkalinize the urine if it is too acidic. The reaction described in answer B is the dissociation of a proton from β-hydroxybutyrate (a ketone body) to form the anion of β-hydroxybutyrate and a proton. This is the reaction that is occurring to bring about the ketoacidosis, and is not the compensatory respiratory alkalosis. Reaction D is the dissociation of water, which cannot buffer the acidosis.

27. The answer is B. Creuzfeldt–Jakob syndrome is a prion disorder, and the infectious agent is a protein. The altered protein forms precipitates in the brain, and shifts the equilibrium of the normal protein to that which will aggregate with the altered protein. The pathologist is concerned that the infectious protein will migrate to his brain and seed the process of aggregation with the normal prion proteins in his brain. Prion disorders are not transmitted by viruses, lipids, bacteria, or any form of nucleic acid.

28. The answer is D. Mad cow disease is a prion disorder, in which a misfolded prion protein in the brain forms aggregrates and precipitates, interfering with normal brain function. Prions can adopt a "stable" conformation, which consists primarily of α-helices, and an aggregation-prone conformation, which consists primarily of β-sheets. Once in the aggregation-prone conformation, the protein aggregates, shifting the equilibrium between structure forms toward the aggregation-prone form. This feeds the aggregation-prone form until the precipitated protein begins to interfere with brain function, and will eventually lead to death. Prion disorders are not due to altered gene expression, viruses, proteolytic cleavage of a prion protein, or to the loss of the nuclear membrane.

29. The answer is C. Free sulfhydryl groups can form disulfide bonds with a cysteine side chain, which may then interfere with the functioning of the protein. Gastric reflux disease is caused by a failure of the gastroesophageal sphincter, allowing stomach contents to travel up the esophagus. Since there is acid in these contents, damage to the esophagus can result. Inhibiting the gastric proton pump, thereby increasing the pH of the stomach contents, would reduce the damage that occurs during reflux due to the increased pH. Inhibiting an intestinal proton pump would not address gastric reflux disease. Methionine residues, though they contain a sulfur atom, do not form disulfide bonds. Free sulfhydryl groups are already reduced, so the drug cannot reduce one further.

30. The answer is B. The family member is exhibiting the symptoms of carbon monoxide (CO) poisoning. CO will bind to hemoglobin, with a higher affinity than oxygen, and decrease oxygen delivery to the tissues. In addition to competing with oxygen for binding to hemoglobin, CO, once bound to hemoglobin, shifts the oxygen-binding curve to the left, stabilizing the "R" state, or oxygenated state, which makes it more difficult for oxygen to be released from hemoglobin in the tissues. Thus, in the presence of CO, oxygen affinity for hemoglobin is actually increased. CO poisoning does not affect the blood flow to the brain.

31. The answer is D. The man has the symptoms of emphysema, due to destruction of lung cells by the protease elastase. Neutrophils in the lung accidentally release elastase as they engulf and destroy inhaled bacteria and other particles, and normally α1-antitrypsin would bind to the elastase and inhibit its activity. In a long-term smoker, however, products from the cigarette smoke oxidize an essential methionine side chain in α1-antitrypsin, rendering it inactive. Thus, over time, noninhibited elastase has been destroying lung tissue until the lung no longer functions properly. Even though the inhibitor will block trypsin activity, the lung damage is the result of increased elastase activity, not trypsin activity. Sulfhydryl groups are not being affected, rather a sulfur in methionine is the target of the cigarette smoke.

32. The answer is E. The student has died from FHC, a thickening of the left ventricle of the heart muscle due to a mutation in β-myosin heavy chain. The exact reason for the hypertrophy, which can be caused by mutations in a variety of sarcomeric proteins, is still unknown. None of the other proteins suggested as answers are muscle sarcomeric proteins. Spectrin is a red blood cell protein, and is not found in the heart. α1-Antitrypsin is a circulating protein synthesized by the liver, and in its absence, emphysema will develop. Collagen is the major structural protein of the body, but there are no mutations in collagen that lead to a greatly hypertrophied heart muscle. A lack of fibrillin leads to Marfan syndrome, which can present with defects in heart valves and the aorta, but not a heart muscle greatly increased in size.

33. **The answer is C.** The patient has Alport syndrome, a mutation in type IV collagen that alters the basement membrane composition of kidney glomeruli. In the absence of a functional basement membrane, the kidneys have difficulty in properly filtering waste products from blood into the urine, and both blood and proteins can enter the urine. Type IV collagen is also important for hearing (it is found in the inner ear) and for the eye. Type IV collagen forms a meshlike structure, which is different from the rodlike structures found in type I collagen, and is found in almost all basement membrane structures. Given sufficient time, the alteration in the basement membrane in the glomeruli will lead to their destruction, and loss of kidney function. A mutation in α1-antitrypsin will lead to emphysema, mutations in spectrin can lead to hereditary spherocytosis, mutations in fibrillin lead to Marfan syndrome, and mutations in β-myosin heavy chain can lead to FHC.

34. **The answer is B.** One treatment for chronic alcoholism is to inhibit the enzyme aldehyde dehydrogenase, which would lead to the accumulation of acetaldehyde if ethanol has been imbibed. Ethanol metabolism, at the first step, converts ethanol to acetaldehyde (the enzyme is alcohol dehydrogenase). Aldehyde dehydrogenase then converts the acetaldehyde to acetic acid, which is eventually converted to acetyl-CoA. The accumulation of acetaldehyde is what initiates the symptoms associated with a hangover, such as headache and nausea. The theory behind the treatment is that if the individual drinks alcohol while on the drug, the buildup of acetaldehyde will make the person feel very uncomfortable, and will lead to a reduction, or cessation, of drinking alcohol. Inhibiting aldehyde dehydrogenase will not lead to elevations of acetic acid, or ethanol, carbon dioxide, or carbon monoxide.

35. **The answer is B.** The patient is exhibiting the signs of osteogenesis imperfecta, brittle bones, as exemplified by various mutations in type 1 collagen, the building blocks of the bones. The aortic regurgitation murmur is also due to a lack of type 1 collagen in the extracellular matrix of the aorta. Mutations in fibrillin give rise to Marfan syndrome, which does exhibit long bones, but not brittle or easily broken bones. Marfan syndrome would also be associated with lens dislocation, which is not occurring in this patient. Mutations in type IV collagen would lead to Alport syndrome, not brittle bones, and there is no mention of kidney/urine problems with the patient. A defect in α1-antitrypsin would lead to emphysema (not brittle bones), and a mutation in β-myosin heavy chain would lead to hypertrophic cardiomyopathy, not brittle bones.

chapter 3

Gene Expression (Transcription), Synthesis of Proteins (Translation), and Regulation of Gene Expression

The main clinical uses of this chapter are to understand, at the molecular level, the basis of genetic mutations, the importance of DNA repair, and infections by certain viruses and bacteria. In addition, protein synthesis and antibiotics that target prokaryotic protein synthesis, as well as toxins, need to be understood. Finally, the use of recombinant DNA technology for the testing and diagnosis of specific diseases is important for understanding the future of diagnostic medicine.

OVERVIEW

- Genetic information is encoded in DNA, which, in eukaryotes, is located mainly in nuclei, with small amounts in mitochondria.
- Genetic information is inherited and expressed. Inheritance occurs by the process of replication. The strands of parental DNA serve as templates for synthesis of copies that are passed on to daughter cells.
- Mutations that result from damage to DNA can lead to genetic alterations, including abnormal cell growth and cancer. Repair mechanisms can correct damaged DNA.
- Recombination of genes promotes genetic diversity.
- Expression of genes requires two steps: transcription and translation. DNA is transcribed to produce messenger RNA (mRNA), which is translated to produce proteins. Ribosomal RNA (rRNA) and transfer RNA (tRNA) participate in the process of translation.
- Proteins are involved in cell structure, and they function as enzymes, determining the reactions that occur in cells. Thus, proteins, the products of genes, determine what cells look like and how they behave.
- Gene expression is regulated by a variety of mechanisms. Only a small fraction of the genome is expressed in any one cell.
- Recombinant DNA technology was developed to study and manipulate genes, and has been used for diagnostics. Gene replacement therapy has been successful in a few cases, with efforts to reduce the side effects being the current focus of investigation.

I. NUCLEIC ACID STRUCTURE

- The monomeric units of nucleic acids are nucleotides; each nucleotide contains a heterocyclic nitrogenous base, a sugar, and phosphate.
- DNA contains the bases adenine (A), guanine (G), cytosine (C), and thymine (T). RNA contains A, G, and C, but has uracil (U) instead of thymine.
- Deoxyribose is present in DNA, whereas RNA contains ribose.
- Polynucleotides consist of nucleosides joined by 3′,5′-phosphodiester bridges. The genetic message resides in the sequence of bases along the polynucleotide chain.
- In DNA, two polynucleotide chains are joined by pairing between their bases (adenine with thymine and guanine with cytosine), and they form a double helix. One chain runs in a 5′ to 3′ direction and the other runs 3′ to 5′.
- DNA molecules in eukaryotes interact with histones to form strands of nucleosomes, which wind into more tightly coiled structures. The term chromatin is used to describe the protein–DNA complexes found on chromosomes, within the nucleus.
- RNA is single-stranded, but the strands loop back on themselves and the bases pair: guanine with cytosine and adenine with uracil. This allows the RNA to form a three-dimensional structure that can be recognized by specific proteins and enzymes.
- mRNA has a cap at the 5′ end and a poly(A) tail at the 3′ end.
- rRNA has extensive base-pairing.
- tRNA forms a cloverleaf structure that contains many unusual nucleotides and an anticodon.

A. The structure of DNA

1. Chemical components of DNA

a. Each polynucleotide chain of DNA contains nucleotides, which consist of a **nitrogenous base** (A, G, C, or T), **deoxyribose,** and **phosphate** (Figs. 3.1 and 3.2).

(1) The bases are the **purines** adenine (A) and guanine (G), and the **pyrimidines** cytosine (C) and thymine (T).

(2) **Phosphodiester bonds** join the 3′-carbon of one sugar to the 5′-carbon of the next sugar (Fig. 3.3).

2. DNA double helix

a. Each DNA molecule is composed of two **polynucleotide chains** joined by hydrogen bonds between the bases (Fig. 3.4).

(1) **Adenine** on one chain forms a base pair with **thymine** on the other chain.

(2) **Guanine** base-pairs with **cytosine**.

(3) The **base sequences** of the two strands are **complementary**. Adenine on one strand is matched by thymine on the other, and guanine is matched by cytosine.

b. The **chains are antiparallel**. One chain runs in a **5′ to 3′** direction and the other chain runs **3′ to 5′** (Fig. 3.5).

c. The double-stranded molecule is twisted to form a **helix** with major and minor grooves (Fig. 3.6).

(1) The **base pairs** that join the two strands are **stacked** like a spiral staircase in the interior of the molecule.

(2) The **phosphate groups** are on the outside of the double helix. Two acidic groups of each phosphate are involved in phosphodiester bonds. The third is free and dissociates its proton at physiologic pH, giving the molecule a **negative charge** (see Fig. 3.3).

(3) The **B form of DNA**, first described by Watson and Crick, is right-handed and contains **10 base pairs per turn**, with each base pair separated by 3.4 Å. Other forms of DNA include the A form, which is similar to the B form but more compact, and the Z form, which is left-handed and has its bases positioned more toward the periphery of the helix.

3. Denaturation, renaturation, and hybridization

a. **Denaturation: Alkali** or **heat** cause the **strands** of DNA to **separate** but do not break phosphodiester bonds.

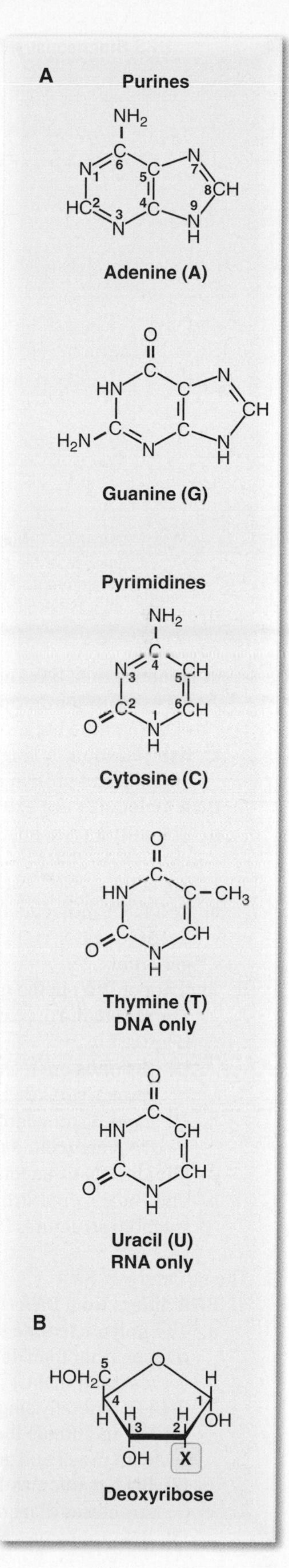

FIGURE 3.1. A. The nitrogenous bases of the nucleic acids. Note the numbering of the purine and pyrimidine rings. Thymine is found in DNA, and uracil only in RNA. **B.** The structure of ribose and deoxyribose, the sugars of the nucleic acids. Ribose has a hydroxyl group on carbon two (indicated as X in the figure), whereas deoxyribose does not (X would be a hydrogen atom for dexoyribose).

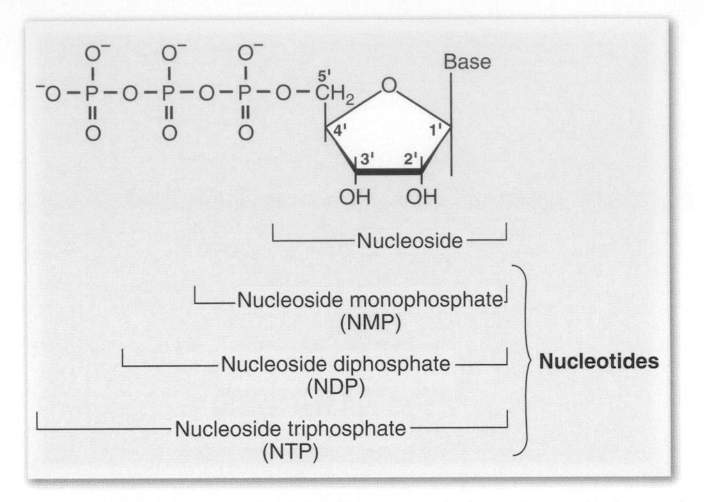

FIGURE 3.2. **Nucleoside and nucleotide structures.** Note the numbering, and the use of the prime symbol, of the carbons of the sugars. The prime symbol is used to distinguish the carbon numbers of the sugars from the carbon and nitrogen numbers of the bases (which do not have the prime symbol).

b. Renaturation: If strands of DNA are separated by heat and then the **temperature** is slowly **decreased** under the appropriate conditions, **base pairs re-form** and complementary strands of DNA come back together.

c. Hybridization: A single strand of DNA pairs with complementary base sequences on another strand of DNA or RNA.

4. DNA molecules are extremely large.

a. The entire chromosome of the bacterium *Escherichia coli* is circular, and contains more than 4×10^6 base pairs. As bacteria lack internal organelles, the DNA is in the cytoplasm of the cell.

b. The DNA molecule in the longest human chromosome is linear, and is over 7.2 cm long. Eukaryotic DNA is found in the nucleus, with the mitochondria also having its own separate genome.

5. Packing of DNA in the nucleus of eukaryotic cells

a. The **chromatin** of eukaryotic cells consists of DNA complexed with histones in **nucleosomes** (Fig. 3.7).

(1) Histones are relatively small, basic proteins with a high content of **arginine** and **lysine**. (Prokaryotes do not have histones.)

(2) Eight histone molecules form an octamer around which approximately 140 base pairs of DNA are wound to form a nucleosome core.

(3) The DNA that joins one nucleosome core to the next is complexed with histone H1.

b. The "beads on a string" nucleosomal structure of chromatin is further compacted to form solenoid structures (helical, tubular coils–see Fig. 3.7).

B. The structure of RNA

1. RNA differs from DNA

a. The polynucleotide structure of **RNA** is similar to DNA except that RNA contains the sugar **ribose** rather than deoxyribose and **uracil** rather than thymine. (A small amount of thymine is present in tRNA.)

b. RNA is generally **single-stranded** (in contrast to DNA, which is double-stranded).

(1) When **strands loop back** on themselves, the bases on opposite sides can pair: adenine with uracil and guanine with cytosine.

(2) RNA molecules have extensive base-pairing, which produces secondary and tertiary structures that are important for RNA function.

(3) RNA molecules recognize DNA and other RNA molecules by base-pairing.

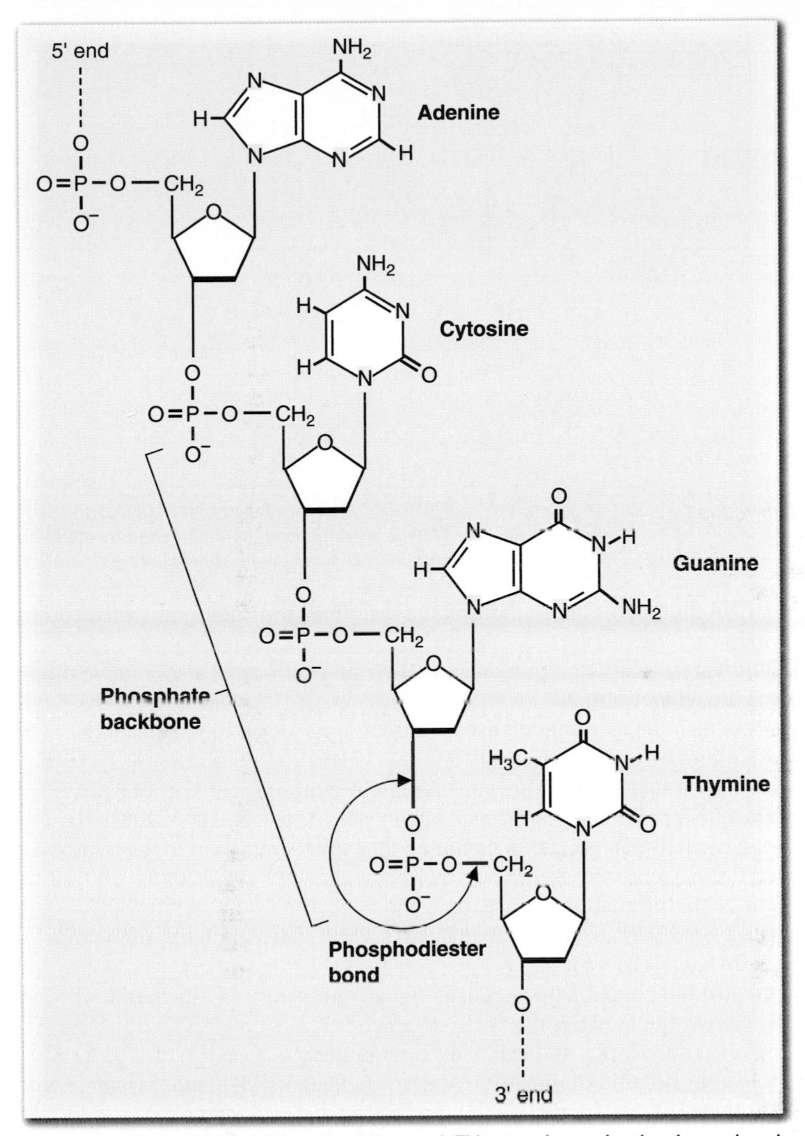

FIGURE 3.3. **A segment of a polynucleotide strand.** This strand contains thymine and exclusively deoxyribose, so it is a segment of DNA. Note the phosphodiester bonds which link the 3′ and 5′ carbons of the sugars.

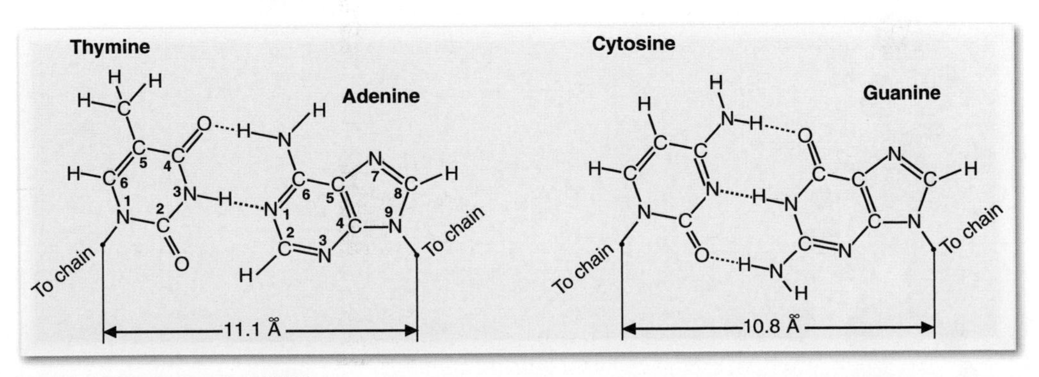

FIGURE 3.4. **The base pairs of DNA.**

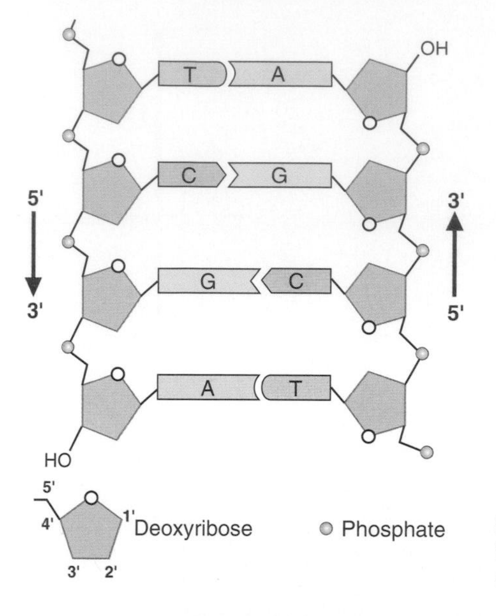

FIGURE 3.5. **Antiparallel strands of DNA.** Note that the strands run in opposite directions, as determined by the hydroxyl groups on carbons 3 and 5 of the deoxyribose (the 3′ and 5′ carbons).

c. Some **RNA molecules** act as **catalysts** of reactions; thus, RNA, as well as protein, can have enzymatic activity.
   (1) **Ribozymes**, usually precursors of rRNA, remove internal segments of themselves, splicing the ends together.
   (2) RNAs also act as **ribonucleases**, cleaving other RNA molecules (e.g., RNase P cleaves tRNA precursors).
   (3) **Peptidyl transferase**, an enzyme in protein synthesis, consists of RNA.

2. **Eukaryotic mRNA** contains a cap structure and a poly(A) tail, and is synthesized in the nucleus.
   a. The **cap** consists of **methylated guanine triphosphate** attached to the hydroxyl group on the ribose at the 5′ end of the mRNA.
      (1) The N7 in the guanine is methylated.
      (2) The 2′-hydroxyl groups of the first and second ribose moieties of the mRNA also may be methylated (Fig. 3.8).
   b. The **poly(A) tail** contains up to 200 adenine (A) nucleotides attached to the hydroxyl group at the 3′ end of the mRNA.

3. **rRNA** contains many loops and extensive base-pairing.
   rRNA molecules differ in their sedimentation coefficients (S). They associate with proteins to form **ribosomes** (Fig. 3.9).
   a. **Prokaryotes** have three types of rRNA: 16S, 23S, and 5S rRNA. They also contain 55 proteins.

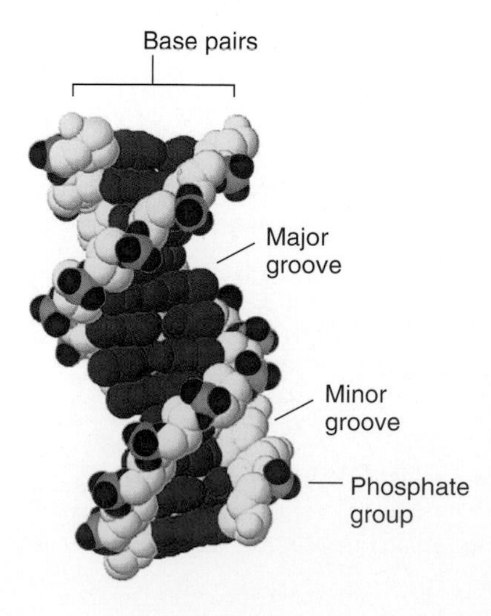

FIGURE 3.6. **The DNA double helix.**

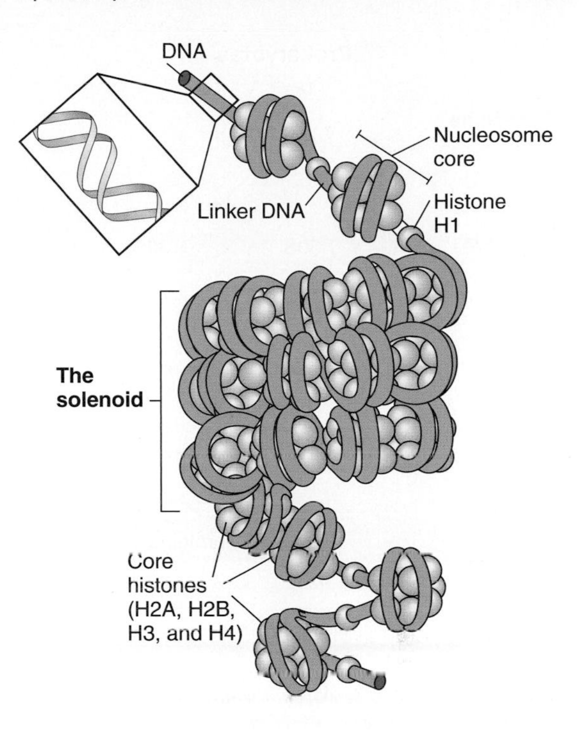

FIGURE 3.7. A polynucleosome, indicating the histone cores and linker DNA. The DNA is depicted in blue, whereas the histones are depicted as light brown spheres.

b. Eukaryotes have four types of cytosolic rRNA: 18S, 28S, 5S, and 5.8S rRNA. The eukaryotic ribosomes have 83 different proteins. Mitochondrial ribosomes are similar to prokaryotic ribosomes.

4. tRNA has a cloverleaf structure and contains modified nucleotides. tRNA molecules are relatively small, containing about 80 nucleotides (Fig. 3.10).

a. In eukaryotic cells, many nucleotides in tRNA are modified.

(1) Modified nucleotides containing **pseudouridine (ψ), dihydrouridine (D),** and **ribothymidine (T)** are present in most tRNAs (see Fig. 3.10).

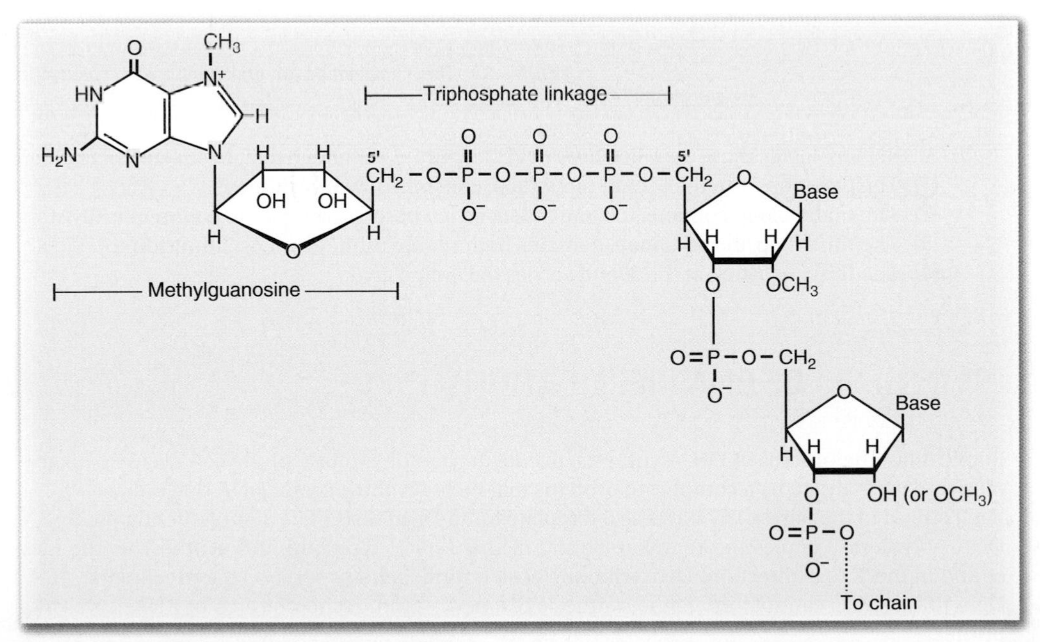

FIGURE 3.8. The cap structure of eukaryotic mRNA.

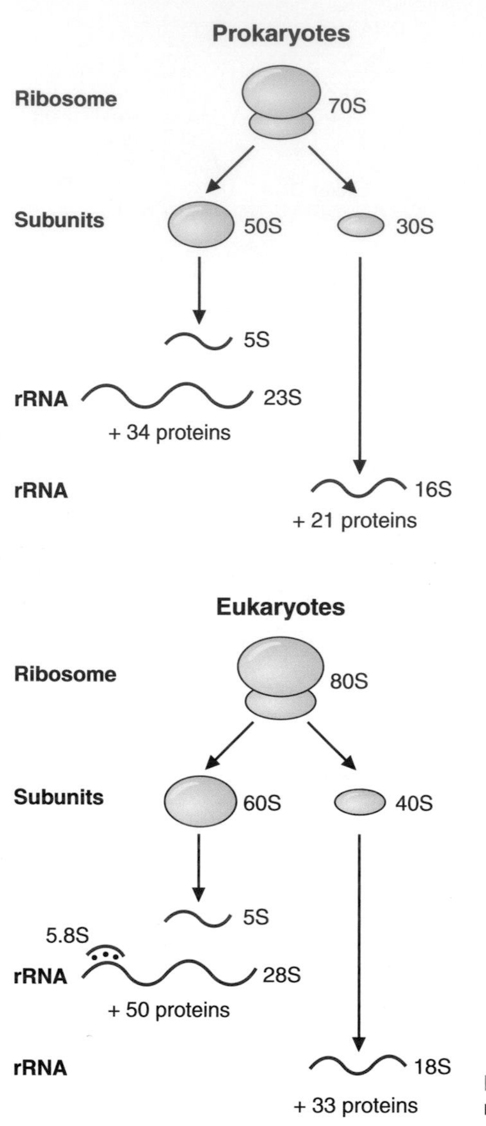

FIGURE 3.9. The composition of prokaryotic and eukaryotic ribosomes.

b. All tRNA molecules have a similar **cloverleaf structure** even though their base sequences differ.

- **(1)** The first loop from the 5′ end, the **D loop**, contains dihydrouridine.
- **(2)** The middle loop contains the **anticodon**, which base-pairs with the codon in mRNA.
- **(3)** The third loop, the **TψC loop**, contains both ribothymidine and pseudouridine.
- **(4)** The **CCA sequence** at the 3′ end carries the amino acid.

II. SYNTHESIS OF DNA (REPLICATION)

- Replication, the process of DNA synthesis, occurs during the S phase of the cell cycle in eukaryotes, and is catalyzed by a complex of proteins that includes the enzyme DNA polymerase.
- Each strand of the parent DNA acts as a template for the synthesis of its complementary strand.
- DNA polymerase copies the template strand in the 3′ to 5′ direction and synthesizes the new strand in the 5′ to 3′ direction. Deoxyribonucleoside triphosphates serve as the precursors.
- DNA polymerase cannot initiate the synthesis of a new strand. A short stretch of RNA serves as a primer.

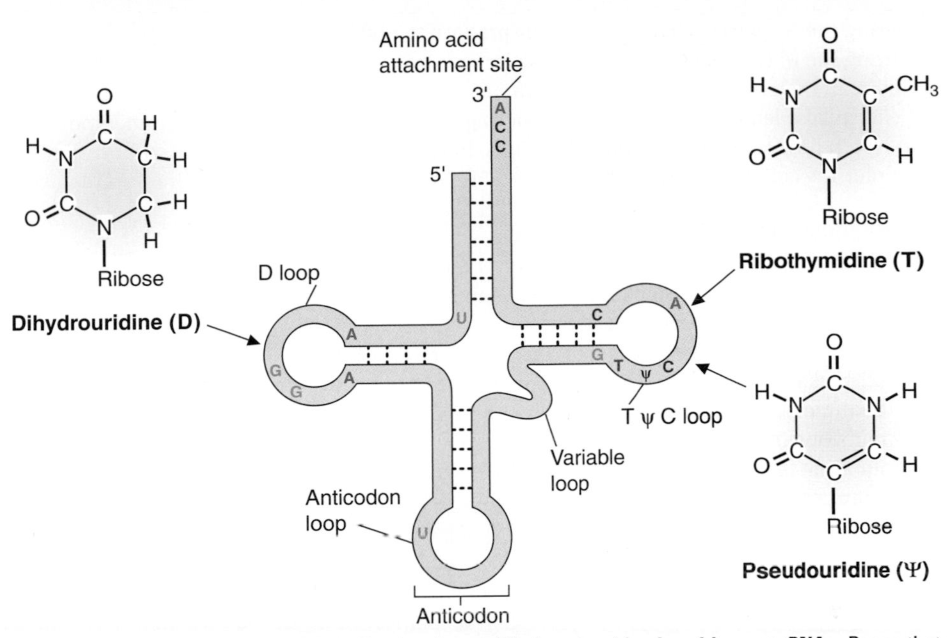

FIGURE 3.10. A generalized view of tRNA, indicating the modified nucleosides found in most tRNAs. Bases that commonly occur in a particular position are indicated by *letters.* Base-pairing in stem regions is indicated by *dashed lines* between strands. Ψ, pseudouridine; *T,* ribothymidine; *D,* dihydrouridine.

- Other proteins and enzymes are required to unwind the parental strands and allow both strands to be copied simultaneously.
- Errors that occur during replication are corrected by enzymes associated with the replication complex.
- In eukaryotic cells, DNA needs to be freed from their associated histones to allow replication to occur.
- Damage that occurs to DNA molecules can be corrected by repair mechanisms, which usually involve removal and replacement of the damaged region with the intact undamaged strand serving as a template.
- DNA molecules can recombine. A portion of a strand from one molecule can be exchanged for a portion of a strand from another molecule.
- Genes can be transposed (moved from one chromosomal site to another).

A. The cell cycle of eukaryotic cells (Fig. 3.11)

1. During the G_1 (first gap) phase, cells **prepare to duplicate** their chromosomes.
2. During the **S** (synthesis) phase, **synthesis of DNA** (replication) occurs.

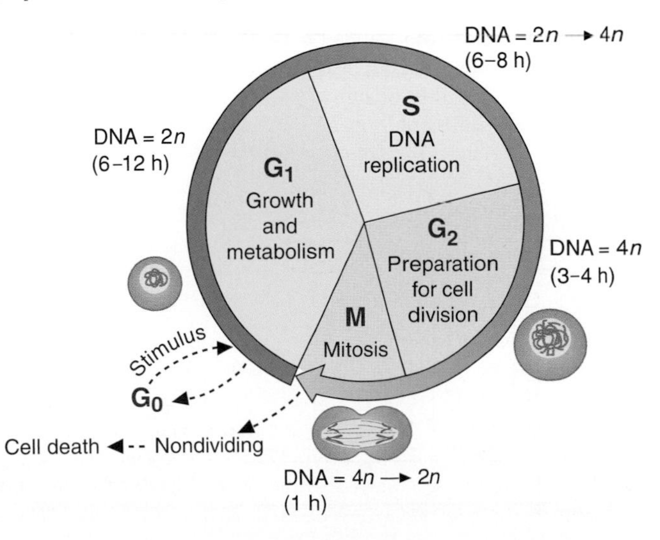

FIGURE 3.11. The cell cycle of eukaryotes.

3. During the G_2 (second gap) phase, cells **prepare to divide**.
4. During the **M** (mitosis) phase, **cell division** occurs.
5. Cells can traverse the cell cycle many times.
6. Cells can also leave the cycle never to divide again, or they can enter a phase (sometimes called G_0) in which they remain for extended periods. In response to an appropriate stimulus, these cells reenter the cell cycle and divide again.

B. Mechanism of replication

1. **Replication** is bidirectional and semiconservative (Fig. 3.12).
   a. **Bidirectional** means that replication begins at a site of origin and simultaneously moves out in both directions from this point.
      (1) Prokaryotes have one site of origin on each chromosome.
      (2) Eukaryotes have multiple sites of origin on each chromosome.
   b. **Semiconservative** means that, following replication, each daughter molecule of DNA contains one intact parental strand and one newly synthesized strand joined by base pairs.
2. **Replication forks** are the sites at which DNA synthesis is occurring.
   The **parental strands** of DNA separate and the helix unwinds ahead of a replication fork (Fig. 3.13).
   a. **Helicases** unwind the helix, and **single-strand binding proteins** hold it in a single-stranded conformation.
   b. **Topoisomerases** act to prevent the extreme supercoiling of the parental helix that would result as a consequence of unwinding at a replication fork.
      (1) Topoisomerases break and rejoin DNA chains.
      (2) **DNA gyrase**, a topoisomerase inhibited by the quinolone family of antibiotics, is found only in prokaryotes.
      (3) A listing of proteins involved in DNA replication is shown in Table 3.1.
3. **DNA polymerases** catalyze the synthesis of DNA.
   a. **Prokaryotes** have three DNA polymerases: **pol I, pol II, and pol III**. Pol III is the replicative enzyme, and pol I is involved in repair and synthesis on the lagging strand.

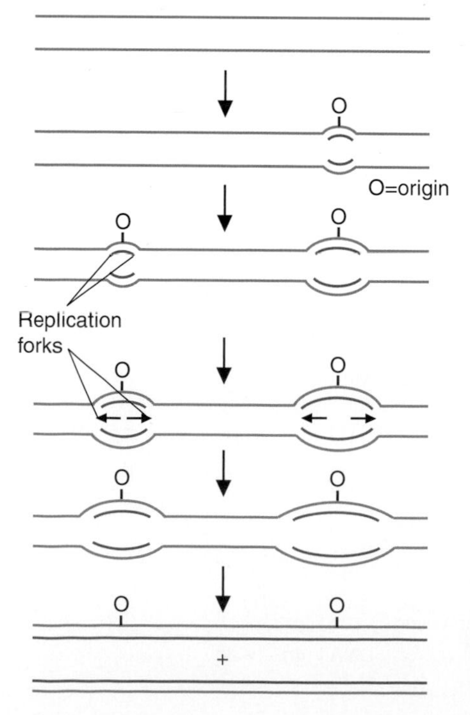

FIGURE 3.12. Replication of a eukaryotic chromosome. *Blue lines* are parental strands. *Red lines* are newly synthesized strands. Synthesis is bidirectional from each point of origin *(O)*, such that there is a replication fork moving in each direction starting from the origin.

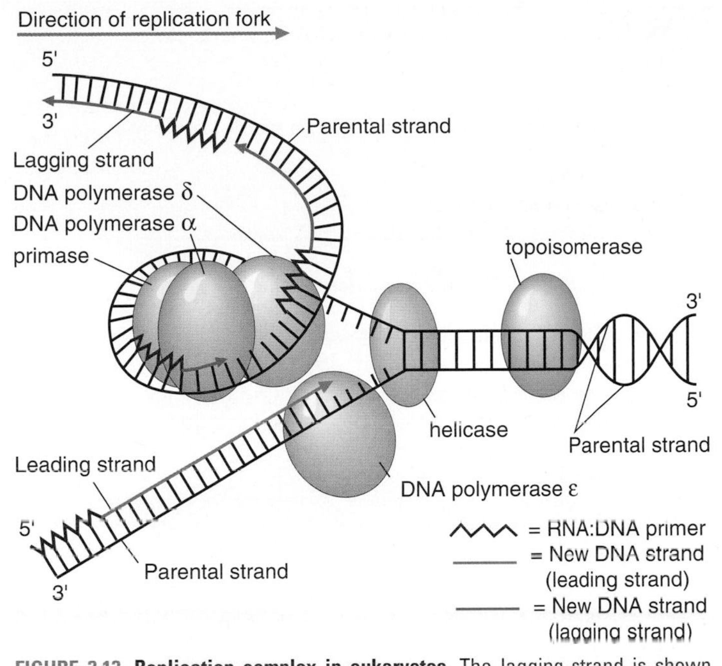

FIGURE 3.13. Replication complex in eukaryotes. The lagging strand is shown looped around the replication complex to demonstrate that all DNA synthesis is in the 5′ to 3′ direction. Single-strand binding proteins (not shown) are bound to the unpaired, single-stranded DNA. Other proteins also participate in this complex (see Table 13.1).

b. **Eukaryotes** have a large number of DNA polymerases, some of which are outlined in Table 3.2. The eukaryotic polymerases are named by Greek letters, and the major ones are α, β, γ, δ, and ε. DNA polymerase α is involved in generating primers for DNA replication. DNA polymerase β is used exclusively for repair. Polymerase δ acts as the lead polymerase on the lagging strand during replication. Polymerase ε is the lead polymerase on the leading strand of DNA, and polymerase γ functions exclusively in mitochondria.
c. **DNA polymerases** can only copy a DNA template in the 3′ to 5′ direction and produce the newly synthesized strand in the 5′ to 3′ direction.
d. **Deoxyribonucleoside triphosphates** (dATP, dGTP, dTTP, and dCTP) are the precursors for DNA synthesis.

table 3.1 Major Proteins Involved in Eukaryotic DNA Replication

DNA polymerases	Add nucleotides to a strand growing 5′ to 3′, copying a DNA template 3′ to 5′
Primase	Synthesizes RNA primers
Helicases	Separate parental DNA strands, i.e., unwind the double helix
Single-strand binding proteins	Prevent single strands of DNA from reassociating
Topoisomerases	Relieve torsional strain on parental duplex caused by unwinding[a]
Enzymes that remove primers	RNase H hydrolyzes RNA of DNA–RNA hybrids Flap endonuclease 1 (FEN1) recognizes "flap" (Unannealed portion of RNA) near 5′-end of primer and cleaves downstream in DNA region of primer; the flap is created by polymerase δ displacing the primer as the Okazaki fragment is synthesized
DNA ligase	Joins, by forming a phosphodiester bond, two adjacent DNA strands that are bound to the same template
PCNA	Enhances processivity of the DNA polymerases; binds to many proteins present at the replication fork

[a]Gyrase is a prokaryotic specific topoisomerase, which is inhibited by the quinolone family of antibiotics.

table 3.2 Functions of Some Eukaryotic DNA Polymerases

Polymerase	Functions[a]	Exonuclease Activity
Pol α	Replication (in a complex with primase and aids in starting the primer) DNA repair	None
Pol β	DNA repair exclusively	None
Pol γ	DNA replication in mitochondria	3′-to-5′
Pol δ	Replication (processive DNA synthesis on lagging strand) DNA repair	3′-to-5′
Pol ε	Replication (processive DNA synthesis on leading strand) DNA repair	3′-to-5′
Pol κ	DNA repair (bypass polymerase)[b]	None
Pol η	DNA repair (bypass polymerase)	None
Pol ξ	DNA repair (bypass polymerase)	None
Pol ι	DNA repair (bypass polymerase)	None

[a]Synthesis of new DNA strands always occurs 5′-to-3′.
[b]Bypass polymerase are able to "bypass" areas of DNA damage and continue DNA replication. Some enzymes are error-free and insert the correct bases; other enzymes are error-prone and insert random bases.

(1) Each precursor pairs with the corresponding base on the template strand and forms a phosphodiester bond with the hydroxyl group on the 3′-carbon of the sugar at the end of the growing chain (Fig. 3.14).

(2) Pyrophosphate is produced and cleaved to two inorganic phosphates.

4. **DNA polymerase** requires a **primer** (Fig. 3.15).
   a. DNA polymerases **cannot initiate** the synthesis of new strands.
   b. **RNA serves as the primer** for DNA polymerase in vivo.
      (1) The RNA primer, which contains about 10 nucleotides, is formed by copying of the parental strand in a reaction initially catalyzed by **DNA primase; polymerase α** then adds a few deoxyribonucleotides to the RNA primer.
   c. **DNA polymerase** adds deoxyribonucleotides to the 3′-hydroxyls of the RNA primers and subsequently to the ends of the growing DNA strands.

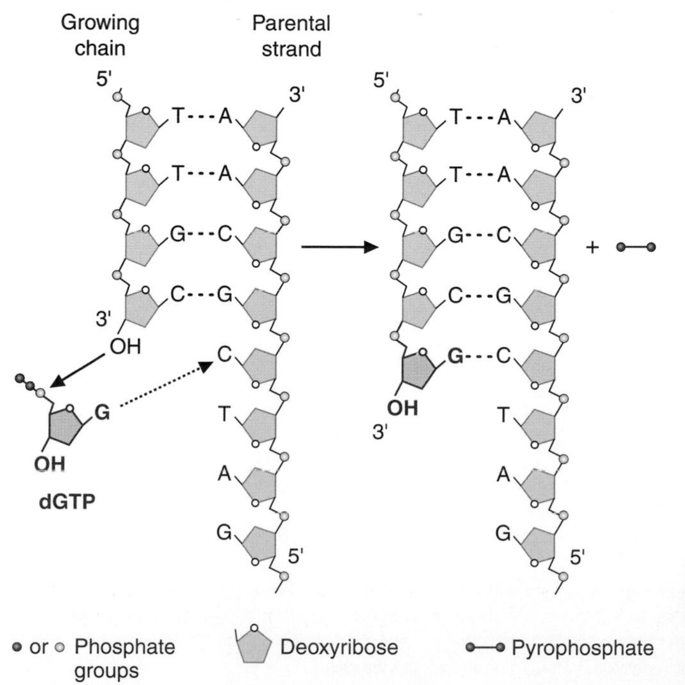

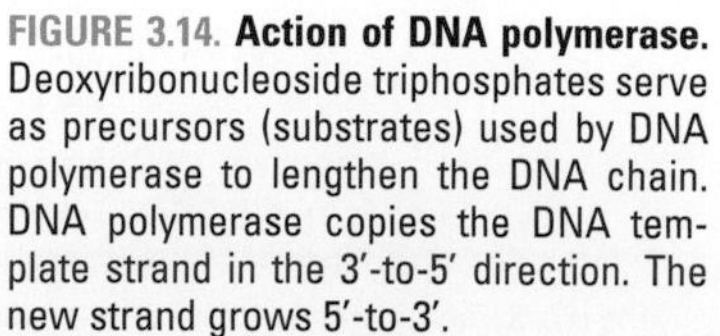

FIGURE 3.14. **Action of DNA polymerase.** Deoxyribonucleoside triphosphates serve as precursors (substrates) used by DNA polymerase to lengthen the DNA chain. DNA polymerase copies the DNA template strand in the 3′-to-5′ direction. The new strand grows 5′-to-3′.

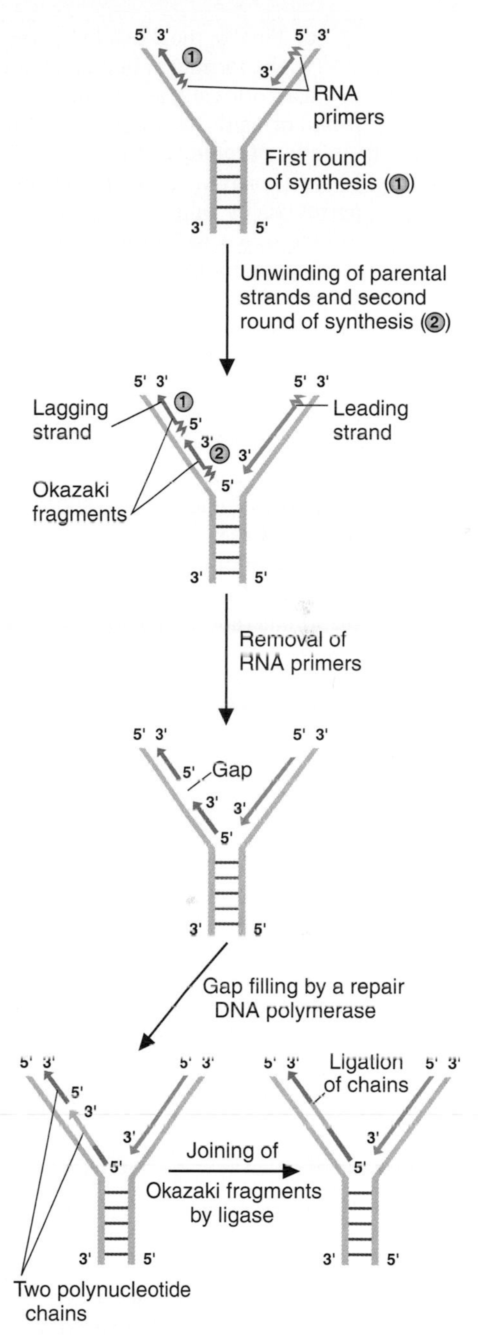

FIGURE 3.15. **Mechanism of DNA synthesis at the replication fork.** Two rounds of polymerase action are shown (❶ and ❷). The number of nucleotides added in each round is much larger than shown; in eukaryotes, about 10 ribonucleotides and 200 deoxyribonucleotides are polymerized on the lagging strand. Synthesis on the leading strand is continuous.

d. DNA parental (template) strands are copied simultaneously at replication forks, although they run in opposite directions.

(1) The **leading strand** is formed by continuous copying of the parental strand that runs 3′ to 5′ **toward** the replication fork.

(2) The **lagging strand** is formed by discontinuous copying of the parental strand that runs 3′ to 5′ **away from** the replication fork.

(a) As more of the helix is unwound, the synthesis of the lagging strand begins from another primer. The short fragments formed by this process are known as **Okazaki fragments**.

(b) The RNA **primers are removed by nucleases** (e.g., RNase H); then the resulting **gaps are filled** with the appropriate deoxyribonucleotides by another DNA polymerase.

(c) Finally, the **Okazaki fragments are joined by DNA ligase**, an enzyme that catalyzes the formation of phosphodiester bonds between two polynucleotide chains.

e. In eukaryotic cells, about 200 deoxyribonucleotides are added to the lagging strand in each round of synthesis, whereas in prokaryotes 1,000 to 2,000 are added.

5. The **fidelity of replication** is very high with an overall error rate of 10^{-9} to 10^{-10}.
   a. **Errors** (insertion of an inappropriate nucleotide) that occur during replication can be **corrected by editing** during the replication process. This proofreading function is performed by a 3′ to 5′ exonuclease activity associated with the polymerase complex.
   b. **Postreplication repair processes** (e.g., mismatch repair) also increase the fidelity of replication.

C. Mutations

Changes in DNA molecules cause mutations. After replication, these changes result in a permanent alteration of the base sequence in the daughter DNA.

1. **Changes causing mutations** include:
   a. **Uncorrected errors** made during replication.
   b. **Damage** that occurs to replicating or nonreplicating DNA caused by oxidative deamination, radiation, or chemicals, resulting in cleavage of DNA strands or chemical alteration or removal of bases.
2. **Types of mutations** include:
   a. **Point mutations** (substitution of one base for another)
   b. **Insertions** (addition of one or more nucleotides within a DNA sequence)
   c. **Deletions** (removal of one or more nucleotides from a DNA sequence)

D. DNA repair (Fig. 3.16)

1. In general, repair involves the **removal** of the segment of DNA that contains a damaged region or mismatched bases, **filling in the gap** by action of a DNA polymerase that uses the undamaged sister strand as a template, and **ligation** of the newly synthesized segment to the remainder of the chain.
2. **Endonucleases, exonucleases**, a **DNA polymerase**, and a **ligase** are required for repair.
   a. **Nucleotide excision repair** involves the removal of a group of **nucleotides** (including the damaged nucleotide) from a DNA strand.
   b. **Base excision repair** involves a specific **glycosylase** that removes a **damaged base** by hydrolyzing an *N*-glycosidic bond, producing an apurinic or apyrimidinic site, which is cleaved and, subsequently, repaired.
   c. **Mismatch repair** involves the removal of the portion of the **newly synthesized strand** of recently replicated DNA that contains a pair of mismatched bases. Bacteria recognize the newly synthesized strand because, in contrast to the parental strand, it has not yet been methylated. The recognition mechanism in eukaryotes is not known.
   d. **Transcription-coupled repair** occurs when the transcription apparatus (synthesizing RNA from DNA) detects DNA damage in the gene being transcribed. Repair enzymes are recruited to the area of damage before transcription is reinitiated.

E. Rearrangements of genes

Several processes produce new combinations of genes, thus promoting genetic diversity. It is also used for the repair of double-strand breaks in DNA (homologous recombination). Nonhomologous end joining (NHEJ) occurs as a mechanism to repair double-strand breaks in DNA.

1. **Recombination** occurs between homologous DNA segments, that is, those that have very similar sequences.
2. **Transposition** involves movement of a DNA segment from one site to a nonhomologous site.
   a. Transposons ("jumping genes") are mobile genetic elements that facilitate the movement of genes.
   b. Certain transposons carry an additional gene, often for antibiotic resistance. The movement of the antibiotic genes between plasmids and the host chromosome can lead to multidrug-resistant bacteria, through the creation of R-plasmids that contain multiple antibiotic-resistance genes.

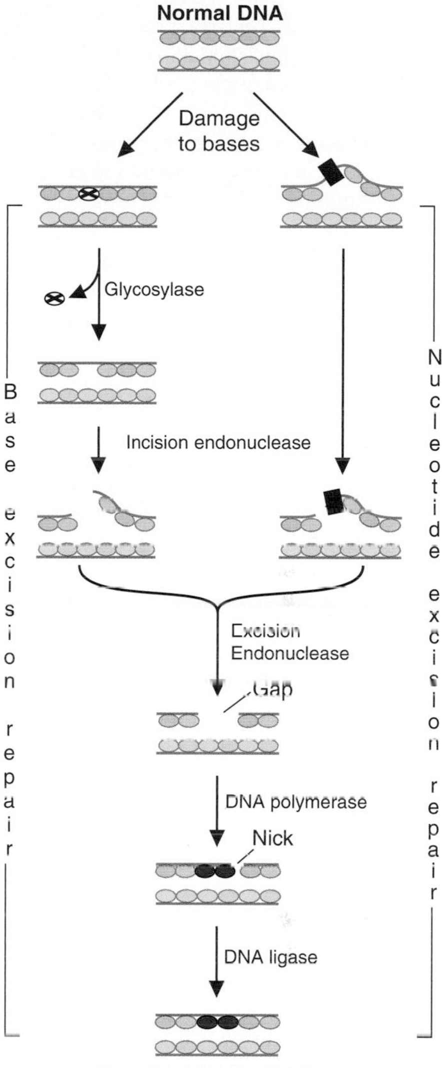

FIGURE 3.16. Base excision and nucleotide excision repair of DNA. Circles indicate normal bases; x's and ■, damaged bases. The actual number of nucleotides removed (the size of the gap) is larger than that shown.

F. **Reverse transcription**

1. Synthesis of DNA from an **RNA template** is catalyzed by **reverse transcriptase**.
2. **Retroviruses** contain RNA as their genetic material.
   a. The retroviral RNA serves as a template for the synthesis of DNA by reverse transcriptase.
   b. The DNA that is generated can be inserted into the genome (chromosomes) of the host cell and be expressed.
3. Certain transposons utilize reverse transcriptase to synthesize DNA copies of the transposon, which can then integrate into the regions of the genome.

CLINICAL CORRELATES Cancer is a group of diseases in which cells are not responsive to the normal restraints on growth. The major causes of cancer are **radiation, chemicals**, and **viruses**. Radiation and chemicals cause **damage to DNA**, which, if not repaired rapidly, produces **mutations** that can result in cancer.

Burning organic material (e.g., cigarettes) produces chemicals such as **benzo(a)pyrene** that covalently bind to the bases in DNA, producing mutations that lead to **lung cancer**. **Ultraviolet (UV) light**, including that from the sun, produces **pyrimidine dimers** in DNA that lead to **skin cancer**. This condition is particularly pronounced in people with **xeroderma pigmentosum (XP)** because their DNA repair system does not

(continued)

function normally. Mutations in other DNA repair systems can lead to **hereditary nonpolyposis colorectal cancer (HNPCC)** (mismatch repair is defective in this disease), **Bloom syndrome** (a defect in a helicase necessary to unwind the DNA strands during replication), **Cockayne syndrome** (a defect in transcription-coupled DNA repair), and **breast cancer** (a defect in repairing single- and double-strand breaks in DNA).

Oncogenes are genes that cause cancer. Their counterparts in normal cells, **proto-oncogenes**, are involved in **normal growth** and **development**. If oncogenes enter cells as a consequence of viral infection or if the normal proto-oncogenes are altered or expressed abnormally, **cancer** can result. Many oncogenes encode proteins that are related to **growth factors**, to receptors for growth factors, to transcription factors, or to proteins whose synthesis is induced by growth factors. Some oncogene products enter the nucleus and **activate genes**. According to the oncogene theory, **viruses cause cancer** by inserting **additional or abnormal copies** of proto-oncogenes into cells or by inserting strong **promoters** into regions that regulate the expression of these genes. Proto-oncogenes can be **amplified**. The gene for the proto-oncogene or its control region may undergo **mutations** owing to radiation or chemicals. Alteration of the product or of the level of expression of a proto-oncogene produces changes in the growth characteristics of cells that can result in cancer.

Chronic myelogenous leukemia (CML) can result from a translocation between chromosomes 9 and 22, which creates a novel protein named bcr-abl. Study of the structure of bcr-abl allowed a drug to be designed, Gleevec, which would bind to and inhibit the unregulated kinase activity of bcr-abl, which led to remission of the cancer.

Cancer can also result from alterations in genes that produce proteins that act as **suppressors of cell growth**. Decreased expression of these **suppressor genes** (e.g., p53, the retinoblastoma gene) results in increased cell growth. MicroRNAs (miRNAs) can be classified as either oncogenes or tumor suppressors, depending on the function of the genes they regulate.

The **treatment** of **cancer** frequently involves **drugs that interfere with DNA synthesis**. For example, 5-fluorouracil (**5-FU**) prevents the conversion of dUMP to dTMP, reducing the level of thymine nucleotides required for DNA synthesis. **Methotrexate** prevents the formation of tetrahydrofolate from its more oxidized precursors. As a result, the formation of both thymine for DNA synthesis and the purines for DNA and RNA syntheses is inhibited. Certain cancers become resistant to methotrexate through amplification of the dihydrofolate reductase gene, the target of methotrexate.

Adriamycin contains a series of rings that intercalate (slip) between the DNA base pairs. When adriamycin is present, DNA cannot act as a template for replication or transcription. **Etoposide** blocks the action of topoisomerase (an enzyme needed for the unwinding of DNA during replication), and will therefore block DNA replication.

III. SYNTHESIS OF RNA (TRANSCRIPTION)

- Transcription, the synthesis of RNA from a DNA template, is catalyzed by RNA polymerase. RNA polymerase copies a DNA template in the 3′ to 5′ direction and synthesizes a single-stranded RNA molecule in the 5′ to 3′ direction. Unlike DNA polymerase, RNA polymerase can initiate the synthesis of new strands, and does not require a primer to do so.
- In eukaryotes, the primary product of transcription is modified and trimmed before it participates in protein synthesis. Transcription occurs in the nucleus, and the final RNA product is exported to the cytoplasm where protein synthesis occurs.
- Eukaryotic mRNA, produced by RNA polymerase II, is capped at the 5′ end and has a poly(A) tail added at the 3′ end. Introns (segments that do not code for protein) are removed, and exons (segments that produce the mature mRNA) are spliced together.
- Eukaryotic rRNA is produced by RNA polymerase I as a 45S precursor that is methylated and cleaved to form three of the rRNAs (18S, 28S, and 5.8S) that appear in ribosomes. The 5S rRNA is produced from a separate gene by RNA polymerase III. The nucleolus is the site of rRNA production and ribosome assembly.
- Eukaryotic tRNA is produced by RNA polymerase III as a precursor that is trimmed at the 5′ and 3′ ends. Introns are removed and exons are spliced together.

- Unusual nucleotides are produced in mature tRNA by posttranscriptional modification of normal nucleotides, and a CCA sequence is added at the 3′ end.
- Eukaryotic RNA must travel from the nucleus to the cytoplasm for translation.
- Bacteria do not contain nuclei, so transcription and translation occur simultaneously. A single RNA polymerase produces mRNA, rRNA, and tRNA in bacteria. Bacterial transcripts (e.g., those from *E. coli*) do not contain introns.

A. RNA polymerase

1. RNA polymerase can **initiate** the synthesis of **new chains.** A primer is not required.
2. The DNA template is copied in the 3′ to 5′ direction, and the RNA chain grows in the 5′ to 3′ direction.
   a. The template strand is also named the antisense strand or the noncoding strand.
3. The DNA strand that is the nontemplate strand will contain the same sequence as the RNA that is transcribed, except that the DNA will contain thymine (T), and RNA uracil (U).
   a. The nontemplate strand is also named the sense strand or the coding strand.
4. **Ribonucleoside triphosphates** (ATP, GTP, UTP, and CTP) serve as the **precursors** for the RNA chain. The process is similar to that for DNA synthesis (see Fig. 3.14).

B. Synthesis of RNA in bacteria

1. The RNA polymerase of *E. coli* contains **four subunits**, $\alpha_2\beta\beta'$, which form the core enzyme, and a fifth subunit, the **sigma factor** (σ), which is required for the initiation of RNA synthesis.
2. Genes contain a **promoter region** to which RNA polymerase binds.
   a. Promoters contain the consensus sequence **TATAAT** (called the Pribnow or TATA box), about 10 bases upstream from (before) the start point of transcription.
      (1) A **consensus sequence** consists of the most commonly found sequence of bases in a given region of all DNAs tested.
   b. **A second consensus sequence** (TTGACA) is usually located upstream from the Pribnow box, about 35 nucleotides (-35) from the start point of transcription.
3. When RNA polymerase binds to a **promoter**, local unwinding of the DNA helix occurs so that the DNA strands partially separate. The polymerase then begins transcription, copying the **template strand**.
   a. As the polymerase moves along the DNA, the next region of the double helix unwinds while the single-stranded region that has already been transcribed rejoins its partner.
   b. Termination occurs in a region in which the transcript forms a hairpin loop that precedes a number of U residues (for **rho-independent** termination).
   c. The ρ **(rho) factor** aids in the termination of some transcripts. This factor binds to the transcript as it is being synthesized, and with the aid of energy obtained from the hydrolysis of ATP, manages to "pull" the transcript off of the template to be released into the cytoplasm.
4. **mRNA** is often produced as a **polycistronic transcript** that is translated as it is being transcribed.
   a. A polycistronic mRNA produces several different proteins during translation, one from each cistron.
   b. *E. coli* mRNA has a short half-life. It is degraded in minutes.
5. **rRNA** is produced as a **large transcript** that is cleaved, producing the 16S rRNA that appears in the 30S ribosomal subunit and the 23S and 5S rRNAs that appear in the 50S ribosomal subunit.
   a. The 30S and 50S ribosomal subunits combine to form the 70S ribosome.
6. **tRNA** is usually produced from **larger transcripts** that are cleaved. One of the cleavage enzymes, RNase P, contains an RNA molecule that acts as a catalyst.

C. Synthesis of RNA in nuclei of eukaryotes

1. **mRNA synthesis** (Fig. 3.17)
   a. Eukaryotic genes that produce mRNA contain a **basal promoter region**. This region binds transcription factors, proteins that bind RNA polymerase II, as well as RNA polymerase II. Promoters contain a number of conserved sequences.
      (1) A **TATA** (Hogness) **box**, containing the consensus sequence TATATAA, is located about 25 base pairs upstream (-25) from the transcription start site.
      (2) A **CAAT box** is frequently found about 70 base pairs upstream from the start site.

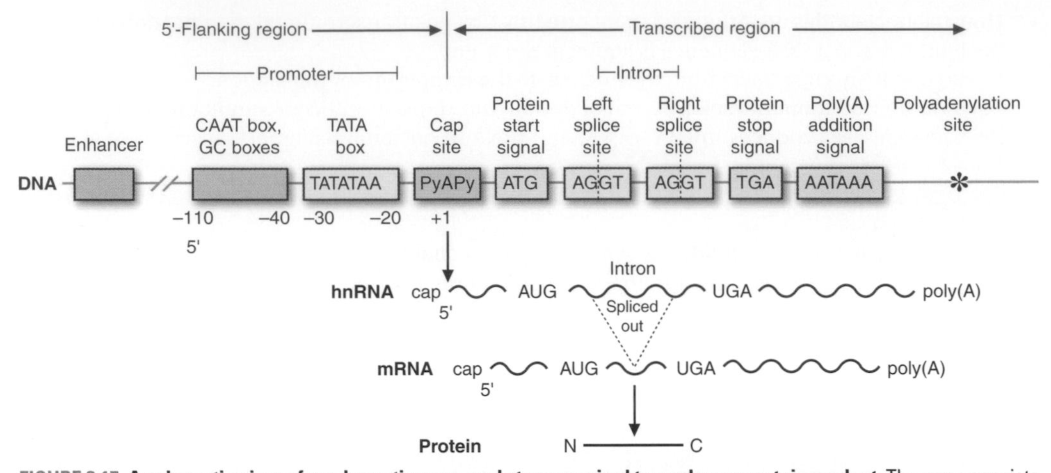

FIGURE 3.17. **A schematic view of a eukaryotic gene, and steps required to produce a protein product.** The gene consists of promoter and transcribed regions. The transcribed region contains introns, which do not contain coding sequences for proteins, and exons, which do carry coding sequences for proteins. The sequences shown are from the non-template strand (the strand of DNA with the same sequence as the RNA which will be produced, except that U will be in RNA and T in DNA). The first RNA form produced is heterogenous nuclear RNA (hnRNA), which contains both intronic and exonic sequences. The hnRNA is modified such that a cap is added at the 5'-end (cap site), and a poly(A) tail is added to the 3'-end. The introns are removed (a process called splicing) to produce the mature mRNA, which leaves the nucleus to direct protein synthesis in the cytoplasm. Py is pyrimidine (C or T). While the TATA box is still included in this figure for historical reasons, only 12.5% of eukaryotic promoters contain this sequence.

(3) **GC-rich regions** (GC boxes) often occur between -40 and -110.

(4) A comparison of eukaryotic and prokaryotic promoters is shown in Fig. 3.18. Note the many elements in the eukaryotic promoter region that have been identified as being important for initiating RNA synthesis, including some DNA regions within the transcribed portion of the gene.

b. **Enhancers** are DNA sequences that function in the **stimulation** of the transcription rate. They can be located thousands of base pairs upstream or downstream from the start site. Other sequences called **silencers** function in the **inhibition** of transcription.

c. **RNA polymerase II** initially produces a large primary transcript called heterogeneous nuclear RNA (**hnRNA**), which contains exons and introns.

(1) **Exons** are sequences within a transcript that appear in the mature **mRNA**.

(2) **Introns** are sequences within the primary transcript that are **removed** and do not appear in the mature mRNA.

d. **Processing of hnRNA** yields mature mRNA, which enters the cytoplasm through nuclear pores.

(1) The **primary transcript** (hnRNA) is capped at its 5′ end as it is being transcribed.

(2) A **poly(A) tail**, 20 to 200 nucleotides in length, is added to the 3′ end of the transcript. The sequence AAUAAA in hnRNA serves as a signal for the cleavage of the hnRNA and addition of the poly(A) tail by poly(A) polymerase. ATP serves as the precursor, and no template is required.

(3) **Splicing** reactions remove introns and connect the exons.

(a) The **splice point** at the **5′ end** of an intron usually has the sequence GU; it is preceded by an **invariant AG at the 3′ end** of the exon adjacent to it. At the **3′ end of the intron,** an **invariant AG** is frequently followed by GU at the 5′ end of the adjacent exon (see Fig. 3.17).

(b) Small nuclear RNAs complexed with protein (**snRNPs**) (e.g., U1 and U2) are involved in the cleavage and splicing process. A **lariat** structure is generated during the splicing reaction.

(4) Some hnRNAs contain 50 or more exons that must be spliced correctly to produce functional mRNA. Other hnRNAs have no introns.

2. **rRNA synthesis** and assembly of ribosomes

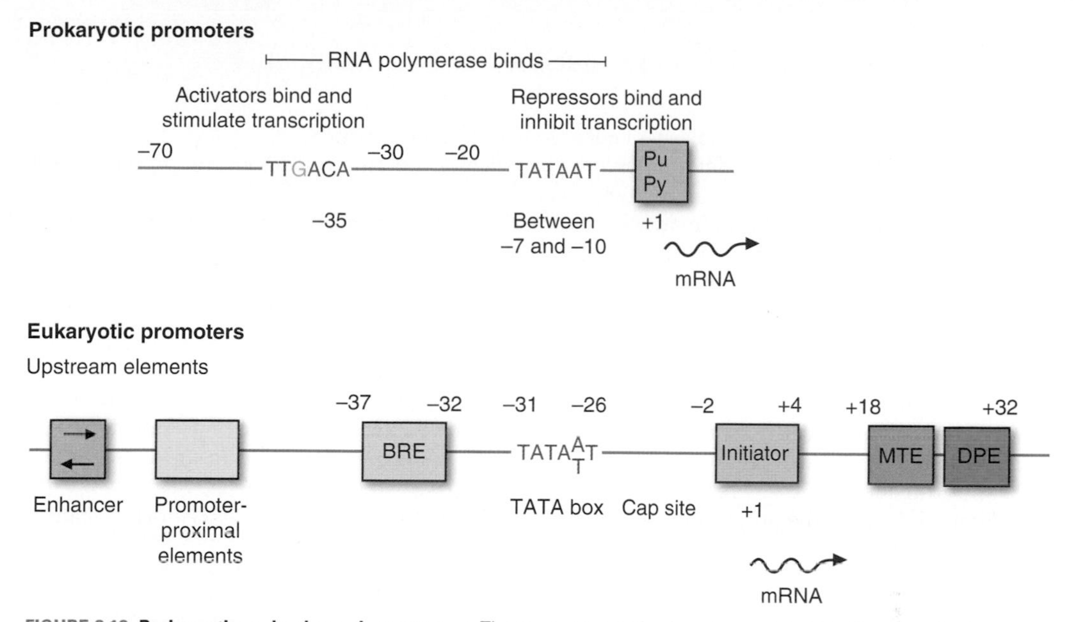

FIGURE 3.18. **Prokaryotic and eukaryotic promoters.** The promoter-proximal region contains binding sites for transcription factors that can accelerate the rate at which RNA polymerase binds to the promoter. BRE, TFIID recognition element, Inr, initiator element; MTE, motif ten element; DPE, downstream promoter element.

a. A **45S precursor** is produced by RNA polymerase I from rRNA genes located in the fibrous region of the **nucleolus.** Many copies of the genes are present, linked together by spacer regions.
b. The **45S precursor** is modified by methylation and undergoes a number of cleavages that ultimately produce 18S rRNA and 28S rRNA; the latter is hydrogen-bonded to a 5.8S rRNA.
c. **18S rRNA** complexes with proteins and forms the 40S ribosomal subunit.
d. The **28S, 5.8S, and 5S rRNAs** complex with proteins and form the 60S ribosomal subunit. 5S rRNA is produced by RNA polymerase III outside of the nucleolus.
e. The **ribosomal subunits migrate** through the nuclear pores into the cytoplasm where they complex with mRNA, forming 80S ribosomes. (Because sedimentation coefficients reflect both shape and particle weight, they are not additive.)
f. rRNA precursors can contain **introns** that are removed during maturation. In some organisms, the enzymatic activity that removes rRNA introns resides in the rRNA precursor. No proteins are required. These autocatalytic RNAs are known as **ribozymes**.

3. **tRNA synthesis**
   a. **RNA polymerase III** is the enzyme that produces tRNA. The promoter is located within the coding region of the gene.
   b. **Primary transcripts** for tRNA are cleaved at the 5′ and 3′ ends.
   c. Some precursors contain **introns** that are removed.
   d. During the processing of tRNA precursors, **nucleotides** are **modified**. Posttranscriptional modification includes the conversion of uridine to pseudouridine (ψ), ribothymidine (T), and dihydrouridine (D). Other unusual nucleotides are also produced.
   e. Addition of the sequence **CCA** to the **3′ end** is catalyzed by nucleotidyl transferase.

CLINICAL CORRELATES

Compounds that inhibit RNA synthesis can be utilized as antibiotics. These **antibiotics** selectively affect bacterial function and have minimal side effects in humans, and are usually selected to treat bacterial infections. **Rifampicin**, which **inhibits the initiation of prokaryotic RNA synthesis**, is used to treat tuberculosis. **α-Amanitin**, derived from the poisonous mushroom *Amanita phalloides,* inhibits **eukaryotic** RNA polymerases, particularly polymerase II. Ingestion of small amounts of α-amanitin initially causes gastrointestinal problems, but can rapidly result in death.

IV. PROTEIN SYNTHESIS (TRANSLATION OF mRNA)

- Translation occurs in the cytoplasm of eukaryotic cells, on ribosomes.
- During translation, mRNA determines the sequence of the amino acids in the protein that is produced.
- mRNA combines with ribosomes, which contain rRNA. Many ribosomes can be attached simultaneously to a single molecule of mRNA, forming a polysome.
- tRNA carries amino acids to the ribosomal site of protein synthesis. The anticodon in each aminoacyl-tRNA combines with the complementary codon in mRNA. A codon is the sequence of three nucleotides in mRNA that specifies a particular amino acid.
- Initiation of a polypeptide chain begins with the amino acid methionine (codon = AUG).
- Subsequently, amino acids are added to the growing polypeptide chain according to the codon sequence in the mRNA. Aminoacyl-tRNAs and GTP provide the energy for chain elongation.
- A protein is synthesized from its N- to its C-terminus, following the codons in the mRNA in the 5′ to 3′ direction.
- When synthesis of the polypeptide is complete, a termination codon (UGA, UAG, or UAA) causes the polypeptide chain to be released.
- Membrane-bound, secreted, and some organelle-targeted proteins are synthesized on the rough endoplasmic reticulum (RER; ribosomes attached to the outer surface of the ER).

A. The genetic code (Table 3.3)

1. The genetic code is the collection of codons that specify all the amino acids found in proteins.
2. A **codon is a sequence of three bases** (triplet) in mRNA (5′ to 3′) that specifies (corresponds to) a particular amino acid. During translation, the successive codons in an mRNA determine the sequence in which amino acids add to the growing polypeptide chain.
   a. **The genetic code is degenerate** (redundant). Each of the 20 common amino acids has at least one codon; many amino acids have numerous codons.
   b. The genetic code is **nonoverlapping** (i.e., each nucleotide is used only once), beginning with a start codon (**AUG**) near the **5′** end of the mRNA and ending with a termination (stop) codon (**UGA, UAG**, or **UAA**) near the 3′ end.
   c. The code is **commaless** (i.e., there are no breaks or markers to distinguish one codon from the next).

table 3.3 The Genetic Code

First Base	Second Base				Third Base
(5′)	U	C	A	G	(3′)
U	Phe	Ser	Tyr	Cys	U
	Phe	Ser	Tyr	Cys	C
	Leu	Ser	Term[a]	Term	A
	Leu	Ser	Term	Trp	G
C	Leu	Pro	His	Arg	U
	Leu	Pro	His	Arg	C
	Leu	Pro	Gln	Arg	A
	Leu	Pro	Gln	Arg	G
A	Ile	Thr	Asn	Ser	U
	Ile	Thr	Asn	Ser	C
	Ile	Thr	Lys	Arg	A
	Met	Thr	Lys	Arg	G
G	Val	Ala	Asp	Gly	U
	Val	Ala	Asp	Gly	C
	Val	Ala	Glu	Gly	A
	Val	Ala	Glu	Gly	G

[a]Term = termination (or stop) codon; the signal to stop translation of a mRNA.

d. The code is **nearly universal**. The same codon specifies the same amino acid in almost all species studied; however, some differences have been found in the codons used in mitochondria.
e. Mitochondrion, the energy-generating organelle of the cell, has its own circular DNA chromosome and protein biosynthetic apparatus. Only a small number of mitochondrial proteins are encoded by the mitochondrial genome; the remainder is encoded by the nuclear genome.
f. The **start codon** (AUG) determines the **reading frame**. Subsequent nucleotides are read in sets of three, sequentially following this codon.

B. Effect of mutations on proteins

1. Mutations in DNA are transcribed into mRNA, and thus can cause changes in the encoded protein.
2. The various types of mutations that occur in DNA have different effects on the encoded protein.
   a. **Point mutations** occur when one base in DNA is replaced by another, altering the codon in mRNA.
      (1) **Silent** mutations do not affect the amino acid sequence of a protein (e.g., CGA to CGG causes no change, since both codons specify arginine).
      (2) **Missense** mutations result in one amino acid being replaced by another (e.g., CGA to CCA causes arginine to be replaced by proline).
      (3) **Nonsense** mutations result in a premature termination of the growing polypeptide chain (e.g., CGA to UGA causes arginine to be replaced by a stop codon).
   b. **Insertions** occur when a base or a number of bases are added to DNA. They can result in a protein with more amino acids than normal.
   c. **Deletions** occur when a base or a number of bases are removed from the DNA. They can result in a protein with fewer amino acids than normal.
   d. **Frameshift mutations** occur when the number of bases added or deleted is not a multiple of three. The reading frame is shifted so that completely different sets of codons are read beyond the point where the mutation starts.

CLINICAL CORRELATES

There are many diseases related to abnormal hemoglobin. Sickle cell anemia results from a **point mutation** (GAG to GTG) that causes valine to replace glutamate at position 6 in the β-globin chain. Hydrophobic interactions between these valine residues on different hemoglobin molecules cause polymerization of sickle cell hemoglobin, which alters the shape of the red blood cells and results in hemolysis. In **hemoglobin Wayne**, deletion of a base causes a **frameshift** that produces the wrong sequence of amino acids in the chain beyond position 127. **Hemoglobin C** is due to a point mutation, also in position 6 of the β-globin chain, which causes a lysine to replace glutamate at this position. This leads to a mild anemia, without sickling. However, HbS/HbC heterozygotes do have substantial sickling taking place, more so than HbA/HbS heterozygotes.

In the **thalassemias** (a group of hemolytic anemias), mutations can affect all steps of RNA metabolism. A thalassemia is an imbalance in the synthesis of the globin proteins; a β-thalassemia is an excess of α subunits, whereas an α-thalassemia is an excess of β subunits. Thalassemias can result from a variety of mutations, some of which are described below. Substitutions in the **TATA box** decrease promoter function. Mutations in **splice junctions** create alternative splice sites. A change in the **polyadenylation site** (AATAAA to AATAGA) results in incorrect processing of the hnRNA, and the abnormal mRNA is degraded. A change from CAG to TAG produces a **stop codon** at position 39 that causes a shortened, nonfunctional protein to be synthesized. These mutations cause insufficient quantities of globin chains to be produced, and an anemia results. Mutations in introns that create a cryptic splice site will also lead to reduced synthesis of one of the globin chains.

C. Formation of aminoacyl-tRNAs (Fig. 3.19)

Amino acids are activated and attached to their corresponding **tRNAs** by highly specific enzymes known as aminoacyl-tRNA synthetases.

1. Each **aminoacyl-tRNA synthetase** recognizes a particular amino acid and the tRNAs specific for that amino acid.

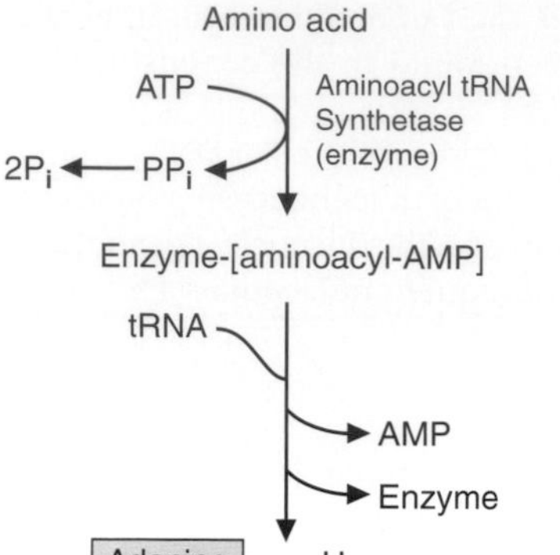

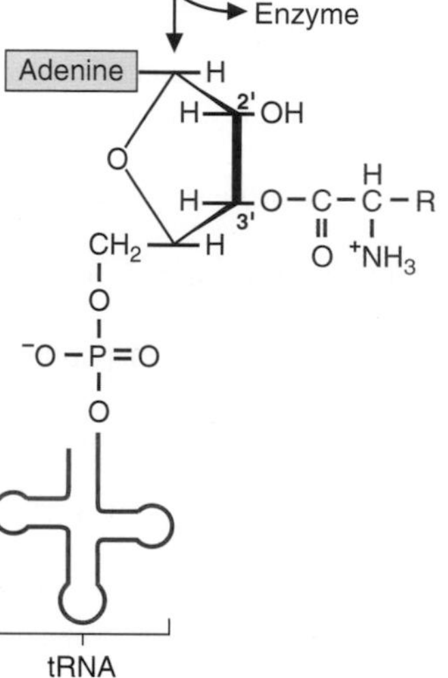

FIGURE 3.19. Formation of aminoacyl-tRNA. The amino acid is covalently linked to either the 3′-hydroxyl of the terminal ribose (as shown) or the 2′-hydroxyl of the terminal ribose.

2. An **amino acid** first reacts with ATP, forming an enzyme (aminoacyl-AMP) complex and pyrophosphate, which is cleaved to 2 P_i.
3. The **aminoacyl-AMP** then **forms an ester** with the 2′- or 3′-hydroxyl of a tRNA specific for that amino acid, producing an aminoacyl-tRNA and AMP.
4. Once an amino acid is attached to a tRNA, insertion of the amino acid into a growing polypeptide chain depends on the codon–anticodon interaction (Fig. 3.20).

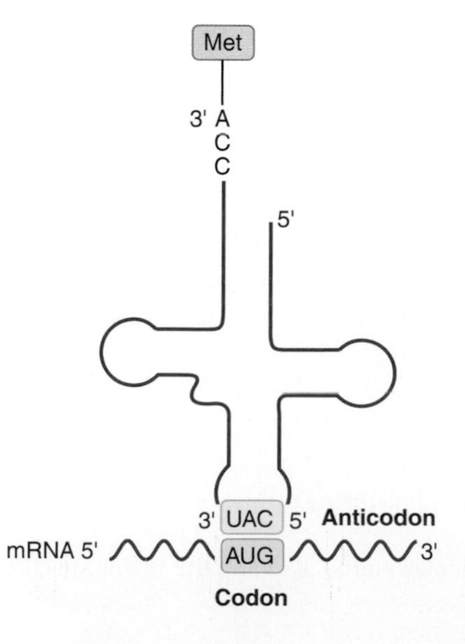

FIGURE 3.20. Antiparallel binding of aminoacyl-tRNA to mRNA.

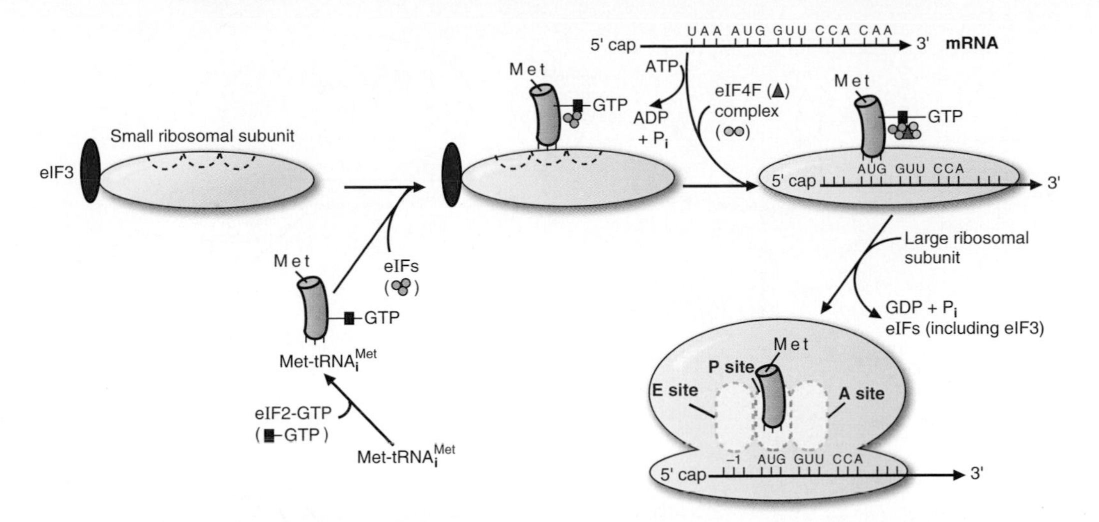

FIGURE 3.21. The initiation reactions of protein synthesis. eIFs are initiation factors in eukaryotes (IFs in prokaryotes). The eukaryotic initiation sequence is shown.

D. Initiation of translation (Fig. 3.21)

1. **In eukaryotes, methionyl-tRNA$_i^{Met}$** binds to the small **ribosomal subunit**. The **5′ cap** of the mRNA binds to the small subunit, and the first AUG codon base-pairs with the anticodon on the methionyl-tRNA$_i^{Met}$. The methionine that initiates protein synthesis is subsequently removed from the N-terminus of the polypeptide.
   a. **In bacteria**, the methionine that initiates protein synthesis is **formylated** and is carried by tRNA$_f^{Met}$.
   b. **Prokaryotes do not contain a 5′ cap** on their mRNA. An mRNA sequence upstream from the translation start site (the Shine–Dalgarno sequence) binds to the 3′ end of 16S rRNA.
2. The **large ribosomal subunit binds**, completing the initiation complex.
   a. The methionyl-tRNA$_i^{Met}$ is bound at the **P** (peptidyl) **site** of the complex.
   b. The **A** (acceptor or aminoacyl) **site** of the complex is unoccupied.
   c. The **E** (ejection) **site** is unoccupied, and is used to remove free tRNA from the ribosome after a peptide bond has been created between the two amino acids carried by tRNA.
3. **Initiation factors, ATP,** and **GTP** are required for the formation of the initiation complex.
   a. The **initiation factors** are designated as IF-1, IF-2, and IF-3 in prokaryotes. In eukaryotes, they are designated as eIF-1, eIF-2, and so on. Seven or more may be present.
   b. The release of the initiation factors involves hydrolysis of GTP to GDP and P_i.
4. The differences in initiation between prokaryotic and eukaryotic cells are summarized in Table 3.4.

table 3.4 Differences between Eukaryotes and Prokaryotes in the Initiation of Protein Synthesis

	Eukaryotes	Prokaryotes
Binding of mRNA to small ribosomal subunit	Cap at 5′-end of mRNA binds eIFs and 40S ribosomal subunit containing tRNA$_i^{Met}$. mRNA is scanned for AUG start codon within the Kozak consensus sequence	Shine–Dalgarno sequence upstream of initiating AUG binds to complementary sequence in 16S rRNA
First amino acid	Methionine	Formyl-methionine
Initiation factors	eIFs (12 or more)	IFs (3)
Ribosomes	80S (40S and 60S subunits)	70S (30S and 50S subunits)

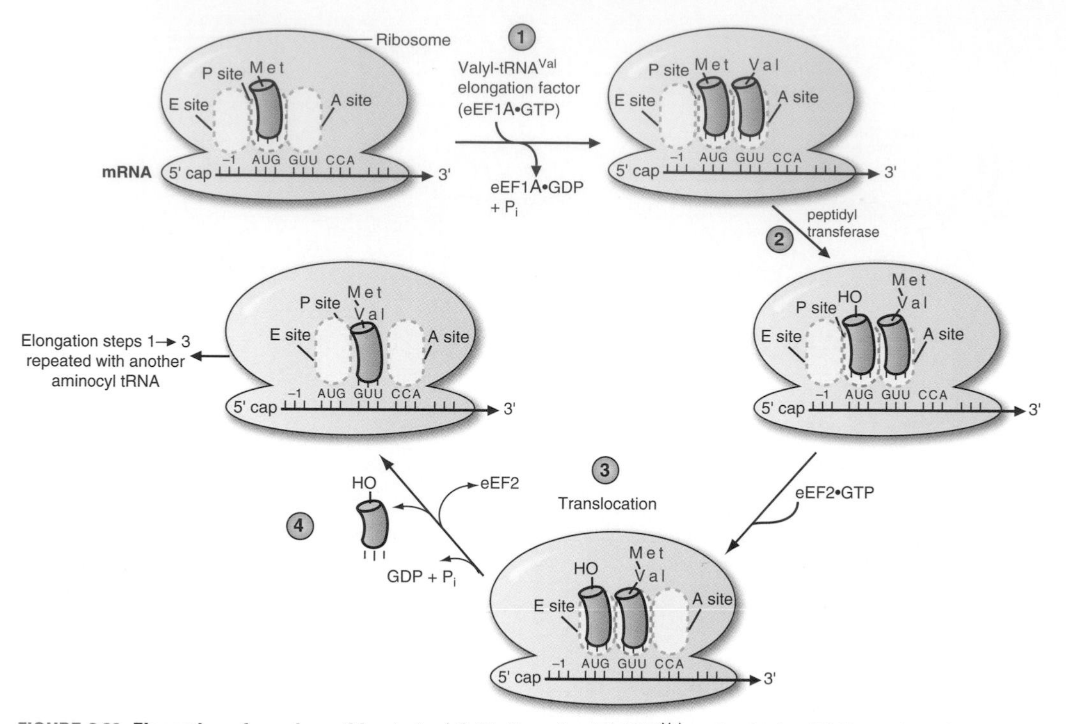

FIGURE 3.22. **Elongation of a polypeptide chain.** (*1*) Binding of valyl-$tRNA^{Val}$ to the A site. (*2*) Formation of a peptide bond. (*3*) Translocation and ejection of the free tRNA. After step 3, step 1 is repeated using the aminoacyl-tRNA for the new codon in the A site. Steps 2 and 3 follow. These three steps keep repeating until termination occurs. eEF, eukaryotic elongation factor.

E. Elongation of polypeptide chains (Fig. 3.22)

The addition of each amino acid to the growing polypeptide chain involves binding of an aminoacyl-tRNA at the A site, formation of a peptide bond, and translocation of the peptidyl-tRNA to the P site.

1. Binding of aminoacyl-tRNA to the A site

- **a.** The **mRNA codon** at the A site determines which aminoacyl-tRNA will bind.
  - **(1)** The **codon and the anticodon** bind by **base-pairing** that is **antiparallel** (see Fig. 3.20).
  - **(2)** Internal methionine residues in the polypeptide chain are added in response to AUG codons. They are carried by $tRNA_m^{Met}$, a second tRNA specific for methionine, which is not used for initiating protein synthesis.
- **b.** An **elongation factor** (EF-Tu in prokaryotes and eEF-1 in eukaryotes) and hydrolysis of GTP are required for binding.

2. Formation of a peptide bond

- **a.** A peptide bond forms between the amino group of the aminoacyl tRNA at the **A site** and the carbonyl of the aminoacyl group attached to the tRNA at the **P site**. The formation of the peptide bond is catalyzed by **peptidyl transferase**, which is an activity of the rRNA in the large ribosomal subunit (an example of a ribozyme).
- **b.** The tRNA at the P site now does not contain an amino acid. It is "uncharged."
- **c.** The growing polypeptide chain is attached to the tRNA in the A site.

3. Translocation of peptidyl-tRNA

- **a.** The peptidyl-tRNA (along with the attached mRNA) moves from the A site to the P site, and the uncharged tRNA moves to the E site before being released from the ribosome. An **elongation factor** (EF-2 in eukaryotes or EF-G in prokaryotes) and the hydrolysis of **GTP** are required for translocation.
- **b.** The next codon in the mRNA is now in the A site.
- **c.** The elongation and translocation steps are repeated until a termination codon moves into the A site.

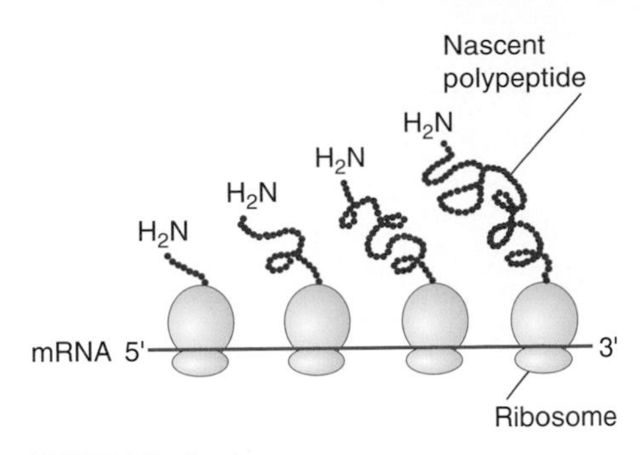

FIGURE 3.23. **A polysome.**

F. Termination of translation

When a termination codon (UGA, UAG, or UAA) occupies the A site, release factors cause the newly synthesized polypeptide to be hydrolyzed from the tRNA and released from the ribosome, and the ribosomal subunits dissociate from the mRNA.

G. Polysomes (Fig. 3.23)

1. More than one ribosome can be attached to a single mRNA at any given time. The complex of mRNA with multiple ribosomes is known as a **polysome**.
2. Each ribosome carries a nascent polypeptide chain that grows longer as the ribosome approaches the 3′ end of the mRNA.

CLINICAL CORRELATES

Compounds that inhibit protein synthesis are also utilized as antibiotics. Streptomycin, tetracycline, chloramphenicol, and erythromycin inhibit protein synthesis **on prokaryotic (70S) ribosomes**, and are used to treat a variety of infections. Because **mitochondria contain 70S-type ribosomes** that function similarly to those in prokaryotic cells, these compounds also **inhibit mitochondrial protein synthesis**. Chloramphenicol is particularly damaging to mitochondrial ribosomes and must be used with caution. **Streptomycin** binds to the 30S ribosomal subunit of prokaryotes and causes misreading of mRNA, thus preventing formation of the initiation complex. **Tetracycline** binds to the 30S ribosomal subunit of prokaryotes and inhibits the binding of aminoacyl-tRNA to the A site. **Chloramphenicol** inhibits the peptidyl transferase activity of the 50S ribosomal subunit of prokaryotes. **Erythromycin** binds to the 50S ribosomal subunit of prokaryotes and prevents translocation. Some inhibitors of protein synthesis are not used clinically but are useful tools for research. **Puromycin** binds at the A site, forms a peptide bond with the growing peptide chain, and prematurely terminates synthesis. It acts in both prokaryotes and eukaryotes. **Cycloheximide** inhibits peptidyl transferase in eukaryotes.

Protein synthesis inhibitors that cause disease have also been identified. **Diphtheria toxin** is produced from phage genes incorporated into the bacterium *Corynebacterium diphtheriae*. The toxin causes diphtheria, a lethal disease of the respiratory tract. The A fragment of the toxin catalyzes the **ADP-ribosylation** of **EF-2**, thus inhibiting translocation in eukaryotes. Cells incorporating the A fragment of the toxin will rapidly die owing to the lack of protein synthesis.

Ricin, a glycoprotein found in the oil of castor beans, is a toxin that inhibits protein synthesis by acting as an *N*-glycosidase, and cleaving a specific adenine base from the 28S rRNA in the large ribosomal subunit, keeping the sugar–phosphate backbone intact. This cleavage inactivates the activity of the ribosome (primarily the initiation factor, and elongation factor binding to the ribosome), thereby inhibiting protein synthesis.

H. Posttranslational processing

After synthesis is completed, **proteins can be modified** by phosphorylation, glycosylation, ADP ribosylation, hydroxylation, and addition of other groups (Fig. 3.24).

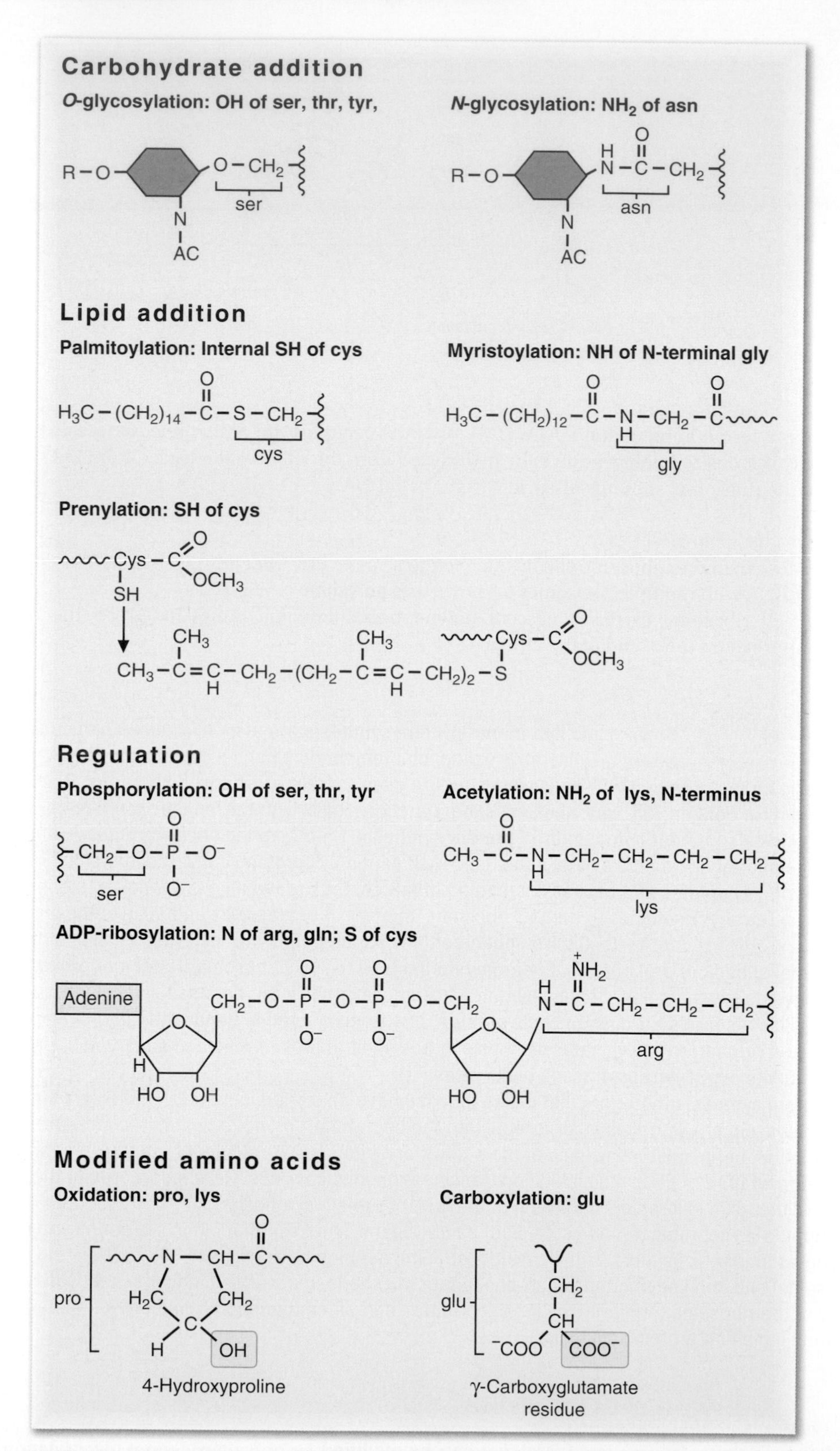

FIGURE 3.24. Posttranslational modifications of amino acids in proteins. The added group is shown in red.

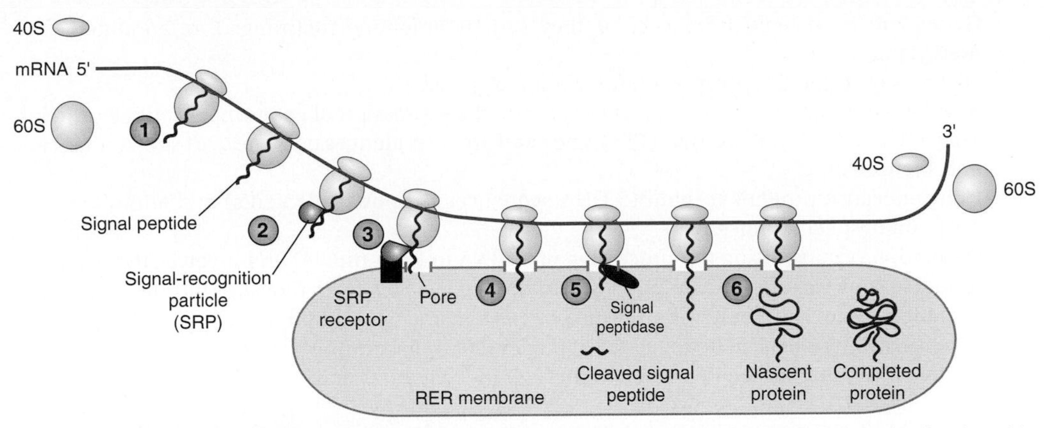

FIGURE 3.25. Synthesis of proteins on the RER. After initiating synthesis in the cytosol the SRP binds to the signal peptide, translocates the ribosome-mRNA complex to the ER, to continue synthesis and to allow the protein to enter the ER lumen as it is synthesized. Once synthesis is complete the ER and Golgi apparatus will appropriately modify and target the protein to its proper location either within or outside the cell.

I. **Synthesis and release of targeted and secretory proteins (Fig. 3.25)**
   1. Certain proteins destined for intracellular organelles are targeted by the inclusion of a particular amino acid sequence within the protein.
      a. **SKL** at the carboxy terminus will target cytoplasmic proteins to the **peroxisomes.**
      b. **KDEL** near the carboxy terminus of proteins in the ER lumen will allow those proteins to remain within the **ER.**
      c. **Highly basic sequences** (such as those containing arginine and lysine, sometimes separated by spacer sequences) will target specific proteins synthesized in the cytoplasm to the **nucleus.**
      d. The addition of **mannose 6-phosphate** to proteins in the ER and Golgi will target those proteins to the **lysosomes.**
   2. **Secretory proteins,** destined for release from the cell, and certain targeted proteins (such as lysosomal proteins) are synthesized on ribosomes attached to the **RER** in eukaryotic cells.
   3. A **hydrophobic signal sequence** at the N-terminus of a secretory protein causes the nascent protein to pass into the lumen of the RER. The signal sequence is cleaved from the N-terminus, and the protein may be glycosylated within the RER.
   4. The protein travels in vesicles to the **Golgi,** where it may be glycosylated further and is packaged in secretory vesicles.
   5. **Secretory vesicles** containing the protein travel from the Golgi to the cell membrane. The protein is released from the cell by **exocytosis.**
   6. If the protein is one to be targeted to an intracellular organelle, the protein binds to specific receptors, which are incorporated into secretory vesicles formed in the Golgi, and the vesicle travels to the target organ to deliver the protein to the organelle.

V. REGULATION OF PROTEIN SYNTHESIS

- Regulation of protein synthesis in prokaryotes occurs mainly at the transcriptional level, and involves genetic units known as operons.
- Operons contain promoter regions where proteins bind and facilitate or inhibit the binding of RNA polymerase.
- When RNA polymerase transcribes the structural genes of an operon, a polycistronic mRNA (i.e., an mRNA that codes for more than one polypeptide) is produced.
- In eukaryotes, regulation of protein synthesis can occur by modification of DNA or at the level of transcription within the nucleus, processing of mRNA in the nucleus, or translation in the cytoplasm.

- Genes can be deleted from cells or they can be amplified, rearranged, or modified (e.g., methylated).
- Histones nonspecifically repress transcription of genes.
- Regulatory elements in DNA sequences control the expression of genes that produce proteins. They include the basal promoter (TATA box and other sequences near the start site), enhancers, and silencers.
- Inducers cause proteins to bind to DNA sequences (response elements) and stimulate transcription of specific genes.
- Regulation occurs during the processing of hnRNA to form mRNA, and involves the use of alternative start sites for transcription, alternative splice sites for removal of introns, alternative polyadenylation sites for addition of the poly(A) tail, and RNA editing.
- Synthesis of proteins can be regulated at the level of translation.
- Synthesis of proteins can be regulated through the degradation of the mRNA by silencing RNA (miRNAs).

A. Regulation of protein synthesis in prokaryotes

1. Relationship of protein synthesis to nutrient supply

a. **Prokaryotes** respond to changes in their supply of **nutrients** in a way that allows them to obtain or conserve energy most efficiently.

(1) Prokaryotes, such as *E. coli*, require a source of **carbon**, which is usually a sugar that is oxidized for energy.

(2) A source of **nitrogen** is also required for the synthesis of amino acids from which structural proteins and enzymes are produced.

b. *E. coli* uses **glucose** preferentially whenever it is available. The enzymes in the pathways for glucose utilization are made **constitutively** (i.e., they are constantly being produced).

c. **If glucose is not present** in the medium but another sugar is available, *E. coli* produces the enzymes and other proteins that allow the cell to derive energy from that sugar. The process by which the synthesis of the enzymes is regulated is called **induction**.

d. **If an amino acid is present** in the medium, *E. coli* does not need to synthesize that amino acid and conserves energy by ceasing to produce the enzymes required for its synthesis. The process by which the synthesis of these enzymes is regulated is called **repression**.

2. Operons

a. An operon is a **set of genes** that are **adjacent** to one another in the genome and are **coordinately controlled**; that is, the genes are either all turned on or all turned off.

b. The **structural genes** of an operon **code** for a series of different **proteins**.

(1) A single **polycistronic mRNA** is transcribed from an operon. This single mRNA codes for all the proteins of the operon.

(2) A series of **start** and **stop codons** on the polycistronic mRNA allows a number of different proteins to be produced at the translational level from the single mRNA.

c. Transcription begins near a **promoter region**, located upstream from the group of structural genes.

d. Associated with the promoter is a short sequence, the **operator**, which determines whether the genes are expressed or not.

e. **Binding of a repressor protein** to the operator region prevents the binding of RNA polymerase to the promoter and **inhibits transcription** of the structural genes of the operon (Fig. 3.26).

(1) Repressor proteins are encoded by regulatory genes, which may be located anywhere in the genome.

3. Induction (Fig. 3.27)

a. Induction is the process whereby an **inducer** (a small molecule) **stimulates** the **transcription** of an operon.

b. The inducer is frequently a sugar (or a metabolite of the sugar), and the proteins produced from the inducible operon allow the sugar to be metabolized.

(1) The inducer binds to the **repressor**, inactivating it.

(2) The **inactive repressor does not bind** to the operator.

(3) **RNA polymerase**, therefore, can **bind** to the promoter and **transcribe** the operon.

(4) The structural **proteins** encoded by the operon are **produced**.

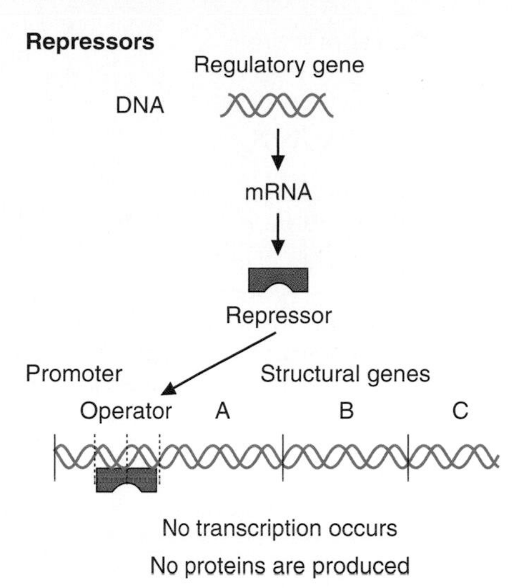

FIGURE 3.26. **Regulation of operons by repressors.** When the repressor protein is bound to the operator, RNA polymerase cannot bind, and transcription therefore does not occur.

c. The lactose (***lac***) operon is **inducible.**
   (1) A metabolite of lactose, **allolactose**, is the inducer.
   (2) Proteins produced by the genes of the *lac* operon allow the cell to oxidize lactose as a source of energy. Gene Z produces a **β-galactosidase**; gene Y, a lactose permease; and gene A, a transacetylase.
   (3) The *lac* operon is induced only in the **absence of glucose**. It exhibits **catabolite repression** (see section below).

4. **Repression**
   a. Repression is the process whereby a corepressor (a small molecule) **inhibits** the **transcription** of an operon (Fig. 3.28).

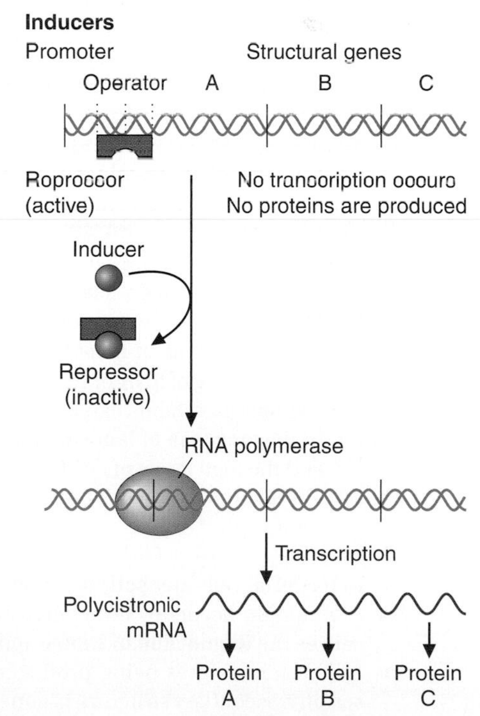

FIGURE 3.27. **An inducible operon (e.g., the *lac* operon).** If the inducer is absent, the repressor is active and binds to the operator, preventing RNA polymerase from binding. Thus, transcription does not occur. If the inducer is present, it binds to and inactivates the repressor, which then does not bind to the operator. Therefore, RNA polymerase can bind and transcribe the structural genes.

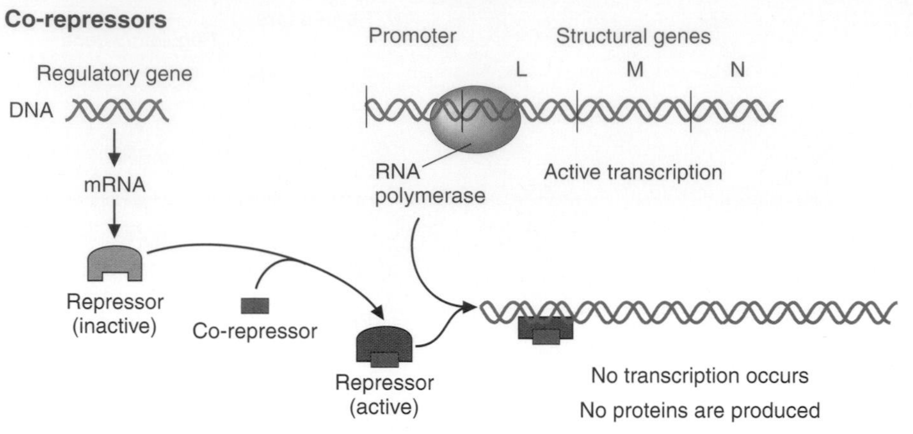

FIGURE 3.28. **A repressible operon.** The repressor is inactive until a small molecule, the co-repressor, binds to it. The repressor-co-repressor complex binds to the operator and prevents transcription.

b. The **corepressor** is usually an amino acid, and the proteins produced from the repressible operon are involved in the synthesis of the amino acid.
  (1) The **corepressor binds to the repressor**, activating it.
  (2) The **active repressor binds to the operator**.
  (3) **RNA polymerase**, therefore, cannot bind to the promoter, and the operon is not transcribed.
  (4) The cell stops producing the structural proteins encoded by the operon.

c. The tryptophan (***trp***) operon is **repressible**.
  (1) **Tryptophan** is the corepressor.
  (2) The proteins encoded by the *trp* operon are involved in the synthesis of tryptophan.
  (3) The *trp* operon is repressed in the presence of tryptophan, since cells do not need to make the amino acid if it is present in the growth medium.

5. Positive control

a. Some operons are turned on by mechanisms that **activate transcription**.

b. When the repressor of the arabinose (*ara*) operon binds arabinose, it changes the conformation and becomes an activator that stimulates the binding of RNA polymerase to the promoter. The operon is then transcribed, and the proteins required for the oxidation of arabinose are produced.

6. Catabolite repression (Fig. 3.29)

a. Cells preferentially use **glucose** when it is available.

b. Some operons (e.g., *lac* and *ara*) are not expressed when glucose is present in the medium. These operons require **cAMP** for their expression.
  (1) Glucose causes cAMP levels in the cells to decrease.
  (2) When **glucose decreases, cAMP levels rise**.
  (3) **cAMP** binds to the catabolite-activator protein **(CAP)**.
  (4) The **cAMP–protein complex** binds to a site near the **promoter** of the operon and facilitates binding of RNA polymerase to the promoter.

c. The ***lac* operon** exhibits catabolite repression.
  (1) In the **presence of lactose** and the **absence of glucose**, the *lac* repressor is inactivated, and the high levels of cAMP facilitate the binding of RNA polymerase to the promoter.
  (2) The **operon is transcribed**, and the proteins that allow the cells to utilize lactose are produced.

7. Attenuation

a. In bacterial cells, **transcription and translation** occur **simultaneously.**

b. Attenuation occurs by a mechanism by which **rapid translation** of the nascent transcript causes the **termination of transcription**.

c. As the transcript is being produced, if ribosomes attach and **rapidly translate** the transcript, a secondary structure is generated in the mRNA that is a **termination signal** for RNA polymerase.

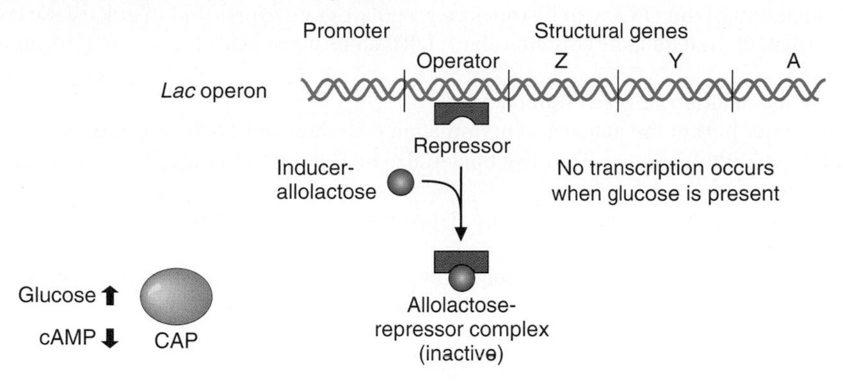

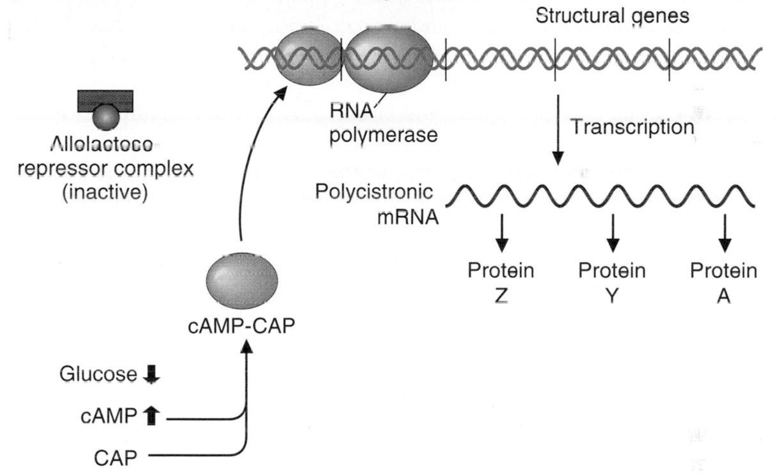

FIGURE 3.29. **Catabolite repression.** The operon is transcribed only when glucose is low. Cyclic adenosine monophosphate (cAMP) is elevated, and the inducer binds to the repressor, inactivating it. Under these conditions, the cAMP-CAP complex forms and binds to the DNA, facilitating the initiation of transcription by RNA polymerase. As shown in this figure, the *lac* operon exhibits catabolite repression.

d. If **translation is slow,** this termination structure does not form, and **transcription continues.**
   (1) Multiple codons for the amino acid are located near the translation start site of the mRNA.
   (2) When cells contain low levels of the amino acid (which is produced by the enzymes encoded by the operon), less aminoacyl-tRNA is available to bind to these codons, and translation slows down.
e. The *trp* operon, as well as other amino acid biosynthetic operons, is regulated by attenuation.

8. **Factors, such as sigma, affect RNA polymerase activity.** These factors bind to the core RNA polymerase and increase its ability to bind to specific promoters.

B. Differences between eukaryotic and prokaryotic cells important for the regulation of gene expression

1. **Eukaryotic cells undergo differentiation,** and the organisms go through various developmental stages.
2. **Eukaryotes contain nuclei.** Therefore, transcription is separated from translation. In prokaryotes, transcription and translation occur simultaneously.

3. **The DNA is complexed with histones in eukaryotes**, but not in prokaryotes. Regulating the association of the DNA with histones can regulate gene expression in eukaryotic cells.
4. The **mammalian genome** contains about **1,000 times more DNA** than *E. coli* (10^9 versus 10^6 base pairs).
5. Most **mammalian** cells are **diploid**.
6. The **major part of the genome** of mammalian cells **does not code for proteins**.
7. **Some eukaryotic genes**, like most bacterial genes, **are unique** (i.e., they exist in one or a small number of copies per genome).
8. **Other eukaryotic genes**, unlike bacterial genes, have **many copies** in the genome (e.g., genes for tRNA, rRNA, histones).
9. Relatively **short, repetitive DNA sequences** are dispersed throughout the eukaryotic genome. They do not code for proteins (e.g., Alu sequences).
10. **Eukaryotic genes contain introns.** Bacterial genes do not.
11. **Bacterial genes** are organized in **operons** (sets that are under the control of a single promoter). **Each eukaryotic gene has its own promoter.**

C. Regulation of protein synthesis in eukaryotes

Regulation can result from changes in genes or from mechanisms that affect transcription, processing and transport of mRNA, mRNA translation, or mRNA stability.

1. **Changes in genes**
   a. **Genes can be lost** (or partially lost) from cells so that functional proteins can no longer be produced (e.g., during differentiation of red blood cells).
   b. **Genes can be amplified.** For example, the drug methotrexate causes hundreds of copies of the gene for the enzyme dihydrofolate reductase to be produced, which results in resistance to the drug.
   c. **Segments of DNA can move** from one location to another on the genome, associating with each other in various ways so that different proteins are produced.
      (1) A number of different potential sequences (or arrangements) occur for various portions of an antibody-producing gene.
      (2) During differentiation of lymphocytes, specific sequences are selected and rearranged so that they are adjacent to each other in the genome and can act as a single transcriptional unit for a specific antibody.
   d. **Modification** of the **bases** in **DNA** affects the **transcriptional activity** of a gene.
      (1) Cytosine can be methylated at its 5 position, which often occurs in CpG islands within promoter regions.
      (2) The greater the extent of methylation, the less readily a gene is transcribed.
         (a) Globin genes are more extensively methylated in nonerythroid cells than in erythroid cells, in which they are expressed.
         (b) Different methylation patterns by males and females form the basis for imprinting (see Chapter 10).
2. **Regulation of the level of transcription**
   a. **Histones,** which are small, basic proteins associated with the DNA of eukaryotes, act as nonspecific repressors.
      (1) **Histone acetyltransferases, or acetylases (HAT or HAC)** will acetylate lysine side chains on histones, which reduces the charge attraction between histones and DNA (Fig. 3.30).
      (2) **Histone deacetylases (HDAC)** will remove the acetate groups from histones, thereby allowing histones to reassociate with the DNA.
      (3) **Heterochromatin** is the tight association of histones and DNA, and represents the transcriptionally inactive areas of the genome.
      (4) **Euchromatin** refers to the transcriptionally active areas of the genome in which histone association with the DNA has been reduced.
   b. The **expression** of specific genes is stimulated by **positive** mechanisms.
   c. **Inducers** (e.g., steroid hormones) enter cells, bind to protein receptors, interact with chromatin in the nucleus, and **activate specific genes** (see Chapter 4).
   d. Some genes have **more than one promoter**. Thus, the promoter that is used can differ under varying physiologic conditions or in different cell types.

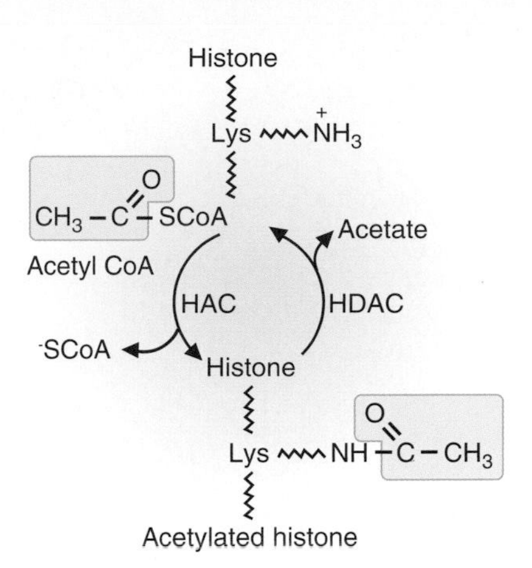

FIGURE 3.30. Histone acetylation. HAC, histone acetylase; HDAC, histone deacetylase.

3. **Chromatin remodeling:** Nucleosome displacement such that transcription can occur
   a. An ATP-driven **chromatin remodeling complex** will bind to the regions of DNA that contain acetylated histones. **Bromodomains** on proteins within the complex recognize the acetylated histones. Once bound, using ATP as an energy source, the complex will move and displace histones to free up an area of DNA for transcription.
   b. **Histone acetylase activity** is often associated with transcription factors that bind to the region of DNA that needs to be transcribed, facilitating the removal of histones from the DNA and binding of the transcription apparatus (see Fig. 3.30).
   c. Figure 3.31 is a representation of the factors involved in regulating the transcription of a gene that contains a TATA sequence, enhancer sequences, and a hormone-response element (HRE). Note that certain proteins (coactivators) do not bind DNA, but rather bind to, and complex with, the DNA-binding proteins.
4. **Regulation during processing and transport of mRNA**
   Regulatory mechanisms that occur during capping, polyadenylation, and splicing can alter the amino acid sequence or the quantity of the protein produced from the mRNA. Editing of mRNA also occurs, and the rate of degradation of mRNA is also regulated.
   a. **Alternative splice sites** can be used to produce different mRNAs.
      (1) The use of different splice sites results in the production of different proteins from the calcitonin gene in the thyroid gland and the brain (Fig. 3.32).
   b. **Alternative polyadenylation sites** can be used to generate different mRNAs.
      (1) Lymphocytes produce a membrane-bound IgM antibody at one stage of development and a soluble form that is secreted at a later stage. The gene for this antibody contains two polyadenylation sites, one after the last two exons (which code for a hydrophobic amino acid sequence) and one before these exons.
      (2) When cleavage and poly(A) addition occur after the last two exons, the antibody contains a hydrophobic region that anchors it in the cell membrane. When polyadenylation occurs at the first site, the antibody lacks the hydrophobic tail and is secreted from the cell.
   c. **mRNAs can be degraded** by nucleases after their synthesis in the nucleus and before their translation in the cytoplasm.
      (1) mRNAs have different half-lives. Some are degraded more rapidly than others.
      (2) Interferon stimulates the synthesis of 2′,5′-oligo(A), which activates a nuclease that degrades mRNA.
   d. **RNA editing** involves the alteration ("editing") of bases in mRNA after transcription (Fig. 3.33).
   e. **Small, interfering RNA (SiRNA)**
      (1) **Gene silencing** can occur through the use of small RNA products (miRNA), which can either block the translation of a target mRNA or induce degradation of the target mRNA.
      (2) **miRNA** molecules are the products of many genes scattered throughout the chromosome, some even located in the introns of the genes they regulate.

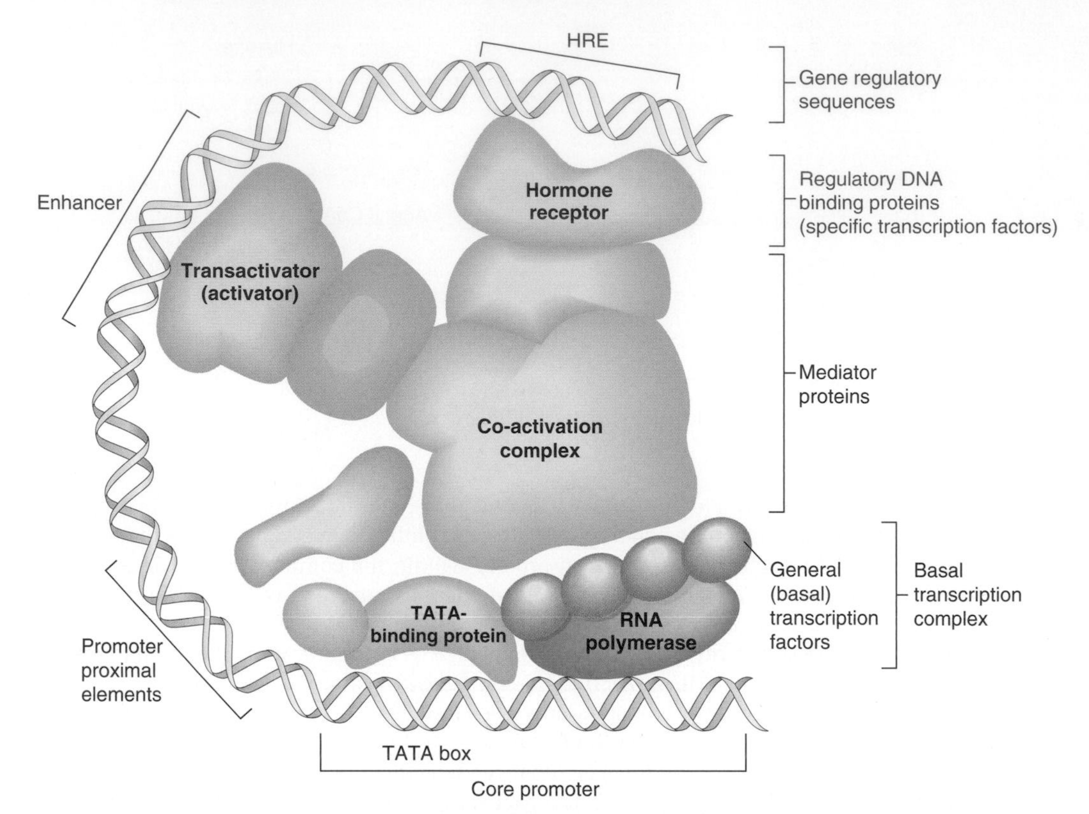

FIGURE 3.31. The gene regulatory control region consists of the promoter region and additional gene regulatory sequences, including enhancers and hormone response elements (HRE). In this case, a promoter containing a TATA box is shown. Gene regulatory proteins that bind directly to DNA (regulatory DNA-binding proteins) are usually called specific transcription factors or transactivators; they may be either activators or repressors of the transcription of specific genes. The specific transcription factors bind mediator proteins (co-activators or co-repressors) that interact with the general transcription factors of the basal transcription complex. The basal transcription complex contains RNA polymerase and associated general transcription factors (TFII factors) and binds, in this case, to the TATA box of the promoter, initiating gene transcription.

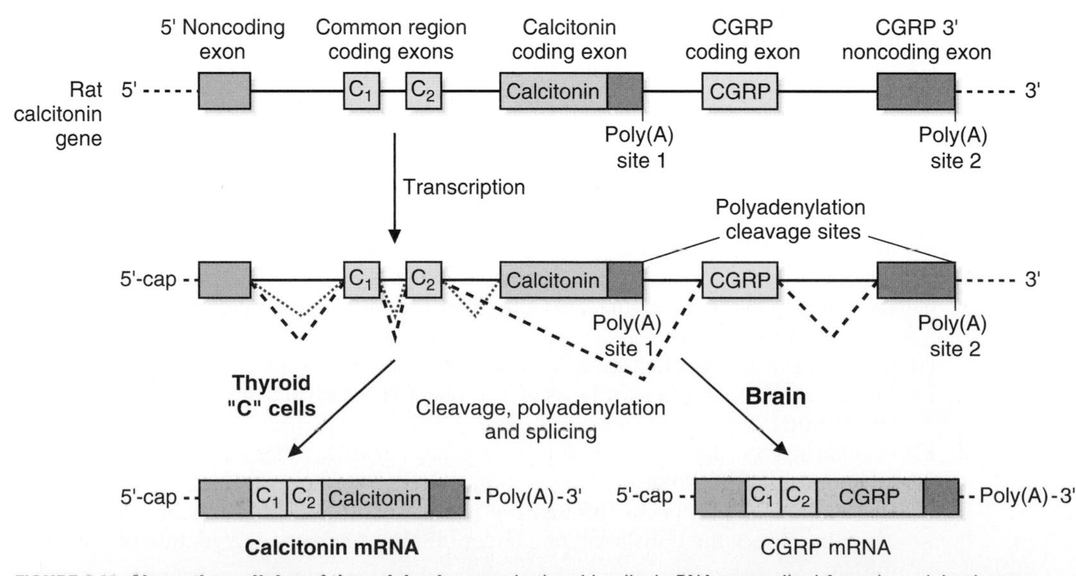

FIGURE 3.32. Alternative splicing of the calcitonin gene. In thyroid cells, hnRNA transcribed from the calcitonin gene is processed to form the mRNA that produces calcitonin. In the brain, the same transcript of this gene is spliced differently. The first polyadenylation site is cleaved out, and a second polyadenylation site is used. The protein product is the calcitonin gene-related protein (CGRP).

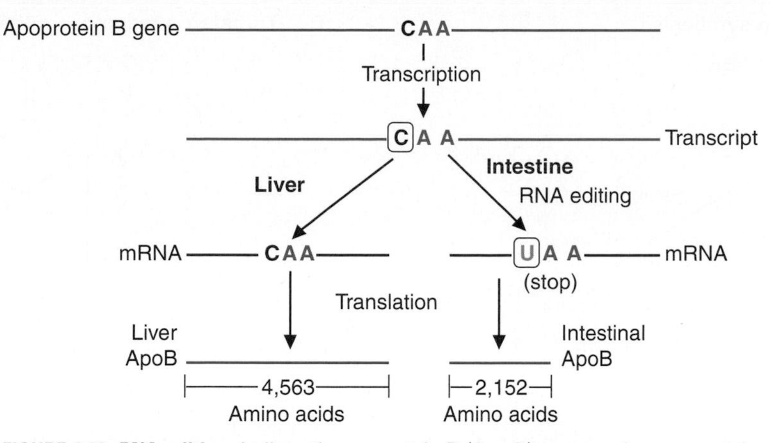

FIGURE 3.33. **RNA editing.** In liver, the apoprotein B (Apo B) gene produces a protein (Apo B-100) that contains more than 4000 amino acids. It is the major apoprotein of very low density lipoprotein (VLDL). In intestinal cells, the same gene produces a protein that contains only 48% of this number of amino acids. This protein (Apo B-48) is the major apoprotein of chylomicrons. "Editing" of mRNA (conversion of a C to a U) generates a stop codon in the intestinal mRNA.

(3) miRNAs are synthesized in the nucleus and processed to form an active molecule that will bind to the target RNA and ablate its expression (Fig. 3.34).

(4) The use of chemically synthesized double-stranded RNA molecules will generate siRNA in the cells, and holds promise as a therapeutic tool of the future.

5. Protein synthesis can be **regulated at the translational level,** during the initiation or elongation reactions.

a. Heme stimulates the synthesis of globin by preventing the phosphorylation and consequent inactivation of eIF-2, a factor involved in the initiation of protein synthesis.

b. Interferon stimulates the phosphorylation of eIF-2, causing inhibition of initiation.

c. Iron-response elements (IREs) in mRNA for ferritin (an iron storage protein) and the transferrin receptor (transferrin carries iron in the circulation, and enters cells via binding to the transferrin receptor) regulate translation of the respective mRNAs. These elements either destabilize the mRNA (transferrin receptor) or allow translation of the mRNA (ferritin) when iron levels are high (Fig. 3.35).

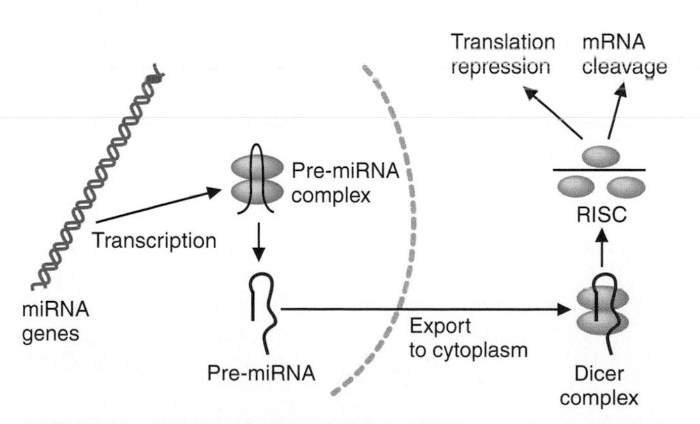

FIGURE 3.34. **miRNA synthesis and action.** miRNA genes are transcribed in the nucleus by RNA polymerase II, generating the primary miRNA, processed to a precursor miRNA (pre-miRNA), and then exported to the cytoplasm. In the cytoplasm the pre-miRNA is further processed by a ribonuclease (Dicer), and resulting double-stranded miRNA is strand selected, with the guide strand (designated in black) entering the RNA-induced silencing complex (RISC). The guide strand of RISC targets the complex to the 3'-untranslated region of the target mRNA, leading to either degradation of the mRNA, or an inhibition of translation.

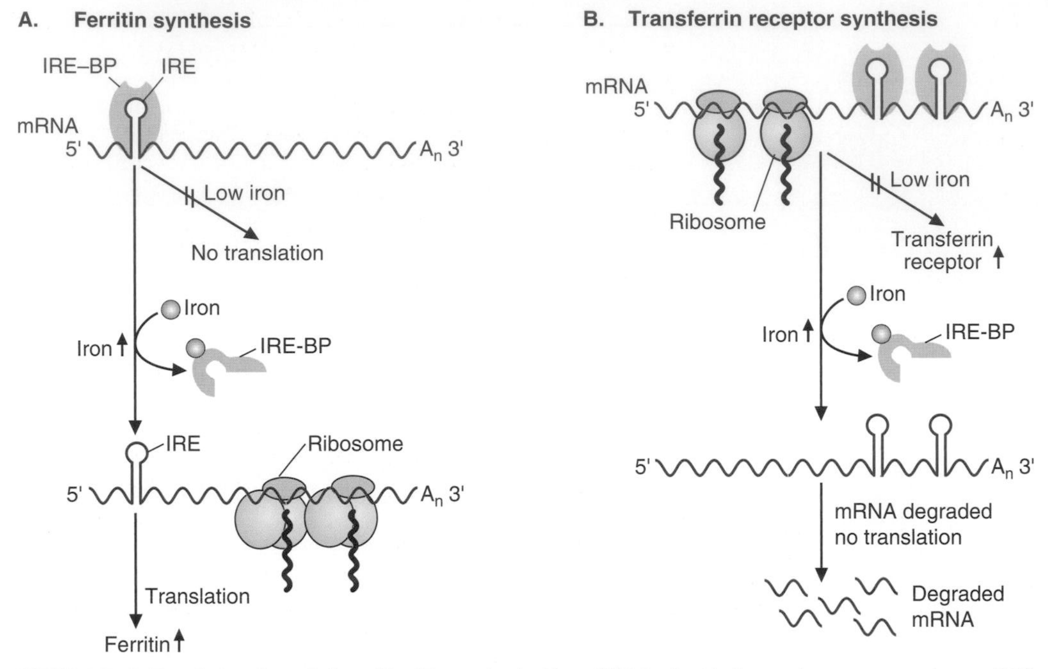

FIGURE 3.35. **A. Translational regulation of ferritin synthesis.** The mRNA for ferritin has an iron response element (IRE). When the iron response-element binding protein IRE-BP does not contain bound iron, it binds to IRE, preventing translation. When IRE-BP binds iron, it dissociates, and the mRNA is translated. **B.** Regulation of degradation of the mRNA for the transferrin receptor. Degradation of the mRNA is prevented by binding of the iron response-element binding protein (IRE-BP) to iron response elements (IRE), which are hairpin loops located at the 3'-end of the transferrin receptor mRNA. When iron levels are high, IRE-BP binds iron and is not bound to the mRNA. The mRNA is rapidly degraded, preventing synthesis of the transferrin receptor.

CLINICAL CORRELATES

Treatment of viral infections. When viruses infect cells, they convert the cells' DNA-, RNA-, and protein-producing machinery to the generation of viral genes and proteins (i.e., to the production of new viruses). Only few drugs are currently effective against viral infections. Azidothymidine (**AZT**), an analog of thymidine, is phosphorylated in the cell and **inhibits retroviral reverse transcriptase** (which is used to make DNA copies of viral RNA) by serving as a **DNA chain terminator**. It has been used to treat **HIV** infections associated with **AIDS**. Other nucleotide analogs, such as dideoxyinosine (**ddI**), also serve as chain terminators. More recently, **inhibitors of the HIV protease** have been produced. These inhibitors prevent the protease from cleaving a polyprotein produced from the viral genome into structural proteins and enzymes required for the assembly of viral particles. A combination of protease inhibitors and DNA chain terminators currently provides the most successful therapy for HIV infections. Viral-infected cells often respond to the virus by producing interferons, which act as antiviral agents, and, in certain cases, as antitumor agents. An interferon-treated cell will exhibit reduced protein synthesis owing to phosphorylation of eIF-2, which complexes with eIF-2B and cannot participate in the initiation of protein synthesis. The inhibition of protein synthesis would reduce the virus's ability to proliferate in interferon-responsive cells.

VI. RECOMBINANT DNA AND MEDICINE

- Newly developed techniques in molecular biology are being used for research, medical diagnosis, and production of therapeutic proteins. They provide hope as future therapy for diseases that are currently considered incurable.

- Restriction enzymes, which cleave within short, specific sequences of DNA, can be used to obtain DNA fragments for study or for insertion into the DNA from other sources. The fusion product is known as chimeric or recombinant DNA.
- Because DNA strands can base-pair with complementary strands of DNA or RNA, a technique known as hybridization has been developed. Labeled DNA can be used as a probe to identify homologous (complementary sequences of) DNA or RNA.
- Gel electrophoresis separates DNA fragments by size.
- The nucleotide sequence of DNA can be determined and used to deduce the amino acid sequence of the protein produced from the DNA.
- Large quantities of DNA can be produced by the polymerase chain reaction (PCR).
- The fragments of DNA obtained, for example, from genomic DNA or DNA copied from mRNA (cDNA), can be amplified by PCR and cloned (i.e., inserted into another organism, where the foreign DNA can be replicated and expressed). The effects of the protein product can then be studied or, in some cases, large quantities of the protein product can be obtained.
- In medicine, recombinant DNA techniques permit the production of specific proteins that are used for therapy or as vaccines. The techniques are also used to diagnose disease, to predict the risk of genetic defects, and to determine parentage or other types of relationships. They have already been used to treat disease (gene therapy).

A. Strategies for obtaining copies of genes or fragments of DNA

1. Short sequences of DNA (**oligonucleotides**) can be synthesized in vitro and used as **primers** for DNA synthesis or as **probes** for the detection of DNA or RNA sequences.
2. **Restriction endonucleases** cleave the DNA into fragments.
   - **a.** Restriction endonucleases recognize short sequences in DNA and cleave both the strands within this region (Fig. 3.36).
   - **b.** Most of the DNA sequences recognized by these enzymes are **palindromes** (i.e., both the strands of DNA have the same base sequence in the 5′ to 3′ direction).
     - **(1)** The enzyme *Eco*R1 cleaves a region between an A and a G on each strand, generating two products.
     - **(2)** The single-stranded regions of the products allow them to reanneal or to recombine with other DNA that has been cleaved by the same restriction endonuclease.
   - **c.** A **DNA fragment**, which contains a **specific gene**, can be isolated from the cellular genome with restriction enzymes. Genes isolated from eukaryotic cells usually contain **introns**, whereas those from bacteria do not.
3. The **mRNA** for a gene can be isolated, and a DNA copy (**cDNA**) can be produced by **reverse transcriptase**. cDNA does not contain introns, nor does it contain the promoter region of a gene, as that region is not transcribed.

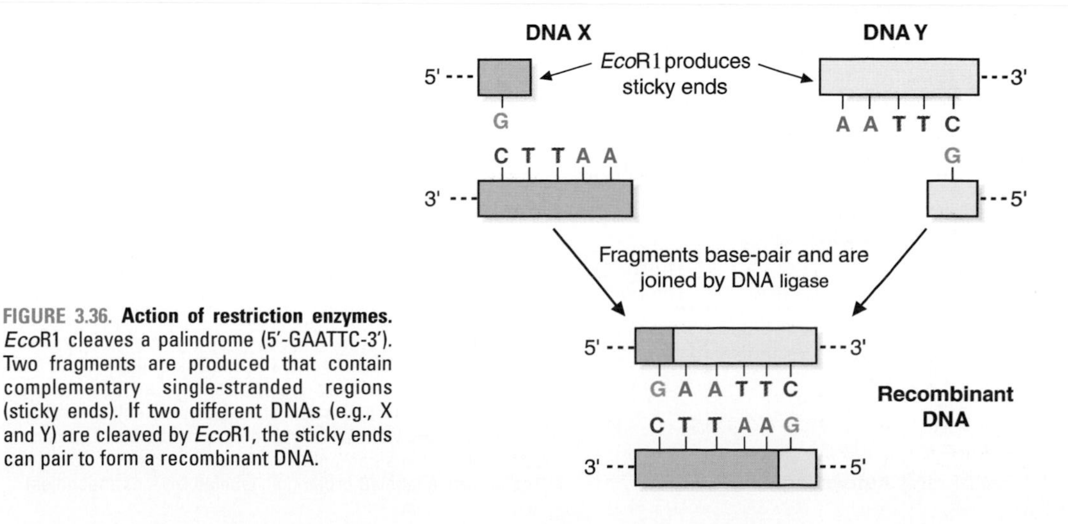

FIGURE 3.36. **Action of restriction enzymes.** *Eco*R1 cleaves a palindrome (5′-GAATTC-3′). Two fragments are produced that contain complementary single-stranded regions (sticky ends). If two different DNAs (e.g., X and Y) are cleaved by *Eco*R1, the sticky ends can pair to form a recombinant DNA.

B. Techniques for identifying DNA sequences

1. Use of probes to detect specific DNA or RNA sequences

- **a.** A **probe** is a single strand of DNA that can **hybridize** (**base-pair**) with a **complementary sequence** on another single-stranded polynucleotide composed of DNA or RNA.
- **b.** The probe must contain a **label** so that it can detect a complementary DNA or RNA. The label may be radioactive (so it can be detected by autoradiography) or a chemical that can be identified, for example, by fluorescence.

2. Gel electrophoresis of DNA

- **a.** Gel electrophoresis **separates** DNA **chains** of varying lengths. Polyacrylamide gels can be used to separate short DNA chains that differ in length by only one nucleotide. Agarose gels separate chains of larger size.
  - **(1)** Because DNA contains negatively charged phosphate groups, it will **migrate** in an electric field **toward the positive electrode**.
  - **(2)** Shorter chains migrate more rapidly through the pores of the gel, so **separation depends on length**.
  - **(3)** DNA **bands** in the gel can be **visualized** by various techniques including staining with dyes (e.g., ethidium bromide) and autoradiography (if the gel contains a radioactive compound, which reacts with a photographic film). **Labeled probes** detect **specific DNA sequences**.
  - **(4)** **Blots** of gels can be made using nitrocellulose paper (Fig. 3.37).
    - **(a)** **Southern blots** are produced when a radioactive **DNA** probe hybridizes with **DNA** on a nitrocellulose blot of a gel.
    - **(b)** **Northern blots** are produced when a radioactive **DNA** probe hybridizes with **RNA** on a nitrocellulose blot of a gel.

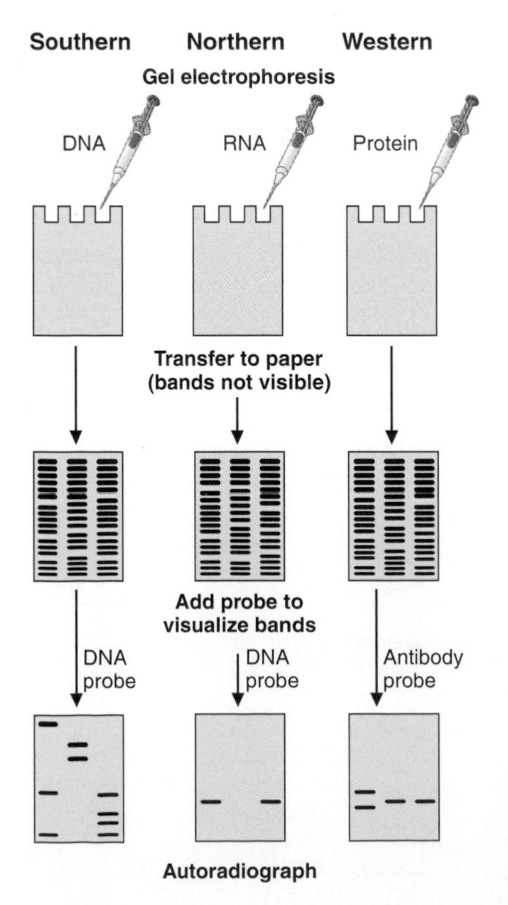

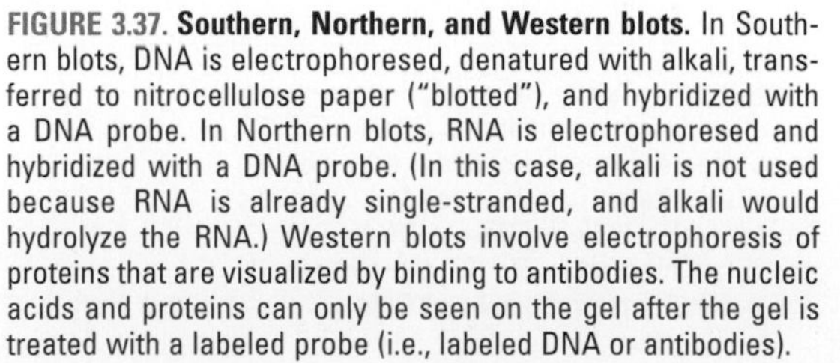
FIGURE 3.37. Southern, Northern, and Western blots. In Southern blots, DNA is electrophoresed, denatured with alkali, transferred to nitrocellulose paper ("blotted"), and hybridized with a DNA probe. In Northern blots, RNA is electrophoresed and hybridized with a DNA probe. (In this case, alkali is not used because RNA is already single-stranded, and alkali would hydrolyze the RNA.) Western blots involve electrophoresis of proteins that are visualized by binding to antibodies. The nucleic acids and proteins can only be seen on the gel after the gel is treated with a labeled probe (i.e., labeled DNA or antibodies).

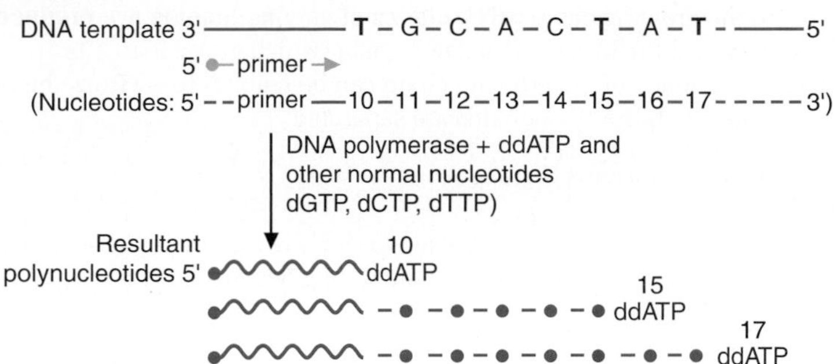

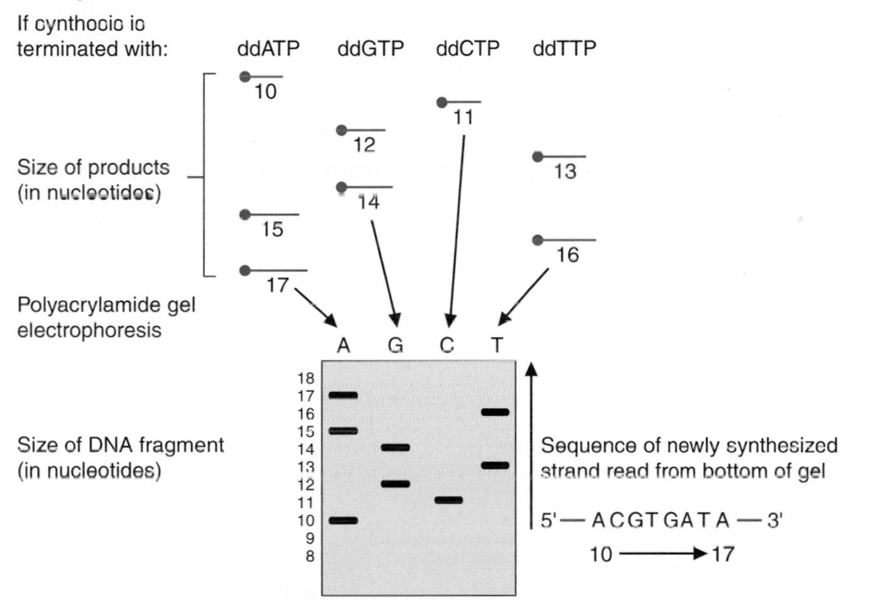

FIGURE 3.38. The Sanger method. A. A reaction mixtures contain one of the dideoxynucleotides, such as ddATP, and some of the normal nucleotide, dATP, which compete for incorporation into the growing polypeptide chain. When a T is encountered on the template strand (position 10), some of the molecules will incorporate a ddATP and the chain will be terminated. Those that incorporate a normal dATP will continue growing until position 15 is reached, where they will incorporate either a ddATP or the normal dATP. Only those that incorporate a dATP will continue growing to position 17. Thus, strands of different length from the 5'-end are produced, corresponding to the position of a T in the template strand. **B.** DNA sequencing by the dideoxynucleotide method. Four tubes are used. Each one contains DNA polymerase, a DNA template hybridized to a primer, plus dATP, dGTP, dCTP, and dTTP. Either the primer or the nucleotides must have a radioactive label, so bands can be visualized on the gel by autoradiography. Only one of the four dideoxyribonucleotides (ddNTPs) is added to each tube. Termination of synthesis occurs where the ddNTP is incorporated into the growing chain. The template is complementary to the sequence of the newly synthesized strand. Automated DNA sequencers utilize fluorescent labeled ddNTPs, and a column to separate the oligonucleotides by size. As samples leave the column their fluorescence is analyzed to determine which base has terminated synthesis of that fragment.

(c) A **Western blot** is a related technique in which **proteins** are separated by gel electrophoresis and probed with **antibodies** that bind a specific protein.

3. **DNA sequencing** by the Sanger dideoxynucleotide method (Fig. 3.38)
   a. **Dideoxynucleotides** are added to solutions in which DNA polymerase is catalyzing the polymerization of a DNA chain.
   b. Because a dideoxynucleotide does not contain a 3′-hydroxyl group, **polymerization of the chain is terminated** wherever a dideoxynucleotide is incorporated into the growing chain.

c. Because the dideoxynucleotide competes with the normal nucleotide for incorporation into the growing chain, **DNA chains of varying lengths are produced**. The shortest chains are nearest the 5′ end of the DNA chain (which grows 5′ to 3′).

d. The sequence of the growing chain can be read (5′ to 3′) from the bottom to the top of the gel on which the DNA chains are separated.

C. Techniques for amplifying DNA sequences

1. PCR

a. PCR is an **in vitro technique** used for rapidly producing large amounts of DNA (Fig. 3.39). It is suitable for clinical or forensic testing because only a very small sample of DNA is required as the starting material.

2. Cloning of DNA

a. DNA from one organism (**"foreign" DNA**, obtained as described earlier) can be **inserted into a DNA vector** and used to **transform** cells from another organism, usually a bacterium, which grows rapidly, replicating the foreign DNA, as well as its own.

b. Large quantities of the foreign DNA can be isolated or, under the appropriate conditions, the DNA can be expressed, and its protein product can be obtained in large quantities.

D. Use of recombinant DNA techniques to detect polymorphisms

Humans differ in their genetic composition. **Polymorphisms** (variations in DNA sequences) occur frequently in the genome both in coding and in noncoding regions. Point mutations cause the simplest type of polymorphisms, but insertions and deletions of varying lengths also occur.

1. Restriction fragment length polymorphism (RFLP)

a. Occasionally, a **mutation** occurs **in a restriction enzyme cleavage site** that is within or tightly linked to a gene. The enzyme can cleave the normal DNA at this site, but not the

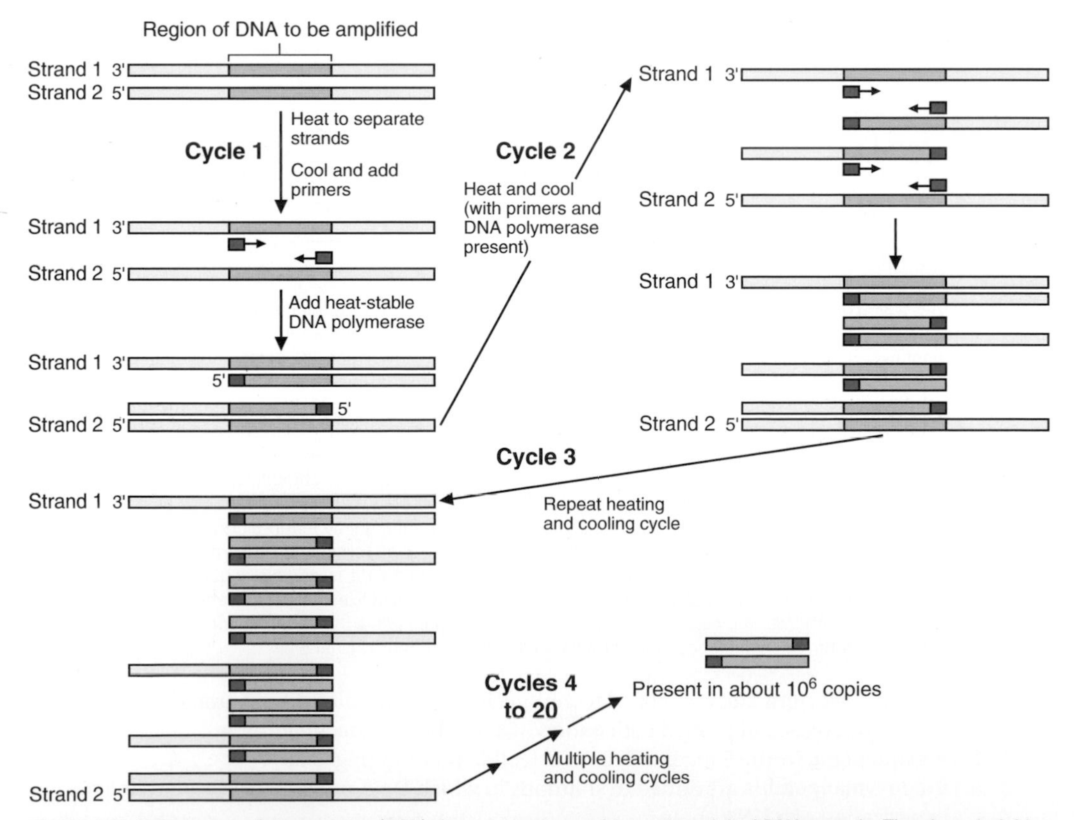

FIGURE 3.39. **Polymerase chain reaction (PCR).** Strand 1 and strand 2 are the original DNA strands. The short dark blue fragments are the primers. After multiple heating and cooling cycles, the original strands remain, but most of the DNA consists of amplified copies of the segment (shown in lighter blue) synthesized by the heat-stable DNA polymerase.

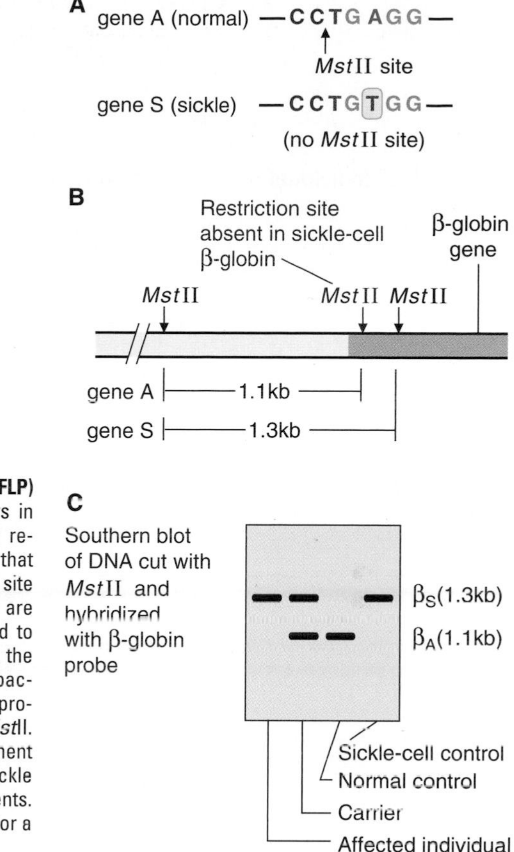

FIGURE 3.40. Restriction fragment length polymorphism (RFLP) caused by loss of a restriction site. If a mutation occurs in a cleavage site for a restriction enzyme, the pattern of restriction fragments differs from normal. **A.** The mutation that causes sickle cell anemia results in the loss of an *Mst*II site in the β-globin gene. **B.** Samples of DNA from individuals are treated with restriction endonucleases and then subjected to electrophoresis on gels. With the Southern blot technique, the restriction fragments on the gel are hybridized with a radioactive cDNA probe for the β-globin gene. The sickle cell allele produces a fragment of 1.3 kilobases (kb) when treated with *Mst*II. A normal allele produces a fragment of 1.1 kb (plus a fragment of 0.2 kb that is not seen on the gel). For a person with sickle cell disease, both alleles produce 1.3-kb restriction fragments. In a normal person, both alleles produce 1.1-kb fragments. For a carrier, both the 1.3- and 1.1-kb fragments are observed.

mutant. Thus, two smaller restriction fragments will be obtained from this region of the normal DNA, compared with only one larger fragment from the mutant (Fig. 3.40).

b. Sometimes, **a mutation creates a restriction site** that is not present in or near the normal gene. In this case, two smaller restriction fragments will be obtained from the mutant, and only one larger fragment from the normal.

c. Normal human DNA has many regions that contain a highly variable number of tandem repeats (**VNTR**). The number of repeats differs from one individual to another (and from one allele to another). Restriction enzymes that cleave on the left and right flanks of a VNTR produce **DNA fragments of variable length**. The length depends on the number of repeats that the DNA contains (Fig. 3.41). The fragments produced by various restriction enzymes from a number of different loci can be used to identify individuals with the accuracy of a fingerprint. Therefore, a technique called **"DNA fingerprinting"** is used to determine parentage or other genetic relationships, or to implicate suspects in criminal cases.

2. Detection of mutations by allele-specific oligonucleotide probes

a. An **oligonucleotide probe** is synthesized that is **complementary** to a region of DNA that contains a **mutation**. A different probe is made for the normal DNA (Fig. 3.42).

b. If the mutant probe binds to a sample of DNA, the sample contains DNA from a mutant allele. If the normal probe binds, the sample contains DNA from a normal allele. If both probes bind, the sample contains DNA from both a mutant and a normal allele (i.e., the person providing the DNA sample is a carrier of the mutation).

3. Testing for mutations by PCR

a. An oligonucleotide complementary to a mutant region is used as a **primer for PCR**. If the primer binds to a DNA sample (i.e., if the sample contains the mutation), amplification of the DNA occurs (i.e., the primer is extended). If the primer does not bind, extension does not occur (i.e., the DNA is normal).

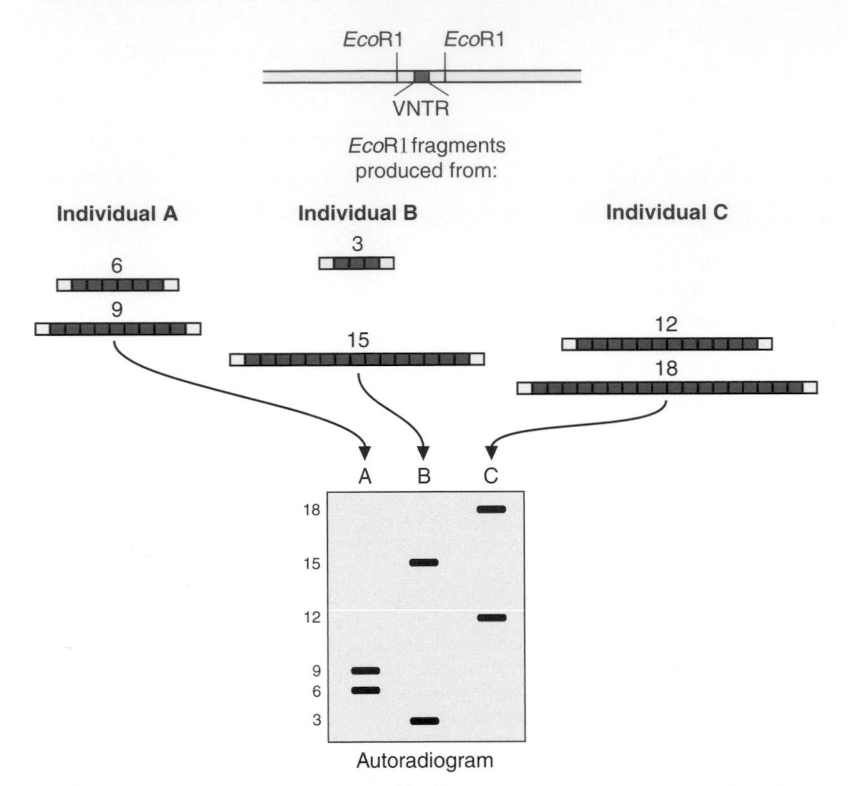

FIGURE 3.41. **Restriction fragments produced from a gene with a variable number of tandem repeats (VNTR).** DNA from three individuals, each with two alleles for this gene and a different number of repeats in each allele, was cleaved, electrophoresed, and treated with a probe for this gene. The length of the fragments depends on the number of repeats that they contain.

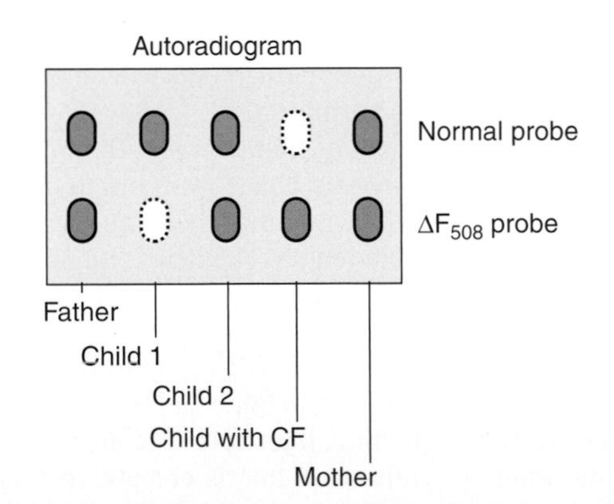

FIGURE 3.42. **The use of oligonucleotide probes to test for cystic fibrosis (CF).** Oligonucleotide probes complementary to the region where a 3-base deletion is located in the CF gene were synthesized. One probe binds only to the mutant (ΔF_{508}), and the other probe binds only to the normal region. DNA was isolated from individuals and a region of the CF gene was amplified by polymerase chain reaction. Two spots were placed on nitrocellulose paper for each person. One spot was treated with the probe for the mutant region of the gene, and the other spot was treated with the probe for the normal region. Dark spots indicate binding of a probe. Only the normal probe binds to the DNA from a normal person, and only the mutant probe binds to the DNA from a person with CF. Both probes bind to DNA from a carrier. In carriers, one allele is normal and the other has the CF mutation.

4. Single nucleotide polymorphisms

a. Single nucleotide polymorphisms (SNPs) have been identified across the human genome as a result of the human genome project. To be considered an SNP, the base change must be present in 1% of the population. SNPs are useful tools for mapping disease genes via positional cloning and in forensic analysis in place of short tandem repeats and the VNTRs. Stringent hybridization conditions (high temperature and low salt) are needed when analyzing for single-base changes between the DNA samples.

E. Alterations in the genetic composition of animals

1. If a **gene** from another organism is **inserted into a fertilized egg**, a **transgenic animal** can be produced. Such animals can be used for research or for other purposes, ranging from the production of human proteins in the milk of transgenic sheep to generation of larger and stronger species.
2. The removal or disruption of genes (**gene knockout**) can be used to develop strains of animals that lack the protein product of the gene. The effects of loss of the protein can then be studied.

F. Mapping of the human genome

The human genome project has led to the sequencing of the human genome, the identification of numerous markers through the genome, such as SNPs, and has identified approximately 25,000 genes in human DNA. The project is now aimed at functional genomics, deducing information about the function of DNA sequences. The project is also addressing the ethical, legal, and societal issues that may arise from the information obtained from the project.

G. Gene therapy

Ultimately, recombinant DNA technology will be used to treat genetic diseases. Already, some diseases (e.g., adenosine deaminase deficiency) have responded to efforts to introduce normal genes into individuals with defective genes. Preventing adverse effects (immune rejection) and promoting long-term expression of the transgene are areas of active research.

H. Proteomics

1. The identification and analysis of all proteins expressed by a given cell under specific conditions constitutes the field of proteomics.
2. The techniques utilized in proteomics (**two-dimensional gel electrophoresis** and **mass-spec** identification of protein fragments) are sensitive enough to allow comparisons between different, but related, cell types (such as a hepatoma cell and a normal liver cell) (Fig. 3.43) and the identification of proteins that differ in expression between the samples being analyzed.

I. Microarrays

1. Screening of thousands of genes simultaneously to determine which alleles of these genes are present in samples obtained from patients, or to compare RNA expression patterns of two different samples.
2. One example of a microarray is "chips," to aid in the diagnosis of infectious disease. The chip contains more than 20,000 distinct oligonucleotides, in an ordered array on the chip, that correspond to vertebrate viruses, bacteria, fungi, and parasites. Patient samples are used as a source of RNA, which is converted to cDNA and used as a probe to bind to the DNA sequences in the chip. Positive hybridization will identify the organism(s) responsible for the patient's symptoms.

CLINICAL CORRELATES Biotechnology is currently being used in the **diagnosis of disease** (e.g., sickle cell anemia, CF, phenylketonuria). Recombinant DNA techniques are used to produce the probes (e.g., cDNA) for **screening human samples**, and they are used to generate large quantities of **proteins for use in therapy** (e.g., human insulin, growth hormone, tissue plasminogen activator, erythropoietin, factor VIII for hemophilia) or as **vaccines** (e.g., hepatitis B). These techniques have been used to introduce normal genes into individuals with defective genes (i.e., in **gene therapy** for inherited disease). Polymorphisms in drug-metabolizing enzymes are also being

(continued)

identified, to determine an effective clinical course for certain individuals. This is accomplished through gene chip microarray experiments.

The identification of particular SNPs have allowed an assessment of risk for an individual acquiring (or being susceptible to) a certain disease. Apolipoprotein E (apoE) has three genes, E2, E3, and E4. Each allele differs by one base, and the protein produced by each allele differs in only one amino acid. Individuals who express one E4 allele are at a greater risk of developing Alzheimer disease than those who do not express the E4 allele. SNP testing is a relatively straightforward procedure to determine which apoE alleles an individual has inherited.

Testing for the presence of HIV in individuals is based on the finding of anti-HIV antibodies in the blood of the tested individuals. This is best accomplished by an initial rapid screening assay (an enzyme-linked immunosorbent assay, ELISA) followed by a Western blot using purified HIV proteins in the gel, and the individual's blood as a source of antibodies for the blot.

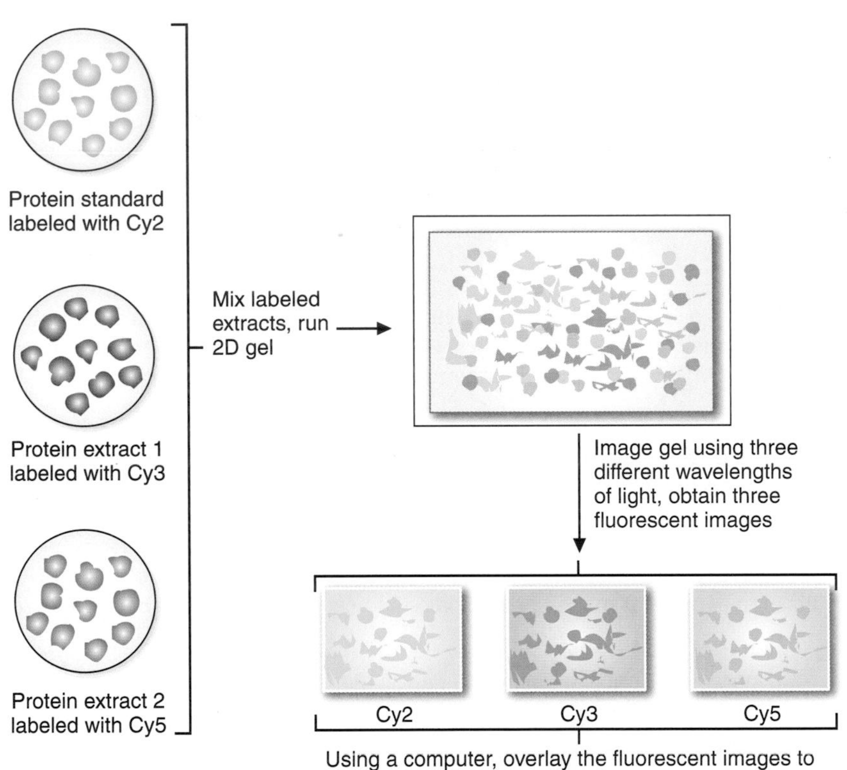

FIGURE 3.43. **Utilizing proteomics to determine if a protein is up- or downregulated.** Proteins from the two different cell types (1 and 2) are isolated and labeled with different fluorescent dyes. The proteins are then separated by two-dimensional gel electrophoresis (the first dimension, or separation, is by charge, and the second dimension is by size), which generates a large number of spots that can be viewed under a fluorescent imaging device, each of these spots corresponding to an individual protein. A computer aligns the spots from the two samples and can determine, by the level of fluorescence expressed at each protein spot, if a protein has been up- or downregulated in one sample compared to the other. Proteins whose expression levels change can then be identified by sensitive techniques involving protein mass spectrometry.

Review Test

Questions 1 to 10 examine your basic knowledge of basic biochemistry and are not in the standard clinical vignette format.

Questions 11 to 35 are clinically relevant, USMLE-style questions.

Basic Knowledge Questions

1. Which one of the following is true for a double-stranded DNA molecule?

	[A]=[T]	[U]=[A]	[G]=[C]	[C]=[T]	**Overall charge**	**Base pairs per one turn of the helix**
A	Yes	No	Yes	No	Negative	10
B	Yes	Yes	No	Yes	Positive	10
C	Yes	No	Yes	No	Negative	12
D	No	Yes	Yes	Yes	Positive	12
E	No	No	No	No	Negative	10
F	No	Yes	Yes	Yes	Positive	12

2. A bacterial mutant grows normally at 32°C but at 42°C accumulates short segments of newly synthesized DNA. Which one of the following enzymes is most likely to be defective at the nonpermissive temperature (the higher temperature) in this mutant?

(A) DNA primase
(B) DNA polymerase
(C) An exonuclease
(D) An unwinding enzyme (helicase)
(E) DNA ligase

3. An RNA produced from a fragment of DNA has the sequence of AAUUGGCU. The sequence of the nontemplate strand in the DNA that gave rise to this sequence is which one of the following?

(A) AGCCAATT
(B) AAUUGGCU
(C) AATTGGCT
(D) TTAACCGA
(E) UUAACCGA

4. Which one of the following changes in the coding region of an mRNA (caused by a point mutation) would result in translation of a protein identical to the normal protein?

(A) UCA → UAA
(B) UCA → CCA
(C) UCA → UCU
(D) UCA → ACA
(E) UCA → GCA

5. Proteins destined for secretion from eukaryotic cells have which of the following in common?

	An N-terminal methionine in the mature protein is:	A signal peptide located at:	Synthesized on which type of ribosome?	Embedded within the ER membrane?
A	Very likely	Carboxy terminus	Rough	Yes
B	Very likely	Amino terminus	Cytoplasmic	No
C	Very likely	Carboxy terminus	Rough	Yes
D	Unlikely	Amino terminus	Rough	No
E	Unlikely	Carboxy terminus	Cytoplasmic	Yes
F	Unlikely	Amino terminus	Cytoplasmic	No

6. Inducible bacterial operons exhibit which of the following properties?

	Inducer binds to the repressor and:	Inducer effect on RNA polymerase binding to the promoter	Repressor produced by:
A	Activates the repressor	Enhances	The polycistronic message
B	Activates the repressor	Enhances	A separate gene
C	Activates the repressor	Inhibits	The polycistronic message
D	Inhibits the repressor	Inhibits	A separate gene
E	Inhibits the repressor	No effect	The polycistronic message
F	Inhibits the repressor	No effect	A separate gene

7. Processes that, in part, can lead to the activation of gene expression in eukaryotes can be best described as which one of the following?

	Methylation of the gene	Formation of polycistronic messages	Histone acetylation levels
A	Increased	Yes	Decreased
B	Decreased	Yes	Increased
C	Increased	Yes	Increased
D	Decreased	No	Decreased
E	Increased	No	Decreased
F	Decreased	No	Increased

8. Gene transcription rates and mRNA levels were determined for an enzyme that is induced by glucocorticoids. Compared with untreated levels, glucocorticoid treatment caused a 10-fold increase in the gene transcription rate and a 20-fold increase in both mRNA levels and enzyme activity. These data indicate that a primary effect of glucocorticoid treatment is to decrease which one of the following?

(A) The activity of RNA polymerase II
(B) The rate of mRNA translation
(C) The ability of nucleases to act on mRNA
(D) The rate of binding of ribosomes to mRNA
(E) The rate of transcription initiation by RNA polymerase II

9. Which region (A to D) of the DNA strands shown could serve as the template for transcription of the region of an mRNA that contains the initial codon for translation of a protein 300 amino acids in length?

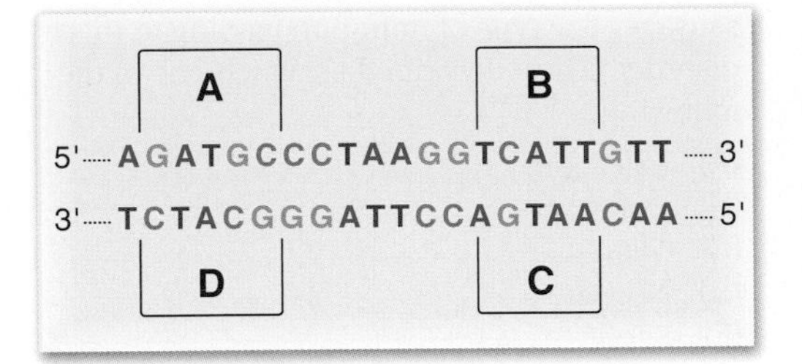

(A) A
(B) B
(C) C
(D) D
(E) None of the indicated areas would suffice.

10. A temperature-sensitive cell line would show early senescence when grown at the nonpermissive temperature, and examination of the chromosomes demonstrated many 3′ overhangs at the ends of the DNA fragments. The defective enzyme, at the nonpermissive temperature, is which one of the following?

(A) Telomerase
(B) DNA ligase
(C) DNA polymerase
(D) A repair DNA polymerase
(E) A helicase

Board-style Questions

11. A family, while on a picnic, picked some wild mushrooms to add to their picnic salad. Shortly thereafter, all the members of the family became ill, with the youngest child showing the most severe symptoms. The family is suffering these effects owing to a primary inability to accomplish which one of the following in their cells and tissues?

(A) Synthesize proteins
(B) Synthesize lipids
(C) Synthesize DNA
(D) Synthesize carbohydrates
(E) Repair damage in DNA

12. A newborn has found to be very photophobic, and his skin burns even with minimal exposure to sunlight, eventually forming skin blisters. Neither parent exhibits this trait, although both are prone to burning when in the sun for a short period of time. As the child grows, he is found to be at average height and weight for his age, and is progressing normally along the developmental guidelines. He is, however, kept inside at all times, and is carefully wrapped if he has to leave the house. Fibroblasts isolated from this child are grown in culture, and in an experiment, exposed to UV light. An analysis of the fibroblast DNA will demonstrate which one of the following?

(A) A preponderance of apurinic sites and apyrimidinic sites
(B) An increase in sister chromatid exchange rate
(C) A preponderance of abnormal base pairs in the DNA
(D) Loss of telomeres within the DNA
(E) An increase in cross-linked bases within the strands of DNA

13. A 15-year-old boy was diagnosed with skin cancer. He had always been sensitive to sunlight, and had remained indoors for most of his life. An analysis of his DNA, from isolated fibroblasts, indicated an increased level of thymine dimers when the cells were exposed to UV light. The boy developed a skin tumor owing to an increased mutation rate, which was caused by which one of the following?

(A) A lack of DNA primase activity
(B) Decreased recombination during mitosis
(C) Increased recombination during mitosis
(D) Loss of base excision repair activity
(E) Loss of nucleotide excision repair activity

14. A 40-year-old male is well controlled on warfarin for a factor V leiden deficiency and recurrent deep vein thrombosis. He presents today with a community-acquired pneumonia, and is placed on erythromycin. Three days later, he develops bleeding and his INR is 8.0 (indicating an increased time for blood clotting to occur, where INR is international normalized ratio). Which of the following best explains why this bleeding occurred?

(A) The erythromycin inhibited cytochrome P450
(B) The erythromycin stimulated cytochrome P450
(C) The causative agent of the pneumonia inhibited vitamin K utilization
(D) The causative agent of the pneumonia stimulated vitamin K utilization
(E) The erythromycin inhibited mitochondrial translation
(F) The erythromycin inhibited mitochondrial transcription

15. Your diabetic patient is using the short-acting insulin lispro to control his blood glucose levels. Lispro is a synthetic insulin formed by reversing the lysine and proline residues on the C-terminal end of the B-chain. This allows for more rapid absorption of insulin from the injection site. The engineering of this drug is an example of which of the following technologies?

(A) Polymorphism
(B) DNA fingerprinting
(C) Site-directed mutagenesis
(D) Repressor binding to a promoter
(E) PCR

16. For the synthesis of lispro insulin (as described in the previous question), which one of the following changes in the coding for the B-chain would be required?

(A) CAAAAA to AAAAAC
(B) CCTAAT to AAACTC
(C) CCGAAG to AAACCA
(D) AAACCA to CCGAAG
(E) AAGCCT to AAACCC

17. A thin, emaciated 25-year-old male presents with purple plaques and nodules on his face and arms, coughing, and shortness of breath. In order to diagnose the cause of his problems most efficiently, you would order which one of the following types of tests?

(A) Southern blot
(B) Northern blot
(C) Western blot
(D) Sanger technique
(E) Southwestern blot

18. A 17-year-old male has large, prominent ears, elongated face, large testicles, hand flapping, low muscle tone, and mild mental retardation. Which type of mutation does his diagnosis represent?

(A) Point
(B) Insertion
(C) Deletion
(D) Mismatch
(E) Silent

19. A young black man was brought to the emergency room (ER) due to severe pain throughout his body. He had been exercising vigorously when the pain started. He has had such episodes about twice a year for the past 10 years. An analysis of the blood shows a reduced blood cell count (anemia), and odd-looking red blood cells that were no longer concave and looked like an elongated sausage. The type of mutation leading to this disorder is best described as which one of the following?

(A) Insertion
(B) Deletion
(C) Missense
(D) Nonsense
(E) Silent

20. A woman has been complaining of a burning sensation when urinating, and a urine culture demonstrated a bacterial infection. The physician placed the woman on ciprofloxacin. Ciprofloxacin will be effective in eliminating the bacteria because it interferes with which one of the following processes?

(A) mRNA splicing
(B) Initiation of protein synthesis
(C) Elongation of protein synthesis
(D) Nucleotide excision repair
(E) DNA replication

21. A 72-year-old man acquired a bacterial infection in the hospital while recuperating from a hip replacement surgery. The staph infection was resistant to a large number of antibiotics, such as amoxicillin, methicillin, and vancomycin, and was very difficult to treat. The bacteria acquired its antibiotic resistance owing to which one of the following? Choose the one best answer.

(A) Spontaneous mutations in existing genes
(B) Large deletions of the chromosome
(C) Transposon activity
(D) Loss of energy production
(E) Alterations in the membrane structure

22. A 42-year-old man is placed on a two-drug regimen to prevent the activation of the tuberculosis bacteria, as his tuberculin skin test (PPD) was positive, but he shows no clinical signs of tuberculosis, and his chest X-ray is negative. One of the drug's mechanism of action is to inhibit which one of the following enzymes?

(A) DNA polymerase
(B) RNA polymerase
(C) Peptidyl transferase
(D) Initiation factor 1 of protein synthesis (IF-1)
(E) Telomerase

23. A 47-year-old woman, who has been on kidney dialysis for the past 7 years, has developed jaundice, fatigue, nausea, a low-grade fever, and abdominal pain. A physical examination indicates a larger-than-normal liver, and blood work demonstrates elevated levels of aspartate aminotransferase (AST) and alanine aminotransferase (ALT). The physician places the patient on two drugs, one of which is a nucleoside analog, geared to inhibit DNA and RNA syntheses. The primary function of the other drug is to do which one of the following?

(A) Inhibit DNA repair in infected cells
(B) Enhance the rate of the elongation phase of protein synthesis
(C) Reduce the rate of initiation of protein synthesis
(D) Inhibit ribosome formation
(E) Promote ribosome formation

24. A 50-year-old female has shortness of breath, cough, and fever for 3 days. She lives with her husband and has no medical problems. Her pulse ox in the office is 89 and her pulse rate is 110. She is admitted for treatment of community-acquired pneumonia, and her intravenous (IV) antibiotic treatment includes levofloxacin. A mutation in which bacterial enzyme would be required for levofloxacin resistance to be observed?

(A) DNA primase
(B) DNA polymerase III
(C) DNA gyrase
(D) DNA ligase
(E) DNA polymerase I

25. An 18-year-old college freshman shares a dorm room with three roommates. One of his roommates has been diagnosed with meningococcal meningitis, caused by the bacteria *Neisseria meningitidis*. The other three roommates are isolated, and treated twice a day with an antibiotic as prophylaxis against this organism, as none of them had received the meningococcal vaccine prior to enrollment. They are told that this antibiotic can give a reddish discoloration of their urine or tears. The reason this drug is effective in killing the bacteria is which one of the following?

(A) DNA synthesis is inhibited.
(B) RNA synthesis is inhibited.
(C) The process of protein synthesis is inhibited.
(D) The bacterial membrane becomes leaky.
(E) ATP generation is reduced.

26. A 38-year-old homeless man who has not received any medical care in the last 20 years presents with 2 days of shortness of breath, chills, fever, drooling, painful swallowing, and a "croupy" cough. A physical examination reveals a bluish discoloration of his skin and a tough, gray membrane adhered to his pharynx. The underlying mechanism through which this disease affects normal cells is which one of the following?

(A) DNA synthesis is inhibited in the target cells.
(B) RNA synthesis is inhibited in the target cells.
(C) The process of protein synthesis is inhibited in the target cells.
(D) The plasma membrane becomes leaky in the target cells.
(E) ATP generation is reduced in the target cells.

27. Disease X has been linked to the creation of a new restriction enzyme site, as indicated in the associated figure. The relevant area of the DNA is shown, along with EcoR1 restriction sites, sizes of fragments obtained, and an area to which a probe is available (the small red box). A family has had DNA obtained to determine if they are carriers for this disorder, and their DNA was digested with EcoR1, and a Southern blot was done using the available probe. A carrier for this disorder would display which bands on the Southern blot?

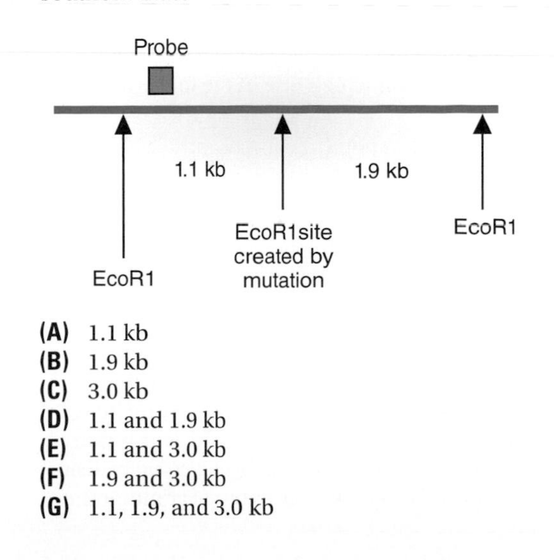

(A) 1.1 kb
(B) 1.9 kb
(C) 3.0 kb
(D) 1.1 and 1.9 kb
(E) 1.1 and 3.0 kb
(F) 1.9 and 3.0 kb
(G) 1.1, 1.9, and 3.0 kb

28. Disease X has been linked to the creation of a new restriction enzyme site, as indicated in the associated figure. A rapid PCR test to determine whether the amplified DNA carries the risk for the disease has been developed. After amplifying this region of the genome with the indicated PCR primers, and treatment of the amplified DNA with the appropriate restriction enzyme, an individual who is a carrier for this disease would express which of the following bands on an ethidium bromide-treated agarose gel?

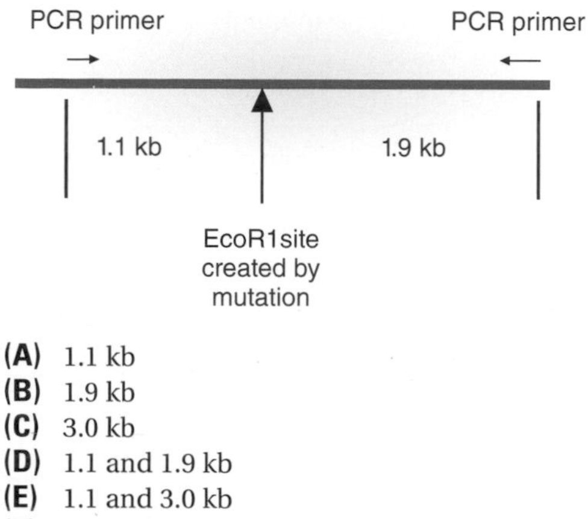

(A) 1.1 kb
(B) 1.9 kb
(C) 3.0 kb
(D) 1.1 and 1.9 kb
(E) 1.1 and 3.0 kb
(F) 1.9 and 3.0 kb
(G) 1.1, 1.9, and 3.0 kb

29. A 4-year-old boy displays a failure to thrive, extreme sensitivity to the sun, hearing loss, severe tooth decay, pigmentary retinopathy, and premature aging. An analysis of fibroblasts from the boy demonstrated extensive DNA damage in cells trying to grow, but minimal damage in quiescent cells, which have a greatly reduced rate of transcription as compared to the growing cells. This child most likely has a defect in which one of the following processes?

(A) Repair of thymine dimers
(B) Base excision repair
(C) Nucleotide excision repair
(D) Mismatch repair
(E) Transcription-coupled DNA repair

30. A 33-year-old man had a screening colonoscopy, and was diagnosed with a right-sided, mucinous colon cancer, with no other lesions or polyps seen. The reason he had a colonoscopy at such an early age is that his father and paternal uncle had colon cancers diagnosed by age 40. His paternal grandmother had ovarian and uterine cancers. A likely defect in the patient is a reduction in the ability to carry out which one of the following processes?

(A) Removal of thymine dimers from the DNA
(B) Inability to remove the base U from DNA
(C) Loss of DNA ligase activity
(D) Inability to correct mismatched bases in newly synthesized DNA
(E) Inability to form a solenoid structure from individual nucleosomes

31. A 10-year-old boy, small for his age in both height and weight with a calculated, projected adult height of less than 5 feet, is photophobic, and develops a "butterfly" rash over his nose and cheeks if exposed to the sun. He has a high-pitched voice, large nose, prominent ears, and has had multiple pneumonias in his childhood. An examination of fibroblasts from this patient demonstrated an increased sister chromatid exchange rate during mitosis as compared to cells from a normal child. The defective enzymatic activity in this child can be traced to which one of the following activities?

(A) A DNA polymerase
(B) An RNA polymerase
(C) A helicase
(D) An exonuclease
(E) An endonuclease

32. An 8-year-old boy has failure to thrive, alopecia totalis, localized scleroderma, a small face and jaw, a "beak" nose, wrinkled skin, and stiff joints. He is determined to have a single-point mutation in a nuclear protein, which is a silent mutation in terms of the primary structure of the protein. How could such a mutation lead to a disease?

(A) Through altering the tertiary structure of the protein
(B) Inhibiting DNA replication
(C) By introducing a premature stop codon into the protein
(D) By creating an alternative splice site in the gene
(E) By creating an alternative start site for transcription in the gene

33. A scientist is studying a novel hepatocyte cell line that cannot produce a nucleolus when the cells are grown at 42°C. When examining cells that have been at 42°C for 96 hours, the scientist finds that the incorporation of ^{14}C-leucine into proteins is greatly reduced as compared to cells grown at 35°C. This is most likely due to which one of the following at the nonpermissive temperature?

(A) Lack of charged tRNA molecules
(B) Inability to form peptide bonds during protein synthesis
(C) Lack of initiation factors
(D) Inability to form mature mRNA
(E) Lack of GTP needed for protein synthesis

34. Use the following figure to answer this question. The gene for CF has been isolated and sequenced. The gel pattern for the DNA sequence of the region that differs from the normal gene in the most common form of CF is shown in the figure. What do the results of this gel indicate about the disease-causing mutation in the altered gene?

(A) It is due to a single nucleotide change.
(B) It is due to an insertion of a small number of bases.
(C) It is due to a deletion of a small number of bases.
(D) It is due to a cytosine deamination.
(E) It is due to a frameshift mutation in the DNA.

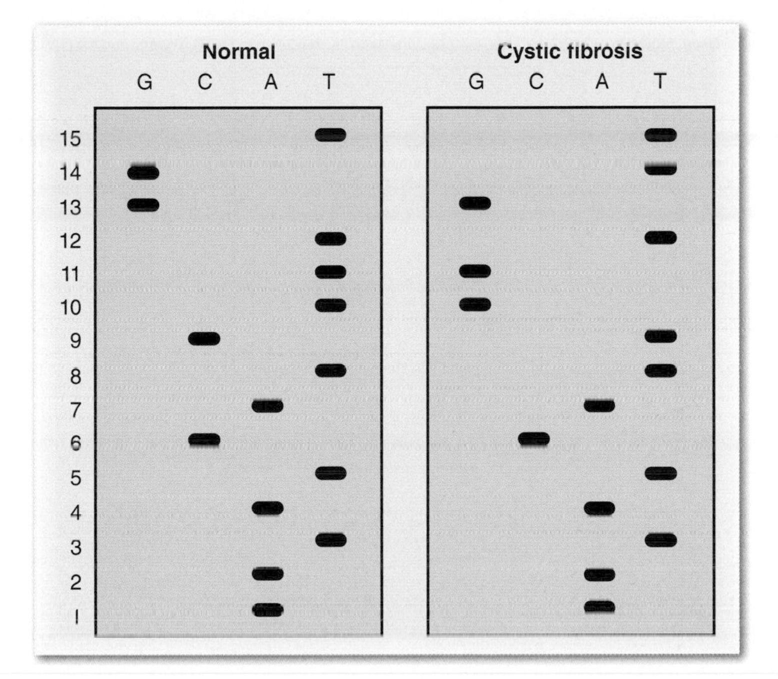

35. Use the following figure to answer this question. Two male infants were born on the same day in the same hospital. Because of concern that the infants had been switched in the hospital nursery, genetic tests based on a DNA restriction fragment that exhibits polymorphism (RFLP) were performed. Blood was drawn from the parents and the infants, the DNA extracted, and PCR performed. The DNA was then treated with the restriction enzyme *BanI*, and the fragments were separated by gel electrophoresis. The results of a Southern blot test are shown in the figure. A radioactive probe was used that bound to a sequence within the *BanI* fragments that exhibited polymorphism. Which of the two infants, C1 or C2, is the genetic offspring of this mother (M) and father (F)?

(A) C1 could be the offspring of these parents.
(B) C2 could be the offspring of these parents.
(C) Both infants could be the offspring of these parents (i.e., this test cannot discriminate).
(D) Either of these infants could be related to the mother, but neither could be related to the father.
(E) Neither infant could be related to this mother or this father.

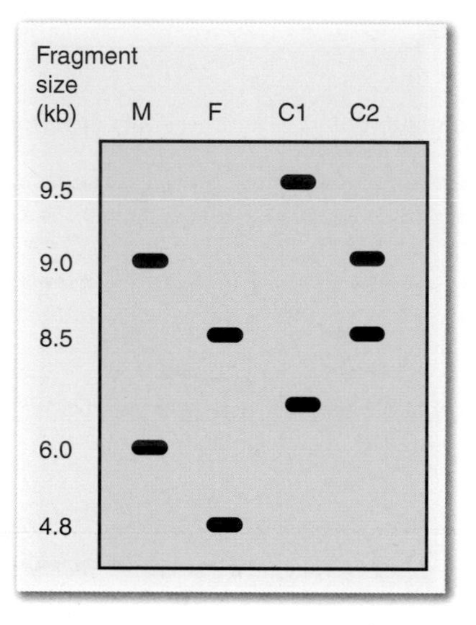

Answers and Explanations

1. **The answer is A.** On a molar basis, DNA contains equal amounts of adenine and thymine and of guanine and cytosine. Uracil is not found in DNA. There are 10 base pairs per turn of the helix, and the overall charge on the molecule is negative, due to the phosphates in the backbone (each phosphodiester bond contains one negative charge).

2. **The answer is E.** The short segments of the newly synthesized DNA that accumulate at 42°C are Okazaki fragments. They are usually joined together by DNA ligase, which most likely exhibits reduced activity at 42°C in this mutant. If the ligase is not functioning, Okazaki fragments would not be joined during replication, so the cells would contain short fragments of the DNA. Endonucleases and exonucleases cleave DNA strands in the middle and at the ends, respectively. They do not join fragments together, nor does DNA polymerase. Unwinding enzymes "unzip" the parental strands, and if these were defective, DNA synthesis most likely would not occur at the nonpermissive temperature, and short DNA fragments would not accumulate.

3. **The answer is C.** The nontemplate strand in the DNA is the same as the coding strand, and will have the same sequence of the RNA that is produced, except that T is in place of U. Both the nontemplate strand and RNA produced will be a complementary sequence to the template strand. As all sequences are written in the 5′ to 3′ direction, unless otherwise specified, an RNA sequence of AAUUGGCU would correspond to a DNA sequence, on the nontemplate strand, of AATTGGCT. The template strand would be the complement of the nontemplate strand, or AGCCAATT (written 5′ to 3′). Note that this sequence is also the complement of the RNA that has been produced. The base U is not found in DNA (so answer choices B and E cannot be correct).

4. **The answer is C.** UCA is a codon for serine. Of the answer choices given, only the codon UCU is also a codon for serine, which would result in a silent mutation (serine would be placed in the protein even though the DNA had been mutated from TGA to TGT). Conversion of UCA to UAA will generate a termination codon, and a truncated protein would be produced. The conversion of UCA to CCA would replace the serine with a proline in the amino acid. Conversion of UCA to ACA results in a threonine being placed in the protein in place of serine, and generation of GCA from UCA would result in alanine being incorporated into the protein in place of the serine. Only the change of UCA to UCU results in the exact same amino acid sequence being produced by the protein.

5. **The answer is D.** Proteins destined for secretion contain a signal sequence at the N-terminal end that causes the ribosomes on which they are being synthesized to bind to the SRP, which transfers the mRNA–ribosome complex to the RER. As they are being produced, they enter the cisternae of the RER, where the signal sequence, including the initial methionine, is removed. It is thus unlikely that the mature protein will contain an N-terminal methionine. Carbohydrate groups can be attached in the RER or the Golgi. Secretory vesicles bud from the Golgi, and the proteins are secreted from the cell by the process of exocytosis. If the proteins have a hydrophobic sequence that embeds in the membrane, they remain attached to the membrane and are not secreted, and become membrane-bound proteins.

6. **The answer is F.** In induction, a regulatory gene produces an active repressor, which is inactivated by binding to the inducer. The inducer prevents binding of the repressor to the operator rather than stimulating the binding of RNA polymerase. The structural genes are coordinately expressed. Transcription yields a single, polycistronic mRNA, which is translated to produce a number of different proteins. The regulatory gene is not a part of the polycistronic message.

7. **The answer is F.** A gene that is methylated is less readily transcribed than the one that is not methylated. Polycistronic mRNAs are only produced in prokaryotes, not in eukaryotic cells.

Histone acetylation will be increased in regions of the chromatin that are being transcribed (euchromatin). Acetylation of histones reduces the positive charges on the proteins, thereby weakening their interaction with the negatively charged phosphates on the DNA.

8. **The answer is C.** If the rate of degradation of the mRNA is not altered by glucocorticoids, then the increase in mRNA levels should reflect the increase in transcription rate. Because the increase in mRNA level is greater than the increase in transcription rate, the glucocorticoids must also be increasing the mRNA stability (i.e., decreasing the rate of degradation by nucleases). The activity of RNA polymerase II is increased (transcription is increased, so transcription initiation is also increased), and the rate of translation (the binding of ribosomes to mRNA) is increased (the enzyme activity is increased), owing to the increased amount of mRNA available.

9. **The answer is B.** Although D contains the sequence 3′-TAC-5′, which produces a start codon (5′-AUG-3′) in the mRNA, there is a sequence (3′-ATT-5′ in the DNA) that would produce a stop codon (5′-UAA-3′) in the mRNA in frame with this start codon. Sequence B, read 3′ to 5′ (from right to left), would produce a start codon in the mRNA transcribed from it. There are no stop codons in this sequence, so it could produce a protein 300 amino acids in length. Sequences A and C do not contain triplets corresponding to the start codon in mRNA.

10. **The answer is A.** Telomerase is defective at the nonpermissive temperature. Owing to the dual requirements of DNA polymerases that they synthesize DNA in the 5′ to 3′ direction, and their need for a template, replicating the ends of linear chromosomes leads to a 3′ overhang after the replication is complete. The overhang is created when the DNA–RNA primer is removed from the template strand, and there is no primer for the DNA polymerase to extend to fill in the gap. Telomerase solves this problem by carrying its own RNA template, and extending the 3′ overhang. After the gap is filled in as best it can, telomere-binding proteins cap the end of the chromosome to protect it from degradation. Lack of DNA ligase would result in a large number of gaps in the phosphodiester backbone (particularly on the lagging strand), but would not affect the telomeres. Lack of DNA polymerase activity would lead to an overall inhibition of DNA replication, and not just affect the telomeres. Lack of helicase activity would also affect the global DNA replication, not just the replication at the telomeres. Lack of repair polymerase would increase the amount of damage in the DNA, but would not specifically target the telomeres.

11. **The answer is A.** The poison in poisonous mushrooms is α-amanitin, an inhibitor of eukaryotic RNA polymerases, primarily RNA polymerase II. As the family ate the mushrooms containing the poison, RNA polymerase II stopped functioning, and mRNA was no longer produced. This led to a lack of protein synthesis. There is no direct effect on the synthesis of lipids, carbohydrate, or DNA, other than replacing the required enzymes due to protein turnover. However, the net effect of α-amanitin poisoning would be to stop protein synthesis, which may then lead to a cessation of lipid or DNA synthesis. α-Amanitin has no direct effect on DNA repair.

12. **The answer is E.** The child has XP, a defect in nucleotide excision repair such that thymine dimers, created by exposure to UV light, cannot be removed from the DNA. XP will not affect the repair of apurinic or apyrimidinic sites (sites missing just the base from DNA, which requires the AP endonuclease for repair). An increase in sister chromatid exchange rates is a finding in Bloom's syndrome, which is a defect in a helicase required for both DNA and RNA syntheses. Patients with Bloom's syndrome are small for their age, unlike those with XP who follow normal developmental milestones. XP does not result in unusual base pairs in DNA, rather the formation of thymine dimers between the adjacent T residues in one strand of DNA. These T residues are still complementary to the A residues in the other strand. XP does not affect the ability of telomerase to extend the ends of the linear chromosomes in the cell.

13. **The answer is E.** The damage to the DNA caused by UV light (pyrimidine dimers) can be repaired by the nucleotide excision repair pathway. In some cases, the missing enzyme is a repair endonuclease. The boy has XP, as determined by the increase in thymine dimers in his DNA after exposure to UV light. Since the dimers cannot be repaired, the DNA polymerase will "guess" when replication occurs across the dimers, increasing the mutation rate of the cells. Eventually, a mutation occurs in a gene that regulates cell proliferation, and a cancer results.

An increase or decrease in mutation rate is not related to the rate of recombination during mitosis, nor to a lack of DNA primase activity (which would lead to reduced DNA synthesis, not inaccurate DNA synthesis). Base excision repair is normal in patients with XP.

14. **The answer is A.** Warfarin is metabolized by a specific subset of induced p450 isozymes. The p450 system is used by cells to modify the xenobiotic (in this case the warfarin) such that it can be more easily excreted. Erythromycin, along with other macrolide antibiotics, inhibits the p450 oxidizing system, which in this case would lead to a higher blood level of warfarin and, therefore, the balance of clotting and bleeding is shifted toward excessive bleeding. A stimulation of p450 production by erythromycin would lead to a lower level of warfarin (due to increased metabolism and loss of warfarin by p450) and the potential of excessive clotting. This effect of p450 is a common drug-drug interaction. The causative agents of community-acquired pneumonia do not affect vitamin K absorption in the small intestine, or distribution throughout the body. Erythromycin does not affect mitochondrial transcription, although it may affect mitochondrial translation. Inhibition of mitochondrial protein synthesis, however, will not alter the inhibition of cytochrome p450 activity, and the increased levels of warfarin present, which may lead to increased bleeding.

15. **The answer is C.** This is a prime example of recombinant DNA technology to manufacture a very useful treatment for human diseases, by using site-directed mutagenesis. As the amino acid and DNA sequences of mature insulin is known, the engineering of lispro required that a proline codon be converted to a lysine codon, and the adjacent lysine codon converted to a proline codon. A polymorphism refers to the differences in DNA sequences amongst individuals in a population at a particular location within the genome. A polymorphism would not result in the synthesis of lispro insulin. DNA fingerprinting is used to identify unknown DNA samples by comparing the polymorphisms present in the sample DNA to a particular individual's DNA. Other than identical twins, everyone's DNA is different, and can be distinguished by fingerprinting using genetic polymorphisms. Repressor binding to a promoter is part of gene regulation in prokaryotypes, and would not be useful in generating the genetic changes necessary to produce lispro insulin. The PCR can amplify a particular DNA segment between two known segments of DNA, but the technique does not lead to the alteration of amino acid sequence between lispro and normal insulin.

16. **The answer is C.** The codons for proline are CCU, CCA, CCG, and CCC. The codons for lysine are AAA and AAG. The normal sequence of these two amino acids in the B-chain of insulin is pro-lys, so examples of this are CCGAAG, or CCCAAA. However, in the genetically engineered lispro variant of insulin, the lysine codon comes first, followed by the options of valine codons. The only answer that does this correctly is answer C. Answer A converts a pro-lys to a lys-asn. Answer B converts a pro-asn to lys-pro. Answer D converts a lys-pro to a pro-lys, and answer E converts a lys-pro to lys-pro.

17. **The answer is C.** The patient has Kaposi's sarcoma and AIDS. The causative agent is HIV, an RNA virus. The Western blot technique is used to identify whether a specific blood sample contains antibodies that will bind to HIV-specific proteins. The HIV proteins are run through a gel, transferred to filter paper, and probed using the sera from the patient. If the patient has antibodies to the HIV proteins, then a positive result will be obtained. A Southern blot is used to identify the DNA, and in this case it is easier to check for the presence of anti-HIV antibodies in the patient's sera. A Northern blot would check for viral RNA, but it is more efficient, and reliable, due to the low levels of viral RNA, to check for anti-HIV proteins instead. The Sanger technique identifies a portion of the DNA chain through sequencing the bases in the DNA, and is not used for determining the HIV status. A Southwestern blot is used to detect DNA binding to proteins, and would not be applicable for AIDS testing.

18. **The answer is B.** The patient has Fragile X syndrome, caused by the expansion of a triplet nucleotide repeat (CGG) within the *FMR1* gene on the X chromosome. The expansion interferes with the normal functioning of this gene product in the brain, leading to the symptoms observed. This is an extreme example of an insertion mutation. A point mutation occurs when only one base is substituted for another, and the change in base results in an amino acid

change in the protein. Deletion is removal of one or more nucleotides from the gene. Mismatch (answer D) repair is a DNA repair process that is utilized when a mismatch is found in the DNA. A silent mutation is one in which a base change in the DNA leads to no change in the corresponding amino acid in the protein (due to degeneracy in the genetic code).

19. **The answer is C.** The patient has sickle cell anemia. Sickle cell anemia arises owing to a substitution of valine (GTG) for glutamate (GAG). This is the definition of a missense mutation (one amino acid replaced by another), and since only one amino acid is replaced, it is also a point mutation. A nonsense mutation is a point mutation that converts a codon to a stop codon and premature termination of the growing peptide chain. A silent mutation is the result of a DNA change that does not change the amino acid sequence of the protein. Since this is a single-nucleotide substitution, it is not due to an insertion or deletion of genetic material.

20. **The answer is E.** The quinolone family of antibiotics (which includes ciprofloxacin) inhibits DNA gyrase, a prokaryotic-specific topoisomerase involved in unwinding the DNA strands for replication to occur. In the absence of gyrase activity, there would be no DNA replication, and the bacteria would not be able to proliferate. The quinolones do not affect eukaryotic topoisomerases. Splicing of hnRNA only occurs in eukaryotic cells. The gyrase is neither involved in nucleotide excision repair, nor in any aspect of protein synthesis.

21. **The answer is C.** Transposons have the ability to move DNA elements from one piece of DNA to another, including antibiotic-resistance genes from R-plasmids to the host chromosome. Thus, over time, as a bacteria obtains plasmids with antibiotic-resistance genes on them, the transposons can move the gene to the bacterial chromosome so it is always expressed by the cell, and no longer requires the plasmid for antibiotic resistance. Alterations in the membrane structure do not occur, nor do large deletions of the bacterial chromosome (antibiotic-resistance genes are not normal components of the bacterial chromosome). Antibiotic resistance is neither due to a loss of energy production, nor to spontaneous mutations in existing genes, as the bacteria do not encode genes that may confer antibiotic resistance to begin with.

22. **The answer is B.** Two drugs are utilized for latent tuberculosis: isoniazid and rifampin. Isoniazid works by blocking the synthesis of mycolic acid, a necessary component of the cell wall of the bacteria that leads to tuberculosis. Rifampin works by inhibiting bacterial RNA polymerase, and blocking the synthesis of new proteins. Neither drug affects DNA polymerase or peptidyl transferase (chloramphenicol is the antibiotic that inhibits bacterial peptidyl transferase activity). Rifampin also has no effect on IF-1.

23. **The answer is C.** The patient has an acute version of hepatitis C infection, which primarily affects the liver and its function. The two-drug treatment for hepatitis C is ribavirin and modified interferon (it is modified so it is more stable). The interferon works by activating a kinase (protein kinase R) that phosphorylates a key initiation factor for protein synthesis, thereby inhibiting the factor from participating in protein synthesis. This leads to a reduction in protein synthesis, and reduced replication of the virus infecting the cells. Interferon does not inhibit DNA repair, enhance the elongation phase of protein synthesis, or affect ribosome formation.

24. **The answer is C.** Levofloxacin is a member of the quinolone family of antibiotics that inhibits bacterial topoisomerases, primarily DNA gyrase (etoposide is the drug that inhibits eukaryotic topoisomerases). Without gyrase activity, the DNA of the bacterial chromosome cannot be unwound properly, and DNA replication would cease, leading to the death of the bacteria. The quinolone family of antibiotics does not directly affect DNA polymerases, DNA ligase, or DNA primase.

25. **The answer is B.** The drug given to prevent Neisseria infection (used prophylaxically), which is common in crowded conditions such as freshman dormitory rooms or military barracks, is rifampin. Rifampin inhibits RNA polymerase, and also exhibits a red color. Loss of rifampin in the urine or tears would give a reddish tint to those fluids. Rifampin does not interfere with DNA synthesis, the bacterial membrane, the process of protein synthesis, or ATP generation by the bacteria.

26. **The answer is C.** The man has contracted diphtheria, and needs the diphtheria antitoxin and then antibiotics to remove the offending organism, *C. diphtheriae*. As the patient has not

received medical care over the past 20 years, he has also missed his diphtheria vaccine, which should be received every 10 years. Diphtheria toxin blocks eukaryotic protein synthesis by phosphorylating an initiation factor, which inhibits protein synthesis in the cells. The toxin does not directly affect DNA or RNA synthesis, nor does it, as a primary target, reduce ATP production by the mitochondria or allow the plasma membrane to become leaky.

27. **The answer is E.** A carrier would have one normal allele (which would generate a 3.0-kb EcoR1 fragment), and one mutated allele (which generates fragments of 1.1 and 1.9 kb). The key to answering this question is that the probe is located within the 1.1-kb fragment that is generated in the mutant allele (it is also within the normal 3.0-kb fragment generated from the normal allele). So when the Southern blot is probed with the given probe, only the 1.1-kb piece of DNA will anneal to the probe and be visible in the mutated allele. Since this section of DNA is also present in the 3.0-kb piece of DNA, a 3.0-kb piece of DNA will also be visible on the Southern blot. If the person had two normal alleles, then only the 3.0-kb piece would be visible–if the person had the disease, then only the 1.1-kb piece would be visible on the Southern blot.

28. **The answer is G.** Amplifying the DNA region between the primers yields a 3.0-kb piece. The DNA from normal alleles will show a 3.0-kb band on an agarose gel after cutting with the appropriate restriction enzyme, as the normal allele does not contain this site within the amplified region. The mutant allele, however, does have this restriction site, such that after amplifying the DNA, and treating with the restriction enzyme, both a 1.1- and 1.9-kb piece will be generated, both of which would be seen on an ethidium bromide–treated gel. Carriers will have one of each allele (normal and mutant), so that a carrier would show three bands on the agarose gel–1.1, 1.9, and 3.0 kb in size. A person with two normal alleles would show only the 3.0-kb fragments, whereas a person with two mutated alleles would show both the 1.1- and 1.9-kb fragments.

29. **The answer is E.** The child is exhibiting the symptoms of Cockayne syndrome, which is due to a defect in transcription-coupled DNA repair. During transcription of genes, if the RNA polymerase notices DNA damage, transcription will stop while the transcription-coupled DNA repair mechanism will correct the DNA damage. This syndrome can be due to mutations in either the *ERCC6* or *ERCC8* gene, and the protein products of both the genes are involved in repairing the DNA of actively transcribed genes. The key to answering the question is the amount of DNA damage in growing cells (which are transcriptionally active) versus the damage in quiescent cells (which express fewer genes). The symptoms described are also unique to individuals with this disorder. The repair of thymine dimers and the processes of base excision repair, nucleotide excision repair, and mismatch repair are all functional in individuals with this disorder.

30. **The answer is D.** The patient has HNPCC, which is due to specific mutations in proteins involved in mismatch repair (mutations in at least four different proteins have been identified that lead to HNPCC). Mismatch repair is not involved in thymine dimer removal, nor base excision repair (the removal of uracil from DNA). HNPCC does not involve a defective DNA ligase, nor does the disease result in defective DNA packaging (solenoid formation) in the nucleus.

31. **The answer is C.** The child has Bloom's syndrome, a DNA synthesis defect due to a defective DNA helicase. The defective helicase leads to an increased mutation rate in the cells, through an unknown mechanism. Cells derived from patients with Bloom's syndrome display a significant increase in recombination events between homologous chromosomes as compared to normal cells (increased sister chromatid exchange rate). Mutations in the helicase increase genomic instability; the normal Bloom's protein suppresses sister chromatid exchange, and helps to maintain genomic stability. Bloom's syndrome is not due to a mutation in either DNA or RNA polymerase, an exonuclease, or an endonuclease.

32. **The answer is D.** The child is expressing the symptoms of Hutchinson–Gilford progeria, a premature aging disease, which is due to a mutation in the *LMNA* gene, which encodes lamin A, a nuclear protein. The most common mutation is C1824T, in which the normal cytosine at position 1,824 of the gene is replaced by a thymine. This is a silent mutation as far as the protein is concerned–G608G. However, the introduction of the T creates a cryptic splice site in the gene, such that as the hnRNA is processed, a lamin A mRNA is created that is missing

150 nucleotides, corresponding to a loss of 50 amino acids near the carboxy terminal of the protein. Under normal conditions, lamin A is farnesylated, which allows the protein to be attached to the endoplasmic reticulum membrane. During processing, the enzyme AMPSTE24 cleaves part of the carboxy terminal, releasing the farnesylated portion of the protein such that lamin A can be transferred to the nucleus, where it is involved in providing a scaffold for the nuclear membrane. In the mutant protein (progerin), the site of cleavage is lost owing to the loss of the C-terminal amino acids, although the site of farnesylation still remains. Thus, the progerin that reaches the nucleus is still bound to the nuclear membrane, distorting the nuclear membrane and contributing to nuclear instability. Chromatin binding to the nuclear membrane is also altered, as are the phosphorylation sites in progerin, which makes it more difficult for the nuclear membrane to dissolve during mitosis. Since this is a silent mutation in the mature protein, the tertiary structure of the protein is not altered, and a premature stop codon has not been introduced into the protein (that would be a nonsense mutation, not a silent mutation). Since the protein amino acid sequence is initially the same, an alternative start site for transcription has not been created, nor does a simple base change lead to an inhibition of DNA replication.

33. The answer is B. The nucleolus is the site within the nucleus at which rRNA is produced, and ribosomal subunits assembled. In the absence of nucleoli, mature ribosome content within the cell will decrease, which will lead to an overall reduction in protein synthesis. One of the functions of the mature ribosomal complex is to catalyze the formation of peptide bonds, using the enzymatic activity within the large ribosomal subunit rRNA (23S in prokaryotes, and 28S in eukaryotes). In nucleoli, rRNA genes are transcribed to produce the 45S rRNA precursor, which is trimmed, modified, and complexed with proteins to form ribosomal subunits. The synthesis of tRNA does not require the nucleolus, and charging reactions occur in the cytoplasm, so it is unlikely that the levels of charged tRNAs will be reduced. The capping, splicing, and polyadenylation of mRNA does not require the nucleolus (or ribosomes), and would proceed normally at the nonpermissive temperature. The only reason for initiation factors to be reduced is if their turnover number is high, and new proteins need to be synthesized to replenish the eIF pool. The lack of nucleoli, for 96 hours, should not affect the ability of the cell to produce energy in the form of ATP and GTP.

34. The answer is C. Sequences are read from the bottom to the top of the gel. In this region, the sequences of the CF and normal genes are identical for the first eight bases. Positions 12 to 15 of the normal gene and 9 to 12 of the CF gene are identical. Therefore, there is a 3-base deletion in the CF gene corresponding to bases 9 to 11 of the normal gene. Since the deletion is a multiple of three (3 bases), there is no frameshift involved. This DNA pattern is indicative of the ΔF508 mutation, in which the codon for a phenylalanine residue is deleted from the gene. This is the most common mutation in CF patients, occurring in about 70% of the individuals with CF.

35. The answer is B. Every chromosome has a homolog. Therefore, there will be two copies of every DNA sequence in the genome. Child C2 could have obtained the 9-kb restriction fragment from this mother, and the 8.5-kb fragment from this father. According to this test, child C1 is not genetically related to either this mother or this father, as neither the mother nor the father contain the 9.5- and 7.0-kb fragments expressed by child C1.

chapter 4

Cell Biology, Signal Transduction, and the Molecular Biology of Cancer

The material in this chapter is important for understanding how cells function and regulate their own proliferation. This is important in understanding cancer and its treatment, as well as disorders that affect specific cellular organelles.

OVERVIEW

- Eukaryotic cells contain a variety of intracellular compartments, each of which has a unique function within the cell.
- **Membranes** separate two aqueous compartments, and can separate intracellular organelles from the cytoplasm or the cytoplasm from the outside environment.
- Membranes consist of **lipids, carbohydrates,** and **proteins**, and one of their roles is to regulate the transport of molecules from one compartment to another.
- **Lysosomes** contain **degradative enzymes** that recycle material the cell no longer requires.
- Mutations in lysosomal enzymes lead to **lysosomal storage diseases**.
- **Mitochondria** are the cells power plant, and **generate ATP** through the oxidation of fuel sources, and reduction of oxygen to water.
- Mitochondrial gene disorders lead to a reduction in energy production, which primarily affects muscle function and the nervous system.
- **Peroxisomes** are cellular compartments devoted to **highly reactive oxidative reactions.** Lack of peroxisomal enzymes, or peroxisomes themselves, will lead to disease.
- The **nucleus** is the site of DNA storage and replication, RNA synthesis, and ribosome assembly. Mutations in nuclear proteins will lead to disease.
- The **endoplasmic reticulum** and **Golgi complex** are utilized for posttranslational modification of proteins, and **sorting of proteins** to their appropriate intracellular or extracellular locations.
- The **cytoskeleton** provides a **scaffold** and framework on which the cell membrane resides.
- Chemical messengers can act in a **paracrine, autocrine**, or **endocrine** manner, but in all cases need to bind to a **specific receptor** to elicit a response from the target cells.
- **Hormone receptors** can be **intracellular** or **transmembrane**; intracellular receptors act as transcription factors, whereas transmembrane receptors initiate a signal transduction cascade that leads to, in part, transcription factor activation.
- Receptors can act through the activation of **kinase activity** (tyrosine kinase or serine-threonine kinase, depending on the receptor), or through the activation of **heterotrimeric GTP-binding proteins** (G-proteins).
- **Second messengers**, induced by chemical messengers binding to their receptors, include **cAMP** and derivatives of **phosphatidylinositol turnover**.

- Uncontrolled cell proliferation (**cancer**) results from mutations in normal cellular genes (**proto-oncogenes**). This can result from DNA damage, which leads to mutations in genes important for the regulation of cell proliferation.
- **Oncogenes** are **gain-of-function** mutations, whereas **tumor suppressor genes** are **loss-of-function** mutations.
- Mutations in growth factor receptors, expression of growth factors, signal transduction proteins, and transcription factors can all lead to uncontrolled cellular proliferation.
- **Apoptosis is programmed cell death**, an orderly process for cell destruction. Mutations that enable apoptosis to be bypassed can lead to uncontrolled cell growth.
- Human cells require multiple mutations in growth-regulatory genes before a tumor develops.

I. COMPARTMENTATION IN CELLS; CELL BIOLOGY AND BIOCHEMISTRY

1. Membranes are lipid structures that separate the contents of the compartment they surround from its environment.

a. Plasma membranes separate the cell from its environment.

b. Organelles have membranes that separate the internal compartment of the organelle from the cytoplasm.

(1) The plasma membrane

(a) Consists of a **lipid bilayer** containing embedded and peripheral proteins (Fig. 4.1).

(b) The major component of membranes is lipids.

(2) The lipids in the plasma membrane are in the form of **phospholipids**, which contain a polar head group attached to two hydrophobic fatty acid tails; the head group faces the aqueous environment, the fatty acid tails the interior of the bilayer (Fig. 4.2)

(a) **Glycerol-based lipids** contain a glycerol backbone, and consist of phosphatidic acid (PA), phosphatidylethanolamine (PE), phosphatidylcholine (PC), phosphatidylserine (PS), phosphatidylglycerol (PG), phosphatidylinositol (PI), and cardiolipin (CL).

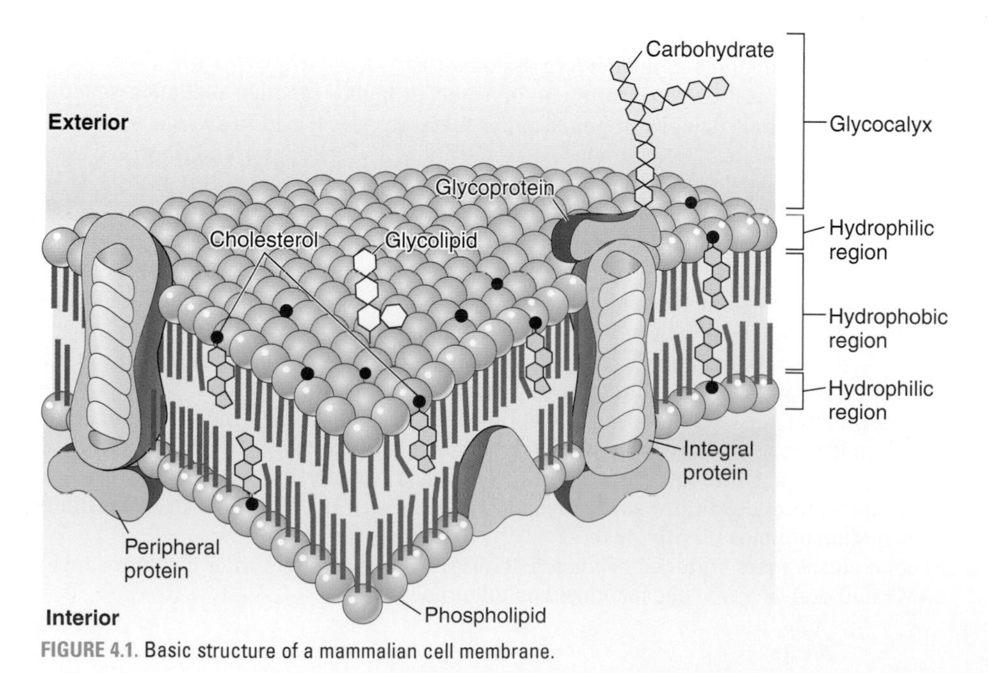

FIGURE 4.1. Basic structure of a mammalian cell membrane.

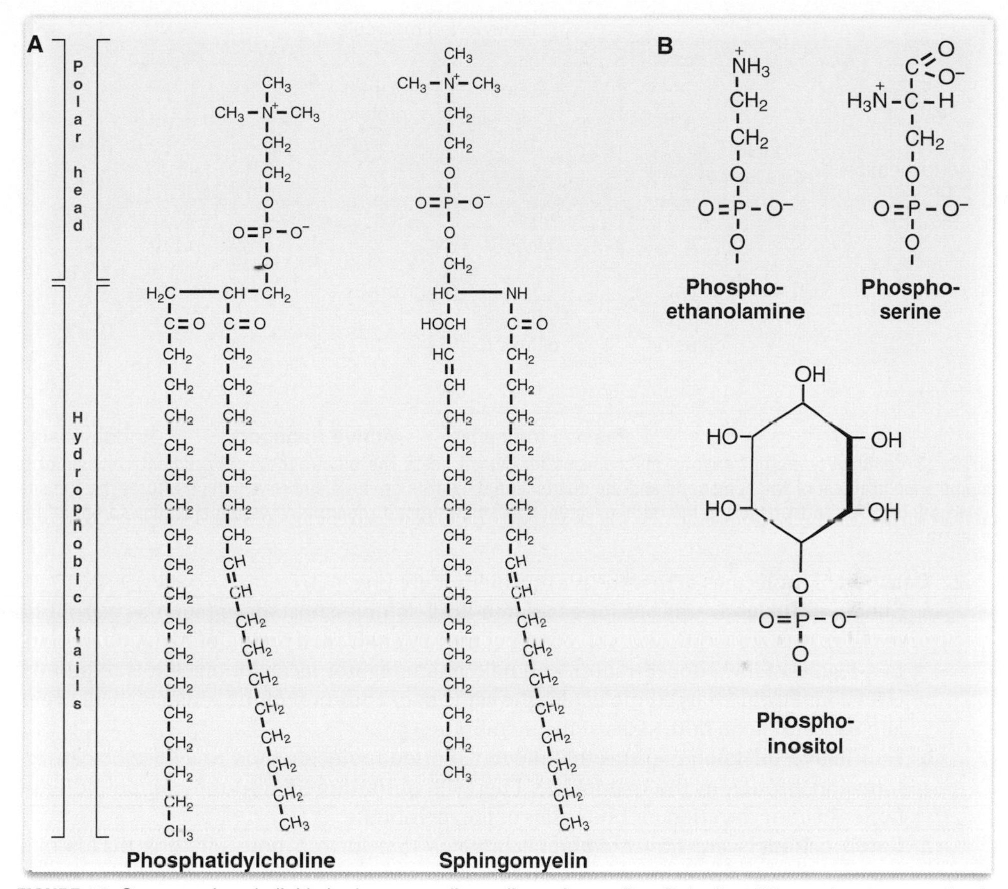

FIGURE 4.2. Common phospholipids in the mammalian cell membrane. Part B depicts different head groups for the glycerol based phospholipids.

(b) The one **sphingosine-based lipid** is sphingomyelin (SM).

(c) **Cholesterol** is present in **eukaryotic** membranes and maintains **membrane fluidity** at a variety of temperatures. Fluidity is also determined by the content of unsaturated fatty acids in the membrane, which are liquids at room temperature, and the chain length of the fatty acids (shorter chains are more fluid than longer chains).

(3) The **embedded proteins** in the plasma membrane function as either **channels or transporters** for the movement of compounds across the membrane, as receptors for the binding of hormones and neurotransmitters, or as structural proteins.

(4) The **peripheral membrane proteins** can be removed from the membrane by ionic agents; these can provide mechanical support to the membrane through the inner membrane skeleton or the cortical skeleton. An example of this is spectrin in the red blood cell membrane.

(5) A third type of membrane proteins is the **glycophosphatidylinositol (GPI) glycan-anchored proteins.** One example of a GPI-anchored protein is the prion protein, present in neuronal membranes. The prion protein can develop an altered pathogenic conformation in both mad cow disease and Creutzfeld–Jakob disease. Other membrane-associated proteins do so through fatty acylation. One example of this class is the ras protein (see Section III, Cancer).

(6) The **plasma membrane glycocalyx** consists of short chains of carbohydrates attached to proteins and lipids which extend in the aqueous media and both protects the cell from digestion and restricts the uptake of hydrophobic molecules.

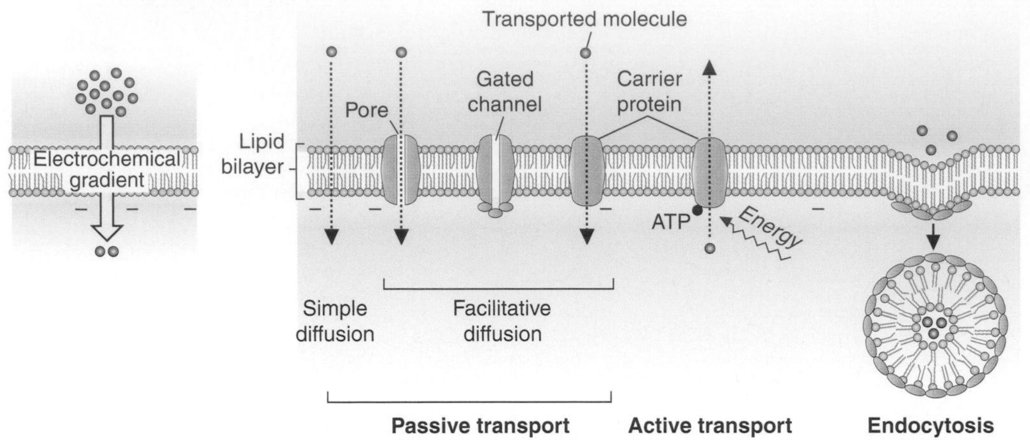

FIGURE 4.3. Common types of transport mechanisms for human cells. The electrochemical gradient consists of the concentration gradient of the compound and the distribution of charge on the membrane, which affects the transport of charged ions. Active transport is the net movement of a solute from a region of low concentration to one of high concentration.

2. **Transport** of molecules across the plasma membrane (Fig. 4.3)
   a. **Simple diffusion** is utilized for gases and lipid-soluble substances (such as steroid hormones). In simple diffusion there is a net movement from a region of high concentration to a region of low concentration, and neither energy nor a carrier protein is required for diffusion. For uncharged molecules the same concentration of the molecule will eventually be reached on both sides of the membrane.
   b. **Facilitative diffusion** requires that the transported molecule bind to a specific carrier or transport protein in the membrane. Energy is not required, and the compound equilibrates in concentration on both sides of the membrane.
   c. **Gated channels** are transmembrane proteins that form a pore for ions that is either opened or closed in response to a stimulus.
      (1) **Voltage-gated channels** respond to a change in voltage across the membrane.
      (2) **Ligand-gated channels** respond to the binding of a ligand to the protein.
      (3) **Phosphorylation-gated channels** respond to a covalent change (phosphorylation) on the protein. The cystic fibrosis transmembrane conductance regulator protein (CFTR) is a chloride channel that provides an example of a ligand-gated channel regulated through phosphorylation.

CLINICAL CORRELATES Inherited mutations in the **CFTR**, in the homozygous state, lead to **cystic fibrosis**. **Chloride transport** is inhibited in a variety of cell types, particularly the lung and pancreas. This leads to an inability to secrete chloride ions from the cell, and the water that would accompany the ions to provide osmotic balance. The net result is thick mucus that dries out and blocks various ducts in the tissues. This leads to problems in the lung (the airways are blocked with a thick mucus lining) and **pancreatic insufficiency**, as pancreatic secretions cannot reach the small intestine owing to blockage of the pancreatic duct. The presence of heterozygotes for CFTR is thought to provide some measure of protection against cholera.

   d. **Active transport** requires energy and transporter proteins
      (1) Compounds can be transported against an electrochemical gradient, whereas in facilitative transport they must go down an electrochemical gradient.
      (2) **Coupled transporters** can carry one compound while another compound travels down its concentration gradient, such as sodium-linked glucose transport (also known as secondary active transport). The sodium is going down its electrochemical gradient and glucose is brought along for the ride. The Na^+-, K^+-ATPase, as indicated in Figure 4.4 , generates the sodium gradient across the membrane.

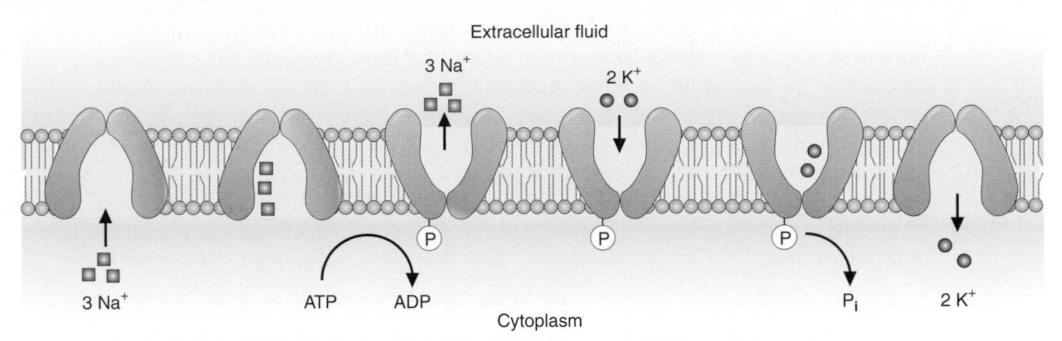

FIGURE 4.4. Active transport by the Na^+, K^+-ATPase. Note that this is an electrogenic transport, as three sodium ions are transported out of the cell in exchange for two potassium ions entering the cell. This creates both a concentration and charge gradient across the membrane.

e. **Vesicular transport** across the plasma membrane
   (1) **Endocytosis** refers to the vesicular transport into the cell
      (a) **Pinocytosis** refers to the condition when the vesicle forms around the fluid containing dispersed molecules.
      (b) **Phagocytosis** refers to the condition when the vesicle forms around particulate matter.
      (c) **Receptor-mediated endocytosis** is the name given to the formation of **clathrin-coated vesicles** that mediate the internalization of membrane-bound receptors in vesicles coated on the intracellular side with subunits of the protein clathrin. Cholesterol enters cells by receptor-mediated endocytosis.
      (d) **Potocytosis** is the name given to endocytosis that occurs via **caveolae** (specific regions of the plasma membrane).
   (2) **Exocytosis** refers to the vesicular transport out of the cell. The release of insulin from the β-cells of the pancreas occurs via exocytosis. Vesicles containing preformed insulin fuse with the plasma membrane, releasing the vesicular contents to the extracellular environment.

3. **Lysosomes**
   a. The lysosomes are the intracellular **organelles of digestion**, which are enclosed by a single membrane.
   b. Enzymes contained by lysosomes include nucleases, phosphatases, glycosidases, esterases and proteases, with **pH optima about 5.5** (Fig. 4.5).

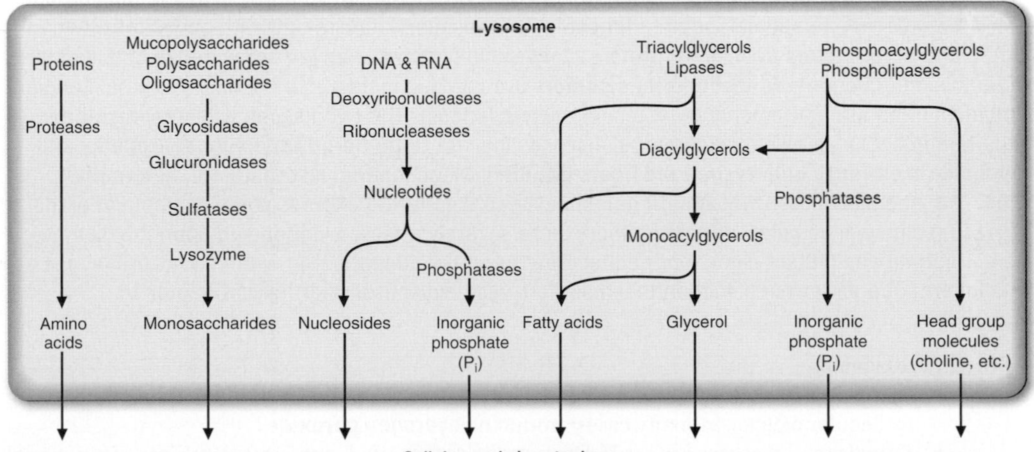

FIGURE 4.5. Lysosomal reactions. These enzymes are active at the acidic pH of the lysosome, and inactive if accidentally released into the cytosol.

c. Lysosomes contain a **vesicular ATPase** (proton pump) that actively transports protons from the cytoplasm into the lysosome to reduce the intralysosomal pH.

d. Lysosomes participate in receptor-mediated endocytosis, as the clathrin-coated vesicle formed by the plasma membrane fuse to form early endosomes, which can mature into late endosomes. The late endosomes will become lysosomes as they accumulate hydrolases, and their intracellular pH is reduced.

e. Lysosomes participate in **autophagy** (self-eating), in which autophagosomes fuse with a lysosome, and the lysosome digests the intracellular contents that were contained in the autophagosome.

CLINICAL CORRELATES **Lysosomal storage diseases** result when a lysosomal enzyme is defective (or a group of enzymes is defective, such as in **I-cell disease**). The lysosome becomes filled with material that cannot be digested, interfering with normal lysosomal function. Examples of specific enzyme defects include **Tay–Sachs disease, Gaucher disease, Pompei disease,** and **Hurler syndrome**. I-cell disease is a problem in targeting lysosomal enzymes to the organelle, and many lysosomal proteins are missing, leading to death in infancy. The targeted proteins are lacking their **mannose-6-phosphate (M6P) signal**, and cannot bind to the M6P receptor for transfer to the lysosomes from the Golgi apparatus. Many of the lysosomal proteins are secreted from the cell, where they cannot do damage since the pH is above 7, and the enzymes have a pH optimum of 5.5.

4. Mitochondria

a. Mitochondria contain most of the enzymes for the pathways of **fuel oxidation and oxidative phosphorylation**, and, thus, generate most of the ATP required by mammalian cells.

b. Mitochondria contain two membranes

(1) The outer membrane contains pores made from proteins called porins and is permeable to molecules with a molecular weight up to about 1000 Da.

(2) The inner membrane is highly impermeable, and supports the establishment of a **proton gradient** across the membrane. Specific transporters are required to shuttle compounds from the cytoplasm to the mitochondria, and vice versa.

c. Mitochondria contain their own circular genome, which encodes 13 different subunits of proteins involved in oxidative phosphorylation. The remainder of the mitochondrial proteins are nuclear encoded, synthesized in the cytoplasm, and transported into the mitochondria.

d. Inherited mutations in mitochondrial DNA lead to muscle, nerve and renal problems.

CLINICAL CORRELATES As the mitochondria contain their own genome, which encodes a number of subunits required for oxidative phosphorylation, along with genes for tRNA, rRNA, and other factors needed for protein synthesis, mutations in any of these genes can lead to defective mitochondria. **Mitochondrial disorders** are due to a mutation in the mitochondrial genome, and most often affect the nervous system or muscle function, the two tissues with very high-energy requirements. Examples of mitochondrial diseases include **Leber hereditary optic neuropathy** and **myocolonic epilepsy with ragged red fibers (MERRF)**. Mitochondrial disorders are **maternally inherited**, and all children who inherit defective mitochondria will express some component of the disease (some will be mildly affected, others more severely affected, depending upon the distribution of normal and mutant mitochondria that were apportioned to the egg when it was produced). The inheritance pattern of mitochondrial disorders will be discussed further in Chapter 10.

5. Peroxisomes

a. Peroxisomes are cytoplasmic organelles that are involved in **oxidative reactions** using molecular oxygen, in many cases producing **hydrogen peroxide**.

b. Peroxisomal diseases are caused by mutations that affect either the synthesis of functional peroxisomal enzymes, the incorporation of these proteins into peroxisomes, or peroxisomal biogenesis.

CLINICAL CORRELATES Peroxisomal disorders include **adrenoleukodystrophy** and **Zellweger syndrome.** Adrenoleukodystrophy refers to several closely related disorders that interfere with **peroxisomal oxidation of very long-chain fatty acids** (mitochondria will oxidize the shorter fatty acids). The very long-chain fatty acids accumulate in the nervous system and adrenal glands, and can interfere with nerve function owing to the disruption of the myelin sheath. **Adrenal function slowly declines** as the very long-chain fatty acids accumulate within that gland. The symptoms include loss of neurologic abilities, degeneration of visual and auditory systems, seizures, and Addison disease (due to chronic adrenal insufficiency). **Zellweger syndrome** is a **peroxisome biogenesis disorder.** When the peroxisomes are not produced properly certain enzymatic reactions cannot take place, including the appropriate formation of myelin in the nervous system. Very long-chain fatty acids accumulate, as does phytanic acid (a plant derivative), which interfere with normal nervous system function. Bile acid and plasmalogen synthesis are also affected, leading to an enlarged liver. The disease is fatal in infants, with death usually occurring in the first 6 months of life.

6. **Nucleus**
   a. The nucleus is the largest subcellular organelle, and houses the **genome** of the cell. **DNA replication, transcription,** and **ribosome assembly** occur within the nucleus.
   b. The nuclear envelope surrounds the nucleus, and consists of an outer and inner membrane joined by pores.
      (1) The **outer membrane** is continuous with the rough endoplasmic reticulum.
      (2) The **inner nuclear** membrane provides more of a permeability barrier.
      (3) RNA and other material that must be exported from the nucleus leave through the pores.
   c. The **nucleolus,** a substructure of the nucleus, is the site of rRNA transcription and processing of ribosome assembly.
7. **Endoplasmic reticulum (ER)**
   a. The ER is a network of membranous materials within the cell consisting of smooth ER (SER), which lacks ribosomes, and rough ER (RER), which is studded with ribosomes.
   b. The SER has many functions, including being the site of synthesis for many large lipids, and the site of cytochrome P450 oxidative enzymes that are used for detoxification and the synthesis of hydrophobic molecules.
   c. The RER is involved in the synthesis of secreted and certain intracellular organelle-targeted proteins (see Chapter 3).
   d. The RER is also the site of the initiation of **posttranslational modifications** to the newly synthesized proteins.
8. **Golgi complex**
   a. The Golgi complex is involved in modifying the proteins produced in the RER and in **sorting and distributing** these proteins to the lysosomes, secretory vesicles, or to the plasma membrane.
   b. **Posttranslational modifications** include complex branched chain oligosaccharide addition, sulfation, and phosphorylation.
9. **Cytoskeleton**
   a. The structure of the cell, the shape of the cell surface, and the arrangement of subcellular organelles are organized by three major protein components, all considered part of the cytoskeleton.
   b. **Microtubules** are responsible for movement of vesicles and organelles.
      (1) The major protein is **tubulin**.

CLINICAL CORRELATES **Hereditary spherocytosis** is due to an inherited mutation in the red cell cytoskeleton. Most often this is due to a deficiency of **spectrin**, an essential component of the red blood cell inner membrane and cytoskeleton. The lack of spectrin leads to a loss of erythrocyte surface area, and since the cells have an altered shape, they are rapidly removed from the circulation by the spleen, leading to an anemia. Other red cell cytoskeletal proteins, which when mutated, can also lead to spherocytosis and include **ankyrin, band 3**, and **protein 4.2**.

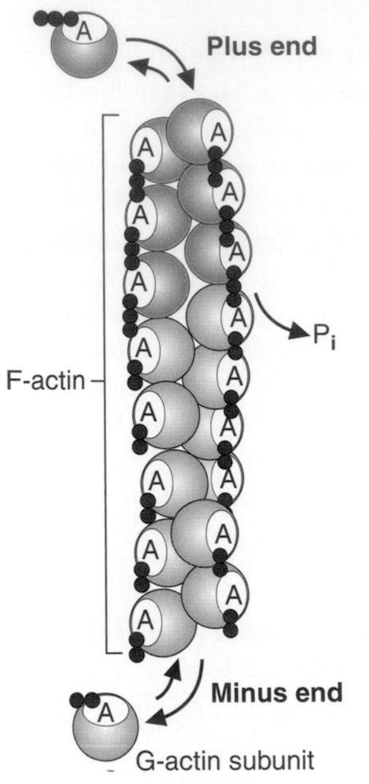

FIGURE 4.6. Actin filaments. The polymer F-actin is assembled from G-actin subunits containing bound ATP. The filament grows at the plus end, and dissociation of actin-GDP occurs at the minus end of the polymer.

(2) The microtubule network begins in the nucleus at the centriole and extends outward to the plasma membrane.

(3) Microtubules consist of polymerized arrays of α- and β-tubulin dimers that form protofilaments organized around a hollow core.

(4) There are three different tubulin polypeptides, α, β, and γ. γ-tubulin is found only in centrosomes and spindle pole bodies.

c. Thin filaments, which form a cytoskeleton, contain primarily **actin**

(1) The actin polymer is F-actin; the monomeric actin is G-actin (Fig. 4.6).

(2) Within F-actin each G-actin contains a bound ATP or ADP that alters the tertiary structure of the actin.

(3) Actin polymers form thin filaments (microfilaments), and are bound to cross-linking proteins associated with the cell surface.

d. Intermediate filaments (IF), which play a structural role, contain primarily fibrous protein polymers that provide structural support to the membranes. There are a wide number of proteins that participate in forming IF.

II. CELL SIGNALING BY CHEMICAL MESSENGERS

1. The general features of chemical messengers are depicted in Figure 4.7.
2. The actions of chemical messengers can be classified as **endocrine**, **paracrine**, and **autocrine**
   - **a.** Endocrine hormones are secreted by a specific cell type, enter the blood, and exert their actions on specific target cells, which may be some distance away.
   - **b.** Paracrine actions are those performed on nearby cells, in which a hormone secreted by a cell acts on the neighboring cells.
   - **c.** Autocrine actions involve a messenger that acts on the cell from which it was secreted.
3. Types of chemical messengers
   - **a.** The nervous system secretes two types of messengers: **small-molecule neurotransmitters** (such as acetylcholine) and **neuropeptides** (normally small peptides between 4 and 35 amino acids in composition).
   - **b.** Endocrine system hormones consist of **polypeptide hormones** (such as insulin and glucagon), **catecholamines** (such as epinephrine), **steroid hormones** (derived from cholesterol, such as estrogen), and **thyroid hormone**.

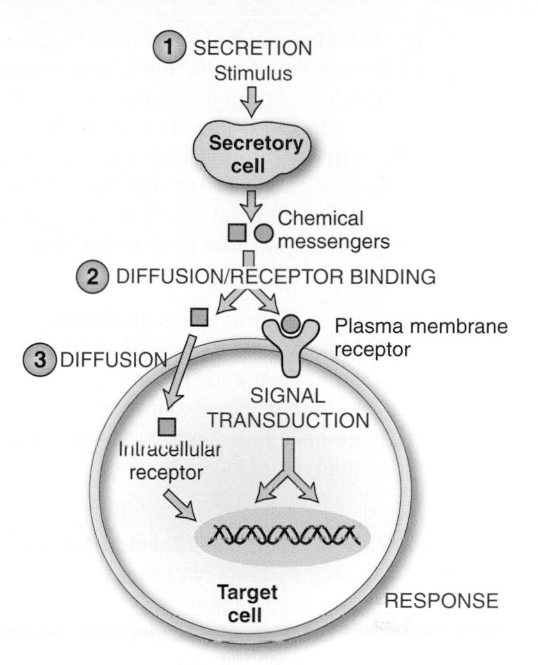

FIGURE 4.7. General features of chemical messengers. 1. Secretion of chemical message. 2. Binding of message to cell surface receptor. 3. Diffusion of a hydrophobic message across plasma membrane and binding to an intracellular receptor.

   c. The immune system utilizes the messengers known as **cytokines**, which are small proteins with an average molecular weight of 20 kDa. There are different classes of cytokines (such as interferons, interleukins, tumor necrosis factors, and colony-stimulating factors), but all are secreted by the cells of the immune system and will induce alterations in gene transcription in the target cells.
   d. The **eicosanoids** are derived from long-chain fatty acids, and consiste of the prostaglandins, thromboxanes, and leukotrienes.
   e. Growth factors are polypeptides that function through the stimulation of cellular proliferation (**hyperplasia**) or cell size (**hypertrophy**).

4. Intracellular versus plasma membrane receptors
   a. **Chemical messenger receptors** can be either **intracellular** or an **embedded membrane protein** (Fig. 4.8).

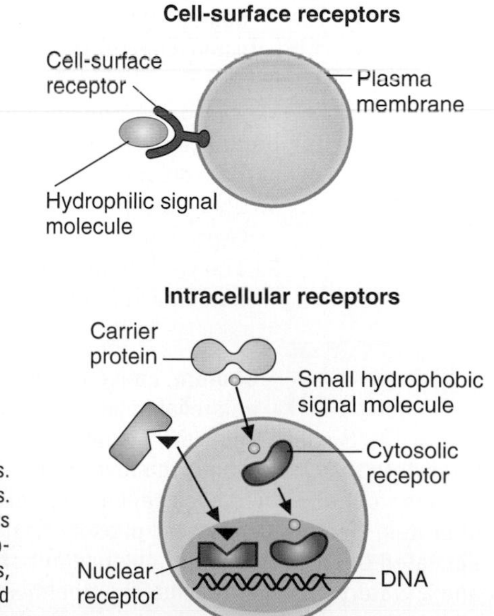

FIGURE 4.8. Intracellular versus plasma membrane receptors. Plasma membrane receptors have extracellular binding domains. Intracellular receptors bind steroid hormones or other messengers that are able to diffuse through the plasma membrane. Such receptors may reside in the cytoplasm and translocate to the nucleus, reside in the nucleus bound to DNA, or reside in the nucleus bound to other proteins.

b. Most intracellular receptors for lipophilic messengers are gene-specific transcription factors.

c. Steroid hormone/thyroid hormone superfamily of receptors

(1) These lipophilic molecules are transported in the blood bound to serum albumin, or to more specific transport proteins, such as steroid-hormone-binding globulin, or thyroid-binding globulin.

(2) Once in the cell the lipophilic messenger binds to its receptor, which will often dimerize (with other intracellular transcription factors) to bind to the promoter-proximal regions of DNA to **alter gene expression**.

CLINICAL CORRELATES **Androgen insensitivity syndrome** (X-linked, nonresponsive to androgens) is due to a mutated androgen receptor, which is encoded by the X chromosome. Males who inherit this mutation from their mothers cannot respond to androgens (such as testosterone) and can be born with ambiguous genitalia. This mutation does not affect female development in the heterozygote condition. There are various levels of phenotypic expression, from mild to severe, depending on the type of mutation inherited.

d. Plasma membrane receptors and signal transduction

(1) Major classes of plasma membrane receptors

(a) Ion-channel receptors.

1. Signal transduction consists of a conformational change when a ligand binds, which allows ion-flow through the channel
2. The **acetylcholine receptor** is an example of an ion-channel receptor.

CLINICAL CORRELATES **Myasthenia gravis** is a neuromuscular disorder due to an **autoimmune** condition that produces antibodies against the **acetylcholine receptor**. The receptor cannot be activated by acetylcholine, and the muscle cell cannot respond to the neurotransmitter, and contraction will not occur. The overlying symptom of this disorder is muscle fatigue. The treatment consists of **acetylcholinesterase inhibitors** (such that acetylcholine is present for an extended period at the neuromuscular junction) and **immunosuppressants** to reduce the production of the autoantibodies targeted against the receptor.

(b) Receptors that are kinases or bind to and activate kinases.

1. Examples of these receptors are shown in Figure 4.9
2. The common feature of this class of receptors is that the **intracellular kinase domain** of the receptor (or the kinase domain of the associated protein) is activated when the messenger binds to the extracellular domain.
3. The signal transduction pathway is propagated downstream through signal transducer proteins that bind to the activated messenger–receptor complex.

(c) Heptahelical receptors (Fig. 4.10) contain seven membrane-spanning domains and are the most common type of plasma membrane receptor.

1. These receptors are **G-protein–coupled receptors (GPCR),** as they work with a G-protein to transmit the signal.
2. The G-proteins lead to a second messenger production, such as cAMP, or alterations in PI-derived molecules.

CLINICAL CORRELATES **Cholera**, caused by a toxin secreted by *Vibrio cholerae,* affects intestinal epithelial cells. The toxin consists of two classes of subunits, designated A and B. The B subunit aids the A subunit in entering the cells, where the A subunit acts to **ADP-ribosylate** the Gαs subunit, inactivating the intrinsic **GTPase activity**. This leads to constitutive activation of adenylate cyclase, and high cAMP levels. The elevated cAMP leads to the activation of protein kinase A, which will phosphorylate a variety of target proteins, including the CFTR. The activated CFTR will transport chloride out of the intestinal epithelial cells, such that water follows and a watery diarrhea results. The treatment is to prevent dehydration. The watery diarrhea usually removes the offending bacteria from the intestinal lumen, so the disease is self-limiting.

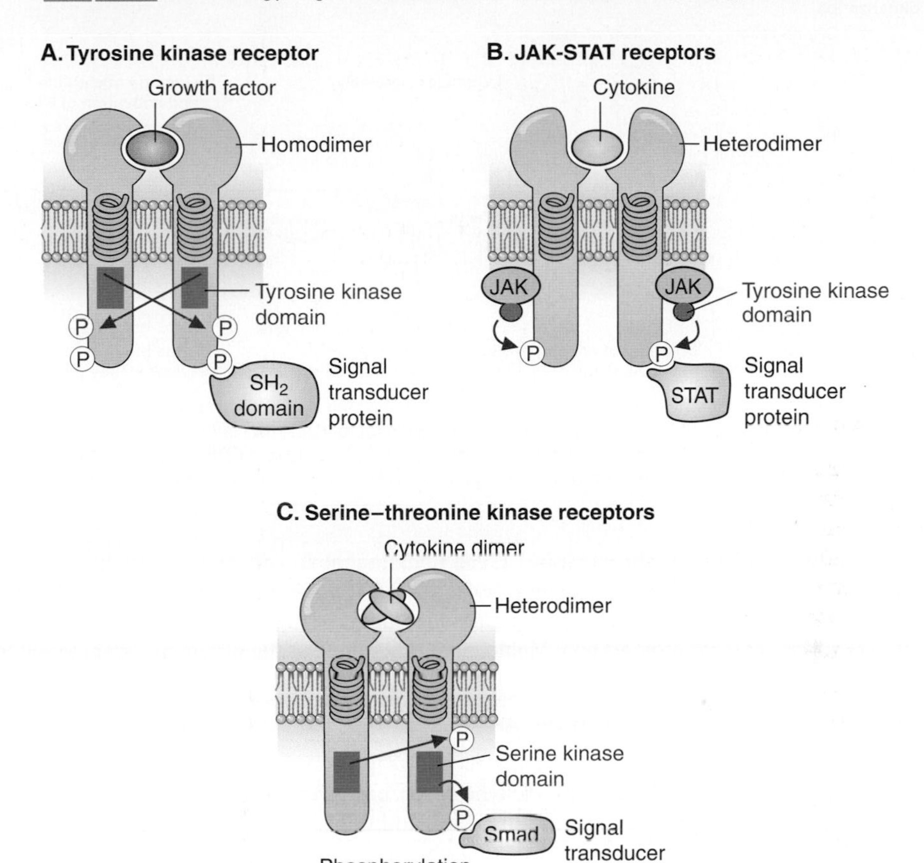

FIGURE 4.9. Receptors that are kinases or that bind kinases. The kinase domains are shown in red, and the phosphorylation sites are indicated with red arrows. **A.** Tyrosine kinase receptors. **B.** JAK-STAT receptors. **C.** Serine-threonine kinase receptors.

(2) Signal transduction through tyrosine kinase receptors
  - **(a)** The activation of **tyrosine kinase receptors** leads to the activation of the **MAP kinase pathway** (Fig. 4.11)
  - **(b)** Once raf is activated, the kinase MEK is activated, which activates the kinase ERK, which leads to an alteration of gene transcription factor activity, thereby either upregulating or downregulating the transcription of many genes involved in cell survival and proliferation

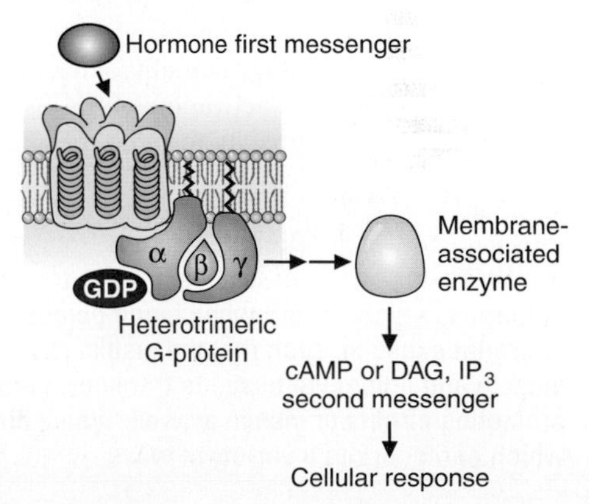

FIGURE 4.10. Serpentine receptors and second messengers. The activated hormone-receptor complex activates a heterotrimeric G-protein and, via stimulation of membrane-bound enzymes, different G-proteins lead to generation of one or more intracellular second messengers.

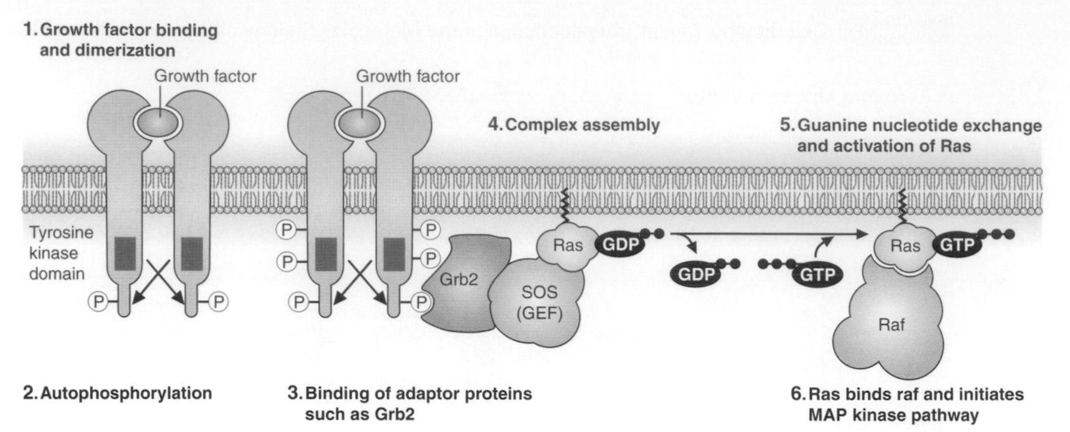

FIGURE 4.11. Signal transduction by tyrosine kinase receptors. 1. Binding and dimerization. 2. Autophosphorylation. 3. Binding of GRB2 and SOS. 4. SOS is a guanine nucleotide exchange factor (GEF) that binds ras, a monomeric G-protein anchored to the plasma membrane. 5. GEF activates the exchange of GTP for bound GDP on ras. 6. Activated ras containing GTP binds the target enzyme raf, thereby activating it and a series of downstream kinases - the MAP kinases.

CLINICAL CORRELATES NF-1 (**neurofibromatosis 1, von Recklinghausen disease**) is due to a mutation in the neurofibromin gene, which is a **GTPase-activating protein (GAP)** for **ras**. Defects in neurofibromin lead to a ras gene product that is active for extended periods of time, leading to abnormal cellular proliferation. Multiple neurofibromas of the nervous system result from inheriting this disorder. The gene product is considered a tumor suppressor gene.

- **(c)** Proteins with SH2 domains recognize phospho-tyrosine residues, and may be specifically targeted to one type of receptor over another
- **(d)** Tyrosine kinase receptors also have additional signaling pathways involving other factors other than the MAP kinase pathway

(3) **Phosphatidylinositol phosphates** in signal transduction

- **(a)** Alterations in PI metabolism can be induced by both tyrosine kinase receptors and heptahelical receptors.
  1. Phosphatidylinositol 4',5'-bisphosphate can be cleaved to generate two intracellular messengers, **diacylglycerol** (DAG) and **inositol trisphosphate** (IP_3)
  2. Phosphatidylinositol 3', 4', 5'-trisphosphate can serve as a plasma-membrane-docking site for signal transduction proteins
  3. The route of generation of these important PI derivatives is shown in Figure 4.12
- **(b)** Phosphoinositide signals are short-lived, as the PI cycle regenerates PI from the signals. This cycle can be interrupted by lithium, which inhibits the phosphatase that converts inositol phosphate to free inositol.

(4) The **insulin receptor** is an example of a receptor that can activate not only the **MAP kinase pathway**, but also the **protein kinase B (Akt) pathway** (Fig. 4.13)

- **(a)** The activation of PI-3-kinase leads to the generation of phosphatidyl inositol 3,4,5-trisphosphate, which is a membrane-binding site for proteins with pleckstrin homology domains, such as PDK1 and PK B (Fig. 4.14)
- **(b)** The activation of PK B (Akt) promotes cell survival (antiapoptotic), as well as propagating some specific insulin effects

CLINICAL CORRELATES **Diabetes** results from alterations in insulin signaling. The insulin receptor is necessary to transmit the signal that insulin has been released by the pancreas, which occurs when blood glucose levels rise. **Type I diabetes** is defined by an inability to produce insulin, such that the insulin receptor is not activated. This leads to an inability of the muscle and adipocyte tissue to transport glucose from the circulation, and reduces the growth-promoting effects of insulin as well. **Type 2 diabetes** is a reduced cellular responsiveness to insulin, which can occur in a variety of ways.

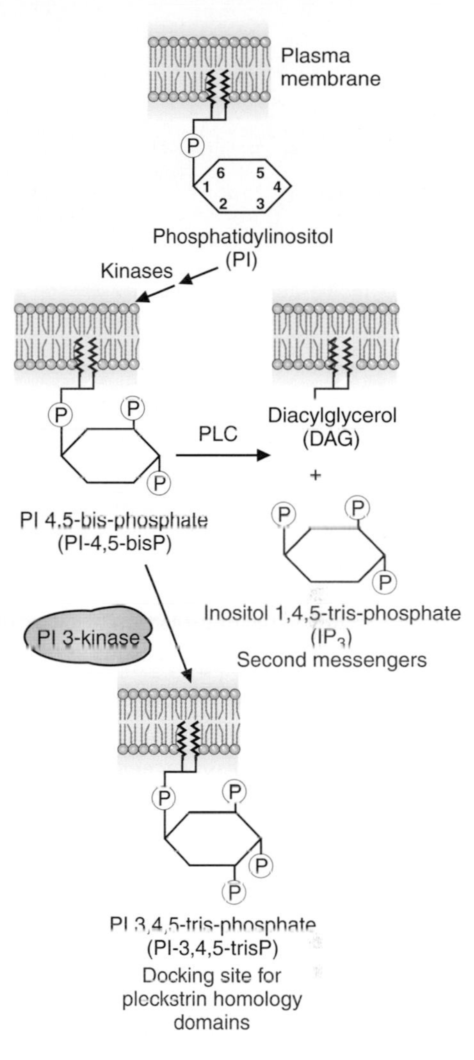

FIGURE 4.12. Major routes for the generation of the phosphoinositide signal molecules. PLC is phospholipase C.

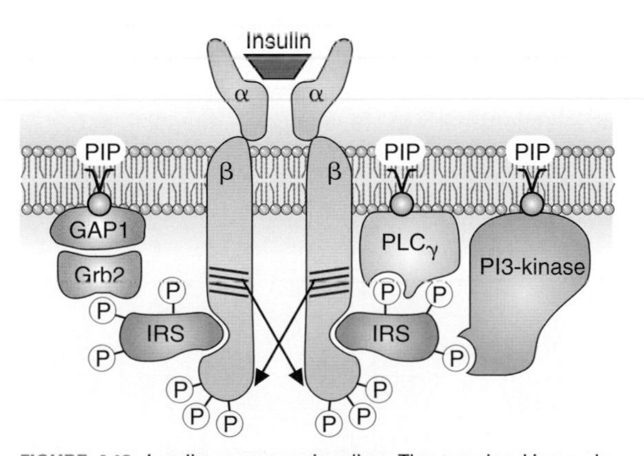

FIGURE 4.13. Insulin receptor signaling. The tyrosine kinase domains are shown in red, and arrows indicate autophosphorylation. The activated receptor binds IRS molecules and phosphorylates IRS at multiple sites, thereby forming binding sites for proteins with SH2 domains, examples being GRB2, phospholipase Cγ, and PI-3-kinase. These associated proteins are also associated with phosphoinositide derivatives (PIP) in the membrane.

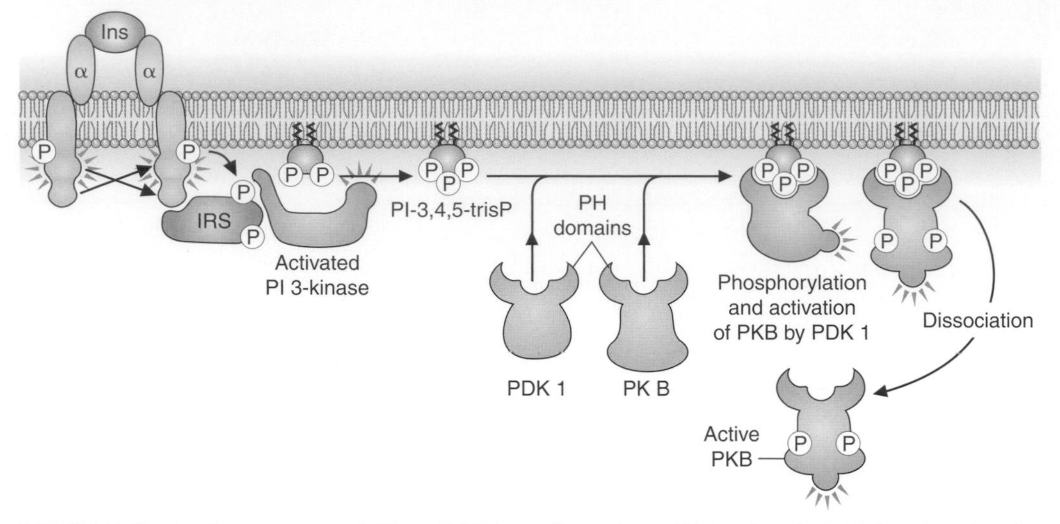

FIGURE 4.14. The insulin receptor protein kinase B (Akt) signaling pathway. PH domains, pleckstrin homology domains; PDK1, phosphoinositide-dependent protein kinase 1; PKB, protein kinase B.

(5) Signal transduction by cytokine receptors: use of **JAK-STAT proteins**

- **(a)** Cytokine receptors do not have an intrinsic kinase domain, but are associated with JAK kinases
- **(b)** When cytokines bind to the receptors dimerization occurs, cross-phosphorylation between JAK kinases occurs, allowing the SH2-domain proteins STAT (signal transducer and activator of transcription) to bind to the receptor–kinase complex
- **(c)** The STATs are phosphorylated, dimerize, and travel to the nucleus to alter gene transcription (Fig. 4.15)
- **(d)** **STAT signaling** is modulated by the **SOCS** (suppressors of cytokine signaling) and **PIAS** (protein inhibitors of activated STAT) family of proteins, some of which are induced by STAT.

CLINICAL CORRELATES **Severe combined immunodeficiency syndrome** (SCID) exists in two types, either as **adenosine deaminase deficiency** (see Chapter 9), or the **X-linked SCID** that is missing a **cytokine receptor subunit** (the defective gene is the IL2RG gene, which stands for interleukin 2 receptor, gamma). This subunit is common to many different cytokine receptors, and when it is defective the immature blood cells cannot appropriately respond to growth and differentiation signals, resulting in the lack of a functional immune system. Individuals inheriting this mutation acquire multiple infections early in life, along with fungal infections. Survival past the first year is rare.

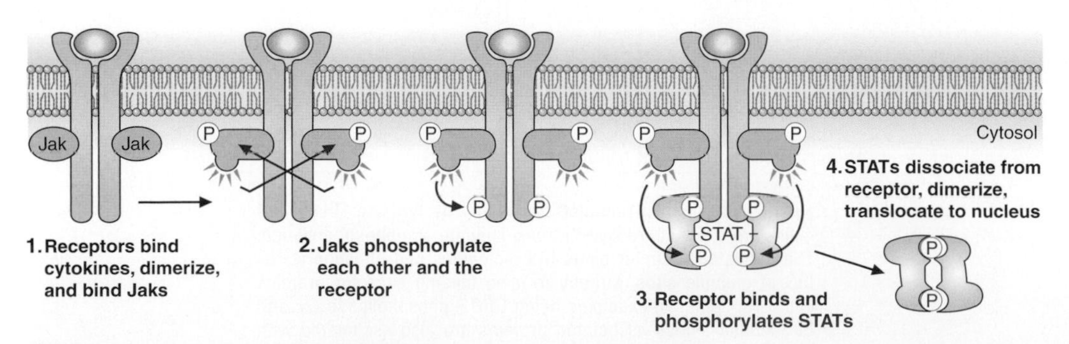

FIGURE 4.15. Steps in cytokine receptor signaling.

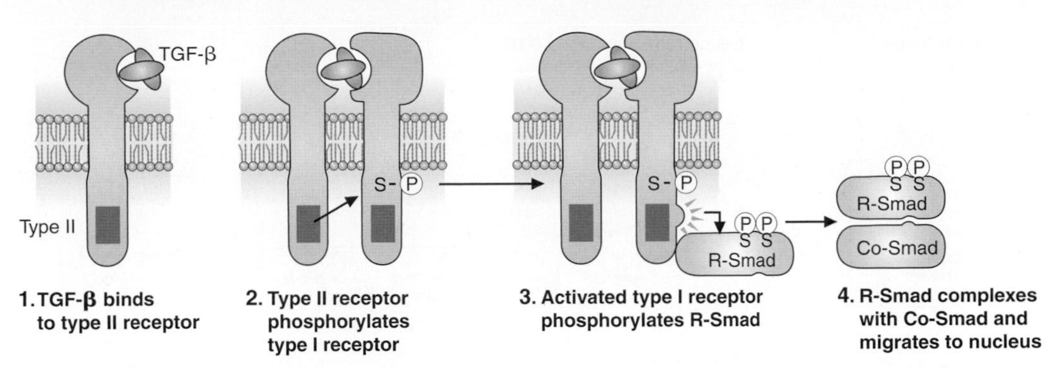

FIGURE 4.16. Serine-threonine receptors and SMAD proteins. The phosphorylated R-SMAD dimerizes with the common SMAD (SMAD 4) and then translocates to the nucleus to either activate or inhibit the expression of target genes.

(6) Receptor serine-threonine kinases

(a) The receptors for the **TGF-β family** of growth factors are serine-threonine kinases (Figure 4.16)

(b) There are **type I and type II receptors**

1. TGF- β initially binds to a type II receptor, which then recruits a type I receptor to the complex
2. The type II receptor phosphorylates, and activates, the kinase activity of the type I receptor.
3. The type I receptor binds a receptor-**SMAD** (R-SMAD), which it phosphorylates on a serine residue
4. The phosphorylated R-SMAD dissociates from the receptor, and dimerizes with SMAD 4, the common SMAD
5. The SMAD complex translocates to the nucleus to alter **gene transcription**
6. One of the genes activated is an inhibitory SMAD, which regulates how long the signal is active (signal termination)

(7) Signal transduction through **heptahelical receptors**

(a) These receptors contain seven membrane-spanning domains.

(b) Ligand binding to the receptor is transduced through the activation of **heterotrimeric G-proteins**, of which there are many classes.

(c) The basic scheme for the activation of heterotrimeric G proteins is illustrated in Figure 4.17. This example depicts a hormone that leads to adenylate cyclase activation.

(d) The α subunits of the heterotrimeric G-proteins, in addition to binding GTP, also have an intrinsic GTPase activity, which self-regulates how long the G-protein is active.

(e) There are five major classes of G-proteins, as indicated in Table 4.1. Note that there are G-proteins that can activate their target, as well as G-proteins that inhibit their target.

(f) cAMP levels are also modulated through the regulation of the **cAMP phosphodiesterase**, the enzyme that converts cAMP to 5'-AMP. Methylxanthines, such as caffeine and theophylline, inhibit the cAMP phosphodiesterase and act by maintaining elevated levels of cAMP.

CLINICAL CORRELATES The bacteria **Bordetalla pertussis** colonizes the lungs and secretes a toxin that enters the lung cells. The toxin is capable of **ADP-ribosylating a Gαi protein**, and blocking the activation of this class of G-proteins. This means that when the cell receives the appropriate signals adenylate cyclase cannot be turned off, and cAMP levels will remain elevated. Protein kinase A is active, and cellular metabolism and signaling is not appropriate for the conditions. This mechanism of action is in contrast to that of cholera toxin, in which the Gαs is continually activated, owing to the ADP-ribosylation modification inhibiting its GTPase activity.

(g) PI signaling through heptahelical receptors occurs through the **Gαq family of G-proteins**. Gαq targets phospholipase Cβ, which hydrolyzes PIP_2 into DAG and IP_3.

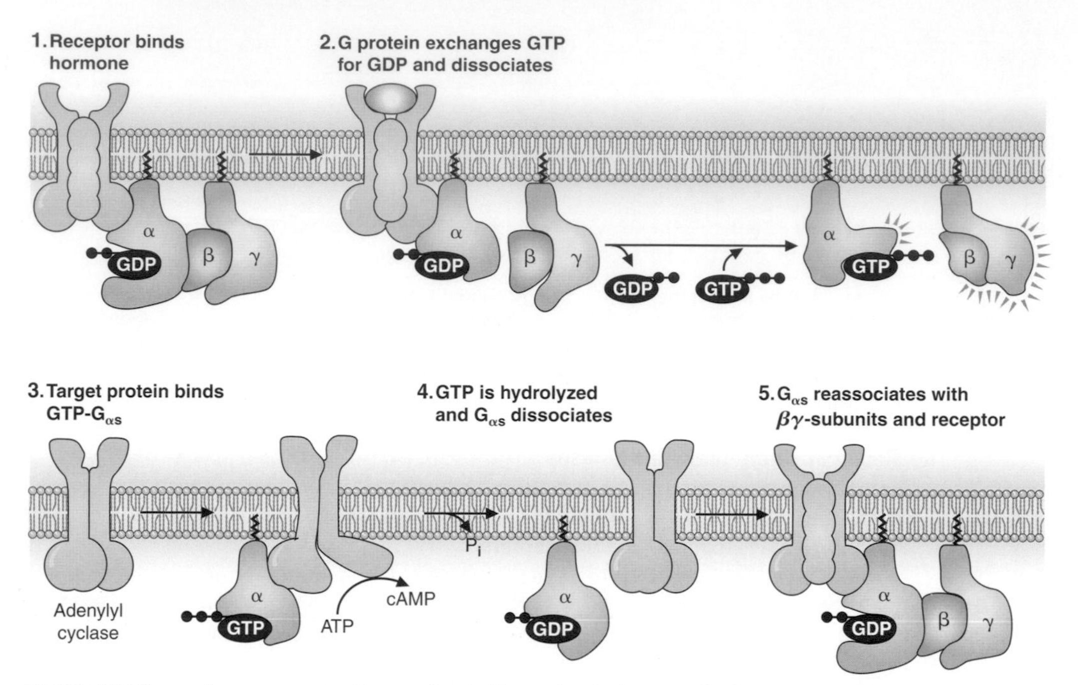

FIGURE 4.17. Serpentine receptors and heterotrimeric G-proteins. In the example shown, the target protein is adenylate cyclase, which produces cAMP. Note that when the G-protein is activated (binds GTP), the α-GTP subunit dissociates from the other two subunits to activate (or inhibit) its target protein.

1. IP_3 binds to the sarcoplasmic or endoplasmic reticulum, which triggers **calcium release** into the cytoplasm. The calcium release activates the enzymes containing the calcium–calmodulin subunit, including a protein kinase.
2. The DAG, which remains membrane-bound, activates **protein kinase C**, which propagates the response by phosphorylating appropriate target proteins.

(8) Receptor activity/number after hormone binding
- **(a)** The receptor number can be reduced by downregulation
- **(b)** The receptor activity may be reduced by phosphorylation or other covalent modifications

(9) The mechanisms of **signal termination** include the following:
- **(a)** The release of the initial chemical messenger is terminated
- **(b)** Receptor desensitization or downregulation occurs

table 4.1 Subunits of Heterotrimeric G-Proteins

G α Subunit	Action	Some Physiologic Uses
α_s; Gα(s)[a]	**S**timulates adenylyl cyclase	Glucagon and epinephrine to regulate metabolic enzymes, regulatory polypeptide hormones to control steroid hormone and thyroid hormone synthesis, and by some neurotransmitters (e.g., dopamine) to control ion channels
$\alpha_{i/o}$; Gα(i/o) (signal also flows through βγ-subunits)	**I**nhibits adenylyl cyclase	Epinephrine; many neurotransmitters, including acetylcholine, dopamine, serotonin
α_t, Gα(t)	Stimulates cGMP phosphodiesterase	Has a role in the transducin pathway, which mediates detection of light in the eye
$\alpha_{q/11}$; Gα(q/11)	Activates phospholipase Cβ	Epinephrine, acetylcholine, histamine, thyroid-stimulating hormone (TSH), interleukin-8, somatostatin, angiotensin
$\alpha_{12/13}$; Gα(12/13)	Activate Rho-GEF (guanine nucleotide exchange factor)	Thromboxane A2, lysophosphatidic acid act as signals to alter cytoskeletal elements

[a]There is a growing tendency to designate the heterotrimeric G-protein subunits without using subscripts.

(c) Protein phosphatases to reverse phosphorylation events are activated
(d) GTPases are activated to block G-protein activity
(e) Phosphodiesterases are activated to reduce cAMP levels
(f) Induced genes are transcribed and translated to produce proteins that antagonize the positive signals initiated by the chemical messenger

III. THE MOLECULAR BIOLOGY OF CANCER

1. Cancer is the term applied to a group of diseases in which cells no longer respond to normal restraints on growth, and are resistant to apoptosis.
2. Cancer is caused by mutations in normal cellular genes, and for every gene that causes cancer (an **oncogene**) there is a corresponding cellular gene called the **proto-oncogene** (Fig. 4.18).

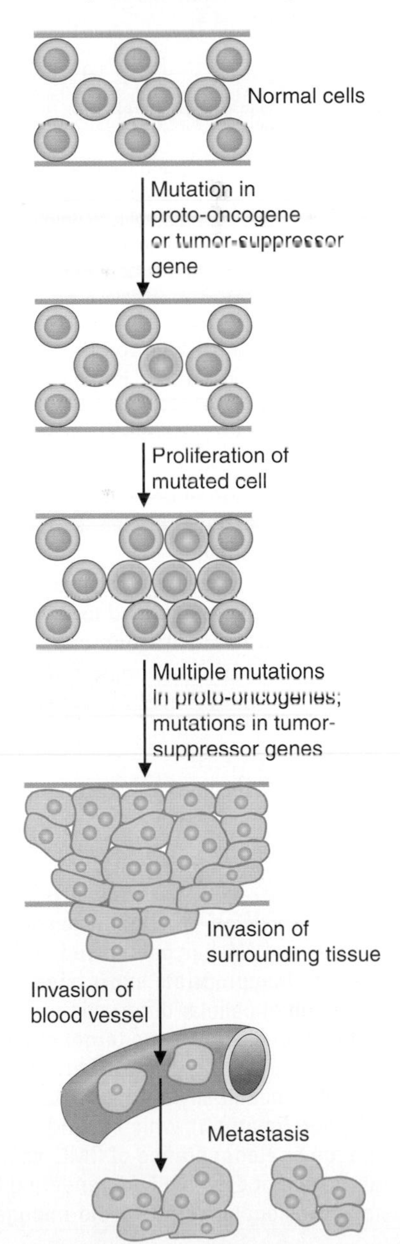

FIGURE 4.18. Development of cancer. Accumulation of mutations in several genes results in transformation. Cancer cells change morphology, proliferate, invade other tissues, and metastasize.

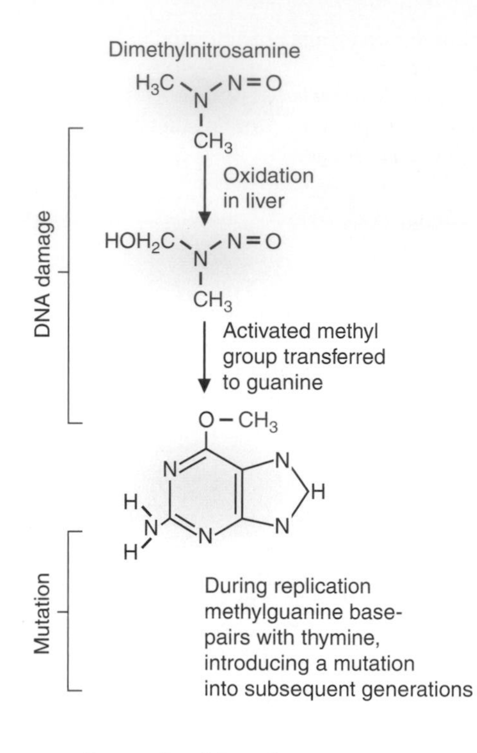

FIGURE 4.19. Mutations in DNA caused by nitrosamines. Nitrosamine metabolites methylate guanine (shown in red in the figure).

3. A **gain-of-function** mutation will lead to an oncogene; a **loss-of-function** mutation, which leads to cancer, is classified as a **tumor suppressor**
4. Tumors can be benign or malignant (infiltrate and destroy surrounding tissue)
5. Tumors can metastasize, separating from the growing mass of the tumor and traveling through the blood or lymph to unrelated organs where they establish new growths of cancer cells
6. **Damage to DNA can lead to mutations**, which lead to cancer
   a. Altering the base sequence of DNA can lead to mutations in proto-oncogenes, leading to cancer
   b. Mutations in the proteins necessary for nucleotide excision repair lead to **xeroderma pigmentosum**, and to skin cancer
   c. Chemical agents can modify DNA and lead to an alteration in the base sequence (Fig. 4.19)
7. Gain-of-function mutations in proto-oncogenes occur via a variety of mechanisms (Fig. 4.20)
   a. Alteration of the **regulation of expression** of the gene
   b. The gene may be **translocated** to another region of the genome, which alters its regulatory pattern
   c. The gene may be **amplified and overexpressed**

CLINICAL CORRELATES

Chronic myelogenous leukemia (CML) results from a **translocation** between chromosomes 9 and 22, and creates a **fusion protein, bcr-abl,** at the location where the DNA has been fused. Abl is a tyrosine kinase that is now under bcr gene control, which leads to **inappropriate expression** of the abl kinase activity. This leads to cell proliferation and an inhibition of cellular differentiation. As this fusion protein is a unique cellular protein, found only in tumor cells, it was the target of study for rational drug design. The analysis of the structure of the bcr-abl protein identified areas where the potential binding of a drug might inhibit the kinase activity–specifically for this kinase, and not for any other kinase. This led to the production of **Gleevec (imatinib),** which bound to the bcr-abl protein, inhibited its activity, and led to remission of the tumor. Reoccurance of CML would occur, and when analyzed, the bcr-abl protein was mutated at the site of drug binding, rendering the kinase resistant to drug inhibition. **Burkitt lymphoma** is due to a translocation of the **proto-oncogene myc**, a transcription factor, which puts expression of myc

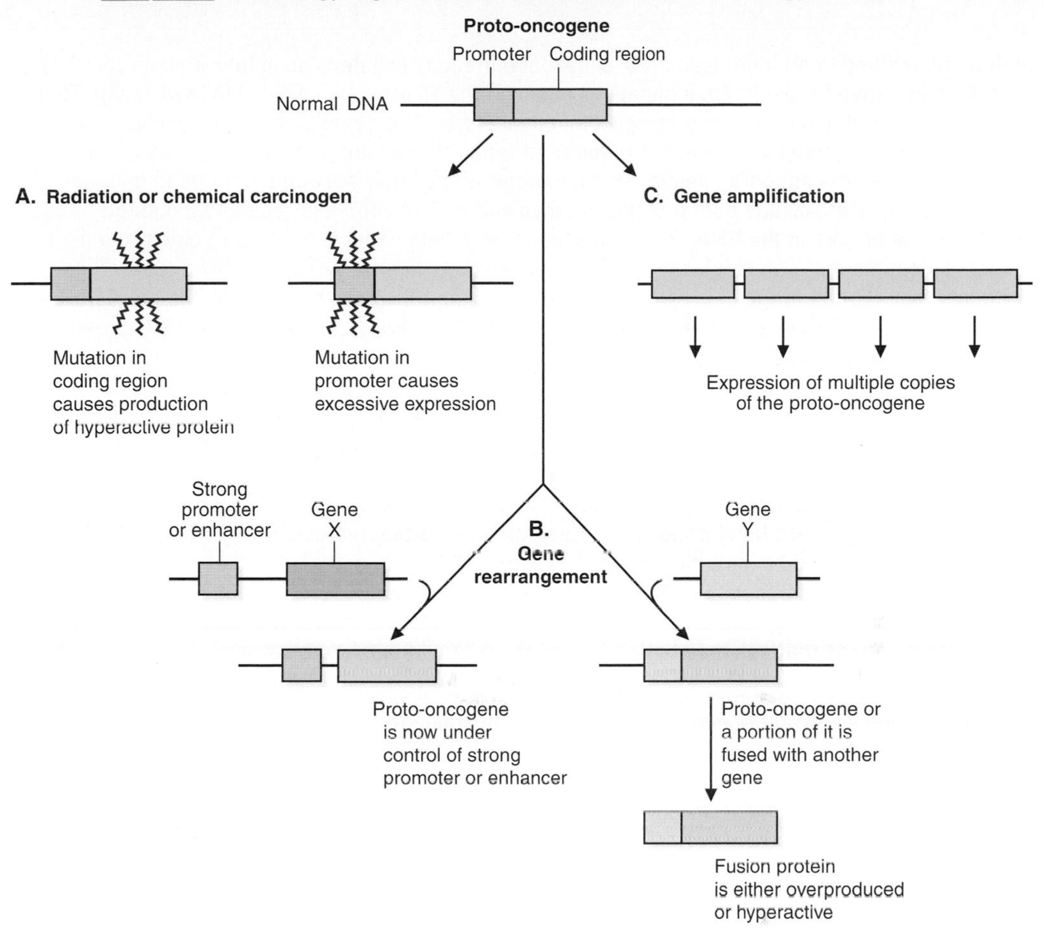

FIGURE 4.20. Transforming mutations in proto-oncogenes.

under a constitutive promoter such that myc is **inappropriately expressed in the cell cycle**, leading to inappropriate cell growth. The most common translocation is between chromosomes 8 and 14 (myc is normally found on chromosome 8, and it is moved to chromosome 14 under the control of immunoglobulin regulatory elements). One form of Burkitt lymphoma is linked to Epstein–Barr virus infection (EBV) in 95% of the cases.

8. **Mutations in DNA repair enzymes**
   a. The enzymes involved in DNA repair are tumor suppressor genes–loss of function results in an increased probability of mutation, and eventual mutation in a proto-oncogene
   b. **BRCA1 and BRCA2** (genes implicated in **breast cancer**) are involved in repairing single- and double-strand breaks in DNA
   c. **HNPCC** (**hereditary nonpolyposis colorectal cancer**) results from mutations in genes involved in mismatch repair

CLINICAL CORRELATES A number of cancers result from mutations in enzymes involved in DNA repair. **Ataxia telangiectasia** results from a mutation in the AT gene, which is required for the production of the ATM protein. The ATM protein kinase transmits the signal that there is damaged DNA to p53, such that the cell cycle can be halted and DNA repair can begin. It is also recruited to the sites of double-strand breaks and signals for other repair enzymes to join it at the site of damage. In the absence of ATM activity, these signals are lost, and the **damaged DNA is**

(continued)

replicated, leading to an increased probability that mutations will develop in future generations of cells. **Fanconi anemia** results from mutations in any of the 15 proteins involved in DNA repair. There are eight core proteins that form a complex that moves from the cytoplasm to the nucleus when DNA damage is detected, and which is required to repair the damage. Mutations in any of these proteins can lead to Fanconi's anemia. **Breast cancer**, of the inherited variety, is due to mutations in either **BRCA1 or BRCA2**, and both proteins are involved in DNA repair, particularly the **single- and double-strand breaks in the DNA.** Breast cancer, in its initial phase, is dependent on estrogen for growth, so agents that can block estrogen's action (such as **tamoxifen**) can be used to treat the disease. A small percentage of breast cancers express the HER2/neu receptor on the cell surface (analogous to the EGF receptor). **Herceptin**, a monoclonal antibody that targets the HER2 protein, has proven to be an effective agent to help target such HER2 overexpressing cells for destruction. **Triple-negative breast cancer cells** (loss of estrogen receptor expression, loss of HER2 expression, and loss of the expression of androgen receptor) are the most difficult to treat, although **PARP inhibitors** have shown promise in treating such cancer cells. **Li–Fraumeni syndrome** occurs when an individual inherits a mutated copy of **p53**. Multiple tumors can develop as the normal allele loses function. This protein is involved in DNA repair, as p53 is the guardian of the genome, as described later in this chapter. **Hereditary colon cancer** can be either **adenomatous polyposis coli** (APC), or HNPCC. APC is initiated by inheriting a mutation in the APC gene, which regulates the transcription factor β-catenin. In the absence of APC, β-catenin is always active, leading to the transcription of growth-promoting genes. HNPCC falls into the class of diseases due to errors in DNA repair, as the four identified genes that can lead to HNPCC are all involved in **mismatch repair**.

9. **Oncogenes**–mutations in genes that control normal growth and division (Table 4.2)
   a. **Signal transduction cascades** (Fig. 4.21)
      (1) Mutations in the genes for growth factors and growth factor receptors can convert the proto-oncogene to an oncogene (overexpression and internal expression of growth factors, and producing receptors that are mutated such that they are always turned on, even in the absence of the ligand).
      (2) Mutations in **signal transduction proteins** can convert proto-oncogenes to oncogenes. The best example of this is a mutation that leads to a loss of GTPase activity within ras. This occurs in a large number of all human tumors.
      (3) Mutations in **transcription factors** can convert proto-oncogenes to oncogenes by altering their normal regulatory functions (misexpressed during the cell cycle, e.g.).

CLINICAL CORRELATES

The **RET proto-oncogene** is a receptor tyrosine kinase. A gain-of-function in RET leads to **multiple endocrine neoplasias** (MEN) types 2A and 2B. The tumors that develop include **pheochromocytoma and medullary thyroid carcinoma**, in addition to other endocrine tumors. **Wilms tumor** is primarily a pediatric disorder that, in 20% of the cases, is due to a mutated **zinc-finger transcription factor**. Additional mutations are also found in **β-catenin** (see colon cancer above), which is important in **regulating the TCF family of transcription factors**. The alteration in the normal developmental expression of genes leads to the kidney tumor being formed.

   b. Oncogenes and the cell cycle
   c. **Cyclins and cyclin-dependent kinases** (CDK) control the progression from one phase of the cell cycle to another
   d. The cyclin–CDK complex is also regulated by phosphorylation and through inhibitory proteins known as **cyclin-dependent kinase inhibitors** (CKI). CKIs slow down cell-cycle progression by binding and inhibiting the cyclin–CDK complexes.
   e. G1/S transition is regulated by two CDKs (cdk4 and cdk6), cyclin D, the retinoblastoma gene product rb, and the E2F family of transcription factors
      (1) Rb binds E2F and prevents new gene transcription.
      (2) When cdk4 and cdk6 bind cyclin D, and are activated, rb is phosphorylated, and dissociates from E2F, allowing new gene transcription and the cell cycle to continue.

table 4.2 Classes of Oncogenes, Mechanism of Activation, and Associated Human Tumors

Class	Proto-Oncogene	Activation Mechanism	Location	Disease
Growth Factors				
Platelet-derived growth-factor β-chain	Sis	Overexpression	Secreted	Glioma Fibrosarcoma
Fibroblast growth factors	int-2	Amplification	Secreted	Breast cancer Bladder cancer Melanoma
	Hst	Overexpression	Secreted	Stomach carcinoma
Growth-factor receptors				
Epidermal growth factor receptor family	erb-B1	Overexpression	Cell membrane	Squamous cell carcinoma of the lung
	erb-B2	Amplification	Cell membrane	Breast, ovarian, lung, stomach cancers
Platelet-derived growth-factor receptor	PDGFR	Translocation	Cell membrane	Chronic myelomonocytic leukemia
Hedgehog receptor	SMO	Point mutation	Cell membrane	Basal cell carcinoma
Signal transduction proteins				
G-proteins	ras	Point mutation	Cytoplasm	Multiple cancers, including lung, colon, thyroid, pancreas, many leukemias
Serine-threonine kinase	akt2	Amplification	Cytoplasm	Ovarian carcinoma
	raf	Overexpression	Cytoplasm	Myeloid leukemia
Tyrosine kinase	abl	Translocation	Cytoplasm	Chronic myeloid leukemia Acute lymphoblastic leukemia
	src	Overexpression	Cytoplasm	Colon carcinoma
Hormone receptors				
Retinoid receptor	RARα	Translocation	Nucleus	Acute promyelocytic leukemia
Transcription factors				
	Hox11	Translocation	Nucleus	Acute T-cell leukemia
	Myc	Translocation Amplification	Nucleus Nucleus	Burkitt lymphoma Neuroblastoma, small-cell carcinoma of the lung
	fos, jun	Phosphorylation	Nucleus	Osteosarcoma, sarcoma
Apoptosis regulators				
	Bcl-2	Translocation	Mitochondria	Follicular B-cell lymphoma
Cell-cycle regulators				
Cyclins	Cyclin D	Amplification	Nucleus	Lymphoma Breast, liver, esophageal cancers
Cyclin-dependent kinase	CDK4	Amplification Point mutation	Nucleus Nucleus	Glioblastoma, sarcoma Melanoma

The table is not meant to be all-inclusive; only examples of each class of gene are presented.

(3) The activated cdk4–cyclin D complex (along with the cdk6–cyclin D complex) is also regulated by CKIs. Loss of function of the CKIs would lead to cancer (a tumor suppressor mechanism).

10. Tumor suppressor genes

a. Loss of function leads to uncontrolled cell growth

b. They are viewed as **recessive oncogenes** as loss of activity of both alleles is required. The oncogenes are considered dominant in action as only one mutated allele is sufficient for tumor cell growth.

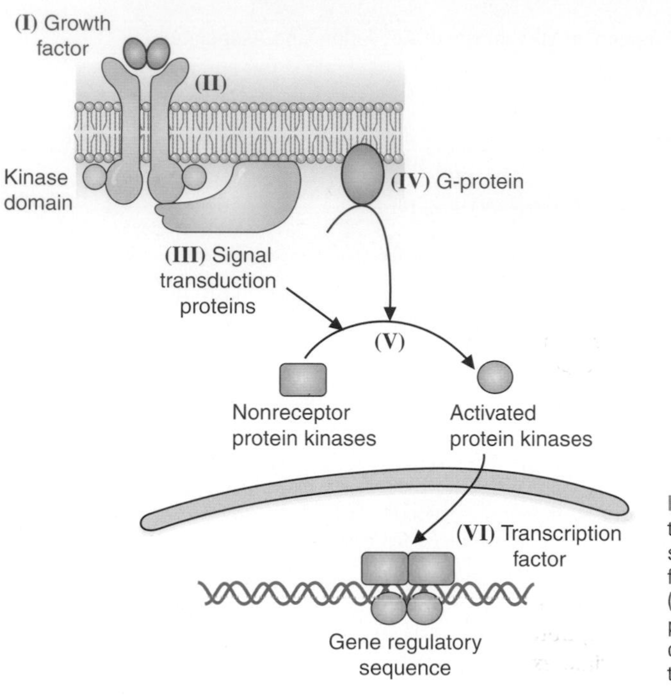

FIGURE 4.21. Proto-oncogene sites for transforming mutations in growth factor signaling pathways. Mutations in growth factors (I) or growth-factor receptors (II), signal transduction proteins (III), G-proteins (IV), non-receptor kinases (V), or transcription factors (VI) can all lead to cancer.

(1) Retinoblastoma gene (rb)
- **(a)** Inheriting a mutated copy of rb, there is a 100% chance of the individual developing retinoblastoma. Loss of rb activity alters the G1/S transition, and allows cells to cycle when they should not be cycling (see above)
- **(b)** Sporadic retinoblastoma acquires two specific mutations in the rb genes during the lifetime of the individual

(2) P53, the guardian of the genome
- **(a)** The protein p53 is a transcription factor that regulates the cell cycle and apoptosis
- **(b)** Loss of p53 activity is found in more than 50% of human tumors
- **(c)** The protein functions by stopping cell-cycle progression when DNA damage is found
  - **1.** When p53 finds DNA damage the synthesis of CKIs is induced, along with proteins to help repair the DNA damage.
  - **2.** If the damage cannot be repaired, p53 activates genes involved in apoptosis, programmed cell death, so the cell dies rather than replicate the damaged DNA.

(3) Ras is regulated by GAPs; loss of GAP activity will keep ras active for extended periods of time, and lead to transformation. Neurofibromatosis is due to a mutation in a GAP protein, NF-1.

(4) The cadherin family of glycoproteins mediates calcium-dependent cell-cell adhesion, and is anchored intracellularly by catenins, which bind to actin filaments.
- **(a)** Catenin proteins have two functions
  - **1.** Anchor cadherins to the cytoskeleton
  - **2.** Act as transcription factors
- **(b)** β-Catenin binds to a complex that contains the regulatory protein APC, which activates β-catenin for degradation.
- **(c)** If APC is inactivated β-catenin levels increase, translocate to the nucleus, and activate myc and cyclin D1 transcription, which leads to cell proliferation. APC is thus a tumor suppressor (loss of function leads to cell proliferation), and mutations in APC are found in most sporadic human colon cancers (Fig. 4.22).
- **(d)** Inherited mutations in APC lead to the most common form of hereditary colon cancer, familial adenomatous polyposis.

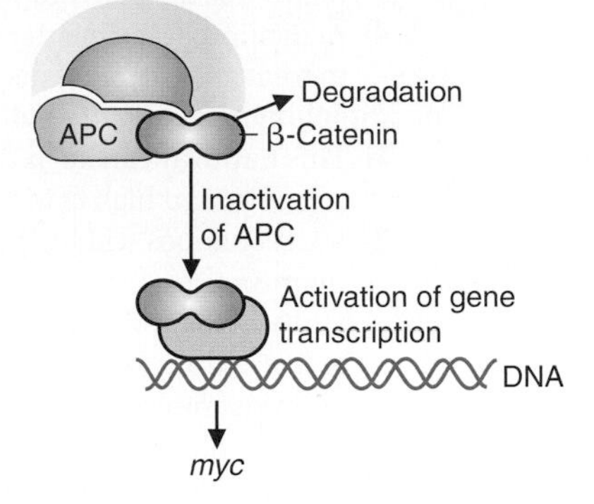

FIGURE 4.22. Role of catenins in regulating gene expression. The APC complex activates β-catenin for proteolytic degradation. If APC is inactivated, β-catenin levels increase.

11. Chromosomal translocations and cancer
- **a.** Certain tumors arise owing to specific chromosomal translocations
- **b.** CML (translocation between chromosomes 9 and 22) creates a novel tyrosine kinase, bcr-abl, which is unregulated and leads to increased cellular proliferation (see prior clinical correlation).
- **c.** Burkitt lymphoma, a translocation between chromosomes 8 and 14, puts the control of the proto-oncogene myc (a transcription factor) under the immunoglobulin promoter, which is constitutive. The inappropriate expression of myc leads to transformation (see prior clinical correlation).
- **d.** Follicular lymphoma results from a translocation between chromosomes 14 and 18, which places the antiapoptotic factor Bcl-2 (see below) under the control of an immunoglobulin promoter. The constant expression of Bcl-2 blocks apoptosis, leading eventually to transformation.

IV. CANCER AND APOPTOSIS

1. Apoptosis is a **regulated energy-dependent sequence of events** via which a cell self-destructs. Unlike necrosis, no inflammatory sequence of events occurs during apoptosis.
   - **a.** The cell shrinks
   - **b.** The chromatin condenses
   - **c.** The nucleus fragments
   - **d.** The cell membrane forms blebs
   - **e.** The cell breaks up into membrane-enclosed apoptotic vesicles
2. Apoptosis consists of an **initiation phase**, a **signal integration phase**, and an **execution phase**
   - **a.** Apoptosis is initiated by external signals that work through death receptors, or deprivation of growth hormones. Loss of mitochondrial integrity can also lead to the initiation of apoptosis.
   - **b.** In the signal integration phase, proapoptotic signals are counterbalanced by antiapoptotic signals to reach a decision as to whether apoptosis should occur
   - **c.** The execution phase is the actual events that occur to allow a cell to begin apoptosis, as first mediated by **caspases**, which are cysteine proteases that cleave bonds next to aspartate residues
     - **(1)** The caspases are present in cells as the inactive zymogens, procaspases, which must be activated before they can cleave target proteins.
     - **(2)** The cell contains both initiator caspases (initiate the activation of caspases) and execution caspases (these are activated by the initiator caspases and actually carry out the destruction of the cell).
   - **d. Death receptor pathway to apoptosis**
     - **(1)** The death receptors are a subset of tumor necrosis factor 1 receptors
     - **(2)** The activated death receptor forms a scaffold that binds two molecules of a procaspase, which cleave each other to form active initiator caspases

(3) The initiator caspases activate execution caspases and Bid (a Bcl-2 family member, see below). Truncated bid will activate the mitochondrial integrity pathway to apoptosis

(4) Activation of the death receptors by drugs (such as infliximab and etanercept) are used to treat autoimmune disorders by inducing apoptosis in the cells of the immune system

e. **Mitochondrial integrity pathway**

(1) This pathway can be initiated by growth-factor withdrawal, cell injury, release of certain steroids, and high cytoplasmic calcium

(2) Such changes will lead to the **release of cytochrome c** from the inner mitochondrial membrane

(3) Once cytochrome c is in the cytoplasm it will bind to **apaf** (proapoptotic protease-activating factor), which binds a caspase, forming an active complex known as the apoptosome, which activates execution caspases

f. Integration of proapoptotic and antiapoptotic signals by the **Bcl-2 family of proteins**

(1) The Bcl-2 family of proteins are decision makers that integrate prodeath and antideath signals to determine if a cell should enter apoptosis (Table 4.3)

(2) The antiapoptotic Bcl-2-type proteins antagonize death signals in either of two ways

(a) Insert into outer mitochondrial membrane to antagonize channel forming pro-apoptotic factors, thereby decreasing the cytochrome c release

(b) May bind apaf to prevent apoptosome formation

(3) Proapoptotic factors fall into two categories, ion-channel–forming and BH-3 domain-only proteins

(a) Ion-channel-forming members dimerize with other BH-3 domain–only members to form a channel in the outer mitochondrial membrane for cytochrome c to leave the mitochondria

(b) The BH-3 domain refers to the protein sequence that allows the proteins to bind to other Bcl-2 family members

(c) BH-3 domain–only members of this family of proteins can only bind to other Bcl-2 family members to activate the prodeath factors

g. **Cancer cells bypass apoptosis**

(1) The Akt pathway allows cells to bypass apoptosis by phosphorylating the proapoptotic factor BAD and inactivating it, thereby promoting cell growth (see Fig. 4.14)

(2) The MAP kinase pathway activates the protein kinase RSK, which also phosphorylates BAD, thereby promoting cell growth

h. MicroRNAs also regulate apoptosis through modulation of mRNA levels for a number of proteins involved in apoptosis, or regulating apoptosis

table 4.3 Examples of Bcl-2 Family Members

Examples of Bcl-2 Family Members
Anti-apoptotic
Bcl-2
Bcl-x
Bcl-w
Proapoptotic *Channel-forming*
Bax
Bak
Bok
Pro-apoptotic *BH3-only*
Bad
Bid
Bim

Roughly 30 Bcl-2 family members are currently known. These proteins play tissue-specific as well as signal pathway–specific roles in regulating apoptosis. The tissue specificity is overlapping. For example, Bcl-2 is expressed in hair follicles, kidney, small intestines, neurons, and the lymphoid system, whereas Bcl-x is expressed in the nervous system and hematopoietic cells.

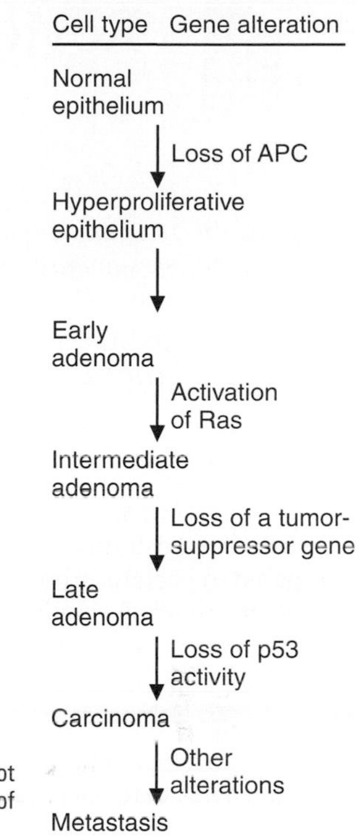

FIGURE 4.23. Possible steps in the development of colon cancer. The changes do not always occur in this order, but the most benign tumors have the lowest frequency of mutations, and the most malignant have the highest frequency.

V. CANCER REQUIRES MULTIPLE MUTATIONS (FIGURE 4.23)

1. Multiple genetic alterations are required to transform normal cells into malignant cells
2. It has been estimated that four to seven mutations are required for normal cells to be transformed

VI. VIRUSES AND HUMAN CANCER

1. **Three RNA retroviruses have been associated with human cancer**
   a. HTLV-1 (adult T-cell leukemia) contains a viral oncogene (tax) without a counterpart in the host human genome
   b. HIV leads to tumor progression due to immunosuppression
   c. Hepatitis C
2. **DNA viruses can also lead to transformation**
   a. Chronic hepatitis B infections lead to hepatocellular carcinoma
   b. Epstein–Barr virus encodes a Bcl-2-like factor, and antagonizes apoptosis in infected cells
   c. Herpes virus (HHV-8) leads to Kaposi sarcoma

Review Test

Questions 1 to 10 examine your basic knowledge and are not in the standard clinical vignette format.

Questions 11 to 35 are clinically relevant, USMLE-style questions.

Basic Knowledge Questions

1. Bacteria grown at 15°C contain a different fatty acid composition in their membranes as compared to bacteria grown at 37°C. Which one of the following would best represent the composition of the fatty acids at these two different temperatures?

(A) Bacteria at the lower temperature would contain a higher percentage of saturated fatty acids than bacteria grown at the higher temperature.
(B) Bacteria grown at the lower temperature would have a higher percentage of long-chain fatty acids than bacteria grown at the higher temperature.
(C) Bacteria grown at the lower temperature would have a higher percentage of unsaturated fatty acids than bacteria grown at the higher temperature.
(D) Bacteria grown at the lower temperature would have an increased level of cholesterol as compared to bacteria grown at the higher temperature.
(E) Bacteria grown at the lower temperature would have a decreased level of cholesterol as compared to bacteria grown at the higher temperature.

2. A 57-year-old pathologist, who had often cut himself while performing autopsies, develops blurred vision, dementia, personality changes, and muscle twitching in a very short period of time. The protein that is leading to these behavioral changes is best described as which one of the following?

(A) A soluble, cytoplasmic protein
(B) A peripheral membrane protein
(C) An embedded membrane protein
(D) A GPI-anchored membrane protein
(E) A secreted protein

3. Many growth factors, upon binding to their receptor, exhibit downregulation, in which the receptor number on the cell surface decreases. This occurs due to which one of the following processes?

(A) Endocytosis
(B) Exocytosis
(C) Pinocytosis
(D) Potocytosis
(E) Phagocytosis

4. Proton gradients across membranes are essential for the functions of which of the following organelles? Choose the one best answer.

(A) Lysosomes
(B) Mitochondria
(C) Nucleus
(D) Lysosomes and mitochondria
(E) Lysosomes and nucleus
(F) Nucleus and mitochondria

5. Assume there is a microRNA that participates in regulating the expression of a particular cyclin kinase inhibitor. How might an alteration in this microRNA lead to uncontrolled cell proliferation?

(A) Overexpression of the microRNA, so it acts as an oncogene.
(B) Reduced expression of the microRNA, so it acts as a tumor suppressor.
(C) A total loss of activity of the microRNA, so it cannot bind to its target mRNA.
(D) A loss of specificity of the microRNA for its target, so different mRNAs are not targeted.
(E) No change in microRNA activity can lead to uncontrolled cell proliferation.

6. The regulation of heterotrimeric G-proteins is similar to the regulation of which one of the following processes?

(A) DNA synthesis
(B) RNA synthesis
(C) Protein synthesis
(D) Active transport
(E) Facilitated transport

7. A message transmitted by which example of a chemical messenger would most likely be negatively affected by a mutation that greatly reduced the fluidity of the plasma membrane?

(A) Cytokine
(B) Steroid hormone
(C) A transforming growth factor
(D) Insulin
(E) Glucagon

8. Li–Fraumeni syndrome results from which one of the following? Choose the one best answer.

(A) Inability to recognize DNA damage
(B) Inability to regulate CDKs
(C) Inability to regulate a tyrosine kinase
(D) Inability to regulate gene transcription
(E) Inability to activate a heterotrimeric G-protein

9. Chromosomal translocations can lead to uncontrolled cell growth due to which one of the following?

(A) Interference with mitosis
(B) Interference with DNA synthesis
(C) Unequal crossing over during mitosis
(D) Inappropriate expression of translocated genes
(E) Loss of gene expression

10. Interference in generalized cytokine signaling can lead to which one of the following disorders?

(A) Adrenoleukodystrophy
(B) SCID
(C) Influenza
(D) Myasthenia gravis
(E) Type 1 diabetes

Board-style Questions

Questions 11 and 12 refer to the following case:

A bodybuilder has gained 50 pounds of muscle over the last 6 months, facilitated by both increased weight lifting and black-market pharmaceutical injection of one substance. He never experienced hypoglycemia during this time frame.

11. The mechanism whereby the illegal substance is entering the muscle cells of the bodybuilder is most likely which one of the following?

(A) Simple diffusion
(B) Active transport
(C) Endocytosis
(D) Facilitative diffusion
(E) Pinocytosis

12. How much energy is needed for the transport of the majority of the illegal substance that the bodybuilder is using for the drug to enter cells?

(A) No energy is required.
(B) One ATP molecule is used for each molecule of the substance transported.
(C) Only a few ATP molecules are being used to open and close the channel through which many substance molecules diffuse.
(D) This is an example of cotransport, in which the energy generates a sodium gradient across the membrane, and it is difficult to calculate an exact energy amount.
(E) The transporter has to be phosphorylated once to allow transport to occur for many solutes.

Questions 13 and 14 are based on the following case:

A 12-year-old boy is admitted to the hospital in ketoacidosis with a blood glucose level of 700 mg/dL (normal fasting levels are between 80 and 100 mg/dL). The boy is shown to have no detectable C-peptide upon further testing.

13. A potential reason for the elevated blood glucose is which one of the following?

(A) A lack of a sodium gradient across cellular membranes
(B) A lack of a calcium gradient across cellular membranes
(C) A lack of a chloride gradient across cellular membranes
(D) A reduced number of glucose transport molecules in the brain membrane
(E) A reduced number of glucose transport molecules in the muscle membrane
(F) A reduce number of glucose transport molecules in the liver membrane

14. The child is treated appropriately such that the glucose levels have been reduced, and he does not become dehydrated. Once glucose is transported into his cells, which organelle is responsible for generating energy from the oxidation of glucose to carbon dioxide and water?

(A) Lysosome
(B) Golgi complex
(C) Mitochondria
(D) Nucleus
(E) Peroxisome

15. A person is diagnosed with group A streptococcal bacteremia. One of the body's major defenses in this type of disease is for eosinophils to phagocytize the bacteria. Once internalized, the bacteria are destroyed by fusing the phagosome with a particular intracellular organelle. Which one of the following would destroy the activity of that organelle such that the bacteria would not be incapacitated?

(A) Inhibiting sodium–potassium ATPase activity
(B) Interfering with mitochondrial protein synthesis
(C) Blocking transport through nuclear pores
(D) Inhibiting a proton-translocating ATPase
(E) Inhibiting a calcium-activated ATPase

16. Lysosomal hydrolases are targeted to the lysosome by the addition of a carbohydrate residue to the protein. An inability to add this carbohydrate leads to a disease in which the lysosomal hydrolases are treated as secreted proteins, and are exported from the cell, rather than taken to the lysosomes. The secreted proteins will have which one of the following effects on the cells and proteins in the circulation?

(A) The blood cells will have their membrane proteins digested.
(B) The blood cells will have their carbohydrates on the cell surface removed.
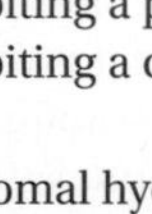
(C) The blood cell membranes will become leaky, leading to the death of the blood cells.
(D) Circulating proteins will be degraded, whereas the blood cells will be protected against the enzymes.
(E) Circulating proteins will be targeted to the spleen for removal.
(F) There will be no effect on the proteins and cells in the circulation.

17. A 25-year-old female presents with intense fear whenever she has to drive her car through a tunnel. She feels faint, sweats profusely, has palpitations, and hyperventilates. She is prescribed diazepam to reduce her symptoms. The type of chemical messenger enhanced by this treatment is best described as which one of the following?

(A) Neuropeptide
(B) Biogenic amine
(C) Large-molecule neurotransmitter
(D) Cytokine
(E) G-protein

18. A 15-month-old girl has been given an MMR immunization. Which of the following chemical messengers is responsible for the body's ability to mount an immune response to this vaccination?

(A) Neuropeptides
(B) Biogenic amines
(C) Steroid hormones
(D) Cytokines
(E) Amino acids

19. A 30-year-old female presents with a 15-pound weight loss over 1 month, heat intolerance, tachycardia, tremor, bilateral exophthalmos, and a mass in the anterior neck. The hormone overproduced in this condition requires which one of the following?

(A) Arachidonic acid
(B) Cholesterol
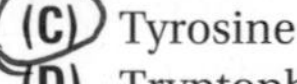
(C) Tyrosine
(D) Tryptophan
(E) Glutamate

20. A 3-year-old boy presents with 3 days of a low-grade fever, joint pain, and a "lacy-"appearing rash on his arms and legs. His rash began on his face and he appeared to have "slapped cheeks." The chemical messenger that caused the symptoms (vasodilatation presenting clinically as a "rash") can be classified as which one of the following?

(A) Cytokine
(B) Neuropeptide
(C) Eicosanoid
(D) Steroid hormone
(E) Amino acid

21. A 62-year-old male has a reddish, rough patch with white scales on the top of his ear. He does not get this treated, and 3 years later it has become an enlarged, raised lesion with a central ulcerated area that will not heal. Of the following, which is the most likely causative factor for this malignancy?

(A) Creation of pyrimidine dimers
(B) Creation of hydroxyl radicals
(C) Oncogenic RNA virus
(D) TNF receptor mutation
(E) Double-strand breaks in the DNA

22. A 32-year-old female has developed breast cancer. Her mother and one maternal aunt had breast cancer and her maternal grandmother had ovarian cancer. Which of the following best describes the mechanism behind this inherited problem?

(A) A tumor suppressor leading to loss of apoptosis
(B) A tumor suppressor leading to an inability to repair DNA
(C) A tumor suppressor leading to a constitutively active MAP kinase pathway
(D) An oncogene leading to loss of apoptosis
(E) An oncogene leading to an inability to repair DNA
(F) An oncogene leading to a constitutively active MAP kinase pathway

23. A 4-year-old boy went to the beach with his parents, and they found some clams, which they later ate for dinner. A few hours later, the boy developed a fever, started vomiting, and had profuse watery diarrhea. After being taken to the emergency room (ER), the boy was treated for potential dehydration, and recovered uneventfully. The root molecular cause of his symptoms was which one of the following?

(A) ADP-ribosylation of a Gαs protein
(B) Phosphorylation of a Gαs protein
(C) Acetylation of a Gαs protein
(D) ADP-ribosylation of a Gαi protein
(E) Phosphorylation of a Gαi protein
(F) Acetylation of a Gαi protein

24. A 4-year-old boy has had multiple episodes of pneumonia, steatorrhea, and has fallen off his normal growth curve. A sweat test was positive for chloride ions. The reason this boy is at risk for repeat episodes of pneumonia is which one of the following?

(A) Elastase destruction of lung cells
(B) Defective α1-antitrypsin activity
(C) Excessive water in the lungs
(D) Dried mucus accumulation in the lungs
(E) Loss of lung cells due to a defect in DNA repair

25. An experimental drug has been added to a eukaryotic cell, and while the drug was designed to interfere with a membrane transport process, the investigators found that in cells treated with the drug the lysosomes quickly turn into inclusion bodies. None of the material directed to the lysosomes for removal was being digested in the lysosome, and remained intact inside the organelle. An analysis of lysosomal contents in drug-treated cells indicated that the full complement of lysosomal enzymes were present in the organelle. Assuming that the drug is targeting just one protein, which one of the following proteins is most likely the target?

(A) An outer membrane protein that allows the lysosomal membrane to become permeable to small molecules
(B) A proton-translocating ATPase in the lysosomal membrane
(C) A chloride pump in the lysosomal membrane
(D) The enzyme that adds mannose-6-phosphate to lysosomal enzymes
(E) The mannose-6-phosphate receptor

26. A 5-year-old boy begins to regress in terms of developmental milestones, particularly neurologically. Shortly thereafter, the child enters a coma, and dies 2 years into the coma. Upon autopsy, the myelin sheath in the brain was found to be abnormal, as it contained a large quantity of very long-chain fatty acids in its phospholipids. The adrenal glands were also abnormal in appearance. The child, at the molecular level, had inherited a mutation that led to an inability to catalyze reactions that occur in which one of the following intracellular organelles?

(A) Lysosomes
(B) Nucleus
(C) Mitochondria
(D) Peroxisomes
(E) Golgi apparatus
(F) Nucleolus

Questions 27 through 29 are based on the following case:

A 42-year-old woman has slowly developed an inability to keep her eyes open at the end of the day. The eyelids droop, despite her best efforts to keep them open. This does not occur first thing in the morning. Further examination shows a generalized muscle weakness as the day progresses.

27. A drug that may help to stabilize this condition would do which one of the following?

(A) Stimulate the production of immune cells
(B) Stimulate the production of epinephrine
(C) Inhibit the production of acetylcholine
(D) Inhibit acetylcholinesterase
(E) Stimulate catechol-O-transferase

28. In addition to the answer to the previous question, a drug that may help to stabilize this condition would do which one of the following?

(A) Stimulate apoptosis
(B) Inhibit apoptosis
(C) Stimulate cell growth
(D) Inhibit cell growth
(E) Induce muscle growth
(F) Inhibit muscle growth

29. In order to diagnose the disease the patient is experiencing, a Western blot was run using the patient's sera as the source of antibodies. The protein run on the gel would need to be which one of the following?

(A) Acetylcholinesterase
(B) Acetylcholine receptor
(C) Epinephrine receptor
(D) Catechol-O-methyltransferase
(E) Glucocorticoid receptor
(F) HMG-CoA reductase

30. A 12-year-old boy is displaying tiredness and lethargy, and is found to have a hypochromic, microcytic anemia. Microscopic examination of the boy's red blood cells demonstrated a spherical shape, rather than concave. The mutation in this child is most often found in a protein located in which part of the red blood cell?

(A) The cytoskeleton
(B) The nucleus
(C) The mitochondria
(D) The endoplasmic reticulum
(E) The plasma membrane

31. A 39-year-old man is brought to the ER for a suspected suicide attempt. He has blurred vision; very dry, hot, red skin; dry mouth; urinary retention; confusion; hallucinations; loss of balance; and tachycardia. Emergency medical technicians (EMTs) found an empty bottle of amitriptyline in his apartment. The date on the bottle was just 1 week ago, yet all the pills were missing. The effects the man is experiencing is due to an inhibition of which of the following processes?

(A) Muscarinic acetylcholine receptor signaling
(B) Nicotinic acetylcholine receptor signaling
(C) GABA signaling
(D) Serotonin signaling
(E) Catecholamine signaling

32. A 21-year-old patient is being evaluated for a major depressive disorder. During the interview, he admits to having several episodes in the past of feeling "on top of the world," able to function very well with only 4 hours of sleep per night, maxing out his credit cards (a very unusual characteristic for him), and "indiscriminate" sexual encounters with multiple partners. The elemental medication most commonly used to treat this patient's disorder will lead to the accumulation of which one of the following compounds?

(A) Inositol
(B) Phosphatidylinositol
(C) Inositol phosphate
(D) Inositol bisphosphate
(E) Inositol trisphosphate
(F) Diacylglycerol

33. On a routine newborn exam, it is noted that the red reflex is absent in one eye. An MRI shows a tumor blocking the retina. Regulation at which phase of the cell cycle would be affected by the mutation that leads to this tumor?

(A) G_0 to G1
(B) G1 to S
(C) S to G2
(D) G2 to M
(E) M to G1
(F) G1 to G_0

34. A 45-year-old man presents with blood in his stool. Workup reveals a stage 3 (Dukes 3) colon carcinoma, with multiple polyps within the colon. A family history reveals that his father and grandfather both had colon cancer in their fifth decade of life. A potential initiating activating event in the development of this tumor is which one of the following?

(A) Loss of β-catenin activity
(B) Activation of β-catenin activity
(C) Loss of transcription factor myc activity
(D) Activation of Bcl-2 activity
(E) Loss of cyclin expression
(F) Gain-of-cyclin expression

35. A 29-year-old female presents with chronic fatigue for the past 9 months. She did have mononucleosis in the past, and blood work reveals a chronic viral infection. An analysis of a liver biopsy indicated that when placed under conditions in which apoptosis should be initiated, the cells continued to grow. The viral infection was most likely caused by which one of the following?

(A) Epstein–Barr virus
(B) Influenza virus
(C) Simian sarcoma virus
(D) Polio virus
(E) Herpes simplex virus

Answers and Explanations

1. **The answer is C.** Membrane fluidity needs to be constant at the two different temperatures for the bacteria to grow. At the lower temperature, a higher percentage of unsaturated fatty acids and short-chain saturated fatty acids would be present as those fatty acids have a lower melting point than long-chain saturated fatty acids. Cholesterol is not found in bacterial membranes.

2. **The answer is D.** The pathologist has obtained a prion disease, which results from the prion protein adopting an alternative conformation and precipitating in neural tissue. The normal prion protein is a GPI-anchored protein.

3. **The answer is A.** Receptor-mediated endocytosis refers to the clustering of receptors over clathrin-coated pits in the inner membrane, and then invagination of the membrane to form an intracellular vesicle that contains the receptor–growth factor complex. Exocytosis is the opposite effect–an intracellular vesicle fuses with the plasma membrane to release its contents into the extracellular space. Pinocytosis refers to endocytosis without the receptors–small particles can be taken into the cell through vesicle formation on the cell surface. Potocytosis refers to receptor-mediated entry into a cell through caveolae, and not through clathrin-coated pits. Phagocytosis refers to the forming of a membrane around a particle, and then the endocytosis of that membrane containing the particle (or bacteria).

4. **The answer is D.** Lysosomes depend on a proton gradient to acidify their intracellular milieu, such that the lysosomal hydrolases will be at their pH optima (around 5.5). Mitochondria require a proton gradient across their inner membrane in order to synthesize ATP via oxidative phosphorylation. The nucleus does not concentrate protons; the intranuclear space has the same pH as the cytoplasm.

5. **The answer is A.** Cyclin kinase inhibitors act as brakes on the cell cycle. If the cyclin kinase inhibitor can be removed from the cell, then the cell cycle could proceed in an uncontrolled fashion. MicroRNAs reduce the amount of protein product formed from the target mRNA. In order to eliminate the production of the cyclin kinase inhibitor, the microRNA would need to be overexpressed, such that all target mRNAs are bound, and translation of the gene product is halted. Reducing the expression of the microRNA would lead to overexpression of the cyclin kinase inhibitor, and more control of the cell cycle. This is also the case if the microRNA lost all activity (overproduction of its target), or lost its specificity (again, overproduction of the target).

6. **The answer is C.** The heterotrimeric G-proteins bind GTP on the α subunit, which activates the subunit. The activation is self-controlled by a built-in GTPase that is present within the α subunit. This activity slowly hydrolyzes the bound GTP to GDP, thereby inactivating the subunit, and allowing the heterotrimer to re-form. Similar events occur in protein synthesis with both initiation factors and elongation factors. These factors are active when GTP is bound to them, and a built-in GTPase activity within these factors limits the length of time they are active. Such control systems are not observed in DNA or RNA synthesis, or in carrier-mediated transport across a cellular membrane.

7. **The answer is B.** Steroid hormones must enter the cell by passive diffusion, and if the membrane is less fluid, it will be more difficult for the hormone to enter the cell to bind to its receptor. Cytokines, transforming growth factors (TGFs), insulin, and glucagon all bind to transmembrane receptors, which transmit their signal to the cytoplasmic portion of the receptor. It is less likely that those conformational signals will be affected by the fluidity of the membrane than the passage of the steroid hormone through the membrane. Decreased membrane fluidity may impair dimerization of the receptors, but the initial binding events should still occur normally.

8. **The answer is A.** Li–Fraumeni syndrome results from an inherited mutation in p53, the guardian of the genome. This protein monitors the DNA for damage, and if damage is found,

acts as a transcription factor to arrest the cell cycle, allow the DNA damage to be repaired, and then to allow the cycle to proceed. If the DNA damage cannot be repaired, then apoptosis is induced so that the cell will not replicate the damaged DNA. P53 does not regulate the CDKs, tyrosine kinases, or G-proteins. Loss of p53 expression will alter gene transcription (repair enzymes will not be induced, nor will apoptosis be induced if the damage cannot be repaired), but does not hinder the normal regulation of gene transcription.

9. **The answer is D.** For most translocations that lead to uncontrolled cell growth, a gene is inappropriately expressed because it has been moved adjacent to a constitutive promoter (such as the myc gene next to the immunoglobulin promoter in Burkitt lymphoma). The dysregulation of cell proliferation does not occur owing to problems with mitosis or DNA replication, nor with crossing over. In most cases, the problem is an increased or inappropriate expression of the translocated gene, and not a loss of gene expression.

10. **The answer is B.** X-linked SCID is due to the lack of a common cytokine receptor subunit (the γ subunit), which affects the ability of a variety of cytokines to transmit signals to hematopoietic cells. Adrenoleukodystrophy is due to the buildup of very long-chain fatty acids, for a variety of reasons. Influenza is due to a virus. Myasthenia gravis is due to the production of autoantibodies directed against the acetylcholine receptor. Type 1 diabetes results from an inability to produce insulin.

11. **The answer is A.** The bodybuilder is injecting (most probably) testosterone, a steroid hormone, which aids in building muscle mass. Steroid hormones are lipid-soluble substances, and cross membranes by simple diffusion. The receptor for steroid hormones is present inside the cell (either the cytoplasm or nucleus), and once the steroid hormone enters the cell, it will bind to the receptor in a saturable manner. Once the concentration of the hormone inside and outside the cells is equal, transport will stop. Active transport refers to using energy to concentrate a solute against its concentration gradient, which is not the case for steroid hormone transport across the membrane. Facilitative diffusion requires a membrane-bound carrier (no energy), but as indicated previously, the carrier (receptor) for these hormones is intracellular. Steroid hormones do not enter cells through either endocytosis or pinocytosis. The fact that the bodybuilder never became hypoglycemic after taking the drug suggests that it was not insulin being injected.

12. **The answer is A.** No energy is needed for simple diffusion, which is the case if this is a steroid hormone. The other answer choices require ATP, which would be necessary for an active transport process, whether it be activation of a channel by phosphorylation, or generation of a gradient across the membrane for cotransport. Simple and facilitated diffusion do not require any energy sources for transport.

13. **The answer is E.** The boy has Type 1 diabetes, and is producing no insulin. One of insulin's effects is to stimulate the translocation of GLUT4 transporters from internal vesicles to the plasma membrane of muscle and fat cells. The increase in glucose transport molecules on the cell surface is important for rapidly reducing blood glucose levels. The GLUT4 transporter is for facilitative diffusion, and is not dependent on an ion gradient across the membrane for effective transport, as are the glucose transporters in the small intestine.

14. **The answer is C.** The mitochondria are organelles of fuel oxidation and ATP generation. Lysosomes contain hydrolytic enzymes that degrade proteins and other large molecules. The Golgi form vesicles for transport of molecules to the plasma membrane and for secretion. The nucleus carries out gene replication and transcription of DNA.

15. **The answer is D.** The phagosomes fuse with lysosomes, where the acidity and digestive enzymes within the lysosomes destroy the contents of the phagosome (in this case, the bacteria within the phagosome). The digestive enzymes have a pH optimum of 5.5, which is maintained within the lysosome through the actions of a proton pump (a proton-translocating ATPase activity). The nucleus and mitochondria are not involved in the lysosomal digestion of phagosome contents. Blocking either the sodium ATPase, potassium ATPase, or a calcium-activated ATPase will greatly affect other organs, but will not affect the ability of the lysosome to degrade its contents.

16. **The answer is F.** Most lysosomal hydrolases have their highest activity near an acidic pH of 5.5 (pH optimum) and little activity in a neutral or basic environment. The intralysosomal pH is maintained near 5.5 by vesicular ATPases, which actively pump protons into the lysosome. The cytosol and other cellular components have a pH near 7.2, and are therefore protected from escaped hydrolases. The pH of the blood is maintained between 7.2 and 7.4, so the escaped lysosomal enzymes will have no activity at that pH, and will not affect the proteins and cells in the circulation. I-cell disease results from the inability to appropriately target lysosomal proteins, and it is a lysosomal storage disease.

17. **The answer is B.** The patient is experiencing an anxiety disorder and panic attacks. These symptoms are often treated with psychotherapy and benzodiazepams. Patients with anxiety disorders have low gamma-aminobutyric acid (GABA). Benzodiazepams, such as diazepam, increase the efficiency of the synaptic transmission of GABA, helping to make any existing GABA more efficacious. This is through the drug binding to GABA receptors such that when GABA binds to the receptor, the response to GABA is greater than in the absence of the drug (one effect is to leave chloride channels open for greater periods of time in response to GABA, thereby depolarizing the membrane and sending an inhibitory signal). GABA is the chief inhibitory neurotransmitter. GABA is a biogenic amine or "small-molecule" neurotransmitter, and is derived from the decarboxylation of glutamate. Neuropeptides are the other type of chemical messenger secreted by the nervous system, and are usually small peptides. Cytokines are small protein messengers of the immune system. G-proteins aid in transmitting the signals induced by proteins that bind to heptahelical receptors (such as the epinephrine or glucagon receptors). GABA does not transmit its signal through a G-protein.

18. **The answer is D.** Cytokines are the messengers of the immune system. Neuropeptides and biogenic amines (small-molecule neurotransmitters) are messengers of the nervous system. Steroid hormones are messengers of the endocrine system. Amino acids (such as glycine and glutamate) can act as mediators within the nervous system. Once the shot is given, immune cells secrete cytokines to induce the synthesis of antibodies against the antigens injected into the girl.

19. **The answer is C.** The patient has hyperthyroidism, or Grave disease, an overproduction of thyroid hormone. Thyroid hormone is derived, in part, from tyrosine, which is iodinated to produce the active forms of thyroid hormone, T3 and T4. Cholesterol is a precursor to steroid hormones. Arachidonic acid is a precursor to eicosanoids (prostaglandins). Tryptophan is a necessity in the production of serotonin, and glutamate is needed to produce GABA. The symptoms described do not occur if there is an overproduction of steroid hormones, eicosanoids, serontonin, or GABA.

20. **The answer is C.** The patient has Fifth disease, a viral illness caused by parvovirus B19. The "slapped cheek" appearance of this rash is very distinctive. The eicosanoids control cellular function in response to injury (in this case, a viral infection). In response to the viral infection, vascular endothelial cells will secrete prostaglandins that act on smooth muscle cells to cause vasodilation, which leads to the reddish appearance of the infected individual. The release of eicosanoids may also be responsible for the fever that sometimes accompanies Fifth disease. Neuropeptides, cytokines, steroid hormones, or amino acids are not responsible for the vasodilation that occurs in this disease.

21. **The answer is A.** This man originally displayed an actinic keratosis that has, over the intervening 3-year period, become a squamous cell carcinoma. Actinic keratosis develops in areas of the skin that are frequently exposed to sunlight, such as the top of the ear. The frequent exposure to UV light led to the creation of pyrimidine dimers in the DNA. If the cells cannot repair the DNA damage rapidly enough, cancerous changes do occur over time. The presentation of actinic keratosis may represent a precancerous state toward squamous cell carcinoma. Removal of the actinic keratosis would have prevented the development of the tumor. Hydroxyl radicals are created by ionizing radiation such as X-rays, not by sunlight. Oncogenic RNA viruses such as HTLV-1 could cause lymphomas or leukemias, but have not been implicated in squamous cell carcinoma. TNF receptor mutations can occur in immune system cells, leading

to apoptosis, which would lead to an immune defect, but not the cancer observed. UV light does not lead to the creation of double-strand breaks in DNA.

22. **The answer is B.** Hereditary breast cancer is due to inherited mutations in either of the tumor suppressor genes *BRCA1* or *BRCA2*. These genes encode proteins that play important roles in DNA repair (primarily single- and double-strand breaks), and it is the loss of this function that predisposes the patient to breast and ovarian cancers. The inability to repair the breaks in the backbone leads to errors during replication, and mutations will develop that eventually lead to a loss of growth control. This is a loss-of-function disorder, which characterizes the genes involved as tumor suppressors. Inheriting one mutated copy of *BRCA1* means that the other, normal copy of *BRCA1* must be lost in a particular cell in order to initiate the disease (loss of heterozygosity). For breast cancer, this occurs 85% of the time (penetrance upon inheriting a *BRCA1* or *BRCA2* mutation). An oncogene is a dominant gene, so only one mutated copy can bring about the disease. *BRCA1* or *BRCA2* mutations do not directly lead to a loss of apoptosis, or to a constitutively active MAP kinase pathway.

23. **The answer is A.** Cholera is caused by *Vibrio cholerae*, found in fecally contaminated food or water and in shellfish. Cholera toxin, which is composed of multiple subunits, utilizes some subunits to allow one particular subunit with enzymatic activity to enter the intestinal epithelial cell. This toxin catalyzes the ADP-ribosylation of Gαs, inhibiting the GTPase activity of the α subunit of the G-protein, leading to constitutive activation of adenylate cyclase, and high cAMP levels. This leads to the activation of ion channels, having potassium, sodium, and chloride ions leave the intestinal epithelial cells into the lumen, along with water, leading to the watery diarrhea. The treatment consists of rehydration with electrolytes. Owing to the volume of water lost, the disease is usually self-limiting, as the bacteria causing the disorder are washed out of the intestine.

24. **The answer is D.** The boy is exhibiting the symptoms of cystic fibrosis, which is due to a mutation in the CFTR. The CFTR is required for chloride transport across the membrane, is activated by phosphorylation by the cAMP-activated protein kinase, and when activated allows chloride to flow down its electrochemical gradient. A defective CFTR also alters the ion composition of mucus, reducing its ability to absorb water through osmosis, leading to the drying of mucus in various ducts and tissues, including the lung cells. The lung cells normally secrete a thin, watery mucus designed to trap small particles, which are moved through the lung so they can be swallowed or removed by coughing. When water cannot leave the lung cells, the mucus dries out, leading to pulmonary dysfunction due to clogged bronchi.

25. **The answer is B.** Lysosomes contain a single membrane (so there is no outer membrane, such as in mitochondria) that contains a proton-translocating ATPase. The ATPase will concentrate protons inside of the lysosome, at the expense of ATP hydrolysis, to acidify the intraorganelle pH such that the lysosomal enzymes will be active. If the intravesicular pH cannot be lowered, the digestive enzymes will be inactive, and no digestion will take place. Targeting a chloride pump in the lysosomal membrane will not affect the activity of the lysosomal enzymes. If the lysosomal enzymes were not marked with a mannose-6-phosphate residue in the Golgi apparatus, they would not be able to bind to the mannose-6-phosphate receptor to be targeted to the lysosomes. If the drug altered either of those proteins (the enzyme responsible for adding the mannose-6-phosphate or the mannose-6-phosphate receptor), then the lysosomal enzymes would not be in the lysosomes, which is not the case. Such a drug would bring about the symptoms of I-cell disease.

26. **The answer is D.** The child has the symptoms of X-linked adrenoleukodystrophy, which is an X-linked disorder with a mutation in the *ABCD1* gene. The *ABCD1* gene is required for the transport of very long-chain fatty acids into the peroxisome for catabolism. In the absence of this activity, the very long-chain fatty acids accumulate, become incorporated into phospholipids, and alter the structure of myelin, leading to the neurological disorders observed. The lysosomes, nucleus, and Golgi apparatus are not involved in very long-chain fatty acid oxidation. The nucleolus is found in the nucleus and is the site of ribosome formation. Mitochondria oxidize fatty acids, but not when they are very long-chain fatty acids (greater than 20 carbons).

In those cases, the initial steps of oxidation occur in the peroxisome, and when the chain length has been reduced, the partially oxidized fatty acid is transferred to the mitochondria to finish the oxidation of the compound.

27. **The answer is D.** The woman has myasthenia gravis, which is due to autoantibodies directed against the acetylcholine receptor. As such, acetylcholine stimulation of muscle cells is decreased, owing to a reduced number of functional acetylcholine receptors at the neuromuscular junction. One way to treat this condition is to inhibit acetylcholinesterase, the enzyme that degrades acetylcholine at the neuromuscular junction. By keeping the levels of acetylcholine high at the junction, there is a greater probability that the receptors that are active are occupied, and the signal from the neuron can be transmitted. Inhibiting the production of acetylcholine would exacerbate the problem, as would stimulating the production of immune cells (more autoantibodies would potentially be generated). Epinephrine is not involved at the neuromuscular junction, and stimulation of catechol-O-transferase is a mechanism to inhibit the action of catecholamines in nonneuronal tissues, and does not contribute to the progression of myasthenia gravis.

28. **The answer is A.** The woman has myasthenia gravis, which is due to an autoimmune disorder in which antibodies directed against the acetylcholine receptor block the ability of acetylcholine to stimulate the muscle cells at the neuromuscular junction. Immunosuppressants can be taken to reduce the autoantibody production. Such drugs work, in part, through the activation of the tumor necrosis factor receptor, which activates apoptosis in the cells, leading to their destruction. Inhibiting apoptosis would exacerbate the problem, as the antibody-producing cells would survive longer and continue to produce the antibodies directed against the acetylcholine receptor. Drugs affecting the muscle would not help with this disorder, as it is a problem unique to the acetylcholine receptor expressed on the muscle surface. Stimulation or inhibition of cell growth does not stop the antibody-producing cells from continuing to make antibodies, and would not be an effective drug target for this disease.

29. **The answer is B.** The woman has myasthenia gravis, which is due to the presence of autoimmune antibodies directed against the acetylcholine receptor in the neuromuscular junction. To confirm the diagnosis by Western blot, a sample of acetylcholine receptor would be run through a polyacrylamide gel, the protein transferred to filter paper, and the filter paper incubated with the patient's sera. If the sera contain antibodies that bind to the acetylcholine receptor, the antibodies will be bound to the filter paper, and then visualized using a secondary antibody linked to a reporter enzyme. Controls would be done to indicate that sera from an individual who did not have myasthenia gravis did not allow for the formation of a band on the Western blot. Running acetylcholinesterase, the epinephrine receptor, catechol-O-methyltransferase, the glucocorticoid receptor, or HMG-CoA reductase on the gel would not allow detection of antibodies against the acetylcholine receptor in the patient's blood sample.

30. **The answer is A.** The boy has hereditary spherocytosis, which is due to a mutation in a red blood cell cytoskeletal protein. The most common mutation is in spectrin, although mutations in ankyrin, band 3, and protein 4.2 can also lead to this phenotype. Owing to the mutation in the cytoskeletal protein, the membrane shape becomes spherical instead of concave. This leads to the removal of the spherical cells by the spleen, leading to both anemia and splenomegaly. Mutations in the proteins in the plasma membrane or the endoplasmic reticulum will not lead to this disorder. Red blood cells do not have a nucleus or mitochondria.

31. **The answer is A.** Many classes of drugs, including some antihistamines (e.g., Benadryl), some antipsychotics (e.g., olanzapine), tricyclic antidepressants (e.g., amitriptyline), and atropine-like drugs (e.g., atropine, scopolamine) have anticholinergic effects or side effects, and function as antagonists to the acetylcholine receptor. Muscarinic acetylcholine receptors act through G-protein activation, whereas nicotinic acetylcholine receptors act as an ion channel, allowing sodium to flow through the receptor once it has been activated. The drug overdose in this case is inhibiting the muscarinic receptors, which occur in the autonomic and central nervous symptoms. This is a very typical case of anticholinergic overdose, with the "classic" symptoms classified as "blind as a bat," "dry as a bone," "red as a beet," "mad as a hatter," and

"hot as a hare." The overdose is not affecting the nicotinic acetylcholine receptors, or receptors for GABA, serotonin, or catecholamines.

32. **The answer is C.** The patient has presented with classic symptoms of bipolar disorder. Lithium is a first-line treatment of bipolar disorder whose mechanism of action is to interrupt the PI cycle by blocking the action of inositol monophosphatases, the enzyme that converts inositiol phosphate to free inositol, such that phosphatidylinositol can be resynthesized from CDP-diacylglycerol and inositol. Through an interruption in the cycle, the key PI-cycle second messengers cannot be continually generated, leading to a reduction in signaling capabilities.

33. **The answer is B.** The child has hereditary retinoblastoma, which is due to an inherited mutation in the rb gene. As the rb gene is a tumor suppressor gene, once loss of heterozygosity occurs, the function of rb in the cell cycle is lost. Rb helps to regulate the E2F family of transcription factors. Once cyclin D is synthesized, and activates a pair of CDKs, rb protein is phosphorylated, which causes it to leave a complex with the E2F factors. The removal of rb from the protein complex activates the E2F proteins, which initiate new gene transcription to allow the cell to transition to the S phase of the cell cycle. In the absence of any functional rb gene product, the transition to S phase is unregulated, and occurs continuously, leading to tumor growth. The rb gene product is not required for any other checkpoints in the cell cycle.

34. **The answer is B.** The patient has hereditary colon cancer, specifically adenomatous polyposis coli, which presents in the fourth or fifth decade of life, with multiple polyps lining the lumen of the colon. The defective protein is APC, which regulates β-catenin activity. The loss of APC function leads to inappropriately activated β-catenin, which is a transcription factor and can stimulate the expression of myc and cyclin D1, to promote cell growth. The inappropriate expression of myc and cyclin D1 due to the loss of APC is an initiating event in tumorigenesis. The APC mutation does not affect BCl-2 activity, which is an antiapoptotic activity. While the loss of APC activity will lead to a gain-of-cyclin expression, that gain is due to the activation of β-catenin, which would be the initiating event for tumor formation.

35. **The answer is A.** Infection by the Epstein–Barr virus will lead to the synthesis of a Bcl-2-like factor that antagonizes apoptosis, and allows the virus-infected cells to survive and continue producing more viruses. This factor is not present in the other viruses listed as potential answers.

chapter 5

Generation of ATP from Metabolic Fuels and Oxygen Toxicity

The main clinical uses of this chapter are to understand the basics of energy production, the function of vitamins, and mechanisms of energy production disruption by toxins or deficiencies. The toxicity of oxygen-derived radicals, and their relationship to disease, is also discussed.

OVERVIEW

- Adenosine triphosphate (ATP) transfers energy from the processes that produce it to those that use it.
- Most carbons of glucose, fatty acids, glycerol, and amino acids are ultimately converted to acetyl-CoA. ATP is produced by these reactions (Fig. 5.1).
- Acetyl-CoA is oxidized in the tricarboxylic acid (TCA) cycle. CO_2 is released, and electrons are passed to NAD^+ and FAD, producing NADH and $FADH_2$.
- NADH and $FADH_2$ transfer the electrons to O_2 via the electron transport chain. Energy from this transfer of electrons is used to produce ATP by the process of oxidative phosphorylation.
- Cofactors, many of which are minerals or compounds produced from vitamins, aid the enzymes that catalyze the reactions of these metabolic pathways.
- Reactive oxygen species can be generated intracellularly, and if their synthesis and utilization are not carefully controlled, cellular damage may result.

I. BIOENERGETICS

- For a biochemical reaction,

$$aA + bB \rightleftharpoons cC + dD$$

the change in free energy (ΔG) is related to the concentrations of the substrates and products and to the change in the standard free energy of the reaction at pH 7 ($\Delta G^{o\prime}$). The change in the standard free energy is determined by the chemical bonds that are being broken and formed.
- Reactions with a negative ΔG proceed spontaneously; those with a positive ΔG do not. If $\Delta G = 0$, the reaction is at equilibrium, and

$$K_{eq} = \frac{[C]^c\,[D]^d}{[A]^a\,[B]^b}$$

where, the substrates and products are at their equilibrium (eq) concentrations. Therefore,

$$\Delta G^{o\prime} = -RT \ln K_{eq}$$

- The change in free energy indicates the direction in which a reaction will proceed, but not the speed of the reaction. Enzymes determine the speed.

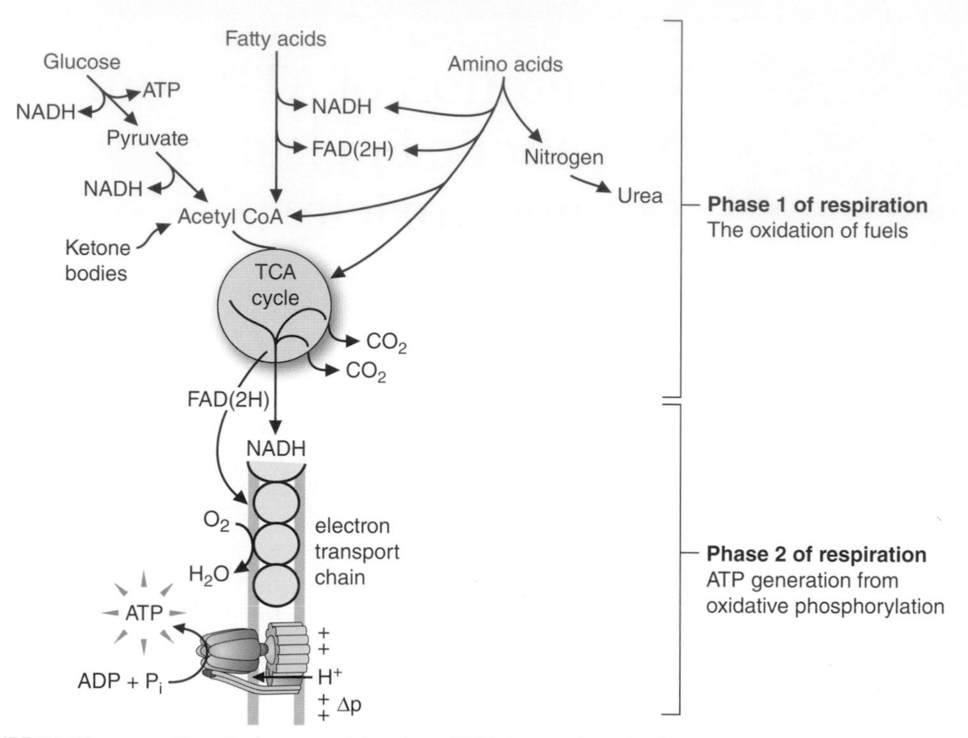

FIGURE 5.1. The generation of adenosine triphosphate (ATP) from fuels in the blood, and cellular respiration. Δp, proton gradient.

A. The change in free energy in biologic systems

1. The **change in free energy** (the energy available to do **useful work** at constant pressure and temperature) is defined by the equation:

$$\Delta G = \Delta H - T\Delta S$$

where, ΔG is the change in free energy; ΔH is the change in enthalpy (heat content); ΔS is the change in entropy (randomness or disorder); and T is the absolute temperature in degrees Kelvin (K).

2. For a biochemical reaction, the change in free energy can be used to predict the **direction** in which the reaction will proceed.
3. For the reaction in which

$$\text{aA} + \text{bB} \rightleftharpoons \text{cC} + \text{dD}$$

(where the upper-case letters symbolize the molecule and the lower-case letters indicate the number of molecules), the free energy change depends on the concentrations of the substrates and products and on the value of the constant $\Delta G^{o\prime}$.

$$\Delta G = \Delta G^{o\prime} + RT\text{In}\frac{[\text{C}]^\text{c}\,[\text{D}]^\text{d}}{[\text{A}]^\text{a}\,[\text{A}]^\text{b}}$$

where, **$\Delta G^{o\prime}$ is the standard free energy change at pH 7** (ignoring the concentration of water); R is the gas constant; T is the absolute temperature; and [ ] means concentration.

4. If ΔG is negative, the reaction will proceed spontaneously with the release of energy. If ΔG is positive, the reaction will not proceed spontaneously. If $\Delta G = 0$, the reaction is at equilibrium, and, although the substrates react to form products and products react to form substrates, there is no net change in the concentrations.

B. The equilibrium constant and the change in free energy

1. At **equilibrium, $\Delta G = 0$** and

$$\Delta G^{o\prime} = -RT\text{In}\frac{[\text{C}]^\text{c}\,[\text{D}]^\text{d}}{[\text{A}]^\text{a}\,[\text{A}]^\text{b}}$$

table 5.1 Additive Nature of Free Energy Changes

Glucose + P_i → glucose-6-P + H_2O	$\Delta G^{o\prime}$ = +3.3 kcal/mol
ATP + H_2O → ADP + P_i	$\Delta G^{o\prime}$ =–7.3 kcal/mol
Sum: glucose + ATP → glucose-6-P + ADP	$\Delta G^{o\prime}$ =–4.0 kcal/mol

2. The **equilibrium constant** (K_{eq})is related to the **concentrations** of the substrates and products **at equilibrium**:

$$K_{eq} = \frac{[C]^c\,[D]^d}{[A]^a\,[B]^b}$$

Therefore, $\Delta G^{o\prime} = -RT \ln K_{eq}$.

a. If $K_{eq} = 1$, $\Delta G^{o\prime} = 0$.
b. If $K_{eq} > 1$, $\Delta G^{o\prime}$ is negative.
c. If $K_{eq} < 1$, $\Delta G^{o\prime}$ is positive.

3. The larger and more negative the $\Delta G^{o\prime}$, the less substrate relative to product is required to produce a negative ΔG (i.e., the more likely the reaction is to proceed spontaneously).
4. For a sequence of reactions that have common intermediates, the **standard free energy changes are additive** (Table 5.1).

C. The relevance of free energy changes to biologic systems

1. The rate of a reaction is not related to its free energy change.
   a. A reaction with a large negative free energy change does not necessarily proceed rapidly.
   b. The **speed** of a reaction depends on the properties of the **enzyme** that catalyzes the reaction.
      (1) An enzyme increases the rate at which a reaction reaches equilibrium. It does not affect K_{eq} (the relative concentrations of the substrates and products at equilibrium).
2. Most biochemical reactions exist in pathways; therefore, other reactions are constantly adding substrates and removing products.
   a. The relative activities of the enzymes that catalyze the individual reactions of a pathway differ.
   b. Some reactions are near equilibrium ($\Delta G = 0$). Their direction can be readily altered by small changes in the concentrations of their substrates or products.
   c. Other reactions are far from equilibrium. Allosteric factors that alter the activity of these enzymes can change the overall flux through the pathway.

II. PROPERTIES OF ADENOSINE TRIPHOSPHATE

- ATP contains the base adenine, the sugar ribose, and three phosphate groups joined to each other by two anhydride bonds.
- ATP is produced from adenosine diphosphate (ADP) and inorganic phosphate (P_i) mainly by the process of oxidative phosphorylation.
- The free energy released when ATP is hydrolyzed is used to drive reactions that require energy.
- ATP can transfer phosphate groups to other compounds such as glucose, forming ADP.
- ADP can accept phosphate groups from compounds such as phosphocreatine, forming ATP.

A. The structure of ATP

ATP consists of the base **adenine**, the sugar **ribose**, and **three phosphate groups** (Fig. 5.2).

1. **Adenosine** (a nucleoside) contains the base adenine linked to ribose.
2. Adenosine monophosphate (**AMP**) is a nucleotide that contains adenosine with a phosphate group esterified to the 5′-hydroxyl of the sugar.

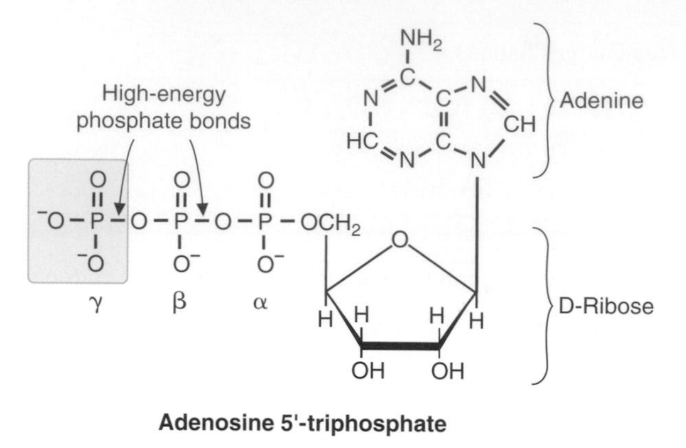

FIGURE 5.2. The structure of adenosine triphosphate (ATP), and the location of the two high-energy bonds in the molecule.

3. **ADP** contains a second phosphate group attached by an anhydride bond.
4. **ATP** contains a third phosphate group.

B. The functions of ATP

ATP plays a central role in **energy exchanges** in the body.

1. ATP is constantly being **consumed and regenerated.**
   a. It is consumed by processes such as muscular contraction, active transport, and biosynthetic reactions.
   b. It is regenerated by the oxidation of foodstuffs.
2. The **free energy** released when ATP is hydrolyzed is used to drive reactions that require energy.
   a. ATP can be hydrolyzed to **ADP** and inorganic phosphate (**P_i**) or to **AMP** and pyrophosphate (**PP_i**). ATP, ADP, and AMP are interconverted by the adenylate kinase reaction (termed myokinase in the muscle).

$$\text{ATP} + \text{AMP} \rightleftharpoons 2\text{ADP}$$

   b. Other nucleoside triphosphates (GTP, UTP, and CTP) are sometimes used to drive biochemical reactions. They can be derived from ATP, and have the same Gibbs free energy of hydrolysis as do the two high-energy bonds in ATP.
3. For **the hydrolysis of ATP to ADP and P_i, $\Delta G^{o\prime} = -7.3$ kcal/mol.**
   a. The anhydride bonds of ATP are often called "high-energy bonds."
   b. **$\Delta G^{o\prime}$ is large**, however, not because a single bond is broken, but because the products of hydrolysis are more stable than ATP.
4. **ATP** can transfer phosphate groups to compounds such as glucose, forming ADP.
5. **ADP** can accept phosphate groups from compounds such as phosphoenolpyruvate, phosphocreatine, or 1,3-bisphosphoglycerate, forming ATP.

III. ELECTRON CARRIERS AND VITAMINS

- Certain cofactors of enzymes are involved in the transfer of electrons from foodstuffs to O_2, a process that generates energy for the production of ATP.
- NAD^+ (derived from the vitamin niacin) and FAD (derived from the vitamin riboflavin) pass electrons to the electron transport chain. In this chain, flavin mononucleotide (FMN) and coenzyme Q (CoQ; ubiquinone) pass the electrons to heme-containing cytochromes, which transfer the electrons to O_2. As a consequence of these processes, ATP is produced.
- Other cofactors involved in deriving energy from food include coenzyme A (CoA; synthesized from the vitamin pantothenate), thiamine pyrophosphate (synthesized from the vitamin thiamine), and lipoic acid.

- Additional cofactors derived from water-soluble vitamins are involved in a variety of metabolic reactions. These cofactors include NADPH (derived from the vitamin niacin), biotin, pyridoxal phosphate (derived from vitamin B_6), tetrahydrofolate (derived from the vitamin folate), vitamin B_{12}, and vitamin C.
- The fat-soluble vitamins (A, D, E, and K) are also involved in metabolism.

A. Major cofactors in the generation of ATP from foodstuffs

As food is oxidized to CO_2 and H_2O, electrons are transferred mainly to nicotinamide adenine dinucleotide (**NAD$^+$**) and flavin adenine dinucleotide (**FAD**).

1. **NAD$^+$** accepts a hydride ion, which reacts with its nicotinamide ring. NAD$^+$ is reduced; the substrate (RH_2) is oxidized; and a proton is released.

$$NAD^+ + RH_2 \rightleftharpoons NADH + H^+ + R$$

   a. NAD$^+$ is frequently involved in oxidizing a hydroxyl group to a ketone.

$$R-\underset{H}{\overset{OH}{\underset{|}{\overset{|}{C}}}}-R_1 + NAD^+ \rightleftharpoons R-\overset{O}{\overset{\|}{C}}-R_1 + NADH + H^+$$

   b. The **nicotinamide ring** of NAD$^+$ is derived from the vitamin **niacin** (nicotinic acid) and, to a limited extent, from the amino acid **tryptophan**.

2. **FAD** accepts two hydrogen atoms (with their electrons). FAD is reduced, and the substrate is oxidized.

$$FAD + RH_2 \rightleftharpoons FADH_2 + R$$

   a. FAD is a better oxidizing agent than NAD$^+$, and is frequently involved in reactions that produce a carbon-carbon double bond.

$$R\text{-}CH_2\text{-}CH_2\text{-}R1 + FAD \rightleftharpoons R\text{-}CH{=}CH\text{-}R1 + FADH_2$$

   b. FAD is derived from the vitamin **riboflavin**.

B. Components of the electron transport chain

The reduced cofactors, NADH and $FADH_2$, transfer electrons to the electron transport chain, which is located in the inner mitochondrial membrane. The chain consists of a number of protein complexes, designated as I through IV.

1. **FMN** receives electrons from NADH in complex I and transfers them through Fe–S centers to coenzyme Q.
   a. FMN is derived from **riboflavin**.
2. **Coenzyme Q** receives electrons from FMN and also through Fe–S centers from $FADH_2$ (such as complex II).
   a. $FADH_2$ is not free in solution like NAD$^+$ and NADH; it is tightly bound to enzymes.
   b. Coenzyme Q can be synthesized in the body. It is not derived from a vitamin.
3. **Cytochromes** in complex III receive electrons from the reduced form of coenzyme Q.
   a. Each cytochrome consists of a **heme** group (Fig. 5.3) associated with a protein.
   b. The **iron** of the heme group is reduced when the cytochrome accepts an electron.

$$Fe^{3+} \rightleftharpoons Fe^{2+}$$

   c. Heme is synthesized from glycine and succinyl-CoA in humans. It is not derived from a vitamin.
4. **Oxygen** (O_2) ultimately receives the electrons at the end of the electron transport chain and is reduced to H_2O (a function of complex IV).

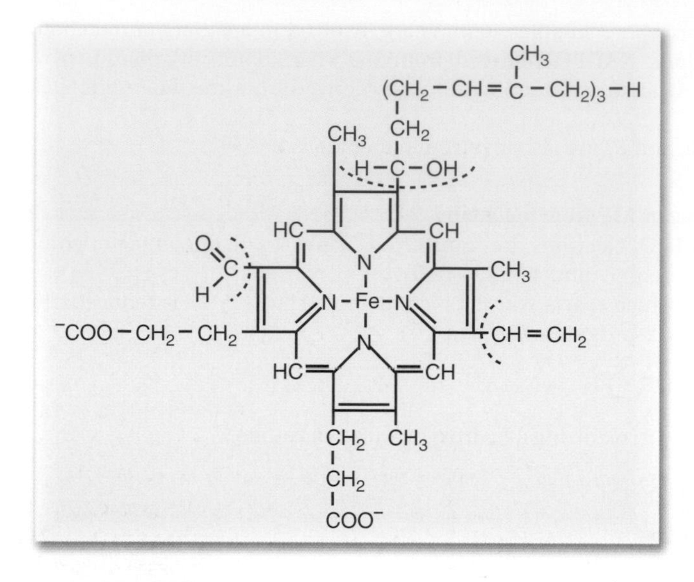

FIGURE 5.3. The general structure of the heme group, which is present in hemoglobin, myoglobin, and the cytochromes b, c, and c_1. Each cytochrome has a heme with a different modification of the side chains (indicated with the dashed lines), resulting in a slightly different reduction potential, and, consequently, a different position in the sequence of electron transfer.

CLINICAL CORRELATES **Iron-deficiency anemia** is due to a lack of iron for heme synthesis, which will lead to reduced oxygen delivery to the tissues. This, in conjunction with reduced heme levels in the electron transfer chain due to the reduced iron levels, can lead to muscle weakness because of an inability to synthesize appropriate amounts of ATP.

C. Coenzyme A

Coenzyme A (CoASH) contains a sulfhydryl group that reacts with carboxylic acids to form **thioesters**, such as acetyl-CoA, succinyl-CoA, and palmitoyl-CoA (Fig. 5.4).

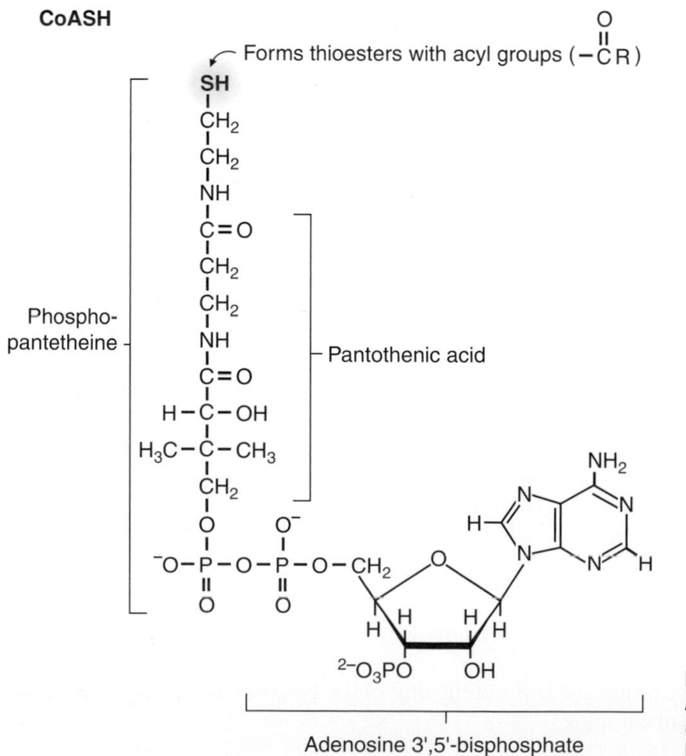

FIGURE 5.4. The structure of coenzyme A. The *arrow* indicates where acyl (e.g., acetyl, succinyl, and fatty acyl) groups bind to form thioesters.

1. The $\Delta G^{o\prime}$ for hydrolysis of the thioester bond is −7.5 kcal/mol (a high-energy bond).
2. Coenzyme A contains the vitamin **pantothenic acid**.
   a. Pantothenic acid is also present in the fatty acid synthase complex.

D. Thiamine and lipoic acid, cofactors for α-keto acid dehydrogenases

1. **α-Keto acid dehydrogenases** catalyze **oxidative decarboxylations** in a sequence of reactions involving thiamine pyrophosphate, lipoic acid, coenzyme A, FAD, and NAD^+.
2. The major α-keto acid dehydrogenases are:
   a. **Pyruvate dehydrogenase,** the enzyme complex that oxidatively decarboxylates pyruvate, forming acetyl-CoA.
   b. **α-Ketoglutarate dehydrogenase,** which catalyzes the conversion of α-ketoglutarate to succinyl-CoA.
   c. The **α-keto acid dehydrogenase** complex involved in the oxidation of the branched chain amino acids.
      (1) Thiamine pyrophosphate (Fig. 5.5A) is involved in the **decarboxylation of α-keto acids.**
         (a) The α-carbon of the α-keto acid becomes covalently linked to thiamine pyrophosphate, and the carboxyl group is released as CO_2.
         (b) Thiamine pyrophosphate is also the cofactor for the enzyme **transketolase** of the pentose phosphate pathway.
         (c) Thiamine pyrophosphate is formed from ATP and the vitamin **thiamine.**
      (2) Lipoic acid participates in the oxidation of the keto group of the decarboxylated α-keto acid (Fig. 5.6).
         (a) After an α-keto acid is decarboxylated, the remainder of the compound is oxidized as it is transferred from thiamine pyrophosphate to lipoic acid, which is reduced in the reaction.

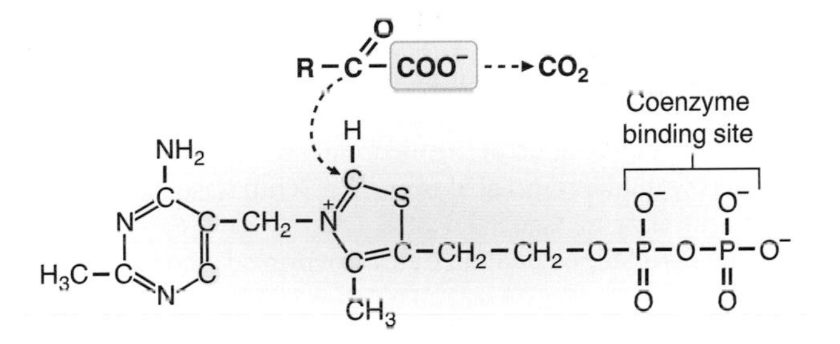

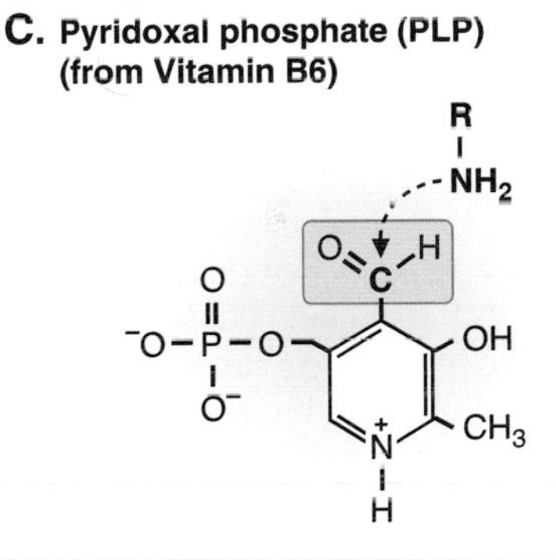

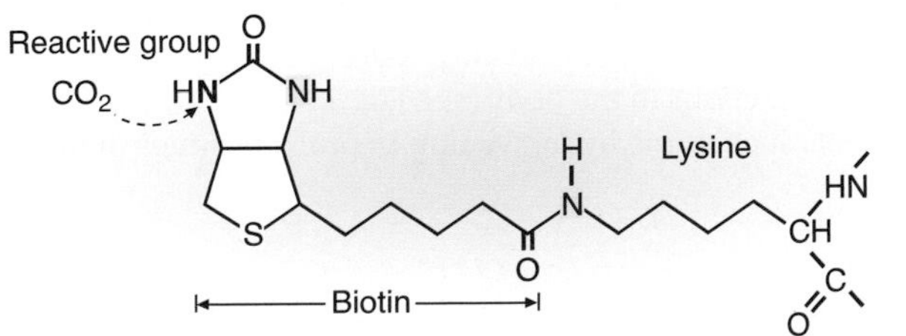

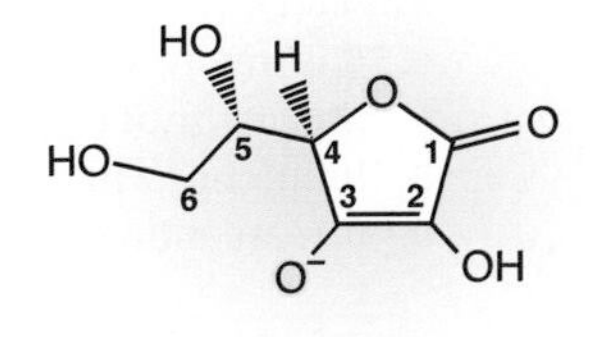

FIGURE 5.5. The structures of thiamine pyrophosphate **(A)**, biotin **(B)**, pyridoxal phosphate **(C)**, and ascorbate **(D)**. *Arrows* indicate the reactive sites. When an α-keto acid binds to thiamine pyrophosphate, the keto group attaches and the carboxyl group is released as CO_2.

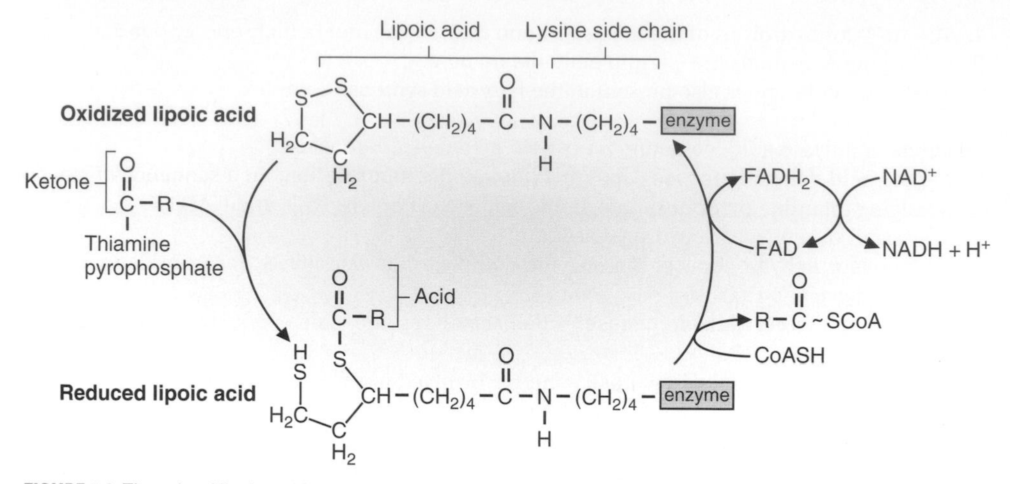

FIGURE 5.6. The role of lipoic acid in oxidative decarboxylation of α-keto acids.

b. The oxidized compound, which forms a thioester with lipoate, is then transferred to the sulfur of coenzyme A.

c. Because there is a limited amount of lipoate in the cell, reduced lipoate must be reoxidized so that it can be reutilized in these types of reactions. It is reoxidized by FAD, which becomes reduced to $FADH_2$ and is subsequently reoxidized by NAD^+.

d. Lipoic acid is not derived from a vitamin.

E. Other cofactors derived from water-soluble vitamins

1. **NADPH** (the reduced form of $NADP^+$) provides reducing equivalents for the synthesis of **fatty acids** and other compounds and for the reduction of **glutathione.**
   a. $NADP^+$ is identical to NAD^+ except that it contains an additional phosphate group.
2. **Biotin** is involved in the **carboxylation** of **pyruvate** (which forms oxaloacetate), **acetyl-CoA** (which forms malonyl-CoA), and **propionyl-CoA** (which forms methylmalonyl-CoA).
   a. The vitamin biotin is covalently linked to a lysyl residue of the enzyme (see Fig. 5.5B).
3. **Pyridoxal phosphate**, an aldehyde, interacts with an amino acid to form a Schiff base. Various products can be generated, depending on the enzyme (see Fig. 5.5C).
   a. Amino acids are **transaminated, decarboxylated**, or **deaminated** in pyridoxal phosphate-requiring reactions.
   b. Pyridoxal phosphate is derived from **vitamin B_6** (pyridoxine).
4. **Tetrahydrofolate** (see Fig. 8.16) **transfers 1-carbon units** (that are more reduced than CO_2) from compounds such as serine to compounds such as dUMP (to form dTMP).
   a. Tetrahydrofolate is synthesized from the **vitamin folate.**
5. **Vitamin B_{12}**, which contains cobalt, is involved in two reactions in the body.
   a. It **transfers methyl groups** from tetrahydrofolate to homocysteine (to form methionine).
   b. It is involved in the **conversion of methylmalonyl-CoA to succinyl-CoA.**
6. **Vitamin C** (ascorbic acid) has at least three functions in the body (see Fig. 5.5D).
   a. It is involved in **hydroxylation reactions**, such as the hydroxylation of prolyl residues in the precursor of collagen.
   b. It functions in the **absorption of iron.**
   c. It is an **antioxidant.**

F. Fat-soluble vitamins (Fig. 5.7)

1. **Vitamin K** is involved in the activation of precursors of prothrombin and other **clotting factors** by carboxylation of glutamate residues. Vitamin K-dependent reactions do not utilize biotin.

A. Vitamin K

Function

CH_3

Blood clotting

B. Vitamin A (retinal)

Vision
Growth
Reproduction

C. Vitamin E

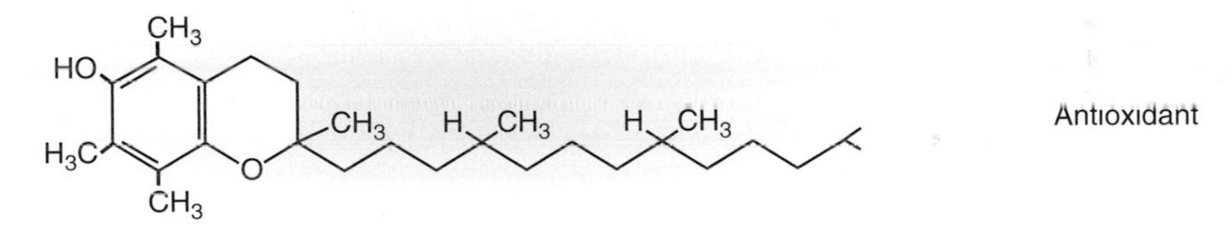

Antioxidant

D. Vitamin D₃

Ca^{2+} uptake from gut and mobilization from bone

FIGURE 5.7. The fat-soluble vitamins and their major functions.

2. **Vitamin A** is necessary for the light reactions of **vision**, for normal **growth** and **reproduction**, and for differentiation and maintenance of **epithelial tissues**.
   a. Δ^{11}-*cis*-Retinal binds to the protein opsin, forming rhodopsin.
      (1) Light causes Δ^{11}-*cis*-retinal in rhodopsin to be converted to all-*trans*-retinal, which dissociates from opsin, causing changes that allow light to be perceived by the brain.
   b. Retinoic acid, the most oxidized form of vitamin A, acts like a steroid hormone (see Chapter 9).
3. **Vitamin E** serves as an **antioxidant.**
   a. It prevents free radicals from oxidizing compounds such as polyunsaturated fatty acids.
   b. A major consequence of preventing free-radical damage is that the integrity of membranes, which contain fatty acid residues in phospholipids, is maintained.
4. **Vitamin D** (as 1,25-dihydroxycholecalciferol) is involved in **calcium metabolism** (see Chapter 9).

table 5.2 Vitamin Deficiencies and Their Manifestations

Fat-soluble Vitamins	Manifestations of Deficiency
Vitamin A	Night blindness, xerophthalmia
Vitamin D	Inadequate bone mineralization, rickets in children
Vitamin E	Reproductive failure, muscular dystrophy, neurologic abnormalities
Vitamin K	Defective blood coagulation
Water-soluble Vitamins	**Manifestations of Deficiency**
Vitamin C	Scurvy
Thiamine	Beriberi
Riboflavin	Oral–buccal cavity lesions
Niacin	Pellagra (diarrhea, dermatitis, dementia, death)
Vitamin B_6 (pyridoxine)	Convulsions, dermatitis, anemia
Folate	Megaloblastic anemia (due to impaired cell division and growth)
Vitamin B_{12}	Megaloblastic anemia, neurologic symptoms (resulting from demyelination)
Biotin	Anorexia, nausea, vomiting, glossitis, alopecia, dry, scaly dermatitis
Pantothenic acid	Listlessness, fatigue, "burning feet" syndrome

CLINICAL CORRELATES **Vitamin deficiencies** (Table 5.2) usually occur because of an insufficient dietary intake or a decreased conversion of the vitamin to its coenzyme derivatives caused by drugs, diseases, or other factors. Decreased absorption from the gut, plasma transport, tissue storage, binding to proteins, or increased excretion may also play a role. Often, multiple deficiencies occur, usually of the water-soluble vitamins, which are stored in limited amounts that may be depleted within weeks. Stores of the fat-soluble vitamins are larger and, therefore, are depleted more slowly. Toxicity due to excessive intake of the fat-soluble vitamins can develop.

IV. TCA CYCLE

- The TCA cycle, also known as the citric acid cycle or the Krebs cycle, is the major energy-producing pathway in the body. The cycle occurs in mitochondria.
- Foodstuffs feed into the cycle as acetyl-CoA and the acetyl-CoA is oxidized to carbon dioxide and water in order to generate energy.
- The cycle also serves in the synthesis of fatty acids, amino acids, and glucose.
- The cycle starts with the 4-carbon compound oxaloacetate, adds two carbons from acetyl-CoA, loses two carbons as CO_2, and regenerates the 4-carbon compound oxaloacetate.
- Electrons are transferred by the cycle to NAD^+ and FAD.
- As the electrons are subsequently passed to O_2 by the electron transport chain, ATP is generated by the process of oxidative phosphorylation.
- ATP is also generated from GTP, produced in one reaction of the cycle by substrate-level phosphorylation.

A. The reactions of the TCA cycle (Fig. 5.8)

1. All the enzymes of the TCA cycle are in the **mitochondrial matrix** except succinate dehydrogenase, which is in the inner mitochondrial membrane.
2. Oxidation of the carbons of acetyl-CoA to carbon dioxide requires capturing eight electrons from the molecule.

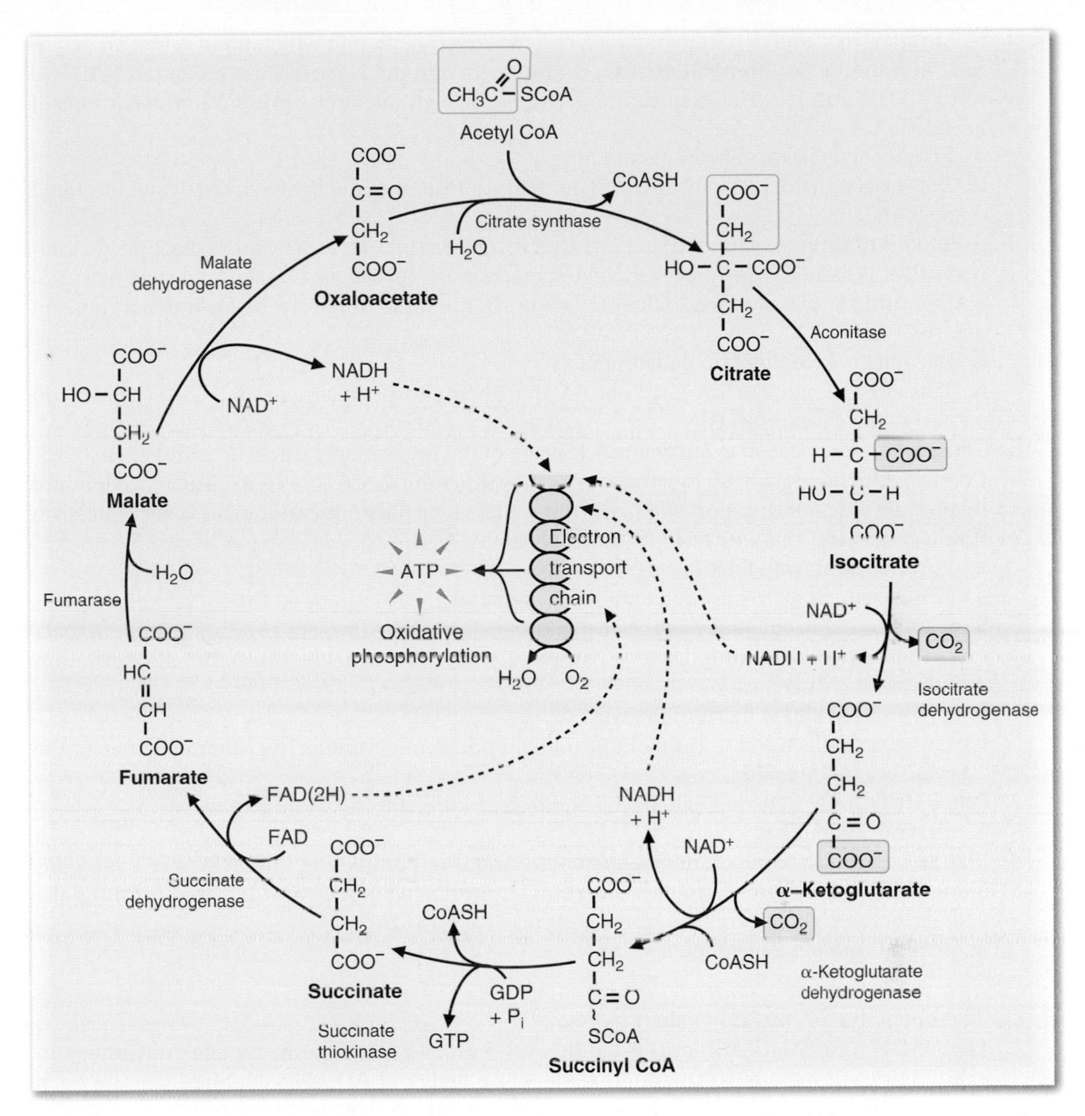

FIGURE 5.8. The TCA cycle. The oxidation–reduction enzymes and coenzymes are shown in red. The entry of the two carbons of acetyl-CoA into the TCA cycle is indicated with the green box. The carbons released as CO_2 are shown with yellow boxes.

a. **Acetyl-CoA** and **oxaloacetate** condense, forming citrate.
   (1) Enzyme: **citrate synthase.**
   (2) Cleavage of the high-energy thioester bond in acetyl-CoA provides the energy for this condensation.
   (3) Citrate (the product) is an inhibitor of this reaction.
b. **Citrate** is isomerized to isocitrate by a rearrangement of the molecule.
   (1) Enzyme: **aconitase.**
   (2) Aconitate serves as an enzyme-bound intermediate.
   (3) Under physiological conditions, this is an unfavorable reaction, favoring citrate formation.

CLINICAL CORRELATES

Fluoroacetate poisoning occurs when fluoroacetate is activated to its CoA derivative, and citrate synthase will condense the fluoroacetyl-CoA with oxaloacetate to produce fluorocitrate, which is a potent inhibitor of aconitase. This leads to the TCA cycle shutting down, reduced energy production, and cell (or organism) death.

3. **Isocitrate** is oxidized to α-ketoglutarate, in a two-step reaction in which there is first an oxidation, and then a decarboxylation. CO_2 is produced, and the electrons are passed to NAD^+ to form NADH and H^+. This step captures two of the eight electrons present in the carbons of acetyl-CoA.
   a. Enzyme: **isocitrate dehydrogenase.**
   b. This key regulatory enzyme of the TCA cycle is allosterically activated by ADP and inhibited by NADH.
4. **α-Ketoglutarate** is converted to succinyl-CoA in an oxidative decarboxylation reaction, mechanistically the same as the pyruvate dehydrogenase reaction. CO_2 is released, and succinyl-CoA, NADH, and H^+ are produced. This step captures another two electrons from the carbons of acetyl-CoA.
   a. Enzyme: **α-ketoglutarate dehydrogenase**.
   b. This enzyme requires five cofactors: thiamine pyrophosphate, lipoic acid, CoASH, FAD, and NAD^+ (see Section III D).
5. **Succinyl-CoA** is cleaved to succinate. Cleavage of the high-energy thioester bond of succinyl-CoA provides energy for the substrate-level phosphorylation of GDP to GTP. Since this does not involve the electron transport chain, it is not an oxidative phosphorylation; however, if electron flow were to stop, this step would also be blocked.
   a. Enzyme: **succinate thiokinase**.
   b. The enzyme is also called **succinyl-CoA synthetase**.
6. **Succinate** is oxidized to fumarate. Succinate transfers two hydrogens together with their electrons to FAD, which forms $FADH_2$. After this step, six of the eight electrons from the carbons in acetyl-CoA have been captured.
   a. Enzyme: **succinate dehydrogenase**.
   b. This enzyme is present in the inner mitochondrial membrane. The other enzymes of the cycle are in the matrix.
7. **Fumarate** is converted to malate by the addition of water across the double bond.
   a. Enzyme: **fumarase**.
8. **Malate** is oxidized, regenerating **oxaloacetate** and thus completing the cycle. Two hydrogens along with their electrons are passed to NAD^+, producing NADH and H^+, and finishing the capture of the eight electrons from the carbons of acetyl-CoA.
   a. Enzyme: **malate dehydrogenase**.

B. Energy production by the TCA cycle

1. The NADH and $FADH_2$ (produced by the cycle) donate electrons to the electron transport chain. For each mole of **NADH**, approximately **2.5 moles of ATP** are generated, and for each mole of **$FADH_2$**, approximately **1.5 moles of ATP** are generated by the passage of these electrons to O_2 (oxidative phosphorylation). In addition, GTP is produced when succinyl-CoA is cleaved. **GTP** produces ATP:

$$(GTP + ADP \rightleftharpoons ATP + GDP)$$

2. The **total energy** generated by one round of the cycle, starting with 1 mole of acetyl-CoA, is approximately **10 moles of ATP**.

C. Regulation of the TCA cycle

The TCA cycle is regulated by the **cell's need for energy** in the form of ATP. The TCA cycle acts in concert with the electron transport chain and the ATP synthase in the inner mitochondrial membrane to produce ATP.

1. The cell has limited amounts of adenine nucleotides (ATP, ADP, and AMP).
2. When ATP is utilized, ADP and inorganic phosphate (P_i) are produced.
3. **When ADP levels are high** relative to ATP–that is, when the cell needs energy–the reactions of the electron transport chain are accelerated. NADH is rapidly oxidized; consequently, the **TCA cycle speeds up**.
   a. One aspect of this process is that ADP allosterically activates isocitrate dehydrogenase.

4. **When the concentration of ATP is high**–that is, when the cell has an adequate energy supply–the electron transport chain slows down, NADH builds up, and consequently **the TCA cycle is inhibited.**
   a. NADH allosterically inhibits isocitrate dehydrogenase. Isocitrate accumulates, and because the aconitase equilibrium favors citrate, the concentration of citrate rises. Citrate inhibits citrate synthase, the first enzyme of the cycle.
   b. High NADH (and low NAD^+) levels also affect the reactions of the cycle that generate NADH, resulting in a slowing down of the cycle by mass action.
      (1) Oxaloacetate is converted to malate when NADH is high and, therefore, less substrate (OAA) is available for the citrate synthase reaction.

D. Vitamins required for reactions of the TCA cycle

1. **Niacin** is used for the synthesis of the nicotinamide portion of **NAD**, which is used in the isocitrate dehydrogenase, α-ketoglutarate dehydrogenase, and malate dehydrogenase reactions.
2. **Riboflavin** is used for the synthesis of **FAD**, which is the cofactor for succinate dehydrogenase. FAD is also required by α-ketoglutarate dehydrogenase.
3. **α-Ketoglutarate dehydrogenase**, a multienzyme complex (see Section III D), contains lipoic acid and four other cofactors that are synthesized from vitamins.
   a. **Thiamine** is used for the synthesis of **thiamine pyrophosphate.**
   b. **Pantothenate** for CoASH.
   c. **Riboflavin** for **FAD.**
   d. **Niacin** for **NAD^+.**

E. Pyruvate dehydrogenase complex

In order for carbons from glucose to enter the TCA cycle, glucose is first converted to pyruvate by glycolysis, then pyruvate forms acetyl-CoA.

1. **Reaction sequence**
   a. Pyruvate dehydrogenase, a multienzyme complex located exclusively in the mitochondrial matrix, catalyzes the oxidative decarboxylation of pyruvate, forming acetyl-CoA.
   b. The reactions catalyzed by the pyruvate dehydrogenase complex are analogous to those catalyzed by the α-ketoglutarate dehydrogenase complex. These enzyme complexes require the same five coenzymes, four of which contain vitamins (see Section IV D 3).
2. **Regulation of pyruvate dehydrogenase**
   a. In contrast to α-ketoglutarate dehydrogenase, pyruvate dehydrogenase exists in a phosphorylated (inactive) form and a dephosphorylated (active) form.
   b. **A kinase** associated with the multienzyme complex phosphorylates the pyruvate decarboxylase subunit, inactivating the pyruvate dehydrogenase complex.
      (1) The products of the pyruvate dehydrogenase reaction, **acetyl-CoA** and **NADH**, activate the kinase, and the substrates, **CoASH** and **NAD^+**, inactivate the kinase. The kinase is also inactivated by ADP.
   c. **A phosphatase** dephosphorylates and activates the pyruvate dehydrogenase complex.
   d. When the concentration of substrates is high, the dehydrogenase is active, and pyruvate is converted to acetyl-CoA. When the concentration of products is high, the dehydrogenase is relatively inactive (Fig. 5.9).

CLINICAL CORRELATES **Arsenic poisoning** is due to the combination of arsenite and arsenate. Arsenite inhibits enzymes and cofactors that contain adjacent free sulfhydryl groups (such as lipoic acid), and will interfere with TCA cycle reactions (pyruvate dehydrogenase and α-ketoglutarate dehydrogenase). Arsenate acts as a phosphate analog and inhibits substrate-level phosphorylation reactions that utilize, at some point, inorganic phosphate as a substrate.

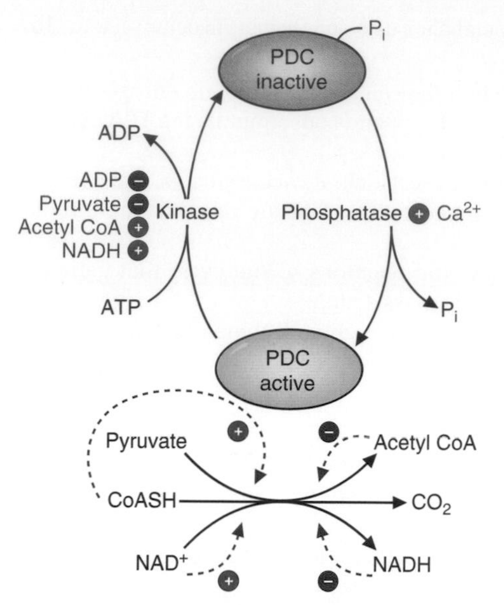

FIGURE 5.9. Regulation of the pyruvate dehydrogenase complex (PDC). The complex contains both PDC kinase and PDC phosphatase.

F. Synthetic functions of the TCA cycle (Fig 5.10)

Intermediates of the TCA cycle are utilized in the fasting state in the liver for the production of **glucose** and in the fed state for the synthesis of **fatty acids**. Intermediates of the TCA cycle are also used to synthesize **amino acids** or to convert one amino acid to another.

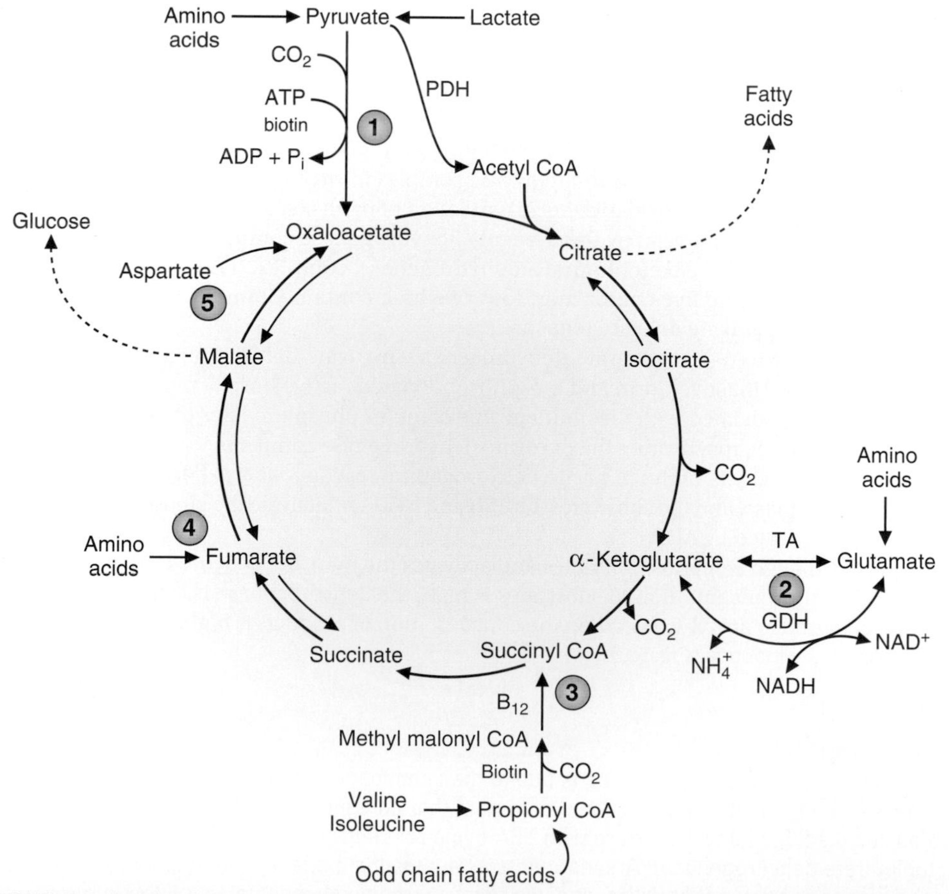

FIGURE 5.10. Anaplerotic and biosynthetic reactions involving the TCA cycle intermediates. Synthetic reactions that form fatty acids and glucose are indicated by *dashed lines*. GDH, glutamate dehydrogenase; PDH, pyruvate dehydrogenase; TA, transamination; ❶ to ❺, anaplerotic reactions.

1. **Anaplerotic reactions** replenish intermediates of the TCA cycle as they are removed for the synthesis of glucose, fatty acids, amino acids, or other compounds.
   a. A key anaplerotic reaction is catalyzed by **pyruvate carboxylase**, which carboxylates pyruvate, forming oxaloacetate.
      (1) Pyruvate carboxylase requires **biotin**, a cofactor that is commonly involved in CO_2-fixation reactions.
      (2) Pyruvate carboxylase, found in liver, brain, and adipose tissue (but not in muscle), is **activated by acetyl-CoA**.
   b. **Amino acids** produce intermediates of the TCA cycle through anaplerotic reactions.
      (1) **Glutamate** is converted to α-ketoglutarate.
      (2) Amino acids that form glutamate include glutamine, proline, arginine, and histidine.
      (3) **Aspartate** is transaminated to form oxaloacetate. Asparagine can produce aspartate.
      (4) **Valine, isoleucine, methionine**, and **threonine** produce propionyl-CoA, which is converted to methylmalonyl-CoA and subsequently to succinyl-CoA, an intermediate of the TCA cycle.
      (5) **Phenylalanine, tyrosine, and aspartate** form fumarate.
2. **Synthesis of glucose**
   a. The synthesis of glucose occurs by the pathway of **gluconeogenesis**, which involves intermediates of the TCA cycle.
   b. As glucose is synthesized, **malate or oxaloacetate** is removed from the TCA cycle and replenished by anaplerotic reactions.
      (1) Pyruvate, produced from lactate or alanine, is converted by pyruvate carboxylase to oxaloacetate, which forms malate.
      (2) Various amino acids that supply carbon for gluconeogenesis are converted to intermediates of the TCA cycle, which form malate and, thus, glucose.
3. **Synthesis of fatty acids**
   a. The pathway for fatty acid synthesis from glucose includes reactions of the TCA cycle.
      (1) From glucose, pyruvate is produced and converted to oxaloacetate (by pyruvate carboxylase) and to acetyl CoA (by pyruvate dehydrogenase).
      (2) Oxaloacetate and acetyl-CoA condense to form **citrate**, which is used for fatty acid synthesis.
      (3) Pyruvate carboxylase catalyzes the anaplerotic reaction that replenishes the TCA cycle intermediates.
4. **Synthesis of amino acids**
   a. Synthesis of amino acids from glucose involves intermediates of the TCA cycle.
      (1) **Glucose** is converted to pyruvate, which forms oxaloacetate, which by transamination forms **aspartate** and, subsequently, **asparagine**.
      (2) Glucose is converted to pyruvate, which forms both oxaloacetate and acetyl-CoA, which condense, forming citrate. Citrate forms isocitrate and then α-ketoglutarate, from which **glutamate, glutamine, proline**, and **arginine** are produced.
5. **Interconversion of amino acids** involves intermediates of the TCA cycle. For example, the carbons of **glutamate** can feed into the TCA cycle at the α-ketoglutarate level and traverse the cycle, forming oxaloacetate, which may be transaminated to **aspartate**.

V. ELECTRON TRANSPORT CHAIN AND OXIDATIVE PHOSPHORYLATION

- ATP is generated as a result of the energy produced when electrons from NADH and $FADH_2$ are passed to molecular oxygen by a series of electron carriers, collectively known as the electron transport chain. The components of the chain include FMN, Fe–S centers, coenzyme Q, and a series of cytochromes (b, c_1, c, and aa_3).
- The energy derived from the transfer of electrons through the electron transport chain is used to pump protons across the inner mitochondrial membrane from the matrix to the cytosolic

side. An electrochemical gradient is generated, consisting of a proton gradient and a membrane potential.

- Protons move back into the matrix through the ATP synthase complex, causing ATP to be produced from ADP and inorganic phosphate.
- ATP is transported from the mitochondrial matrix to the cytosol in exchange for ADP (the ATP–ADP antiport system).
- The oxidation of 1 mole of NADH generates approximately 2.5 moles of ATP, whereas the oxidation of 1 mole of $FADH_2$ generates approximately 1.5 moles of ATP.
- Because energy generated by the transfer of electrons through the electron transport chain to O_2 is used in the production of ATP, the overall process is known as oxidative phosphorylation.
- Electron transport and ATP production occur simultaneously and are tightly coupled.
- NADH and $FADH_2$ are oxidized only if ADP is available for conversion to ATP (i.e., if ATP is being utilized and converted to ADP).

CLINICAL CORRELATES **Malignant hyperthermia** can result from inhalation anesthetics. The major inhalation anesthetics (halothane, ether, and methoxyflurane) trigger a reaction in susceptible people that results in the uncoupling of oxidative phosphorylation from electron transport. ATP production decreases; heat is generated; and the temperature rises markedly. The TCA cycle is stimulated, and excessive CO_2 production leads to respiratory acidosis.

A. Overview of the electron transport chain

1. **NADH and $FADH_2$** are produced by glycolysis, β-oxidation of fatty acids, the TCA cycle, and other oxidative reactions. NADH and $FADH_2$ pass electrons to the components of the electron transport chain, which are located in the inner mitochondrial membrane.
2. **NADH** freely diffuses from the matrix to the membrane, whereas **$FADH_2$** is tightly bound to enzymes that produce it within the inner mitochondrial membrane.
3. **Mitochondria** are separated from the cytoplasm by two membranes. The soluble interior of a mitochondrion is called the **matrix**. The matrix is surrounded by the inner membrane, which contains infoldings known as **cristae**.
   a. The **transfer of electrons** from NADH to O_2 occurs in three stages, each of which involves a large protein complex in the inner mitochondrial membrane.
   b. Each complex uses the energy from electron transfer to **pump protons** to the cytosolic side of the membrane.
   c. An **electrochemical potential** or proton-motive force is generated.
      (1) The electrochemical potential consists of both a membrane potential and a pH gradient.
      (2) The cytosolic side of the membrane is more acidic (i.e., has a higher $[H^+]$) than the matrix.
4. The inner mitochondrial membrane is impermeable to protons. The **protons** can **reenter** the matrix only **through appropriate carriers. One of these is the ATP synthase** complex (the F_0-F_1/ATPase), **causing ATP to be generated**.
   a. The ATP synthase complex contains proteins (F_0) that form a channel in the inner mitochondrial membrane, through which the protons can flow, and a stalk that is attached to an ATP-synthesizing head (F_1) that projects into the matrix.
5. During the transfer of electrons through the electron transport chain, **some** of the **energy is lost as heat**.
6. The electron transport chain has a large negative $\Delta G^{o\prime}$, and thus the electrons flow from NADH (or $FADH_2$) toward O_2.

CLINICAL CORRELATES **Cyanide poisoning** is due to cyanide binding to Fe^{3+} in cytochrome aa_3. As a result, O_2 cannot receive electrons; respiration is inhibited; energy production is halted; and death occurs rapidly.

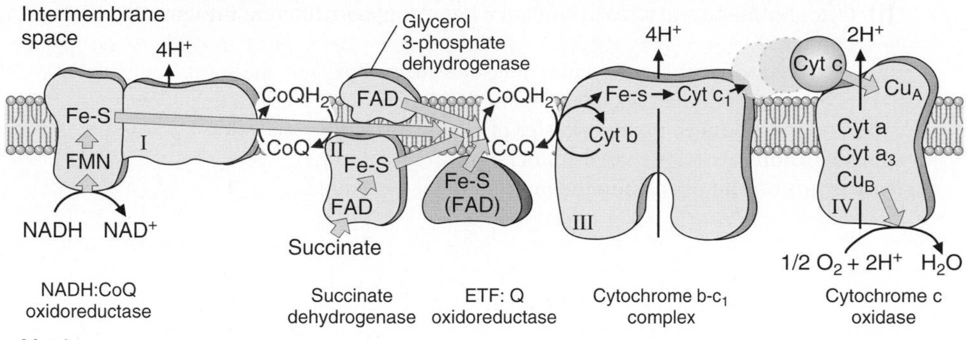

FIGURE 5.11. Components of the electron transport chain. *Heavy arrows* indicate the flow of electrons. CoQ, coenzyme Q (ubiquinone); Cyt, cytochrome; Fe–S, iron–sulfur centers; FMN, flavin mononucleotide.

B. The three major stages of electron transport (Fig. 5.11)

1. Transfer of electrons from NADH to coenzyme Q

a. **NADH** passes electrons via the **NADH dehydrogenase complex (complex I)** to FMN. The complex is also known as the **NADH:CoQ oxidoreductase.**

   (1) NADH is produced by the α-ketoglutarate dehydrogenase, isocitrate dehydrogenase, and malate dehydrogenase reactions of the TCA cycle, by the pyruvate dehydrogenase reaction that converts pyruvate to acetyl-CoA, by β-oxidation of fatty acids, and by other oxidation reactions.

   (2) NADH produced in the mitochondrial matrix diffuses to the inner mitochondrial membrane where it passes electrons to FMN, which is tightly bound to a protein.

b. **FMN** passes the electrons through a series of iron–sulfur (Fe–S) complexes to **coenzyme Q**, which accepts electrons one at a time, forming first the semiquinone and then ubiquinol.

c. The energy produced by these electron transfers is used to pump protons to the cytosolic side of the inner mitochondrial membrane.

d. As the protons flow back into the matrix through the pores in the ATP synthase complex, ATP is generated.

2. Transfer of electrons from coenzyme Q to cytochrome c

a. **Coenzyme Q** passes electrons through Fe–S centers to **cytochromes b and c_1**, which transfer the electrons to **cytochrome c**. The protein complex involved in these transfers is called **complex III, or the cytochrome b-c_1 complex.** The complex is also known as **CoQ:C1 oxidoreductase**.

   (1) These cytochromes each contain heme as a prosthetic group but have different apoproteins.

   (2) In the **ferric (Fe^{3+})** state, the heme iron can accept one electron and be reduced to the **ferrous (Fe^{2+})** state.

   (3) Because the cytochromes can only carry one electron at a time, two molecules in each cytochrome complex must be reduced for every molecule of NADH that is oxidized.

b. The energy produced by the transfer of electrons from coenzyme Q to cytochrome c is used to pump protons across the inner mitochondrial membrane.

c. As the protons flow back into the matrix through the pores in the ATP synthase complex, ATP is generated.

d. **Electrons from $FADH_2$**, produced by reactions such as the oxidation of succinate to fumarate, **enter** the electron transport chain **at complex II, which contains succinate dehydrogenase**. Complex II will transfer electrons to coenzyme Q,without the associated proton pumping across the inner mitochondrial membrane (see Fig. 5.11).

3. Transfer of electrons from cytochrome c to oxygen

a. **Cytochrome c** transfers electrons to the **cytochrome aa_3** complex, which transfers the electrons to molecular **oxygen**, reducing it to water. **Cytochrome oxidase (complex IV)** catalyzes this transfer of electrons.

(1) Cytochromes a and a_3 each contain a heme and two different proteins that each contain **copper**.

(2) Two electrons are required to reduce one atom of oxygen; therefore, for each NADH that is oxidized, one-half of O_2 is converted to H_2O.

b. The energy produced by the transfer of electrons from cytochrome c to oxygen is used to pump protons across the inner mitochondrial membrane.

c. As the protons flow back into the matrix, ATP is generated.

CLINICAL CORRELATES **Lactic acidosis** is an accumulation of lactic acid in the blood such that the acid overwhelms the buffering capabilities of the blood, leading to a reduced blood pH. This is a life-threatening situation that requires rapid treatment to bring the blood pH back to normal. One cause of lactic acidosis is hypoxia. Under anaerobic conditions, the TCA cycle and oxidative phosphorylation cannot occur, and glycolysis (see Chapter 6) must provide all energy to the cell. In order for glycolysis to run effectively in the absence of oxygen, the pyruvate produced is reduced to lactate by NADH, and the lactate accumulates. An inherited deficiency in the pyruvate dehydrogenase complex can also lead to lactic acidosis, as pyruvate would no longer be converted to acetyl-CoA efficiently, and the excess pyruvate will be metabolized to lactate in order to allow glycolysis to run efficiently. An inherited mutation in pyruvate carboxylase would also lead to an accumulation of pyruvate, and excess lactate being formed.

C. ATP production

1. The production of ATP is coupled to the transfer of electrons through the electron transport chain to O_2. The overall process is known as **oxidative phosphorylation**.
2. Protons flow down their electrochemical gradient through the membrane-bound ATP synthase. The flow of protons through the ATPase allows the enzyme to synthesize ATP.
3. The exact amount of ATP that is generated by this process has not been unequivocally established, but current thought indicates that for each pair of electrons that enters the chain from NADH, 10 protons are pumped out of the mitochondria. As it takes four protons to flow through the ATPase to synthesize one ATP, 2.5 moles (10 divided by 4) of ATP can be generated from 1 mole of NADH.
   a. For every mole of NADH that is oxidized, 0.5 moles of O_2 is reduced to H_2O and approximately 2.5 moles of ATP are produced.
   b. For every mole of $FADH_2$ that is oxidized, approximately **1.5 moles of ATP** are generated because the electrons from $FADH_2$ enter the chain via coenzyme Q, bypassing the NADH dehydrogenase step. Electrons entering via complex II lead to the extrusion of 6 protons per pair of electrons, instead of the 10 protons per pair of electrons starting at complex I. This is because complex II does not extrude protons as electrons flow through the complex.

D. The ATP–ADP antiport

ATP produced within mitochondria is transferred to the cytosol in exchange for ADP by a transport protein in the inner mitochondrial membrane known as the adenine nucleotide translocase (ANT). The energy for their transport is the proton gradient, as ATP has four negative charges (leaving the matrix), and ADP contains three negative charges (entering the matrix).

E. Inhibitors of electron transport and oxidative phosphorylation

1. **Agents that act on components of the electron transport chain**
   a. If there is a block at any point in the electron transport chain, all carriers before the block will accumulate in their reduced states, whereas those after the block will accumulate in their oxidized states. As a result, O_2 will not be consumed; ATP will not be generated; and the TCA cycle will slow down owing to the accumulation of NADH.
   (1) **Rotenone**, a fish poison, complexes with complex I, causing NADH to accumulate. It does not block the transfer of electrons to the chain from $FADH_2$.

(2) **Antimycins** (antibiotics) block the passage of electrons through the cytochrome b-c_1 complex (complex III).

(3) **Cyanide and carbon monoxide**, poisons commonly used for suicide, combine with cytochrome oxidase (complex IV) and block the transfer of electrons to O_2.

CLINICAL CORRELATES **An acute myocardial infarction** is due to a reduction in blood flow to a specific region of the heart. Coronary arteries frequently become narrow because of **atherosclerotic plaques**. If **coronary occlusions** occur, regions of heart muscle may be deprived of blood flow and, therefore, of oxygen for prolonged periods of time. **Lack of oxygen** causes inhibition of the processes of electron transport and oxidative phosphorylation, which results in a decreased production of ATP. **Heart muscle**, suffering from a lack of energy required for contraction and maintenance of membrane integrity, becomes **damaged**. **Enzymes** from the damaged cells (including the MB fraction of creatine kinase) **leak into the blood**. If the damage is relatively mild, the person may recover. If heart function is severely compromised, death may result.

2. **Inhibitors of ATP synthesis**
   a. Because the synthesis of ATP and electron transport are coupled, if the ATP synthase complex is inhibited or if an adequate supply of ADP is not available, ATP synthesis will be inhibited, O_2 will not be consumed, the components of the electron transport chain will accumulate in their reduced states, and the TCA cycle will slow down.
   b. Atractyloside will inhibit the ANT. Therefore, ATP synthesis will stop owing to a lack of ADP in the mitochondrial matrix.
   c. Agents such as dicyclohexylcarbodiimide (DCCD) and oligomycin block the proton pore of the ATPase, preventing ATP synthesis and blocking oxidative phosphorylation.

CLINICAL CORRELATES **AIDS treatments** may have unintended consequences. **Azidothymidine** (AZT), which interferes with reverse transcriptase activity, can also act as an inhibitor of mitochondrial DNA polymerase. Under these conditions, AZT treatment can lead to the depletion of mitochondrial DNA in cells, leading to muscle weakness due to an inability to produce sufficient amounts of ATP.

3. **Uncoupling agents**
   a. Agents such as **dinitrophenol** are ionophores that allow protons from the cytosol to reenter the matrix without going through the pore in the ATP synthase complex. Thus, they uncouple electron transport and ATP production.
   b. Uncouplers **increase** the rate of O_2 consumption, **electron transport**, the **TCA** cycle, and CO_2 production.
   c. **ATP production decreases** because the proton gradient across the inner mitochondrial membrane is dissipated.
   d. The **energy** generated by the increased rate of respiration (electron transport and O_2 consumption) is **lost as heat**.
4. A summary of the electron transfer chain and oxidative phosphorylation inhibitors is presented in Table 5.3.

CLINICAL CORRELATES **Oxidative phosphorylation disorders** exhibit a variety of symptoms. Mitochondria contain their own genome, and mutations in mitochondrial DNA can lead to disease. The major diseases are outlined in Table 5.4. The genetics of mitochondrial inheritance are discussed in Chapter 10.

table 5.3 Inhibitors of Oxidative Phosphorylation

Inhibitor	Site of Inhibition
Rotenone, amytal	Transfer of electrons from complex I to coenzyme Q.
Antimycin C	Transfer of electrons from complex III to cytochrome c.
Carbon monoxide (CO)	Transfer of electrons from complex IV to oxygen.
Cyanide (CN)	Transfer of electrons through complex IV to oxygen.
Atractyloside	Inhibits the ANT.
Oliogmycin	Inhibits proton flow through the F_0 component of the ATP synthase.
Dinitrophenol	An uncoupler; facilitates proton transfer across the inner mitochondrial membrane.
Valinomycin	A potassium ionophore; facilitates potassium ion transfer across the inner mitochondrial membrane.

VI. OXYGEN TOXICITY AND FREE-RADICAL INJURY

- Oxygen can accept single electrons to create highly reactive oxygen radicals.
- Oxygen can form the **reactive oxygen species (ROS), superoxide (O_2^-), hydrogen peroxide (H_2O_2), and the hydroxyl radical (HO·).**

table 5.4 Examples of OXPHOS Diseases Arising from mtDNA Mutations

Syndrome	Characteristic Symptoms	mtDNA Mutation
I. mtDNA rearrangements in which genes are deleted or duplicated		
Kearns–Sayre syndrome	Onset before 20 y of age, characterized by ophthalmoplegia, atypical retinitis pigmentosa, mitochondrial myopathy, and one of the following: cardiac conduction defect, cerebellar syndrome, or elevated CSF proteins.	Deletion of contiguous segments of tRNA and OXPHOS polypeptides, or duplication mutations consisting of tandemly arranged normal mtDNA and a mtDNA with a deletion mutation.
Pearson syndrome	Systemic disorder of oxidative phosphorylation that predominantly affects bone marrow.	Deletion of contiguous segments of tRNA and OXPHOS polypeptides, or duplication mutations consisting of tandemly arranged normal mtDNA and a mtDNA with a deletion mutation.
II. mtDNA point mutations in tRNA or ribosomal RNA genes		
MERRF (**m**yoclonic **e**pilepsy and **r**agged **r**ed **f**iber disease)	Progressive myoclonic epilepsy, a mitochondrial myopathy with ragged red fibers, and a slowly progressive dementia. Onset of symptoms: late childhood to adult.	$tRNA^{Lys}$
MELAS (mitochondrial **m**yopathy, **e**ncephalomyopathy, **l**actic **a**cidosis, and **s**trokelike episodes)	Progressive neurodegenerative disease characterized by strokelike episodes first occurring between 5 and 15 y of age and a mitochondrial myopathy.	80%–90% mutations in $tRNA^{Leu}$
III. mtDNA missense mutations in OXPHOS polypeptides		
Leigh disease (subacute necrotizing encephalopathy)	Mean age of onset, 1.5–5 y; clinical manifestations include optic atrophy, ophthalmoplegia, nystagmus, respiratory abnormalities, ataxia, hypotonia, spasticity, and developmental delay or regression.	7%–20% of cases have mutations in F_0 subunits of F_0-F_1/ATPase.
LHON (**L**eber **h**ereditary **o**ptic **n**europathy)	Late onset, acute optic atrophy.	90% of European and Asian cases result from mutation in NADH dehydrogenase.

- ROS can be generated in a controlled manner (enzymatically), or in a noncontrolled manner (nonenzymatically).
- Oxygen radicals can be deadly to cells, through initiation of free-radical damage to lipids and proteins.
- **Reactive nitrogen–oxygen species (RNOS)** are generated during metabolism, and are also deadly to cells.
- Protection against ROS and RNOS include enzymes, such as **superoxide dismutase (SOD)** and **glutathione peroxidase.**
- **Vitamins C and E** protect against radical damage.

A. Oxygen and the generation of ROS

1. Oxygen is a **biradical** (contains two single unpaired electrons in distinct orbitals).
2. Oxygen can accept four electrons to reduce it to water (Fig. 5.12).
   a. Addition of one electron generates **superoxide.**
   b. Addition of a second electron generates **hydrogen peroxide.**
   c. Addition of a third electron generates water and the **hydroxyl radical.**
   d. Addition of the last electron generates the second molecule of water.

B. Characteristics of ROS

1. Reactive free radicals extract electrons from other compounds to complete their own orbitals, thereby initiating free-radical chain reactions.
2. Hydroxyl radical is the most potent ROS. Chain reactions initiated by hydroxyl radicals lead to the formation of **lipid peroxides and organic radicals.**
3. Superoxide is also highly reactive, but its range of action is limited by its reduced solubility.
4. Hydrogen peroxide is not a radical, but easily generates hydroxyl radicals through interactions with **transition metals**, such as iron (+2) and copper (+1).
5. RNOS are derived principally from the free-**radical nitric oxide (NO)**, which combines with oxygen or superoxide to produce additional RNOS (Table 5.5).

C. Major sources of primary ROS in the cell

1. **Coenzyme Q** generates superoxide as a by-product of the electron transport chain (Fig. 5.13).
2. **Oxidases, oxygenases**, and **peroxidases**

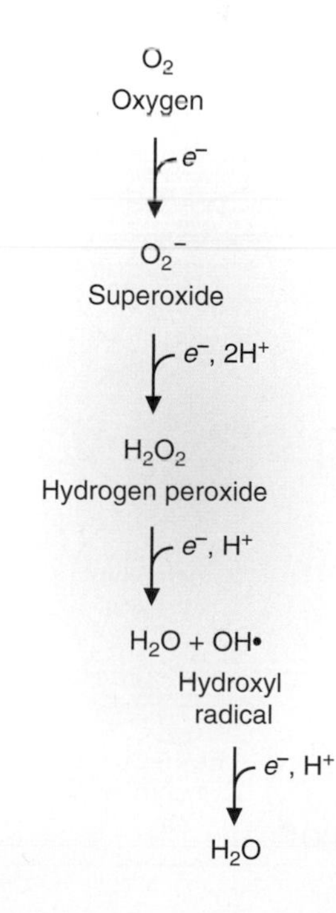

FIGURE 5.12. Reduction of oxygen by four one-electron steps. Hydrogen peroxide, the half-reduced form of oxygen, has accepted two electrons and is therefore not an oxygen radical.

table 5.5 Reactive Oxygen Species (ROS) and Reactive Nitrogen–Oxygen Species (RNOS)

Reactive Species	Properties
O_2^-, superoxide anion	Produced by the electron transport chain and at other sites. Cannot diffuse far from the site of origin. Generates other ROS.
H_2O_2, hydrogen peroxide	Not a free radical, but can generate free radicals by reaction with a transition metal (e.g., Fe^{2+}). Can diffuse into and through cell membranes.
OH•, hydroxyl radical	The most reactive species in attacking biologic molecules. Produced from H_2O_2 in the Fenton reaction in the presence of Fe^{2+} or Cu^+.
RO•, R•, R–S, organic radicals	Organic free radicals (R denotes the remainder of the compound). Produced from ROH, RH (e.g., at the carbon of a double bond in a fatty acid), or RSH OH• attack.
RCOO•, peroxyl radical	An organic peroxyl radical, such as that occurs during lipid degradation (also denoted LOO•).
HOCl, hypochlorous acid	Produced in neutrophils during the respiratory burst to destroy invading organisms. Toxicity is through halogenation and oxidation reactions. Attacking species is OCl^-.
$O_2^{\downarrow\uparrow}$, singlet oxygen	Oxygen with antiparallel spins. Produced at high oxygen tensions from absorption of UV light. Decays so fast that it is probably not a significant in vivo source of toxicity.
NO, nitric oxide	RNOS. A free-radical produced endogenously by nitric oxide synthase. Binds to metal ions. Combines with O_2 or other oxygen-containing radicals to produce additional RNOS.
$ONOO^-$, peroxynitrite	RNOS. A strong oxidizing agent that is not a free radical. It can generate NO_2 (nitrogen dioxide), which is a radical.

a. Most cellular oxidases bind oxygen and transfer single electrons to oxygen via a metal cofactor. Free-radical intermediates of these reactions may be accidentally released.

b. **Cytochrome P450 enzymes** are a major source of free radicals leaked from reactions.

(1) Cytochrome P450 enzymes attempt to detoxify xenobiotics by solubilizing them through oxidation reactions.

(2) Induction of cytochrome P450 enzymes by alcohol, drugs, or chemical toxicants lead to an increased generation of free radicals, and cellular injury.

c. Hydrogen peroxide and lipid peroxides are generated enzymatically as reaction products in peroxisomes, mitochondria, and the endoplasmic reticulum.

3. Ionizing radiation has sufficient energy that the radiation can split water into hydroxyl and hydrogen radicals, leading to radiation damage to the skin, DNA base alterations, cancer, and cell death.

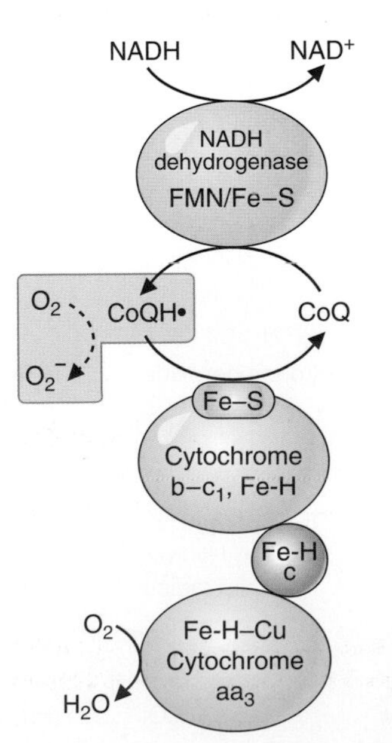

FIGURE 5.13. Generation of superoxide by coenzyme Q in the electron transport chain. In the process of transporting electrons to oxygen, some of the electrons escape when CoQH· accidently interacts with oxygen to form superoxide. Fe–H represents the Fe–heme center of the cytochromes.

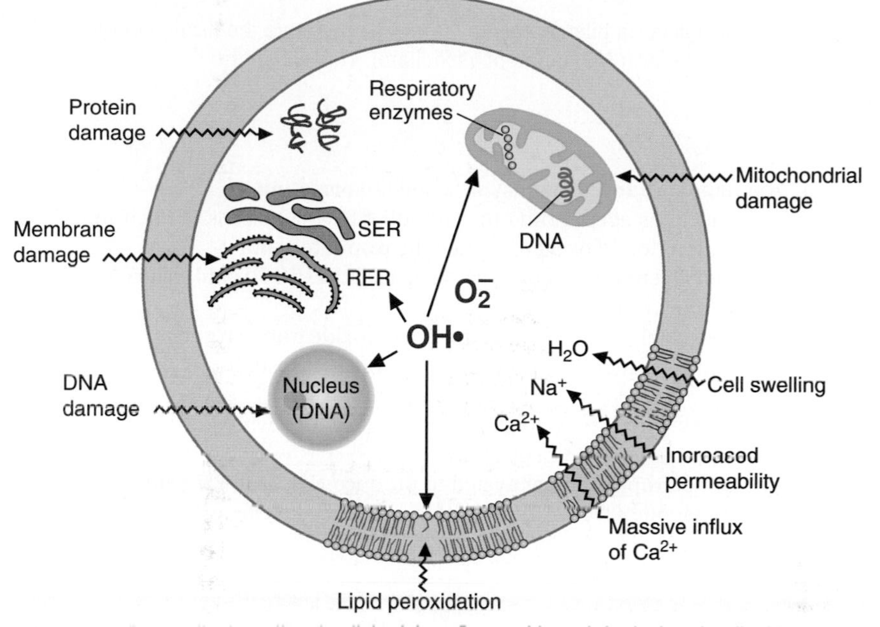

FIGURE 5.14. Free radical–mediated cellular injury. Superoxide and the hydroxyl radical initiate lipid peroxidation in the cellular, mitochondrial, nuclear, and endoplasmic reticulum membranes. The increase in cellular permeability results in an influx of calcium, which causes further mitochondrial damage. The cysteine sulfhydryl groups and other amino acid residues on proteins are oxidized and degraded. Nuclear and mitochondrial DNA can be oxidized, resulting in strand breaks and other types of damage. RNOS have similar effects.

D. ROS reactions with cellular components (Fig. 5.14)

1. Chain reactions that form lipid free radicals and lipid peroxides in membranes make a major contribution to ROS-induced injury. This leads to membrane damage in all organelles.
2. In proteins, the amino acids **proline, histidine, arginine, cysteine, and methionine** are susceptible to hydroxyl-mediated attack and oxidative damage. This can lead to protein fragmentation or cross-linking, aggregate formation, and loss of protein activity.
3. Oxygen-derived free radicals are also a major source of **DNA damage**, such as base alterations or single- and double-strand breaks in the backbone.

CLINICAL CORRELATES **Age-related macular degeneration** is a leading cause of blindness, which is due to, over time, oxidative damage to the retinal pigment epithelium, such that the brain no longer processes light correctly.

E. Nitric oxide and reactive nitrogen–oxygen species

1. **Nitric acid synthase** produces **nitric oxide** (NO), which is a free radical. NO, while an important second messenger in various physiological processes, can also be toxic.
   a. NO can exert direct toxic effects by combining with iron-containing compounds that also have single electrons.
   b. When NO is present in high concentrations, it can lead to the formation of other dangerous RNOS.
2. RNOS include **peroxynitrite** (formed from NO and superoxide), **peroxynitrous acid, nitrogen dioxide**, and **nitronium ion**, all of which can lead to protein damage and loss of protein function.

Nitroglycerin pills are given for **angina** because the nitroglycerin will decompose to form NO, a potent vasodilator. This will increase blood flow to the heart, and reduce the pain.

F. Formation of free radicals during phagocytosis and inflammation

1. In response to infectious agents and other stimuli, phagocytic cells of the immune system exhibit a rapid consumption of oxygen, termed the respiratory burst.
2. The respiratory burst generates ROS, **hyperchlorous acid (HOCl)**, and RNOS in order to destroy the phagocytosed materials.
   a. **NADPH oxidase** catalyzes the formation of superoxide from oxygen and NADPH (Fig. 5.15). The superoxide generated is released into the phagolysosome and is converted into other species of ROS.
   b. **Myeloperoxidase** generates hypochlorous acid (HOCl), which is a powerful oxidant and destroys the contents of the phagosome.
   c. When human neutrophils are activated to produce NO, NADPH oxidase is also activated, and the combination of NO and superoxide generates RNOS.

CLINICAL CORRELATES

Chronic granulomatosis disease can result from genetic defects in NADPH oxidase. A defective NADPH oxidase means that superoxide production during the respiratory burst is greatly compromised, and the patient exhibits enhanced susceptibility to bacterial and fungal infections. There is also a dysregulation of normal inflammatory responses.

G. Cellular defenses against oxygen toxicity (Fig. 5.16)

1. Antioxidant scavenging enzymes
   a. **SOD** produces hydrogen peroxide and oxygen from two molecules of superoxide. SOD is found in mitochondria.

CLINICAL CORRELATES

Amyotrophic lateral sclerosis (ALS; Lou Gehrig disease) can be due to several causes. One form of ALS is familial (inheritable), and the genetic defect has been traced to a form of SOD. This suggests that oxidative damage over a long period of time will eventually lead to muscle failure.

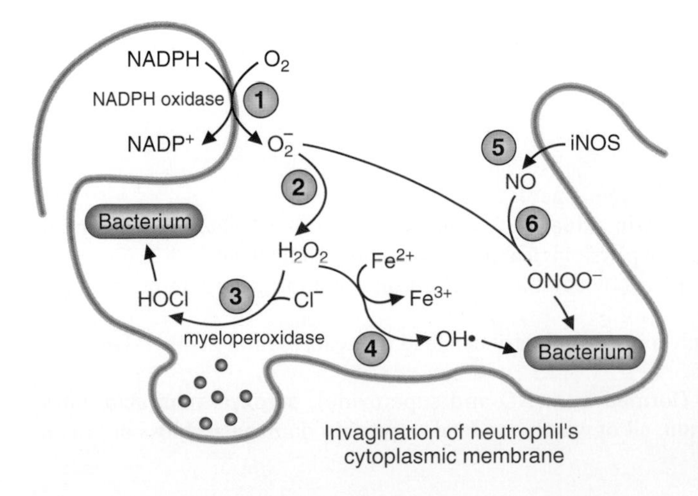

FIGURE 5.15. Production of ROS during the phagocytic burst by activated neutrophils. *(1)*—generation of superoxide by NADPH oxidase. *(2)* through *(6)*—conversion of superoxide into other ROS species (and RNOS). The result is an attack on the membranes and other components of phagocytosed cells and eventual lysis.

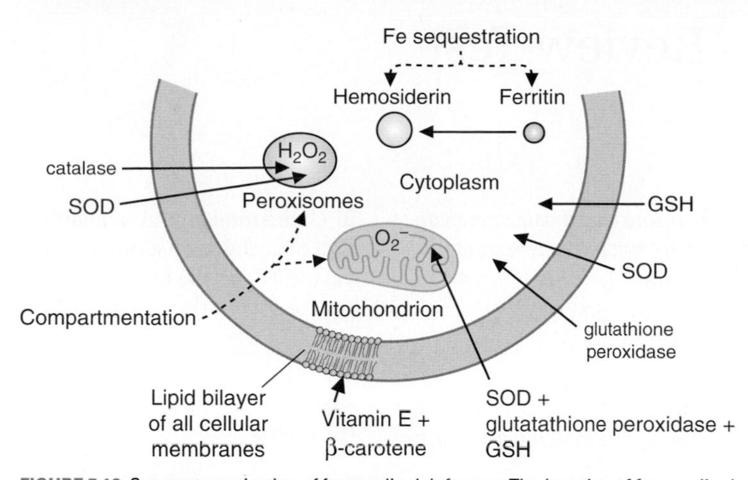

FIGURE 5.16. Compartmentization of free-radical defenses. The location of free-radical defense enzymes (shown in red) matches the type and amount of ROS generated in each subcellular compartment. Another form of compartmentization involves the sequesteration of iron, which is stored as mobilizable iron in ferritin. Excess iron is stored in nonmobilizable hemosiderin deposits. Glutathione (GSH) is a nonenzymatic antioxidant.

b. Catalase converts two molecules of hydrogen peroxide into two molecules of water and molecular oxygen. Catalase is present in peroxisomes and cells that utilize a respiratory burst.

c. Glutathione peroxidase and **glutathione reductase** utilize the protective tripeptide glutathione (γ-glutamyl-cysteine-glycine) to protect against oxidative damage.

(1) Glutathione peroxidase utilizes two molecules of reduced GSH, plus hydrogen peroxide, to generate oxidized glutathione (GSSG) and two molecules of water. This enzyme requires the metal selenium, and accounts for the majority of the daily selenium requirement in the diet.

(2) Glutathione reductase will reduce GSSG into reduced glutathione (GSH) using NADPH as a source of electrons.

2. Nonenzymatic antioxidants (free-radical scavengers)

a. Vitamin E, which is acquired through the lipid fractions of liver, egg yolks, and cereal.

b. Vitamin C (ascorbic acid).

c. Carotenoids (derivatives of β-carotene).

d. Flavonoids, which are acquired in red wine, green tea, and chocolate.

Review Test

Questions 1 to 10 examine your basic knowledge of biochemistry and are not in the standard clinical vignette format.

Questions 11 to 35 are clinically relevant, USMLE-style questions.

Basic Knowledge Questions

Refer to the following equation when answering Questions 1 and 2.

Consider the reaction catalyzed by fumarase: fumarate + $H_2O \rightleftharpoons$ malate

1. When measured in the absence of fumarase, the $\Delta G^{o\prime}$ for this reaction is 0 kcal/mol (neglecting any terms associated with H_2O). The equilibrium constant for this reaction would therefore be which one of the following?

(A) 0
(B) 0.5
(C) 1.0
(D) 10.0
(E) 50.0

2. Fumarase was added to a solution that initially contained 20 μM fumarate. After the establishment of equilibrium, the concentration of malate was which one of the following?

(A) 2 μM
(B) 5 μM
(C) 10 μM
(D) 20 μM
(E) 50 μM

Questions 3 and 4 require reference to the following reactions and associated values when answering them.

Reaction	Approximate $\Delta G^{o\prime}$ (kcal/mol)
Acetate + 2 $O_2 \rightarrow$ 2 CO_2 + 2 H_2O	-243
NADH + H^+ + ½ $O_2 \rightarrow NAD^+$ + H_2O	-53
$FADH_2$ + ½ $O_2 \rightarrow$ FAD + H_2O	-41
GTP $\rightarrow$ GDP + P_i	-8
ATP $\rightarrow$ ADP + P_i	-8

3. Of the total energy available from the oxidation of acetate, what percentage is transferred via the TCA cycle to NADH, $FADH_2$, and GTP?

(A) 38%
(B) 42%
(C) 81%
(D) 86%
(E) 100%

4. What percentage of the energy available from the oxidation of acetate is converted to ATP?

(A) 3%
(B) 30%
(C) 40%
(D) 85%
(E) 100%

5. A genetic mutation caused the cellular concentration of an enzyme to increase 100-fold for a biochemical reaction. Therefore, the equilibrium constant for the reaction catalyzed by the enzyme would change in which one of the following ways?

(A) It would decrease two-fold.
(B) It would remain the same.
(C) It would increase in proportion to the enzyme concentration.
(D) It would change inversely with the enzyme concentration.
(E) It would decrease 100-fold.

6. Consider the section of the TCA cycle in which isocitrate is converted to fumarate. This segment of the TCA cycle can be best described by which one of the following?

(A) These reactions yield 5 moles of high-energy phosphate bonds per mole of isocitrate.
(B) These reactions require a coenzyme synthesized in the human from niacin (nicotinamide).
(C) These reactions are catalyzed by enzymes located solely in the mitochondrial membrane.
(D) These reactions produce 1 mole of CO_2 for every mole of isocitrate oxidized.
(E) These reactions require GTP to drive one of the reactions.

7. In the TCA cycle, a role for thiamine pyrophosphate is which one of the following?

(A) To accept electrons from the oxidation of pyruvate and α-ketoglutarate
(B) To accept electrons from the oxidation of isocitrate
(C) To form a covalent intermediate with the α-carbon of α-ketoglutarate
(D) To form a thioester with the sulfhydryl group of CoASH
(E) To form a thioester with the sulfhydryl group of lipoic acid

8. Which one of the following is a property of pyruvate dehydrogenase?

(A) The enzyme contains only one polypeptide chain.
(B) The enzyme requires thiamine pyrophosphate as a cofactor.
(C) The enzyme produces oxaloacetate from pyruvate.
(D) The enzyme is converted to an active form by phosphorylation.
(E) The enzyme is activated when NADH levels increase.

9. Which one of the following components of the electron transport chain only accepts electrons, and does not donate them?

(A) Cytochrome b
(B) Oxygen
(C) Coenzyme Q
(D) FMN
(E) Cytochrome C1

10. Which one of the following tissues of the eye relies almost solely on anaerobic metabolism instead of the TCA/electron transport cycle?

(A) Cornea
(B) Lens
(C) Ciliary muscle
(D) Retina
(E) All of the tissues of the eye use only anaerobic glycolysis as an energy source.

Board-style Questions

Questions 11 and 12 are based on the following case:

A 43-year-old female has been on a "grapefruit and potatoes" diet for several months in an effort to lose weight. She now complains of a rash covering most of her body, a large, beefy tongue, nausea and diarrhea, and some confusion.

11. Which one of the following cofactors or enzyme complexes would be most affected by this condition?

(A) The concentration of NAD^+
(B) The concentration of FAD
(C) The concentration of coenzyme Q
(D) The functioning of the FMN components of complex I
(E) The functioning of the cytochrome-containing components of complex III

12. To reverse the symptoms described in the patient, a diet high in which one of the following should be recommended?

(A) Green, leafy vegetables
(B) Whole grains and meat
(C) Citrus fruits
(D) Orange and yellow vegetables
(E) Chocolate cake

Questions 13 and 14 are based on the following case:

An alcoholic presents with swelling and fissuring of the lips, cracking at the angles of the mouth, red eyes, and an oily, scaly rash of his scrotum.

13. Which one of the following cofactors of enzyme complexes would be most affected by this condition?

(A) The concentration of NAD^+
(B) The concentration of $NADP^+$
(C) The concentration of coenzyme Q
(D) The functioning of the FMN components of complex I
(E) The functioning of the cytochrome-containing components of complex III

14. Which of the following foods would best help reverse the symptoms described in the above patient?

(A) Broccoli
(B) Carrots
(C) Grapefruits
(D) Wheat
(E) Chocolate cake

15. A firefighter is brought to the emergency room (ER) from the scene of a fire complaining of headaches, weakness, confusion, and difficulty in breathing. His skin and mucous membranes appear very pink/red. The causative agent of these symptoms inhibits electron transport and oxidative phosphorylation by which one of the following mechanisms?

(A) Uncoupling of electron transport and phosphorylation
(B) Combining with NADH dehydrogenase
(C) Combining with cytochrome oxidase
(D) Inhibiting an adequate supply of ADP
(E) Combining with coenzyme Q

16. A patient is undergoing an appendectomy under general anesthesia (succinylcholine and an inhaled anesthetic) when she begins to develop muscle rigidity, tachycardia, and hyperthermia. Which of the following best describes the mechanism of this process?

(A) Uncoupling of electron transport and phosphorylation
(B) Inhibition of NADH dehydrogenase
(C) Inhibition of cytochrome oxidase
(D) Inhibiting an adequate supply of ADP
(E) Combining with coenzyme Q

17. A patient is in septic shock and his tissues are poorly perfused and oxygenated. The major end product of glucose metabolism in these tissues will be an accumulation of which one of the following?

(A) Pyruvate
(B) Acetyl-CoA
(C) Lactate
(D) Urea
(E) Citrate

18. Which one of the following cell types cannot utilize the TCA cycle or electron transport chain?

(A) Brain
(B) Red blood cells
(C) Liver
(D) Kidney
(E) Heart

19. Metformin is the standard first-line oral medication for Type 2 diabetes. The use of the drug has the potential side effect of lactic acidosis. Which of the following explains why this lactic acid buildup is rarely seen clinically?

(A) The red blood cells utilize the lactate as fuel.
(B) The renal cell utilizes the lactate as fuel.
(C) The cardiac muscle cells utilize the lactate as fuel.
(D) The large, voluntary muscle groups utilize the lactate as fuel.
(E) The lactate directly enters the TCA cycle to be oxidized.

20. In which one of the following scenarios should a patient with Type 2 diabetes, who is taking metformin for glucose control, be advised to discontinue the metformin owing to an increased risk of lactic acidosis resulting from continuation of the metformin?

(A) Severe anemia
(B) Early pyelonephritis
(C) Severe loss of cardiac tissue from a myocardial infarction
(D) Severe tear of the quadriceps muscle
(E) Significant weight gain

21. The most potent ROS is which one of the following?

(A) Hydrogen peroxide
(B) Superoxide
(C) Hydroxyl radical
(D) Nitric oxide
(E) Coenzyme Q

22. Using Internet sources, a patient has developed his own diet plan in an attempt to prevent macular degeneration. For breakfast every morning, he has poached eggs, orange and carrot juice, and red wine. Which of the following in his diet can potentially protect against the patient developing macular degeneration?

	Carotenoids	Flavonoids	Vitamin C	Vitamin D	Vitamin E
A	Yes	Yes	Yes	No	Yes
B	Yes	No	Yes	Yes	Yes
C	Yes	Yes	No	No	Yes
D	No	No	No	Yes	Yes
E	No	Yes	No	No	No
F	No	No	Yes	Yes	No

23. Which of the following medications, given chronically, could create a physiologic response that would be a major source of free-radical production?

(A) Ciprofloxacin
(B) Isoniazide
(C) Cimetidine
(D) Ketaconazole
(E) None of these drugs has the potential to increase free-radical formation.

24. A contestant on a TV reality show, in which the contestants had to survive off the land for an extended period of time, developed recurrent diarrhea, dermatitis, and had trouble remembering things. These symptoms could be brought about due to the lack of which one of the following in the contestant's diet?

(A) Niacin
(B) Thiamine
(C) Riboflavin
(D) Vitamin C
(E) Vitamin D

25. A 40-year-old chronic alcoholic enters the hospital because of a variety of symptoms, including loss of feeling in his hands and feet, nystagmus, and difficulty with his balance when walking. This patient would have difficulty catalyzing which one of the following reactions?

(A) α-Ketoglutarate dehydrogenase
(B) Succinate dehydrogenase
(C) Fumarase
(D) Malate dehydrogenase
(E) Pyruvate carboxylase

26. A scientist has developed a drug that, when added to eukaryotic cells, leads to elevated lactate levels. An analysis of mitochondrial contents also demonstrated elevated α-ketoglutarate levels in drug-treated cells. This drug may be interfering with a reaction that requires which one of the following vitamins?

(A) Biotin
(B) Vitamin K
(C) Pantothenate
(D) Ascorbate
(E) Pyridoxine

27. A man presents to the ER with an elevated temperature, sweats, and increased rate of breathing. He had been spraying insecticide and accidentally inhaled some of the poison. Using the insecticide on cultured cells, it was demonstrated that the rate of oxygen consumption by the cells was much greater than in the absence of the compound. This drug is acting most like which one of the following?

(A) Dintrophenol
(B) Cyanide
(C) Carbon monoxide
(D) Rotenone
(E) Atractyloside

28. A patient suffering a heart attack was brought to the ER. An atherosclerotic plaque had blocked a major coronary artery, preventing blood from reaching a region of her heart. As a result, in cells of the affected heart muscle, there was an increase in which one of the following?

(A) The rate of CO_2 production
(B) The rate of electron transport by the electron transport chain
(C) The concentration of ADP
(D) The proton gradient across the inner mitochondrial membrane
(E) The rate of O_2 consumption

29. A scientist is conducting an experiment on isolated mitochondria in a buffered solution, in the presence of ADP. At time = 0, oxygen and succinate were added to the mitochondrial suspension, and oxygen consumption measured as a function of time. At various times after starting the experiment (labeled as 1, 2, 3, and 4 in the figure shown), different chemicals were added to the solution.

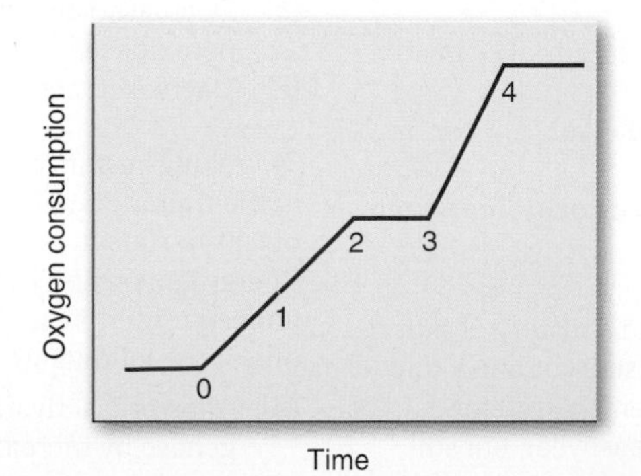

Which one of the following correctly describes which chemicals were added at times 1, 2, 3, and 4?

	Time 1	Time 2	Time 3	Time 4
A	Antimycin	Oligomycin	Cyanide	Carbon monoxide
B	Antimycin	Cyanide	Dinitrophenol	Oligomycin
C	Rotenone	Oligomycin	Dinitrophenol	Cyanide
D	Rotenone	Dinitrophenol	Oligomycin	Carbon monoxide
E	Dinitrophenol	Rotenone	Oligomycin	Cyanide
F	Dinitrophenol	Cyanide	Oligomycin	Dinitrophenol

30. A 52-year-old man suddenly collapsed at work, and was administered CPR by some co-workers until the ambulance arrived. Upon arrival at the hospital, blood work demonstrated an increased troponin I level. Tissue plasminogen activator was administered to the patient, but further damage to the affected organ resulted from this treatment. This may have happened owing to which one of the following?

(A) Release of cytochrome a from the mitochondria
(B) Increased generation of oxygen-derived radicals
(C) Elevated lactic acid levels
(D) Inhibition of the TCA cycle due to plasminogen activator administration
(E) Uncoupling of oxidation from phosphorylation

31. A 23-year-old man was diagnosed as HIV positive 3 years ago. He has been on a variety of anti-HIV drugs since then, including AZT, along with some dideoxy compounds and HIV protease inhibitors. He recently developed muscle weakness, to the point where he had difficulty walking. This complication has come about due to which one of the following?

(A) AZT inhibition of nuclear RNA polymerase
(B) AZT inhibition of nuclear DNA polymerase
(C) AZT inhibition of mitochondrial RNA polymerase
(D) AZT inhibition of mitochondrial DNA polymerase
(E) AZT uncoupling of oxidative phosphorylation
(F) AZT induced release of cytochrome c from the mitochondria

32. A 4-year-old boy has had a history of skin infections, pneumonia, nausea, vomiting, and abdominal pain. He has been on antibiotics prophylactically for the past year, but still contracts various sort of infections, both bacterial and fungal. A physical exam demonstrated hepatosplenomegaly. The boy most likely has inherited a mutation that prevents which one of the following reactions?

(A) Hydrogen peroxide conversion to hydroxyl radicals
(B) Superoxide conversion to hydrogen peroxide and oxygen
(C) Oxidized glutathione to reduced glutathione
(D) The formation of hydrochloric acid
(E) The formation of superoxide

33. A 42-year-old man had experienced cramping and stiffness of muscles, coupled with severe muscle weakness in his left leg. Over time, the muscle weakness becomes more severe, and his speech has become slurred, and he has to use a wheelchair to get around. He developed swallowing difficulties, and upon diagnosis was told he had less than 2 years to live. A family history indicated that the man's father had begun to experience mild muscle weakness before he died in an automobile accident. The man's symptoms can be best explained at the molecular level by which one of the following?

(A) Increased activity of NADPH oxidase
(B) Increased activity of catalase
(C) Elevated levels of hydrogen peroxide
(D) Elevated levels of superoxide
(E) Decreased activity of glutathione peroxidase
(F) Decreased activity of myeloperoxidase

34. A long-distance runner is training, and part of the training requires running multiple sets of 800-m runs in order to increase her stamina. Energy generation by the TCA cycle is enhanced under these conditions owing to which one of the following?

(A) Allosteric activation of isocitrate dehydrogenase by increased NADH

(B) Allosteric activation of fumarase by increased ADP
(C) A rapid decrease in the concentration of 4-carbon intermediates
(D) Product inhibition of citrate synthase
(E) Stimulation of the flux through a number of enzymes by a decreased NADH/NAD^+ ratio

35. A man presents to the ER after ingesting an insecticide. His respiration rate is very low. Information from the Poison Control Center indicates that this particular insecticide binds to and completely inhibits cytochrome c. Therefore, which one of the following would occur in this man's mitochondria?

(A) Coenzyme Q would be in the oxidized state
(B) Cytochromes a and a_3 would be in the reduced state
(C) The rate of ATP synthesis would be approximately zero
(D) The rate of CO_2 production would be increased
(E) The rate of oxygen consumption would be increased

Answers and Explanations

1. **The answer is C.** If $\Delta G^{o\prime} = 0$, then$-RT \ln K_{eq} = 0$, since $\Delta G^{o\prime} = -RT \ln K_{eq}$. For$-RT \ln K_{eq}$ to be equal to 0, the ln K_{eq} must be 0, which means that $K_{eq} = 1$ (the natural log of 1 = 0).

2. **The answer is C.** From the answer to Question 1, we know that $K_{eq} = 1 =$ [Malate]/[Fumarate] $= X/(20–X)$. Therefore, $(20–X) = X$, $20 = 2X$, and $X = 10$ μM.

3. **The answer is D.** In the TCA cycle, each turn of the cycle produces 3 NADH, 1 $FADH_2$, and 1 GTP. Each NADH releases 53 kcal/mol; the 3 NADH thus yield 159 kcal/mol of energy. $FADH_2$ releases 41 kcal/mol, and GTP 8 kcal/mol. The energy captured is 159 + 41 + 8, or 208 kcal/mol. The total energy available is 243 kcal/mol, so the fraction of energy captured is 208/243, or 86%.

4. **The answer is B.** About 10 ATP (7.5 from NADH, 1.5 from $FADH_2$, and 1 from GTP) are produced by the TCA cycle (10×8 kcal = 80 kcal). The percentage of the total energy available from oxidation of acetate that is converted to ATP is 80/243, or 33%.

5. **The answer is B.** An enzyme increases the rate at which a reaction reaches equilibrium but does not change the concentration of the reactants and products at equilibrium; that is, the K_{eq} is not affected by an enzyme, so a change in enzyme concentration will have no effect on the K_{eq}.

6. **The answer is B.** In the conversion of isocitrate to fumarate, 2 CO_2, 2 NADH (which contains niacin), 1 GTP, and 1 $FADH_2$ are produced. A total of approximately 7.5 ATP are generated. The enzymes for these reactions are all located in the mitochondrial matrix except succinate dehydrogenase, which is an inner mitochondrial membrane protein. GTP is not required in any of the reactions but is produced in the conversion of succinyl-CoA to succinate.

7. **The answer is C.** Thiamine pyrophosphate is involved in the making and breaking of carbon-carbon bonds. It is a necessary cofactor for oxidative decarboxylation reactions, in which a carbon-carbon bond is broken and carbon dioxide is released. Mechanistically, thiamine pyrophosphate forms a covalent intermediate with the α-carbon of an α-keto acid substrate, which, in the TCA cycle, is α-ketoglutarate. Thiamine pyrophosphate is not involved in redox reactions, or in thioester formation.

8. **The answer is B.** Pyruvate dehydrogenase converts pyruvate to acetyl-CoA. It contains multiple subunits: a dehydrogenase component that oxidatively decarboxylates pyruvate, a dihydrolipoyl transacetylase that transfers the acetyl group to coenzyme A, and a dihydrolipoyl dehydrogenase that reoxidizes lipoic acid. Thiamine pyrophosphate, lipoic acid, coenzyme A, NAD^+, and FAD serve as cofactors for these reactions. In addition, a kinase is present that phosphorylates and inactivates the decarboxylase component. Acetyl-CoA and NADH activate this kinase, thus inactivating pyruvate dehydrogenase. A phosphatase dephosphorylates the decarboxylase subunit, thereby reactivating pyruvate dehydrogenase.

9. **The answer is B.** Under physiological conditions, oxygen is the terminal electron acceptor in the electron transport chain, and water will not donate electrons to other substrates to regenerate oxygen. The cytochromes, FMN, and coenzyme Q both accept and donate electrons during the course of electron flow through the electron transport chain.

10. **The answer is B.** Aerobic metabolism requires an O_2 supply. Oxygen is usually obtained from the blood, which is circulating through the blood vessels. However, transparent tissue cannot have an extensive network of blood vessels since these would create opacities that would block the transmission of light. The cornea is exposed to air and gets its oxygen by diffusion from air. The lens has no capillaries and is not exposed to the air, so it utilizes anaerobic metabolism. Glucose and lactate diffuse from and into aqueous and vitreous humor. The ciliary muscle and

retina have extensive blood vessel systems, and can carry out oxidative phosphorylation in order to generate energy.

11. **The answer is A.** This patient has the classic symptoms of pellegra, a vitamin B_3 (niacin) deficiency. NAD^+ is derived from niacin. Pellagra leads to the four Ds-dermatitis, dementia, diarrhea, and death. Riboflavin is the precursor for both FAD and FMN. Coenzyme Q is synthesized from acetyl-CoA, and its levels would not be affected as much as those of NAD^+. Heme is synthesized from succinyl-CoA and glycine, and a reduction in heme levels would lead to an anemia and not the symptoms as described for this patient.

12. **The answer is B.** While green, leafy vegetables are rich in other B vitamins, whole grains, meats, fish, and liver are the best sources of niacin. Citrus fruits are high in vitamin C. Orange and yellow vegetables are high in vitamin A. Chocolate cake is high in flavonoids, an antioxidant, fats, and carbohydrates.

13. **The answer is D.** This patient has vitamin B_2 (riboflavin) deficiency, ariboflavinosis, as indicated by the symptoms displayed by him. Both FAD and FMN require vitamin B_2 to be produced. NAD^+ and $NADP^+$ are derived from niacin. Coenzyme Q is derived from acetyl-CoA, and vitamin B_2 is not needed in the synthesis of the heme ring, which is derived from succinyl-CoA and glycine.

14. **The answer is A.** Dark green vegetables, especially broccoli, meats, and dairy products are all high in riboflavin. Carrots are high in vitamin A, grapefruits in vitamin C, and whole grains in niacin. Chocolate cake is high in flavonoids, an antioxidant, fats, and carbohydrates.

15. **The answer is C.** The symptoms experienced by the firefighter could be caused by either cyanide or carbon monoxide, both of which inhibit cytochrome c oxidase. Both carbon monoxide and cyanide are by-products of fuel oxidation, and would be generated during a fire. The firefighter most likely inhaled smoke that contained one or both of these compounds. Both agents will block the reduction of oxygen to water, thereby halting the electron transfer chain and oxidative phosphorylation. Neither agent is an uncoupler, nor do they block the ANT (so ADP levels will not be decreased). Rotenone, a fish poison, complexes with NADH dehydrogenase (complex I) to inhibit electron flow from complex I to coenzyme Q. Neither cyanide nor carbon monoxide will bind to coenzyme Q and block its ability to either accept or donate electrons.

16. **The answer is A.** The patient is experiencing malignant hyperthermia, which is similar in symptoms to the uncoupling of the electron transfer chain from ATP synthesis. Succinylcholine and several inhaled anesthetics can act as uncouplers of electron transport in susceptible individuals. An inhibition of complex I (NADH dehydrogenase), or cytochrome oxidase, would block both electron flow and ATP synthesis, and muscle rigidity and hyperthermia would not result. The same is true if coenzyme Q were prevented from accepting and donating electrons. An uncoupler will not lead to a decrease in ADP levels as ATP cannot be synthesized under these conditions, and ADP levels would be expected to increase.

17. **The answer is C.** Without oxygen, aerobic metabolism cannot function, so the electron transfer chain will stop (without oxygen, cytochrome oxidase cannot remove electrons from the chain to reduce oxygen to water). ATP synthesis in the mitochondria will stop because of the coupling of oxidation and phosphorylation. The TCA cycle will stop because of the accumulation of NADH in the mitochondria (as the electron transfer chain is fully reduced, owing to the lack of oxygen, NADH cannot donate electrons to a reduced complex I), and NADH inhibits key enzymes of the TCA cycle. However, the tissues still need energy, and so they use anaerobic glycolysis to generate ATP. The end product, pyruvate, is converted to lactate to regenerate NAD^+ such that glycolysis can continue. An accumulation of lactate (which is a metabolic dead end) can lead to lactic acidosis in the patient.

18. **The answer is B.** Cellular respiration occurs in the mitochondria. Red blood cells lack mitochondria, and can only utilize anaerobic glycolysis for energy production. The brain, liver, kidney, and heart cells contain mitochondria, and can carry out oxidative phosphorylation, as well as anaerobic glycolysis.

19. The answer is C. Metformin can increase glucose uptake by tissues and lead to increased lactate formation. In addition, metformin, through unknown mechanisms, appears to block lactate uptake by the liver (this could occur because of the reduced gluconeogenesis occurring in the liver in the presence of metformin, as lactate is a key substrate for gluconeogenesis). The cardiac muscles, with their massive amount of mitochondria, will utilize lactate as fuel and can overcome a lactate buildup from therapeutic doses of metformin. It is only in the rare case (about 1 in 10,000) that metformin treatment will lead to lactic acidosis. Lactate does not enter the TCA cycle to be oxidized, as it first has to be converted to pyruvate, then to acetyl-CoA before it can enter the cycle. Red blood cells generate lactate, but do not utilize it as a fuel. Kidney cells, and other muscles, do not utilize lactate to an appreciable extent as a fuel, as compared to the heart muscle.

20. The answer is C. Metformin tends to increase circulating blood lactate levels owing to reduced use of lactate by the liver for gluconeogenesis, which is inhibited in the presence of metformin. The excess lactate, however, can be used by the heart as an energy source, which reduces circulating lactate levels. If the loss of cardiac muscle cells (and their mitochondria) is significant due to a myocardial infarction, the elevated lactate due to metformin use will accumulate since it is no longer being metabolized by the heart. This could lead to lactic acidosis, which may be fatal. Lactate can also accumulate in renal failure; however, none of the other conditions listed will lead to renal failure. Pyelonephritis usually does not lead to renal failure. Torn muscles may cause myoglobinuria, and this may lead to renal failure, but it would be an uncommon outcome. Red blood cells produce lactate, and so anemia would cause reduced lactate formation. A significant weight gain would not be the cause to stop taking metformin.

21. The answer is C. The hydroxyl radical is the most potent ROS. It creates chain reactions that produce lipid peroxides and organic radicals. Hydrogen peroxide is not a radical, but it can generate hydroxyl radicals through interactions with transition metals. Superoxide is a potent ROS, but its potency is diminished by its reduced solubility. NO and coenzyme Q are not ROS, although NO is a radical, and will give rise to reactive nitrogen–oxygen species (RNOS).

22. The answer is A. Macular degeneration can come about through oxidative damage to the macula, so the patient is trying to increase his intake of antioxidant compounds to protect against the generation of ROS. There are no sources of vitamin D in the patient's diet. Carotenoids, found in the carrot juice, are antioxidants. Flavonoids, another potent antioxidant, are found in wine. Orange juice contains vitamin C, another antioxidant. Vitamin E, a very potent antioxidant, is found in egg yolks. The role of vitamin D as an antioxidant is controversial, as its primary role is to maintain calcium homeostasis, although there are some reports of it being utilized as an antioxidant. In either event, none of the patient's dietary sources is providing for vitamin D.

23. The answer is B. Ciprofloxacin, cimetidine, and ketaconazole all inhibit cytochrome P450. Their mode of detoxification does not require the actions of cytochrome P450. Isoniazide, however, induces cytochrome P450 formation as a means of oxidizing the drug to remove it from the body. Cytochrome P450 enzymes are a major source of free-radical production that can occur when electrons are accidentally leaked from reactions and react with molecular oxygen.

24. The answer is A. The contestant has the symptoms of pellagra, which is characterized by the four Ds (i.e., diarrhea, dermatitis, dementia, and eventual death). Pellagra is due to a lack of niacin in the diet. A thiamine deficiency will lead to beriberi; a riboflavin deficiency to ariboflavinosis; a vitamin C deficiency to scurvy; and a vitamin D deficiency to rickets. Only pellagra would yield the symptoms observed by the patient.

25. The answer is A. The patient is exhibiting the symptoms of beriberi, due to a vitamin B_1 deficiency. Thiamine (B_1) is required for oxidative decarboxylation reactions, such as those catalyzed by pyruvate dehydrogenase and α-ketoglutarate dehydrogenase. α-Ketoglutarate dehydrogenase requires thiamine (as thiamine pyrophosphate), lipoic acid, CoASH, FAD, and NAD^+. Succinate dehydrogenase only requires FAD, fumarase has no cofactor requirement, malate dehydrogenase requires NAD^+, and pyruvate carboxylase requires biotin.

26. **The answer is C.** The elevation of lactate, as well as α-ketoglutarate in the TCA cycle, suggests a defect in reactions that catalyze oxidative decarboxylations. Pyruvate would accumulate if pyruvate dehydrogenase was defective, and the increase of concentration of pyruvate would increase the production of lactate. Five cofactors are needed for oxidative decarboxylation reactions, and they are NAD^+, FAD, lipoic acid, thiamine pyrophosphate, and coenzyme A (which is derived from pantothenic acid). The drug appears to be acting by blocking the conversion of pantothenic acid to coenzyme A. Biotin is used for carboxylation reactions, as is vitamin K. Ascorbate is utilized for proline hydroxylation, and vitamin B_6 helps to catalyze a variety of reactions involving amino acids.

27. **The answer is A.** The insecticide is acting as an uncoupler of oxidation and phosphorylation, reducing ATP synthesis while increasing oxygen uptake (increased rate of breathing). As the uncoupler dissipates the proton gradient, the energy used to generate a gradient is lost as heat (leading to the sweating and increased temperature observed in the worker). Of the drugs listed as possible answers, only dinitrophenol (DNP) acts as an uncoupler. Cyanide and carbon monoxide interfere with cytochrome oxidase (complex IV) and block electron transfer to oxygen. Rotenone blocks electron flow from complex I to coenzyme Q, and atractyloside inhibits the ANT, leading to reduced ADP in the mitochondria and a cessation of oxidative phosphorylation. Oxygen consumption would not continue in the presence of atractyloside owing to the loss of ATP synthesis, and the coupling of oxygen consumption and ATP production.

28. **The answer is C.** A lack of blood flow decreased the flow of O_2 to the heart, which slowed down the electron transport chain. The reduction in the rate of the electron transfer chain will lead to an increase in intramitochondrial NADH levels, which slows down the TCA cycle. A slowdown of the TCA cycle leads to a reduction in carbon dioxide production. ATP levels within the mitochondria would also drop as the ATP was transported into the cytoplasm in exchange for ADP. Since oxidative phosphorylation was blocked, the ADP that enters the mitochondria cannot be converted back into ATP, leading to an accumulation of ADP under these conditions.

29. **The answer is C.** At time = 0, succinate is donating electrons to the electron transfer chain through complex II. Succinate is converted to fumarate, and the $FADH_2$ generated donates the electrons to coenzyme Q via complex II. The addition of a chemical at time = 1 does not affect oxygen consumption, and the rate of oxygen consumption is the same as before adding the chemical. Rotenone inhibits at complex I, and would not affect electrons being donated through complex II. Antimycin blocks the transfer of electrons from complex III to complex IV, and would be expected to block electron transfer, and oxygen consumption, from electrons donated from complex II. Dinitrophenol is an uncoupler, and would be expected to increase the rate of oxygen consumption when added, as the uncoupler dissipates the proton gradient and makes it easier for the electron transfer chain to pump protons out of the mitochondria. At time = 2, electron flow stops, but it restarts at time = 3, and at a faster rate. Thus, at time = 3, an uncoupler (dinitrophenol) is added. The fact that electron flow can start again, after stopping, suggests that the chemical added at time = 2 inhibited phosphorylation, as phosphorylation and oxygen consumption are coupled. Inhibiting phosphorylation, via oligomycin, will block oxygen consumption without inhibiting any steps in the electron transfer chain. This is why oxygen consumption can resume again once the uncoupler is added. At time = 4, all electron flow stops, and that can be due to either the addition of cyanide or carbon monoxide, both of which inhibit at complex IV.

30. **The answer is B.** The patient is experiencing ischemic-reperfusion injury. The patient has suffered a heart attack, in which a region of the heart muscle has become anaerobic due to a lack of oxygen delivery. In these cells, the mitochondrial electron transfer chain is reduced, since the terminal electron acceptor (oxygen) is missing. When tPA is administered to dissolve the clot that is causing the blockage of oxygen delivery to the damaged heart muscle, it allows oxygen to rapidly reenter the damaged cells. In these cells, due to the hypoxia, coenzyme Q is fully reduced in the mitochondrial membrane, and the sudden influx of oxygen leads to some accidental electron transfers to oxygen, generating the superoxide radical. This leads to increased radical damage to the already damaged heart muscle, and further damage to the tissue.

Cytochrome a is not released from mitochondria (there may be release of cytochrome c if the damage is sufficient, which would initiate apoptosis in the heart cells). Lactic acid levels are elevated owing to anaerobic glycolysis being used to generate energy, but the high lactate levels are not contributing to further damage to the heart muscle. tPA does not have a direct effect on the activity of the TCA cycle enzymes, and is not an uncoupler.

31. **The answer is D.** AZT is a DNA chain terminator, with greatest affinity for the viral reverse transcriptase. However, mitochondrial DNA polymerase will also recognize and utilize AZT as a substrate, thereby interfering with mitochondrial DNA replication, and mitochondrial division. For some patients, after prolonged AZT treatment, the reduction in mitochondrial function results in the symptoms observed in this patient. Removal of the offending agent will usually reverse this complication. AZT does not affect RNA polymerase, nor is it an uncoupler of oxidative phosphorylation. It itself does not induce the release of cytochrome c from the mitochondria (a signal to start apoptosis), although if sufficient mitochondria are damaged within a cell, the cell might decide to undergo apoptosis.

32. **The answer is E.** The boy has the symptoms of chronic granulomatosis, which in the inherited condition is due to a defect in a component of NADPH oxidase. NADPH oxidase is responsible for the respiratory burst in neutrophils, and produces superoxide from oxygen and NADPH. The superoxide formed helps to destroy the invading bacteria and fungi. In the absence of this activity, the body has difficulty protecting itself from bacterial and fungal infections. The conversion of hydrogen peroxide to hydroxyl radicals is not enzyme-catalyzed, and so the loss of such conversion cannot be inherited. Mutations in SOD (which catalyzes the conversion of superoxide to hydrogen peroxide and oxygen) can lead to ALS, but not chronic granulomatosis. The loss of glutathione reductase can lead to hemolytic anemia, but not the observed symptoms. Neutrophils do not produce hydrochloric acid. The stomach does however, and a loss of acid production can lead to digestive disorders.

33. **The answer is D.** The man is experiencing the inherited form of ALS, which is most often due to an inactivating mutation in SOD. Lack of SOD activity would lead to elevated levels of superoxide, which would then, in an unknown manner, lead to a loss of motor neuron activity. Inherited mutations in NADPH oxidase, catalase, glutathione peroxidase, or myeloperoxidase have not been correlated with ALS.

34. **The answer is E.** During aerobic exercise, the muscle needs to generate ATP via oxidative phosphorylation. For this to occur, NADH levels will decrease (an increase in NADH levels would inhibit the TCA cycle enzymes), leading to a decrease in the NADH/NAD^+ ratio. While citrate is an inhibitor of citrate synthase, if this were to occur during exercise, the TCA cycle would stop, not push forward. The activity of isocitrate dehydrogenase is inhibited (not activated) by NADH, and fumarase is not a regulated enzyme. If TCA cycle intermediates were to decrease in concentration, then the cycle would slow down as well, the opposite of what is needed under conditions in which energy needs to be generated, such as during exercise.

35. **The answer is C.** If cytochrome c cannot function, all the components of the electron transport chain between it and O_2 remain in the oxidized state, and the components of the chain before cytochrome c are reduced. The electron transport chain will not function; O_2 will not be consumed; a proton gradient will not be generated; and ATP will not be produced. NADH will not be oxidized, thereby increasing the NADH/NAD^+ ratio. Owing to this increased ratio, the TCA cycle will slow down and, therefore, CO_2 production will decrease.

chapter 6

Carbohydrate Metabolism

The major clinical uses of the material presented in this chapter are understanding glucose and carbohydrate metabolism, the rationale behind the treatment of diabetes and other carbohydrate disorders, and the formation of glucose-containing compounds used in all body tissues.

OVERVIEW

- Dietary carbohydrates include starch, sucrose, lactose, and indigestible fiber.
- The major product of digestion of carbohydrates is glucose, but some galactose and fructose are also produced.
- Glucose is a major fuel source that is oxidized by cells for energy. After a meal, it is converted to glycogen or to triacylglycerols and stored.
- Glucose is also converted to compounds such as proteoglycans, glycoproteins, and glycolipids.
- When glucose enters cells, it is converted to glucose-6-phosphate, which is a pivotal compound in several metabolic pathways.
  - The major fate of glucose-6-phosphate is to enter the pathway of glycolysis, which produces pyruvate and generates NADH and ATP.
  - Glucose-6-phosphate can be converted to glucose-1-phosphate and then to UDP-glucose, which is used for the synthesis of glycogen or compounds such as the proteoglycans.
  - Glucose-6-phosphate can also enter the pentose phosphate pathway, which produces NADPH (for reactions such as the biosynthesis of fatty acids) and ribose-5-phosphate for nucleotide production.
- Fructose and galactose are converted to intermediates in the pathways by which glucose is metabolized.
- Glycogen is the major storage form of carbohydrate in animals. The largest stores are in muscle and liver.
  - Muscle glycogen is used to generate ATP for muscle contraction.
  - Liver glycogen is used to maintain blood glucose levels during fasting or exercise.
- The maintenance of blood glucose levels is a major function of the liver.
  - The liver produces glucose by glycogenolysis and gluconeogenesis.

I. CARBOHYDRATE STRUCTURE

- Carbohydrates are compounds that contain at least three carbon atoms, a number of hydroxyl groups, and usually an aldehyde or ketone group. They may contain phosphate, amino, or sulfate groups.
- In the body, monosaccharides, the simplest carbohydrates, are usually of the D series.

- Monosaccharides form rings that usually contain five or six members and are called furanoses and pyranoses, respectively. The hydroxyl group on the anomeric carbon (the carbonyl carbon) may be in either the α or the β configuration.
- Monosaccharides are joined by *O*-glycosidic bonds to form disaccharides, oligosaccharides, and polysaccharides.
- Nucleotides contain *N*-glycosidic bonds.
- Monosaccharides can be oxidized to the corresponding acids or reduced to the corresponding polyols.

A. Monosaccharides

1. Nomenclature

a. The simplest monosaccharides have the formula $(CH_2O)_n$. Those with three carbons are called **trioses**; four, **tetroses**; five, **pentoses**; and six, **hexoses**.

b. They are called **aldoses** or **ketoses**, depending on whether their most oxidized functional group is an aldehyde or a ketone (Fig. 6.1).

2. D and L sugars

a. The configuration of the asymmetric carbon atom farthest from the aldehyde or ketone group determines whether a monosaccharide belongs to the D or L series. In the D form, the hydroxyl group is on the right; in the L form, it is on the left (see Fig. 6.1).

b. An **asymmetric carbon** atom has **four different chemical groups** attached to it.

c. **Sugars of the D** series, which are related to D-glyceraldehyde, are the most common in nature (Fig. 6.2).

3. Stereoisomers, enantiomers, and epimers

a. **Stereoisomers** have the same chemical formula but differ in the position of the hydroxyl groups on one or more of their asymmetric carbons.

b. **Enantiomers** are stereoisomers that are mirror images of each other (see Fig. 6.1).

c. **Epimers** are stereoisomers that differ in the position of the hydroxyl group at only one asymmetric carbon. For example, D-glucose and D-galactose are epimers that differ at carbon 4 (see Fig. 6.2).

4. Ring structures of carbohydrates

a. Although **monosaccharides** are often drawn as straight chains (Fischer projections), they exist mainly as ring structures in which the aldehyde or ketone group has reacted with a hydroxyl group in the same molecule (Fig. 6.3).

b. **Furanose** and **pyranose** rings contain five and six members, respectively, and are usually drawn as Haworth projections (see Fig. 6.3).

c. The **hydroxyl group on the anomeric carbon** may be in the α or β configuration. In the **α configuration**, the hydroxyl group on the anomeric carbon is on the right in the Fischer projection and below the plane of the ring in the Haworth projection. In the **β configuration**, it is on the left in the Fischer projection and above the plane in the Haworth projection (Fig. 6.4).

d. In solution, **mutarotation occurs**. The α and β forms equilibrate via the straight-chain aldehyde form (see Fig. 6.4).

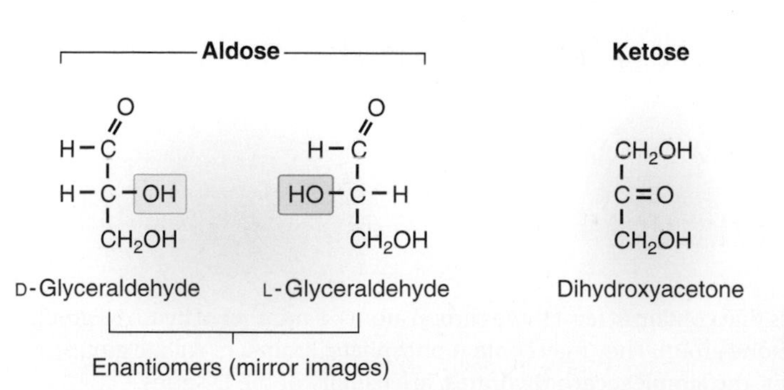

FIGURE 6.1. Examples of trioses, the smallest monosaccharides.

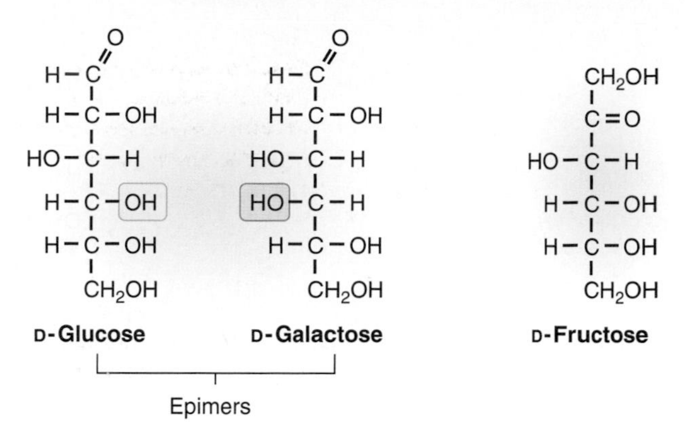

FIGURE 6.2. The common hexoses of the D configuration.

B. Glycosides

1. Formation of glycosides

a. Glycosidic bonds form when the **hydroxyl** group on the anomeric carbon of a monosaccharide reacts with an -OH or -NH group of another compound.

b. **α-Glycosides** or **β-glycosides** are produced depending on the position of the atom attached to the anomeric carbon of the sugar.

2. *O*-Glycosides

a. **Monosaccharides** can be linked via *O*-glycosidic bonds to another monosaccharide, forming *O*-glycosides.

b. **Disaccharides** contain two monosaccharides. Sucrose, lactose, and maltose are the common disaccharides (Fig. 6.5).

c. **Oligosaccharides** contain up to approximately 12 monosaccharides (see Section II B and C).

d. **Polysaccharides** contain more than 12 monosaccharides; for example, glycogen, starch, and glycosaminoglycans (to be further discussed below).

3. *N*-Glycosides

a. Monosaccharides can be linked via *N*-glycosidic bonds to compounds that are not carbohydrates. Nucleotides contain *N*-glycosidic bonds (see Chapter 3).

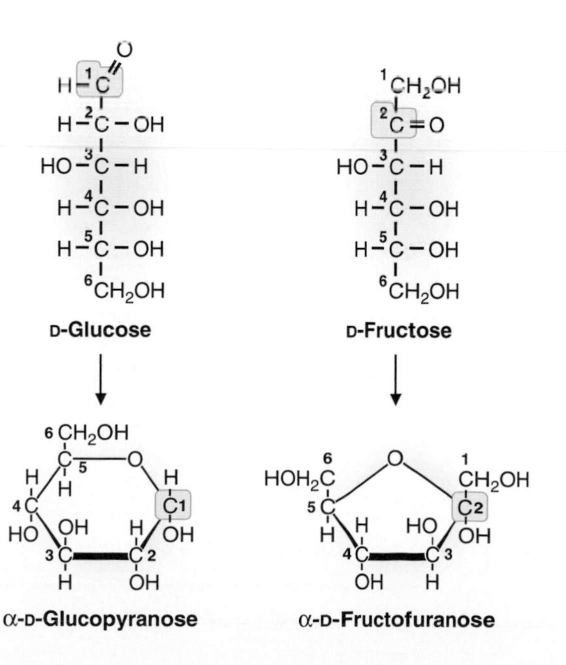

FIGURE 6.3. Furanose and pyranose rings formed from fructose and glucose. The anomeric carbons are highlighted.

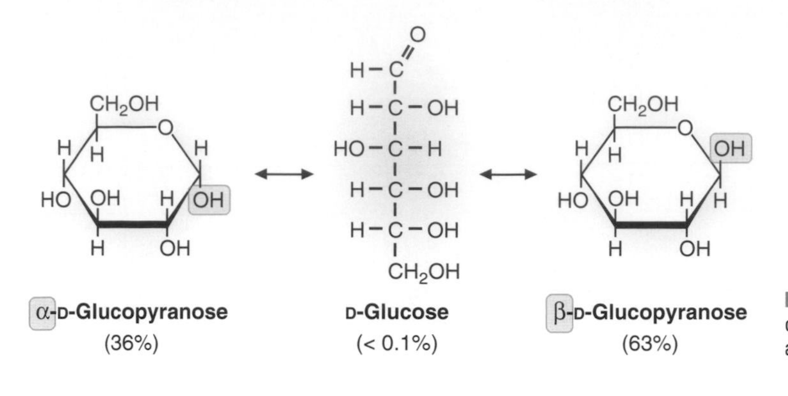

FIGURE 6.4. Mutarotation of glucose in solution. The percentage of each form is indicated.

C. Derivatives of carbohydrates

1. **Phosphate groups** can be attached to carbohydrates.
   a. Glucose and fructose can be phosphorylated on carbons 1 and 6.
   b. Phosphate groups can link sugars to nucleotides, as in UDP-glucose.
2. **Amino groups**, which are often acetylated, can be linked to sugars (e.g., glucosamine and galactosamine).
3. **Sulfate groups** are often found on sugars (e.g., chondroitin sulfate and other glycosaminoglycans).

D. Oxidation of carbohydrates

1. **Oxidized forms**
   a. The anomeric carbon of an aldose (C1) can be oxidized to an acid. Glucose forms **gluconic acid** (gluconate). **6-Phosphogluconate** is an intermediate in the pentose phosphate pathway.
   b. **Carbon 6 of a hexose** can be oxidized to a uronic acid.

CLINICAL CORRELATES

Oxidation of glucose by glucose oxidase (a highly specific test for glucose) is used by clinical and other laboratories to measure the amount of glucose in solution.

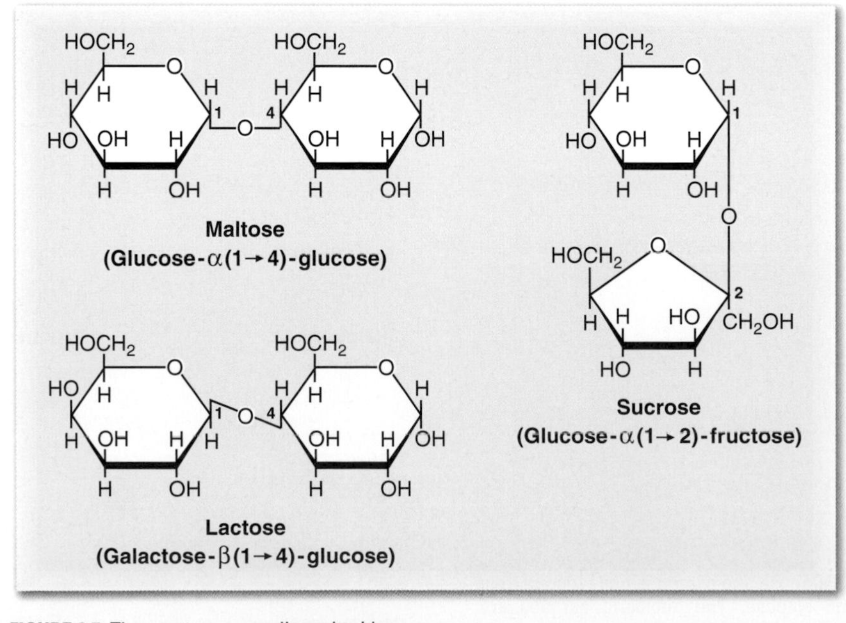

FIGURE 6.5. The most common disaccharides.

(1) Uronic acids are found in glycosaminoglycans of proteoglycans (see Fig. 6.8).
(2) Glucose forms **glucuronic acid**. Conjugation with glucuronic acid makes lipid compounds more water soluble (e.g., bilirubin diglucuronide).

2. **Test for reducing sugars**
   a. Reducing sugars contain a free anomeric carbon that can be oxidized.
      (1) When the anomeric carbon is oxidized, another compound is reduced. If the reduced product of this reaction is colored, the intensity of the color can be used to determine the amount of the reducing sugar that has been oxidized.
      (2) This reaction is the basis of the reducing-sugar test, which is used by clinical laboratories. The test is not sugar specific. Aldoses such as glucose give a positive test result. Ketoses such as fructose are also reducing sugars because they form aldoses under test conditions.

E. Reduction of carbohydrates

1. The aldehyde or ketone group of a sugar can be reduced to a hydroxyl group, forming a **polyol** (polyalcohol).
2. Glucose is reduced to **sorbitol** and galactose to **galactitol**.

F. Glycosylation of proteins

Aldehyde groups of sugars **nonenzymatically** form Schiff bases with amino groups of proteins. Subsequently, Amadori rearrangements form stable covalent interactions.

CLINICAL CORRELATES A glycosylated fraction of hemoglobin, **HbA_{1c}**, is normally 4% to 6% of the total hemoglobin, but increases when red blood cells are exposed to high levels of blood glucose. Measurement of HbA_{1c} levels can determine glycemic control in individuals with diabetes. The American Diabetes Association uses the range of 5.7% to 6.4% to define "prediabetes."

II. PROTEOGLYCANS, GLYCOPROTEINS, AND GLYCOLIPIDS

- Proteoglycans consist of long linear chains of glycosaminoglycans attached to a core protein. Each chain is composed of a repeating disaccharide that is usually negatively charged and contains a hexosamine and a uronic acid. Sulfate groups are often present.
  - The glycosaminoglycans are synthesized from UDP-sugars.
- Glycoproteins contain smaller polysaccharide chains that are usually branched.
  - In addition to glucose, galactose, and their amino derivatives, glycoproteins contain mannose, L-fucose, and *N*-acetylneuraminic acid (NANA).
  - For O-linked chains, the polysaccharide grows by the sequential addition of monosaccharide units from UDP-sugars to serine or threonine residues in a protein.
  - For N-linked chains, branched carbohydrates are synthesized on dolichol phosphate and transferred to the amide nitrogen of an asparagine residue in a protein.
- Glycolipids are members of the class of sphingolipids.
  - The carbohydrate portion is synthesized from UDP-sugars that add to the hydroxymethyl group of ceramide and then sequentially to the nonreducing end of the chain.
  - *N*-acetylneuraminic acid (derived from CMP-NANA) often forms branches from the main chain.
  - Proteoglycans, glycoproteins, and glycolipids are synthesized in the endoplasmic reticulum (ER) and Golgi complex; they are degraded by the action of lysosomes.

A. Proteoglycans are found in the extracellular matrix or ground substance of connective tissue, synovial fluid of joints, vitreous humor of the eye, secretions of mucus-producing cells, and in cartilage.

1. Structure of proteoglycans

a. Proteoglycans consist of a core protein with long unbranched polysaccharide chains (**glycosaminoglycans**) attached. The overall structure resembles a bottle brush.

b. These chains are composed of **repeating disaccharide units**, which usually contain a **uronic acid** and a **hexosamine**. The uronic acid is generally D-glucuronic or L-iduronic acid.

c. The amino group of the hexosamine is usually **acetylated**, and **sulfate** groups are often present on carbons 4 and 6.

d. A xylose and two galactose residues connect the chain of repeating disaccharides to the core protein.

2. Synthesis of proteoglycans

a. The **protein** is synthesized on the ER.

b. **Glycosaminoglycans** are produced by the addition of sugars to serine or threonine residues of the protein. UDP-sugars serve as the precursors.

c. In the ER and the Golgi, the glycosaminoglycan chains grow by sequential addition of sugars to the nonreducing end.

(1) **Sulfate groups**, donated by 3'-phosphoadenosine 5'-phosphosulfate (PAPS), are added after the hexosamine is incorporated into the chain.

(2) Because of the uronic acid and sulfate groups, the glycosaminoglycans are **negatively charged**, causing the chains to be heavily hydrated.

d. **Proteoglycans** are **secreted** from the cell.

e. Proteoglycans can associate noncovalently with **hyaluronic acid** (a glycosaminoglycan) forming large aggregates, which act as molecular sieves, that can be penetrated by small but not by large molecules.

3. Degradation of proteoglycans by lysosomal enzymes

a. Because proteoglycans are located outside the cell, they are taken up by **endocytosis**. The endocytic vesicles fuse with lysosomes.

b. **Lysosomal enzymes** specific for each monosaccharide remove the sugars, one at a time, from the nonreducing end of the chain.

c. **Sulfatases** remove the sulfate groups before the sugar residue is hydrolyzed.

B. Glycoproteins serve as enzymes, hormones, antibodies, and structural proteins. They are found in extracellular fluids and in lysosomes and are attached to the cell membrane. They are involved in cell-cell interactions.

1. Structure of glycoproteins

a. The **carbohydrate** portion of glycoproteins differs from that of proteoglycans in that it is **shorter** and often **branched** (Fig. 6.6).

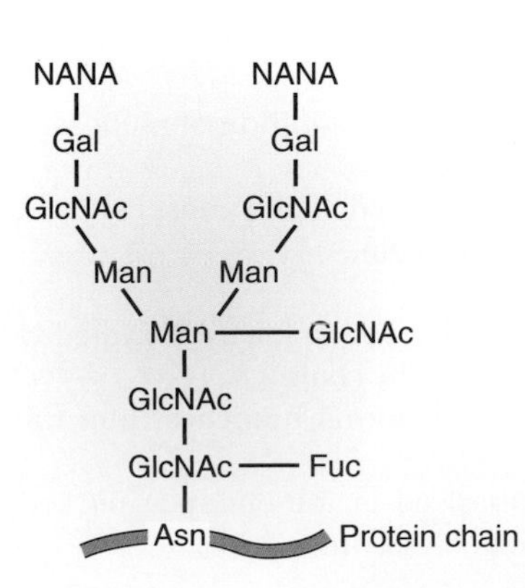

FIGURE 6.6. An example of the carbohydrate moiety of a glycoprotein. Note that, in this case, the carbohydrate is attached to an asparagine (N-linked). NANA, *N*-acetylneuraminic acid; Gal, galactose; GlcNAc, *N*-acetylglucosamine; Man, mannose; Fuc, fucose.

(1) Glycoproteins contain mannose, L-fucose, and NANA in addition to glucose, galactose, and their amino derivatives. NANA is a member of the class of sialic acids.

(2) The antigenic determinants of the ABO and Lewis blood group substances are sugars at the ends of these carbohydrate branches.

b. The carbohydrates are attached to the protein via the hydroxyl groups of **serine and threonine** residues or the amide N of **asparagine**.

2. **Synthesis of glycoproteins**

a. The protein is synthesized on the ER. In the ER and the Golgi, the **carbohydrate chain** is produced by the sequential addition of monosaccharide units to the nonreducing end. UDP-sugars, GDP-mannose, GDP-L-fucose, and CMP-NANA act as precursors.

b. For O-linked glycoproteins, the initial sugar is added to a **serine or threonine** residue in the protein and the carbohydrate chain is then elongated.

c. **Dolichol phosphate** is involved in the synthesis of N-linked glycoproteins in which the carbohydrate moiety is attached to the amide N of asparagine.

(1) Dolichol phosphate, a long-chain alcohol containing about 20 5-carbon isoprene units, can be synthesized from acetyl-CoA.

(2) Sugars are added sequentially to dolichol phosphate, which is associated with the membrane of the ER.

(3) The branched polysaccharide chain is transferred to an amide N of an asparagine residue in the protein.

(4) In the ER and the Golgi, sugars are removed from the chain and other sugars are added.

d. **Glycoproteins** are **segregated** into lysosomes within the cell, **attached** to the cell membrane, or **secreted** by the cell.

(1) **Lysosomal enzymes** are glycoproteins. A mannose phosphate residue targets these glycoproteins to lysosomes. An inability to add this mannose-6-phosphate modification to the protein results in the secretion of the protein from the cell, a condition known as **I-cell disease**.

(2) When a glycoprotein is attached to the **cell membrane**, the carbohydrate portion extends into the extracellular space and a hydrophobic segment of the protein is anchored in the membrane.

CLINICAL CORRELATES

I-cell disease is a disorder in the targeting of lysosomal proteins to the lysosomes (the formal name for I-cell disease is mucolipidosis type II). The presence of **mannose-6-phosphate** on the carbohydrate chain of lysosomal proteins is recognized by a mannose-6-phosphate receptor for proper sorting of these proteins from the Golgi apparatus to the lysosome. Adding the mannose-6-phosphate to the precursors of lysosomal proteins is a two-step process, and mutations in the first enzyme of this process are defective in I-cell disease. Lysosomal proteins are secreted instead of being delivered to the lysosomes. Lysosomal function is lost, and death most often results within the first 10 years of life. The lysosomes become engorged with nondigested material and resemble **inclusion bodies**, for which the disease is named.

3. **Degradation of glycoproteins**

a. **Lysosomal enzymes** specific for each monosaccharide remove sugars sequentially from the nonreducing ends of the chains.

C. Glycolipids

1. Glycolipids (or **sphingolipids**) are derived from the lipid **ceramide** (see Fig. 7.3). This class of compounds includes cerebrosides and gangliosides. Some bacterial toxins and viruses use glycolipids as receptors.

a. **Cerebrosides** are synthesized from ceramide and UDP-sugars.

b. **Gangliosides** have *N*-acetylneuraminic acid residues (derived from CMP-NANA) branching from the linear oligosaccharide chain.

table 6.1 Defective Enzymes in the Mucopolysaccharidoses

Disease	Enzyme Deficiency	Accumulated Products
Hunter	Iduronate sulfatase	Heparan sulfate, dermatan sulfate
Hurler + Scheie	α-L-iduronidase	Heparan sulfate, dermatan sulfate
Maroteaux–Lamy	*N*-acetylgalactosamine sulfatase	Dermatan sulfate
Mucolipidosis VII	β-Glucuronidase	Heparan sulfate, dermatan sulfate
Sanfilippo A	Heparan sulfamidase	Heparan sulfate
Sanfilippo B	*N*-acetylglucosaminidase	Heparan sulfate
Sanfilippo D	*N*-acetylglucosamine 6-sulfatase	Heparan sulfate

These disorders share many clinical features, although there are significant variations between disorders, and even within a single disorder, based on the amount of residual activity remaining. In most cases, multiple organ systems are affected (with bone and cartilage being a primary target). For some disorders, there is significant neuronal involvement, leading to mental retardation.

2. Glycolipids are found in the **cell membrane**, with the carbohydrate portion extending into the extracellular space.
3. They are degraded by **lysosomal enzymes**.

CLINICAL CORRELATES A **deficiency of lysosomal enzymes** results in the inability to degrade the carbohydrate portions of proteoglycans and sphingolipids, leading to the **mucopolysaccharidoses** and **sphingolipidoses (gangliosidoses)**, respectively. Partially digested products accumulate in lysosomes. Tissues become engorged with these "residual bodies," and their function is impaired. These diseases, which include Hunter's and Hurler's mucopolysaccharidoses and Tay–Sachs' and Gaucher's gangliosidoses, are often **fatal.** Table 6.1 lists the mucopolysaccharidoses. The sphingolipidoses are described in more detail in Chapter 7.

III. DIGESTION OF CARBOHYDRATES

- The major dietary carbohydrates are starch, sucrose, and lactose.
- In the mouth, salivary α-amylase acts on starch, cleaving α-1,4 linkages between glucose residues.
- In the intestine, pancreatic α-amylase continues the digestion of starch.
- Enzymes associated with the brush border of intestinal epithelial cells digest sucrose, lactose, and the products generated from starch by α-amylase.
- The final products of carbohydrate digestion–glucose, fructose, and galactose–are absorbed by intestinal epithelial cells and then secreted from these cells to enter the blood.

A. Dietary carbohydrates (mainly starch, sucrose, and lactose) constitute about 50% of the calories in the average diet in the United States.

1. **Starch**, the storage form of carbohydrate in plants, is similar in structure to glycogen (Fig. 6.7).
   a. Starch contains amylose (long, unbranched chains with glucose units linked α-1,4) and amylopectin (α-1,4-linked chains with α-1,6-linked branches). Amylopectin has fewer branches than glycogen.
2. **Sucrose** (a component of table sugar and fruit) contains glucose and fructose residues linked via their anomeric carbons (see Fig. 6.5).
3. **Lactose** (milk sugar) contains galactose linked β-1,4 to glucose (see Fig. 6.5).

B. Digestion of dietary carbohydrates in the mouth (Fig. 6.8)

In the mouth, **salivary α-amylase** cleaves starch by breaking α-1,4 linkages between glucose residues within the chains (see Fig. 6.10). Dextrins (linear and branched oligosaccharides) are the major products that enter the stomach.

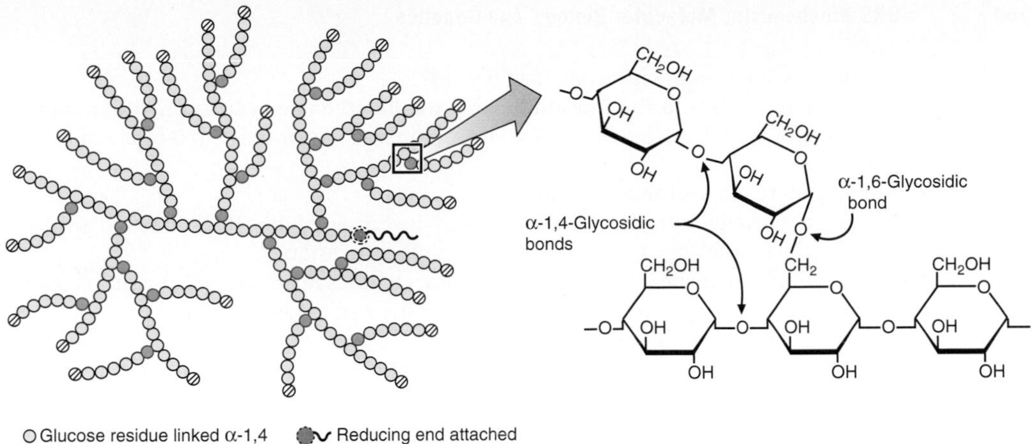

FIGURE 6.7. Glycogen structure. Glycogen is composed of glucosyl units linked by α-1,4 and α-1,6 glycosidic linkages.

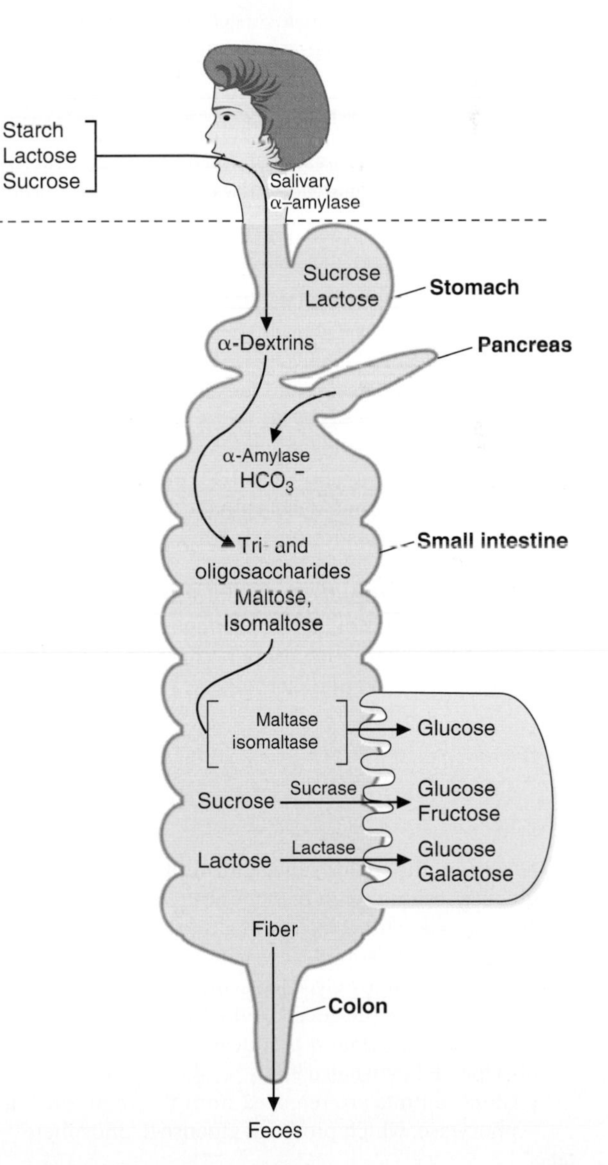

FIGURE 6.8. An overview of carbohydrate digestion. Starch is digested by salivary and pancreatic α-amylases and intestinal cell maltase and isomaltase. Sucrose and lactose are digested by intestinal enzymes. Subsequent metabolic reactions occur after the sugars are absorbed.

C. Digestion of carbohydrates in the intestine (see Fig. 6.8)

The stomach contents pass into the intestine where the **bicarbonate** secreted by the pancreas neutralizes the stomach acid, raising the pH into the optimal range for the action of the intestinal enzymes.

1. **Digestion by pancreatic enzymes** (see Fig. 6.8)
   a. The pancreas secretes an **α-amylase** that acts in the lumen of the small intestine and, like salivary amylase, cleaves α-1,4 linkages between glucose residues.
   b. The products of pancreatic α-amylase are the disaccharides maltose and isomaltase, trisaccharides, and small oligosaccharides containing α-1,4 and α-1,6 linkages.
2. **Digestion by enzymes of intestinal cells**
   a. **Complexes of enzymes**, produced by intestinal epithelial cells and located in their **brush borders**, continue the digestion of carbohydrates (see Fig. 6.8).
      (1) **Glucoamylase** (an **α-glucosidase**) and other **maltases** cleave glucose residues from the nonreducing ends of oligosaccharides and also cleave the α-1,4 bond of maltose, releasing the two glucose residues.
      (2) **Isomaltase** cleaves α-1,6 linkages, releasing glucose residues from branched oligosaccharides.
      (3) **Sucrase** converts sucrose to glucose and fructose.
      (4) **Lactase** (a β-galactosidase) converts lactose to glucose and galactose.

CLINICAL CORRELATES **Intestinal lactase deficiency** is a common condition in which **lactose cannot be digested** and is oxidized by bacteria in the gut, which produce gas, and cause bloating and watery diarrhea. This can also occur through a loss of intestinal epithelial cells due to viral gastroenteritis.

D. Carbohydrates that cannot be digested

Indigestible polysaccharides are part of the **dietary fiber** that passes through the intestine into the feces. For example, because enzymes produced by human cells cannot cleave the β-1,4 bonds of cellulose, this polysaccharide is indigestible.

E. Absorption of glucose, fructose, and galactose

Glucose, fructose, and galactose, the final products generated by digestion of dietary carbohydrates, are absorbed by intestinal epithelial cells.

1. They are transported into the cells on **transport proteins**, moving down a concentration gradient.
2. Glucose also moves into cells on a transport protein that carries sodium ions in addition to the monosaccharide. This is **a secondary active transport process.**
3. The sugars are then passed into the blood using facilitative transporters on the serosal side of the intestinal epithelial cells.

IV. GLYCOGEN STRUCTURE AND METABOLISM

- Glycogen, the major storage form of carbohydrate in animals, consists of chains of α-1,4-linked glucose residues with branches that are attached by α-1,6 linkages.
- Glycogen is synthesized from glucose (Fig. 6.9).
  - UDP-glucose supplies the glucose moieties, which are added to the nonreducing ends of a glycogen primer by glycogen synthase.
  - Branches are produced by the branching enzyme, glucosyl 4:6 transferase.
- Glycogen degradation produces glucose-1-phosphate as the major product, but free glucose is also formed (see Fig. 6.9).
  - Glucose units are removed from the nonreducing ends of glycogen chains by glycogen phosphorylase, which produces glucose-1-phosphate.

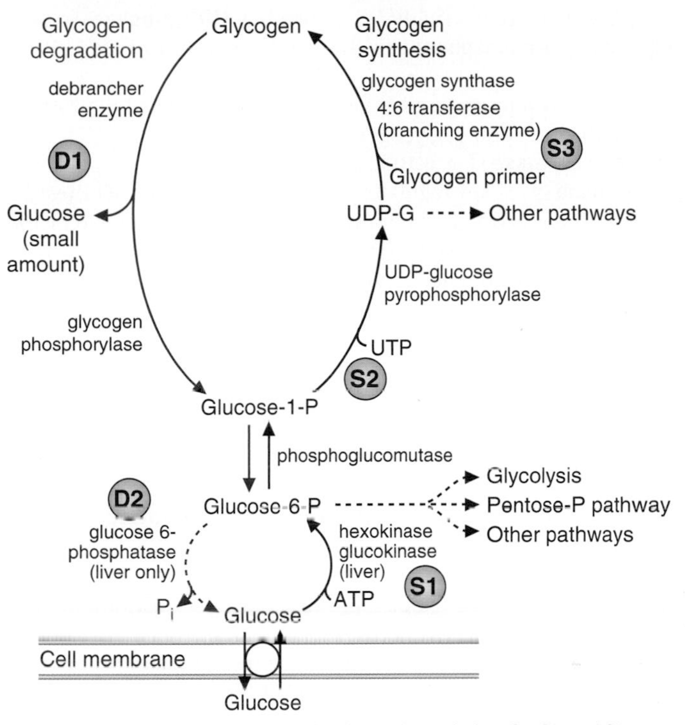

FIGURE 6.9. The scheme of glycogen synthesis and degradation. S1, S2, and S3 represent the steps involved in the synthesis of glycogen. D1 and D2 represent the steps involved in glycogen degradation. UDP-G represents UDP-glucose, an activated nucleotide sugar.

  - Three of the four glucose units at a branch point are moved by a glucosyl 4:4 transferase to the nonreducing end of another chain.
  - The remaining glucose unit that is linked α-1,6 at the branch point is released as free glucose by an α-1,6-glucosidase.
- Liver glycogen is used to maintain blood glucose levels during fasting or exercise.
  - Its breakdown is stimulated by glucagon and by epinephrine via a mechanism that involves cyclic adenosine monophosphate (cAMP) (see Chapter 3).
- Muscle glycogen is utilized to generate ATP for muscle contraction.
  - Epinephrine, via cAMP, stimulates muscle glycogen breakdown.

A. Glycogen structure

Glycogen is a large, **branched polymer** consisting of D-glucose residues (see Fig. 6.7).

1. The **linkages** between glucose residues are **α-1,4** except at branch points where the linkage is **α-1,6**. Branching is more frequent in the interior of the molecule and less frequent at the periphery, the average being an α-1,6 branch every 8 to 10 residues.
2. One glucose unit, located at the reducing end of each glycogen molecule, is attached to the protein **glycogenin**.
3. The glycogen molecule branches like a tree and has **many nonreducing ends** at which addition and release of glucose residues occur during synthesis and degradation, respectively.

B. Glycogen synthesis

UDP-glucose is the **precursor** for glycogen synthesis.

1. **Synthesis of UDP-glucose**
   a. Glucose enters cells and is phosphorylated to glucose-6-phosphate by **hexokinase** (or by **glucokinase** in the liver). ATP provides the phosphate group.
   b. **Phosphoglucomutase** converts glucose-6-phosphate to glucose-1-phosphate.

c. Glucose-1-phosphate reacts with UTP, forming **UDP-glucose** in a reaction catalyzed by **UDP-glucose pyrophosphorylase**. Inorganic pyrophosphate (PP_i) is released in this reaction.
   (1) **PP_i** is cleaved by a pyrophosphatase to 2 P_i. This removal of product helps to drive the process in the direction of glycogen synthesis.

2. **Action of glycogen synthase** (Fig. 6.10)
   a. **Glycogen synthase** is the key regulatory enzyme for glycogen synthesis. It transfers glucose residues from UDP-glucose to the nonreducing ends of a glycogen primer.
      (1) UDP is released and reconverted to UTP by reaction with ATP.
   b. The primers, which are attached to glycogenin, are glycogen molecules that were partially degraded in liver during fasting or in muscle and liver during exercise.
3. **Formation of branches** (see Fig. 6.10)
   a. When a chain contains 11 or more glucose residues, an **oligomer**, six to eight residues in length, is removed from the nonreducing end of the chain. It is **reattached** via an **α-1,6 linkage** to a glucose residue within an α-1,4-linked chain.
   b. These branches are formed by the branching enzyme, a **glucosyl 4:6 transferase**, that breaks an α-1,4 bond and forms an α-1,6 bond.
   c. The new branch points are at least four residues and an average of 7 to 11 residues from previously existing branch points.
4. **Growth of glycogen chains**
   a. Glycogen synthase continues to add glucose residues to the nonreducing ends of the newly formed branches as well as to the ends of the original chains.
   b. As the chains continue to grow, additional branches are produced by the branching enzyme.

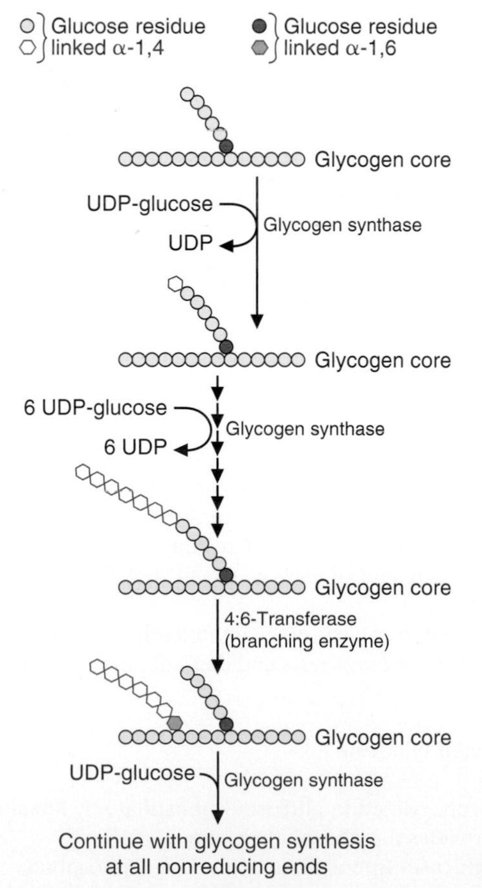

FIGURE 6.10. Glycogen synthesis.

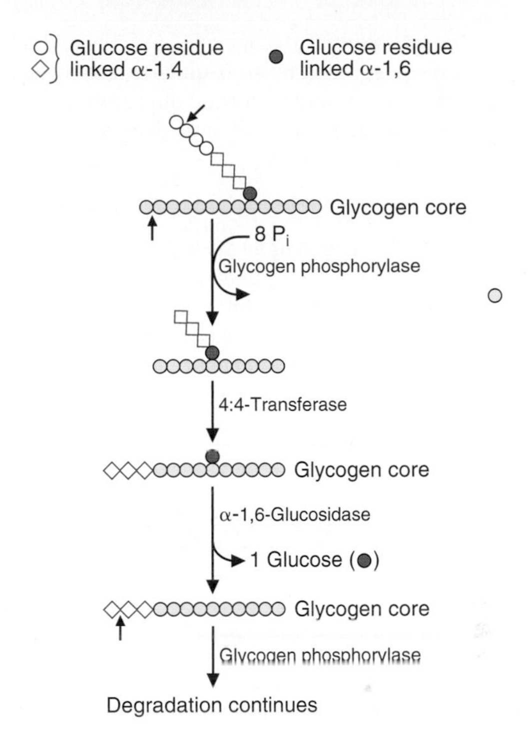

FIGURE 6.11. Glycogen degradation.

C. Glycogen degradation (Fig. 6.11)

1. Action of glycogen phosphorylase

a. Glycogen phosphorylase, the key regulatory enzyme for glycogen degradation, removes glucose residues, one at a time, from the nonreducing ends of glycogen molecules.

b. Phosphorylase uses inorganic phosphate (P_i) to cleave α-1,4 bonds, producing **glucose-1-phosphate**.

c. Phosphorylase can act only until it is four glucose units from a branch point.

2. Removal of branches

a. The four units remaining at a branch are removed by the **debranching enzyme**, which has both glucosyl 4:4 transferase and α-1,6-glucosidase activity.

(1) Three of the four glucose residues that remain at the branch point are removed as a trisaccharide and attached to the nonreducing end of another chain by a **4:4 transferase**, which cleaves an α-1,4 bond and forms a new α-1,4 bond.

(2) The last glucose unit at the branch point, which is linked α-1,6, is hydrolyzed by **α-1,6-glucosidase**, forming free glucose.

3. Degradation of glycogen chains

a. The **phosphorylase/debranching process is repeated**, generating glucose-1-phosphate and free glucose in about a 10:1 ratio that reflects the length of the chains in the outer region of the glycogen molecule.

4. Fate of glucosyl units released from glycogen (see Fig. 6.9)

a. In the liver, glycogen is degraded to maintain blood glucose levels.

(1) Glucose-1-phosphate is converted by **phosphoglucomutase** to glucose-6-phosphate.

(2) Inorganic phosphate is released by **glucose 6-phosphatase**, and free glucose enters the blood. This enzyme also acts in gluconeogenesis (see Section VI A 3).

b. In muscle, glycogen is degraded to provide energy for contraction.

(1) Phosphoglucomutase converts glucose-1-phosphate to glucose-6-phosphate, which enters the pathway of **glycolysis** and is converted either to lactate or to CO_2 and H_2O, generating ATP.

(2) Muscle does not contain glucose-6-phosphatase and, therefore, does not contribute to the maintenance of blood glucose levels.

D. Lysosomal degradation of glycogen

Glycogen is also degraded by an α-**glucosidase** located in lysosomes. Lysosomal degradation is not necessary for maintaining normal blood glucose levels. A lack of this enzyme activity leads to a fatal glycogen storage disease, **Pompe disease**.

CLINICAL CORRELATES In the **glycogen storage diseases, glycogen accumulates** primarily in the liver or muscle, or both. Enzyme deficiencies occur mainly in glycogen degradation or conversion to glucose. Because different forms of the enzymes (isozymes) occur in the liver and muscle, one tissue may be affected, but not the other. In the **liver**, glycogen storage diseases result in hepatomegaly and conditions ranging from mild hypoglycemia to liver failure. In **muscle**, they cause problems ranging from difficulty in performing strenuous exercise to cardiorespiratory failure (Table 6.2). Some of these disorders are fatal at an early age.

table 6.2 Glycogen Storage Diseases

Type	Enzyme Affected	Primary Organ Involved	Manifestations[a]
0	Glycogen synthase	Liver	Hypoglycemia, hyperketonemia, failure to thrive, early death.
I[b]	Glucose 6-phosphatase (Von Gierkes disease)	Liver	Enlarged liver and kidney, growth failure, severe fasting hypoglycemia, acidosis, lipemia, gout, thrombocyte dysfunction.
II	Lysosomal α-glucosidase (Pompe disease): may see clinical symptoms in childhood, juvenile, or adult life stages, depending on the nature of the mutation.	All organs with lysosomes	Infantile form: early-onset progressive muscle hypotonia, cardiac failure, death before age 2 y. Juvenile form: later-onset myopathy with variable cardiac involvement. Adult form: limb-girdle muscular dystrophy-like features. Glycogen deposits accumulate in lysosomes.
III	Amylo-1,6-glucosidase (debrancher): form IIIa is the liver and muscle enzymes, form IIIb is a liver-specific form, and IIIc a muscle-specific form	Liver, skeletal muscle, heart	Fasting hypoglycemia; hepatomegaly in infancy in some myopathic features. Glycogen deposits have short outer branches.
IV	Amylo-4,6-glucosidase (branching enzyme) (Andersen disease)	Liver	Hepatosplenomegaly; symptoms may arise from a hepatic reaction to the presence of a foreign body (glycogen with long outer branches). Usually fatal.
V	Muscle glycogen phosphorylase (McArdle's disease) (expressed as either adult or infantile form)	Skeletal muscle	Exercise-induced muscular pain, cramps, and progressive weakness, sometimes with myoglobinuria.
VI[c]	Liver glycogen phosphorylase (Her's disease) and its activating system (includes mutations in liver phosphorylase kinase and liver PKA)	Liver	Hepatomegaly, mild hypoglycemia; good prognosis.
VII	Phosphofructokinase 1 (Tarui syndrome)	Muscle, red blood cells	As in type V; in addition, enzymopathic hemolysis.
XI	GLUT2 (glucose/galactose transporter); Fanconi–Bickel syndrome	Intestine, pancreas, kidney, liver	Glycogen accumulation in liver and kidney; rickets, growth retardation, glucosuria.

[a]All of these diseases except type 0 are characterized by increased glycogen deposits.
[b]Glucose 6-phosphatase is composed of several subunits that also transport glucose, glucose- 6-phosphate, phosphate, and pyrophosphate across the endoplasmic reticulum membranes. Therefore, there are several subtypes of this disease, corresponding to defects in the different subunits. Type Ia is a lack of glucose 6-phosphatase activity; type Ib is a lack of glucose-6-phosphate translocase activity; type Ic is a lack of phosphotranslocase activity; type Id is a lack of glucose translocase activity.
[c]Glycogen storage diseases IX (hepatic phosphorylase kinase) and X (hepatic protein kinase A) have been reclassified to VI, which now refers to the hepatic glycogen phosphorylase activating system.

Sources: Parker PH, Ballew M and Greene HL. Nutritional management of glycogen storage disease. *Annu Rev Nutr.* 1993;13:83–109. Copyright © 1993 by Annual Reviews, Inc.; Shin YS. Glycogen storage disease: clinical, biochemical and molecular heterogeneity. *Semin Ped Neurol.* 2006;13:115–120; Bayraktar Y. Glycogen storage diseases: new perspectives. World J. Gastroenterol. 2007;13:2541-2553.

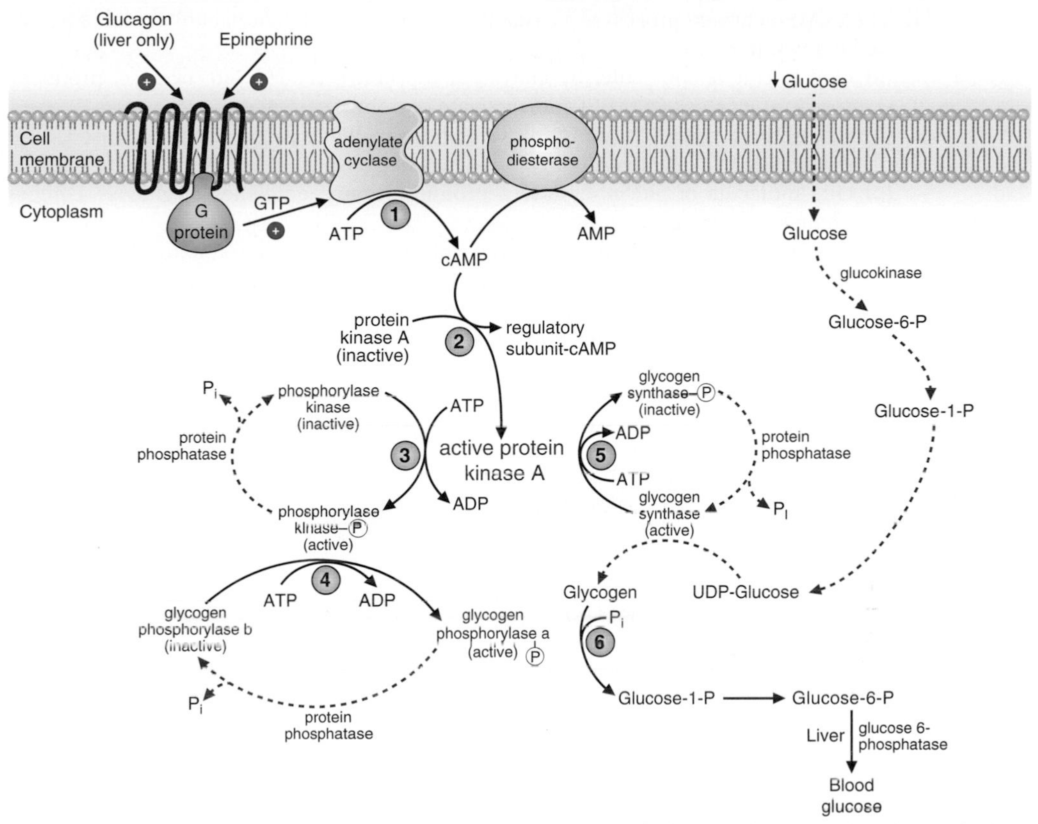

FIGURE 6.12. Hormonal regulation of glycogen synthesis and degradation. *Solid lines* indicate reactions that predominate when glucagon or epinephrine is elevated. Steps 1 through 6, indicated by *circled numbers*, correspond to Section IV E 1–6 in the text. The *red dashed lines* indicate reactions that are decreased in the livers of fasting individuals (they are also promoted in the presence of insulin). Note that protein kinase A phosphorylates both phosphorylase kinase and glycogen synthase.

E. Regulation of glycogen degradation (Fig. 6.12)

1. **Hormones** that use **3',5'-cyclic AMP** (cAMP) as a second messenger stimulate a mechanism, resulting in the phosphorylation of enzymes.
2. Glycogen degradation is stimulated, and synthesis is inhibited when the enzymes of glycogen metabolism are phosphorylated.
   a. **Glucagon** acts on liver cells and **epinephrine** (adrenaline) acts on both liver and muscle cells to stimulate glycogen degradation.
      (1) These hormones via G-proteins activate **adenylate cyclase** in the cell membrane, which converts ATP to cAMP (see Figs. 4.10 and 4.17).
      (2) Adenylate cyclase is also called adenyl or adenylyl cyclase.
   b. cAMP **activates protein kinase A** (see Fig. 6.12), which consists of two regulatory and two catalytic subunits. cAMP binds to the regulatory (inhibitory) subunits, releasing the catalytic subunits in an active form.
   c. **Protein kinase A** phosphorylates **glycogen synthase**, causing it to be less active, thus decreasing the glycogen synthesis.
   d. **Protein kinase A** phosphorylates **phosphorylase kinase**.
   e. **Phosphorylase kinase** phosphorylates **phosphorylase b**, converting it to its active form, phosphorylase a.
   f. **Phosphorylase a** cleaves glucose residues from the nonreducing ends of glycogen chains, producing glucose-1-phosphate, which is oxidized or, in the liver, converted to blood glucose.
   g. **The cAMP cascade**

(1) The cAMP-activated process is a cascade in which the initial **hormonal signal is amplified** many times.

(a) One hormone molecule, by activating the enzyme adenylate cyclase, produces many molecules of cAMP, which activate protein kinase A.

(b) One active protein kinase A molecule phosphorylates many phosphorylase kinase molecules, which convert many molecules of phosphorylase b to phosphorylase a.

(c) One molecule of phosphorylase a produces many molecules of glucose-1-phosphate from glycogen.

(d) The net result is that one hormone molecule can generate tens of thousands of molecules of glucose-1-phosphate, which form glucose 6-phosphate. Oxidation of glucose-6-phosphate generates hundreds of thousands of molecules of ATP.

h. Additional regulatory mechanisms in muscle

(1) In addition to cAMP-mediated regulation, **adenosine monophosphate (AMP)** and **Ca^{2+}** stimulate glycogen breakdown in muscle.

(a) **Phosphorylase b** is activated by the rise in **AMP**, which occurs during muscle contraction by the following reactions:

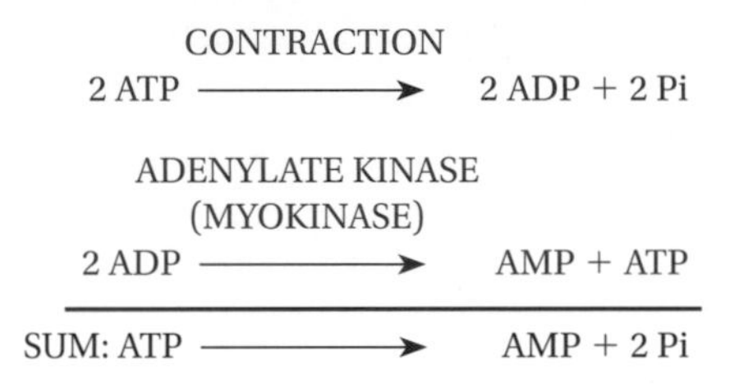

(b) **Phosphorylase kinase** is activated by **Ca^{2+}**, which is released from the sarcoplasmic reticulum during muscle contraction.

1. Ca^{2+} binds to **calmodulin**, which serves as a subunit of phosphorylase kinase. The conformational change induced by calcium binding to calmodulin is sufficient to activate the nonphosphorylated form of phosphorylase kinase.

F. Regulation of glycogen synthesis (see Fig. 6.12)

Insulin, which is elevated after a meal, stimulates the synthesis of glycogen in liver and muscle.

1. Factors that promote glycogen synthesis in the liver

a. In the fed state, glycogen degradation decreases because **glucagon** is **low**, and the cAMP cascade is not activated.

(1) cAMP is converted to AMP by a cell membrane **phosphodiesterase.**

(2) As **cAMP decreases**, the regulatory subunits rejoin the catalytic subunits of **protein kinase A**, and the enzyme is **inactivated.**

(3) **Dephosphorylation** of phosphorylase kinase and phosphorylase a causes these enzymes to be inactivated. **Insulin** causes the activation of the **phosphatases** that dephosphorylate these enzymes.

(a) A key phosphatase is protein phosphatase I (PP-1).

(b) PP-1 is regulated by a protein inhibitor, which is activated by phosphorylation by protein kinase A. The inhibitor, when phosphorylated, binds to and inhibits PP-1 activity.

(c) The PP-1:inhibitor complex allows for slow hydrolysis of the phosphorylated inhibitor by PP-1. When the inhibitor is dephosphorylated, it no longer has affinity for PP-1 and falls out of the complex, leading to a fully active PP-1.

b. **Glycogen synthesis** is promoted by the activation of **glycogen synthase** and by the increased concentration of glucose, which enters liver cells from the hepatic portal vein.

(1) The inactive, phosphorylated form of glycogen synthase is dephosphorylated, causing the enzyme to become active. **Insulin** causes the activation of the **phosphatase** that catalyzes this reaction.

2. Factors that promote glycogen synthesis in muscle

a. After a meal, muscle will have low levels of cAMP, AMP, and Ca^{2+} if it is not contracting and epinephrine is low. Consequently, muscle glycogen degradation will not occur.

b. **Insulin stimulates glycogen synthesis** by mechanisms similar to those in the liver.
c. In addition, **insulin stimulates the transport of glucose** into muscle cells, providing increased substrate for glycogen synthesis.

CLINICAL CORRELATES

Insulinomas and glucagonomas are rare neuroendocrine tumors of the pancreas that can episodically release large amounts of either insulin or glucagon, respectively. Insulinomas will lead to **hypoglycemia**, due to the stimulation of glucose transport into the muscle and fat cells, particularly if the insulin is released under fasting conditions. Glucagonomas will lead to **hyperglycemia**, as the liver is instructed to release glucose via glycogenolysis and gluconeogenesis in the presence of glucagon.

V. GLYCOLYSIS

- Glycolysis is the pathway by which glucose is converted to pyruvate. It occurs in the cytosol of all cells of the body.
- In the initial reactions, a hexose is phosphorylated twice by ATP and then cleaved to yield two triose phosphates.
  - Glucose is phosphorylated to glucose-6-phosphate, which is isomerized to fructose-6-phosphate.
  - Fructose-6-phosphate is phosphorylated by the key regulatory enzyme, phosphofructokinase-1. The product is fructose-1,6-bisphosphate, which is cleaved, forming two triose phosphates.
- In the second sequence of reactions, the triose phosphates produce ATP.
- Overall, glycolysis produces ATP, NADH, and pyruvate.
  - ATP is produced directly by reactions catalyzed by phosphoglycerate kinase and pyruvate kinase.
  - Although NADH produced in the cytosol cannot directly enter mitochondria, reducing equivalents can be shuttled into this organelle, where they generate ATP.
  - Pyruvate can enter mitochondria and be converted to acetyl-CoA, which is oxidized by the tricarboxylic acid (TCA) cycle, generating additional ATP.
  - Pyruvate can also be converted to oxaloacetate (OAA) by a reaction that replenishes the intermediates of the TCA cycle, and it can be reduced to lactate or transaminated to alanine.

A. Transport of glucose into cells

1. Glucose travels across the cell membrane on a **transport protein**.
2. **Insulin stimulates** glucose transport into **muscle** and **adipose** cells by causing glucose transport proteins (GLUT4) within cells to move to the cell membrane (Table 6.3).
3. Insulin does not significantly stimulate the transport of glucose into tissues such as liver, brain, and red blood cells.

table 6.3 Effect of Insulin on Glucose Transport Systems of Various Tissues

Tissues	Insulin Effect on Glucose Transport
Liver	0
Brain	0
Red blood cells	0
Adipose	+
Muscle	+

0 indicates no effect; + indicates stimulation.

CLINICAL CORRELATES **A GLUT1 deficiency** can have serious consequences. The GLUT1 transporter translocates glucose across the blood–brain barrier. When one allele is defective, the rate of glucose entry into the nervous system is insufficient for the cells' needs, leading to seizures, developmental delays, and microcephaly. The treatment consists of a ketogenic diet, one high in fat, in order to produce ketone bodies as an alternative energy source for the nervous system.

B. Reactions of glycolysis (Fig. 6.13)

1. **Glucose** is converted to glucose-6-phosphate in a reaction that uses ATP and produces ADP.
   a. Enzymes: **hexokinase** in all tissues and, in the liver and pancreas, **glucokinase**. Both of these enzymes are subject to regulatory mechanisms.

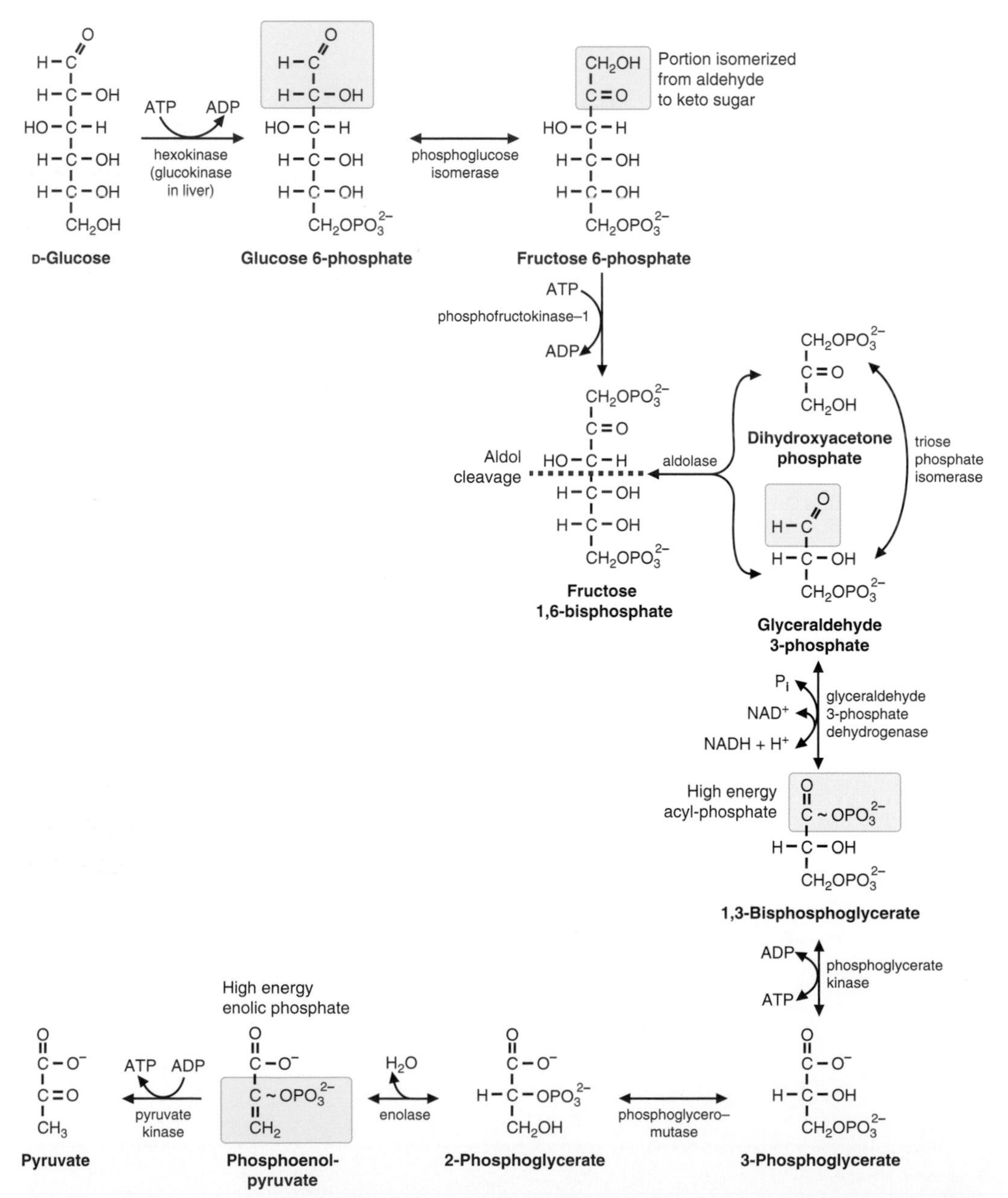

FIGURE 6.13. The reactions of glycolysis. These reactions occur in the cytosol. High-energy phosphates are indicated by the *red squiggles*.

2. **Glucose-6-phosphate** is isomerized to fructose-6-phosphate.
   a. Enzyme: **phosphoglucose isomerase**
3. **Fructose-6-phosphate** is phosphorylated by ATP, forming fructose-1,6-bisphosphate and ADP. This reaction is the first committed step in glycolysis.
   a. Enzyme: **phosphofructokinase 1** (PFK1)
   b. PFK1 is regulated by a number of effectors.
4. **Fructose-1,6-bisphosphate** is cleaved to form the triose phosphates, glyceraldehyde-3-phosphate and dihydroxyacetone phosphate (DHAP).
   a. Enzyme: **aldolase**
5. **DHAP** is isomerized to glyceraldehyde-3-phosphate.
   a. Enzyme: **triose phosphate isomerase**
      (1) *Note: The net result of reactions 1 through 5 is that 2 moles of glyceraldehyde-3-phosphate are formed from 1 mole of glucose, at the expense of two high-energy bonds.*
6. **Glyceraldehyde-3-phosphate** is oxidized by NAD^+ and reacts with inorganic phosphate to form 1,3-bisphosphoglycerate and NADH + H^+.
   a. Enzyme: **glyceraldehyde-3-phosphate dehydrogenase**.
   b. The aldehyde group of glyceraldehyde-3-phosphate is oxidized to a carboxylic acid, which forms a high energy anhydride with inorganic phosphate.
   c. A cysteine residue at the active site is essential for this reaction to proceed.
7. **1,3-Bisphosphoglycerate** reacts with ADP to produce 3-phosphoglycerate and **ATP**.
   a. Enzyme: **phosphoglycerate kinase**
8. **3-Phosphoglycerate** is converted to 2-phosphoglycerate by transfer of the phosphate group from carbon 3 to carbon 2.
   a. Enzyme: **phosphoglyceromutase**
9. **2-Phosphoglycerate** is dehydrated to phosphoenolpyruvate (PEP), which creates a high-energy enol phosphate.
   a. Enzyme: **enolase**
10. **Phosphoenolpyruvate** reacts with ADP to form **pyruvate** and **ATP** in the last reaction of glycolysis.
    a. Enzyme: **pyruvate kinase**
    Pyruvate kinase is more active in the fed state than in the fasting state.
11. The summary of the reactions of the glycolytic pathway is that glucose + 2 NAD^+ + 2 P_i + 2 ADP yields 2 pyruvate + 2 NADH + 4 H^+ + 2 ATP + 2 H_2O.

CLINICAL CORRELATES **Deficiency of pyruvate kinase** causes decreased production of ATP from glycolysis. Red blood cells have insufficient ATP for their membrane pumps, and a **hemolytic anemia** results, although oxygen delivery to tissues is not necessarily affected. As phosphoenolpyruvate accumulates, it is converted to 2-phosphoglycerate, which leads to increased levels of 2,3-bisphosphoglycerate in the red blood cells. The elevated levels of 2,3-bisphosphoglycerate promote oxygen release from hemoglobin in the tissues to an extent that is greater than in the presence of normal 2,3-bisphosphoglycerate levels.

C. **Special reactions in red blood cells**
1. In red blood cells, 1,3-bisphosphoglycerate can be converted to **2,3-bisphosphoglycerate**, a compound that decreases the affinity of hemoglobin for oxygen.
2. 2,3-Bisphosphoglycerate is dephosphorylated to form inorganic phosphate and 3-phosphoglycerate, an intermediate that reenters the glycolytic pathway.

D. **Regulatory enzymes of glycolysis (Fig. 6.14)**
1. **Hexokinase** is found in most tissues, and is geared to provide glucose-6-phosphate for ATP production even when blood glucose levels are low.
   a. Hexokinase has a **low** K_m for glucose (about 0.1 mM). Therefore, it is working near its maximum rate (V_{max}), even at fasting blood glucose levels (about 5 mM).

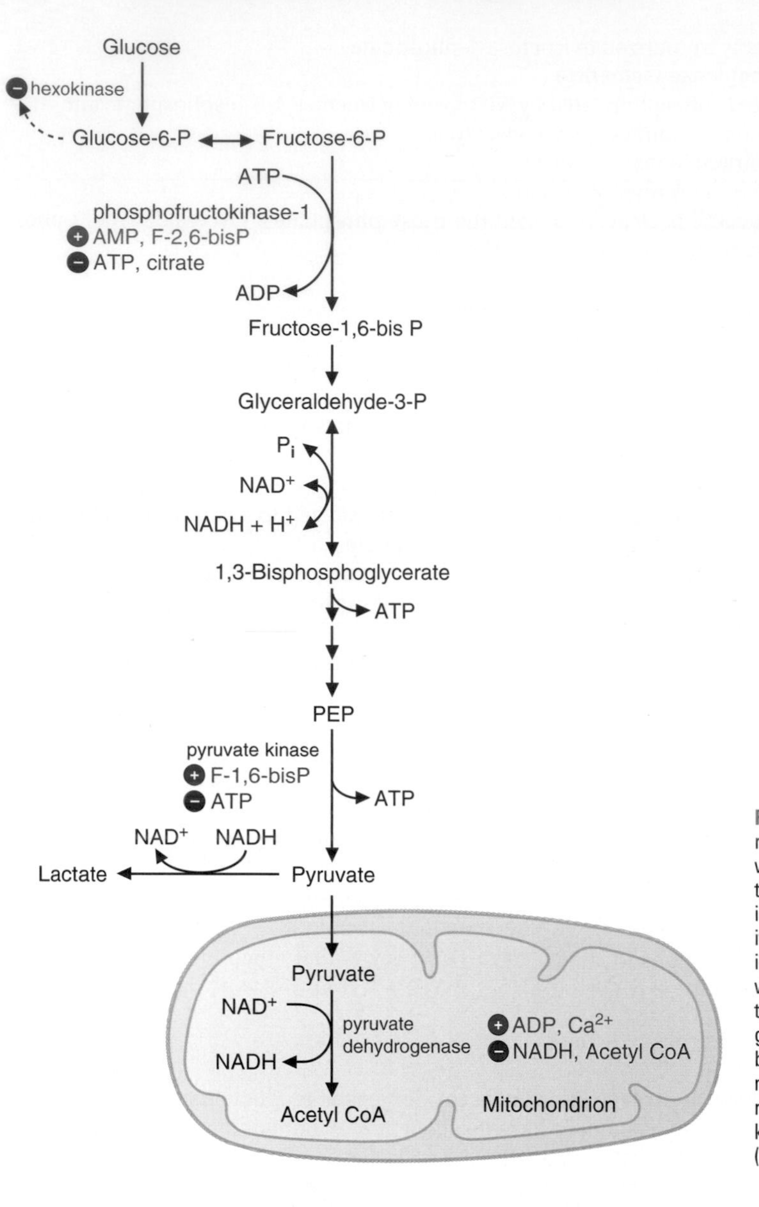

FIGURE 6.14. The major sites of regulation in the glycolytic pathway. Hexokinase and PFK1 are the major regulatory enzymes in skeletal muscle. The activity of pyruvate dehydrogenase in the mitochondria determines whether pyruvate is converted to lactate or acetyl-CoA. In liver, glucokinase, PFK1 (activated by F-2,6-P), and pyruvate kinase are the key enzymes. The regulation shown for pyruvate kinase occurs only for the liver (L) isozyme.

b. Hexokinase is **inhibited** by its product, **glucose-6-phosphate**. Therefore, it is most active when glucose-6-phosphate is being rapidly utilized.

2. **Glucokinase** is found in the **liver and pancreas** and functions at a significant rate only after a meal.
   a. Glucokinase has a **high K_m** for glucose (about 6 mM). Therefore, it is very **active after a meal** when glucose levels in the hepatic portal vein are high, and it is relatively inactive during fasting when glucose levels are low.
   b. Glucokinase is **induced** when insulin levels are high.
   c. Glucokinase is not inhibited by its product, glucose-6-phosphate, at physiologic concentrations.
   d. Glucokinase is regulated by a glucokinase regulatory protein, which binds to glucokinase at low glucose concentrations and sequesters glucokinase in the nucleus. When glucose levels increase, the glucokinase is brought back to the cytoplasm and released from the regulatory protein.
3. **PFK1** is regulated by several factors. It functions at a rapid rate in the liver when blood glucose levels are high or in cells such as muscle when there is a need for ATP.

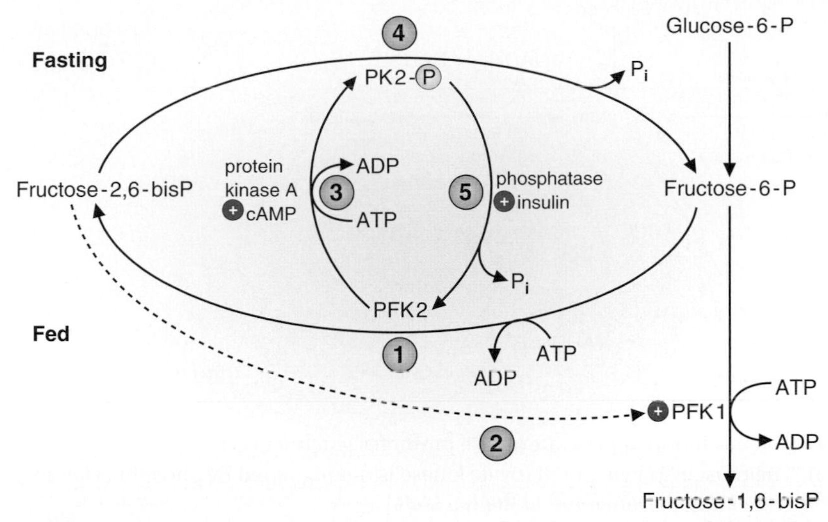

FIGURE 6.15. The regulation of F-2,6-P levels in the liver. F-2,6-P is an activator of phosphofructokinase 1 (PFK1), which converts fructose-6-phosphate to fructose 1,6-bisphosphate. Phosphofructokinase 2 (PFK2) acts as a kinase in the fed state and as a phosphatase during fasting. It regulates the cellular levels of fructose 2,6-bisphosphate. The *circled numbers* correspond with (1) to (5) under Section V D 3 a in the text.

a. PFK1 is **activated by fructose 2,6-bisphosphate (F-2,6-P)**, an important regulatory mechanism in the liver (Fig. 6.15).

(1) After a meal, F-2,6-P is formed from fructose-6-phosphate by **phosphofructokinase 2 (PFK2)**.

(2) F-2,6-P activates PFK1, and **glycolysis** is **stimulated**. The liver is using glycolysis to produce fatty acids for triacylglycerol synthesis.

(3) **In the fasting state** (when glucagon is elevated), **PFK2** is phosphorylated by **protein kinase A**, which is activated by cAMP.

(4) Phosphorylated PFK2 converts F-2,6-P to fructose-6-phosphate. F-2,6-P levels fall, and **PFK1** is **less active.**

(5) **In the fed state, insulin** causes **phosphatases** (such as PP-1) to be stimulated. A phosphatase dephosphorylates PFK2, causing it to become more active in forming F-2,6-P from fructose-6-phosphate. F-2,6-P levels rise, and **PFK1** is **more active.**

(6) Thus, **PFK2** acts as a **kinase** (in the **fed state** when it is dephosphorylated) and as a **phosphatase** (in the **fasting state** when it is phosphorylated). PFK2 catalyzes two different reactions.

(7) The muscle isozyme of PFK2 is not regulated by phosphorylation, although the heart isozyme is, and in the heart the kinase activity of PFK2 is activated upon phosphorylation (the opposite of what occurs in the liver).

b. PFK1 is **activated by AMP**, an important regulatory mechanism in **muscle** (see Fig. 6.14).

(1) In muscle during **exercise**, AMP levels are high and ATP levels are low.

(2) Glycolysis is promoted by a more active PFK1, and ATP is generated.

c. PFK1 is **inhibited** by **ATP** and **citrate**, the important regulatory mechanisms in **muscle.**

(1) When ATP is high, the cell does not need ATP, and glycolysis is inhibited.

(2) High levels of citrate indicate that adequate amounts of substrate are entering the TCA cycle, and that intramitochondrial levels of NADH and ATP are high. Therefore, glycolysis slows down.

4. Pyruvate kinase

a. Pyruvate kinase is **activated** by **fructose 1,6-bisphosphate** and **inhibited** by **alanine** and by **phosphorylation** in the liver **during fasting** when glucagon levels are high (see Fig. 6.14).

(1) Glucagon via cAMP activates **protein kinase A**, which phosphorylates and inactivates pyruvate kinase.

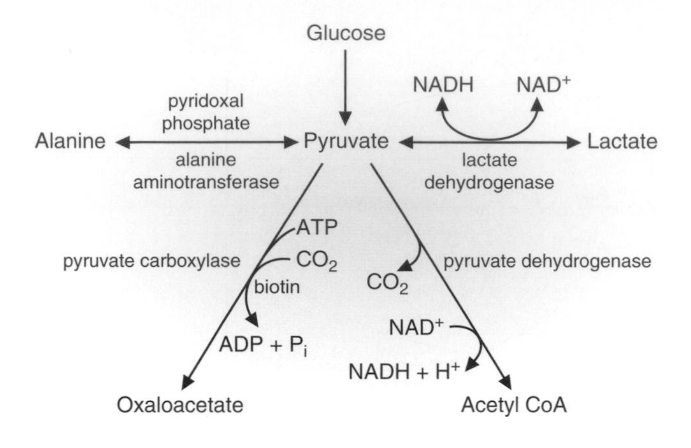

FIGURE 6.16. The fate of pyruvate.

(2) The inhibition of pyruvate kinase promotes gluconeogenesis.

(3) The muscle isozyme of pyruvate kinase is not regulated by phosphorylation.

b. Pyruvate kinase is **activated in the fed state.**

(1) Insulin stimulates phosphatases that dephosphorylate and activate pyruvate kinase.

E. The fate of pyruvate (Fig. 6.16)

1. Conversion to lactate

a. Pyruvate can be reduced in the cytosol by NADH, forming **lactate**, and regenerating NAD^+. The enzyme is lactate dehydrogenase (LDH).

(1) **NADH**, which is produced by glycolysis, **must be reconverted to NAD^+** so that carbons of glucose can continue to flow through glycolysis. This is particularly important under **anaerobic conditions.**

CLINICAL CORRELATES An increase of lactate levels in the blood causes an **acidosis (lactic acidosis).** This condition can result from **hypoxia** or **alcohol ingestion**. Lack of oxygen slows down the electron transport chain, resulting in increased NADH levels. High NADH levels cause more than normal amounts of pyruvate to be converted to lactate. High NADH levels from alcohol metabolism also cause increased conversion of pyruvate to lactate. **Thiamine deficiency,** which is common in alcoholics, decreases pyruvate dehydrogenase activity, causing pyruvate to accumulate and form lactate. Thiamine deficiency also slows down the TCA cycle at the α-ketoglutarate dehydrogenase step. This and other conditions that slow down the TCA cycle can also produce a **lactic acidosis**.

(2) **LDH** converts pyruvate to lactate. LDH consists of four subunits that can be either of the muscle (M) or the heart (H) type.

(a) Five isozymes occur (MMMM, MMMH, MMHH, MHHH, and HHHH), which can be separated by electrophoresis.

(b) Different tissues have different mixtures of these isozymes.

(3) Lactate is released by tissues (e.g., red blood cells or exercising muscle) and is used by the liver for gluconeogenesis or by tissues such as the heart and kidney where it is converted to pyruvate and oxidized for energy.

(4) The **LDH** reaction is **reversible.**

2. Conversion to acetyl-CoA

a. Pyruvate can enter mitochondria and be converted by **pyruvate dehydrogenase** to acetyl-CoA, which can enter the TCA cycle.

3. Conversion to oxaloacetate

a. Pyruvate can be converted to oxaloacetate by **pyruvate carboxylase**, an enzyme found in tissues such as the liver and brain, and recently been shown to be present in muscle as well.

b. This reaction serves to replenish the intermediates of the TCA cycle.

4. Conversion to alanine

a. Pyruvate can be transaminated to form the amino acid **alanine**.

F. Generation of ATP by glycolysis

1. Production of ATP and NADH in the glycolytic pathway

a. Overall, when 1 mole of glucose is converted to 2 moles of pyruvate, 2 moles of ATP are used in the process, and 4 moles of ATP are produced, for a net yield of 2 moles of ATP. In addition, 2 moles of cytosolic NADH are generated.

2. Energy generated by conversion of glucose to lactate

a. If the NADH generated by glycolysis is used to reduce pyruvate to lactate, the net yield is 2 moles of ATP per mole of glucose converted to lactate.

3. Energy generated by conversion of glucose to CO_2 and H_2O (Fig. 6.17)

a. When glucose is oxidized completely to CO_2 and H_2O, approximately 30 or 32 moles of ATP are generated.

(1) Two moles of ATP and 2 moles of NADH are generated from the conversion of 1 mole of glucose to 2 moles of pyruvate.

(2) The 2 moles of pyruvate enter the mitochondria and are converted to 2 moles of acetyl-CoA, producing 2 moles of NADH, which generate approximately 5 moles of ATP by oxidative phosphorylation.

(3) The 2 moles of acetyl-CoA are oxidized in the TCA cycle, generating approximately 20 moles of ATP.

(4) NADH, produced in the cytosol by glycolysis, cannot directly cross the mitochondrial membrane. Therefore, the electrons are passed to the mitochondrial electron transport chain by two shuttle systems.

(a) Glycerol phosphate shuttle (Fig. 6.18)

1. Cytosolic DHAP is reduced to glycerol-3-phosphate by NADH.

2. Glycerol-3-phosphate reacts with an FAD-linked dehydrogenase in the inner mitochondrial membrane. DHAP is regenerated and reenters the cytosol.

3. Each mole of $FADH_2$ that is produced generates approximately 1.5 moles of ATP via oxidative phosphorylation.

4. Because glycolysis produces 2 moles of NADH per mole of glucose, approximately **3 moles of ATP are produced by this shuttle.**

(b) Malate aspartate shuttle (see Fig. 6.19)

1. Cytosolic oxaloacetate is reduced to malate by NADH. The reaction is catalyzed by cytosolic malate dehydrogenase.

2. Malate enters the mitochondrion and is reoxidized to oxaloacetate by the mitochondrial malate dehydrogenase, generating NADH in the matrix.

3. Oxaloacetate cannot cross the mitochondrial membrane. In order to return carbon to the cytosol, oxaloacetate is transaminated to aspartate, which can be

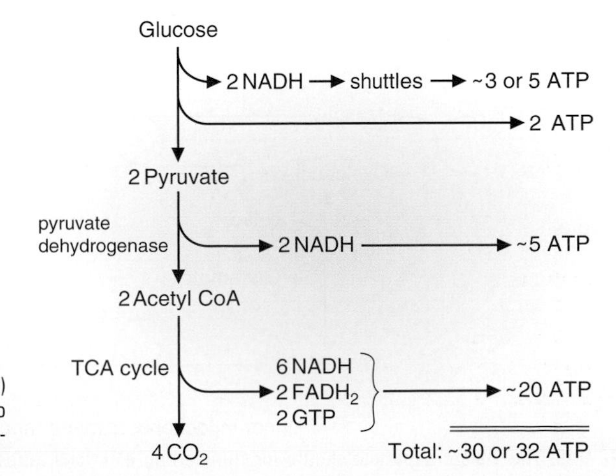

FIGURE 6.17. Adenosine triphosphate (ATP) produced by the conversion of glucose to CO_2. The ATP produced by oxidative phosphorylation is approximate (indicated by ~).

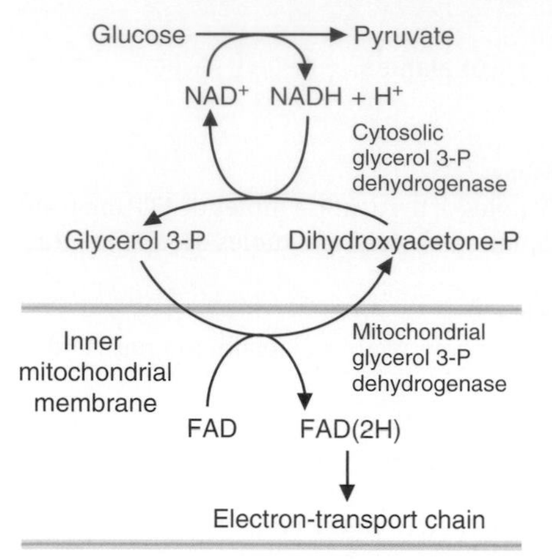

FIGURE 6.18. The glycerol-3-phosphate shuttle for transferring electrons across the inner mitochondrial membrane. DHAP, dihydroxyacetone phosphate.

transported into the cytosol and reconverted to oxaloacetate by another transamination reaction.

4. Note that the transporter for malate requires the opposite movement of α-ketoglutarate; for aspartate the opposite movement of glutamate is required. These are exchange transporters across the inner mitochondrial membrane.
5. In the mitochondrial matrix, each mole of **NADH** generates approximately 2.5 moles of ATP via oxidative phosphorylation.
6. Because glycolysis produces 2 moles of NADH per mole of glucose, approximately **5 moles of ATP are produced by this shuttle**.

4. Maximal ATP production

a. Overall, when 1 mole of glucose is oxidized to CO_2 and H_2O, approximately 30 moles of ATP are produced if the glycerol phosphate shuttle is used, or 32 moles if the malate aspartate shuttle is used.

VI. GLUCONEOGENESIS (FIG. 6.20)

- Gluconeogenesis, which occurs mainly in the liver, is the synthesis of glucose from compounds that are not carbohydrates.
- The major precursors for gluconeogenesis are lactate, amino acids (which form pyruvate or TCA cycle intermediates), and glycerol (which forms DHAP). Even-chain length fatty acids do not produce any net glucose.

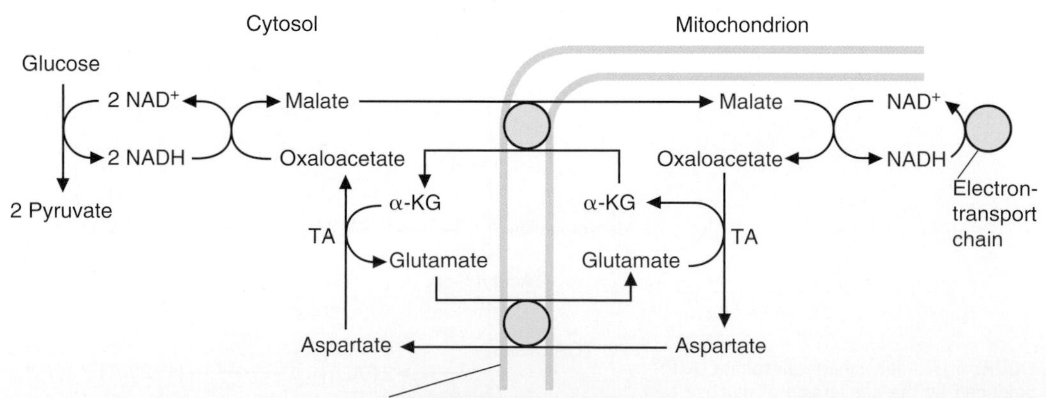

FIGURE 6.19. The malate aspartate shuttle for transferring electrons across the inner mitochondrial membrane. The process is described in the text. α-KG, α-ketoglutarate.

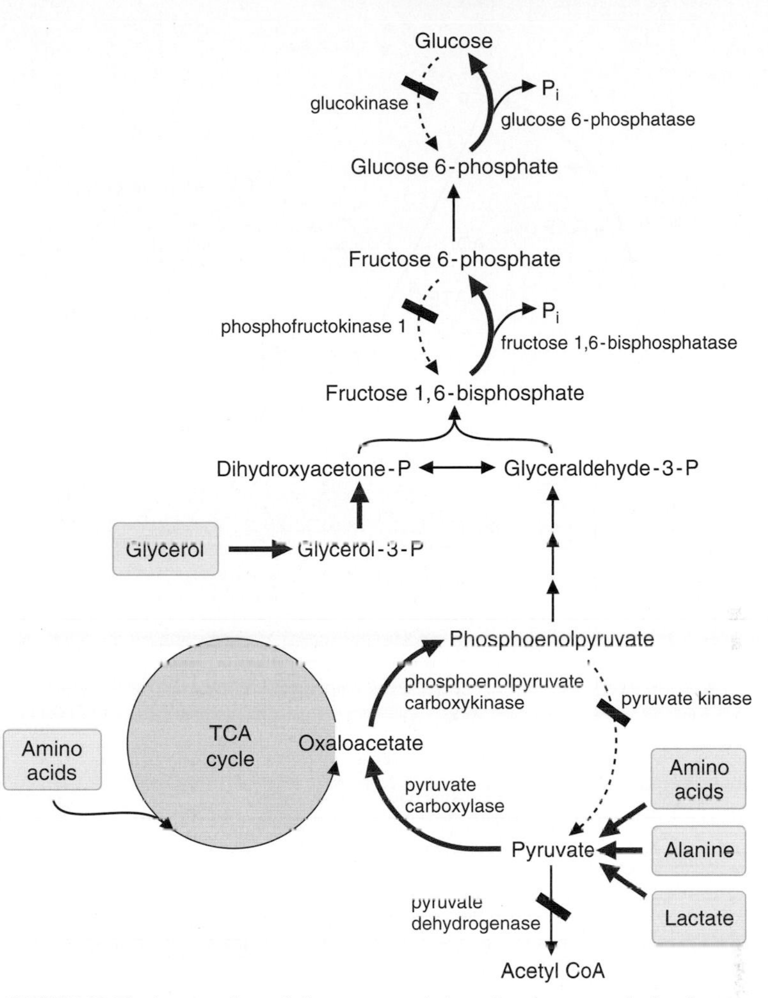

FIGURE 6.20. The key reactions of gluconeogenesis from the precursors alanine, lactate, and glycerol. *Heavy arrows* indicate steps that differ from those of glycolysis. *Broken arrows* are glycolytic reactions that are inhibited (**I**) under conditions in which gluconeogenesis is occurring.

- Gluconeogenesis involves several enzymatic steps that do not occur in glycolysis; thus, glucose is not generated by a simple reversal of glycolysis.
- Pyruvate carboxylase converts pyruvate to oxaloacetate in the mitochondrion. Oxaloacetate is converted to malate or aspartate, which travels to the cytosol and is reconverted to oxaloacetate.
- Phosphoenolpyruvate carboxykinase converts oxaloacetate to phosphoenolpyruvate. Phosphoenolpyruvate forms fructose 1,6-bisphosphate by reversal of the steps of glycolysis.
- Fructose 1,6-bisphosphatase converts fructose 1,6-bisphosphate to fructose-6-phosphate, which is converted to glucose-6-phosphate.
- Glucose-6-phosphatase converts glucose-6-phosphate to free glucose, which is released into the blood.
- Gluconeogenesis occurs under conditions in which pyruvate dehydrogenase, pyruvate kinase, PFK1, and glucokinase are relatively inactive. The low activity of these enzymes prevents futile cycles from occurring and ensures that, overall, pyruvate is converted to glucose.
- The synthesis of 1 mole of glucose from 2 moles of lactate requires energy equivalent to about 6 moles of ATP.

A. Reactions of gluconeogenesis

1. Conversion of pyruvate to phosphoenolpyruvate (Fig. 6.21)

a. In the liver, pyruvate is converted to phosphoenolpyruvate.

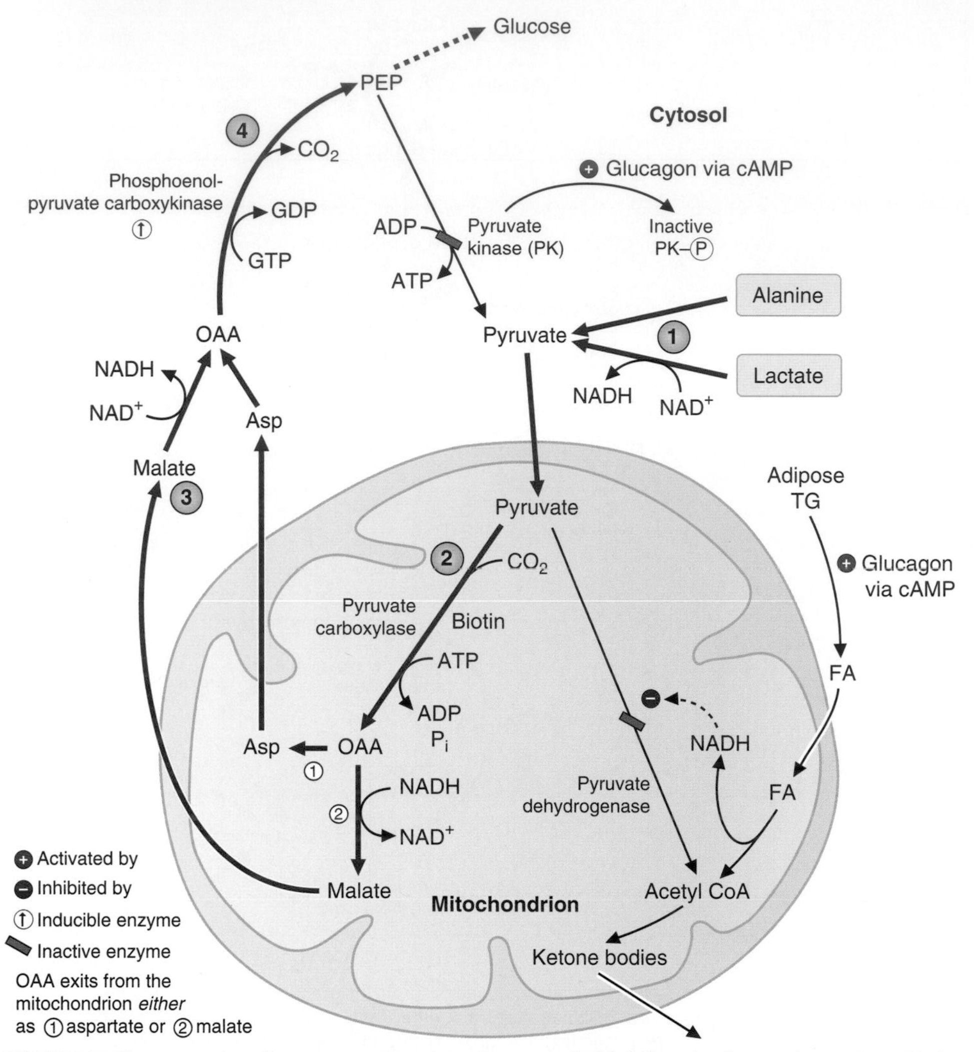

FIGURE 6.21. The conversion of pyruvate to phosphoenolpyruvate (PEP). Follow the diagram by starting with the precursors alanine and lactate. FA, fatty acid; OAA, oxaloacetate; TG, triacylglycerol.

(1) Pyruvate (produced from lactate, alanine, and other amino acids) is first converted to oxaloacetate by **pyruvate carboxylase**, a mitochondrial enzyme that requires biotin and ATP.

(a) Oxaloacetate cannot directly cross the inner mitochondrial membrane. Therefore, it is converted to malate or to aspartate, which can cross the mitochondrial membrane and be reconverted to oxaloacetate in the cytosol.

(2) Oxaloacetate is decarboxylated by **phosphoenolpyruvate carboxykinase** to form phosphoenolpyruvate. This reaction requires GTP.

(3) Phosphoenolpyruvate is converted to fructose 1,6-bisphosphate by reversal of the glycolytic reactions (Fig. 6.22).

2. Conversion of fructose 1,6-bisphosphate to fructose-6-phosphate (see Fig. 6.22)

a. Fructose-1,6-bisphosphate is converted to fructose-6-phosphate in a reaction that releases inorganic phosphate and is catalyzed by fructose-**1,6-bisphosphatase**.

b. Fructose-6-phosphate is converted to glucose 6- phosphate by the same isomerase used in glycolysis.

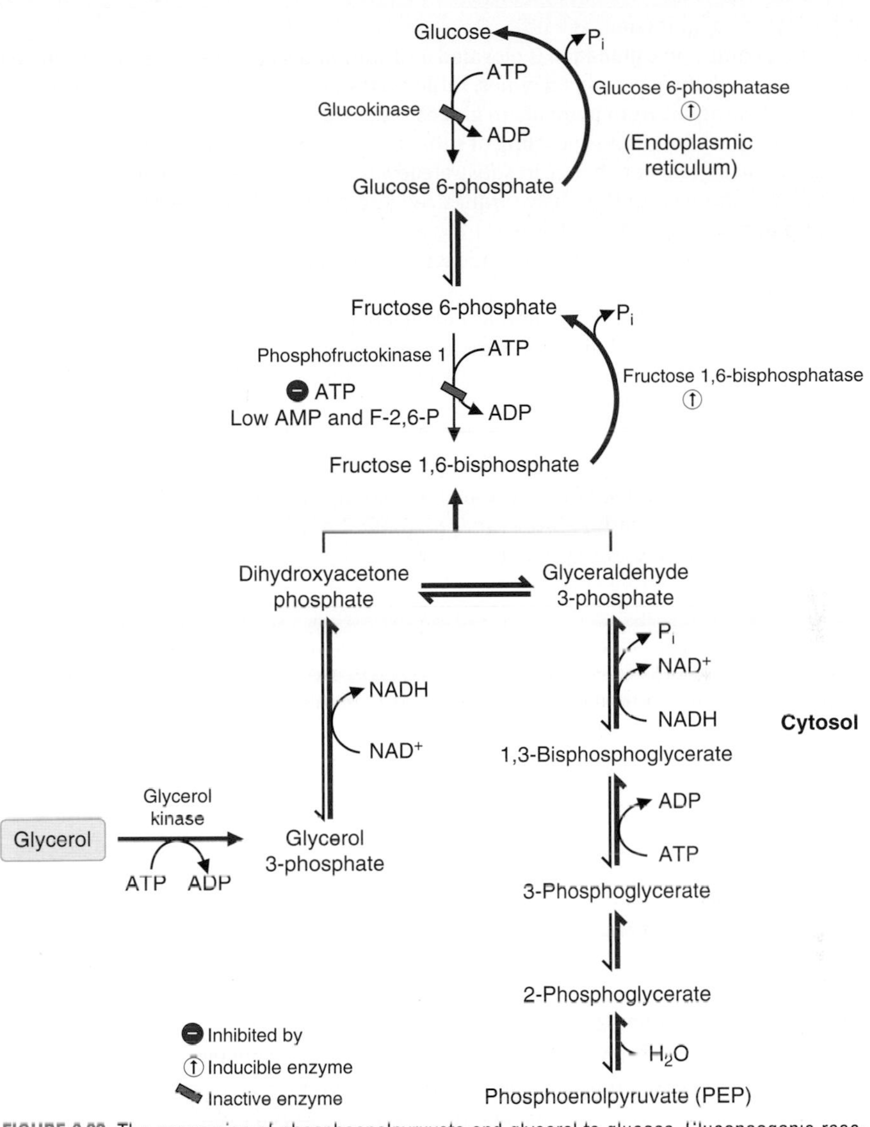

FIGURE 6.22. The conversion of phosphoenolpyruvate and glycerol to glucose. Gluconeogenic reactions are indicated by the *red arrows*. F-2,6-P, fructose-2,6-bisphosphate.

3. Conversion of glucose-6-phosphate to glucose

   a. Glucose-6-phosphate releases inorganic phosphate, which produces free glucose that enters the blood. The enzyme is **glucose 6-phosphatase**.

   b. Glucose-6-phosphatase is involved both in **gluconeogenesis and glycogenolysis** (see Fig. 6.12).

CLINICAL CORRELATES **Hypoglycemia (low blood sugar)** is caused by the inability of the liver to maintain blood glucose levels. It can result from excessive insulin, excessive cellular uptake of glucose, or an impairment of glycogenolysis or gluconeogenesis. Hypoglycemia is caused by liver disease, insulin-secreting tumors, and administration of inappropriately high doses of insulin or sulfonylureas. **Excessive alcohol ingestion** also can cause hypoglycemia. Metabolism of alcohol increases levels of NADH in the liver, which inhibit gluconeogenesis. Extremely rare mutations have been found in both phosphoenolpyruvate carboxykinase and fructose 1,6-bisphosphatase. Mutations in either enzyme will lead to hypoglycemia; inherited mutations in phosphoenolpyruvate carboxykinase will lead to an early death.

B. Regulatory enzymes of gluconeogenesis

1. Under fasting conditions, **glucagon** is elevated and stimulates gluconeogenesis. Because of the changes in the activity of certain enzymes, futile cycles are prevented from occurring, and the overall flow of carbon is from pyruvate to glucose (see Figs. 6.21 and 6.22).
2. A futile cycle is the continuous recycling of substrates and products with the net consumption of energy and no significant change in substrate levels. This is also known as substrate cycling, as the same substrate is continuously synthesized and degraded (recycled).
   a. **Pyruvate dehydrogenase** (see Fig. 6.21)
      (1) Decreased insulin and increased glucagon stimulate the **release of fatty acids** from adipose tissue.
      (2) **Fatty acids** travel to the liver and **are oxidized**, producing acetyl-CoA, NADH, and ATP, which cause inactivation of pyruvate dehydrogenase.
      (3) Because **pyruvate dehydrogenase** is relatively **inactive**, pyruvate is converted to oxaloacetate, not to acetyl-CoA.
   b. **Pyruvate carboxylase**
      (1) Pyruvate carboxylase, which converts pyruvate to oxaloacetate, is **activated by acetyl-CoA** (which is generated from fatty acid oxidation within the mitochondria).
      (2) Note that pyruvate carboxylase is active in both the fed and fasting states.
   c. **Phosphoenolpyruvate carboxykinase (PEPCK)**
      (1) PEPCK is an **inducible** enzyme.
      (2) **Transcription** of the gene encoding PEPCK is stimulated by binding of proteins (CREB, for cyclic AMP response element–binding protein) that are phosphorylated in response to cAMP and by binding of glucocorticoid–protein complexes to regulatory elements in the gene.
      (3) Increased production of PEPCK mRNA leads to increased translation, resulting in higher PEPCK levels in the cell.
   d. **Pyruvate kinase**
      (1) **Glucagon**, via cAMP and protein kinase A, causes pyruvate kinase to be phosphorylated and **inactivated**.
      (2) Because pyruvate kinase is relatively inactive, phosphoenolpyruvate formed from oxaloacetate is not reconverted to pyruvate but, in a series of steps, forms fructose-1, 6-bisphosphate, which is converted to fructose-6-phosphate.
   e. **Phosphofructokinase 1** (see Fig. 6.22)
      (1) PFK1 is relatively **inactive** because the concentrations of its activators, AMP and F-2,6-P, are low and its inhibitor, ATP, is relatively high.
   f. **Fructose 1,6-bisphosphatase**
      (1) The level of **F-2,6-P**, an inhibitor of fructose 1,6-bisphosphatase, is **low** during fasting. Therefore, fructose 1,6-bisphosphatase is **more active**.
      (2) Fructose 1,6-bisphosphatase is also **induced** in the fasting state.
   g. **Glucokinase**
      (1) **Glucokinase** is relatively **inactive** because it has a **high K_m** for glucose, and under conditions that favor gluconeogenesis, the glucose concentration is low. Therefore, free glucose is not reconverted to glucose-6-phosphate.

C. Precursors for gluconeogenesis

Lactate, amino acids, and glycerol are the major precursors for gluconeogenesis in humans.

1. **Lactate** is oxidized by NAD^+ in a reaction catalyzed by LDH to form pyruvate, which can be converted to glucose (see Fig. 6.21).
   a. The sources of lactate include red blood cells and exercising muscle.
2. **Amino acids** for gluconeogenesis come from degradation of muscle protein.
   a. Amino acids are released directly into the blood from muscle, or carbons from amino acids are converted to alanine and glutamine and released.
      (1) **Alanine** is also formed by transamination of pyruvate that is derived by the oxidation of glucose.
      (2) **Glutamine** is converted to alanine by tissues such as gut and kidney.
   b. Amino acids travel to the liver and provide carbon for gluconeogenesis. Quantitatively, **alanine is the major gluconeogenic amino acid**.
   c. Amino acid **nitrogen** is converted to **urea**.

3. **Glycerol**, which is derived from **adipose** triacylglycerols, reacts with ATP to form glycerol-3-phosphate, which is oxidized to DHAP and converted to glucose (see Fig. 6.22).

D. Role of fatty acids in gluconeogenesis

1. **Even-chain fatty acids**
   - a. Fatty acids are oxidized to acetyl-CoA, which enters the TCA cycle.
   - b. For every two carbons of acetyl-CoA that enter the TCA cycle, two carbons are released as CO_2. Therefore, there is **no net synthesis of glucose from acetyl-CoA.**
   - c. The pyruvate dehydrogenase reaction is irreversible, thus acetyl-CoA cannot be converted to pyruvate.
   - d. Although even-chain fatty acids do not provide carbons for gluconeogenesis, β-oxidation of fatty acids provides **ATP** that drives gluconeogenesis.
2. **Odd-chain fatty acids**
   - a. The three carbons at the carbonyl-end of an odd-chain fatty acid are converted to propionate. **Propionate** enters the TCA cycle as succinyl-CoA, which forms **malate**, an intermediate in glucose formation (see Fig. 6.21).

E. Energy requirements for gluconeogenesis

1. **From pyruvate** (see Figs. 6.21 and 6.22)
   - a. Conversion of pyruvate to oxaloacetate by pyruvate carboxylase requires one ATP.
   - b. Conversion of oxaloacetate to phosphoenolpyruvate by phosphoenolpyruvate carboxykinase requires one GTP (the equivalent of one ATP)
   - c. Conversion of 3-phosphoglycerate to 1,3-bisphosphoglycerate by phosphoglycerate kinase requires one ATP.
   - d. Since 2 moles of pyruvate are required to form 1 mole of glucose, **6 moles of high-energy phosphate are required for the synthesis of 1 mole of glucose.**
2. **From glycerol** (see Fig. 6.22)
   - a. Glycerol enters the gluconeogenic pathway at the DHAP level.
     - (1) Conversion of glycerol to glycerol-3-phosphate, which is oxidized to DHAP, requires one ATP.
     - (2) Since 2 moles of glycerol are required to form 1 mole of glucose, 2 moles of high-energy phosphate are required for the synthesis of 1 mole of glucose.

VII. FRUCTOSE AND GALACTOSE METABOLISM

- Although glucose is the most abundant monosaccharide derived from the diet, fructose and galactose are usually obtained in significant quantities, mainly from sucrose and lactose.
- After fructose and galactose enter the cells, they are phosphorylated on carbon 1 and converted to intermediates in the pathways of glucose metabolism.
- Fructose is metabolized mainly in the liver, where it is converted to fructose-1-phosphate and cleaved to produce DHAP and glyceraldehyde, which is phosphorylated to glyceraldehyde-3-phosphate. These two triose phosphates are intermediates of glycolysis.
- Fructose can be produced from sorbitol, which is generated from glucose.
- Galactose is phosphorylated to galactose-1-phosphate, which reacts with UDP-glucose. The products are glucose-1-phosphate and UDP-galactose, which is epimerized to UDP-glucose. The net result is that galactose is converted to the glucose moieties of UDP-glucose and glucose-1-phosphate, intermediates in the pathways of glucose metabolism.
  - UDP-galactose is used in the synthesis of glycoproteins, glycolipids, and proteoglycans.
  - UDP-galactose reacts with glucose in the mammary gland to form the milk sugar lactose.
- Galactose can be reduced to galactitol.

A. Metabolism of fructose

The major dietary source of fructose is the disaccharide sucrose in table sugar and fruit, but it is also present as the monosaccharide in corn syrup, which is used as a sweetener.

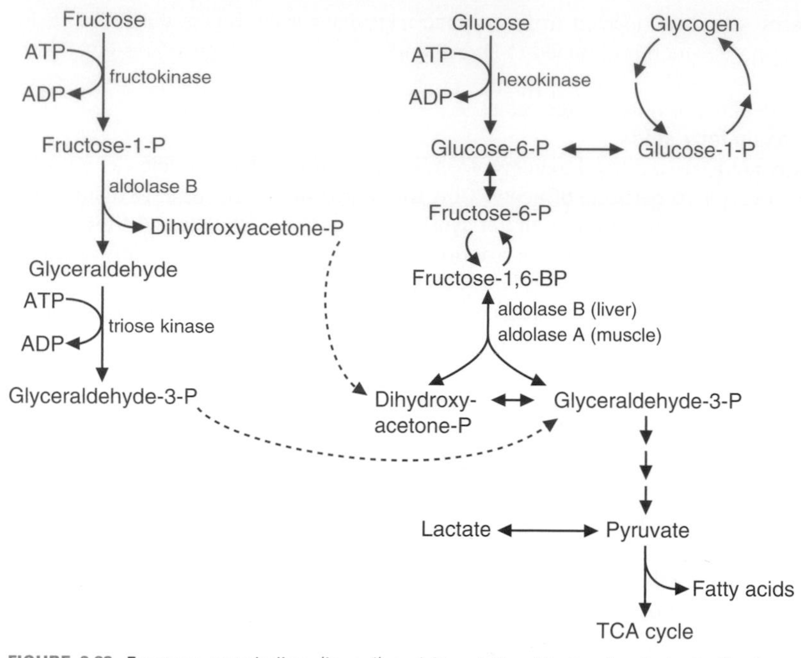

FIGURE 6.23. Fructose metabolism (in red) and its relationship to glycolysis. In the liver, aldolase B cleaves both fructose-1-phosphate F-1-P) in the pathway for fructose utilization and fructose 1,6-bisphosphate in the pathway for glycolysis.

1. **Conversion of fructose to glycolytic intermediates** (Fig. 6.23)
   a. **Fructose** is metabolized mainly in the **liver** where it is converted to pyruvate or, under fasting conditions, to glucose.
      (1) **Fructose** is phosphorylated by ATP to form fructose-1-phosphate. The enzyme is **fructokinase.**
      (2) **Fructose-1-phosphate** is cleaved by **aldolase B** to form DHAP and glyceraldehyde, which is phosphorylated by ATP to form glyceraldehyde-3-phosphate. DHAP and glyceraldehyde-3-phosphate are intermediates of glycolysis. (Aldolase B is the same liver enzyme that cleaves fructose-1,6-bisphosphate in glycolysis.)
   b. In tissues other than liver, the major fate of fructose is phosphorylation by hexokinase to form fructose-6-phosphate, which enters glycolysis. Hexokinase has an affinity for fructose about one-twentieth of that for glucose.
2. **Production of fructose from glucose (the polyol pathway)**
   a. **Glucose** is reduced to sorbitol by **aldose reductase**, which reduces the aldehyde group to an alcohol.
   b. **Sorbitol** is then reoxidized at carbon 2 by sorbitol dehydrogenase to form fructose.
   c. **Fructose**, derived from glucose in seminal vesicles, is the major energy source for sperm cells.

CLINICAL CORRELATES

There are two disorders associated with fructose metabolism. **Fructokinase is deficient** in **essential fructosuria**; therefore, fructose cannot be metabolized as rapidly as normal. Blood fructose levels rise, and fructose appears in the urine. The condition is **benign**. The more serious disorder is hereditary fructose intolerance (HFI). In HFI, a**ldolase B**, the primary liver isozyme of the glycolytic enzyme aldolase, is defective. Under such conditions, aldolase B still functions normally in glycolysis (and other isozymes of aldolase may participate in glycolysis) but not in fructose metabolism (only aldolase B can cleave fructose- 1-phosphate). Fructose-1-phosphate accumulates and inhibits glucose production, causing severe **hypoglycemia** if fructose is ingested. Dietary fructose (found mainly in sucrose) must be avoided.

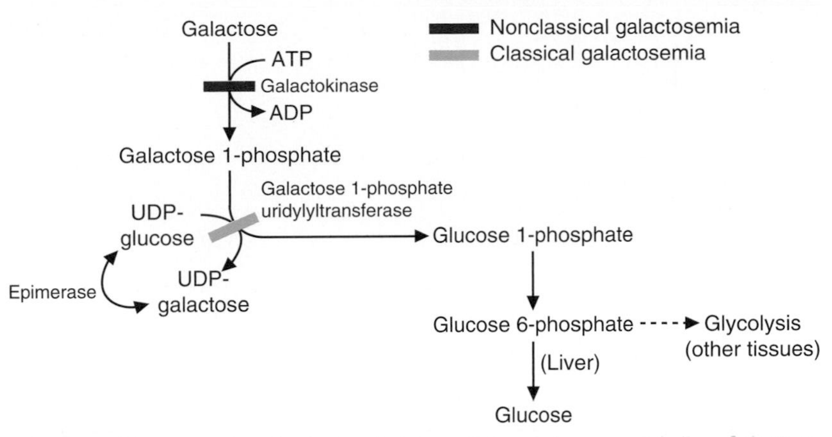

FIGURE 6.24. The conversion of galactose to intermediates of glucose metabolism. Galactose-1-phosphate uridylyl transferase is deficient in classical galactosemia, and galactokinase in nonclassical galactosemia.

B. Metabolism of galactose

The disaccharide **lactose**, found in milk or milk products, is the major dietary source of galactose.

1. **Conversion of galactose** to intermediates of glucose pathways (Fig. 6.24)
   a. **Galactose** is phosphorylated by ATP to galactose-1-phosphate. The enzyme is **galactokinase**.
   b. **Galactose-1-phosphate** reacts with UDP-glucose and forms glucose 1-phosphate and UDP-galactose. The enzyme is **galactose 1-phosphate uridylyl transferase**.
   c. **UDP-galactose** is epimerized to UDP-glucose in a reaction that is readily reversible. The enzyme is **UDP-glucose epimerase**.
   d. Repetition of the above-mentioned three reactions results in conversion of galactose to **UDP-glucose** and **glucose-1-phosphate**.
      (1) In the **liver**, these glucose derivatives are converted to blood glucose during fasting or to glycogen after a meal. In various tissues, the glucose-1-phosphate forms glucose-6-phosphate and feeds into glycolysis.

CLINICAL CORRELATES Disorders associated with galactose metabolism give rise to **galactosemias**. The appearance of high concentrations of galactose in the blood after lactose ingestion may be due to a **galactokinase deficiency** or to a **uridylyl transferase deficiency**. In both conditions, galactose accumulates and is reduced to **galactitol**, which causes **cataracts**. **Uridylyl transferase deficiency** is more severe, causing elevation of galactose-1-phosphate, which inhibits phosphoglucomutase, interfering with glycogen synthesis and degradation. **Hypoglycemia** can occur after ingestion of galactose. Dietary galactose (found mainly in milk and milk products, but also in "artificial sweeteners" and as a "filler" in some medications) must be avoided.

2. **Other fates of UDP-galactose** (Fig. 6.25)
   a. UDP-galactose can be produced either from galactose or from glucose via UDP-glucose and an epimerase.
      (1) UDP-galactose supplies galactose moieties for the synthesis of **glycoproteins, glycolipids**, and **proteoglycans**.
         (a) The enzyme that adds galactose units to the growing polysaccharide chains is **galactosyl transferase**.
      (2) UDP-galactose reacts with glucose in the **lactating mammary gland** to produce the milk sugar **lactose**.
         (a) The modifier protein, **α-lactalbumin**, binds to galactosyl transferase, lowering its K_m for glucose so that glucose adds to galactose (from UDP-galactose), forming lactose.

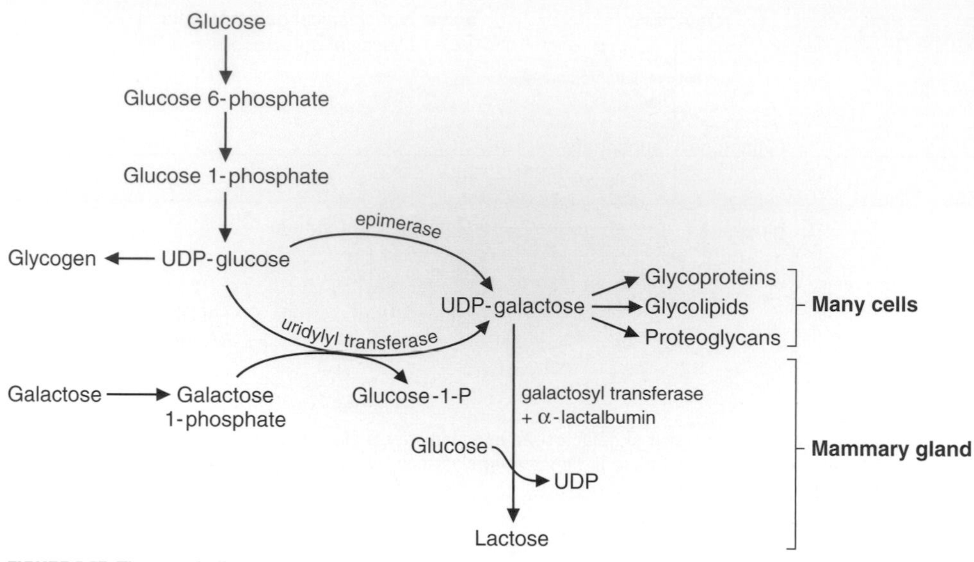

FIGURE 6.25. The metabolism of uridine diphosphate galactose (UDP-galactose). UDP-galactose can be produced from dietary glucose or galactose.

3. **Conversion of galactose to galactitol**
   a. Aldose reductase reduces the aldehyde of galactose to an alcohol, forming galactitol. Galactitol is a C4 epimer of sorbitol.

VIII. PENTOSE PHOSPHATE PATHWAY

- In the irreversible oxidative reactions of the pathway, one carbon of glucose-6-phosphate is released as CO_2; NADPH is generated; and ribulose-5-phosphate is produced (Fig. 6.26).
  - NADPH is used for reductive biosynthesis (particularly of fatty acids) and for protection against oxidative damage (e.g., by reduction of glutathione).
  - Ribulose-5-phosphate provides ribose-5-phosphate for nucleotide biosynthesis or generates pentose phosphates, which enter the nonoxidative portion of the pathway.
- In the reversible nonoxidative reactions, pentose phosphates produced from ribulose-5-phosphate are converted to the glycolytic intermediates fructose-6-phosphate and glyceraldehyde-3-phosphate.
- Because the nonoxidative reactions are reversible, they can be used to generate ribose-5-phosphate for nucleotide synthesis from intermediates of glycolysis.

A. Reactions of the pentose phosphate pathway

1. **The oxidative reactions** (Fig. 6.27)
   a. **Glucose-6-phosphate** is converted to 6-phosphogluconolactone, and $NADP^+$ is reduced to NADPH + H^+.
      (1) Enzyme: **glucose-6-phosphate dehydrogenase**
   b. **6-Phosphogluconolactone** is hydrolyzed to 6-phosphogluconate.
      (1) Enzyme: **gluconolactonase**
   c. **6-Phosphogluconate** undergoes an oxidation, followed by a decarboxylation. CO_2 is released, and a second NADPH + H^+ is generated from $NADP^+$. The remaining carbons form ribulose-5-phosphate.
      (1) Enzyme: **6-phosphogluconate dehydrogenase**

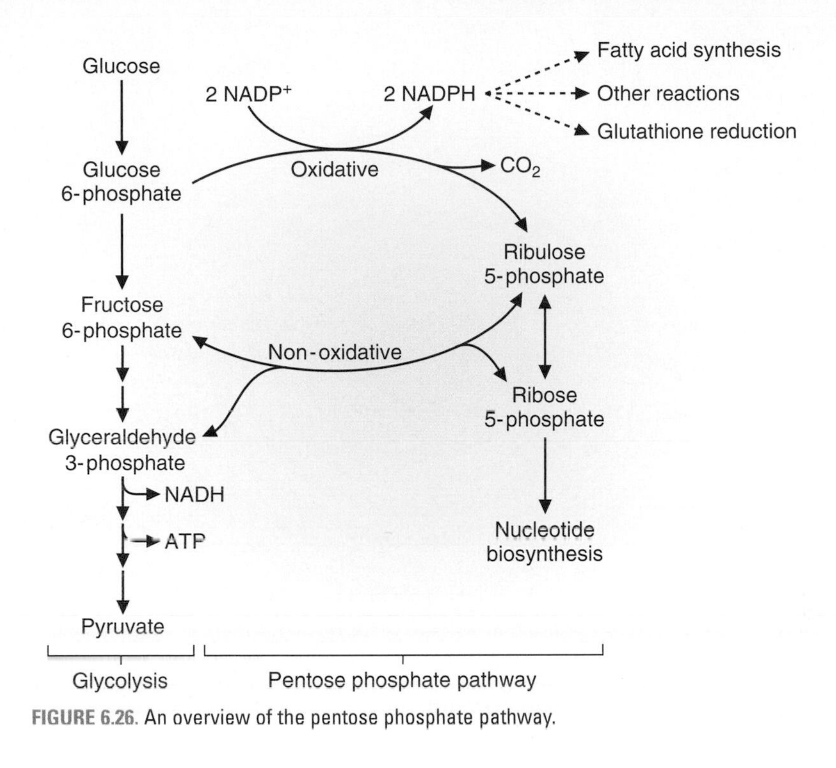

FIGURE 6.26. An overview of the pentose phosphate pathway.

CLINICAL CORRELATES A deficiency of **glucose-6-phosphate dehydrogenase** causes insufficient amounts of NADPH to be produced under certain conditions (e.g., when anti-malarial drugs are being used). As a result, glutathione is not adequately reduced and, in turn, is not available to reduce compounds that are produced by the metabolism of these drugs. Red blood cells lyse, and a **hemolytic anemia** can occur. This is an X-linked disorder, and individuals who inherit this mutation have protection against malaria.

2. **The nonoxidative reactions** (see Fig. 6.26)
   a. **Ribulose-5-phosphate** is isomerized to ribose-5-phosphate or epimerized to **xylulose-5-phosphate**.
   b. **Ribose-5-phosphate** and **xylulose-5-phosphate** undergo reactions, catalyzed by **transketolase** and **transaldolase**, that transfer carbon units, ultimately forming fructose 6-phosphate and glyceraldehyde-3-phosphate.
      (1) **Transketolase**, which requires **thiamine pyrophosphate**, transfers two-carbon units.
      (2) **Transaldolase** transfers three-carbon units.
3. **Overall reactions of the pentose phosphate pathway** (Fig. 6.28)

$$3 \text{ Glucose-6-P} + 6 \text{ NADP}^+ \rightarrow 3 \text{ ribulose-5-P} + 3 \text{ CO}_2 + 6 \text{ NADPH}$$

$$3 \text{ Ribulose-5-P} \rightarrow 2 \text{ xylulose-5-P} + \text{ribose-5-P}$$

$$2 \text{ Xylulose-5-P} + \text{ribose-5-P} \rightarrow 2 \text{ fructose-6-P} + \text{glyceraldehyde-3-P}$$

B. Functions of NADPH (see Fig. 6.26)

1. The pentose phosphate pathway produces NADPH for **fatty acid synthesis**. Under these conditions, the fructose-6-phosphate and glyceraldehyde-3-phosphate generated in the pathway reenter glycolysis.

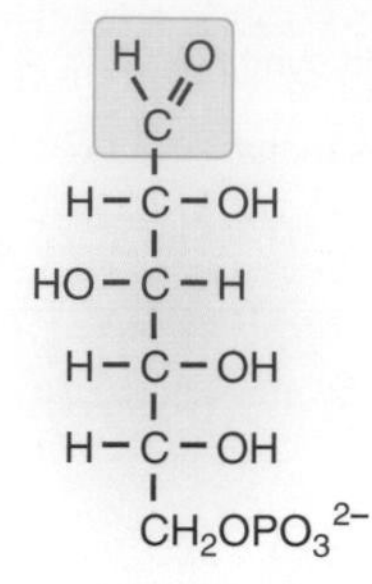

Glucose 6-phosphate

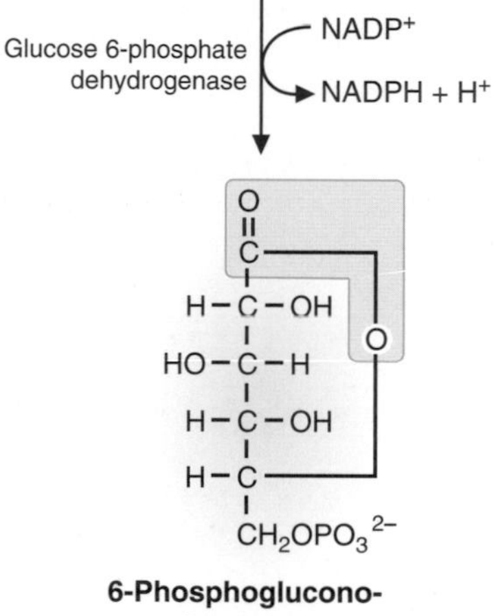

6-Phosphoglucono-δ-lactone

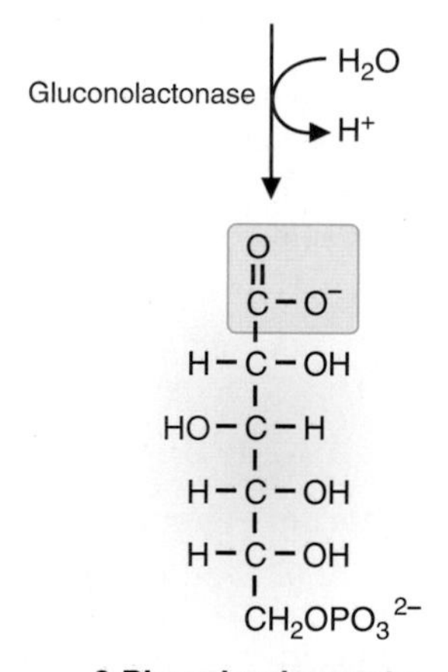

6-Phosphogluconate

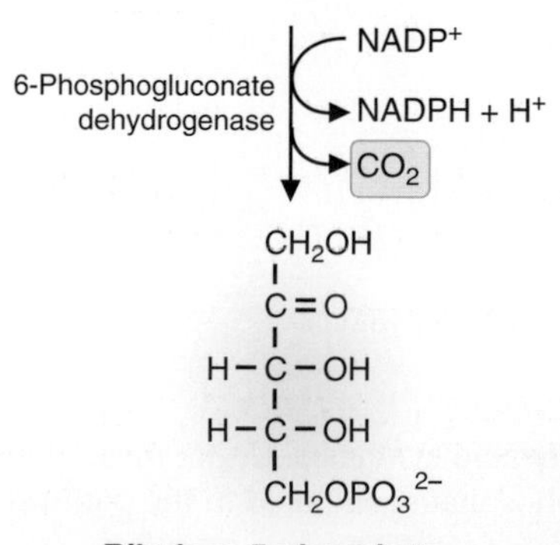

Ribulose 5-phosphate

FIGURE 6.27. The oxidative reactions of the pentose phosphate pathway. These reactions are irreversible. Deficiency of glucose-6-phosphate dehydrogenase can result in hemolytic anemia. Carbon 1 of glucose-6-phosphate is oxidized to an acid and then released as carbon dioxide in an oxidation followed by a decarboxylation. Each of the oxidation steps generates NADPH.

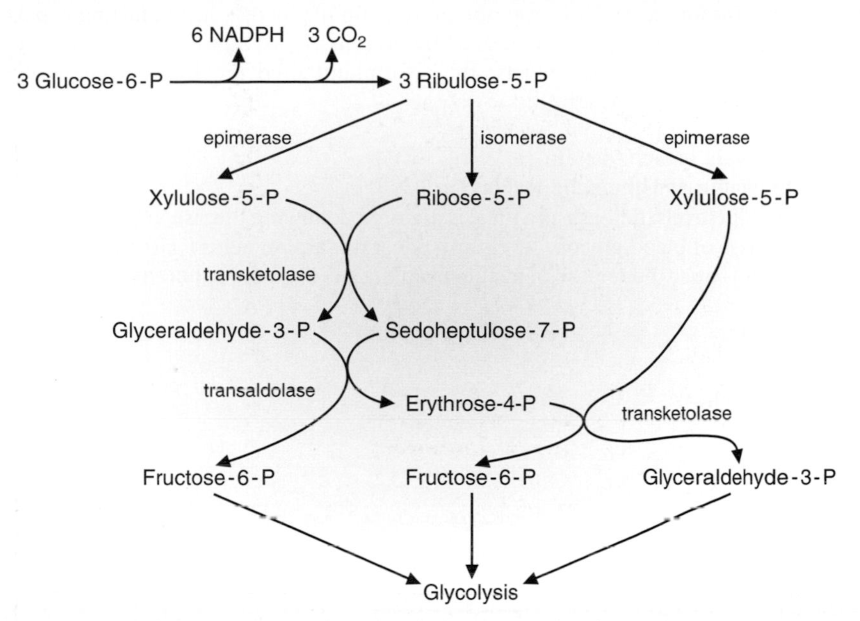

FIGURE 6.28. A balanced sequence of reactions in the pentose phosphate pathway

2. NADPH is also used to **reduce glutathione** (γ-glutamylcysteinylglycine).
   a. Glutathione helps to prevent oxidative damage to cells by reducing hydrogen peroxide (H_2O_2) (see Chapter 4).
   b. Glutathione is also used to transport amino acids across the membranes of certain cells by the γ-glutamyl cycle.

C. **Generation of ribose-5-phosphate (see Fig. 6.26)**
   1. When **NADPH levels are low**, the oxidative reactions of the pathway can be used to generate ribose-5-phosphate for nucleotide biosynthesis.
   2. When **NADPH levels are high**, the reversible nonoxidative portion of the pathway can be used to generate ribose-5-phosphate for nucleotide biosynthesis from fructose-6-phosphate and glyceraldehyde-3-phosphate.

IX. MAINTENANCE OF BLOOD GLUCOSE LEVELS

- Blood glucose levels are maintained within a very narrow range, although the nature of the diet varies widely and the normal person eats periodically during the day and fasts between meals and at night. Even under circumstances when a person does not eat for extended periods of time, blood glucose levels decrease only slowly.
- The major hormones that regulate blood glucose are insulin and glucagon.
- After a meal, blood glucose is supplied by dietary carbohydrate.
- During fasting, the liver maintains blood glucose levels by the processes of glycogenolysis and gluconeogenesis.
  - Within the first few hours of fasting, glycogenolysis is primarily responsible for maintaining blood glucose levels.
  - As a fast progresses and glycogen stores decrease, gluconeogenesis becomes an important additional source of blood glucose.
  - After about 30 hours, when liver glycogen stores are depleted, gluconeogenesis becomes the only source of blood glucose.

- All cells use glucose for energy; however, the production of glucose during fasting is particularly important for tissues such as the brain and red blood cells.
- During exercise, blood glucose levels are also maintained by liver glycogenolysis and gluconeogenesis.

A. Blood glucose levels in the fed state

1. Changes in insulin and glucagon levels (Fig. 6.29)

a. Blood insulin levels increase as a meal is digested, following the rise in blood glucose levels.

(1) Increases of blood glucose levels and of certain amino acids (particularly arginine and leucine) cause the release of insulin from the β cells of the **pancreas**.

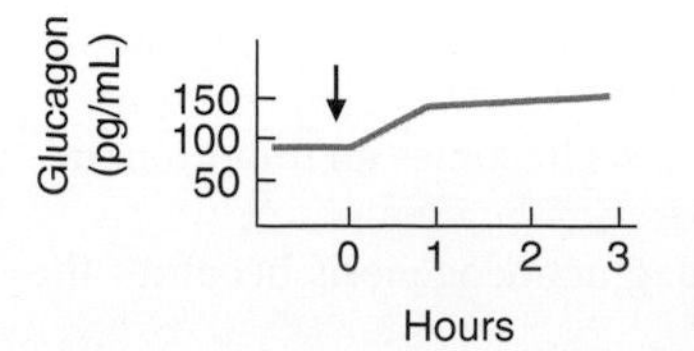

FIGURE 6.29. Changes in blood glucose, insulin, and glucagon levels in response to a glucose or a protein meal.

b. Blood glucagon levels change depending on the content of the meal. A high-carbohydrate meal causes glucagon levels to decrease. A high-protein meal causes glucagon to increase (see Fig. 6.29).

(1) On a normal mixed diet, glucagon will remain relatively constant after a meal while insulin increases.

2. Fate of dietary glucose in the liver

a. Glucose is **oxidized** for energy. Excess glucose is converted to **glycogen** and to the **triacylglycerols** of very low density lipoprotein (**VLDL**).

(1) The enzyme **glucokinase** has a **high K_m** for glucose (about 6 mM), thus its velocity increases after a meal when glucose levels are elevated. On a high-carbohydrate diet, glucokinase is **induced**.

(2) **Glycogen synthesis** is promoted by insulin, which leads to the activation of PP-1, the phosphatase that dephosphorylates and activates glycogen synthase.

(3) The **synthesis of triacylglycerols** is also stimulated. The triacylglycerols are converted to VLDL and released into the blood.

3. Fate of dietary glucose in peripheral tissues

a. All cells oxidize glucose for energy.

(1) Insulin stimulates the **transport** of glucose into **adipose** and **muscle** cells.

(2) In **muscle**, insulin stimulates the synthesis of **glycogen**.

(3) Adipose cells convert glucose to the **glycerol** moiety for synthesis of triacylglycerols.

4. Return of blood glucose to fasting levels

a. The **uptake of dietary glucose** by tissues (particularly liver, adipose, and muscle) causes blood glucose to decrease.

b. By 2 hours after a meal, blood glucose has returned to the fasting level of 5 mM or 80 to 100 mg/dL.

B. Blood glucose levels in the fasting state (Fig. 6.30)

1. Changes in insulin and glucagon levels

a. During fasting, insulin levels decrease and glucagon levels increase.

b. These hormonal changes promote **glycogenolysis** and **gluconeogenesis** in the liver so that blood glucose levels are maintained.

2. Stimulation of glycogenolysis

a. Within a few hours after a meal, as **glucagon** levels increase, glycogenolysis is stimulated and begins to supply glucose to the blood (see Fig. 6.12).

3. Stimulation of gluconeogenesis

a. By 4 hours after a meal, the liver is supplying glucose to the blood via gluconeogenesis and glycogenolysis (Fig. 6.31).

b. Regulatory mechanisms prevent futile cycles from occurring and promote the conversion of gluconeogenic precursors to glucose (see Figs. 6.21 and 6.22).

4. Stimulation of lipolysis (see Fig. 6.30)

a. During fasting, the **breakdown of adipose triacylglycerols** is stimulated, and fatty acids and glycerol are released into the blood.

b. Fatty acids are **oxidized** by certain tissues and converted to **ketone bodies** by the liver. The ATP and NADH produced by β-oxidation of fatty acids promote gluconeogenesis.

c. Glycerol is a source of carbon for gluconeogenesis in the liver.

5. Relative roles of glycogenolysis and gluconeogenesis in maintaining blood glucose levels (see Fig. 6.31)

a. Glycogenolysis is stimulated as blood glucose falls to the fasting level after a meal. It is the main source of blood glucose for the next 8 to 12 hours.

b. Gluconeogenesis is stimulated within a few (4) hours after a meal, and supplies an increasingly larger share of blood glucose as the fasting state persists.

c. By 20 hours of fasting, **gluconeogenesis and glycogenolysis** are approximately **equal** as sources of blood glucose.

d. As liver glycogen stores become depleted, **gluconeogenesis predominates.**

e. By about 30 hours of fasting, liver glycogen is depleted, and thereafter, **gluconeogenesis** is the **only source** of blood glucose.

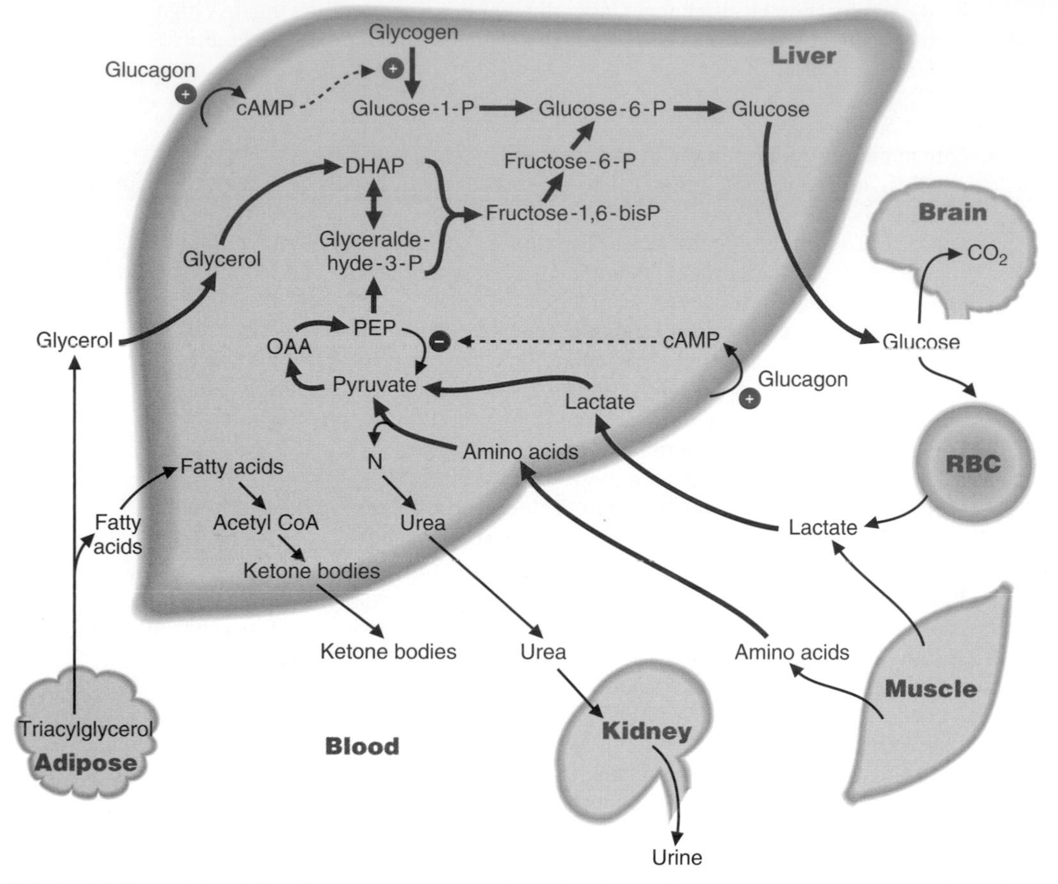

FIGURE 6.30. Tissue interrelationships in glucose production during fasting. Trace the precursors lactate, amino acids, and glycerol to blood glucose.

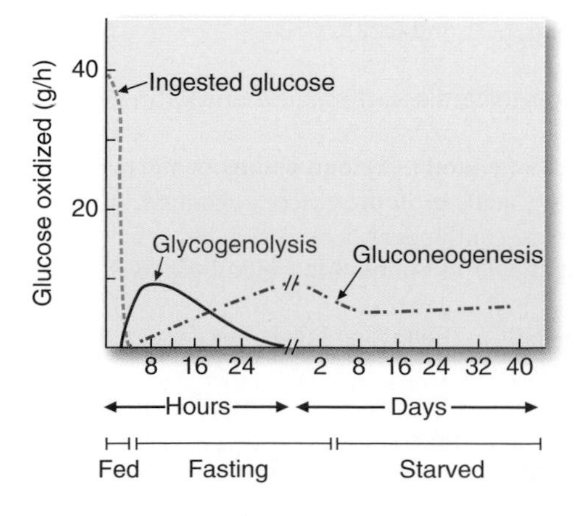

FIGURE 6.31. Sources of blood glucose in fed, fasting, and starved states. Note that the scale changes from hours to days. (Modified from Hanson RW, Mehlman MA. (eds). *Gluconeogenesis: Its Regulation in Mammalian Species.* New York, NY: John Wiley & Sons; 1976:518. Copyright ©1976 by John Wiley & Sons, Inc. Reprinted by permission of John Wiley & Sons, Inc.)

CLINICAL CORRELATES

Diabetes mellitus occurs via a variety of mechanisms. **High blood glucose levels** occur because of either a **deficiency of insulin** (Type 1, formerly insulin-dependent diabetes mellitus [IDDM]) or decreased secretion or an **inability of tissues to respond** to insulin (Type 2, formerly noninsulin-dependent diabetes mellitus [NIDDM]). If diabetes mellitus is untreated, the body responds as if it is starving. Fuel stores are degraded in the face of high blood glucose levels, and **ketoacidosis** may occur, particularly in Type 1. Many metabolic pathways are affected. Exposure of red blood cells to glucose results in glycosylation of hemoglobin. An increase in the **HbA1c** fraction above 6% of the total hemoglobin is an indication that a diabetic patient's blood glucose levels have been elevated during the last 6 to 8 weeks.

Drugs used to treat Type 2 diabetes include sulfonylurea compounds, which increase insulin secretion by the pancreas, and metformin, which functions through the activation of the AMP-activated protein kinase in the liver and muscle. Activation of the AMP-activated protein kinase leads to a reduction of gluconeogenesis, and translocation of GLUT4 transporters in the muscle from intracellular vesicles to the cell surface. The overall effect is a lowering of blood glucose levels, without increasing insulin secretion. Drugs that inhibit intestinal glucoamylase (a brush border enzyme that will split di- and trisaccharides) can also reduce blood glucose levels after a meal. However, under these conditions, flatulence and diarrhea can result owing to the sugars being metabolized by the colonic bacteria (as in lactase deficiency).

Maturity onset of diabetes in the young (MODY) is a group of glucose-regulation disorders that are difficult to classify as either Type 1 or Type 2 diabetes. Many forms of MODY are inherited, and there are now six known genes, which, if mutated, can lead to MODY. A mutation in pancreatic glucokinase and hepatocyte nuclear factor-1-alpha (HNF-1-alpha) are the most commonly mutated genes giving rise to MODY. The symptoms expressed by the individuals vary according to the gene mutated, but have in common elevated blood glucose levels under fasting conditions.

C. Blood glucose levels during prolonged fasting (starvation)

Even after 5 to 6 weeks of starvation, blood glucose levels are still in the range of 65 mg/dL. Changes in fuel utilization by various tissues prevent blood glucose levels from decreasing abruptly during prolonged fasting.

1. The levels of ketone bodies rise in the blood, and the **brain uses ketone bodies** for energy, decreasing its utilization of blood glucose.
2. The rate of **gluconeogenesis** and, therefore, of **urea** production by the liver **decreases**.
3. **Muscle protein is spared.** Less muscle protein is used to provide amino acids for gluconeogenesis.

D. Blood glucose levels during exercise

During exercise, blood glucose levels are maintained by essentially the same mechanisms that are used during fasting.

1. **Use of endogenous fuels**
   a. As the exercising muscle contracts, **ATP** is utilized.
   b. ATP is regenerated initially from **creatine phosphate**.
   c. **Muscle glycogen** is oxidized to produce ATP. AMP activates phosphorylase b, and Ca^{2+}-calmodulin activates phosphorylase kinase. The hormone epinephrine causes the production of cAMP, which stimulates glycogen breakdown. (see Fig. 6.12).
2. **Use of fuels from the blood**
   a. As blood flow to the exercising muscle increases, **blood glucose** and **fatty acids** are taken up and **oxidized by muscle**. An AMP-activated protein kinase in muscle will stimulate glucose uptake by muscle in the absence of insulin.
   b. As blood glucose levels begin to decrease, the **liver**, by the processes of **glycogenolysis** and **gluconeogenesis**, acts to maintain blood glucose levels.

Review Test

Questions 1 to 10 examine your basic knowledge of basic biochemistry and are not in the standard clinical vignette format.

Questions 11 to 35 are clinically relevant, USMLE-style questions.

Basic Knowledge Questions

1. After digestion of a piece of cake that contains flour, milk, and sucrose as its primary ingredients, the major carbohydrate products entering the blood are which one of the following? Choose the one best answer.

(A) Glucose
(B) Fructose and galactose
(C) Galactose and glucose
(D) Fructose and glucose
(E) Glucose, fructose, and galactose

2. The immediate degradation of glycogen under normal conditions gives rise to which one of the following?

(A) More glucose than glucose-1-phosphate
(B) More glucose-1-phosphate than glucose
(C) Equal amounts of glucose and glucose-1-phosphate
(D) Neither glucose nor glucose-1-phosphate
(E) Only glucose-1-phosphate

3. Phosphorylase kinase can be best described by which one of the following?

	Activated in response to:	Target of enzyme activity:	Active in presence of caffeine?	Required substrate for enzymatic activity
A	Insulin	Glycogen phosphorylase	Yes	ATP
B	Insulin	Glycogen phosphorylase	Yes	GTP
C	Insulin	Branching enzyme	No	ATP
D	Epinephrine	Branching enzyme	No	GTP
E	Epinephrine	Glycogen phosphorylase	Yes	ATP
F	Epinephrine	Glycogen phosphorylase	Yes	GTP

4. In an embryo with a complete deficiency of pyruvate kinase, how many net moles of ATP are generated in the conversion of 1 mole of glucose to 1 mole of pyruvate?

(A) 0
(B) 1
(C) 2
(D) 3
(E) 4

5. Which one of the following is a regulatory mechanism employed by muscle for glycolysis?

(A) Inhibition of PFK1 by AMP
(B) Inhibition of hexokinase by its product
(C) Activation of pyruvate kinase when glucagon levels are elevated
(D) Inhibition of aldolase by fructose 1,6-bisphosphate
(E) Inhibition of glucokinase by F-2,6-P

6. Caffeine, a methyl xanthine, has been added to a variety of cell types. Which one of the following would be expected in various cell types treated with caffeine and epinephrine?

(A) Decreased activity of liver protein kinase A
(B) Decreased activity of muscle protein kinase A
(C) Increased activity of liver pyruvate kinase
(D) Decreased activity of liver glycogen synthase
(E) Decreased activity of liver glycogen phosphorylase

7. Which one of the following occurs during the conversion of pyruvate to glucose during gluconeogenesis?

(A) Biotin is required as a cofactor.
(B) The carbon of CO_2, added in one reaction, appears in the final product.
(C) Energy is utilized only in the form of GTP.

(D) All of the reactions occur in the cytosol.
(E) All of the reactions occur in the mitochondrion.

8. A common intermediate in the conversion of glycerol and lactate to glucose is which one of the following?

(A) Pyruvate
(B) Oxaloacetate
(C) Malate
(D) Glucose-6-phosphate
(E) Phosphoenolpyruvate

9. The pentose phosphate pathway generates which one of the following?

(A) NADH, which may be used for fatty acid synthesis.
(B) Ribose-5-phosphate, which may be used for the biosynthesis of ATP.
(C) Pyruvate and fructose 1,6-bisphosphate by the direct action of transaldolase and transketolase.
(D) Xylulose-5-phosphate by one of the oxidative reactions.
(E) Glucose from ribose-5-phosphate and CO_2.

10. Which one of the following metabolites is used by all cells for glycolysis, glycogen synthesis, and the hexose monophosphate shunt pathway?

(A) Glucose-1-phosphate
(B) Glucose-6-phosphate
(C) UDP-glucose
(D) Fructose-6-phosphate
(E) Phosphoenolpyruvate

Board-style Questions

11. A patient has a genetic defect that causes intestinal epithelial cells to produce disaccharidases of much lower activity than normal. Compared with a normal person, after eating a bowl of milk and oatmeal sweetened with table sugar, this patient will have higher levels of which one of the following? Choose the one best answer.

(A) Maltose, sucrose, and lactose in the stool
(B) Starch in the stool
(C) Galactose and fructose in the blood
(D) Glycogen in the muscles
(E) Insulin in the blood

12. An infant, who was nourished by a synthetic formula, had a sugar in the blood and urine. This compound gave a positive reducing-sugar test but was negative when measured with glucose oxidase. Treatment of blood and urine with acid (which cleaves glycosidic bonds) did not increase the amount of reducing sugar measured. Which of the following compounds is most likely to be present in this infant's blood and urine?

(A) Glucose
(B) Fructose
(C) Sorbitol
(D) Maltose
(E) Lactose

13. A 3-year-old girl has been a fussy eater since being weaned, particularly when fruit is part of her diet. She would get cranky, sweat, and display dizziness, and lethargy, after eating a meal with fruit. Her mother noticed this correlation, and as long as fruit was withdrawn from her diet the child did not display such symptoms. The problems the girl exhibits when eating fruit is most likely due to which one of the following?

(A) Decreased levels of fructose in the blood
(B) Elevated levels of glyceraldehyde in liver cells
(C) High levels of sucrose in the stool
(D) Elevated levels of fructose-1-phosphate in liver cells
(E) Decreased levels of fructose in the urine

14. A 1-year-old child, on a routine well child visit, was discovered to have cataract formation in both eyes. Blood work demonstrated elevated galactose and galactitol levels. In order to determine which enzyme might be defective in the child, which intracellular metabolite should be measured?

(A) Galactose
(B) Fructose
(C) Glucose
(D) Galactose-1-phosphate
(E) Fructose-1-phosphate
(F) Glucose-6-phosphate

15. A pregnant woman who has a lactase deficiency and cannot tolerate milk in her diet is concerned that she will not be able to produce milk of sufficient caloric value to nourish her baby. The best advice to her is which one of the following?

(A) She must consume pure galactose in order to produce the galactose moiety of lactose.
(B) She will not be able to breastfeed her baby because she cannot produce lactose.

(C) The production of lactose by the mammary gland does not require the ingestion of milk or milk products.
(D) She can produce lactose directly by degrading α-lactalbumin.
(E) A diet rich in saturated fats will enable her to produce lactose.

16. A chronic alcoholic has recently had trouble with their ability to balance, becomes easily confused, and displays nystagmus. An assay of which of the following enzymes can determine a biochemical reason for these symptoms?

(A) Isocitrate dehydrogenase
(B) Transaldolase
(C) Glyceraldehyde-3-phosphate dehydrogenase
(D) Transketolase
(E) Glucose-6-phosphate dehydrogenase

17. In a glucose tolerance test, an individual in the basal metabolic state ingests a large amount of glucose. If an individual displays a normal response, this ingestion results in which one of the following?

(A) Enhanced glycogen synthase activity in liver
(B) An increased ratio of phosphorylase a to phosphorylase b in the liver
(C) An increased rate of lactate formation by erythrocytes
(D) Inhibition of PP-1 activity in the liver
(E) Increased activity of CREB

18. A 3-month-old infant was cranky and irritable, became quite lethargic between feedings, and began to develop a potbelly. A physical exam demonstrated an enlarged liver, while blood work taken between feedings demonstrated elevated lactate and uric acid levels, as well as hypoglycemia. This child most likely has a mutation in which one of the following enzymes?

(A) Liver glycogen phosphorylase
(B) Glycogen synthase
(C) Glucose 6-phosphatase
(D) Muscle glycogen phosphorylase
(E) Pyruvate kinase

19. A 16-year-old patient with Type 1 diabetes mellitus was admitted to the hospital with a blood glucose level of 400 mg/dL. (The reference range for blood glucose is 80 to 100 mg/dL.) One hour after an insulin infusion was begun, her blood glucose level had decreased to 320 mg/dL. One hour later, it was 230 mg/dL. The patient's glucose level decreased because the infusion of insulin led to which one of the following?

(A) The stimulation of the transport of glucose across the cell membranes of the liver and brain
(B) The stimulation of the conversion of glucose to glycogen and triacylglycerol in the liver
(C) The inhibition of the synthesis of ketone bodies from blood glucose
(D) The stimulation of glycogenolysis in the liver
(E) The inhibition of the conversion of muscle glycogen to blood glucose

Questions 20 and 21 are based on the following case:

A patient presented with a bacterial infection that produced an endotoxin that was found, after extensive laboratory analysis, to inhibit phosphoenolpyruvate carboxykinase.

20. The patient would have very little glucose produced from which one of the following gluconeogenic precursors?

(A) Alanine
(B) Glycerol
(C) Even-chain fatty acids
(D) Phosphoenolpyruvate
(E) Fructose

21. Administration of a high dose of glucagon to this patient 2 to 3 hours after a high-carbohydrate meal would result in which one of the following?

(A) A substantial increase in blood glucose levels
(B) A decrease in blood glucose levels
(C) Have little effect on blood glucose levels

22. Administration of a high dose of glucagon to this patient 30 hours after a high-carbohydrate meal would result in which one of the following?

(A) A substantial increase in blood glucose levels
(B) A decrease in blood glucose levels
(C) Have little effect on blood glucose levels

23. A 65-year-old patient complains of occasional swelling, pain, and a scraping sensation in the knees. X-rays show a narrowing of the joint space. The patient only wants to take "natural" oral products to help reverse this condition. The products available for consumption are examples of which one of the following types of compounds?

(A) Proteoglycan
(B) Polyol
(C) Glycolipid
(D) Disaccharide
(E) Glycoprotein

Questions 24 and 25 refer to the following case:

A woman undergoing chemotherapy for breast cancer has developed bloating, diarrhea, and excess gas whenever she drinks milk. She never had this problem before.

24. Which one of the following best describes the mechanism causing the symptoms in the above-mentioned patient?

(A) Chemotherapy damage to the salivary gland
(B) Chemotherapy damage of the pancreas
(C) Chemotherapy damage of the brush border of the intestine
(D) Cancer infiltration into the small intestine
(E) Cancer infiltration into the pancreas
(F) Cancer infiltration into the salivary gland

25. The symptoms the woman is experiencing is due to a reduced synthesis of which one of the following enzymes?

(A) Sucrase
(B) Lactase
(C) Amylase
(D) Isomaltase
(E) Trehalase

Questions 26 through 28 refer to the following case:

A 50-year-old male with Type 2 diabetes is taking glipizide to help control his blood sugar levels. On one day he could not remember if he had taken the medication, so he accidently took a second dose of the drug. Two hours later, he suddenly develops irritability, tremors, tachycardia, and lightheadedness.

26. The patient is experiencing which one of the following due to his drug overdose?

(A) Hyperglycemia
(B) Hypoglycemia
(C) Lactic acidosis
(D) Ketoacidosis
(E) Hyperammonemia

27. The symptoms the patient is experiencing are caused by which one of the following hormones?

(A) Insulin
(B) Glucagon
(C) Epinephrine
(D) Glucocorticosteroids
(E) Testosterone

28. In response to the overdose of glipizide, the patient has released hormones that will lead to glucose being released by the liver. This occurs through an initial activation of which one of the following liver enzymes?

(A) Adenylate cyclase
(B) Protein kinase A
(C) Glycogen synthase
(D) Phosphorylase kinase
(E) Glycogen phosphorylase

29. A patient with Type 1 diabetes self-injected insulin prior to their evening meal, but then was distracted and forgot to eat. A few hours later, the individual fainted, and after the paramedics arrived they did a STAT blood glucose level and found it to be 45 mg/dL. The blood glucose level was so low because which one of the following tissues assimilated most of it under these conditions?

(A) Brain
(B) Liver
(C) Red blood cells
(D) Adipose tissue
(E) Intestinal epithelial cells

30. A patient has been diagnosed with Type 1 diabetes in their late teens and is being treated with exogenous insulin, but a second physician is not convinced that the patient has Type 1 diabetes, but rather has Type 2 diabetes. A measurement of which one of the following would allow the physician to determine which diagnosis is correct?

(A) Insulin levels
(B) C-peptide levels
(C) Glucagon levels
(D) Epinephrine levels
(E) HbA1c levels

31. A 3-month-old infant, who was experiencing seizures, was diagnosed with a GLUT1 deficiency, resulting in reduced glucose uptake into the brain. As a result, which one of the following substrates was providing energy for the brain?

(A) Lactate
(B) Amino acids
(C) Fatty acids
(D) Glycerol
(E) Ketone bodies

Questions 32 and 33 refer to the following case:

A 50-year-old male with a history of coronary artery disease presents with episodes of lightheadedness, tremors, palpitations, hunger, headache, weakness, and confusion. He is fine between episodes. He had such an episode at his last doctor's visit, at which time blood was drawn for various analyses. The lab results revealed high insulin, high C-peptide, and low blood glucose levels.

32. Which of the following would be most consistent with his symptoms and lab values?

(A) Insulinoma
(B) Pheochromocytoma
(C) Exogenous insulin injection
(D) Carcinoid tumor
(E) Liver cancer

33. In order to treat the patient's symptoms during an episode, which one of the following would be safest to administer?

(A) Epinephrine
(B) Glucagon
(C) Insulin
(D) Amylin
(E) Testosterone

34. Patients with diabetes frequently report changing visual acuities when their glucose levels are chronically high. Which of the following could explain the fluctuating acuity with high blood glucose levels?

(A) Increased sorbitol in the lens
(B) Decreased fructose in the lens
(C) Increased oxidative phosphorylation in the lens
(D) Macular degeneration
(E) Increased galactitol in the lens

35. A couple and their two sons were going to visit Panama in the summer, and obtained drugs from friends (who had these leftover from their trip the year before) to help combat the possibility of acquiring the malarial parasite while in that country. The family members took one drug every day while visiting, then, once they arrived back home, they had to continue the drug treatment for an additional week. During the trip, one of the sons complained of being tired; after the family returned home, he was even more tired and complained of pain in his upper abdomen. He was taken to the emergency department where it was determined that he was anemic. Careful examination demonstrated a slight yellowing in the whites of his eyes. In the presence of the drug, the boy had difficulty in carrying out which one of the following reactions?

(A) The synthesis of heme
(B) The conversion of oxidized glutathione to reduced glutathione
(C) The absorption of iron, reducing hemoglobin synthesis
(D) The conversion of superoxide to oxygen
(E) The conversion of hydrogen peroxide to oxygen

Answers and Explanations

1. **The answer is E.** The cake contains starch, lactose (milk sugar), and sucrose (table sugar). Digestion of starch produces glucose. Lactase cleaves lactose to galactose and glucose, and sucrase cleaves sucrose to fructose and glucose. Thus, the intestinal epithelial cells will absorb from the intestinal lumen, and then secrete into the blood, glucose, galactose, and fructose. The intestinal epithelial cells will not use these sugars as an energy source.

2. **The answer is B.** Phosphorylase produces glucose-1-phosphate from glucose residues linked α-1,4. Free glucose is produced from α-1,6-linked residues at branch points by an α-1, 6-glucosidase activity of the debranching enzyme. Degradation of glycogen produces glucose-1-phosphate and glucose in about a 10:1 ratio, which is the ratio of the α-1,4 linkages to α-1,6 linkages.

3. **The answer is E.** Glucagon in the liver and epinephrine in both the liver and muscle cause cAMP levels to rise, activating protein kinase A. Protein kinase A phosphorylates and activates phosphorylase kinase, which in turn phosphorylates and activates phosphorylase. These phosphorylation reactions require ATP. Branching enzyme is not a substrate for phosphorylase kinase. Phosphodiesterase inhibitors, such as caffeine, keep cAMP elevated, which allows protein kinase A to be active, which keeps phosphorylase kinase active, and in its phosphorylated form.

4. **The answer is A.** Normally, 1 mole of ATP is used to convert 1 mole of glucose to 1 mole of glucose-6-phosphate and a second to convert 1 mole of fructose-6-phosphate to the bisphosphate. Two triose phosphates are produced by cleavage of fructose-1,6-bisphosphate. As the two triose phosphates are converted to pyruvate, four ATPs are generated; two by phosphoglycerate kinase and two by pyruvate kinase. Net, two ATPs are produced. If pyruvate kinase is completely deficient, two less ATPs will be produced, and thus the net ATP production will be zero. It is unlikely that the embryo would survive with a complete deficiency of this enzyme.

5. **The answer is B.** Hexokinase is inhibited by its product, glucose-6-phosphate. PFK1 is activated by AMP and F-2,6-P. F-2,6-P does not inhibit glucokinase, nor is glucokinase present in the muscle. Aldolase is not inhibited by its substrate, fructose 1,6-bisphosphate. Pyruvate kinase is inactivated by glucagon-mediated phosphorylation in the liver, but not in the muscle. The muscle isozyme of pyruvate kinase is not a substrate for protein kinase A. In addition, muscle cells do not respond to glucagon as they do not express glucagon receptors.

6. **The answer is D.** If the phosphodiesterase that degrades cAMP were inhibited (an effect of caffeine) in the presence of epinephrine, cAMP levels would be elevated. Protein kinase A would become more active in the liver and muscle; pyruvate kinase would become less active in the liver; and glycogen synthase activity would be decreased in both muscle and liver. Phosphorylase activity would be increased in both muscle and liver owing to constant phosphorylation by phosphorylase kinase, which is activated by protein kinase A.

7. **The answer is A.** In the mitochondria, CO_2 is added to pyruvate to form oxaloacetate. The enzyme is pyruvate carboxylase, which requires biotin and ATP. Oxaloacetate leaves the mitochondrion as malate or aspartate and is regenerated in the cytosol. Oxaloacetate is converted to phosphoenolpyruvate by a reaction that utilizes GTP and releases the same CO_2 that was added in the mitochondrion. The remainder of the reactions occur in the cytosol.

8. **The answer is D.** The only intermediate included on the list that the pathway of gluconeogenesis from glycerol has in common with the pathway of gluconeogenesis from lactate is glucose-6-phosphate. Glycerol enters gluconeogenesis as DHAP. Therefore, it bypasses the other compounds (pyruvate, oxaloacetate, malate and phosphoenolpyruvate) through which the carbons of lactate must pass on its pathway to glucose synthesis.

9. **The answer is B.** In the oxidative reactions of the pentose phosphate pathway, glucose is converted to ribulose-5-phosphate and CO_2, with the production of NADPH. These reactions are not reversible. Ribose-5-phosphate and xylulose-5-phosphate are formed from ribulose-5-phosphate by two of the nonoxidative reactions of the pathway. Ribose-5-phosphate is used for biosynthesis of nucleotides such as ATP. A series of reactions catalyzed by transketolase and transaldolase produce the glycolytic intermediates fructose-6-phosphate and glyceraldehyde-3-phosphate. Glucose is produced by gluconeogenesis in humans, and not directly by the hexose monophosphate shunt pathway.

10. **The answer is B.** Glucose-6-phosphate is common to all pathways. It can be converted to glucose-1-phosphate for glycogen synthesis or go directly into the pentose phosphate pathway, or proceed through fructose-6-phosphate in glycolysis. UDP-glucose is formed from glucose-1-phosphate and can be used to form glycogen, lactose, glycoproteins, and glycolipids.

11. **The answer is A.** In this patient, starch will be digested by salivary and pancreatic α-amylases to small oligosaccharides and maltose, but a lower than normal amount of glucose will be produced because of the deficiency of the brush border disaccharidases, which have maltase, isomaltase, sucrase, and lactase activity. Sucrose and lactose will not be cleaved. There will be more maltose, sucrose, and lactose in the stool and less monosaccharides in the blood and tissues. Insulin levels will be lower than normal, due to the reduced levels of glucose entering the blood. Muscle glycogen will not increase since there is less glucose in the circulation, and insulin, which is required for glucose entry into the muscle, may not be secreted under these conditions.

12. **The answer is B.** Fructose gives a positive result in a reducing-sugar test and a negative result in a glucose oxidase test. It is a monosaccharide, and, so, is not cleaved by acid. Glucose gives a positive test result with the enzyme glucose oxidase. Sorbitol has no aldehyde or ketone group, and, thus, cannot be oxidized in the reducing-sugar test. Maltose and lactose are disaccharides that undergo acid hydrolysis, which doubles the amount of reducing sugar. This infant probably has benign fructosuria or the more dangerous condition, HFI. A galactose oxidase test would rule out the possibility that the sugar was galactose.

13. **The answer is D.** The patient has HFI, which is due to a mutation in aldolase B. Sucrose would still be cleaved by sucrase, thus it would not increase in the stool. Fructose would not be metabolized normally, therefore it would be elevated in the blood and urine. Aldolase B would not cleave fructose 1-phosphate, thus its levels would be elevated and the product, glyceraldehyde, would not be produced.

14. **The answer is D.** The child has a form of galactosemia. The elevated galactitol enters the lens of the eye, and is trapped. The difference in osmotic pressure across the lens of the eye leads to cataract formation. Galactose is phosphorylated by galactokinase to galactose 1-phosphate, which reacts with UDP-glucose in a reaction catalyzed by galactose-1-phosphate uridylyl transferase to form UDP-galactose and glucose 1-phosphate. An epimerase converts UDP-galactose to UDP-glucose. Deficiencies in either galactokinase (nonclassical) or galactose-1-phosphate uridylyl transferase (classical) result in galactosemia, with elevated levels of galactose and galactitol (reduced galactose) in the blood. An intracellular measurement of galactose-1-phosphate can allow a definitive diagnosis to be obtained (such levels would be nonexistent if the defect were in galactokinase, and the levels would be greatly elevated if the galactose-1-phosphate uridylyl transferase enzyme were defective).

15. **The answer is C.** The woman will be able to breastfeed her baby because she can produce lactosc from amino acids and other carbohydrates. She will not have to eat pure galactose, or even lactose, to do so. Glucose, which can be provided by gluconeogenesis or obtained from the diet, can be converted to UDP-galactose (glucose → glucose-6-phosphate → glucose-1-phosphate → UDP-glucose → UDP-galactose). UDP-galactose reacts with free glucose to form lactose. α-Lactalbumin is a protein that serves as the modifier of galactosyl transferase, which catalyzes this reaction. The amino acids of α-lactalbumin can be used to produce glucose, but the immediate products of α-lactalbumin degradation are not lactose. Carbohydrates cannot be synthesized from fats.

16. **The answer is D.** The patient has the symptoms of beriberi, which is due to a thiamine deficiency. Of the enzymes listed, transketolase would be less active because it requires thiamine pyrophosphate as a cofactor. The other enzymes listed do not require cofactors except for the three dehydrogenases, which require either NAD^+ or $NADP^+$, depending on the enzyme.

17. **The answer is A.** After ingestion of glucose the insulin:glucagon ratio increases, the cAMP phosphodiesterase is activated, cAMP levels drop, and protein kinase A is inactivated. This leads to the activation of glycogen synthase by PP-1. The ratio of phosphorylase a to phosphorylase b is decreased by PP-1 as well, thus glycogen degradation decreases. Red blood cells continue to use glucose and form lactate at their normal rate as glucose is the sole energy source for such cells. CREB is also inactivated under these conditions, thereby reducing the levels of PEPCK (via transcriptional regulation) within the cell.

18. **The answer is C.** The child has the symptoms of von Gierke's disease, which is due to a lack of glucose 6-phosphatase activity. In this disorder, neither liver glycogen nor gluconeogenic precursors (e.g., alanine and glycerol) can be used to maintain normal blood glucose levels. The last step (conversion of glucose-6-phosphate to glucose) is deficient for both glycogenolysis and gluconeogenesis. Muscle glycogen cannot be used to maintain blood glucose levels because muscle does not contain glucose 6-phosphatase. A defective liver glycogen phosphorylase (Her's disease) will not affect the ability of the liver to raise blood glucose levels by gluconeogenesis. In addition, the lack of liver glycogen phosphorylase does not lead to lactic and uric acid accumulation, although mild fasting hypoglycemia can be observed. Defects in liver glycogen synthase (type 0 glycogen storage disease) will lead to an early death, with hypoglycemia and hyperketonemia observed. Muscle does not contribute to blood glucose levels, so a defect in muscle glycogen phosphorylase (McArdle's disease) will not lead to the observed symptoms, but will lead to exercise intolerance. A defect in pyruvate kinase will lead to hemolytic anemia, but not the other symptoms observed in the patient.

19. **The answer is B.** Blood glucose decreases because insulin stimulates the transport of glucose into muscle and adipose cells and stimulates the conversion of glucose to glycogen and triacylglycerols in the liver. Ketone bodies are not made from blood glucose. During fasting, when the liver is producing ketone bodies, it is also synthesizing glucose. Carbon for ketone body synthesis comes from fatty acids. Insulin stimulates glycogen synthesis, not glycogenolysis. Muscle glycogen is not converted to blood glucose.

20. **The answer is A.** Phosphoenolpyruvate carboxykinase converts oxaloacetate to phosphoenolpyruvate. It is a gluconeogenic enzyme required for the conversion of amino acid carbons and lactate (but not phosphoenolpyruvate or glycerol) to glucose. Acetyl-CoA from the oxidation of fatty acids is not converted to glucose. Fructose can be converted to glucose without the need for PEPCK activity (fructose to fructose-1-phosphate, fructose-1-phosphate to DHAP and glyceraldehyde, glyceraldehyde to glyceraldehyde-3-phosphate, then the production of fructose-1,6-bisphosphate from DHAP and glyceraldehyde-3-phosphate, loss of phosphate to fructose-6-phosphate, isomerization to glucose-6-phosphate, then loss of phosphate to produce glucose).

21. **The answer is A.** By 2 to 3 hours after a high-carbohydrate meal, the patient's glycogen stores would be filled. Glucagon would stimulate glycogenolysis, and blood glucose levels would rise. Gluconeogenesis would still be impaired, but since glycogen levels are high, the liver would be able to export significant amounts of glucose.

22. **The answer is C.** Thirty hours after a meal, liver glycogen is normally depleted, and blood glucose level is maintained solely by gluconeogenesis after this time. However, in this case, a key gluconeogenic enzyme is inhibited by an endotoxin. Therefore, gluconeogenesis will not occur at a normal rate and glycogen stores will be depleted more rapidly than normal. Blood glucose levels will not change significantly if glucagon is administered after 30 hours of fasting.

23. **The answer is A.** The patient has osteoarthritis, and wants to use glucosamine/chondroitin sulfate to provide cushioning in the joint. These molecules are proteoglycans, which consist of long, linear chains of glycosaminoglycans attached to a core protein. Each chain is composed

of repeating disaccharides, but disaccharides are, by definition, only 2 sugars. A polyol is a polyalcohol. A glycolipid is a sphingolipid, and does not contribute to joint stability. The typical glycoproteins are not found in the joints, nor are they available as oral supplements as the proteoglycans are.

24. The answer is C. Chemotherapy targets rapidly growing cells. The outer cells of the intestinal lining (brush border) are rapidly growing cells and are commonly affected by chemotherapy. Lactase is found in the brush border. Cancer metastases to the small bowel would not disrupt the entire small intestine. Cancer infiltration to the pancreas or salivary gland would also not affect lactase activity.

25. The answer is B. Lactase converts lactose (milk sugar) to glucose and galactose. In the absence of lactase activity (the chemotherapy is destroying the rapidly growing cells, such as the intestinal epithelial cells, where lactase is found), the lactose enters the colon, where the bacterial flora metabolize it to produce gases and acids. The gases produce flatulence, and the acids lead to an osmotic imbalance that drives water to leave the colonic epithelium and enter the lumen of the colon, leading to the diarrhea. Sucrase converts sucrose (table sugar) to glucose and fructose. Amylase helps digest plant starches. Isomaltase releases glucose residues from branched oligosaccharides. Trehalase splits trehalose, which is glucose α-1,1-glucose (a disaccharide).

26. The answer is B. The patient has become hypoglycemic due to excessive release of insulin from the pancreas. Glipizide (glucotrol) is a sulfonylurea drug that stimulates insulin release from the pancreas. If taken in excess, the insulin will promote fat and muscle cells to take up glucose from the circulation, leading to hypoglycemia and insufficient blood glucose levels for normal brain function. Lactic acidosis may result from such an overdose, but it would be secondary to the hypoglycemic symptoms observed. Elevated ammonia levels would not occur, as glipizide does not alter amino acid metabolism. The high levels of insulin released by the drug would inhibit fatty acid release from the adipocytes, and therefore the precursors for ketone body synthesis are not available, and ketoacidosis would not occur.

27. The answer is C. The patient is having a hypoglycemic attack. Glipizide is a sulfonylurea that stimulates insulin release from the pancreas and can cause hypoglycemia. Glucagon, epinephrine, and glucocorticosteroids are all released to raise blood glucose levels. The symptoms observed in the patient are side effects of epinephrine, acting in the autonomic nervous system. Testosterone levels would not be altered in the presence of glipizide.

28. The answer is A. Epinephrine, via binding to its receptor, activates a Gs-protein, which binds to and activates adenylate cyclase. Adenylate cyclase will convert ATP to cAMP. As cAMP levels increase, the cAMP binds to the regulatory subunits of protein kinase A, allowing the regulatory subunits to be released from the catalytic subunits. This activates protein kinase A, which then phosphorylates both glycogen synthase and phosphorylase kinase. Glycogen phosphorylase is activated by phosphorylation by phosphorylase kinase. Thus, of the events listed, activation of adenylate cyclase is the initial event.

29. The answer is D. Insulin stimulates glucose transport into muscle and adipose cells through mobilization of GLUT4 transporters from internal vesicles to the cell surface. Insulin does not significantly stimulate glucose transport into tissues such as liver, brain, or RBCs, which utilize different variants of the glucose transporters. Only GLUT4 is insulin-responsive.

30. The answer is B. The major difference between Type 1 and Type 2 diabetes is the ability of the body to produce endogenous insulin. Patients with Type 1 diabetes do not produce insulin, whereas patients with Type 2 diabetes do produce insulin, but have difficulty responding to the insulin. When insulin is synthesized as preproinsulin, it is then modified and the C-peptide is removed from the molecule, to produce active insulin. Persons with Type 1 diabetes would be lacking C-peptide (exogenous insulin that is injected also lacks C-peptide), whereas persons with Type 2 diabetes would be producing C-peptide. The levels of glucagon and epinephrine would be similar in both types of diabetes. Since the patient is on insulin already, measuring the level of mature insulin in the blood would be unhelpful. HbA1c levels measure glycemic control over the past 6 weeks, and are usually elevated in both types of diabetes. Measuring HbA1c would not enable one to distinguish between Type 1 and Type 2 diabetes in this patient.

31. The answer is E. The brain can only use glucose or ketone bodies as an energy source. Even though the heart can use lactate for energy, the brain does not do so. If glucose levels are low, the only available substrate would be ketone bodies. Fatty acids will not cross the blood–brain barrier and are not a good energy source for the brain. The liver will convert fatty acids to ketone bodies for use by the brain. Amino acids are a good source of carbon for gluconeogenesis, but the brain does not oxidize amino acids at an appreciable rate. Glycerol cannot be used by the brain as an energy source as the brain lacks glycerol kinase, a necessary enzyme in the metabolism of glycerol. The treatment for a GLUT1 deficiency is a ketogenic diet–one high in fats such that ketone bodies are continuously generated to provide fuel for the brain.

32. The answer is A. The symptoms and lab results are classic for insulinoma. An insulinoma is a tumor of the pancreatic β cells that episodically releases large amounts of insulin. At those times, the patient experiences the symptoms of hypoglycemia. A pheochromocytoma is a tumor of the adrenal gland that episodically releases epinephrine and norepinephrine throughout the body. A pheochromocytoma would not lead to hypoglycemia (epinephrine stimulates the liver to export glucose), or high insulin or C-peptide levels. If the patient were injecting insulin, the C-peptide should be low, as exogenous insulin lacks the C-peptide. Neither a liver tumor nor a carcinoid tumor would release insulin to the blood.

33. The answer is B. Glucagon is the major catabolic hormone that counters insulin's effects. It can raise blood glucose through stimulation of gluconeogenesis and glycogenolysis. Adding insulin would exacerbate the metabolic situation, as excessive insulin is causing the problem. Amylin suppresses glucagon action, and would not overcome the effects of the high insulin levels. Epinephrine can help counter insulin and raise blood glucose, but would be dangerous in a patient with known coronary artery disease (CAD) and palpitations.

34. The answer is A. Fluctuating levels of sugars and sugar alcohols in the lens can cause fluctuating visual acuity. With high blood glucose, there would be increased levels of sorbitol in the lens. The lens does not contain mitochondria and cannot use the TCA cycle/electron transport chain to generate energy. Galactitol causes the same problems as sorbitol, but galactitol is derived from galactose, whereas sorbitol is produced from glucose. The patient has high glucose levels, so galactitol would not be expected to accumulate in the lens. Macular degeneration affects the retina, but in this case, it is the lens that is the affected tissue. Reducing fructose levels in the lens would reduce sorbitol levels, which would ease the visual acuity problem, not make it occur.

35. The answer is B. The boy lacks glucose-6-phosphate dehydrogenase activity (an X-linked disorder) and, in response to the drug (most likely primaquine), has developed a hemolytic anemia due to an inability to regenerate reduced glutathione to protect red blood cell membranes from oxidative damage. The yellow in the eyes is due to a buildup of bilirubin, as the released hemoglobin from the red blood cells cannot be adequately metabolized by the liver, and converted to the more soluble diglucuronide form. The abdominal pain may be due to bilirubin stones being formed in the gall bladder. Glucose-6-phosphate dehydrogenase produces NADPH in the red blood cells, which is required for glutathione reductase, the enzyme that converts oxidized glutathione to reduced glutathione. The drug does not block heme synthesis, the absorption of iron, or affect radical oxygen species metabolism (superoxide dismutase or catalase).

chapter 7 Lipid and Ethanol Metabolism

The major clinical uses of this chapter are understanding the basics of lipid disorders and treatment, understanding obesity and weight loss, and understanding the rationale of medication that targets eicosanoids.

OVERVIEW

- Lipids are a diverse group of compounds that are related by their insolubility in water.
- Membranes contain lipids, particularly phosphoglycerides, sphingolipids, and cholesterol.
- Triacylglycerols, which provide the body with its major source of energy, are obtained from the diet or synthesized mainly in the liver. They are transported in the blood as lipoproteins and are stored in adipose tissue (Fig. 7.1A).
- The major classes of blood lipoproteins include chylomicrons, very low-density lipoprotein (VLDL), intermediate-density lipoprotein (IDL), low-density lipoprotein (LDL), and high-density lipoprotein (HDL).
- Chylomicrons are produced in intestinal cells from dietary lipid, and VLDL is produced in the liver, mainly from dietary carbohydrate.
- The triacylglycerols of chylomicrons and VLDL are hydrolyzed in the blood by lipoprotein lipase to fatty acids and glycerol. In adipose cells, the fatty acids are converted to triacylglycerols and stored.
- IDL consists of the remains of VLDL after digestion of some of the triacylglycerols. IDL can either be endocytosed by liver cells and digested by lysosomal enzymes or converted to LDL by further digestion of triacylglycerols.
- LDL undergoes endocytosis and lysosomal digestion, both in the liver and in the peripheral tissues.
- Chylomicron remnants are endocytosed by the liver.
- Cholesterol travels through the blood as a component of the blood lipoproteins. Cholesterol is synthesized in most cells of the body. The key regulatory enzyme is hydroxymethylglutaryl (HMG)-CoA reductase. Cholesterol is a component of cell membranes. In the liver, cholesterol is converted to bile salts, and it forms steroid hormones in endocrine tissues.
- HDL transfers proteins (including an activator of lipoprotein lipase, apoC_II) to chylomicrons and VLDL. HDL also picks up cholesterol from peripheral tissues and from other blood lipoproteins. This cholesterol ultimately returns to the liver.
- During fasting, fatty acids (derived from adipose triacylglycerol stores) are oxidized by various tissues to produce energy (see Fig. 7.1B). In the liver, fatty acids are converted to ketone bodies, which are oxidized by tissues such as muscle and kidney.
- Eicosanoids (prostaglandins, thromboxanes, and leukotrienes) are derived from polyunsaturated fatty acids.

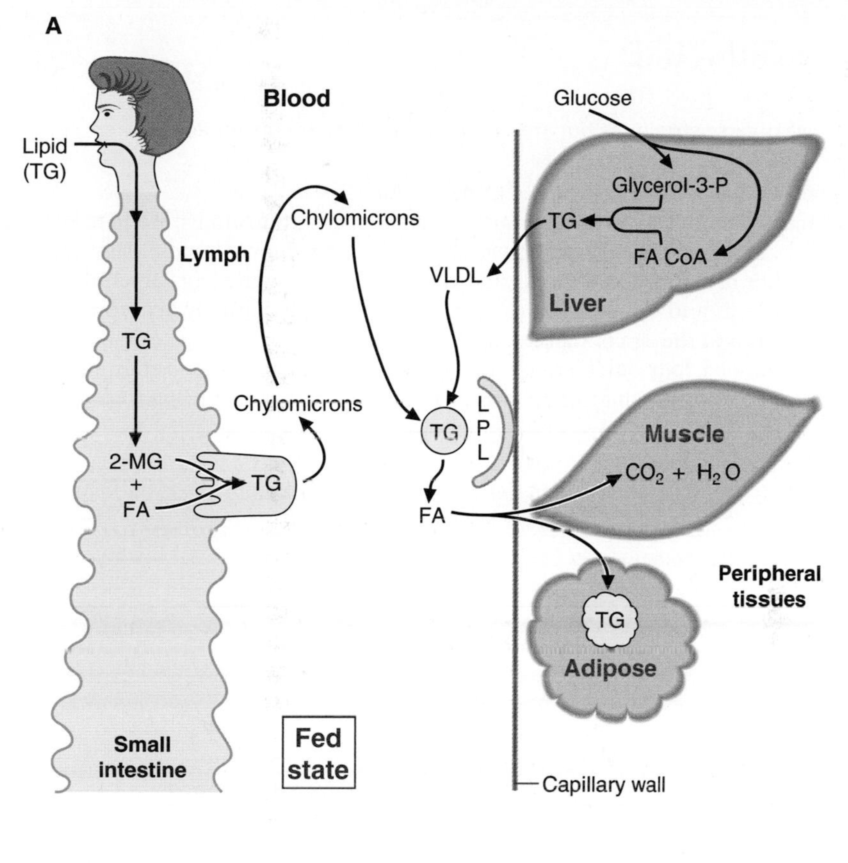

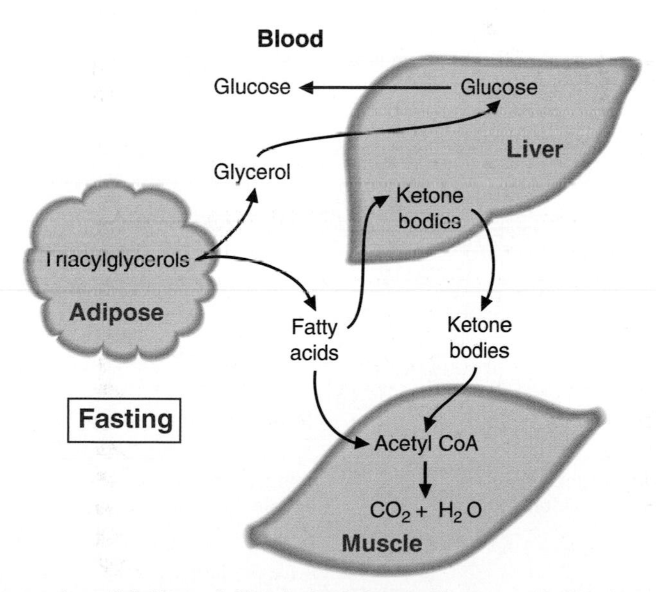

FIGURE 7.1. **A. An overview of lipid metabolism in the fed state**. FA, fatty acid; HDL, high-density lipoprotein; LPL, lipoprotein lipase; 2-MG, 2-monoacylglycerol; TG, triacylglycerol; *circled* TG, triacylglycerols of VLDL and chylomicrons; VLDL, very low-density lipoprotein. **B.** An overview of lipid metabolism in the fasting state.

I. LIPID STRUCTURE

- Lipids have diverse structures but are similar in that they are insoluble in water.

A. Fatty acids exist freely or esterified to glycerol (Fig. 7.2).

1. In humans, fatty acids usually have an **even number** of carbon atoms, are **16 to 20 carbon atoms** in length, and may be **saturated** or **unsaturated** (contain double bonds). They are described by the number of carbons and the positions of the double bonds (e.g., arachidonic acid, which has 20 carbons and 4 double bonds, is 20:4, $\Delta^{5,8,11,14}$). All naturally occurring fatty acids have double bonds in the cis configuration.
2. **Polyunsaturated fatty acids** are often classified according to the position of the first double bond from the ω-end (the carbon farthest from the carboxyl group; e.g., ω-3 or ω-6).

B. Monoacylglycerols (monoglycerides), diacylglycerols (diglycerides), and triacylglycerols (triglycerides) contain one, two, and three fatty acids esterified to glycerol, respectively.

C. Phosphoglycerides contain fatty acids esterified to positions 1 and 2 of the glycerol moiety and a phosphoryl group at position 3 (e.g., phosphocholine).

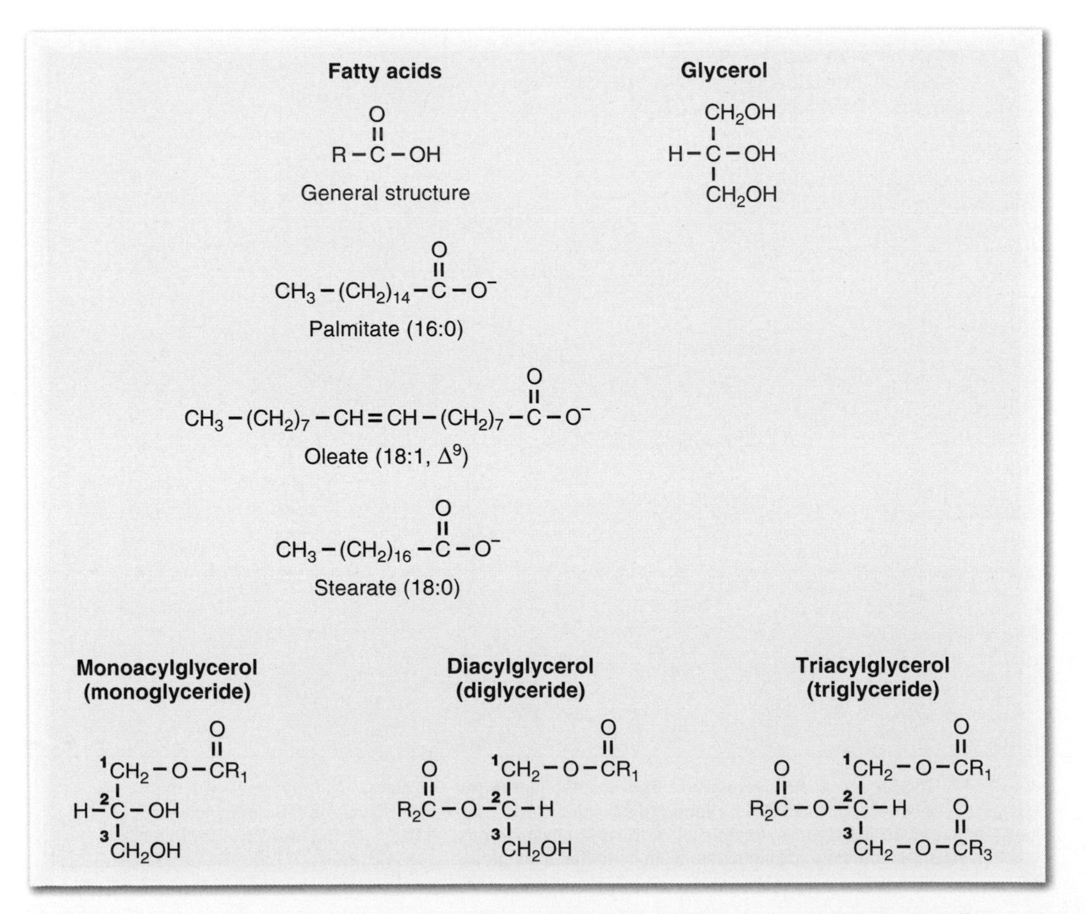

FIGURE 7.2. The structures of fatty acids, glycerol, and the acylglycerols. R indicates a linear aliphatic chain. Fatty acids are identified by the number of carbons and the number of double bonds and their positions (e.g., 18:1, Δ^9).

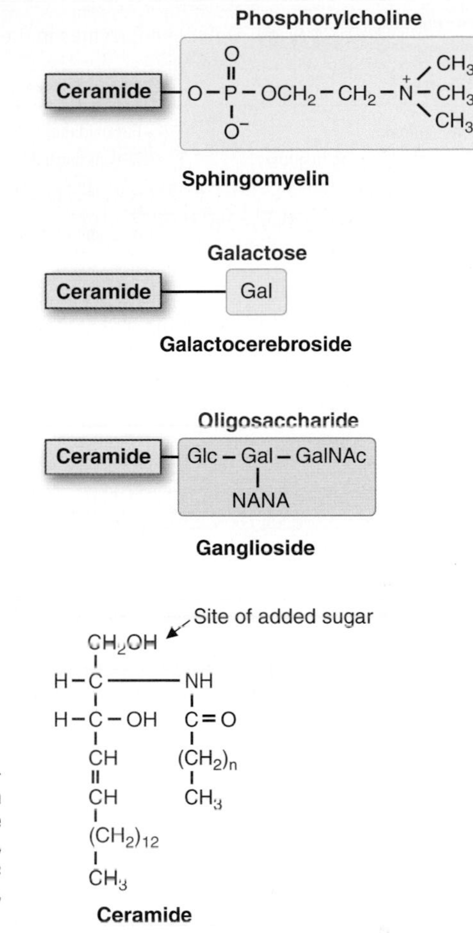

FIGURE 7.3. **Sphingolipids, derivatives of ceramide.** The structure of ceramide is shown at the bottom of the figure. The portion of ceramide shown in red is sphingosine. Different groups are added to the hydroxyl portion of ceramide to form sphingomyelin, cerebrosides, and gangliosides. NANA, *N*-acetylneuraminic acid, also called sialic acid; Glc, glucose; Gal, galactose; GalNac, *N*-acetylgalactosamine.

D. Sphingolipids contain ceramide with a variety of groups attached (Fig. 7.3).

1. **Sphingomyelin** contains phosphocholine.
2. **Cerebrosides** contain a sugar residue.
3. **Gangliosides** contain a number of sugar residues, one of which is sialic acid.

E. Cholesterol contains four rings and an aliphatic side chain (see Fig. 7.11).
Bile salts and steroid hormones are derived from cholesterol (see Fig. 7.12).

F. Prostaglandins and leukotrienes are derived from polyunsaturated fatty acids such as arachidonic acid.

G. The fat-soluble vitamins include vitamins A, D, E, and K (see Fig. 5.7).

CLINICAL CORRELATES Sphingolipids are normally degraded by lysosomal enzymes. If these enzymes are deficient, partially degraded sphingolipids accumulate in cells, compromising cell function; death may result. An α-galactosidase is deficient in **Fabry's disease**, a β-glucosidase in **Gaucher's disease**, a sphingomyelinase in **Neimann–Pick disease**, and a hexosaminidase in **Tay–Sachs disease**. These diseases are known as the **sphingolipidoses**, or **gangliosidoses**. The sphingolipidoses are summarized in Table 7.1.

table 7.1 Defective Enzymes in the Sphingolipidoses (Gangliosidoses)

Disease	Enzyme Deficiency	Accumulated Lipid
Fucosidosis	α-Fucosidase	Cer–Glc–Gal–GalNAc–Gal:Fuc H-isoantigen
Generalized gangliosidosis	G_{M1}-β-galactosidase	Cer–Glc–Gal(NeuAc)–GalNAc:Gal G_{M1} ganglioside
Tay–Sachs disease	Hexosaminidase A	Cer–Glc–Gal(NeuAc):GalNAc G_{M2} ganglioside
Tay–Sachs variant or Sandhoff's disease	Hexosaminidases A and B	Cer–Glc–Gal–Gal:GalNAc globoside plus G_{M2} ganglioside
Fabry's disease	α-Galactosidase	Cer–Glc–Gal:Gal globotriaosylceramide
Ceramide lactoside lipidosis	Ceramide lactosidase (β-galactosidase)	Cer–Glc:Gal ceramide lactoside
Metachromatic leukodystrophy	Arylsulfatase A	Cer–Gal:OSO_3^{3-}sulfogalactosylceramide
Krabbe's disease	β-Galactosidase	Cer:Gal galactosylceramide
Gaucher's disease	β-Glucosidase	Cer:Glc glucosylceramide
Niemann–Pick disease	Sphingomyelinase	Cer:P–choline sphingomyelin
Farber's disease	Ceramidase	Acyl:sphingosine ceramide

NeuAc, *N*-acetylneuraminic acid; Cer, ceramide: Glc, glucose; Gal, galactose; Fuc, fucose. The colon indicates the bond that cannot be broken owing to the enzyme deficiency associated with the disease.

II. MEMBRANES

- The cell (plasma) membrane is a fluid mosaic of lipids and proteins.
- The proteins serve as transporters, enzymes, receptors, and mediators that allow extracellular compounds, such as hormones, to exert intracellular effects.

A. Membrane structure

1. **Membranes** are composed mainly of lipids and proteins (see Fig. 4.1).
2. **Phosphoglycerides** are the major membrane lipids, but sphingolipids and cholesterol are also present.
   a. **Phospholipids** form a bilayer, with their hydrophilic head groups interacting with water on both the extracellular and intracellular surfaces, and their hydrophobic fatty acyl chains in the central portion of the membrane.
3. **Peripheral proteins** are embedded at the periphery; **integral proteins** span from one side to the other.
4. **Carbohydrates** are attached to proteins and lipids on the exterior side of the cell membrane. They extend into the extracellular space.
5. **Lipids and proteins** can **diffuse laterally** within the plane of the membrane. Therefore, the membrane is a fluid mosaic.

B. Membrane function

1. Membranes serve as **barriers** that separate the contents of a cell from the external environment or the contents of organelles from the remainder of the cell.
2. The **proteins** in the cell membrane have many functions.
   a. Some are involved in the **transport** of substances across the membrane.
   b. Some are **enzymes** that catalyze biochemical reactions.
   c. Those on the exterior surface can function as **receptors** that bind external ligands such as hormones or growth factors.
   d. Others are **mediators** that aid the ligand–receptor complex in triggering a sequence of events (e.g., G-proteins); as a consequence, **second messengers** (e.g., cAMP) that alter metabolism are produced inside the cell. Therefore, an external agent, such as a hormone, can elicit intracellular effects without entering the cell.

III. DIGESTION OF DIETARY TRIACYLGLYCEROL

- The major dietary fat is triacylglycerol, which is obtained from the fat stores of the plants and animals in the food supply.
- The dietary triacylglycerols, which are water-insoluble, are emulsified by bile salts and digested in the small intestine to fatty acids and 2-monoacylglycerols. These digestive products are resynthesized to triacylglycerols in intestinal epithelial cells and are secreted in chylomicrons via the lymph into the blood.
- Medium- and short-chain fatty acids are sufficiently soluble to pass through the intestinal epithelial cells and to enter the circulation without being incorporated into triglycerides.

A. Dietary triacylglycerols are digested in the small intestine by a process that requires bile salts and secretions from the pancreas (Fig. 7.4).

1. **Bile salts** are synthesized in the liver from cholesterol and are secreted into the bile. They pass into the intestine, where they emulsify the dietary lipids.
2. The **pancreas** secretes digestive enzymes and bicarbonate, which neutralizes stomach acid, raising the pH into the optimal range for the digestive enzymes.

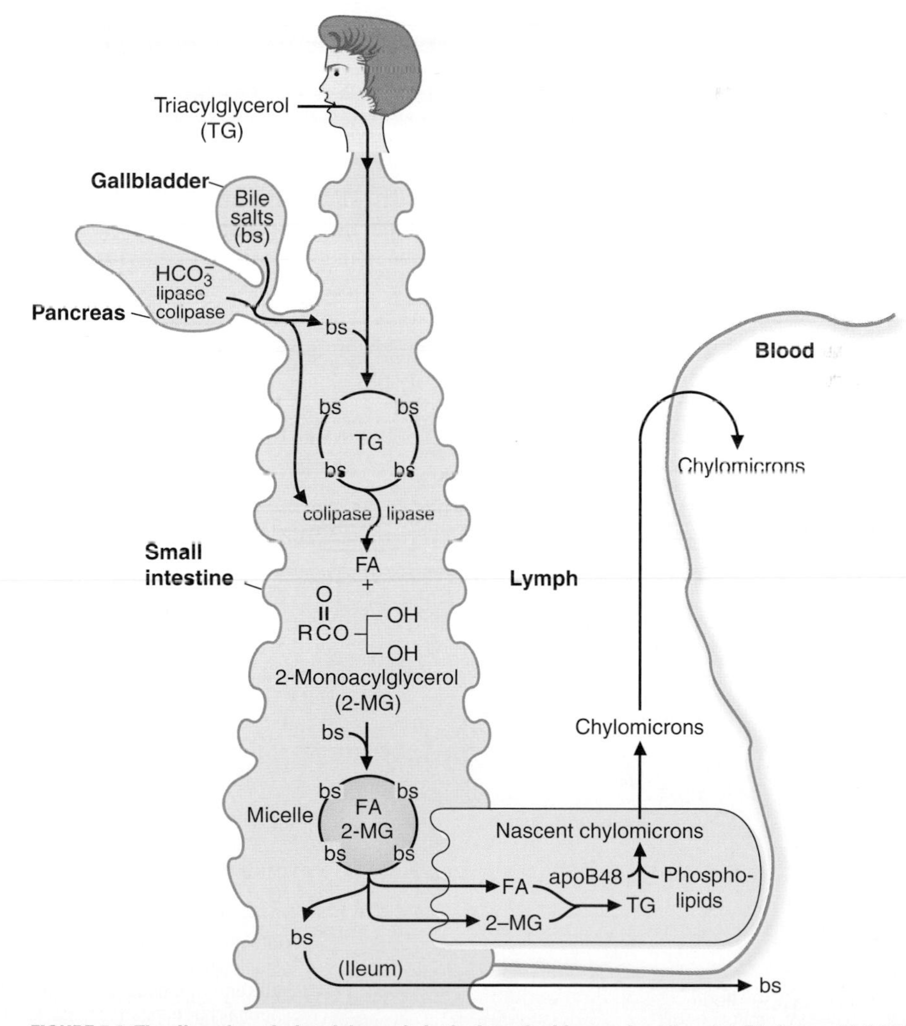

FIGURE 7.4. The digestion of triacylglycerols in the intestinal lumen. bs, bile salts; FA, fatty acid; 2-MG, 2-monoacylglycerol; TG, triacylglycerols.

3. **Pancreatic lipase**, with the aid of colipase, digests the triacylglycerols to 2-monoacylglycerols and free fatty acids, which are packaged into micelles. The micelles, which are tiny microdroplets emulsified by bile salts, also contain other dietary lipids such as cholesterol and the fat-soluble vitamins.
4. The **micelles** travel to the microvilli of the intestinal epithelial cells, which absorb the fatty acids, 2-monoacylglycerols, and other dietary lipids.
5. The **bile salts are resorbed**, recycled by the liver, and secreted into the gut during subsequent digestive cycles (Fig. 7.5).

B. Synthesis of chylomicrons

1. In intestinal epithelial cells, the **fatty acids** from micelles are **activated** by fatty acyl-CoA synthetase (thiokinase) to form fatty acyl-CoA.
2. A **fatty acyl-CoA** reacts with a 2-monoacylglycerol to form a **diacylglycerol**. Then another fatty acyl-CoA reacts with the diacylglycerol to form a **triacylglycerol**.
3. The triacylglycerols pass into the lymph packaged in **nascent (newborn) chylomicrons**, which eventually enter the blood.
4. Medium-chain fatty acids do not need to be incorporated into triglycerides in order to enter the circulation. These fatty acids are sufficiently soluble that they enter the blood directly, and are taken up by target organs as an energy source.

CLINICAL CORRELATES **Blockage of the bile duct** caused by problems such as cholesterol-containing gallstones or duodenal or pancreatic tumors can lead to an inadequate concentration of bile salts in the intestine. The digestion and absorption of dietary lipids are diminished. Diseases that affect the pancreas, such as cystic fibrosis and alcoholism, can lead to a decrease in bicarbonate and digestive enzymes in the intestinal lumen. (Bicarbonate is required to raise the intestinal pH so that bile salts and digestive enzymes can function.) If dietary fats are not adequately digested, **steatorrhea** can result. **Malabsorption of fats** can lead to **caloric deficiencies** and **lack of fat-soluble vitamins** and **essential fatty acids**.

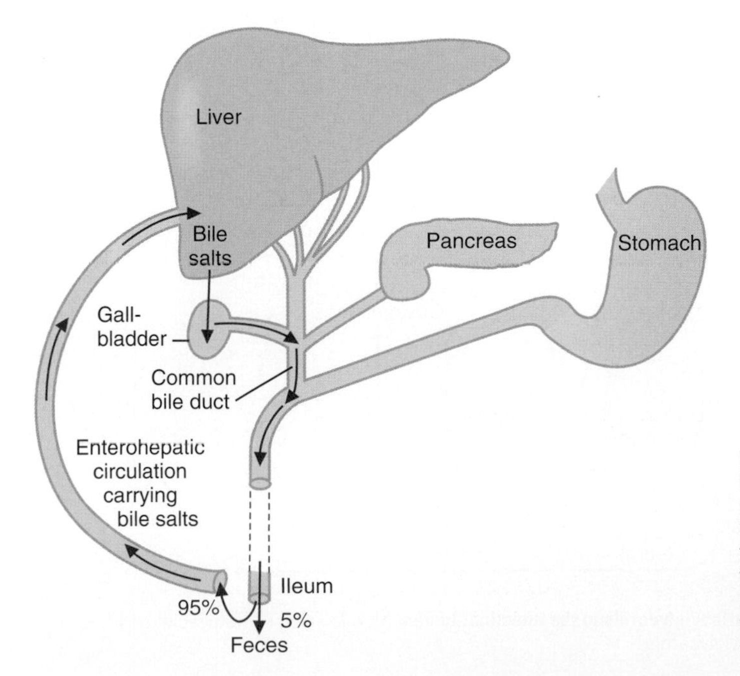

FIGURE 7.5. Recycling of bile salts. Bile salts are synthesized in the liver, stored in the gallbladder, secreted into the small intestine, resorbed in the ileum, and returned to the liver via the enterohepatic circulation. Under normal circumstances, 5% or less of the bile acids in the intestinal lumen are excreted in the stool.

IV. FATTY ACID AND TRIACYLGLYCEROL SYNTHESIS

- Lipogenesis, the synthesis of fatty acids and their esterification to glycerol to form triacylglycerols, occurs mainly in the liver in humans, with dietary carbohydrate as the major source of carbon.
- The de novo synthesis of fatty acids from acetyl-CoA occurs in the cytosol on the fatty acid synthase complex.
- Acetyl-CoA, derived mainly from glucose, is converted by acetyl-CoA carboxylase to malonyl-CoA.
- The growing fatty acyl chain on the fatty acid synthase complex is elongated, two carbons at a time, by the addition of the 3-carbon compound, malonyl-CoA, which is subsequently decarboxylated. With each 2-carbon addition, the growing chain, which initially contains a β-keto group, is reduced in a series of steps that require NADPH.
  - NADPH is produced by the pentose phosphate pathway and by the reaction catalyzed by malic enzyme.
- Palmitate, the product released by the fatty acid synthase complex, is converted to a series of other fatty acyl-CoAs by elongation and desaturation reactions.
- The fatty acyl-CoA combines with glycerol-3-phosphate in the liver to form triacylglycerols by a pathway in which phosphatidic acid serves as an intermediate.
- The triacylglycerols, packaged in VLDL, are secreted into the blood.

A. Conversion of glucose to acetyl-CoA for fatty acid synthesis (Fig. 7.6)

1. **Glucose** enters liver cells and is converted via glycolysis to pyruvate, which enters mitochondria.
2. **Pyruvate** is converted to acetyl-CoA by pyruvate dehydrogenase and to **oxaloacetate (OAA)** by pyruvate carboxylase.
3. Because acetyl-CoA cannot directly cross the mitochondrial membrane and enter the cytosol to be used for the process of fatty acid synthesis, acetyl-CoA and oxaloacetate condense to form **citrate**, which can cross the mitochondrial membrane.
4. In the cytosol, **citrate is cleaved** to oxaloacetate and acetyl-CoA by citrate lyase, an enzyme that requires ATP and is induced by insulin.
   a. **Oxaloacetate** from the citrate lyase reaction is reduced in the cytosol by NADH, producing NAD^+ and **malate**. The enzyme is cytosolic malate dehydrogenase.
   b. In a subsequent reaction, **malate** is converted to pyruvate, NADPH is produced, and CO_2 is released. The enzyme is the **malic enzyme** (also known as decarboxylating malate dehydrogenase or $NADP^+$-dependent malate dehydrogenase).
      (1) **Pyruvate** reenters the mitochondrion and is reutilized.
      (2) **NADPH** supplies the reducing equivalents for reactions that occur on the fatty acid synthase complex.
         (a) **NADPH** is produced not only by the **malic enzyme** but also by the **pentose phosphate pathway**.
5. **Acetyl-CoA** (from the citrate lyase reaction or from other sources) supplies carbons for the fatty acid synthesis in the cytosol.

B. Synthesis of fatty acids by the fatty acid synthase complex (Fig. 7.7)

1. **Fatty acid synthase** is a multienzyme complex located in the cytosol. It has two identical subunits with seven catalytic activities.
2. This enzyme contains a **phosphopantetheine residue**, derived from the vitamin pantothenic acid, and a **cysteine residue**; both contain sulfhydryl groups that can form thioesters with acyl groups. The growing fatty acyl chain moves from one to the other of these sulfhydryl residues as it is elongated.
   a. **Addition of 2-carbon units**
      (1) Initially, **acetyl-CoA** reacts with the phosphopantetheinyl residue and then the acetyl group is transferred to the cysteinyl residue. This acetyl group provides the **ω-carbon** of the fatty acid produced by the fatty acid synthase complex.
      (2) A malonyl group from **malonyl-CoA** forms a **thioester** with the phosphopantetheinyl sulfhydryl group.

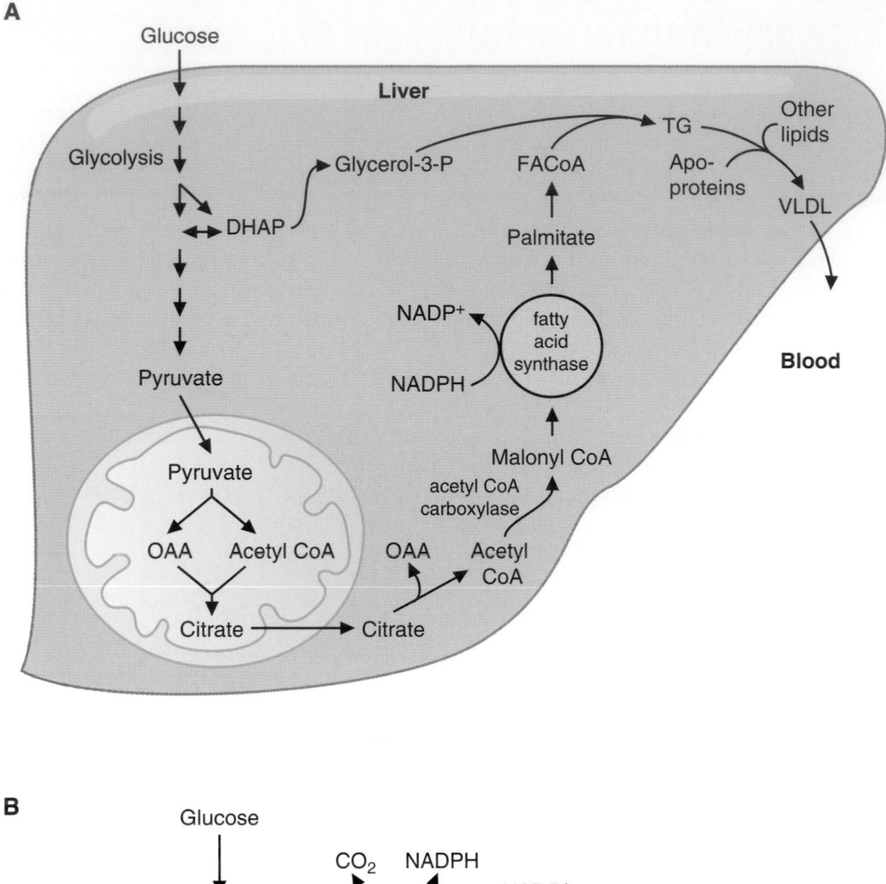

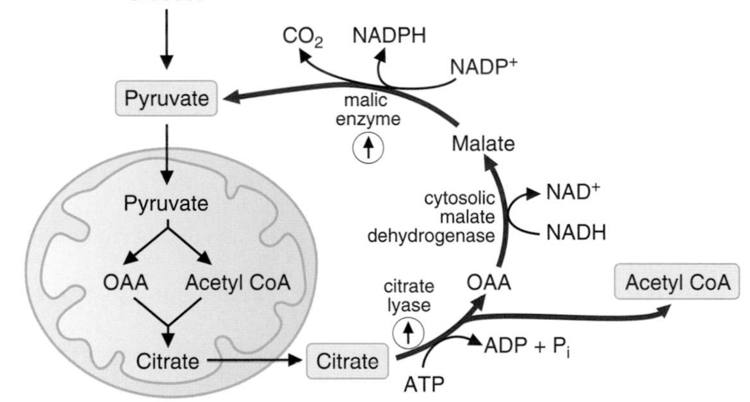

FIGURE 7.6. **Lipogenesis, the synthesis of triacylglycerols (TGs) from glucose. A.** In humans, the synthesis of fatty acids from glucose occurs mainly in the liver. Fatty acids (FAs) are converted to TG, packaged into VLDL, and secreted into the circulation. **B.** Citrate provides acetyl-CoA for fatty acid synthesis, as well as initiating a pathway for NADPH production via malic enzyme. OAA, oxaloacetate.

(a) **Malonyl-CoA** is formed from acetyl-CoA by a carboxylation reaction that requires **biotin** and ATP.

(b) The enzyme is **acetyl-CoA carboxylase**, a regulatory enzyme that is inhibited by phosphorylation, activated by dephosphorylation and by citrate, and induced by insulin. The enzyme that phosphorylates acetyl-CoA carboxylase is the **AMP-activated protein kinase** (not protein kinase A).

(3) The **acetyl group** on the fatty acid synthase complex condenses with the malonyl group; the CO_2 that was added to the malonyl group by acetyl-CoA carboxylase is released; and a β-**ketoacyl group**, containing four carbons, is produced.

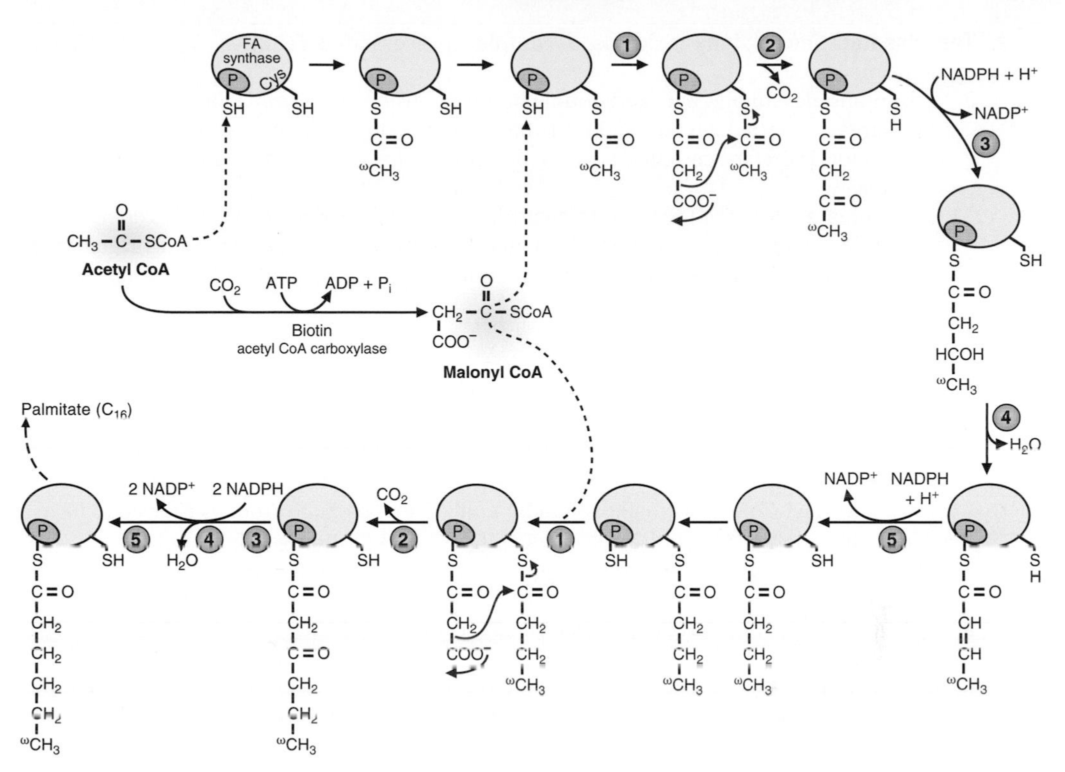

FIGURE 7.7. **Fatty acid synthesis.** Malonyl-CoA provides the 2-carbon units that are added to the growing fatty acyl chain. The addition and reduction steps (1–5) are repeated until palmitic acid is produced. *cys*-SH, a cysteinyl residue; P, a phosphopantetheinyl group attached to the fatty acid synthase complex.

b. Reduction of the β-ketoacyl group

(1) The β-keto group is **reduced** by NADPH to a β-hydroxy group.

(2) Then **dehydration** occurs, producing an **enoyl group** with the double bond between carbons 2 and 3.

(3) Finally, the double bond is reduced by NADPH, and a **4-carbon acyl group** is generated.

(a) The **NADPH** for these reactions is produced by the **pentose phosphate pathway** and by the **malic enzyme**.

c. Elongation of the growing fatty acyl chain

(1) The acyl group is transferred to the cysteinyl sulfhydryl group, and **malonyl-CoA** reacts with the phosphopantetheinyl group. Condensation of the acyl and malonyl groups occurs with the release of CO_2, followed by the three reactions that reduce the β-keto group. The chain is now longer by two carbons.

(2) This sequence of reactions repeats until the growing chain is 16 carbons in length.

(3) **Palmitate**, a 16-carbon saturated fatty acid, is the final product released by hydrolysis from the fatty acid synthase complex.

C. Elongation and desaturation of fatty acids

Palmitate can be elongated and desaturated to form a **series of fatty acids.**

1. The elongation of long-chain fatty acids occurs on the endoplasmic reticulum, by reactions similar, but not identical, to those that occur on the fatty acid synthase complex.

a. Malonyl-CoA provides the 2-carbon units that add to palmitoyl-CoA or to longer-chain fatty acyl-CoAs.

b. Malonyl-CoA condenses with the carbonyl group of the fatty acyl residue and CO_2 is released.

c. The β-keto group is reduced by NADPH to a β-hydroxy group, dehydration occurs, and a double bond is formed, which is reduced by NADPH.

2. **The desaturation of fatty acids** is a complex process that requires O_2, NADPH, and cytochrome b_5.
   a. In humans, desaturates may add double bonds at positions 5, 6, and 9 of a fatty acyl-CoA.
      (1) Plants can introduce double bonds between carbon 9 and the ω-carbon, but animals cannot. Therefore, certain unsaturated fatty acids from plants are required in the human diet.
      (2) **Linoleate** (18:2, $\Delta^{9,12}$) and **α-linolenate** (18:3, $\Delta^{9,12,15}$) are the major sources of the essential fatty acids required in the human diet. They are used for the synthesis of **arachidonic acid** and other polyunsaturated fatty acids from which eicosanoids (e.g., prostaglandins) are produced.

D. Synthesis of triacylglycerols (Fig. 7.8)

1. **In intestinal epithelial** cells, triacylglycerol synthesis occurs by a different pathway than in other tissues (see Section III B). This triacylglycerol becomes a component of chylomicrons. Ultimately, the fatty acyl groups are stored in adipose triacylglycerols.
2. **In the liver and adipose tissue**, glycerol-3-phosphate provides the glycerol moiety that reacts with two fatty acyl-CoAs to form **phosphatidic acid**. The phosphate group is cleaved to form a diacylglycerol, which reacts with another fatty acyl-CoA to form a triacylglycerol.
   a. **The liver** can use glycerol to produce glycerol-3-phosphate by a reaction that requires ATP and is catalyzed by glycerol kinase.
   b. **Adipose tissue**, which **lacks glycerol kinase**, cannot generate glycerol-3-phosphate from glycerol.
   c. **Both liver and adipose tissue** can convert glucose, through glycolysis, to dihydroxyacetone phosphate (**DHAP**), which is reduced by NADH to glycerol-3-phosphate.
   d. **Triacylglycerol** is **stored in adipose tissue.**
   e. In the **liver**, triacylglycerol is incorporated into **VLDL**, which enters the blood. Ultimately, the fatty acyl groups are stored in adipose triacylglycerols.

E. Synthesis of phospholipids

1. The phospholipids are synthesized from phosphatidic acid.
2. The head groups are added to phosphatidic acid via either of the two mechanisms (Fig. 7.9).
   a. Phosphatidylcholine and phosphatidylethanolamine are produced by **head group activation**.
   b. Phosphatidyl serine is formed by a **head group substitution** of serine for ethanolamine in phosphatidylethanolamine.
   c. Phosphatidylinositol, phosphatidylglycerol, and cardiolipin are formed by **activating phosphatidic acid** to CDP-DAG (CDP-diacylglycerol).
   d. The fatty acids at positions 1 and 2 do not have to be the same; in many cases, the fatty acid at position 2 is unsaturated, whereas the fatty acid at position 1 is saturated.

CLINICAL CORRELATES **Dipalmitoylphosphatidylcholine** serves as the major component of lung surfactant in adults, allowing the lungs to function normally. This phospholipid develops in the fetus after week 30 of gestation. Premature infants do not have an adequate amount of this phospholipid. As a result, **acute respiratory distress syndrome** can occur.

F. Regulation of triacylglycerol synthesis from carbohydrate

1. The synthesis of triacylglycerols from carbohydrate occurs in the liver in the **fed state**.
2. The **key regulatory enzymes** in the pathway are activated, and a high-carbohydrate diet causes their induction.
   a. The glycolytic enzymes **glucokinase**, **phosphofructokinase 1**, and **pyruvate kinase** are active (see Chapter 6 for mechanisms).
   b. **Pyruvate dehydrogenase** is dephosphorylated and active.
   c. **Pyruvate carboxylase** is activated by acetyl-CoA.

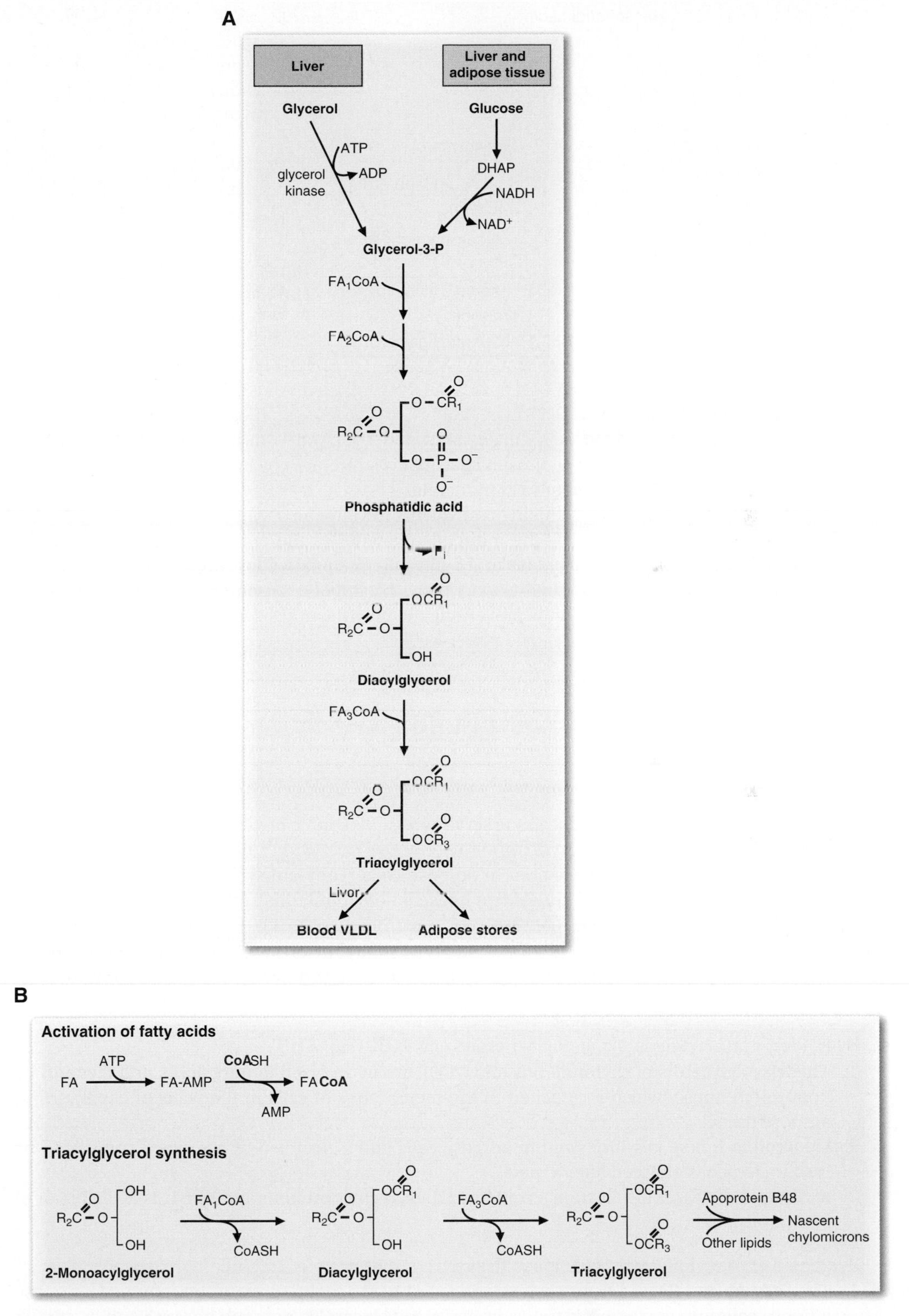

FIGURE 7.8. The synthesis of triacylglycerols in (A) liver, adipose tissue, and **(B)** intestinal cells. DHAP, dihydroxyacetone phosphate; R, aliphatic chain of a fatty acid; VLDL, very low-density lipoprotein.

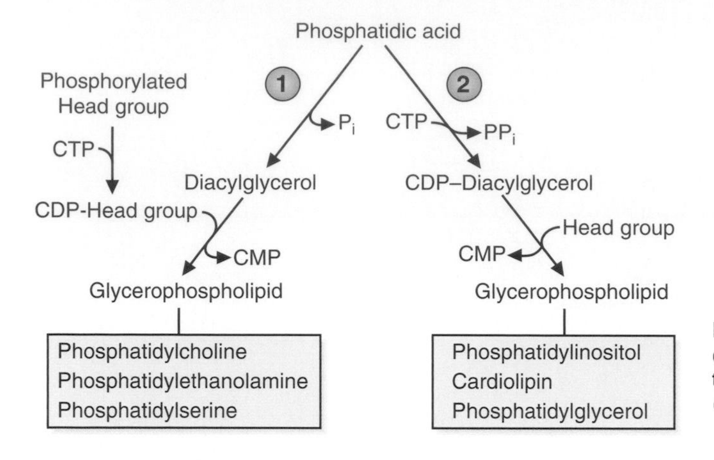

FIGURE 7.9. **Strategies for the addition of the head group to phosphatidic acid to form glycerophospholipids.** In both cases, CTP is used to drive the reaction.

d. **Citrate lyase** is inducible.
e. **Acetyl-CoA carboxylase** is induced, activated by citrate, and converted to its active, dephosphorylated state by a phosphatase that is stimulated by insulin.
f. The **fatty acid synthase complex** is inducible.

3. **NADPH**, which provides the reducing equivalents for fatty acid synthesis, is produced by the inducible **malic enzyme** and by the inducible enzymes of the pentose phosphate pathway, **glucose-6-phosphate dehydrogenase** and **6-phosphogluconate dehydrogenase.**
4. **Malonyl-CoA**, the product of acetyl-CoA carboxylase, **inhibits carnitine acyltransferase I** (carnitine palmitoyltransferase I), thus preventing the newly synthesized fatty acids from entering mitochondria and undergoing β-oxidation (see Fig. 7.15).

V. FORMATION OF TRIACYLGLYCEROL STORES IN ADIPOSE TISSUE

- The triacylglycerol stores of adipose tissue serve as a major source of fuel for the human body. The average 70-kg man has about 15 kg of fat.
- After a meal, triacylglycerols are stored in adipose cells. They are synthesized from fatty acids (derived mainly by the action of lipoprotein lipase on chylomicrons and VLDL) and from a glycerol moiety (derived from glucose).
- The storage of triacylglycerols in adipose tissue is mediated by insulin, which stimulates adipose cells to secrete lipoprotein lipase and to take up glucose, the source of glycerol (via the formation of DHAP) for triacylglycerol synthesis.

A. Hydrolysis of triacylglycerols of chylomicrons and VLDL (Fig. 7.10)

1. The triacylglycerols of chylomicrons and VLDL are hydrolyzed to **fatty acids** and **glycerol** by lipoprotein lipase, which is attached to the membranes of cells in the walls of capillaries in adipose tissue.
2. **Lipoprotein lipase** is synthesized in adipose cells and is secreted by a process stimulated by insulin, which is elevated after a meal.
   a. **Apoprotein C_{II}**, which is transferred from HDL to chylomicrons and VLDL, is an activator of lipoprotein lipase.

B. Synthesis of triacylglycerols in adipose tissue

1. The **fatty acids** released from chylomicrons and VLDL by lipoprotein lipase are taken up by adipose cells and converted to triacylglycerols, but glycerol is not used because adipose tissue lacks glycerol kinase (see Fig. 7.8).
   a. The **transport of glucose** into adipose cells is **stimulated by insulin**, which is elevated after a meal.

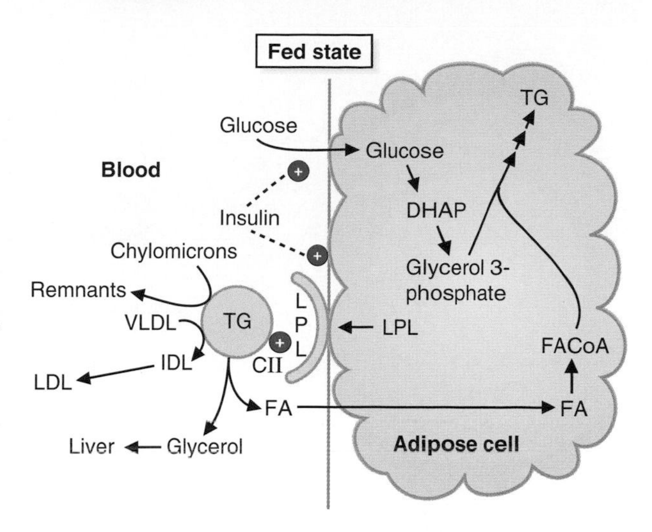

FIGURE 7.10. **The formation of triacylglycerol stores in adipose tissue in the fed state.** Note that insulin stimulates both the transport of glucose into adipose cells and the synthesis and secretion of lipoprotein lipase (LPL) from the cells. Apoprotein C-II activates LPL. DHAP, dihydroxyacetone phosphate; FA, fatty acid; LPL, lipoprotein lipase; ⊕, stimulated by; *circled* TG, triacylglycerol of chylomicrons and VLDL; TG, triacylglycerol.

b. Glucose is converted to **DHAP**, which is reduced by NADH to form **glycerol-3-phosphate**, which is used to produce the glycerol moiety.

2. The triacylglycerols are stored in large fat globules in adipose cells.

VI. CHOLESTEROL AND BILE SALT METABOLISM

- Although cholesterol is synthesized in most tissues of the body where it serves as a component of cell membranes, it is produced mainly in the liver and intestine.
- Cholesterol and cholesterol esters are transported in blood lipoproteins.
- All the carbons of cholesterol are derived from acetyl-CoA.
- The key intermediates in cholesterol biosynthesis are HMG-CoA, mevalonic acid, isopentenyl pyrophosphate, and squalene. The major regulatory enzyme is HMG-CoA reductase.
- In the liver, bile salts are formed from cholesterol by hydroxylation of the sterol ring, oxidation of the side chain, and conjugation of the carboxylic acid group with glycine or taurine.
- The bile salts are stored in the gallbladder and released during a meal to aid in lipid digestion. Ninety-five percent of the bile salts are resorbed and recycled.
- The sterol ring cannot be degraded. It is excreted intact, mainly as unresorbed bile salts.
- Cholesterol is stored in tissues as cholesterol esters.
- In certain endocrine tissues, cholesterol is converted to steroid hormones.
- A cholesterol precursor can be converted to 1,25-dihydroxycholecalciferol, the active form of vitamin D_3.

A. Cholesterol is synthesized from cytosolic acetyl-CoA by a sequence of reactions (Fig. 7.11).

1. **Glucose** is a major source of carbon for acetyl-CoA. Acetyl-CoA is produced from glucose by the same sequence of reactions used to produce cytosolic acetyl-CoA for fatty acid biosynthesis (see Fig. 7.6).
2. **Cytosolic acetyl-CoA** forms acetoacetyl-CoA, which condenses with another acetyl-CoA to form HMG-CoA.
   a. Acetyl-CoA undergoes similar reactions in the mitochondrion, where HMG-CoA is used for ketone body synthesis.
3. **Cytosolic HMG-CoA**, a key intermediate in cholesterol biosynthesis, is reduced in the endoplasmic reticulum to mevalonic acid by the regulatory enzyme HMG-CoA reductase.
   a. **HMG-CoA reductase** is inhibited by cholesterol. In the liver, it is also inhibited by bile salts and is induced when blood insulin levels are elevated. It is also regulated by phosphorylation by the AMP-activated protein kinase.

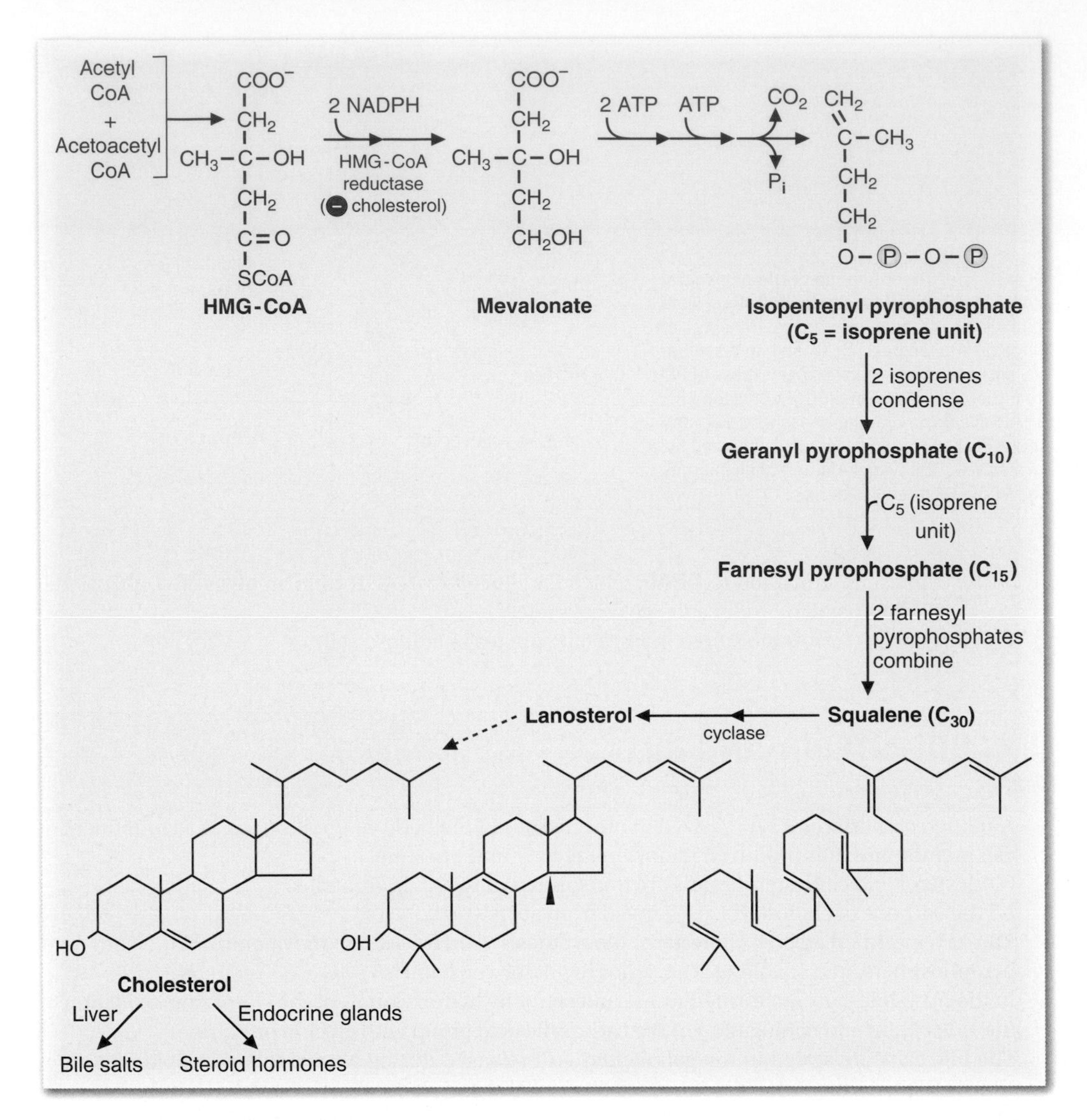

FIGURE 7.11. Cholesterol biosynthesis. HMG-CoA, hydroxymethylglutaryl-CoA; ⊖, inhibited by; *circled* P, phosphate group.

4. **Mevalonic acid** is phosphorylated and decarboxylated to form the 5-carbon (C-5) isoprenoid, isopentenyl pyrophosphate.
5. Two **isopentenyl pyrophosphate** units condense, forming a C-10 compound, geranyl pyrophosphate, which reacts with another C-5 unit to form a C-15 compound, farnesyl pyrophosphate.
6. **Squalene** is formed from two C-15 units and then oxidized and cyclized, forming lanosterol.
7. **Lanosterol** is converted to **cholesterol** in a series of steps.
8. The **ring structure** of cholesterol **cannot be degraded** in the body. The bile salts in the feces are the major form in which the steroid nucleus is excreted.

B. Bile salts are synthesized in the liver from cholesterol (Fig. 7.12).

1. An **α-hydroxyl group** is added to carbon 7 of cholesterol. A **7α-hydroxylase**, which is inhibited by bile salts, catalyzes this rate-limiting step (see Fig. 7.12A).
2. The **double bond** is **reduced** and **further hydroxylations** occur, resulting in two compounds. One has α-hydroxyl groups at positions 3 and 7, and the other at positions 3, 7, and 12.

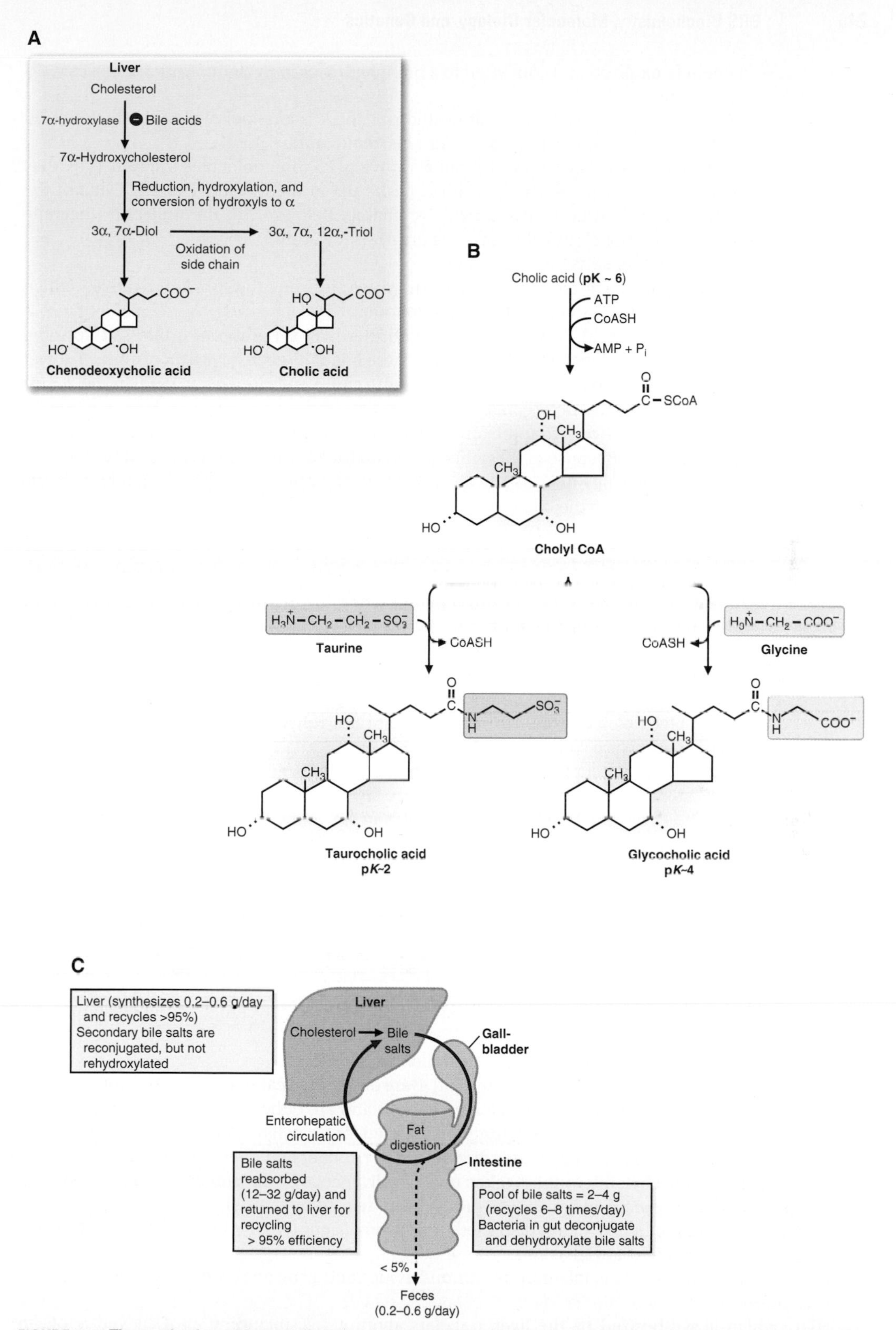

FIGURE 7.12. The synthesis and fate of bile salts. A. The synthesis of bile salts. Two sets of bile salts are generated, one with α-hydroxyl groups at positions 3 and 7 (the chenocholate series) and the other with α-hydroxyl groups at positions 3, 7, and 12 (the cholate series). **B.** Conjugation of bile salts, which lowers the pK_a of the bile salts, making them more effective detergents. **C.** An overview of bile salt metabolism.

3. The **side chain is oxidized** and converted to a branched, 5-carbon chain, containing a carboxylic acid at the end.
   a. The bile acid with hydroxyl groups at positions 3 and 7 is **chenocholic acid**. The bile acid with hydroxyl groups at positions 3, 7, and 12 is **cholic acid**.
   b. These bile acids each have a **p*K*** of about **6**. Above pH 6, the molecules are salts (i.e., they ionize and carry a negative charge). At pH 6 (the pH in the intestinal lumen), half of the molecules are ionized and carry a negative charge. Below pH 6, the molecules become protonated, and their charge decreases as the pH is lowered.
4. **Conjugation of the bile salts** (see Fig. 7.12B)
   a. The bile salts are activated by ATP and coenzyme A, forming their CoA derivatives, which can form conjugates with either **glycine** or **taurine**.
   b. **Glycine**, an amino acid, forms an amide with the carboxyl group of a bile salt, forming **glycocholic acid or glycochenocholic acid**. These bile salts each have a **p*K*** of about **4**, lower than the unconjugated bile salts, so they are more completely ionized at pH 6 in the gut lumen and serve as better detergents.
   c. **Taurine**, which is derived from the amino acid cysteine, forms an amide with the carboxyl group of a bile salt. Because of the sulfite group on the taurine moiety, the **taurocholic** and **taurochenocholic acids** have a **p*K*** of about **2**. They ionize very readily in the gut and are the best detergents among the bile salts.
5. **Fate of the bile salts** (see Fig. 7.12C)
   a. Cholic acid, chenocholic acid, and their conjugates are known as the primary bile salts. They are made in the liver and secreted via the **bile** through the **gallbladder** into the **intestine**, where, because they are amphipathic (contain both hydrophobic and hydrophilic regions), they aid in **lipid digestion**.
   b. In the intestine, bile salts can be **deconjugated** and **dehydroxylated** (at position 7) **by intestinal bacteria**.
   c. Bile salts are **resorbed** in the ileum and return to the liver, where they can be reconjugated with glycine or taurine. However, they are not rehydroxylated. Those that lack the 7α-hydroxyl group are called secondary bile salts.
   d. The **liver recycles** about 95% of the bile salts each day; 5% are lost in the feces.

C. Steroid hormones are synthesized from cholesterol, and 1,25-dihydroxycholecalciferol (active vitamin D_3) is synthesized from a precursor of cholesterol (see Chapter 9).

VII. BLOOD LIPOPROTEINS

- The blood lipoproteins serve to transport water-insoluble triacylglycerols and cholesterol from one tissue to another.
- The major carriers of triacylglycerols are chylomicrons and VLDL.
- The triacylglycerols of the chylomicrons and VLDL are digested in capillaries by lipoprotein lipase. The fatty acids that are produced are taken up by cells and are either oxidized for energy or converted to triacylglycerols and stored. The glycerol is used for triacylglycerol synthesis or converted to DHAP and oxidized for energy, either directly or after conversion to glucose in the liver.
- The remnants of the chylomicrons are taken up by the liver cells by the process of endocytosis and are degraded by lysosomal enzymes. The products are reused by the cell.
- VLDL is converted to IDL, which is degraded by lysosomal action in the liver or converted to LDL by further digestion of triacylglycerols.
- LDL, produced from IDL, is taken up by various tissues and provides cholesterol, which the tissues utilize.
- HDL, which is synthesized by the liver, transfers apoproteins, including apoC-II and apoE, to chylomicrons and VLDL.
- HDL picks up cholesterol from the cell membranes or from other lipoproteins. Cholesterol is converted to cholesterol esters by the lecithin:cholesterol acyltransferase (LCAT) reaction. Some

of this cholesterol ester is transferred to other lipoproteins. The cholesterol ester is carried by these lipoproteins or by HDL to the liver and hydrolyzed to free cholesterol, which is used for the synthesis of VLDL or converted to bile salts.

CLINICAL CORRELATES **Atherosclerosis** involves the formation of **lipid-rich plaques** in the intima of arteries. The plaques begin as fatty streaks containing foam cells, which initially are macrophages filled with oxidized LDL. These early lesions develop into fibrous plaques that can occlude an artery and cause a **myocardial infarct** or a **cerebral infarct**. The formation of these plaques is often associated with abnormalities in plasma lipoprotein metabolism. In contrast to the other lipoproteins, HDL has a protective effect.

A. Composition of the blood lipoproteins (Table 7.2)

The major components of lipoproteins are triacylglycerols, cholesterol, cholesterol esters, phospholipids, and proteins. The protein components (called apoproteins) are designated A, B, C, and E (Table 7.3).

1. **Chylomicrons** are the least dense of the blood lipoproteins because they have the most triacylglycerol and the least protein.
2. **VLDL** is more dense than chylomicrons but still has a high content of triacylglycerol.
3. **IDL**, which is derived from VLDL, is more dense than VLDL and has less than one-half the amount of triacylglycerol.
4. **LDL** has less triacylglycerol and more protein and, therefore, is more dense than the IDL from which it is derived. LDL has the highest content of cholesterol and its esters.
5. **HDL** is the most dense lipoprotein. It has the lowest triacylglycerol and the highest protein content.

B. Metabolism of chylomicrons (Fig.7.13)

1. Chylomicrons are **synthesized in intestinal epithelial cells**. Their triacylglycerols are derived from dietary lipid, and their major apoprotein is apoB-48.
2. Chylomicrons travel through the lymph into the blood. **ApoC-II**, the activator of lipoprotein lipase, and **apoE** are transferred to nascent chylomicrons **from HDL**, and mature chylomicrons are formed.

table 7.2 Characteristics of the Major Lipoproteins

Lipoprotein	Density Range (g/mL)	Particle Diameter (mm) Range	Electrophoretic Mobility	Lipid (%)[a] TG	Chol	PL	Function
Chylomicrons	0.930	75–1,200	Origin	80–95	2–7	3–9	Deliver dietary lipids
Chylomicron remnants	0.930–1.006	30–80	Slow pre-β				Return dietary lipids to the liver
VLDL	0.930–1.006	30–80	Pre-β	55–80	5–15	10–20	Deliver endogenous lipids
IDL	1.006–1.019	25–35	Slow pre-β	20–50	20–40	15–25	Return endogenous lipids to the liver; precursor of LDL
LDL	1.019–1.063	18–25	β	5–15	40–50	20–25	Deliver cholesterol to cells
HDL_2	1.063–1.125	9–12	α	5–10	15–25	20–30	Reverse cholesterol transport
HDL_3	1.125–1.210	5–9	α				Reverse cholesterol transport
Lip(a)	1.050–1.120	25	Pre-β				

TG, triacylglycerols; Chol, the sum of free and esterified cholesterol; PL, phospholipid; VLDL, very low-density lipoprotein; IDL, intermediate-density lipoprotein; LDL, low-density lipoprotein; HDL, high-density lipoprotein.

[a]The remaining percent composition is composed of apoproteins.

table 7.3 Characteristics of the Major Apoproteins

Apoprotein	Primary Tissue Source	Molecular Mass (daltons)	Lipoprotein Distribution	Metabolic Function
ApoA-I	Intestine, liver	28,016	HDL (chylomicrons)	Activates LCAT; structural component of HDL
ApoA-II	Liver	17,414	HDL (chylomicrons)	Uncertain; may regulate the transfer of apoproteins from HDL to other lipoprotein particles
ApoA-IV	Intestine	46,465	HDL (chylomicrons)	Uncertain; may be involved in the assembly of HDL and chylomicrons
ApoB-48	Intestine	264,000	Chylomicrons	Assembly and secretion of chylomicrons from small bowel
ApoB-100	Liver	540,000	VLDL, IDL, LDL	VLDL assembly and secretion; structural protein of VLDL, IDL, and LDL; ligand for LDL receptor
ApoC-I	Liver	6,630	Chylomicrons, VLDL, IDL, HDL	Unknown; may inhibit the hepatic uptake of chylomicron and VLDL remnants
ApoC-II	Liver	8,900	Chylomicrons, VLDL, IDL, HDL	Cofactor activator of lipoprotein lipase (LPL)
ApoC-III	Liver	8,800	Chylomicrons, VLDL, IDL, HDL	Inhibitor of LPL; may inhibit the hepatic uptake of chylomicrons and VLDL remnants
ApoE	Liver	34,145	Chylomicron remnants, VLDL, IDL, HDL	Ligand for binding of several lipoproteins to the LDL receptor, to the LDL receptor–related protein (LRP), and possibly to a separate apoE receptor
Apo(a)	Liver		Lipoprotein "little" a (Lp(a))	Unknown; consists of apoB-100 linked by a disulfide bond to apoprotein (a)

3. In peripheral tissues, particularly adipose and muscle, the triacylglycerols are **digested by lipoprotein lipase.**
4. The chylomicron remnants interact with the receptors on liver cells and are taken up by **endocytosis**. The contents are degraded by **lysosomal enzymes**, and the products (amino acids, fatty acids, glycerol, cholesterol, and phosphate) are released into the cytosol and reutilized.

C. Metabolism of VLDL (see Fig. 7.13B)

1. **VLDL** is synthesized in the **liver**, particularly after a high-carbohydrate meal. It is formed from triacylglycerols that are packaged with cholesterol, apoproteins (particularly apoB-100), and phospholipids, and it is released into the blood.
2. In **peripheral tissues**, particularly adipose and muscle, VLDL triacylglycerols are **digested by lipoprotein lipase**, and VLDL is converted to IDL.

CLINICAL CORRELATES

Familial lipoprotein lipase (LPL) deficiency is characterized by very high levels of circulating triglycerides (hypertriglyceridemia), due to the triglycerides in chylomicrons remaining in the circulation as they cannot be digested by the missing LPL activity. Patients present with recurrent abdominal pain (pancreatitis), the presence of xanthomas, and hepatosplenomegaly. The treatment consists of reducing fat consumption in the diet to less than 15% of total calories, or about 20 g of fat a day. This will greatly reduce chylomicron synthesis, and dramatically reduce the levels of circulating triglycerides. An assay for LPL involves measuring LPL activity in the blood of an individual treated with heparin; LPL is associated with capillary walls through binding to the glycosaminoglycan heparin, so circulating heparin can compete with the surface-bound heparin and release the bound LPL from the surface. If, in the presence of active apolipoprotein C-II (the activator of LPL), minimal activity is found, then an LPL deficiency can be diagnosed. **Familial apolipoprotein C-II (apoC-II) deficiency** is a very rare condition with the same symptoms as LPL deficiency. ApoC-II is the activator of LPL on the capillary surface, and in the absence of apoC-II activity, LPL activity is greatly reduced, leading to chylomicronemia. ApoC-II deficiency can be distinguished from LPL deficiency by the assay mentioned above for LPL activity in heparinized plasma. By adding functional apoC-II to the assay mixture, one can determine whether the LPL activity is still able to be activated by apoC-II. If it is, then the defect is in the apoC-II.

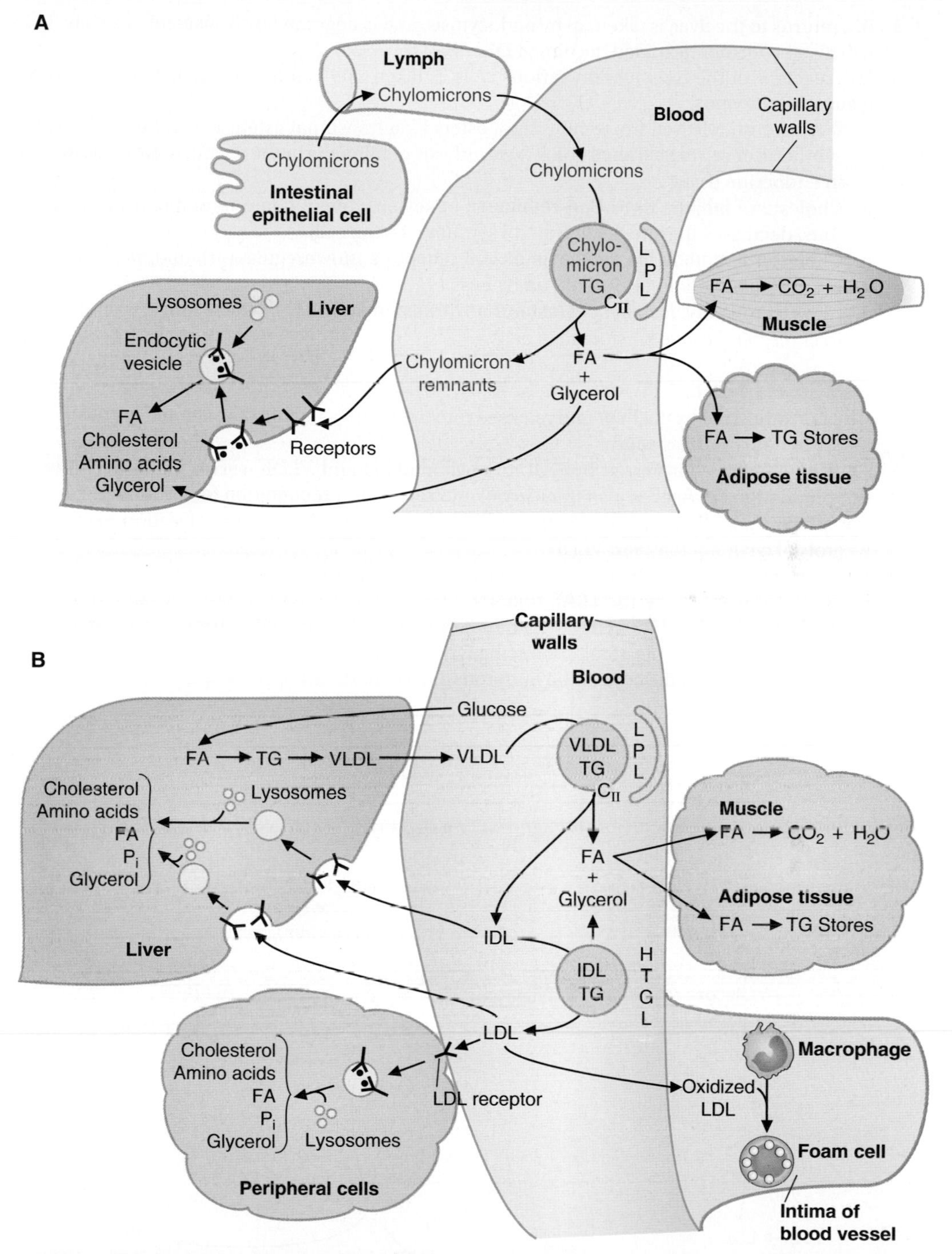

FIGURE 7.13. **The metabolism of chylomicrons and VLDL. A.** The fate of chylomicrons. **B.** The fate of VLDL. FA, fatty acids; P_i, inorganic phosphate; HTGL, hepatic triglyceride lipase.

3. **IDL** returns to the liver, is taken up by endocytosis, and is degraded by **lysosomal enzymes**. IDL can also be further degraded, forming LDL.
4. **LDL** reacts with the receptors on various cells, is taken up by endocytosis, and is digested by **lysosomal enzymes**.
   a. **Cholesterol**, released from cholesterol esters by a lysosomal esterase, can be used for the synthesis of cell **membranes** or for the synthesis of bile salts in the liver or **steroid hormones** in endocrine tissue.
   b. Cholesterol **inhibits HMG-CoA reductase** (a key enzyme in cholesterol biosynthesis) and, thus, decreases the rate of cholesterol synthesis by the cell.
   c. Cholesterol **inhibits the synthesis of LDL receptors** (downregulation), and, thus, reduces the amount of cholesterol taken up by cells.
   d. Cholesterol **activates acyl:cholesterol acyltransferase (ACAT)**, which converts cholesterol to cholesterol esters for storage in cells.

D. Metabolism of HDL (Fig. 7.14)

1. **HDL** is synthesized by the **liver** and released into the blood as small, disk-shaped particles. The major **protein** of **HDL** is **apoA**.
2. **ApoC-II**, which is transferred by HDL to chylomicrons and VLDL, serves as an **activator of lipoprotein lipase**. **ApoE** is also transferred and serves as a **recognition factor** for **cell surface receptors**. ApoC-II and apoE are transferred back to HDL following the digestion of triacylglycerols of chylomicrons and VLDL.
3. **Cholesterol**, obtained by **HDL** from the cell membranes or from other lipoproteins, is converted to **cholesterol esters** by the **LCAT reaction**, which is activated by apoA_I. A fatty acid from position 2 of lecithin (phosphatidylcholine), a component of HDL, forms an ester with the 3-hydroxyl group of cholesterol, producing lysolecithin and a cholesterol ester. As cholesterol esters accumulate in the core of the lipoprotein, HDL particles become spheroids.

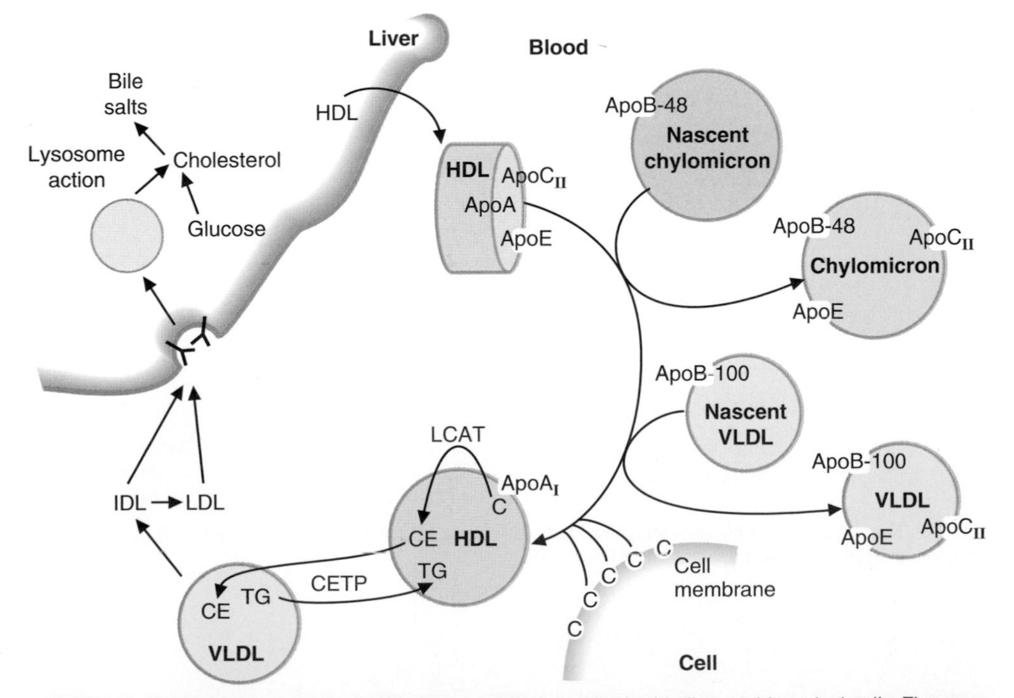

FIGURE 7.14. The functions and fate of HDL. Nascent HDL is synthesized in liver and intestinal cells. The steps are described in the text. C, cholesterol; CE, cholesterol ester; LCAT, lecithin:cholesterol acyltransferase; PL, phospholipid; TG, triacylglycerol.

CLINICAL CORRELATES **Familial LCAT deficiency** will lead to increased blood cholesterol levels. LCAT is the enzyme that esterifies cholesterol in HDL particles (through removal of a fatty acid at position 2 of phosphatidylcholine, and esterification of the cholesterol with this fatty acid). The inability of HDL to accept cholesterol from the tissues leads to elevated free cholesterol in the blood, which tends to deposit in specific tissues–the cornea, kidneys, and erythrocytes. The major complications of this disorder are **renal failure, anemia, and corneal opacities. Tangier disease** is a codominant disorder resulting in a greatly **reduced level of HDL** in the circulation. The characteristic feature of this disease is orange tonsils, due to the buildup of lipid within the tonsils. Owing to the loss of the protective effect of HDL, these individuals are also subject to **premature coronary heart disease**. The mutation is in the ABC1 protein, which is responsible for transporting cholesterol from the cells to the HDL particle. There is no treatment for this disorder.

4. **HDL transfers cholesterol esters** to other lipoproteins in exchange for various lipids. Cholesterol ester transfer protein (CETP) mediates this exchange. HDL and other lipoproteins carry the cholesterol esters back to the liver.
5. **HDL particles** and other lipoproteins are taken up by the liver by endocytosis and **hydrolyzed by lysosomal enzymes**.
6. Cholesterol, released from cholesterol esters, can be packaged by the liver in VLDL and released into the blood or converted to bile salts and secreted into the bile.

CLINICAL CORRELATES In the **hyperlipidemias**, the blood levels of cholesterol or triacylglycerols, or both, are elevated resulting from overproduction of lipoproteins or defects in various stages of their degradation. Elevations of blood lipid levels (particularly LDL) are associated with a high incidence of heart attacks and strokes. In **familial hypercholesterolemia**, cellular receptors for LDL are defective. Therefore, LDL is not taken up at a normal rate by the cells and degraded by lysosomal enzymes. The consequent increase of blood LDL (which contains a large amount of cholesterol and cholesterol ester) is associated with **xanthomas** (lipid deposits often found under the skin) and **coronary artery disease**. The treatment may involve diets low in saturated fat and cholesterol, HMG-CoA reductase inhibitors (e.g., lovastatin), bile acid binding resins, and nicotinic acid (niacin). In **hypertriglyceridemia** due to deficiencies in LPL or apoC-II (the lipoprotein lipase activator), triacylglycerol levels rise markedly because of decreased degradation of VLDL and chylomicrons. These deficiencies are associated with characteristic xanthomas and an intolerance to fatty foods. Low-fat diets may be effective (see below). In **diabetes mellitus (DM)**, VLDL levels are often elevated, which results in high blood triacylglycerol levels. Cholesterol may also be elevated. In diabetes, elevated VLDL levels result from deranged carbohydrate and lipid metabolism caused by decreased insulin levels (Type 1 DM) or insulin resistance (Type 2 DM). The consumption of trans–fatty acids (trans fats) is also detrimental to overall lipid health. Trans fats (which occur when polyunsaturated fatty acids are partially hydrogenated in order to increase their shelf life; the act of reducing some of the double bonds in the fatty acids results in some trans double bonds being created. Recall that virtually all naturally occurring unsaturated fatty acids are of the cis configuration.) raise LDL levels, and reduce HDL levels, through an ill-defined mechanism. The recommended amount of trans fat consumption is no more than 1% of total calories consumed, which covers the small amount of trans fats found in our foods.

VIII. FATE OF ADIPOSE TRIACYLGLYCEROLS

- During fasting, fatty acids and glycerol are released from adipose triacylglycerol stores and serve as a source of fuel for other tissues.
- Insulin falls and glucagon rises during fasting, causing the activation of hormone-sensitive lipase by a cAMP-dependent mechanism. The hormone-sensitive lipase initiates the conversion of adipose triacylglycerols to fatty acids and glycerol, which are released into the blood.

- Fatty acids are transported in the blood complexed with albumin, taken up by various tissues, and oxidized for energy. In the liver, fatty acids are converted to ketone bodies, and glycerol is converted to glucose. These fuels serve as energy sources for other tissues.

A. Lipolysis of adipose triacylglycerols

1. In the **fasting state**, lipolysis of adipose triacylglycerols occurs.
2. Insulin levels decrease and glucagon levels rise, **stimulating lipolysis**. (Epinephrine and other hormones promote lipolysis by the same mechanism.)
   a. **cAMP** levels rise, and protein kinase A is activated.
   b. **Protein kinase A** phosphorylates and thus activates the hormone-sensitive lipase of adipose tissue.
3. The **hormone-sensitive lipase** initiates lipolysis, and fatty acids and glycerol are released from adipose cells.

B. Fate of fatty acids and glycerol

1. **Fatty acids** are carried on albumin in the blood.
   a. In tissues such as **muscle** and **kidney**, fatty acids are oxidized for energy.
   b. In the **liver**, fatty acids are converted to **ketone bodies** that are oxidized by tissues such as muscle and kidney. During starvation (after fasting has lasted for about 3 or more days), the brain uses ketone bodies for energy.
2. **Glycerol** is used by the liver as a source of carbon for **gluconeogenesis**, which produces glucose for tissues such as the brain and red blood cells.

IX. FATTY ACID OXIDATION

- Fatty acids, which are the major source of energy in the human body, are oxidized mainly by β-oxidation.
- Prior to oxidation, long-chain fatty acids are activated, forming fatty acyl-CoA, which is transported into mitochondria by a carnitine carrier system.
- The process of β-oxidation occurs in mitochondria. In the four steps that produce $FADH_2$ and NADH, two carbons are cleaved from a fatty acyl-CoA and are released as acetyl-CoA. This series of steps is repeated until an even-chain fatty acid is completely converted to acetyl-CoA.
- ATP is obtained when $FADH_2$ and NADH interact with the electron transport chain or when acetyl-CoA is oxidized further.
- In tissues such as skeletal and heart muscle, acetyl-CoA enters the tricarboxylic acid (TCA) cycle and is oxidized to CO_2 and H_2O. In the liver, acetyl-CoA is converted to ketone bodies.
- β-Oxidation is regulated by the mechanisms that control oxidative phosphorylation (i.e., by the demand for ATP).
- Fatty acids also undergo α- and ω-oxidation and peroxisomal oxidation.

A. Activation of fatty acids

1. In the cytosol of the cell, long-chain fatty acids are activated by **ATP** and **coenzyme A**, and **fatty acyl-CoA** is formed (Fig. 7.15). Short-chain fatty acids are activated in mitochondria.
2. The **ATP** is converted to **AMP and pyrophosphate** (PP_i), which is cleaved by pyrophosphatase to two inorganic phosphates (2 P_i). Because two high-energy phosphate bonds are cleaved, the equivalent of two molecules of ATP is used for fatty acid activation.

B. Transport of fatty acyl-CoA from the cytosol into mitochondria

1. **Fatty acyl-CoA** from the cytosol reacts with **carnitine** in the outer mitochondrial membrane, forming fatty acylcarnitine. The enzyme is **carnitine acyltransferase I** (CAT I), which is also called carnitine palmitoyltransferase I (CPT I). **Fatty acylcarnitine** passes to the inner membrane, where it **re-forms to fatty acyl-CoA**, which enters the matrix. The second enzyme is carnitine acyltransferase II (CAT II).

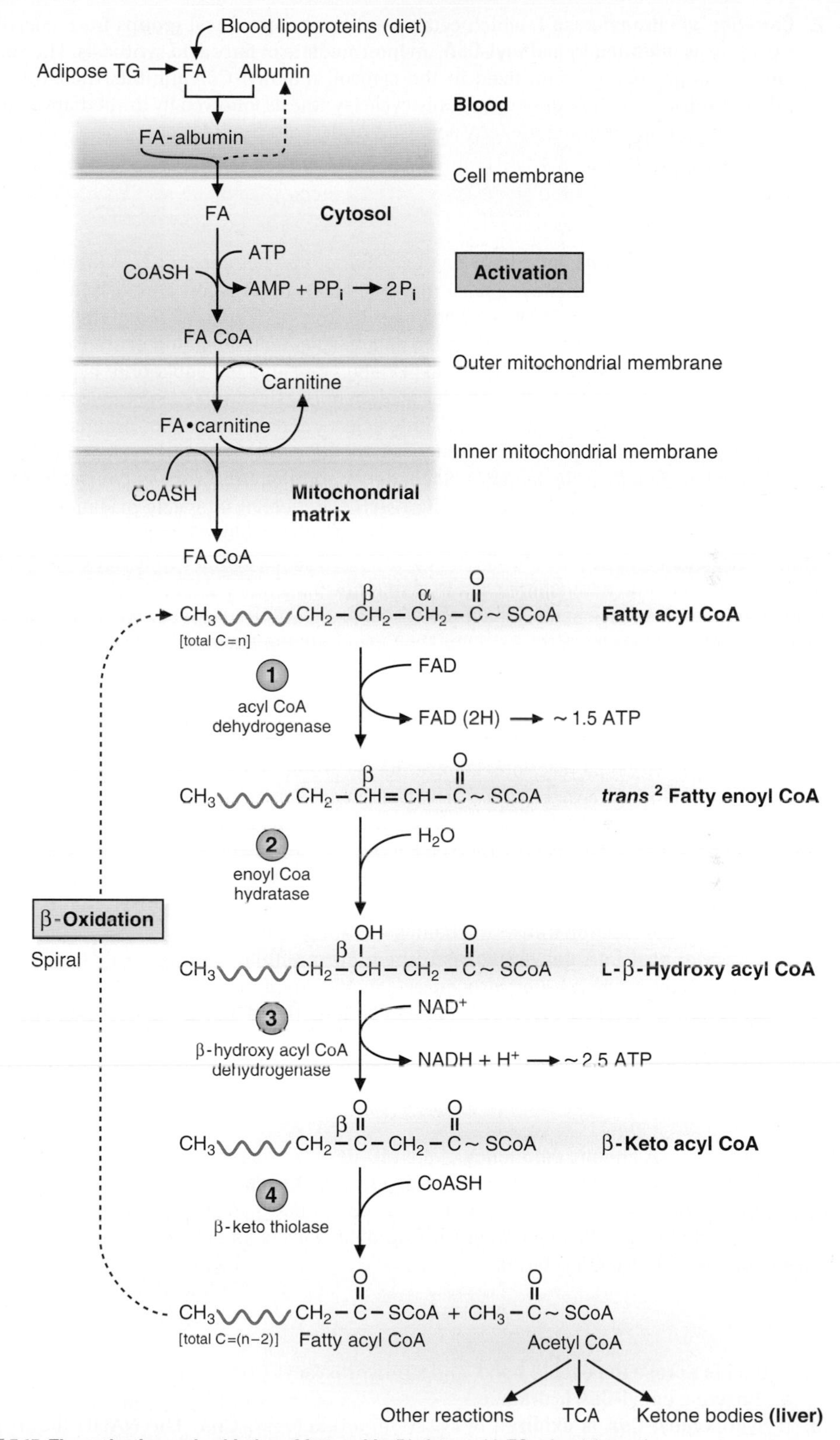

FIGURE 7.15. The activation and oxidation of fatty acids. FA, fatty acid; TG, triacylglycerol.

2. **Carnitine acyltransferase I**, which catalyzes the transfer of acyl groups from coenzyme A to carnitine, is **inhibited by malonyl-CoA**, an intermediate in fatty acid synthesis. Therefore, when fatty acids are being synthesized in the cytosol, malonyl-CoA inhibits their transport into mitochondria and, thus, prevents a futile cycle (synthesis followed by immediate degradation).
3. Inside the mitochondrion, the fatty acyl-CoA undergoes **β-oxidation**.

CLINICAL CORRELATES **Carnitine deficiency** can be either primary or secondary. A **primary carnitine deficiency** results from an inability to transport carnitine (a nonessential compound) into the cells that need it–liver and muscle. This results in reduced fatty acid oxidation, and in the case of muscle, exercise intolerance and muscle damage during exercise occurs, leading to myobloginuria. In the liver, lack of fatty acid oxidation can lead to **hypoketotic hypoglycemia**–low blood glucose levels (due to lack of energy for gluconeogenesis) coupled with below-normal levels of ketone bodies (due to the deficiency in fatty acid oxidation). The major organs and systems involved include the cardiac muscle (cardiomyopathy), the central nervous system (not enough fuel), and the skeletal muscle (muscle damage). **Secondary carnitine deficiency** is caused by other metabolic disorders (such as a carnitine acyltransferase II mutation, or fatty acid oxidation disorders). Acylcarnitine derivatives can accumulate within tissues and the blood in a secondary carnitine deficiency. The accumulation of long-chain acylcarnitines is toxic, and can lead to a sudden cardiac arrest. The accumulation of organic acids, from defects in amino acid metabolism, can also lead to carnitine depletion, as these acids, which are formed from CoA derivatives, are often transferred to carnitine as a means to remove the accumulating acid from the body.

C. β-Oxidation of even-chain fatty acids

β-Oxidation (in which all reactions involve the β-carbon of a fatty acyl-CoA) is a spiral consisting of four sequential steps, the first three of which are similar to those in the TCA cycle between succinate and oxaloacetate. These steps are repeated until all the carbons of an even-chain fatty acyl-CoA are converted to acetyl-CoA (see Fig. 7.15).

1. **FAD accepts hydrogens** from a fatty acyl-CoA in the first step. A double bond is produced between the α- and β-carbons, and an enoyl-CoA is formed. The $FADH_2$ that is produced interacts with the electron transport chain, generating ATP.
   a. Enzyme: **acyl-CoA dehydrogenase** (there are multiple variants of this enzyme, such as short-chain acyl-CoA dehydrogenase [SCAD], medium-chain acyl-CoA dehydrogenase [MCAD], long-chain acyl-CoA dehydrogenase [LCAD], and very long-chain acyl-CoA dehydrogenase [VLCAD].)

CLINICAL CORRELATES A **genetic deficiency** of the **MCAD** of β-oxidation prevents the normal use of fatty acids as fuels. **Fasting hypoglycemia** results, and dicarboxylic acids, produced by ω-oxidation, are excreted in the urine, as are acylglycines (glycine will conjugate with dicarboxylic acids to aid in their excretion). MCAD deficiency is an autosomal recessive disease with a frequency of 1/15,000 live births.

2. **H_2O adds across the double bond**, and a β-hydroxyacyl-CoA is formed.
   a. Enzyme: **enoyl-CoA hydratase**
3. **β-Hydroxyacyl-CoA is oxidized** by NAD^+ to a β-ketoacyl-CoA. The NADH that is produced interacts with the electron transport chain, generating ATP.
   a. Enzyme: **L-3-hydroxyacyl-CoA dehydrogenase** (which is specific for the L-isomer of the β-hydroxyacyl-CoA)

CLINICAL CORRELATES

Jamaican vomiting sickness is caused by a toxin (hypoglycin) from the unripe fruit of the akee tree. This toxin inhibits an acyl-CoA dehydrogenase of β-oxidation; consequently, more glucose must be oxidized to compensate for the decreased ability to use fatty acids as a fuel, and severe **hypoglycemia** can occur. ω-Oxidation of fatty acids is increased, and dicarboxylic acids are excreted in the urine. Unwary children are usually the victims of this frequently fatal disease.

4. The **bond between the α- and β-carbons** of the β-ketoacyl-CoA is cleaved by a **thiolase** that requires coenzyme A. Acetyl-CoA is produced from the two carbons at the carboxyl end of the original fatty acyl-CoA, and the remaining carbons form a fatty acyl-CoA that is two carbons shorter than the original.
   a. Enzyme: **β-ketothiolase**
5. The shortened **fatty acyl-CoA repeats** these four steps. Repetitions continue until all the carbons of the original fatty acyl-CoA are converted to acetyl-CoA.
   a. The 16-carbon palmitoyl-CoA undergoes seven repetitions.
   b. In the last repetition, a 4-carbon fatty acyl-CoA (butyryl-CoA) is cleaved to two acetyl-CoAs.
6. **Energy is generated** from the products of β-oxidation.
   a. When one palmitoyl-CoA is oxidized, seven $FADH_2$, seven NADH, and eight acetyl-CoA are formed.
      (1) The seven $FADH_2$ each generate approximately 1.5 ATP, for a total of about 10.5 ATP.
      (2) The seven NADH each generate about 2.5 ATP, for a total of about 17.5 ATP.
      (3) The eight acetyl-CoA can enter the TCA cycle, each producing about 10 ATP, for a total of about 80 ATP.
      (4) From the oxidation of palmitoyl-CoA to CO_2 and H_2O, a total of about 108 ATP are produced.
   b. The **net ATP** produced from palmitate that enters the cell from the blood is about 106 because palmitate must undergo activation (a process that requires the equivalent of 2 ATP) before it can be oxidized (108 ATP − 2 ATP = 106 ATP).
   c. **The oxidation of other fatty acids** will yield different amounts of ATP.

D. Oxidation of odd-chain and unsaturated fatty acids

1. **Odd-chain fatty acids** produce acetyl-CoA and propionyl-CoA.
   a. These fatty acids repeat the four steps of the β-oxidation spiral, producing **acetyl-CoA** until the last cleavage when the three remaining carbons are released as propionyl CoA.
   b. **Propionyl-CoA**, but not acetyl-CoA, can be converted to glucose. (See Chapter 6, Section VI D 2; and Fig. 8.7.)
2. **Unsaturated fatty acids**, which comprise about half the fatty acid residues in human lipids, require enzymes in addition to the four that catalyze the repetitive steps of the β-oxidation spiral. The reaction pathway differs depending on whether the double bond is at an even- or odd-numbered carbon position.
   a. **β-Oxidation** occurs until a double bond of the unsaturated fatty acid is near the carboxyl end of the fatty acyl chain.
      (1) If the double bond originated at an odd carbon number (such as 3, 5, 7, etc.), an isomerase will convert the eventual cis-Δ 3 to a trans-Δ 2 fatty acid (Fig.7.16).
      (2) If the double bond originated at an even carbon number (such as 4, 6, 8, etc.), the eventual trans-Δ 2, cis-Δ 4 fatty acid will be reduced by a 2,4-dienoyl-CoA reductase, which requires NADPH and generates a *trans*-Δ 3-acyl-CoA and $NADP^+$. The isomerase will convert the trans-Δ 3 fatty acyl-CoA to a trans-Δ 2 fatty acyl-CoA to allow β-oxidation to continue.
   b. ATP yield for unsaturated fatty acids
      (1) If the double bond originated at an odd carbon position, then compared to a fully saturated fatty acid of the same carbon length, there will be 1.5 ATP less for each unsaturation at the odd carbon position, due to one less $FADH_2$ being produced for each unsaturation.

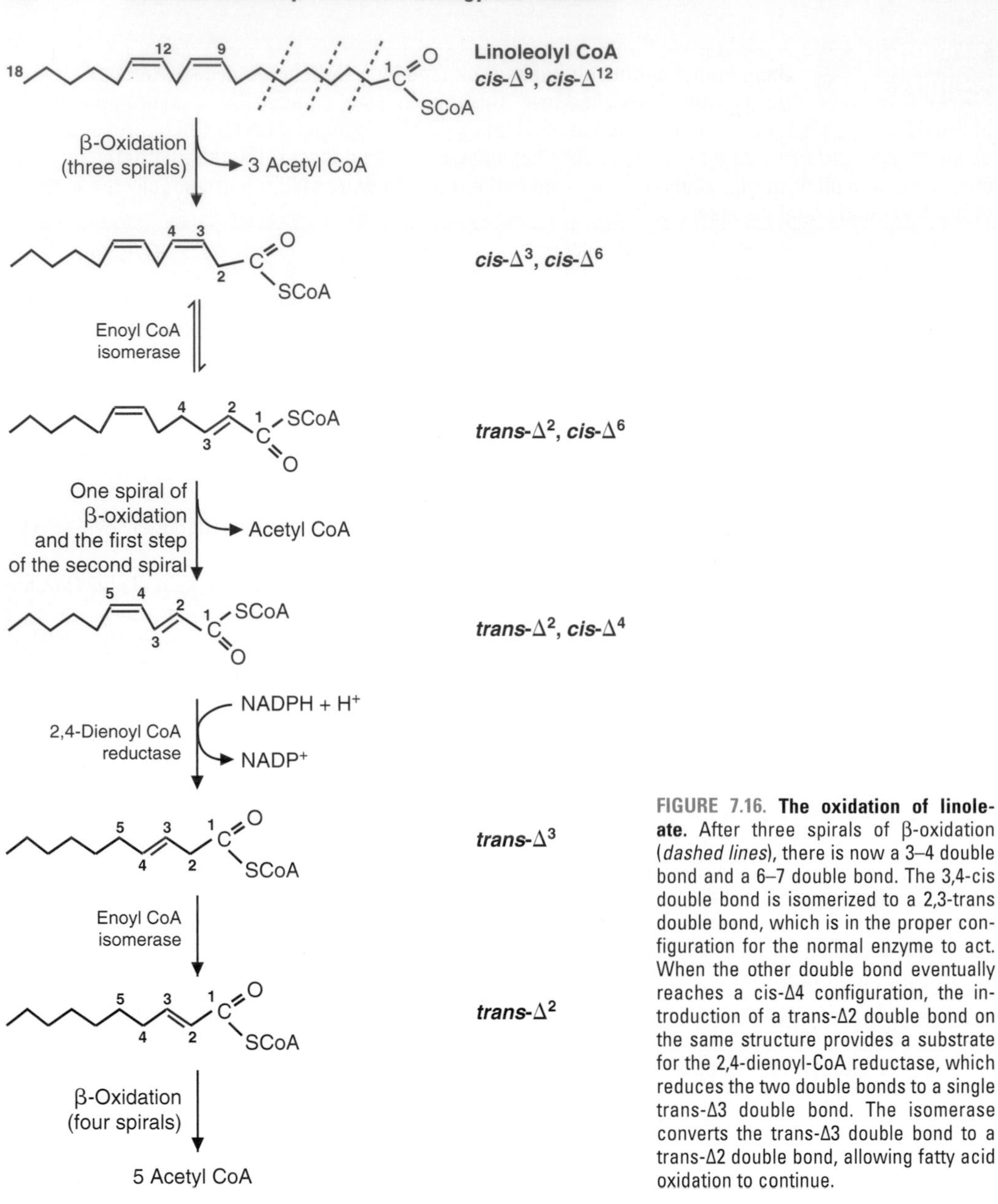

FIGURE 7.16. **The oxidation of linoleate.** After three spirals of β-oxidation (*dashed lines*), there is now a 3–4 double bond and a 6–7 double bond. The 3,4-cis double bond is isomerized to a 2,3-trans double bond, which is in the proper configuration for the normal enzyme to act. When the other double bond eventually reaches a cis-Δ4 configuration, the introduction of a trans-Δ2 double bond on the same structure provides a substrate for the 2,4-dienoyl-CoA reductase, which reduces the two double bonds to a single trans-Δ3 double bond. The isomerase converts the trans-Δ3 double bond to a trans-Δ2 double bond, allowing fatty acid oxidation to continue.

(2) If the double bond originated at an even-numbered carbon position, then compared to an equivalent length fully saturated fatty acid, there is one less NADH equivalent (or 2.5 ATP) produced, due to the use of NADPH in the step catalyzed by the 2,4-dienoyl-CoA reductase.

E. ω-Oxidation of fatty acids

1. The ω (omega)-**carbon** (the methyl carbon) of fatty acids is oxidized to a carboxyl group in the endoplasmic reticulum.
2. β-Oxidation can then occur in mitochondria at this end of the fatty acid as well as from the original carboxyl end. **Dicarboxylic acids** are produced.

F. **Oxidation of very long-chain fatty acids in peroxisomes**

1. The process differs from β-oxidation in that **molecular O_2** is used by VLCAD, **hydrogen peroxide** (H_2O_2) is formed, and **$FADH_2$** is not generated at the first step of β-oxidation in the peroxisomes.
2. The shorter-chain fatty acids that are produced travel to mitochondria, where they undergo β-oxidation, generating **ATP**.

G. **α-Oxidation of fatty acids in peroxisomes**

1. **Branched chain fatty acids** are oxidized at the α-carbon (mainly in brain and other nervous tissue), and the carboxyl carbon is released as CO_2. Branches can interfere with the normal β-oxidation pathway, most often at the acyl-CoA dehydrogenase step.
2. The fatty acid is thus degraded by one carbon initially, and then two carbons at a time. Both acetyl-CoA and propionyl-CoA are products if the branches are methyl groups.

CLINICAL CORRELATES

Peroxisomal disorders include **adrenoleukodystrophy** and **Zellweger syndrome.** Adrenoleukodystrophy is an X-linked disorder that affects the **transport of very long-chain fatty acids into the peroxisomes** for initial oxidation events. The loss of this activity leads to the accumulation of very long-chain fatty acids, which appear to target the adrenal glands and the myelin sheath for destruction, through incorporation into the membrane lipids surrounding those structures. Children who inherit this mutation will experience cognitive deficiencies, nervous system deterioration, seizures, visual impairment, and may develop Addison's disease, a loss of adrenal gland function. Zellweger syndrome is a **peroxisome biogenesis disorder**, which is one of the luekodystrophies. The lack of peroxisomes leads to the buildup of very long-chain fatty acids, an inability to degrade branched fatty acids (such as phytanic acid), and gives rise to Zellweger syndrome, neonatal adrenoleukodystrophy, and infantile refsum disease. Myelin structure is altered owing to the accumulation of these fatty acids, particularly **phytanic** acid. Patient symptoms include an enlarged liver, mental retardation, and seizures. Infants with Zellweger syndrome lack appropriate muscle strength and may be unable to move or suck because of their weakened muscles.

X. KETONE BODY SYNTHESIS AND UTILIZATION

- The ketone bodies, acetoacetate and β-hydroxybutyrate, serve as a source of fuel. They are synthesized mainly in liver mitochondria whenever fatty acid levels are high in the blood.
- Fatty acids are activated in liver cells and converted to acetyl-CoA, generating ATP. As NADH and ATP levels rise, acetyl-CoA accumulates.
- Acetyl-CoA reacts with acetoacetyl-CoA to form HMG-CoA, which is cleaved to form acetoacetate.
- Acetoacetate can be reduced to a second ketone body, 3-hydroxybutyrate (β-hydroxybutyrate), by NADH.
- Acetone is produced by spontaneous (nonenzymatic) decarboxylation of acetoacetate.
- The liver cannot use ketone bodies because it lacks the thiotransferase enzyme that activates acetoacetate.
- Ketone bodies are used as fuels by tissues such as muscle and kidney. During starvation (after about 3 to 5 days of fasting), the brain also oxidizes ketone bodies.
- Ketone bodies enter cells, where 3-hydroxybutyrate is oxidized to form acetoacetate in a reaction that produces NADH.
- Acetoacetate, obtained directly from the blood or produced from 3-hydroxybutyrate, is activated to acetoacetyl-CoA by reacting with succinyl-CoA. Acetoacetyl-CoA is cleaved by β-ketothiolase to two acetyl-CoAs, which enter the TCA cycle and are oxidized to CO_2 and H_2O, generating ATP.

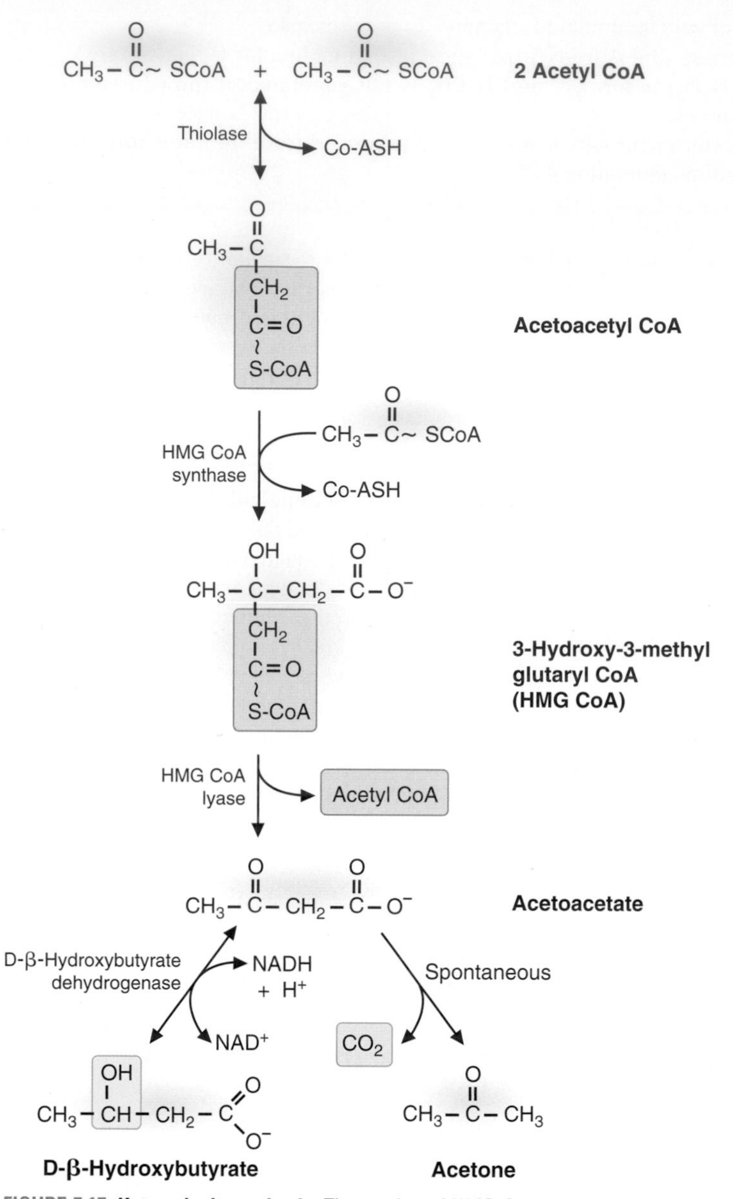

FIGURE 7.17. **Ketone body synthesis.** The portion of HMG-CoA shown in the *tinted box* is released as acetyl-CoA, and the remainder of the molecule forms acetoacetate. Acetoacetate is reduced to β-hydroxybutyrate if NADH levels are high, and the spontaneous decarboxylation of acetoacetate forms acetone.

A. **The synthesis of ketone bodies (Fig. 7.17) occurs in liver mitochondria when fatty acids are in high concentration in the blood (during fasting, starvation, or as a result of a high-fat diet).** **β-Oxidation** produces NADH and ATP and results in the accumulation of acetyl-CoA. The liver is producing glucose, using oxaloacetate, so there is decreased condensation of acetyl-CoA with oxaloacetate to form citrate.

1. **Two molecules of acetyl-CoA** condense to produce acetoacetyl-CoA. This reaction is catalyzed by thiolase or an isoenzyme of **thiolase**.
2. Acetoacetyl-CoA and acetyl-CoA form HMG-CoA in a reaction catalyzed by HMG-CoA synthase.
3. **HMG-CoA** is cleaved by HMG-CoA lyase to form acetyl-CoA and acetoacetate.

4. **Acetoacetate** can be reduced by an NAD-requiring dehydrogenase (3-hydroxybutyrate dehydrogenase) to **3-hydroxybutyrate**. This is a reversible reaction.
5. Acetoacetate is also spontaneously **decarboxylated**, in a nonenzymatic reaction, forming **acetone** (the source of the odor on the breath of ketotic diabetics).
6. The **liver** lacks succinyl-CoA-acetoacetate-CoA transferase (a thiotransferase) and so it **cannot use ketone bodies**. Therefore, acetoacetate and 3-hydroxybutyrate are released into the blood by the liver.

CLINICAL CORRELATES If a patient with Type 1 **diabetes** who has failed to take insulin is suffering from an illness or is subjected to stress, blood glucose may rise markedly, and **diabetic ketoacidosis** may occur. Decreased insulin and elevated glucagon levels cause adipose tissue to release increased amounts of fatty acids, which are converted to ketone bodies by the liver. Decarboxylation of acetoacetate produces acetone, which gives a characteristic odor to the patient's breath. Ketone body levels can become extremely high, causing a **metabolic acidosis** that, if not treated rapidly and effectively, can lead to **coma and death**.

B. Utilization of ketone bodies

1. When ketone bodies are released from the liver into the blood, they are taken up by peripheral tissues such as **muscle and kidney**, where they are oxidized for energy.
   a. During **starvation**, ketone bodies in the blood increase to a level that permits entry into **brain** cells, where they are oxidized.
2. **Acetoacetate** can enter cells directly, or it can be produced from the oxidation of 3-hydroxybutyrate by 3-hydroxybutyrate dehydrogenase. NADH is produced by this reaction and can generate ATP.
3. Acetoacetate is activated by reacting with succinyl-CoA to form **acetoacetyl-CoA** and succinate. The enzyme is succinyl-CoA-acetoacetate-CoA transferase (a thiotransferase).
4. Acetoacetyl-CoA is cleaved by **thiolase** to form two acetyl-CoAs, which enter the TCA cycle and are oxidized to CO_2 and H_2O.
5. **Energy is produced** from the oxidation of ketone bodies.
   a. One acetoacetate produces two acetyl-CoAs, each of which can generate about 10 ATP, or a total of about 20 ATP via the TCA cycle.
   b. However, the activation of acetoacetate results in the generation of one less ATP because GTP, the equivalent of ATP, is not produced when succinyl-CoA is used to activate acetoacetate. (In the TCA cycle, when succinyl-CoA forms succinate, GTP is generated.) Therefore, the oxidation of acetoacetate produces a net yield of only 19 ATP.
   c. When **3-hydroxybutyrate** is oxidized, 2.5 additional ATP are formed because the oxidation of 3-hydroxybutyrate to acetoacetate produces NADH.

XI. PHOSPHOLIPID AND SPHINGOLIPID METABOLISM

- Phospholipids and sphingolipids are the major components of cell membranes. They are amphipathic molecules; that is, one portion of the molecule is hydrophilic and associates with H_2O, and another portion contains the hydrocarbon chains derived from fatty acids, which are hydrophobic and associate with lipids (see Figs. 7.2 and 7.3).
- Phosphoglycerides (the major phospholipids) contain glycerol, fatty acids, and phosphate. The phosphate is esterified to choline, serine, ethanolamine, or inositol.
- The phosphoglycerides are synthesized via a number of pathways.
- The degradation of phosphoglycerides involves phospholipases, which are each specific for one of the ester linkages of the phosphodiester bonds.
- The sphingolipids include sphingomyelin (which contains phosphocholine) and the cerebrosides and gangliosides (which contain sugar residues). These compounds are the major components of cell membranes in nervous tissue.

- The sphingolipids are synthesized from ceramide, which is produced from serine and palmitoyl-CoA.
- During degradation, the phosphocholine and sugar units of the sphingolipids are removed by lysosomal enzymes.

A. Synthesis and degradation of phosphoglycerides

The phosphoglycerides are synthesized by a process similar in its initial steps to triacylglycerol synthesis (glycerol-3-phosphate combines with two fatty acyl-CoAs to form **phosphatidic acid**) (see Fig. 7.9).

1. Synthesis of phosphatidylinositol

a. **Phosphatidic acid** reacts with CTP to form CDP-diacylglycerol, which reacts with inositol to form phosphatidylinositol.

b. **Phosphatidylinositol** can be further phosphorylated to form phosphatidylinositol 4,5-bisphosphate, which is cleaved in response to various stimuli to form the compounds inositol 1,4,5-trisphosphate (IP_3) and diacylglycerol (DAG), which serve as second messengers (see Chapter 9).

2. Synthesis of phosphatidylethanolamine, phosphatidylcholine, and phosphatidylserine (Fig. 7.18)

a. **Phosphatidic acid** releases inorganic phosphate, and diacylglycerol is produced. **Diacylglycerol** reacts with compounds containing cytosine nucleotides to form **phosphatidylethanolamine** and **phosphatidylcholine**.

(1) Phosphatidylethanolamine

(a) Diacylglycerol reacts with CDP-ethanolamine to form phosphatidylethanolamine.

(b) Phosphatidylethanolamine can also be formed by the decarboxylation of phosphatidylserine.

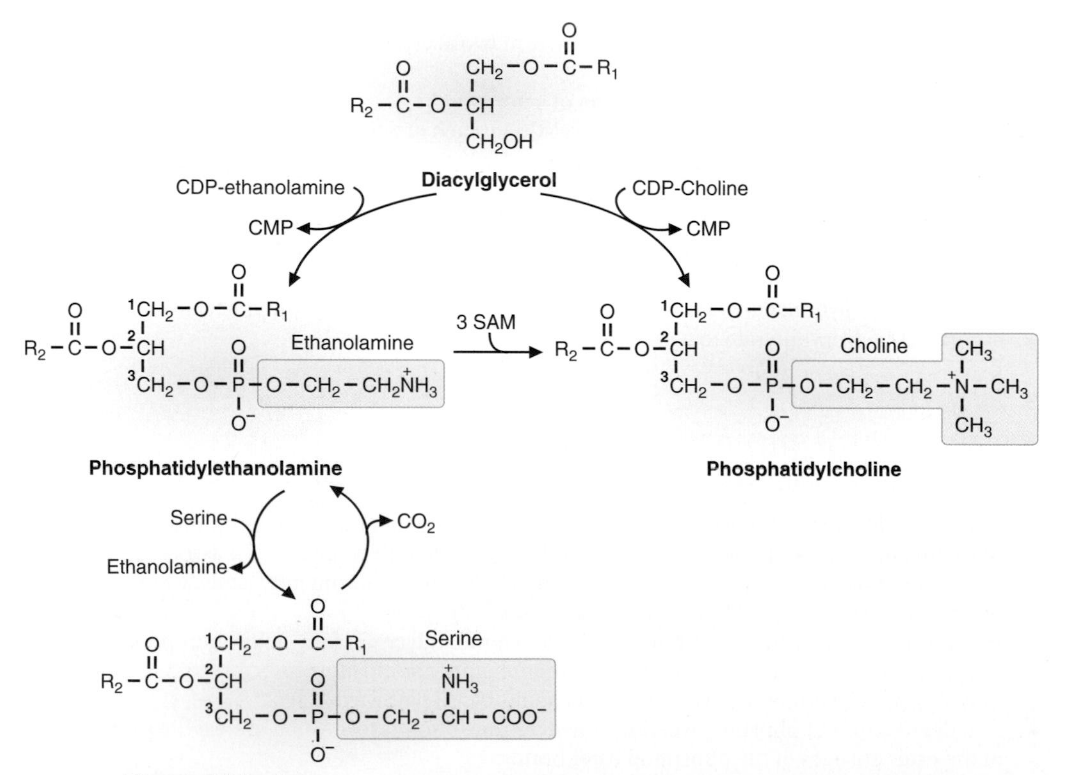

FIGURE 7.18. The synthesis of major phospholipids. SAM, *S*-adenosylmethionine, a methyl group donor for many biochemical reactions

(2) **Phosphatidylcholine**

(a) Diacylglycerol reacts with CDP-choline to form **phosphatidylcholine (lecithin)**.

(b) Phosphatidylcholine can also be formed by the methylation of phosphatidylethanolamine. *S*-Adenosylmethionine (SAM) provides the methyl groups.

1. In addition to being an important component of cell membranes and the blood lipoproteins, phosphatidylcholine provides the fatty acid for the synthesis of cholesterol esters in HDL by the **LCAT reaction** and, as the dipalmitoyl derivative, serves as **lung surfactant**. If choline is deficient in the diet, phosphatidylcholine can be synthesized de novo from glucose (see Fig. 7.18).

(3) **Phosphatidylserine**

(a) Phosphatidylserine is formed when phosphatidylethanolamine reacts with serine, which replaces the ethanolamine moiety (see Fig. 7.18).

3. **Degradation of phosphoglycerides**

a. Phosphoglycerides are hydrolyzed by **phospholipases**.

b. Phospholipase A_1 releases the fatty acid at position 1 of the glycerol moiety; phospholipase A_2 releases the fatty acid at position 2; phospholipase C releases the phosphorylated base (e.g., choline) at position 3; and phospholipase D releases the free base.

B. **Synthesis and degradation of sphingolipids (Fig. 7.19)**

Sphingolipids are derived from **serine** rather than glycerol.

1. **Serine** condenses with **palmitoyl-CoA** in a reaction in which the serine is decarboxylated by a pyridoxal phosphate–requiring enzyme.
2. The product is converted to a derivative of **sphingosine**.
3. A fatty acyl-CoA forms an amide with the nitrogen, and the resulting compound is **ceramide**.
4. The hydroxymethyl moiety of ceramide combines with various compounds to form **sphingolipids**.
   a. **Phosphatidylcholine** reacts with ceramide to form **sphingomyelin**.
   b. **UDP-galactose**, or **UDP-glucose**, reacts with ceramide to form **galactocerebrosides** or **glucocerebrosides**.
   c. A series of sugars can add to ceramide, UDP-sugars serving as precursors. **CMP-NANA** (*N*-acetylneuraminic acid, a sialic acid) can form branches from the carbohydrate chain. These ceramide-oligosaccharide compounds are **gangliosides**.
5. Sphingolipids are degraded by **lysosomal enzymes**. A loss of one of these enzymes can lead to a sphingolipidosis (see Table 7.1).

CLINICAL CORRELATES

Understanding the pathways of lipid metabolism has allowed various drugs to be developed to attempt to control lipid levels in humans. **Statins** reduce cholesterol levels through an inhibition of **HMG-CoA reductase**; the reduced intracellular cholesterol levels leads to the upregulation of the **LDL receptor**, which removes LDL (and its cholesterol) from the circulation, thereby resulting in a reduction of circulating cholesterol. Individuals with nonfunctional LDL receptors (homozygous) would not be helped by statin treatment. **Bile acid sequesterants** work by binding to the bile acids in the intestine and interfering with the enterohepatic circulation, as the bile acid–drug combination is excreted in the feces rather than recycling the bile acid. This forces intracellular cholesterol levels to be lowered, as more cholesterol has to be converted to bile acids for digestion still to operate properly. Drugs such as ezetimide interfere with **cholesterol absorption** in the small intestine, such that dietary cholesterol is excreted in the feces. Such drugs, in combination with statins, may be more effective than statins alone in reducing circulating cholesterol levels. **Fibrates** act by **reducing triglyceride levels** and, in some cases, elevating HDL levels. The fibrates activate the transcription factor PPARα, which regulates the genes involved in fatty acid and triglyceride synthesis in the liver.

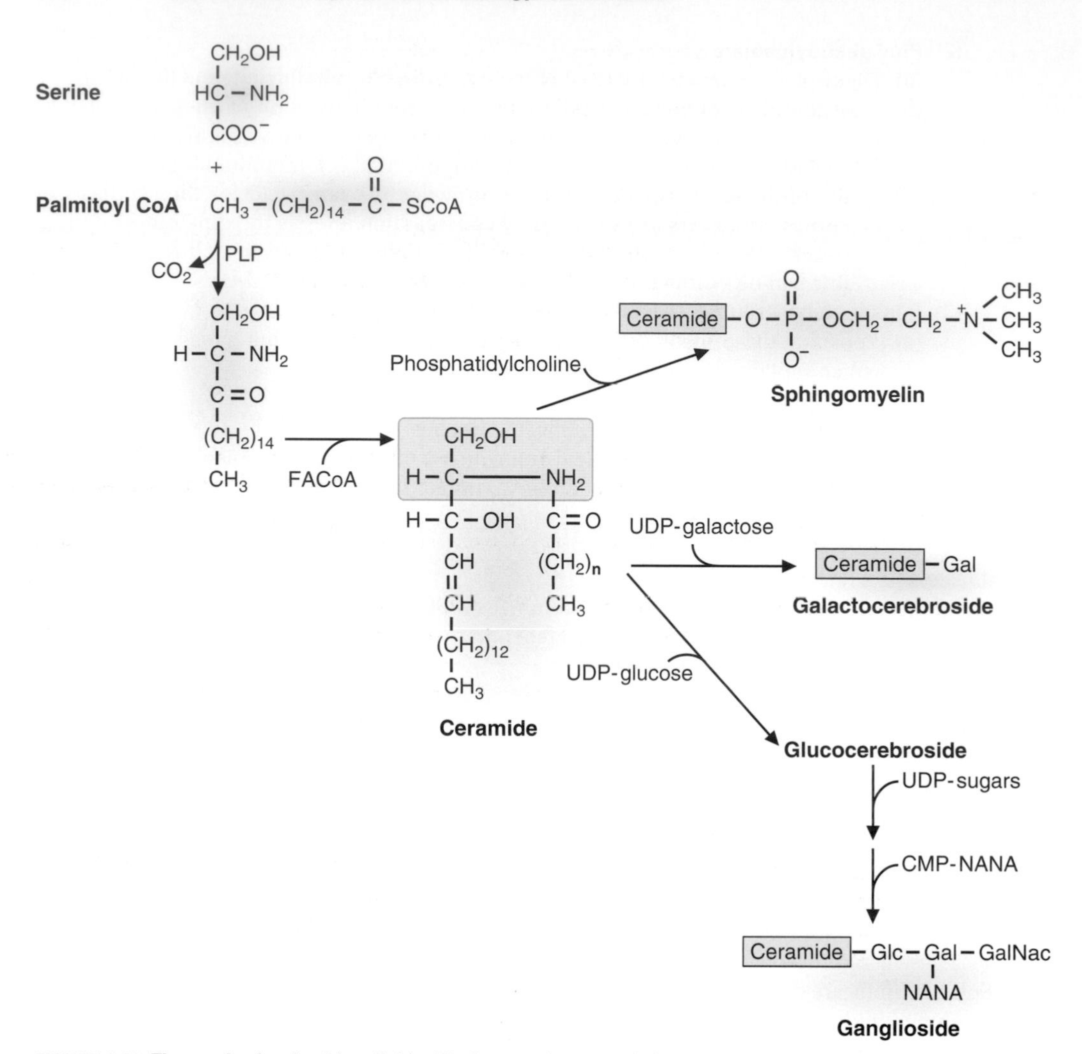

FIGURE 7.19. **The synthesis of sphingolipids.** FA, fatty acyl groups derived from fatty acids; Gal, galactose; GalNAc, *N*-acetylgalactosamine; Glc, glucose; NANA, *N*-acetylneuraminic acid; PLP, pyridoxal phosphate. The *dashed box* contains the portion of ceramide derived from serine.

XII. METABOLISM OF THE EICOSANOIDS

- The eicosanoids (prostaglandins, thromboxanes, and leukotrienes) are synthesized from polyunsaturated fatty acids (e.g., arachidonic acid). These fatty acids are released from membrane phospholipids by phospholipase A_2, which is inhibited by glucocorticoids and other steroidal anti-inflammatory agents.
- For prostaglandin synthesis, the polyunsaturated fatty acid is cyclized and oxidized by a cyclooxygenase, which is inhibited by aspirin and the nonsteroidal anti-inflammatory agents. Further oxidations and rearrangements occur that produce a series of prostaglandins, including the prostacyclins.
- Thromboxanes are produced from certain prostaglandins.
- Leukotrienes are produced from arachidonic acid by a pathway that differs from that for prostaglandin synthesis.

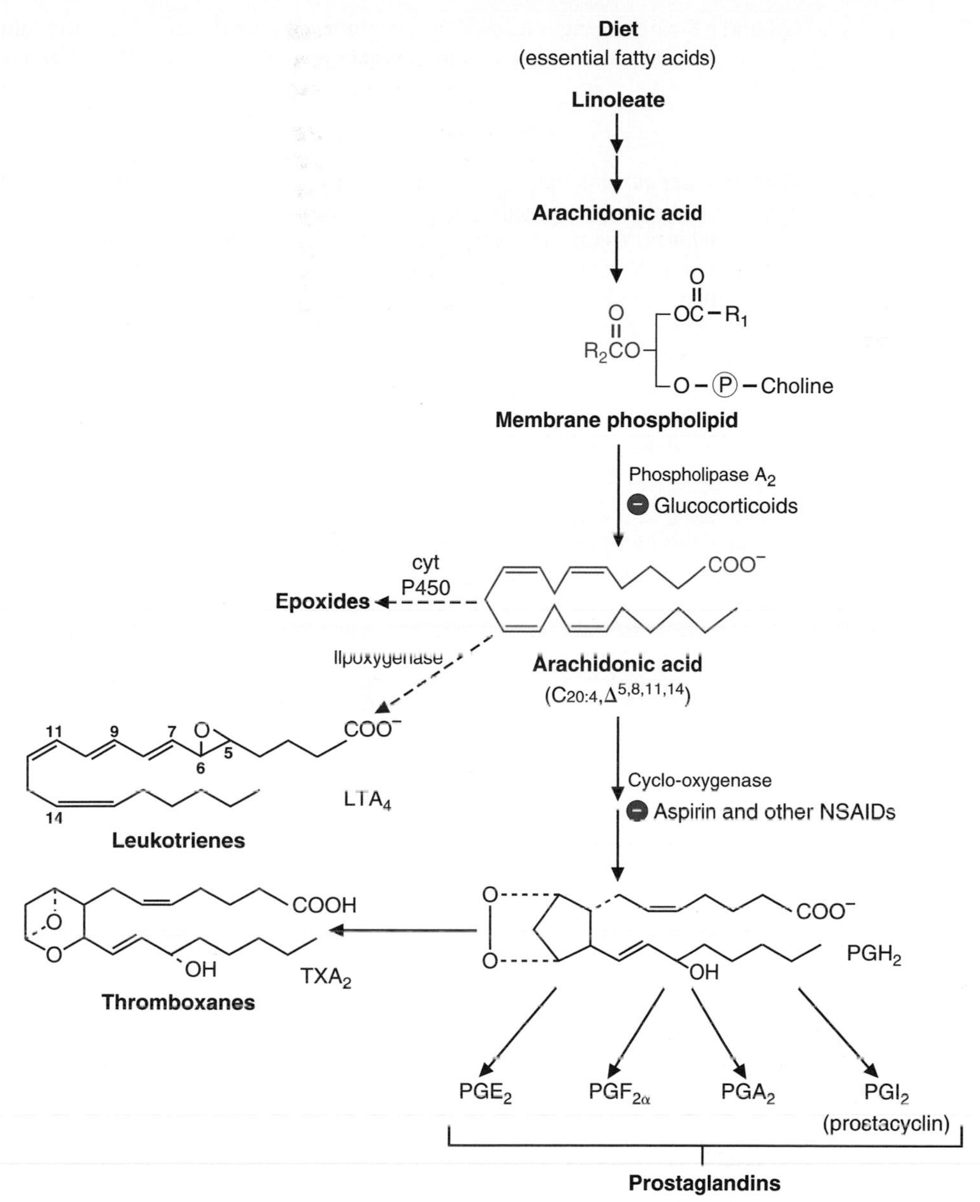

FIGURE 7.20. **Prostaglandins, thromboxanes, and leukotrienes.** LT, leukotriene; PG, prostaglandin; TX, thromboxane. For each of the classes of prostaglandins (H, E, F, A), the *ring* contains hydroxyl and keto groups at different positions, and the subscript refers to the number of double bonds in the nonring portion. The class with two double bonds is derived from arachidonate. Other classes (with one or three double bonds) are derived from other polyunsaturated fatty acids. NSAIDS, nonsteroidal anti-inflammatory drugs; –, inhibits.

A. Prostaglandins, prostacyclins, and thromboxanes (Fig.7.20)

1. **Polyunsaturated fatty acids** containing 20 carbons and three to five double bonds (e.g., arachidonic acid) are usually esterified to position 2 of the glycerol moiety of phospholipids in cell membranes. These fatty acids require **essential fatty acids** such as dietary linoleic acid (18:2, $\Delta^{9,12}$) for their synthesis.
2. The polyunsaturated fatty acid is cleaved from the membrane phospholipid by **phospholipase A_2**, which is inhibited by the steroidal anti-inflammatory agents.

3. Oxygen is added and a 5-carbon ring is formed by a **cyclooxygenase** that produces the initial prostaglandin, which is converted to other classes of **prostaglandins** and to the **thromboxanes**.

CLINICAL CORRELATES **Nonsteroidal anti-inflammatory drugs** (NSAIDs), such as aspirin and ibuprofen, inhibit the cyclooxygenase involved in prostaglandin synthesis. These drugs reduce pain, inflammation, and fever associated with the action of the prostaglandins. Aspirin irreversibly acetylates the enzyme in platelets, inhibiting thromboxane (TXA_2) formation, thus reducing platelet aggregation for the life span of the platelet. Because platelets turn over rapidly, the daily ingestion of small doses of aspirin is often recommended to inhibit platelet aggregation (thrombus formation) that, in conjunction with atherosclerotic plaques, often precipitates heart attacks. There are two forms of cyclooxygenase, **COX1 and COX2**. Aspirin and many nonsteroidal anti-inflammatory drugs affect both, but **COX2-specific drugs**, such as celecoxib, are reversible inhibitors that only affect COX2, the enzyme induced during inflammatory events.

a. The **prostaglandins** have a multitude of effects that differ from one tissue to another and include inflammation, pain, fever, and aspects of reproduction. These compounds are known as **autocoids** because they exert their effects primarily in the tissue in which they are produced.
b. Certain **prostacyclins** (PGI_2), produced by vascular endothelial cells, **inhibit platelet aggregation**, whereas certain **thromboxanes** (TXA_2) **promote platelet aggregation**.

4. Inactivation of the prostaglandins occurs when the molecule is oxidized from the carboxyl and ω-methyl ends to form **dicarboxylic acids** that are excreted in the urine.

B. Leukotrienes

Arachidonic acid, derived from membrane phospholipids, is the major precursor for the synthesis of the leukotrienes.

1. In the first step, oxygen is added by lipoxygenases, and a family of linear molecules, hydroperoxyeicosatetraenoic acids (**HPETEs**), is formed.
2. A series of compounds, comprising the family of leukotrienes, is produced from these HPETEs. The leukotrienes are involved in **allergic reactions**. Leukotrienes also contribute to the symptoms of asthma by acting as bronchoconstricting agents, narrowing the airway, and making it more difficult to breathe.

XIII. ETHANOL METABOLISM

A. Ethanol is both lipid- and water-soluble.

B. Greater than 80% of the absorbed ethanol is metabolized in the liver.

1. One pathway of ethanol metabolism is through **alcohol and acetaldehyde dehydrogenase** (Fig. 7.21).
2. The other route is through the liver **microsomal ethanol oxidizing system** (MEOS), which requires a cytochrome P450–containing enzyme (CYP2E1) (Fig. 7.22).
3. The ethanol-metabolizing enzymes exist as a family of isozymes, and individual variations in isozyme expression can determine an individual's tolerance to alcohol.
   a. There are at least seven different forms of medium-chain alcohol dehydrogenase.
   b. There are at least three different forms of aldehyde dehydrogenase.
      (1) ALDH2 is the mitochondrial version of aldehyde dehydrogenase and exhibits a low K_m for its substrate.
      (2) ALDH2*2 is a common allelic variant with a greatly increased K_m and reduced V_{max} as compared to ALDH2.

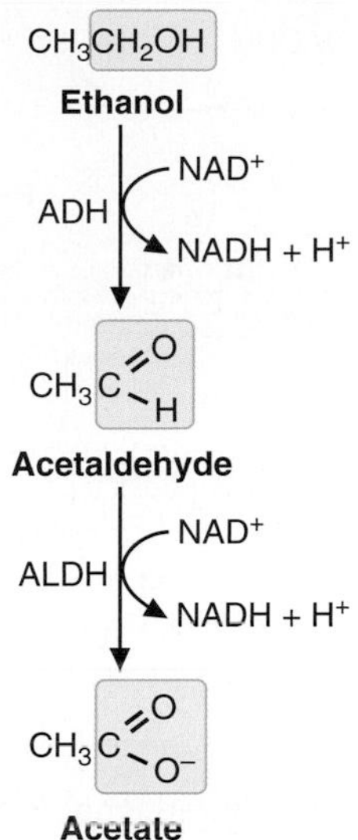

FIGURE 7.21. **The pathway of ethanol metabolism.** ADH, alcohol dehydrogenase; ALDH, acetaldehyde dehydrogenase.

(3) **Acetaldehyde accumulation** causes nausea and vomiting.
  (a) Homozygotes for ALDH2*2 have a very low tolerance for alcohol, due to the rapid accumulation of acetaldehyde.
  (b) Inhibition of aldehyde dehydrogenase by drugs is a treatment for alcoholism (disulfiram).

4. The acetate generated by ethanol metabolism in the liver can be converted to acetyl-CoA for energy use by the liver, or secreted into the circulation for use by skeletal muscles.
5. The MEOS system (Fig. 7.23)
   a. MEOS has a much higher K_m for ethanol than alcohol dehydrogenase, and is induced by ethanol.
      (1) Many cytochrome P450 enzymes are induced by substrate, which gives rise to tolerance.
      (2) Drugs are metabolized via cytochrome P450 enzymes, and as the rate of clearance of the drug increases, higher doses of the drug are required to obtain the same effect (tolerance).
      (3) Ethanol can inhibit certain P450 systems, which can lead to adverse ethanol drug interactions, particularly when tolerance is involved.
   b. MEOS is utilized when ethanol concentrations are high.

C. Toxic effects of ethanol

1. Alcohol-induced liver disease includes **fatty liver, alcohol-induced hepatitis, and cirrhosis.**
   a. Fatty liver results from ethanol inhibition of fatty acid oxidation, resulting in fatty acid buildup in the liver.

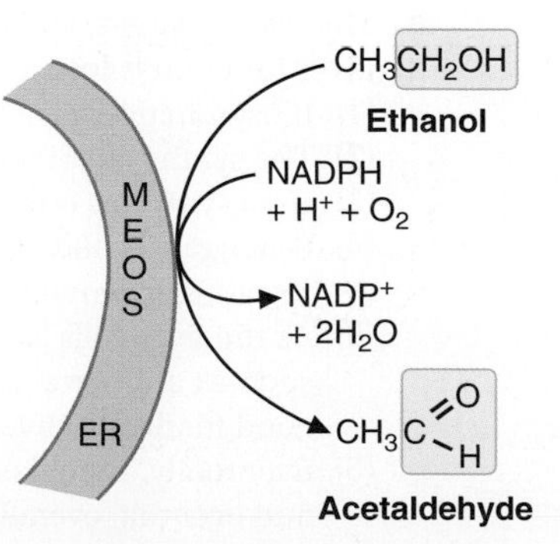

FIGURE 7.22. **The reaction catalyzed by the microsomal ethanol oxidizing system (MEOS) in the endoplasmic reticulum (ER).**

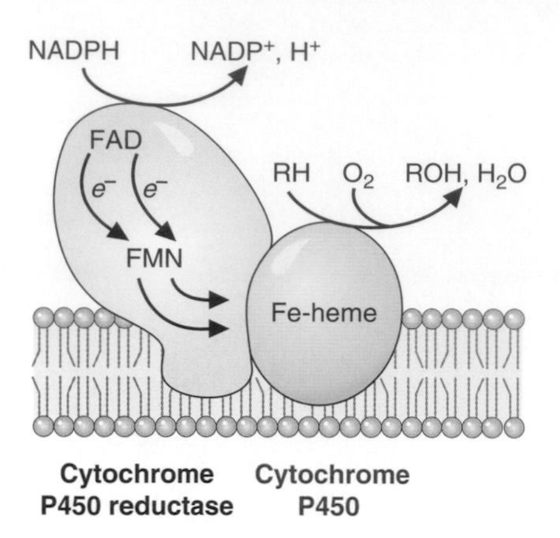

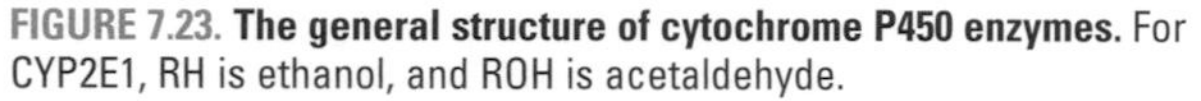
FIGURE 7.23. **The general structure of cytochrome P450 enzymes.** For CYP2E1, RH is ethanol, and ROH is acetaldehyde.

b. Alcohol-induced hepatitis results from acetaldehyde and free-radical generation from ethanol metabolism in the liver (via the MEOS oxidation pathway).
c. Cirrhosis occurs as an accumulation of damage to the hepatocytes, leading to fibrosis and loss of liver function.

2. Acute effects of ethanol arise from the increased **NADH/NAD$^+$ ratio** due to ethanol metabolism.
   a. Alterations in fatty acid metabolism occur as fatty acid oxidation is inhibited by the high levels of NADH. Fatty acids accumulate in the liver, produce triacylglycerols, and increase the production of VLDL. The export of VLDL is diminished in chronic alcoholics, leading to a fatty liver, due to an impairment in protein synthesis due to chronic liver dysfunction.
   b. **Alcohol-induced ketoacidosis** occurs because of the high levels of acetyl-CoA produced from both ethanol metabolism and fatty acid oxidation. The high NADH inhibits the TCA cycle, leading to ketone body formation. The tissues, however, are using acetate as fuel instead of the ketone bodies, which leads to the ketoacidosis.
   c. **Lactic acidosis, hyperuricemia, and hypoglycemia** occur owing to the high NADH levels in the liver. The high NADH levels convert pyruvate to lactate, and the elevated lactate interferes with the excretion of uric acid by the kidney. Hypoglycemia also results from the elevated NADH levels owing to the diversion of gluconeogenic precursors away from gluconeogenesis. Due to the high NADH levels lactate is not converted to pyruvate, malate is not converted to oxaloacetate, and glycerol-3-phosphate is not converted to DHAP.
3. **Acetaldehyde toxicity** (Fig. 7.24) leads to alcohol-induced hepatitis and free-radical damage (through acetaldehyde binding to free radical–defense enzymes) (see Fig. 7.24, steps 1 and 2.
4. **Free-radical formation** is enhanced during ethanol intoxication owing to the induction of MEOS (see Fig. 7.24, steps 3, 4, and 5).
5. **Hepatic cirrhosis and loss of liver function** may ultimately occur owing to **chronic ethanol intoxication.**
   a. The liver enlarges, and increases its fat content and collagen content.
   b. Liver function is lost, and normal metabolic pathways are lost.
      (1) Biosynthetic and detoxification pathways are lost.
      (2) The synthesis of blood proteins is reduced.
      (3) Urea synthesis is reduced, resulting in hyperammonemia.
      (4) Conjugation and excretion of bilirubin (the product of heme degradation) are reduced, resulting in jaundice.
      (5) As the liver cells lose their ability to maintain their membranes, liver-specific enzymes, such as aminotransferases (AST and ALT), will be measurable in the blood, and are a good marker for liver damage.
      (6) Eventually, fibroblasts infiltrate the liver and produce collagen, leading to liver fibrosis, and eventual overall liver failure.

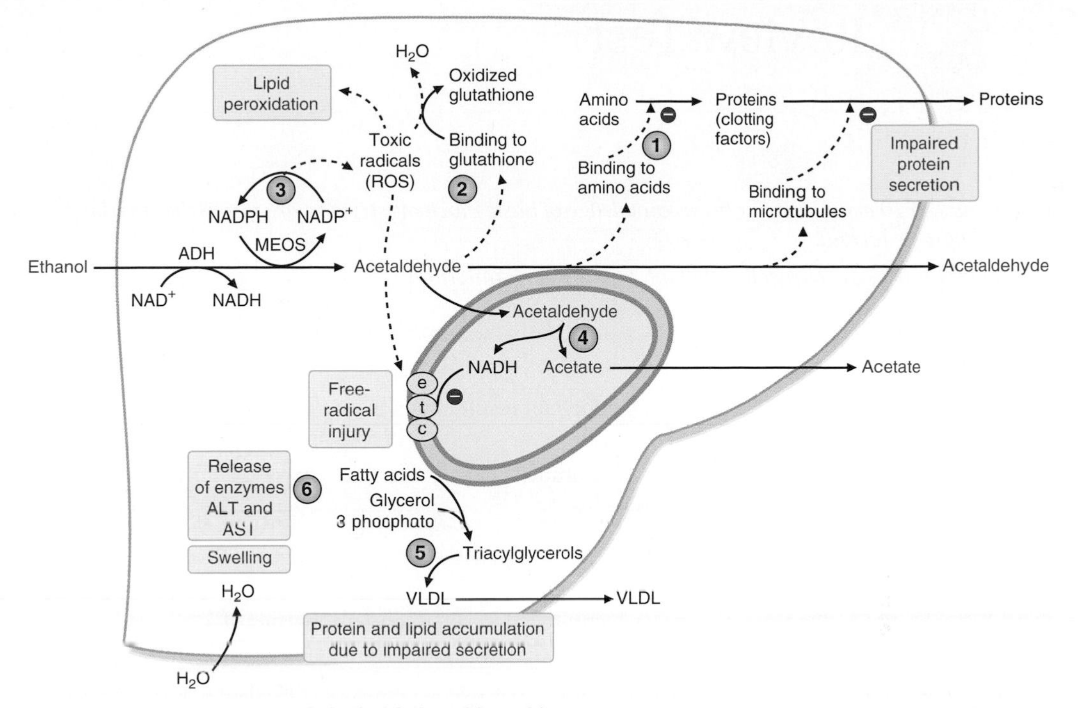

FIGURE 7.24. The development of alcohol-induced hepatitis. Note the effects that acetaldehyde has on protein secretion, lipid peroxidation, free-radical injury, liver swelling and release of enzymes, and impaired VLDL secretion.

Review Test

Questions 1 to 10 examine your basic knowledge of basic biochemistry and are not in the standard clinical vignette format.

Questions 11 to 35 are clinically relevant, USMLE-style questions.

Basic Knowledge Questions

1. A deficiency of pancreatic exocrine secretion can result in which one of the following?

(A) An increased pH in the intestinal lumen
(B) An increased absorption of fat-soluble vitamins
(C) A decreased formation of bile salt micelles
(D) Increased levels of blood chylomicrons
(E) Decreased amounts of fat in the stool

2. Choose the one best answer that most accurately describes some properties of acetyl-CoA carboxylase.

	Required cofactor	Intracellular location	Allosteric modifier	Enzyme that catalyzes a covalent modification
A	Biotin	Mitochondrial	Citrate	PKA
B	Biotin	Cytoplasmic	Citrate	AMP-activated protein kinase
C	Thiamine	Mitochondrial	Acetyl-CoA	PKA
D	Thiamine	Cytoplasmic	Acetyl-CoA	AMP-activated protein kinase
E	None	Mitochondrial	Malonyl-CoA	PKA
F	None	Cytoplasmic	Malonyl-CoA	AMP-activated protein kinase

3. The synthesis of fatty acids from glucose in the liver is best described by which one of the following?

(A) The pathway occurs solely in the mitochondria.
(B) It requires a covalently bound derivative of pantothenic acid.
(C) It requires NADPH derived solely from the pentose phosphate pathway.
(D) The pathway is primarily regulated by isocitrate.
(E) The pathway does not utilize a carboxylation reaction.

4. Which one of the following best describes the synthesis of triglyceride in adipose tissue?

	Source of fatty acids	Source of backbone	Requires coenzyme A	Requires lipoprotein lipase	Requires 2-monoacylglycerol
A	VLDL	Glycerol	Yes	No	Yes
B	Chylomicrons	Glycerol	No	Yes	No
C	VLDL and chylomicrons	DHAP	Yes	No	Yes
D	VLDL and chylomicrons	DHAP	Yes	No	No
E	Chylomicrons	DHAP	No	Yes	Yes
F	VLDL	Glycerol	No	Yes	No

5. Which one of the following sequences places the lipoproteins in the order of most dense to least dense?

(A) HDL/VLDL/chylomicrons/LDL
(B) HDL/LDL/VLDL/chylomicrons
(C) LDL/chylomicrons/HDL/VLDL
(D) VLDL/chylomicrons/LDL/HDL
(E) LDL/chylomicrons/VLDL/HDL

6. Which one of the following best represents fasting conditions?

	Activity of hormone-sensitive lipase	Fate of glycerol	VLDL production	Modification of acetyl-CoA carboxylase	Ketone body production
A	Inactive	Glycolysis	High	Dephosphorylated	No
B	Active	Glycolysis	High	Phosphorylated	Yes
C	Inactive	Glycolysis	High	Dephosphorylated	No
D	Active	Gluconeogenesis	Low	Phosphorylated	No
E	Inactive	Gluconeogenesis	Low	Dephosphorylated	Yes
F	Active	Gluconeogenesis	Low	Phosphorylated	Yes

Questions 7 through 10 are based on the following scenario:

A molecule of palmitic acid, attached to carbon 1 of the glycerol moiety of a triacylglycerol, is ingested and digested. It passes into the blood, is stored in a fat cell, and ultimately is oxidized to CO_2 and H_2O in a muscle cell. Choose the molecular complex in the blood in which the palmitate residue is carried from the first site to the second in each of the four questions that follow. An answer may be used once, more than once, or not at all.

7. From the lumen of the gut to the surface of the gut epithelial cell
8. From the gut epithelial cell to the blood
9. From the intestine through the blood to a fat cell
10. From a fat cell to a muscle cell

(A) VLDL
(B) Chylomicron
(C) Fatty acid–albumin complex
(D) Bile salt micelle
(E) LDL

Board-style Questions

Questions 11 through 13 are based on the following case:

A 6-month-old baby was doing well until he developed viral gastroenteritis and was unable to tolerate oral feeding for 2 days. He is admitted to the hospital with encephalopathy, cardiomegally and heart failure, poor muscle tone, and hypoketotic hypoglycemia. Blood work did not detect any medium-chain dicarboxylic acids.

11. Once this baby is diagnosed and treated, his diet will need to be very restricted. Theoretically, which one of the following fatty acids will he be able to consume and metabolize?

(A) An 8-carbon fatty acid
(B) A 14-carbon fatty acid
(C) A 20-carbon fatty acid
(D) Only unsaturated fatty acids, regardless of chain length
(E) Only saturated fatty acids, regardless of chain length

12. Which one of the following foods or supplements would be allowable on the above patient's restricted diet?

(A) Coconut oil
(B) Tuna
(C) Walnuts
(D) Spinach
(E) Oleic acid supplements

13. Dietary supplementation of which one of the following would be beneficial to this patient?

(A) Pantothenic acid
(B) Niacin
(C) Riboflavin
(D) Carnitine
(E) Thiamine

Questions 14 and 15 are based on the following case:

A 50-year-old male patient has high cholesterol levels and is placed on lovastatin. He is counseled to stop drinking his usual glass of grapefruit juice every morning.

14. Which of the following may occur when someone taking lovastatin chronically consumes grapefruit juice?

(A) Cholesterol levels increase
(B) Muscle pain and discomfort
(C) Steatorrhea
(D) Acid reflux
(E) A decrease in HDL levels

15. Given that grapefruit juice interferes with lovastatin action, which one of the following best explains this interaction?

(A) Grapefruit juice inhibits the cytochrome p450 enzyme which modifies lovastatin for excretion.
(B) Grapefruit juice stimulates the cytochrome p450 enzyme which modifies lovastatin for excretion.
(C) Grapefruit juice is a competitive inhibitor of lovastatin binding to cholesterol.
(D) Grapefruit juice is a competitive inhibitor of HMG-CoA reductase.
(E) Grapefruit juice reduces the maximal velocity of HMG-CoA reductase.

16. Aspirin is used in small daily doses to help prevent heart attacks and/or strokes. Aspirin can be used in this fashion because it inhibits which one of the following?

(A) Prostaglandin synthesis
(B) Thromboxane synthesis
(C) Arachidonic acid synthesis
(D) Leukotriene synthesis
(E) Linolenic acid synthesis

17. A patient with a hyperlipoproteinemia would most likely benefit from a low-carbohydrate diet if the lipoproteins that are elevated in the blood belong to which class of lipoproteins? Choose the one best answer.

(A) Chylomicrons
(B) VLDL
(C) LDL
(D) HDL
(E) Chylomicrons and VLDL
(F) VLDL and LDL
(G) LDL and HDL

18. An individual has been determined to have hypertriglyceridemia, with a triglyceride level of 350 mg/dL (normal is <150 mg/dL). The patient decides to reduce this level by keeping his caloric intake the same, but switching to a low-fat, low-protein, high-carbohydrate diet. Three months later, after sticking faithfully to his diet, his triglyceride level was 375 mg/dL. This increase in lipid content is being caused by which component of his new diet?

(A) Phospholipids
(B) Triglycerides
(C) Amino acids
(D) Carbohydrates
(E) Cholesterol

19. An alcoholic who went on a weekend binge without eating any food was found to have severe hypoglycemia. Hypoglycemia occurred because the metabolism of ethanol prevented the production of blood glucose from which one of the following? Choose the one best answer.

(A) Glycogen
(B) Lactate
(C) Glycerol
(D) Oxaloacetate
(E) Lactate, glycerol, and oxaloacetate

20. A man has just received his fourth DUI citation. The judge orders an alcohol dependency program complete with a medication that makes him have nausea and vomiting if he drinks alcohol while taking the medication. The drug-induced illness is due to the buildup of which one of the following?

(A) Ethanol
(B) Acetaldehyde
(C) Acetate
(D) Acetyl-CoA
(E) Acetyl phosphate

21. A 52-year-old man, after suffering a heart attack, was put on 81 mg of aspirin daily by his cardiologist. The purpose of this treatment is to reduce the levels of which one of the following?

(A) Cytokines
(B) Leukotrienes
(C) Thromboxanes
(D) Cholesterol
(E) Triglycerides

22. Remembering the distribution and solubility of ethanol, after several drinks with an evening meal, in which of the following tissues would you find the LEAST amount of alcohol?

(A) Brain
(B) Liver
(C) Fatty tissue
(D) Central cornea

Questions 23 through 25 are based on the following case:

A 14-year-old girl with Type 1 diabetes has had viral gastroenteritis for 5 days, and she has been vomiting, been nauseous, and had trouble taking fluids by mouth. Because she was not eating, she did not take any insulin during her illness. She becomes weak and confused and is taken to the emergency room (ER) by her parents. The ER doctor notices a fruity odor to her breath, hyperventilation, and a blood glucose level of 600 mg/dL.

23. A blood pH measurement is taken. You would expect this value to be which one of the following?

(A) 6.75
(B) 7.15
(C) 7.40
(D) 7.65
(E) 8.00

24. The patient is hyperventilating because of which one of the following?

(A) The low pH of the blood
(B) The elevated pH of the blood
(C) The increased glucagon/insulin ratio in the blood
(D) Lack of fluids in the body
(E) Difficulty in breathing due to the lack of food

25. The fruity odor noticed by the ER physician is due to which one of the following?

(A) Oxidation of acetoacetate
(B) Reduction of acetoacetate
(C) Conversion of acetoacetate to acetoacetyl-CoA
(D) Decarboxylation of acetoacetate
(E) Carboxylation of acetoacetate

26. A 2-day-old infant born at 32 weeks gestation has had breathing difficulties since birth and is currently on a respirator and 100% oxygen. These difficulties occur due to which one of the following?

(A) An inability of the lung to contract to exhale
(B) An inability of the lung to expand when taking in air
(C) An inability of the lung to respond to insulin
(D) An inability of the lung to respond to glucagon
(E) An inability of the lung to produce energy

27. An 8-month-old baby girl had normal growth and development for the first few months, but then progressively deteriorated with deafness, blindness, atrophied muscle, inability to swallow, and seizures. Early on in the diagnosis of the child, it was noticed that a cherry red macula was present in both eyes. Considering the child in the above case, measurement of which one of the following would enable one to determine whether the mutation were in the hex A or hex B gene?

(A) GM1
(B) GM2
(C) Globoside
(D) Glucocerebroside
(E) Ceramide

28. A patient with high blood cholesterol levels was treated with lovastatin. This drug lowers blood cholesterol levels due primarily to which one of the following?

(A) Inhibition of absorption of dietary cholesterol
(B) Inhibition of lipoprotein lipase in adipose tissue
(C) Inhibition of citrate lyase in the liver
(D) Induction of LDL receptors in the liver and peripheral tissues
(E) Inhibition of HMG-CoA reductase in the liver and peripheral tissues

29. A patient is taking a statin primarily to reduce the risk of atherosclerosis. A potential problem of statin treatment is which one of the following?

(A) Reduced intracellular cholesterol
(B) Reduced synthesis of coenzyme Q
(C) Reduced synthesis of intracellular phospholipids
(D) Reduced intestinal absorption of cholesterol
(E) Reduced levels of chylomicrons

30. A person with an intestinal infection caused by a proliferation of bacteria in the gut would most likely have an increase in which one of the following?

(A) The synthesis of bile salts in the liver
(B) The amount of conjugated bile salts in the intestine

(C) The absorption of dietary lipid by intestinal cells
(D) Body stores of fat-soluble vitamins
(E) Body stores of water-soluble vitamins

31. A 28-year-old man was found to have elevated cholesterol levels of 325 mg/dL on a routine checkup. His father died of a heart attack at the age of 42, and also had greatly elevated cholesterol levels throughout his life. The man's physician placed him on lovastatin, and his cholesterol levels dropped to 170 mg/dL. The nature of the elevated cholesterol in this patient is most likely due to a mutation in which one of the following proteins?

(A) HMG-CoA reductase
(B) Protein kinase A
(C) ACAT
(D) LDL receptor
(E) Lipoprotein lipase

32. A person with a known hyperlipoproteinemia went in for a scheduled checkup. The lab values revealed a blood cholesterol level of 360 mg/dL (the recommended level is below 200 mg/dL) and a blood triglyceride (triacylglycerol) level of 140 mg/dL (the recommended level is below 150 mg/dL). The hyperlipoproteinemia is most likely due to which one of the following?

(A) A decreased ability for receptor-mediated endocytosis of LDL
(B) A decreased ability to degrade the triacylglycerols of chylomicrons
(C) An increased ability to produce VLDL
(D) An inactive lipoprotein lipase
(E) A decreased ability to convert VLDL to IDL

33. A 4-month-old child exhibited extreme tiredness, irritable moods, poor appetite, and fasting hypoglycemia associated with vomiting and muscle weakness. Blood work showed elevated levels of free fatty acids, but low levels of acylcarnitine. A muscle biopsy demonstrated a significant level of fatty acid infiltration in the cytoplasm. The most likely molecular defect in this child is in which one of the following enzymes?

(A) Medium-chain acyl-CoA dehydrogenase
(B) Carnitine transporter
(C) Acetyl-CoA carboxylase
(D) Carnitine acyltransferase II
(E) HMG-CoA synthase

34. Type 1 diabetes mellitus is caused by a decreased ability of the β cells of the pancreas to produce insulin. A person with Type 1 diabetes mellitus who has neglected to take insulin injections will exhibit which one of the following?

(A) Increased fatty acid synthesis from glucose in liver
(B) Decreased conversion of fatty acids to ketone bodies
(C) Increased stores of triacylglycerol in adipose tissue
(D) Increased production of acetone
(E) Increased glucose transport into muscle cells

35. A premature infant, when born, had low Apgar scores and was having difficulty breathing. The NICU physician injected a small amount of a lipid mixture into the child's lungs, which greatly reduced the respiratory distress the child was experiencing. In addition to proteins, a key component of the mixture was which one of the following?

(A) Sphingomyelin
(B) A mixture of gangliosides
(C) Triacylglycerol
(D) Phosphatidylcholine
(E) Prostaglandins E and F

Answers and Explanations

1. **The answer is C.** The pancreas produces bicarbonate (which neutralizes stomach acid) and digestive enzymes (including pancreatic lipase and colipase, enzymes that degrade dietary lipids). The decreased production of bicarbonate will lead to a decrease of intestinal pH. Lower levels of pancreatic lipase will result in the decreased digestion of dietary triacylglycerols, which will lead to the formation of fewer bile salt micelles. The reduced pH will also interfere with the ability of the bile salts to effectively form micelles. Intestinal cells will have less substrate for chylomicron formation, and less fat-soluble vitamins will be absorbed. More dietary fat will be excreted in the feces.

2. **The answer is B.** Biotin is required for the acetyl-CoA carboxylase reaction in which the substrate, acetyl-CoA, is carboxylated by the addition of CO_2 to form malonyl-CoA. This reaction occurs in the cytosol. Malonyl CoA provides the 2-carbon units that add to the growing fatty acid chain on the fatty acid synthase complex. As the growing chain is elongated, malonyl-CoA is decarboxylated. Citrate is an allosteric activator of the enzyme, and the enzyme is inhibited by phosphorylation by the AMP-activated protein kinase.

3. **The answer is B.** The synthesis of fatty acids from glucose occurs in the cytosol, except for the mitochondrial reactions in which pyruvate is converted to citrate (pyruvate to oxaloacetate, pyruvate to acetyl-CoA, and oxaloacetate and acetyl-CoA condense to form citrate). Biotin is required for the conversion of pyruvate to oxaloacetate (a carboxylation reaction), which combines with acetyl-CoA to form citrate. Biotin is also required by acetyl-CoA carboxylase. Citrate, not isocitrate, is a key regulatory compound for acetyl-CoA carboxylase. Pantothenic acid is covalently bound to the fatty acid synthase complex as part of a phosphopantetheinyl residue. During the reduction reactions on the synthase complex, the growing fatty acid chain is attached to this residue. NADPH, produced by the malic enzyme as well as by the pentose phosphate pathway, provides the reducing equivalents.

4. **The answer is D.** Fatty acids, cleaved from the triacylglycerols of chylomicrons and VLDL by the action of lipoprotein lipase, are taken up by adipose cells and react with coenzyme A to form fatty acyl-CoA. The lipoprotein lipase is not required to synthesize triglyceride within the adipocyte. Glucose is converted via DHAP to glycerol-3-phosphate, which reacts with fatty acyl-CoA to form phosphatidic acid. Adipose tissue lacks glycerol kinase and cannot use glycerol to directly form glycerol-3-phosphate. After inorganic phosphate is released from phosphatidic acid, the resultant diacylglycerol reacts with another fatty acyl-CoA to form a triacylglycerol, which is stored in the adipose cells. (2-Monoacylglycerol is an intermediate for triglyceride synthesis only in intestinal cells, and is not produced in the adipocyte.)

5. **The answer is B.** Because chylomicrons contain the most triacylglycerol, they are the least dense of the blood lipoproteins. Because VLDL contains more protein than chylomicrons, it is more dense than chylomicrons, but less dense than LDL. Because LDL is produced by the degradation of the triacylglycerols of VLDL, LDL is denser than VLDL. HDL is the most dense of the blood lipoproteins. It has the most protein and the least triacylglycerol (see Tables 7.2 and 7.3).

6. **The answer is F.** During fasting, the hormone-sensitive lipase of adipose tissue is activated by a mechanism involving increased glucagon (and decreased insulin), cAMP, and protein kinase A. Phosphorylation of hormone-sensitive lipase activates the enzyme. Triacylglycerols are degraded, and fatty acids and glycerol are released into the blood. In the liver, glycerol is converted to glucose by gluconeogenesis and fatty acids are oxidized to produce ketone bodies. These fuels are released into the blood and supply energy to various tissues. During fasting, the

liver does not produce significant quantities of VLDL. Fatty acid synthesis is reduced owing to the phosphorylation and inactivation of acetyl-CoA carboxylase by the AMP-activated protein kinase.

7. **The answer is D.** A palmitate residue attached to carbon 1 of a dietary triacylglycerol is released by pancreatic lipase and carried from the intestinal lumen to the gut epithelial cell in a bile salt micelle, which will allow absorption of the fatty acid by the intestinal epithelial cell.

8. **The answer is B.** Palmitate is absorbed into the intestinal cell and utilized to synthesize a triacylglycerol, which is packaged in a nascent chylomicron and secreted via the lymph into the blood.

9. **The answer is B.** The chylomicron, containing the palmitate, matures in the blood by accepting proteins from HDL. It travels to a fat cell. VLDL is the particle made in the liver with endogenous triglyceride.

10. **The answer is C.** The chylomicron triacylglycerol is digested by lipoprotein lipase, and the palmitate enters a fat cell and is stored as triacylglycerol. It is released as free palmitate and carried, complexed with albumin, to a muscle cell, where it is oxidized.

11. **The answer is A.** This baby has primary carnitine deficiency, an autosomal recessive disorder. The lack of medium-chain dicarboxylic acids in the blood rules out an MCAD deficiency. He is unable to transport blood-borne carnitine into the muscle and liver, thereby blocking fatty acid oxidation in those tissues. Carnitine is required to transfer most fatty acids from the cytoplasm to the matrix of the mitochondria. However, short- and medium-chain fatty acids (up to 10 or 12 carbons) are sufficiently water-soluble such that they can enter cells and be transferred into the mitochondria in the absence of carnitine. Once inside the mitochondria, an acyl-CoA synthetase will activate the fatty acid to an acyl-CoA such that β-oxidation can occur. The transfer is not affected whether the fatty acid is saturated or unsaturated; the chain length is the determining factor. Dietary restriction of long-chain fatty acids is essential to treat this disorder and alleviate the symptoms. The patient was doing well while feeding on a regular schedule because of the carbohydrate in the diet. Once the child had an extended fast, and needed to oxidize fatty acids for energy, the symptoms of carnitine deficiency became apparent. The hypoketotic hypoglycemia is a strong indication that the problem is in fatty acid oxidation.

12. **The answer is A.** The patient has a primary carnitine deficiency and can only metabolize medium-chain fatty acids. Coconut oil is high in medium-chain saturated fatty acids. Tuna and certain nuts are high in very long-chain fatty acids and omega-3 fatty acids. Spinach is a good source of ALA (alpha-linolenic acid), and omega-6 fatty acids. Oleic acid is a cis-$\Delta 9$ C18:1 fatty acid, and would not be metabolized in a child lacking carnitine in the cells.

13. **The answer is D.** In many cases of primary carnitine deficiency, increasing the blood levels of carnitine is sufficient to allow some transport of carnitine into cells such that fatty acid oxidation can occur. While pantothenic acid (part of coenzyme A), niacin (the precursor for NAD^+), and riboflavin (needed for FAD) are required for fatty acid oxidation, the rate-limiting step in these patients is the transport of the fatty acids from the cellular cytoplasm to the matrix of the mitochondria.

14. **The answer is B.** Grapefruit juice contains furanocoumarins, which inhibit the cytochrome P450 complex CYP3A4. This complex modifies various statins for rapid excretion from the body. Thus, in the presence of grapefruit juice, statin levels will be higher than expected. This will lead to prolonged inhibition of HMG-CoA reductase and a reduction of cholesterol levels (with minimal effect on HDL levels), but will also increase the probability of side effects from statin treatment, one of which is muscle pain and weakness. The grapefruit juice plus statin will not lead to gastric reflux or steatorrhea.

15. **The answer is A.** Grapefruit juice contains furanocoumarins, which inhibit the cytochrome P450 enzyme CYP3A4 that prepares statins for excretion. Grapefruit juice does not act as

an inhibitor of HMG-CoA reductase. Lovastatin binds to HMG-CoA reductase, but not to cholesterol.

16. **The answer is B.** Thromboxanes promote platelet aggregation, and aspirin blocks this function through reducing the synthesis of thromboxanes. This decreases the chances of a clot forming in a coronary artery (MI) or in the artery that feeds the brain (CVA). Aspirin also inhibits prostaglandin synthesis, but this is an anti-inflammatory property. Leukotrienes are involved in allergies and asthma, and their synthesis requires lipoxygenase, which is not inhibited by aspirin. Arachidonic acid is derived from linoleic acid, and that synthesis (fatty acid elongation) is not inhibited by aspirin. Linolenic acid is an essential fatty acid, and cannot be synthesized by humans.

17. **The answer is B.** VLDL is produced mainly from dietary carbohydrate, LDL is produced from VLDL, and chylomicrons contain primarily dietary triacylglycerol. Elevated HDL levels are desirable and are not considered to be a lipid disorder. HDL also contains low levels of triglyceride. A low-carbohydrate diet would be expected to reduce the level of circulating VLDL due to reduced fatty acid and triglyceride synthesis in the liver.

18. **The answer is D.** Dietary glucose is the major source of carbon for synthesizing fatty acids in humans. In a high-carbohydrate diet, excess carbohydrates are converted to fat (fatty acids and glycerol) in the liver, packaged as VLDL, and sent into the circulation for storage in the fat cells. The new diet has reduced dietary lipids, which lower chylomicron levels, but the excess carbohydrate in the diet is leading to increased VLDL synthesis and elevated triglyceride levels. Dietary amino acids are usually incorporated into proteins, particularly in a low-protein diet.

19. **The answer is E.** Ethanol metabolism (which produces high NADH levels) does not prevent glycogen degradation. In fact, glycogen stores would be rapidly depleted under these conditions because of decreased gluconeogenesis. Lactate is converted to pyruvate during gluconeogenesis. The pyruvate–lactate equilibrium greatly favors lactate when NADH is high. Thus, alanine and lactate are prevented from producing glucose. Lactate levels are elevated, and a lactic acidosis can result. Glycerol normally enters gluconeogenesis by forming glycerol-3-phosphate, which is oxidized to DHAP. High NADH levels prevent this oxidation. Aspartic acid is converted to oxaloacetate (via transamination), as do other amino acid degradation products that enter the TCA cycle (α-ketoglutarate, succinyl-CoA, fumarate). However, the high NADH levels favor malate formation from oxaloacetate, reducing the amount of oxaloacetate available for gluconeogenesis (through the phosphoenolpyruvate carboxykinase reaction). Thus, the three major gluconeogenic precursors (alanine, glycerol, and lactate) do not form glucose because of the high NADH levels, and as glycogen stores are depleted, hypoglycemia results.

20. **The answer is B.** The court-ordered medication is disulfiram. Disulfiram inhibits aldehyde dehydrogenase, which greatly reduces the amount of acetaldehyde that is converted to acetate. This causes an accumulation of acetaldehyde, which is the substance responsible for the symptoms of a "hangover," including nausea and vomiting. Alcohol dehydrogenase reduces ethanol to acetaldehyde. Acetyl-CoA synthetase converts acetate to acetyl-CoA.

21. **The answer is C.** Platelet aggregation is often a determining factor in heart attacks. Thromboxane A_2, produced by platelets, promotes platelet aggregation when clotting is required, and inhibition of thromboxane A_2 synthesis by aspirin reduces the potential for inappropriate clot formation, and further heart attacks. Thromboxane A_2 is produced from arachidonic acid by the action of cyclooxygenase, the enzyme covalently modified and irreversibly inhibited by aspirin. Leukotrienes are also synthesized from arachidonic acid, but utilize lipoxygenase in their synthesis, which is not inhibited by aspirin. Cholesterol, triglyceride, and cytokine synthesis do not require cyclooxygenase activity.

22. **The answer is D.** Ethanol is both water- and lipid-soluble. It is easily absorbed from the gastrointestinal tract and is distributed throughout the body via the blood stream. It is mostly

metabolized in the liver, so the level would be high in this tissue. It is lipid-soluble, so it would be found in fatty tissue. It has many central effects in the brain, so it easily passes the blood–brain barrier. The central cornea has no arterial supply. The only way alcohol could accumulate in the central cornea would be through diffusion into the aqueous humor and then into the central cornea, a slower and less-efficient system. Therefore, the tissue with the lowest level of alcohol would be the central cornea.

23. The answer is B. The patient is exhibiting the symptoms of diabetic ketoacidosis. Normal blood pH is in the range of 7.40. Diabetic ketoacidosis reduces the blood pH since ketone bodies accumulate and produce acid, which the blood has trouble buffering. A mild ketoacidosis would reduce the pH to about 7.25; one in which the patient exhibits neurological changes (weak and confused) would lower the pH even further to 7.15. Life-threatening diabetic ketoacidosis would be a pH of 7.0. Answers D and E are incorrect because they represent an alkalization of the blood, which does not occur when acids accumulate.

24. The answer is A. Hyperventilation is the body's way to try and raise the lowered blood pH by exhaling carbon dioxide rapidly. Carbon dioxide will form carbonic acid and a proton in the blood; as the carbon dioxide is exhaled, the acid and proton will associate so that the carbonic acid can form carbon dioxide and water. This will decrease the proton concentration in the blood, and raise the pH. The hyperventilation is not due to the altered hormonal ratios in the blood, the lack of fluids, or the lack of food.

25. The answer is D. The fruity odor is due to acetone, which is being exhaled. The acetone is derived from the spontaneous decarboxylation of acetoacetate (one of the ketone bodies) to acetone within the blood and tissues.

26. The answer is B. The baby has respiratory distress syndrome, due to an inability to produce surfactant, a hydrophobic molecule that is secreted by the type II cells in the lung and coats the airways, reducing surface tension during contraction, and allowing relatively easy expansion of the lung during inhalation. This is due to the lungs not yet producing surfactant, which contains a few proteins and a large amount of dipalmitoylphosphatidyl choline. Respiratory distress syndrome is not related to insulin or glucagon response by the lung, or the ability of the lung cells to generate energy.

27. The answer is C. The child is exhibiting the symptoms of either Tay–Sachs or Sandhoff's disease, both of which are sphingolipidoses. The hex A gene codes for hexosaminidase A, whereas the hex B gene codes for hexosaminidase B. The hex A protein consists of two A and two B subunits, and cleaves only GM2. The hex B protein is a B tetramer, and cleaves both GM2 and globoside. In Tay–Sachs disease, a loss of hex A activity, globoside degradation is normal as the hex B protein is normal. The loss of hex B activity affects both hex A (since two subunits are of the B variant) and hex B (tetramer) activity, and globoside will accumulate in Sandhoff's disease, but not in Tay–Sachs disease.

28. The answer is D. The class of drugs known as the statins (e.g., lovastatin) lower blood cholesterol levels through the induction of LDL receptor expression on the liver and peripheral tissue cell surface. Statins directly inhibt HMG-CoA reductase, a key regulatory enzyme in cholesterol biosynthesis, which reduces intracellular cholesterol levels. The reduction of intracellular cholesterol leads to the induction of LDL receptors, as the cells now need to obtain their cholesterol from the circulation. Ezetimibe inhibits the intestinal absorption of cholesterol. Statins do not inhibit lipoprotein lipase or citrate lyase.

29. The answer is B. Statins inhibit HMG-CoA reductase, thereby reducing intracellular cholesterol levels. The reduced intracellular cholesterol induces the expression of the LDL receptor, which binds LDL and internalizes it within the cell, thereby reducing circulating cholesterol levels, and reducing the risk of developing atherosclerosis. The inhibition of HMG-CoA reductase reduces mevalonate production, which is the precursor for isoprene biosynthesis. Isoprenes are required for the synthesis of coenzyme Q, as well as dolichol phosphate. Thus, taking statins may reduce endogenous coenzyme Q levels. HMG-CoA reductase does not play a role in intestinal cholesterol absorption, nor will it affect the production of chylomicrons, which carry dietary (exogenous) cholesterol throughout the body.

30. The answer is A. Bacteria in the intestine deconjugate and dehydroxylate bile salts, converting them to secondary bile salts. Therefore, the bile salts become less water-soluble and less effective as detergents, less readily absorbed, and more likely to be excreted in the feces than recycled by the liver. Fewer micelles would be produced, so less dietary lipid (including the fat-soluble vitamins) would be absorbed. Because fewer bile salts would return to the liver, more bile salts would be synthesized. Bile salts inhibit the 7α-hydroxylase that is involved in their synthesis. In addition, the person's food intake might decrease, which would augment some of the effects noted above. Charges on bile salts will not affect the uptake of water-soluble vitamins. Since the bacteria in the gut deconjugate and dehydroxylate the bile salts, the amount of conjugated bile salts in the intestine will decrease.

31. The answer is D. The patient most likely has familial hypercholesterolemia due to a mutated LDL receptor. Lovastatin is an inhibitor of HMG-CoA reductase. HMG-CoA reductase inhibitors cause cells to decrease the rate of cholesterol synthesis. Lower cellular levels of cholesterol cause a decreased conversion of cholesterol to cholesterol esters (by the ACAT reaction) for storage and an increased production of LDL receptors. An increased number of receptors will cause more LDL to be taken up by cells from the circulation and degraded by lysosomes. Thus, blood cholesterol levels will decrease. Blood triacylglycerol levels will also decrease but not to a great extent because LDL contains only small amounts of triacylglycerol. The patient most likely is heterozygous for a mutation in the LDL receptor. If the patient were homozygous for such a mutation, then lovastatin would not reduce circulating cholesterol levels, as there would be no functional LDL receptors to upregulate. The patient does not have a mutation in HMG-CoA reductase, as that is the target of lovastatin, and if an HMG-CoA reductase mutation were to reduce HMG-CoA reductase activity, then lovastatin would have little to no effect. A mutation in protein kinase A, rendering it inactive, would not affect intracellular cholesterol synthesis (as it is the AMP-activated protein kinase that regulates the HMG-CoA reductase activity). An inactivating mutation in ACAT would not lead to a reduction of circulating cholesterol in response to lovastatin. Inactivating mutations in lipoprotein lipase lead to hypertriglyceridemia, and do not respond to lovastatin.

32. The answer is A. Of the blood lipoproteins, LDL contains the highest concentration of cholesterol and lowest concentration of triacylglycerols. Elevation of blood LDL levels (the result of decreased endocytosis of LDL) would result in high blood cholesterol levels and relatively normal triacylglycerol levels. A decreased ability to degrade the triacylglycerols of chylomicrons or to convert VLDL to IDL, as well as an increased ability to produce VLDL, would all result in elevated triacylglycerol levels. An inactive lipoprotein lipase would keep chylomicron and VLDL levels elevated, so an increased triglyceride level would have been seen if this were the defect.

33. The answer is B. The child has the symptoms of primary carnitine deficiency. Carnitine cannot be transported from the blood into the liver and muscle, and fatty acid oxidation in those tissues is severely impaired. The inability to utilize fatty acids for energy give rise to muscle weakness, and an accumulation of fatty acids can occur within the muscle tissue. The inability of the liver to oxidize fatty acids will lead to fasting hypoglycemia as there is insufficient energy for gluconeogenesis. An MCAD deficiency would not show fatty infiltration in the muscle, nor elevated levels of free fatty acids (the presence of medium-chain dicarboxylic acids, or acylglycines, would be observed instead). A defect in carnitine acyltransferase II would result in acylcarnitines in the circulation. An HMG-CoA synthase deficiency would not allow ketone body formation, and would not present with these symptoms. A lack of acetyl-CoA carboxylase would greatly reduce the fatty acid content within the fat cell, as endogenous fatty acids would not be able to be synthesized from acetyl-CoA.

34. The answer is D. Decreased insulin levels cause fatty acid synthesis to decrease and glucagon levels to increase. Adipose triacylglycerols are degraded, and fatty acids are released. They are converted to ketone bodies in the liver, and a ketoacidosis can occur. Nonenzymatic decarboxylation of acetoacetate forms acetone, which causes the odor associated with diabetic ketoacidosis. Insulin is required for efficient glucose transport into muscle cells.

35. **The answer is D.** The premature infant is experiencing respiratory distress syndrome, which is caused by a deficiency of lung surfactant. The lung cells do not begin to produce surfactant until near birth, and premature infants frequently are not producing sufficient surfactant to allow the lungs to expand and contract as needed. The surfactant is composed of a number of hydrophobic proteins and dipalmitoylphosphatidylcholine. Sphingomyelin, gangliosides, triglyceride, and prostaglandins are not components of the surfactant. The phosphatidylcholine content of the surfactant is 85% of the total lipids associated with the complex.

chapter 8

Nitrogen Metabolism–Amino Acids, Purines, Pyrimidines, and Products Derived from Amino Acids

The major clinical uses of this chapter are understanding protein, nitrogen, and amino acid metabolisms. Products made from protein metabolism include enzymes, DNA, RNA, and hormones. Abnormalities or deficiencies lead to a plethora of disease processes (anemias, liver disease, renal disease, inborn errors, etc).

CLINICAL CORRELATES Many important steps in nitrogen metabolism occur in the liver. If **liver disease** is severe enough, urea production can be compromised. Blood urea nitrogen **(BUN) levels decrease**, and the levels of the toxic compound **ammonia increase**. Because the liver is normally involved in converting bilirubin to the diglucuronide that is excreted in the bile, in liver disease, the levels of **bilirubin increase** in the body and **jaundice** can occur. When the liver cells are damaged, enzymes such as aspartate aminotransferase (**AST**) and alanine aminotransferase (**ALT**) leak into the blood. Acetaminophen poisoning occurs in the liver; excess acetaminophen is metabolized by a cytochrome P450 enzyme into a toxic intermediate, which can be detoxified by glutathione, a tripeptide of γ-glutamyl-cysteinyl-glycine. The treatment for acetaminophen poisoning includes the administration of *N*-acetylcysteine, to boost the synthesis of glutathione in order to continue to detoxify the toxic intermediate.

OVERVIEW

- Nitrogen is obtained mainly from protein in the diet, which is digested to amino acids by the combined action of proteases produced by the stomach, pancreas, and intestinal epithelial cells.
- Amino acids are absorbed by intestinal epithelial cells, pass into the blood, and are taken up by other cells of the body.
- Amino acids are used by cells for the synthesis of proteins, which is a dynamic process; proteins are constantly being synthesized and degraded.
- After nitrogen is removed from amino acids, the carbon skeletons can be oxidized for energy.
- The nitrogen of amino acids is converted to urea in the liver and ultimately excreted by the kidney.
- Although urea is the major nitrogenous excretory product, nitrogen is also excreted as NH_4^+, uric acid, and creatinine.

- In the fed state, the liver can convert amino acid carbons to fatty acids and glycerol, which form the triacylglycerols of very low density lipoprotein (VLDL).
- During fasting, muscle protein is degraded and supplies amino acids to the blood, and the liver converts amino acid carbons to glucose or ketone bodies.
- The essential amino acids (histidine, isoleucine, leucine, lysine, methionine, phenylalanine, threonine, tryptophan, and valine) are required in the diet. Arginine and increased amounts of histidine are required during periods of growth.
- The nonessential amino acids can be synthesized in the body.
  - The carbons of 10 of the nonessential amino acids can be derived from glucose. However, the synthesis of cysteine requires sulfur from the essential amino acid methionine.
  - Tyrosine, the 11th nonessential amino acid, is produced by the hydroxylation of the essential amino acid phenylalanine.
- Amino acids are used for the synthesis of many nitrogen-containing compounds such as the purine and pyrimidine bases, heme, creatine, nicotinamide, serotonin, thyroxine, epinephrine, melanin, and sphingosine.

I. PROTEIN DIGESTION AND AMINO ACID ABSORPTION

- Proteins are converted to amino acids by digestive enzymes.
- Many of the digestive proteases are produced and secreted as inactive zymogens. They are converted to their active forms by the removal of a peptide fragment in the lumen of the digestive tract.
- The digestion of proteins begins in the stomach, where pepsin converts dietary proteins into smaller polypeptides.
- In the lumen of the small intestine, proteolytic enzymes produced by the pancreas (trypsin, chymotrypsin, elastase, and the carboxypeptidases) cleave the polypeptides into oligopeptides and amino acids.
- The digestive enzymes produced by the intestinal epithelial cells (aminopeptidases, dipeptidases, and tripeptidases) cleave the small peptides to amino acids.
- Amino acids, the final products of protein digestion, are absorbed through intestinal epithelial cells and enter the blood.

A. Digestion of proteins (Fig. 8.1)

1. The 70 to 100 g of **protein consumed** each day and an equal or larger amount of protein that enters the digestive tract as **digestive enzymes** or in **sloughed-off cells** from the intestinal epithelium are converted to amino acids by the **digestive enzymes**.
2. In the **stomach**, pepsin is the major proteolytic enzyme. It cleaves proteins to smaller polypeptides.
   - a. **Pepsin** is produced and secreted by the chief cells of the stomach as the inactive zymogen **pepsinogen**.
   - b. **HCl**, produced by the parietal cells of the stomach, causes a conformational change in pepsinogen that enables it to cleave itself (autocatalysis), forming pepsin.
   - c. **Pepsin** has a **broad specificity** but tends to cleave peptide bonds in which the carboxyl group is contributed by the aromatic amino acids or by leucine.
3. In the **intestine**, the partially digested material from the stomach encounters **pancreatic secretions**, which include bicarbonate and a group of proteolytic enzymes.
   - a. **Bicarbonate** neutralizes the stomach acid, raising the pH of the contents of the intestinal lumen into the optimal range for the digestive enzymes to act.
   - b. **Endopeptidases** from the pancreas cleave peptide bonds within protein chains.
     - (1) **Trypsin** cleaves peptide bonds in which the carboxyl group is contributed by **arginine** or **lysine**.
       - (a) Trypsin is secreted as the inactive zymogen **trypsinogen**.

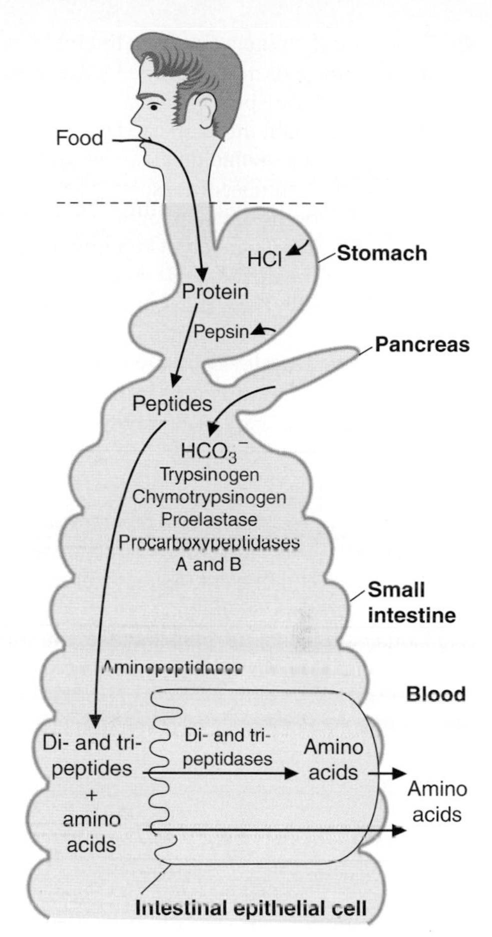

FIGURE 8.1. The digestion of proteins. The proteolytic enzymes trypsin, chymotrypsin, elastase, and the carboxypeptidases are produced as zymogens (the [pro] and [ogen], in red, accompanying the enzyme name) that are activated by cleavage after they enter the gastrointestinal lumen. Pepsinogen is produced within the stomach and is activated within the stomach (to pepsin) as the pH drops due to HCl secretion.

(b) Trypsinogen is cleaved to form trypsin by the enzyme **enteropeptidase** (enterokinase), which is produced by the intestinal cells. Trypsinogen may also undergo autocatalysis by trypsin.

(2) **Chymotrypsin** usually cleaves peptide bonds in which the carboxyl group is contributed by the **aromatic amino acids** or by **leucine**. **Chymotrypsinogen**, the inactive zymogen, is cleaved to form chymotrypsin by trypsin.

(3) **Elastase** cleaves at the carboxyl end of amino acid residues with small, uncharged side chains such as alanine, glycine, or serine. **Proelastase**, the inactive zymogen, is cleaved to elastase by trypsin.

c. **Exopeptidases** from the pancreas (carboxypeptidases A and B) cleave one amino acid at a time from the C-terminal end of the peptide.

(1) The carboxypeptidases are produced as **procarboxypeptidases**, which are cleaved to the active form by trypsin.

(2) **Carboxypeptidase A** cleaves **aromatic** amino acids from the C-terminal end of peptides.

(3) **Carboxypeptidase B** cleaves the **basic** amino acids, lysine and arginine, from the C-terminal end of peptides.

d. **Proteases** produced by the **intestinal epithelial cells** complete the conversion of dietary proteins to amino acids.

(1) **Aminopeptidases** are exopeptidases produced by the intestinal cells that cleave one amino acid at a time from the N-terminal end of peptides.

(2) **Dipeptidases** and **tripeptidases** associated with the intestinal cells produce amino acids from dipeptides and tripeptides.

B. Transport of amino acids from the intestinal lumen into the blood

1. Amino acids are absorbed by the intestinal epithelial cells and released into the blood by two types of transport systems.
2. At least eight different carrier proteins transport different groups of amino acids.
   a. **Sodium–amino acid carrier system**
      (1) A number of transport systems involve the uptake by the cell of a **sodium ion** and an **amino acid** by the same carrier protein on the luminal surface (examples are systems A, ASC, N, and B).
      (2) The **sodium ion** is pumped from the cell into the blood by the Na^+–K^+ ATPase, while the **amino acid** travels down its concentration gradient into the blood.
         (a) Thus, the transport of amino acids from the intestinal lumen to the blood is driven by the hydrolysis of ATP (secondary active transport).
   b. **The L-system of amino acid transport is not dependent on sodium for cotransport.** The L-system transports branched-chain and aromatic amino acids, and will be exploited in a treatment for phenylketonuria (PKU).

CLINICAL CORRELATES

Defective membrane-transport systems for amino acids result in decreased absorption of amino acids from the intestine and resorption of amino acids by the kidney (and thus increased excretion in the urine). In **cystinuria**, the transport of cysteine is defective. Cysteine is synthesized in the body and oxidized to cystine, which can crystallize, forming kidney stones. In **Hartnup disease**, the transport of neutral amino acids is defective, resulting in **deficiencies of essential amino acids** because they are not absorbed from the diet. As tryptophan is one of the amino acids affected in Hartnup disease, transient episodes of symptoms similar to **niacin deficiency** may occur, as tryptophan can be converted to the nicotinamide ring portion of NAD.

II. ADDITION AND REMOVAL OF AMINO ACID NITROGEN

- When amino acids are synthesized, nitrogen is added to the carbon precursors.
- When amino acids are oxidized to produce energy, the nitrogen is removed and converted mainly to urea.
- Nitrogen is transferred from one amino acid to another by transamination reactions, which always involve two different pairs of amino acids and their corresponding α-keto acids.
  - Glutamate and α-ketoglutarate usually serve as one of the pairs.
  - Pyridoxal phosphate (PLP) is the cofactor.
- Nitrogen is removed as ammonium ions from glutamate by glutamate dehydrogenase; from glutamine by glutaminase; from histidine by histidase; from serine and threonine by a dehydratase; and from asparagine by asparaginase. Ammonium ions are also removed from amino acids by the purine nucleotide cycle.
- Glutamate is a pivotal compound in amino acid metabolism.

A. Transamination reactions (Fig. 8.2)

Transamination involves the **transfer of an amino group** from one amino acid (which is converted to its corresponding α-keto acid) to an α-keto acid (which is converted to its corresponding amino acid). Thus, the nitrogen from one amino acid appears in another amino acid.

1. The enzymes that catalyze transamination reactions are known as **transaminases or aminotransferases.**
2. **Glutamate** and **α-ketoglutarate** are often involved in transamination reactions, serving as one of the amino acid/α-keto acid pairs.
3. Transamination reactions are readily reversible, and can be used in the **synthesis** or the **degradation** of amino acids.
4. Most amino acids participate in transamination reactions. **Lysine** is an exception; it **is not transaminated.**
5. **PLP** serves as the cofactor for transamination reactions. PLP is derived from vitamin B_6.

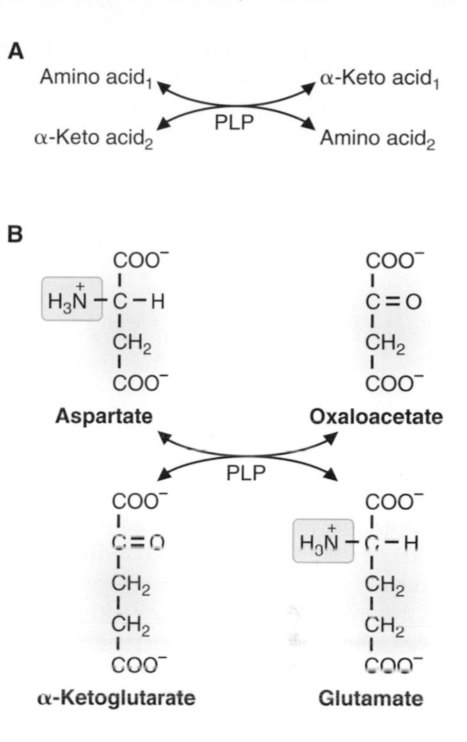

FIGURE 8.2. **Transamination. A.** Generalized reaction. **B.** The AST reaction. PLP, pyridoxol phosphate.

CLINICAL CORRELATES Vitamin B_6 is required for the formation of PLP, an important cofactor in nitrogen metabolism. **Deficiencies of vitamin B_6** are caused by a lack of the vitamin in the diet or by the administration of drugs such as isoniazid (for tuberculosis), which interfere with its metabolism. The synthesis of neurotransmitters, NAD, and heme is decreased, resulting in **neurologic** and **pellagra-like symptoms** and **anemia**.

B. Removal of amino acid nitrogen as ammonia

A number of amino acids undergo reactions in which their nitrogen is released as ammonia or ammonium ion (NH_4^+).

1. **Glutamate dehydrogenase** catalyzes the oxidative deamination of glutamate. Ammonium ion is released, and α-ketoglutarate is formed. The glutamate dehydrogenase reaction, which is readily reversible, requires either NAD^+ or $NADP^+$.
2. **Histidine** is deaminated by histidase to form NH_4^+ and urocanate.
3. **Serine** and **threonine** are deaminated by serine dehydratase, which requires PLP. Serine is converted to pyruvate, and threonine to α-ketobutyrate; NH_4^+ is released.
4. The amide groups of **glutamine** and **asparagine** are released as ammonium ions by hydrolysis. Glutaminase converts glutamine to glutamate and NH_4^+. Asparaginase converts asparagine to aspartate and NH_4^+.
5. The **purine nucleotide cycle** serves to release NH_4^+ from amino acids, particularly in muscle.
   a. Glutamate collects nitrogen from other amino acids and transfers it to aspartate by a transamination reaction.
   b. Aspartate reacts with inosine monophosphate (IMP) to form adenosine monophosphate (AMP) and generate fumarate.
   c. NH_4^+ is released from AMP, and IMP is re-formed.

C. The role of glutamate

1. **Glutamate provides nitrogen for the synthesis** of many amino acids.
   a. NH_4^+ provides the nitrogen for amino acid synthesis by reacting with α-ketoglutarate to form glutamate in the glutamate dehydrogenase reaction.

b. Glutamate transfers nitrogen by transamination reactions to α-keto acids to form their corresponding α-amino acids.

2. Glutamate plays a key role in removing nitrogen from amino acids.
   a. Glutamate collects nitrogen from other amino acids by means of transamination reactions.
   b. The nitrogen of glutamate is released as NH_4^+ via the glutamate dehydrogenase reaction.
   c. NH_4^+ and aspartate provide nitrogen for urea synthesis via the urea cycle. Aspartate obtains its nitrogen from glutamate by transamination of oxaloacetate.

III. UREA CYCLE

- Ammonia, which is very toxic in humans, is converted to urea, which is nontoxic, very soluble, and readily excreted by the kidneys.
- Urea is formed in the urea cycle from NH_4^+, CO_2, and the nitrogen of aspartate. The cycle occurs mainly in the liver. Nitrogen is transported from the tissues to the liver in the forms of alanine (primarily muscle) and glutamine.
- NH_4^+, CO_2, and ATP react to form carbamoyl phosphate. The enzyme, carbamoyl phosphate synthetase I, is activated by *N*-acetylglutamate.
- Carbamoyl phosphate reacts with ornithine to form citrulline.
- Citrulline reacts with aspartate to form argininosuccinate, which releases fumarate to form arginine.
- The cleavage of arginine by arginase releases urea and regenerates ornithine.
- The enzymes of the urea cycle are induced if a high-protein diet is consumed for several days.
- When the nitrogen of amino acids is converted to urea in the liver, their carbon skeletons are converted either to glucose (in the fasting state) or to fatty acids (in the fed state).

CLINICAL CORRELATES

Nitrogenous excretory products are removed from the body mainly in the urine. In **renal failure**, these products are retained. **BUN, creatinine**, and **uric acid levels rise**.

A. Transport of nitrogen to the liver

Ammonia is **very toxic**, particularly to the central nervous system.

1. The concentration of ammonia and ammonium ions in the blood is normally very low. ($NH_3 + H^+ \leftrightarrow NH_4^+$.)
2. Ammonia travels to the **liver** from other tissues, mainly in the form of **alanine and glutamine**. It is released from amino acids in the liver by a series of transamination and deamination reactions.
3. Ammonia is also produced **by bacteria in the gut** and travels to the liver via the hepatic portal vein.

B. Reactions of the urea cycle (Fig. 8.3)

NH_4^+ and **aspartate** provide the nitrogen that is used to produce **urea**, and CO_2 provides the carbon. Ornithine serves as a carrier that is regenerated by the cycle.

1. **Carbamoyl phosphate** is synthesized in the first reaction from NH_4^+, CO_2, and two ATP. Inorganic phosphate and two ADP are also produced.
   a. Enzyme: **carbamoyl phosphate synthetase I**, which is located in mitochondria and is activated by *N*-acetylglutamate.
2. **Ornithine** reacts with carbamoyl phosphate to form citrulline. Inorganic phosphate is released.
   a. Enzyme: **ornithine transcarbamoylase**, which is found in mitochondria. The product, citrulline, is transported to the cytosol in exchange for cytoplasmic ornithine.
3. **Citrulline** combines with aspartate to form argininosuccinate in a reaction that is driven by the hydrolysis of ATP to AMP and inorganic pyrophosphate.
   a. Enzyme: **argininosuccinate synthetase**

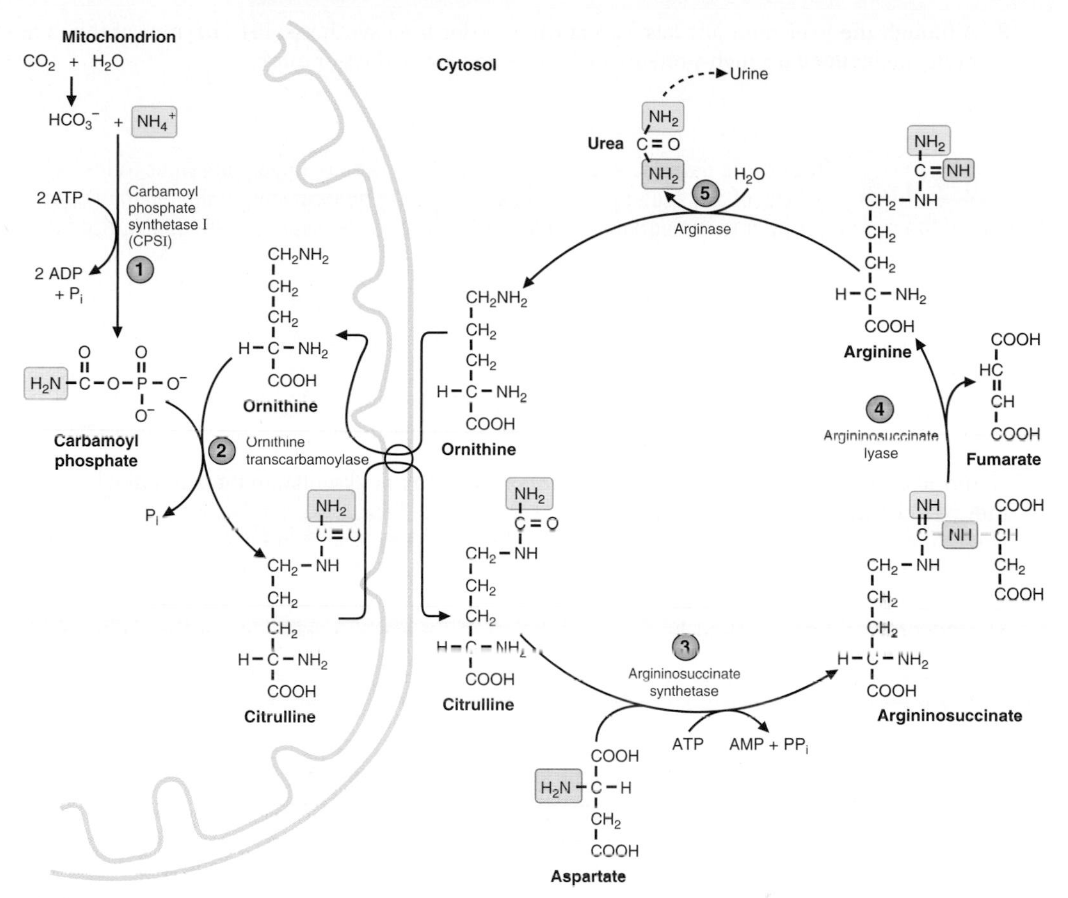

FIGURE 8.3. The urea cycle. The *dashed boxes* indicate the nitrogen-containing groups from which urea is formed. The *numbers* correspond to the steps described in the text in Section III B.

4. **Argininosuccinate** is cleaved to form arginine and fumarate.
   a. Enzyme: **argininosuccinate lyase.** This reaction occurs in the cytosol.
      (1) The carbons of fumarate, which are derived from the aspartate added in reaction 3, can be converted to malate.
      (2) In the fasting state in the liver, malate can be converted to glucose or to oxaloacetate, which is transaminated to regenerate the aspartate required for reaction 3.
5. **Arginine** is cleaved to form urea and regenerate ornithine.
   a. Enzyme: **arginase**, which is located primarily in the liver and is inhibited by ornithine.
6. **Urea** passes into the blood and is excreted by the kidneys.
   a. The urea excreted each day by a healthy adult (about 30 g) accounts for about 90% of the nitrogenous excretory products.
7. **Ornithine** is transported back into the mitochondrion (in exchange for citrulline) where it can be used for another round of the cycle.
   a. When the cell requires additional **ornithine**, it is synthesized from glucose via glutamate (see Fig. 8.7).
   b. **Arginine** is a nonessential amino acid in adults. It is synthesized from glucose via ornithine and the first four reactions of the urea cycle.

C. Regulation of the urea cycle

1. ***N*-Acetylglutamate** is an activator of carbamoyl phosphate synthetase I, the first enzyme of the urea cycle.
   a. **Arginine** stimulates *N*-acetylglutamate synthase, which catalyzes the synthesis of *N*-acetylglutamate from acetyl-CoA and glutamate.

2. Although the liver normally has a great capacity for urea synthesis, the enzymes of the urea cycle are **induced** if a high-protein diet is consumed for 4 days or more.

CLINICAL CORRELATES **Urea cycle defects** result from mutations in the urea cycle enzymes. A loss of activity of any urea cycle enzyme will lead to **hyperammonemia**, along with a buildup of the precursor prior to the enzymatic defect. The urea cycle disorders are summarized in Table 8.1.

IV. SYNTHESIS AND DEGRADATION OF AMINO ACIDS

- Of the 20 amino acids commonly found in proteins, 11 are not essential in the adult diet because they can be synthesized in the body.
- Ten of the nonessential amino acids contain carbon skeletons that can be derived from glucose (Fig. 8.4).
- Tyrosine is produced by the hydroxylation of the essential amino acid phenylalanine.
- The major products obtained by the degradation of the carbon skeletons of the amino acids are pyruvate, intermediates of the tricarboxylic acid (TCA) cycle, acetyl-CoA, and acetoacetate (Fig. 8.5).

A. Synthesis of amino acids

Messenger RNA contains codons for 20 amino acids. Eleven of these amino acids can be synthesized in the body. The carbon skeletons of 10 of these amino acids can be derived from **glucose**.

1. **Amino acids derived from the intermediates of glycolysis** (Fig. 8.6)
   a. The intermediates of glycolysis serve as the precursors for serine, glycine, cysteine, and alanine.
      (1) **Serine** can be synthesized from the glycolytic intermediate 3-phosphoglycerate, which is oxidized, transaminated by glutamate, and dephosphorylated.
      (2) **Glycine** and **cysteine** can be derived from serine.

table 8.1 Inherited Defects in Urea Cycle Enzymes

Defective enzyme	Metabolites that accumulate in the blood/urine	Comments
Carbamoyl phosphate synthetase I	Ammonia	A very rare disorder, no treatment exists, and the disorder is fatal.
Ornithine transcarbamoylase	Ammonia, orotic acid	An X-linked disorder, most common of the inherited disorders in the urea cycle. Orotic acid accumulates owing to the carbamoyl phosphate produced in the mitochondria diffusing into the cytoplasm and activating pyrimidine synthesis.
Argininosuccinate synthetase	Ammonia, citrulline	The second most common urea cycle defect; as with an OTC deficiency, mental retardation can result if not rapidly treated.
Argininosuccinate lyase	Ammonia, argininosuccinate	The ammonia accumulation is not as severe as the other defects as two nitrogens have been fixed into argininosuccinate, which is now excreted. Arginine is now an essential amino acid.
Arginase	Ammonia, arginine	As with an argininosuccinate lyase deficiency, the ammonia accumulation is not as severe as previous defects owing to nitrogens being fixed into arginine, which is excreted.

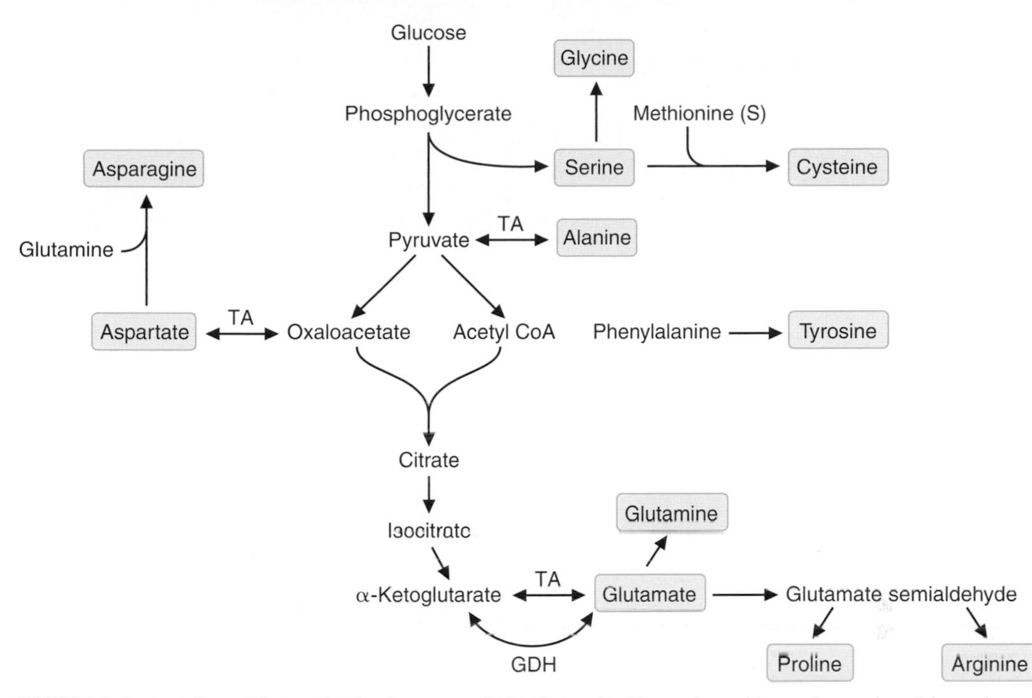

FIGURE 8.4. **An overview of the synthesis of nonessential amino acids**. Ten amino acids can be produced from glucose via the intermediates of glycolysis or the TCA cycle. The 11th nonessential amino acid, tyrosine, is synthesized by the hydroxylation of the essential amino acid phenylalanine. Only the sulfur of cysteine comes from the essential amino acid methionine; its carbons and nitrogen come from serine. GDH, glutamate dehydrogenase; TA, transamination.

(a) **Glycine** can be produced from serine by a reaction in which a methylene group is transferred to tetrahydrofolate (FH_4) (a cofactor that carries 1-carbon groups at various levels of oxidation).

(b) **Cysteine** derives its carbon and nitrogen from serine. The essential amino acid **methionine** supplies the sulfur.

(3) **Alanine** can be derived by the transamination of pyruvate.

CLINICAL CORRELATES **Primary hyperoxaluria type 1** results from a defect in glycine transaminase, resulting in an accumulation of **glyoxylate** (glycine transaminase catalyzes the transamination of glyoxylate to glycine, or the reverse reaction, glycine to glyoxylate). Glyoxylate will be oxidized to oxalate, which is sparingly soluble and will form calcium salts in the kidney, precipitating and leading to stone formation. **Renal failure** can occur in this disorder.

2. **Amino acids derived from TCA cycle intermediates** (see Fig. 8.4)
   a. **Aspartate** can be derived from oxaloacetate by transamination.
      (1) **Asparagine** is produced from aspartate by amidation.
   b. **Glutamate** is derived from α-ketoglutarate by the addition of NH_4^+ via the glutamate dehydrogenase reaction or by transamination. **Glutamine, proline, and arginine** can be derived from glutamate (Fig. 8.7).
      (1) **Glutamine** is produced by the amidation of glutamate.
      (2) Proline and arginine can be derived from **glutamate semialdehyde**, which is formed by the reduction of glutamate.
         (a) **Proline** can be produced by the cyclization of glutamate semialdehyde.
         (b) **Arginine**, via three reactions of the urea cycle, can be derived from ornithine, which is produced by the transamination of glutamate semialdehyde.

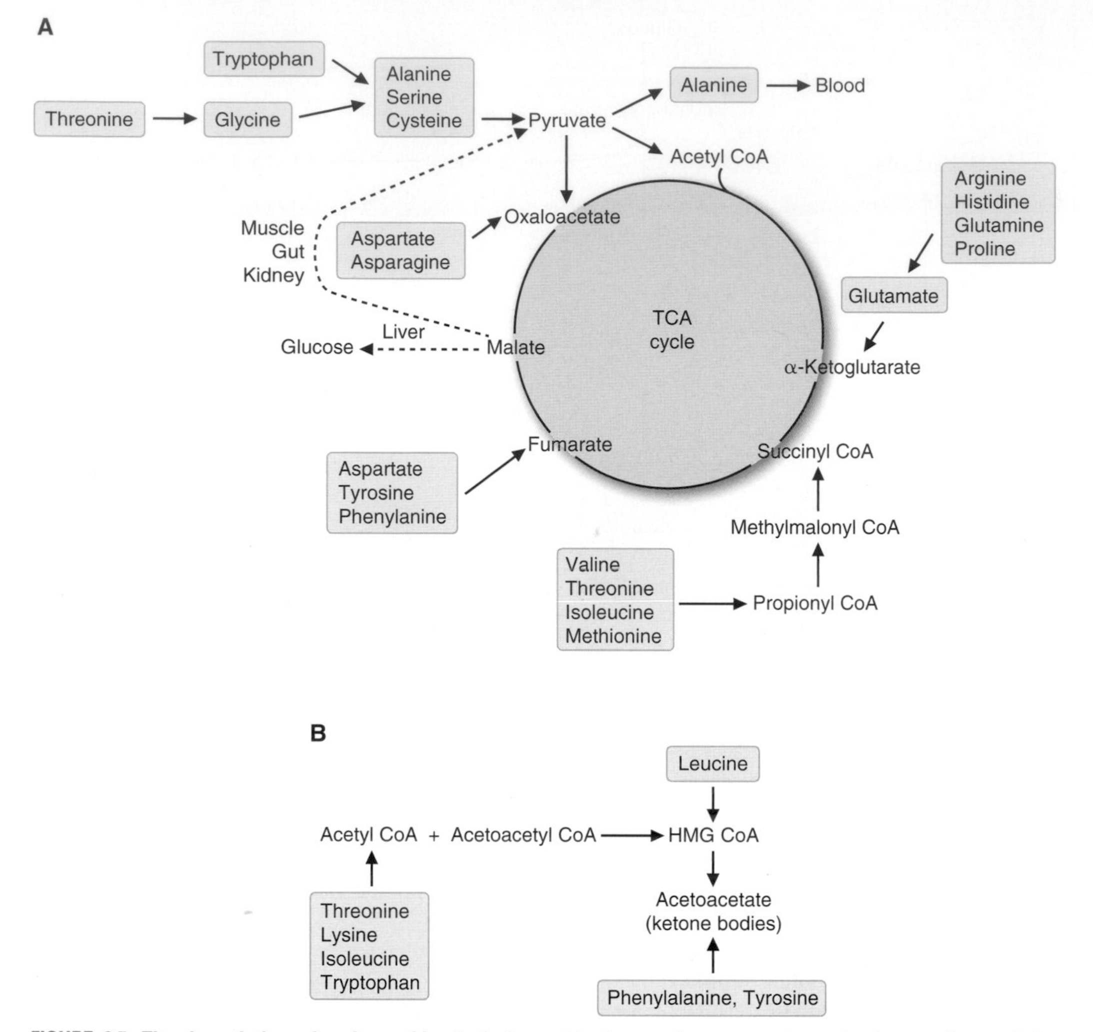

FIGURE 8.5. The degradation of amino acids. A. Amino acids that produce pyruvate or the intermediates of the TCA cycle. These amino acids are considered glucogenic because their carbons can produce glucose in the liver. **B.** Amino acids that produce acetyl-CoA or ketone bodies. These amino acids are considered ketogenic. HMG-CoA, hydroxymethylglutaryl-CoA.

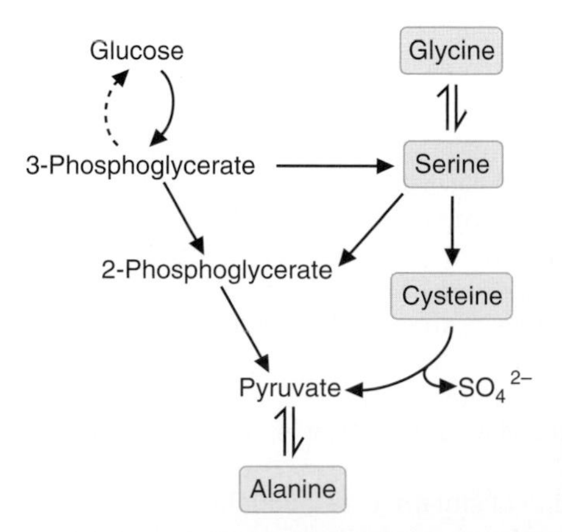

FIGURE 8.6. Amino acids derived from the intermediates of glycolysis. These amino acids can be synthesized from glucose and can be reconverted to glucose in the liver.

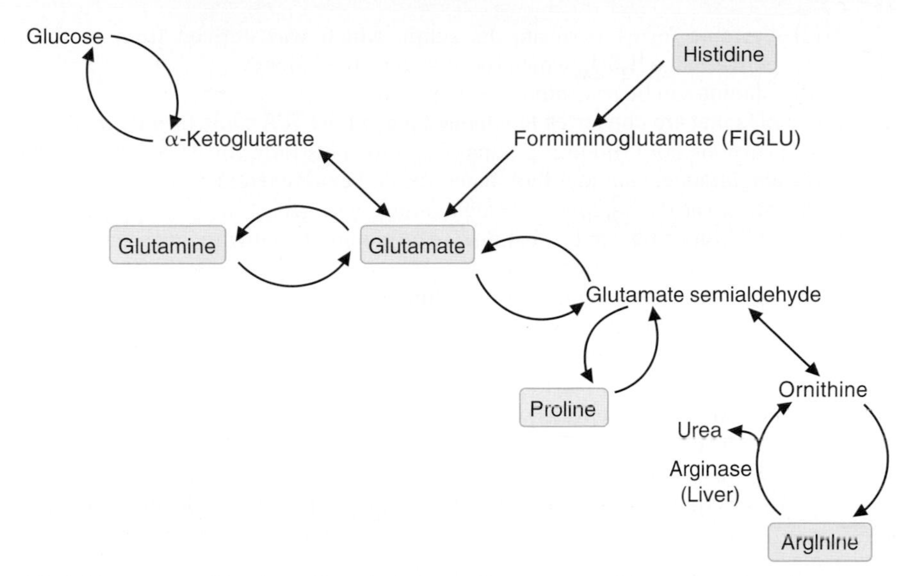

FIGURE 8.7. **Amino acids related through glutamate.** These amino acids contain carbons that can be converted to glutamate, which can be converted to glucose in the liver. All of these amino acids except histidine can be synthesized from glucose.

3. **Tyrosine**, the 11th nonessential amino acid, is synthesized by the hydroxylation of the essential amino acid phenylalanine in a reaction that requires tetrahydrobiopterin, which is a required cofactor for ring hydroxylation reactions.

B. Degradation of amino acids

1. When the carbon skeletons of amino acids are degraded, the major products are pyruvate, intermediates of the TCA cycle, acetyl-CoA, and acetoacetate (see Fig. 8.5).
2. Amino acids that form pyruvate or intermediates of the TCA cycle in the liver are **glucogenic** (or gluconeogenic); that is, they provide carbon for the synthesis of glucose (see Fig. 8.5A).
3. Amino acids that form acetyl-CoA or acetoacetate are **ketogenic**; that is, they form ketone bodies (see Fig. 8.5B).
4. Some amino acids (isoleucine, tryptophan, phenylalanine, and tyrosine) are both glucogenic and ketogenic.

CLINICAL CORRELATES **Pellagra** is caused by a deficiency of the vitamin **niacin** or of **tryptophan**. Niacin is required for the production of NAD and NADP. These compounds can also be generated from tryptophan. Pellagra results in the four Ds: **dermatitis, diarrhea, dementia, and death**.

a. **Amino acids that are converted to pyruvate** (see Fig. 8.6).
   (1) The amino acids that are synthesized from the intermediates of glycolysis (serine, glycine, cysteine, and alanine) are degraded to form pyruvate.
      (a) **Serine** is converted to 2-phosphoglycerate, an intermediate of glycolysis, or directly to pyruvate and NH_4^+ by serine dehydratase, an enzyme that requires PLP.
      (b) **Glycine**, in a reversal of the reaction utilized for its synthesis, reacts with methylene-FH_4 to form serine.
         1. Glycine also reacts with FH_4 and NAD^+ to produce CO_2 and NH_4^+ (glycine cleavage enzyme).
         2. Glycine can be converted to glyoxylate, which can be oxidized to CO_2 and H_2O or converted to oxalate.

(c) **Cysteine** forms pyruvate. Its sulfur, which was derived from methionine, is converted to $\mathbf{H_2SO_4}$, which is excreted by the kidneys.

(d) **Alanine** can be transaminated to pyruvate.

b. **Amino acids that are converted to intermediates of the TCA cycle** (see Fig. 8.5).

(1) The carbons from the four groups of amino acids form the TCA cycle intermediates **α-ketoglutarate, succinyl-CoA, fumarate,** and **oxaloacetate.**

(a) Amino acids that form α-ketoglutarate (see Fig. 8.7).

1. **Glutamate** can be deaminated by glutamate dehydrogenase or transaminated to form α-ketoglutarate.
2. **Glutamine** is converted by glutaminase to glutamate with the release of its amide nitrogen as NH_4^+.
3. **Proline** is oxidized so that its ring opens, forming glutamate semialdehyde, which is oxidized to glutamate.
4. **Arginine** is cleaved by arginase in the liver to form urea and ornithine. Ornithine is transaminated to glutamate semialdehyde, which is oxidized to glutamate.
5. **Histidine** is converted to formiminoglutamate (FIGLU). The formimino group is transferred to FH_4, and the remaining five carbons form glutamate.

CLINICAL CORRELATES

In **histidinemia**, histidase, which converts histidine to urocanate, is defective. The earlier cases were reported to be associated with mental retardation, but more recently, deleterious consequences have not been observed. The defect is an autosomal recessive disorder, and is now considered benign.

(b) **Amino acids that form succinyl-CoA** (Fig. 8.8)

1. Four amino acids are converted to **propionyl-CoA**, which is carboxylated in a biotin-requiring reaction to form methylmalonyl-CoA, which is rearranged to form succinyl-CoA in a reaction that requires vitamin B_{12} (seen previously in the metabolism of odd-chain number fatty acids).

a) **Threonine** is converted by a dehydratase to NH_4^+ and α-ketobutyrate, which is oxidatively decarboxylated to propionyl-CoA.
In a different set of reactions, threonine is converted to glycine and acetyl-CoA.

b) **Methionine** provides **methyl groups** for the synthesis of various compounds; its sulfur is incorporated into **cysteine**; and the remaining carbons form **succinyl-CoA**.

i. Methionine and ATP form ***S*-adenosylmethionine (SAM)**, which donates a methyl group and forms homocysteine.

ii. **Homocysteine** is reconverted to methionine by accepting a methyl group from the FH_4 pool via vitamin B_{12}.

iii. **Homocysteine** can also react with serine to form **cystathionine.** The cleavage of cystathionine produces cysteine, NH_4^+, and α-ketobutyrate, which is converted to propionyl-CoA.

CLINICAL CORRELATES

In **homocystinuria**, the cystathionine synthase is defective. Therefore, homocysteine does not react with serine to form cysteine (see Fig. 8.8). The homocysteine that accumulates is oxidized to homocystine and excreted in the urine. Some cases respond to increased doses of vitamin B_6, which forms PLP, the cofactor for the synthase enzyme. Other causes of elevated levels of homocyst(e)ine (homocysteine or homocystine) in the blood or urine are deficiencies of the enzyme methionine synthase (that converts homocysteine to methionine) or dietary deficiencies of the vitamin cofactors for this enzyme (folate and B_{12}). Elevated levels of homocyst(e)ine have been associated with increased risk of **coronary artery disease**, although this is controversial.

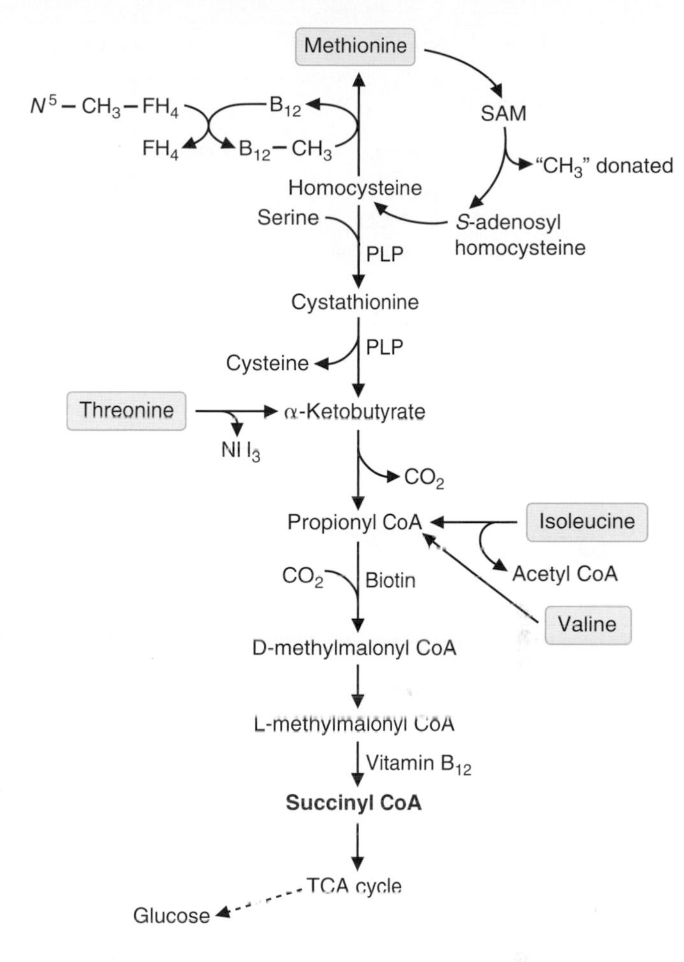

FIGURE 8.8. **Amino acids that can be converted to succinyl-CoA**. The amino acids methionine, threonine, isoleucine, and valine, which form succinyl-CoA via methylmalonyl-CoA, are all essential. Because succinyl-CoA can form glucose, these amino acids are glucogenic. The carbons of serine are converted to cysteine and do not form succinyl-CoA by this pathway. A defect in cystathionine synthase causes homocystinuria. A defect in cystathionase causes cystathionuria. PLP, pyridoxal phosphate; SAM, *S*-adenosylmethionine; B_{12}-CH_3, methylcobalamin; N^5-CH_3-FH_4, N^5-methyltetrahydrofolate.

c) **Valine and isoleucine**, two of the three branched-chain amino acids, form succinyl-CoA (see Fig. 8.8).
   i. The degradation of all three branched-chain amino acids begins with a **transamination** followed by an **oxidative decarboxylation** catalyzed by the branched-chain α-keto acid dehydrogenase complex (Fig. 8.9). This enzyme, like pyruvate dehydrogenase and α-ketoglutarate dehydrogenase, requires thiamine pyrophosphate, lipoic acid, coenzyme A, FAD, and NAD^+.
   ii. **Valine** is eventually converted to succinyl-CoA via propionyl-CoA and methylmalonyl-CoA.
   iii. **Isoleucine** also forms succinyl-CoA after two of its carbons are released as acetyl-CoA.

CLINICAL CORRELATES

In **maple syrup urine disease (MSUD)**, the enzyme complex that decarboxylates the transamination products of the branched-chain amino acids (the α-keto acid dehydrogenase) is defective (see Fig. 8.9). Valine, isoleucine, and leucine accumulate. The urine has the odor of maple syrup. **Mental retardation** occurs. MSUD is more difficult to treat than PKU owing to the need to titrate three essential amino acids in the diet.

(c) Amino acids that form fumarate

1. Three amino acids (phenylalanine, tyrosine, and aspartate) are converted to fumarate (see Fig. 8.5).

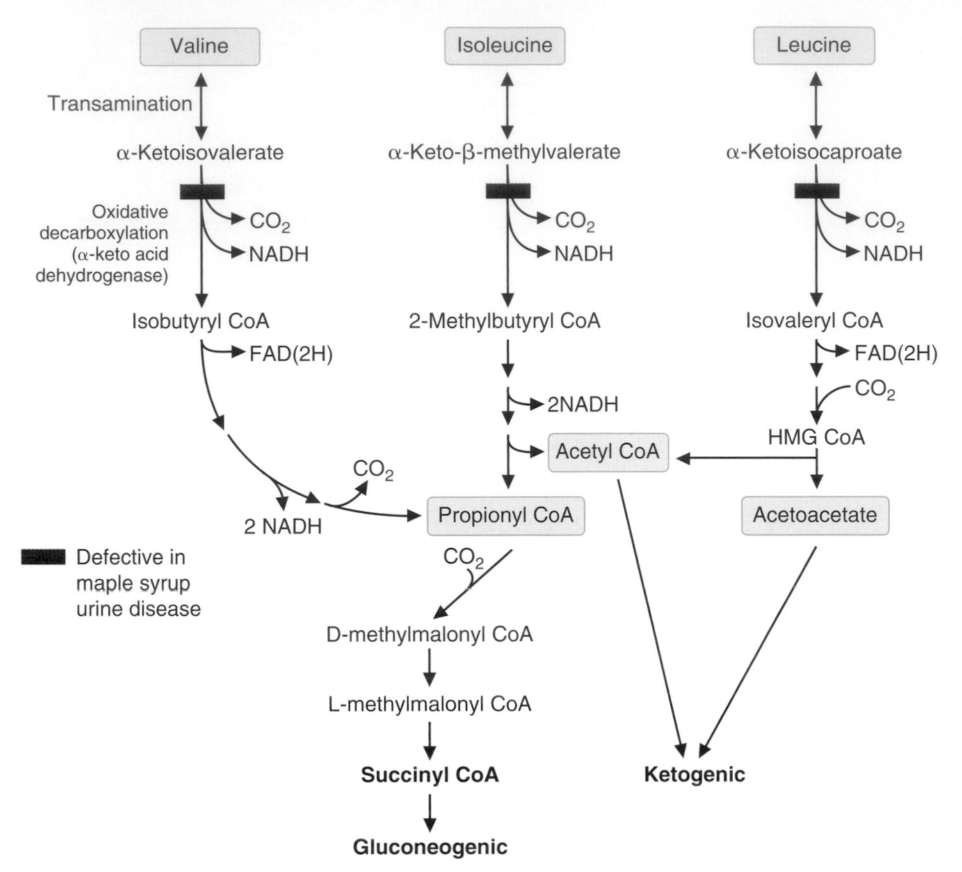

FIGURE 8.9. The degradation of the branched-chain amino acids. Valine forms propionyl-CoA. Isoleucine forms propionyl-CoA and acetyl-CoA. Leucine forms acetoacetate and acetyl-CoA.

a) **Phenylalanine** is converted to **tyrosine** by phenylalanine hydroxylase in a reaction requiring tetrahydrobiopterin and O_2 (Fig. 8.10).
b) **Tyrosine**, obtained from the diet or by hydroxylation of phenylalanine, is converted to homogentisic acid. The aromatic ring is opened and cleaved, forming **fumarate** and **acetoacetate**.
c) **Aspartate** is converted to fumarate via reactions of the **urea cycle** and the **purine nucleotide cycle**.
   Aspartate reacts with IMP to form AMP and fumarate in the purine nucleotide cycle.

CLINICAL CORRELATES There are a number of disorders associated with phenylalanine and tyrosine metabolism. In **PKU**, the conversion of phenylalanine to tyrosine is defective. Some cases of PKU are due to defects in phenylalanine hydroxylase and some to defects in the synthesis of **tetrahydrobiopterin**. Phenylalanine accumulates, and is converted to compounds such as the **phenylketones**, which give the urine a musty odor. **Mental retardation** occurs. PKU is treated by restriction of phenylalanine in the diet. In **alcaptonuria**, homogentisic acid, a product of phenylalanine and tyrosine metabolism, accumulates because homogentisate oxidase is defective (see Fig. 8.10). Homogentisic acid auto-oxidizes and the products polymerize, forming dark-colored pigments, which accumulate in various tissues and are sometimes associated with a **degenerative arthritis**. Defects in tyrosine metabolism can lead to **tyrosinemia, types I and II**. Tyrosinemia, type I, is due to a defect in fumarylacetoacetate hydrolase (splits fumarylacetoacetate

(continued)

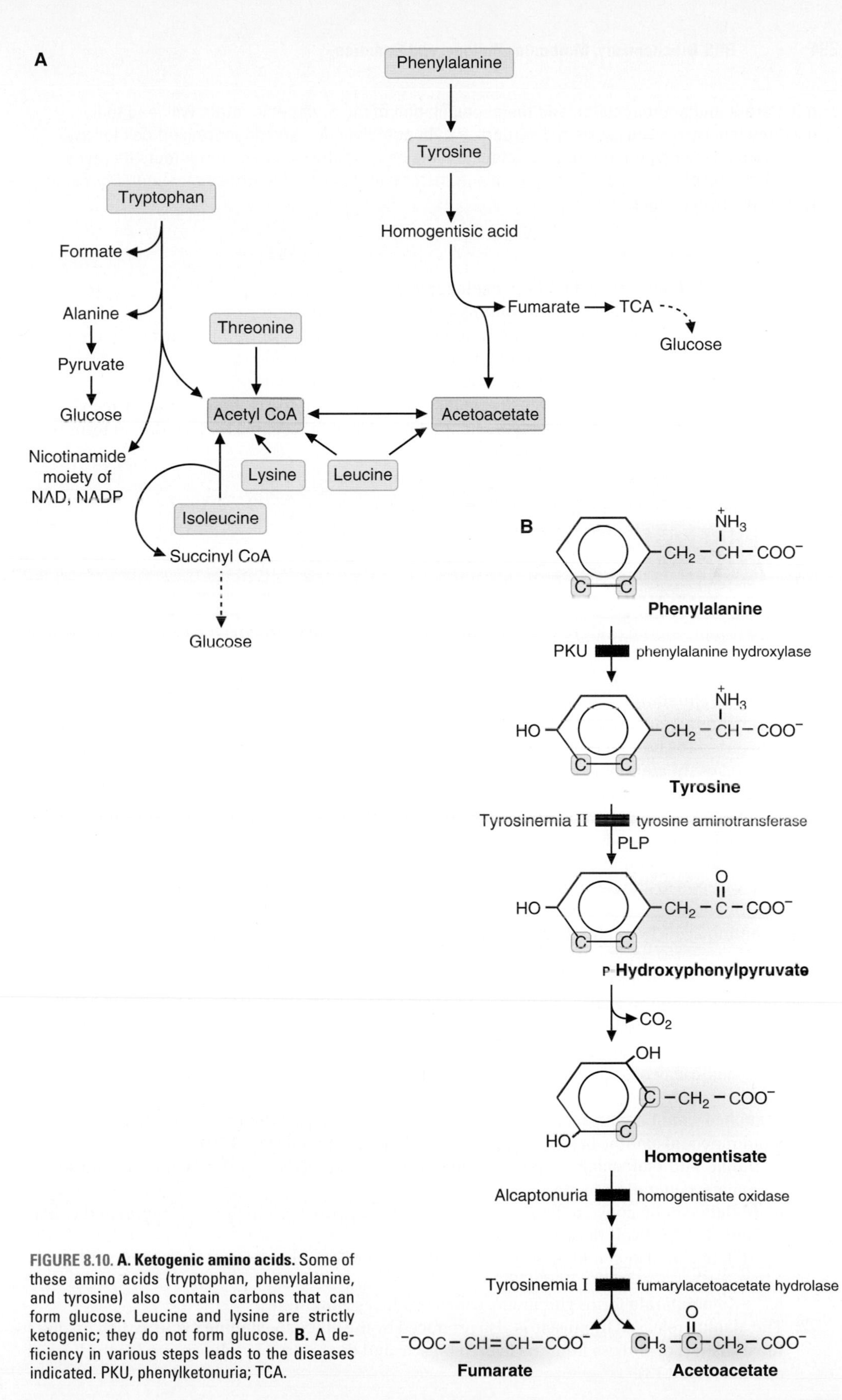

FIGURE 8.10. A. Ketogenic amino acids. Some of these amino acids (tryptophan, phenylalanine, and tyrosine) also contain carbons that can form glucose. Leucine and lysine are strictly ketogenic; they do not form glucose. **B.** A deficiency in various steps leads to the diseases indicated. PKU, phenylketonuria; TCA.

into fumarate and acetoacetate), and the accumulation of fumarylacetoacetate will lead to liver and kidney failure, nervous system disorders, a cabbage-like odor, and an increased risk for liver cancer. Tyrosinemia type II is due to a defect in tyrosine aminotransferase, and affects the eyes, skin, and mental development. Tyrosine aminotransferase catalyzes the transamination of tyrosine to 4-hydroxyphenylpyruvate.

(d) Amino acids that form oxaloacetate (see Fig. 8.5)

1. **Aspartate** is transaminated to form oxaloacetate.
2. **Asparagine** loses its amide nitrogen as NH_4^+, forming aspartate in a reaction catalyzed by asparaginase.

c. Amino acids that are converted to acetyl-CoA or acetoacetate (see Fig. 8.10)

(1) Four amino acids (**lysine, threonine, isoleucine**, and **tryptophan**) can form acetyl-CoA, and **phenylalanine** and **tyrosine** form acetoacetate. **Leucine** is degraded to form both acetyl-CoA and acetoacetate.

V. INTERRELATIONSHIPS OF VARIOUS TISSUES IN AMINO ACID METABOLISM

- In the fed state, amino acids released by the digestion of dietary proteins travel to the liver, where they are used for the synthesis of proteins, glucose, and triglyceride (Fig. 8.11A).
- During fasting, amino acids from muscle protein are released into the blood, predominantly as alanine and glutamine (see Fig. 8.11B).
- The branched-chain amino acids are oxidized by muscle to produce energy. Some of the carbons are converted to glutamine and alanine. Alanine is also produced by the glucose–alanine cycle.
- The gut takes up glutamine from the blood and converts it to alanine, citrulline, and ammonia, which are released.
- The kidney takes up glutamine from the blood and releases ammonia into the urine and alanine and serine into the blood.
- The liver takes up alanine and other amino acids from the blood and converts the nitrogen to urea and the carbons to glucose and ketone bodies, which are released into the blood and oxidized by tissues for energy.
- Thus, amino acids from muscle protein serve as a source of energy for many other tissues.

A. Amino acid metabolism in muscle

During **fasting**, amino acids are released from muscle protein. Some of the **amino acids** are **partially oxidized** in muscle and their remaining carbons are converted to **alanine** and **glutamine**. Thus, although about 50% of the amino acids released into the blood from muscle are alanine and glutamine, these two amino acids constitute much less than 50% of the total amino acid residues in muscle protein.

1. **Branched-chain amino acids** are oxidized in muscle. Some of their carbons are converted to glutamine and alanine before they are released into the blood (Fig. 8.12).
   - **a. Valine** and **isoleucine** produce succinyl-CoA, which feeds into the TCA cycle and forms malate, generating energy.
   - **b. Malate** can be converted by the malic enzyme to pyruvate, which is transaminated to alanine or oxidatively decarboxylated to acetyl-CoA.
   - **c.** Malate can also continue around the TCA cycle to α-ketoglutarate, generating additional energy.
   - **d. α-Ketoglutarate** forms glutamate, which produces glutamine.
2. The **alanine** released by muscle is also produced by the **glucose–alanine cycle**, which involves the transport of glucose from the liver to muscle and the return of carbon atoms to the liver as alanine (Fig. 8.13).

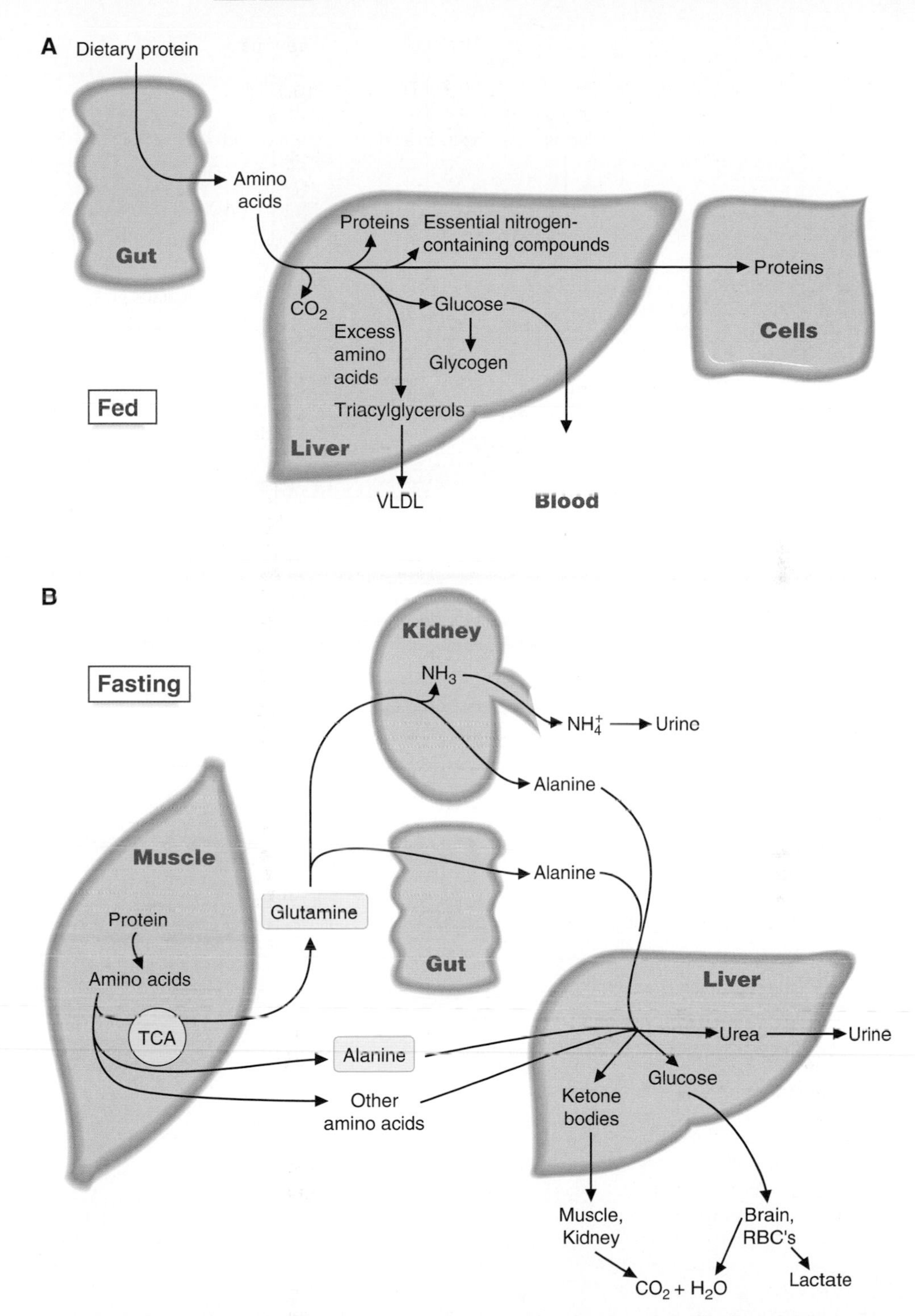

FIGURE 8.11. **A. Interrelationships of various tissues in amino acid metabolism**. **B.** During fasting, muscle releases amino acids, particularly alanine and glutamine. Glutamine is converted by the kidney to alanine and serine or by the gut to alanine. Alanine is taken up by the liver. The liver converts the carbons of alanine and other amino acids to glucose or to ketone bodies and the nitrogens to urea. Glucose and ketone bodies are oxidized by other tissues. Muscle can oxidize glucose, forming alanine, which is reconverted to glucose in the liver (the glucose–alanine cycle). RBCs, red blood cells; TCA.

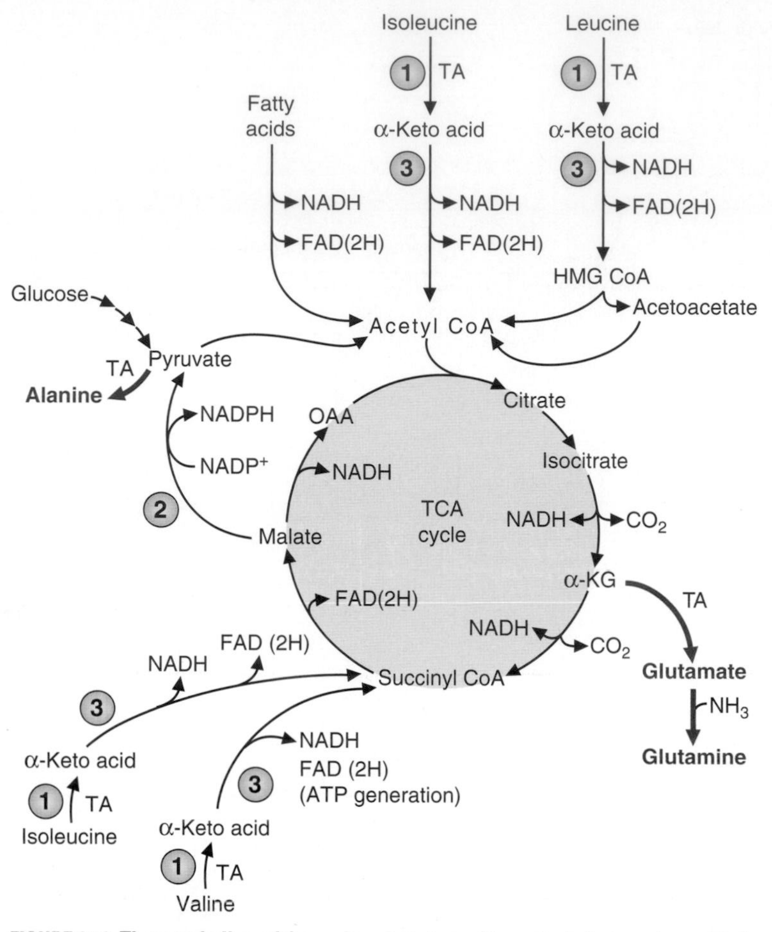

FIGURE 8.12. The metabolism of the carbon skeletons of branched-chain amino acids in skeletal muscle. (1) The first step in the metabolism of these amino acids is transamination (TA). (2) Carbon from valine and isoleucine enters the TCA cycle as succinyl-CoA and is converted to pyruvate by malic enzyme. (3) The oxidative pathways generate NADH and $FADH_2$ even before the carbon skeletons enter the TCA cycle. The carbon skeletons can also be converted to glutamate and alanine, shown in red, for nitrogen transport to the liver.

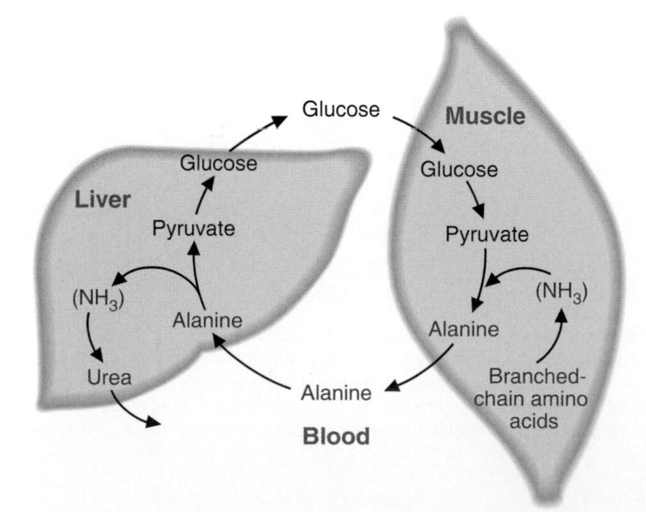

FIGURE 8.13. The glucose–alanine cycle. The pathway for the transfer of the amino group from branched-chain amino acids in skeletal muscle to urea in the liver is shown in red.

a. Glucose is oxidized in muscle to **pyruvate**, producing energy.
b. Pyruvate can be transaminated to **alanine**, which travels to the liver carrying nitrogen for urea synthesis and carbon for gluconeogenesis.

B. Amino acid metabolism in the gut (Fig. 8.14)

The gut takes up **glutamine** and releases alanine, citrulline, and ammonia.

1. In the gut, **glutamine** is converted to **NH_4^+** and glutamate, which forms α-ketoglutarate.
2. **α-Ketoglutarate** is converted to malate, generating energy. In the cytosol, malate is decarboxylated by the malic enzyme to form pyruvate, which is transaminated to **alanine**.
3. **Glutamate** is also converted to ornithine, which forms **citrulline**.

C. Amino acid metabolism in the kidney (Fig. 8.15)

The kidney takes up glutamine, which is deaminated by glutaminase, forming ammonia and glutamate, which is converted to alanine and serine.

1. **Ammonia** (NH_3) is released into the **urine**, where it forms NH_4^+, **buffering** the hydrogen ions produced by phosphoric acid, sulfuric acid (produced from cysteine), and various metabolic acids (e.g., lactic acid and the ketone bodies, acetoacetic acid and ß-hydroxybutyric acid).
2. **Alanine** and **serine** produced from glutamate are released into the **blood.**
   a. **Glutamate** is deaminated or transaminated to form α-ketoglutarate, which enters the TCA cycle and is converted to malate.
   b. **Malate** enters the cytosol and is oxidized to oxaloacetate, which is converted by phosphoenolpyruvate carboxykinase to phosphoenolpyruvate (PEP).
   c. **Phosphoenolpyruvate** feeds into glycolysis and forms alanine and serine, which are released into the blood.

D. Amino acid metabolism in the liver (see Fig. 8.11)

The liver takes up alanine, serine, and other amino acids from the blood and converts their **nitrogen to urea** and their **carbons to glucose or to ketone bodies**, which are released into the blood and oxidized by other tissues.

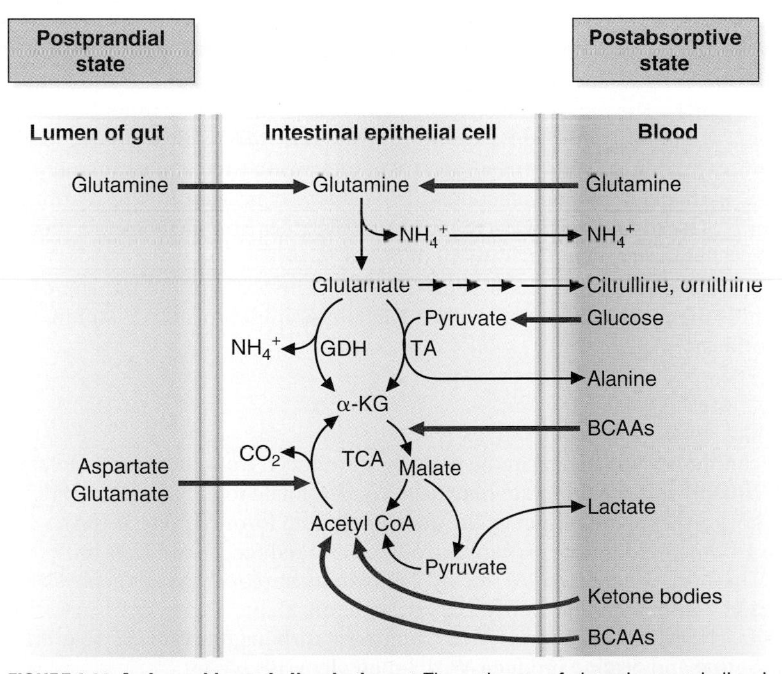

FIGURE 8.14. Amino acid metabolism in the gut. The pathways of glutamine metabolism in the gut are the same whether it is supplied by the diet or from the blood (postabsorptive state). The cells of the gut also metabolize aspartate, glutamate, and branched-chain amino acids. αKG, α-ketoglutarate; GDH, glutamate dehydrogenase; TA, transaminase.

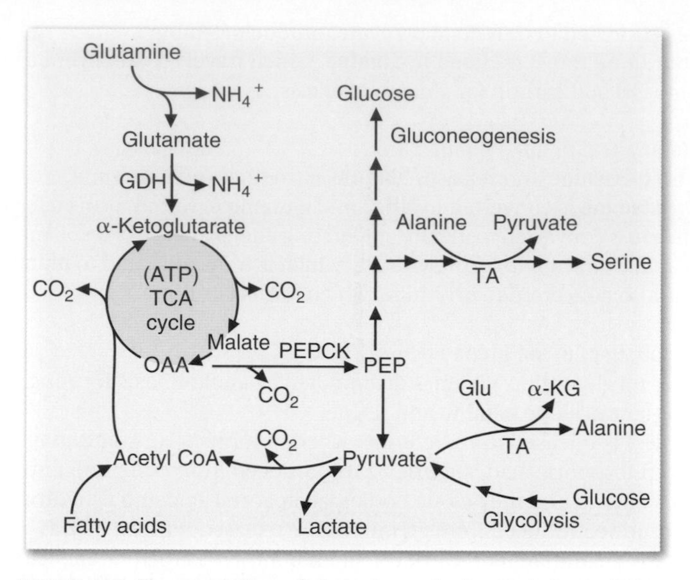

FIGURE 8.15. **The metabolism of glutamine and other fuels in the kidney.** GDH, glutamate dehydrogenase; OAA, oxaloacetate; PEPCK, phosphoenolpyruvate carboxykinase; TA, transaminase.

VI. TETRAHYDROFOLATE, VITAMIN B_{12}, AND *S*-ADENOSYLMETHIONINE

- The groups containing a single carbon atom can be transferred from one compound to another.
- The 1-carbon groups at lower levels of oxidation than CO_2 are transferred by FH_4, vitamin B_{12}, and SAM. (CO_2 is transferred by biotin.)
- FH_4, which is produced from the vitamin folate, obtains 1-carbon groups from serine, glycine, histidine, formaldehyde, and formate. The 1-carbon groups are oxidized and reduced while they are attached to FH_4. The most reduced form, methyl-FH_4, however, cannot be oxidized.
- The 1-carbon groups carried by FH_4 are transferred to dUMP to form dTMP; to glycine to form serine; to purine precursors to form C2 and C8 of the purine ring; and to vitamin B_{12}.
- Vitamin B_{12} is involved in two reactions in the body. It is used in the rearrangement of the methyl group of methylmalonyl-CoA to form succinyl-CoA, and it transfers a methyl group from 5-methyl-FH_4 to homocysteine to form methionine.
- SAM, which is produced from methionine and ATP, is involved in the transfer of methyl groups to compounds that form creatine, phosphatidylcholine, epinephrine, melatonin, and methylated polynucleotides.

A. Tetrahydrofolate

1. The nature of FH_4 and its derivatives

a. **FH_4** cannot be synthesized in the body. It is produced from the vitamin folate.

(1) **NADPH** and **dihydrofolate reductase** convert folate to dihydrofolate (**FH_2**), which undergoes a second reduction by the same enzyme to form **FH_4** (Fig. 8.16A).

b. The **1-carbon groups** of FH_4 can be oxidized and reduced (see Fig. 8.16B).

(1) The most reduced form, N^5-methyl-FH_4, cannot be reoxidized under physiologic conditions.

2. Sources of 1-carbon groups carried by FH_4

a. Serine, glycine, histidine, and formate transfer 1-carbon groups to FH_4 (Fig. 8.17, steps 1 to 4).

(1) **Serine** and **glycine** produce N^5,N^{10}-methylene-FH_4.

(a) **Serine** transfers a 1-carbon group to FH_4, and is converted to glycine in a reversible reaction. Because serine can be derived from glucose, this 1-carbon group can be obtained from dietary carbohydrate.

(b) When **glycine** transfers a 1-carbon unit to FH_4, NH_4^+ and CO_2 are produced.

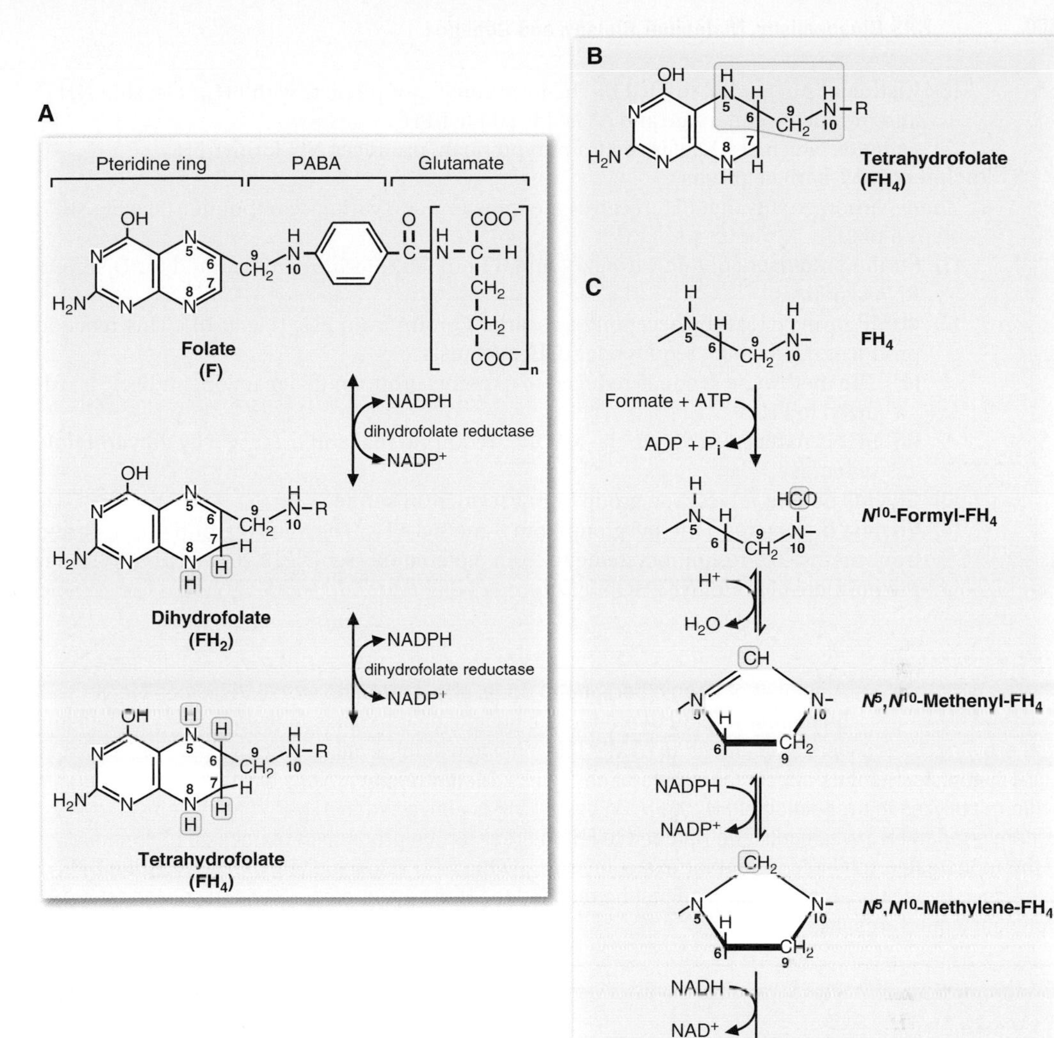

FIGURE 8.16. **Tetrahydrofolate (FH_4).** **A.** Reduction of folate by dihydrofolate reductase. **B.** The structure of FH_4. **C.** One-carbon groups carried by FH_4. The 1-carbon groups are indicated by yellow boxes. Only atoms 5, 6, 9, and 10 of FH_4 are shown.

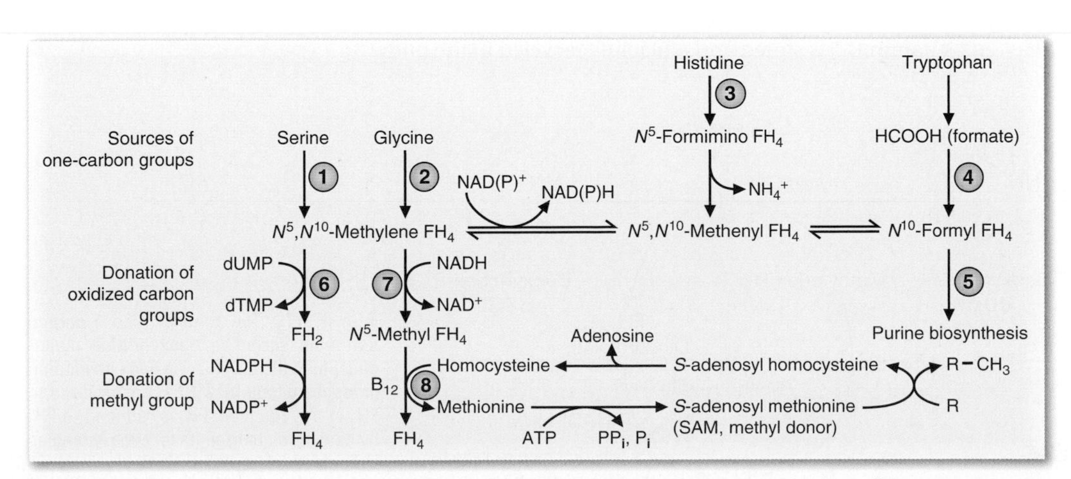

FIGURE 8.17. The sources of carbon (1 to 4) for the FH_4 pool and the recipients of carbon (5 to 8) from the pool. FH_2, dihydrofolate; FH_4, tetrahydrofolate.

(2) **Histidine** is degraded to FIGLU. The formimino group reacts with FH_4, releasing NH_4^+ and producing glutamate and N^5,N^{10}-methenyl-FH_4.

(3) **Formate**, which can be derived from tryptophan, produces N^{10}-formyl-FH_4.

3. **Recipients of 1-carbon groups**

a. The 1-carbon groups that FH_4 receives are transferred to various compounds (see Fig. 8.17, steps 5 to 8).

(1) **Purine precursors** obtain carbons 2 and 8 from FH_4. Purines are required for DNA and RNA synthesis.

(2) **dUMP** forms dTMP by accepting a 1-carbon group from FH_4 (Fig. 8.18). This reaction produces the **thymine** required for **DNA synthesis**.

(a) The methylene group is reduced to a methyl group in this reaction, and FH_4 is oxidized to FH_2.

(b) FH_2 is reduced to FH_4 in the NADPH-requiring reaction catalyzed **by dihydrofolate reductase**.

(3) **Glycine** obtains a 1-carbon group from FH_4 to form **serine**.

(4) **Vitamin B_{12}** obtains a methyl group from 5-methyl-FH_4. The methyl group is transferred from methyl-B_{12} to homocysteine to form **methionine** (see Fig. 8.20, reaction 8).This is the only fate of 5-methyl-FH_4.

CLINICAL CORRELATES A number of **chemotherapeutic drugs affect nitrogen metabolism** (see Fig. 8.18). **5-Fluorouracil (5-FU)** produces a nucleotide that binds to thymidylate synthetase and prevents the conversion of dUMP to dTMP. This inhibition of the synthesis of thymine **prevents DNA synthesis** and thus affects the proliferation of cells. **Methotrexate** inhibits dihydrofolate reductase, which catalyzes the reduction of FH_2 to FH_4. When dUMP is converted to dTMP, N^5,N^{10}-methylene-FH_4 is converted to FH_2, which must be reduced to FH_4 in order for the production of thymine to continue. If the reductase is inhibited by methotrexate, thymine synthesis is also inhibited, thus **preventing DNA synthesis**. In addition, dietary folate cannot be reduced to FH_4 by dihydrofolate reductase, and thus a deficiency of FH_4 results.

B. Vitamin B_{12}

1. **Source of vitamin B_{12}**

a. Vitamin B_{12} is produced by microorganisms, but not by plants.

b. Animals obtain vitamin B_{12} from their intestinal flora, from bacteria in their food supply, or by consuming the tissues of other animals.

c. **Intrinsic factor**, produced by gastric parietal cells, is required for the absorption of vitamin B_{12} by the intestine.

d. Vitamin B_{12} is stored and efficiently recycled in the body.

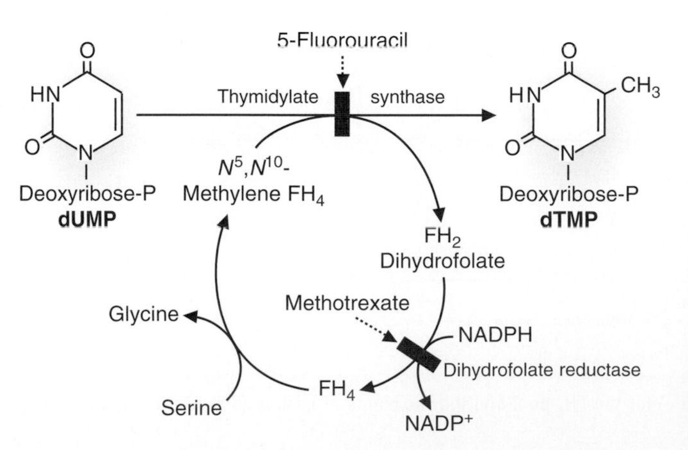

FIGURE 8.18. The transfer of a 1-carbon unit from serine to deoxyuridine monophosphate (dUMP) to form deoxythymidine monophosphate (dTMP). FH_4 is oxidized to FH_2 in this reaction. FH_2 is reduced to FH_4 by dihydrofolate reductase. The rectangles indicate the steps at which the antimetabolites methotrexate and 5-fluorouracil (5-FU) act.

CLINICAL CORRELATES

Folate and vitamin B_{12} deficiencies result in decreased DNA and RNA synthesis, which causes a **megaloblastic anemia**. FH_4 is directly involved in providing carbon for thymine (and thus DNA) synthesis and for the synthesis of the purines (see Fig. 8.23), which are required for both DNA and RNA synthesis. In a folate deficiency, normal cell proliferation cannot occur, and megaloblasts develop. If vitamin B_{12}, which accepts methyl groups from 5-methyl-FH_4, is deficient, 5-methyl-FH_4 accumulates. Since this compound cannot be reconverted to other FH_4 derivatives, a folate deficiency is secondarily produced (the methyl trap theory). Vitamin B_{12} also is involved in converting methylmalonyl-CoA to succinyl-CoA (see Fig. 8.8). Folate deficiencies have also been associated with neural tube defects at birth. Once this correlation was established, women who were considering getting pregnant were encouraged to take supplemental folic acid, which has significantly reduced the incidence of neural tube defects in newborns. In a **vitamin B_{12} deficiency**, methylmalonic acid is excreted in the urine, and, in addition to a megaloblastic anemia, **neurologic symptoms**, which cannot be alleviated by the administration of folate, occur. They are due to demyelination of nerves, and thus are irreversible. The **absence of intrinsic factor**, which is produced by the stomach and necessary for vitamin B_{12} absorption in the intestine, causes **pernicious anemia**. The accumulation of methylmalonyl-CoA, as in a B_{12} deficiency, will also lead to an accumulation of its precursor, propionyl-CoA. At high-enough propionyl-CoA levels, methylcitrate will accumulate, due to propionyl-CoA replacing acetyl-CoA in the citrate synthase reaction. Elevated methylcitrate is indicative of either a B_{12} deficiency or a lack of activity of propionyl-CoA carboxylase. Methylcitrate can inhibit citrate synthase and block the TCA cycle from functioning if present at sufficient levels, and lead to ketosis and hypoglycemia.

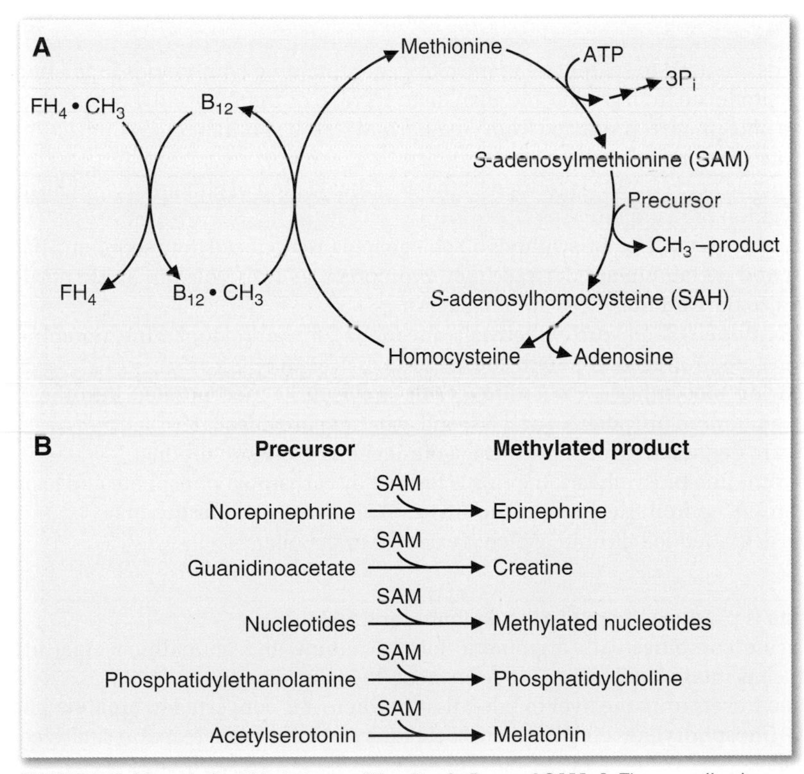

FIGURE 8.19. The relationship between FH_4, vitamin B_{12}, and SAM. A. The overall scheme. **B.** Some specific reactions that require SAM as a methylating agent.

2. **Functions of vitamin B_{12}**
   a. Vitamin B_{12} contains **cobalt** in a corrin ring that resembles a porphyrin.
      (1) **Vitamin B_{12}** is the cofactor for methylmalonyl-CoA mutase, which catalyzes the rearrangement of **methylmalonyl-CoA to succinyl-CoA** (see Fig. 8.8).
         (a) This reaction is involved in the production of succinyl-CoA from valine, isoleucine, threonine, methionine, thymine, and the propionyl-CoA formed by the oxidation of fatty acids with an odd number of carbons.
      (2) **Vitamin B_{12}** is involved in the transfer of methyl groups from FH_4 to **homocysteine to form methionine** (see Fig. 8.8).

C. *S*-Adenosylmethionine

1. SAM is synthesized from **methionine** and **ATP**.
2. **Methyl groups** are supplied by SAM for the following conversions (Fig. 8.19):
   a. Guanidinoacetate to **creatine**
   b. Phosphatidylethanolamine to **phosphatidylcholine**
   c. Norepinephrine to **epinephrine**
   d. Acetylserotonin to **melatonin**
   e. Polynucleotides to **methylated polynucleotides**
3. When SAM transfers its methyl group to an acceptor, *S*-adenosylhomocysteine is produced.
4. *S*-Adenosylhomocysteine releases adenosine to form homocysteine, which obtains a methyl group from vitamin B_{12} to form methionine. Methionine reacts with ATP to regenerate SAM (see Figs. 8.8 and 8.17).

VII. SPECIAL PRODUCTS DERIVED FROM AMINO ACIDS

- Amino acids are used to synthesize many nitrogen-containing compounds in the body.
- Creatine is produced from glycine, the guanidinium group of arginine, and the methyl group of SAM.
  - Creatine phosphate is produced from creatine and ATP. It spontaneously cyclizes to form creatinine, which is excreted in the urine.
- γ-Aminobutyric acid (GABA) is formed by the decarboxylation of glutamate, and histamine by the decarboxylation of histidine.
- Ceramide, which is used for the synthesis of sphingolipids, is produced from serine and palmitoyl-CoA.
- Serotonin and melatonin are derived from tryptophan, as is the nicotinamide portion of NAD^+ (which is also derived from the vitamin niacin).
- Thyroid hormone, 3,4-dihydroxyphenylalanine (dopa), melanin, dopamine, norepinephrine, and epinephrine are produced from tyrosine.
- During purine biosynthesis, the entire glycine molecule is incorporated into the growing ring structure; glutamine provides N3 and N9; and aspartate provides N1.
  - Purines are degraded to form uric acid, a nitrogenous excretory product.
- During pyrimidine biosynthesis, the ring is formed by carbamoyl phosphate and aspartate.
- Heme is produced from glycine and succinyl-CoA via a series of porphyrins.
  - Heme is degraded to bilirubin, which is excreted in the bile.

A. Creatine

1. **Creatine** is produced from glycine, arginine, and SAM.
   a. Glycine combines with arginine to form ornithine and guanidinoacetate in the kidney, which is methylated by SAM to form creatine in the liver.
2. **Creatine** travels from the liver to other tissues where it is converted to **creatine phosphate**.
   a. ATP phosphorylates creatine to form creatine phosphate in a reaction catalyzed by **creatine kinase**.
      (1) **Muscle** and **brain** contain large amounts of **creatine phosphate.**
      (2) **Creatine phosphate** provides a small reservoir of high-energy phosphate that readily regenerates ATP from ADP. It plays a particularly important role during the early stages of exercise in muscle, where the largest quantities of creatine phosphate are found.
      (3) Creatine also transports high-energy phosphate from mitochondria to actomyosin fibers.

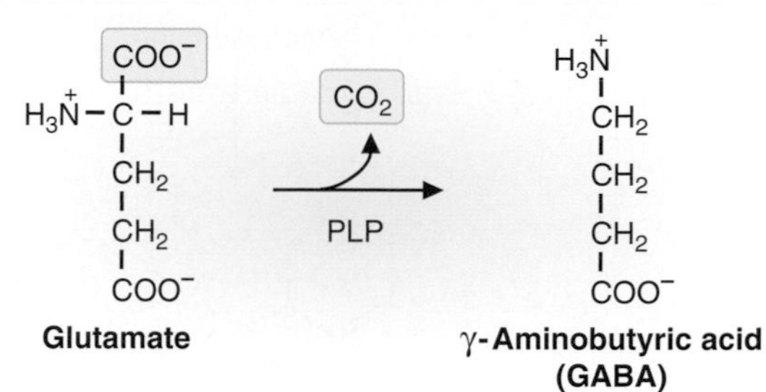

FIGURE 8.20. The decarboxylation of glutamate to form γ-aminobutyric acid (GABA). PLP, pyridoxal phosphate.

3. Creatine phosphate spontaneously cyclizes, forming **creatinine**, which is **excreted by the kidney.**
   a. The amount of **creatinine** excreted per day depends on the **body muscle mass** and **kidney function**, and is constant at about 15 mmol for the average person.

B. Products formed by amino acid decarboxylations

Amines are produced by the decarboxylation of amino acids in reactions that use **PLP** as a cofactor.

1. **GABA**, an inhibitory neurotransmitter, is produced by the decarboxylation of **glutamate** (Fig. 8.20).
2. **Histamine** is produced by the decarboxylation of **histidine**.
   a. Histamine causes vasodilation and bronchoconstriction. In the stomach, it stimulates the secretion of HCl.
3. The initial step in **ceramide** formation involves the condensation of **palmitoyl-CoA** with **serine**, which undergoes a simultaneous decarboxylation.
   a. Ceramide forms the sphingolipids (e.g., sphingomyelin, cerebrosides, gangliosides) (see Fig. 7.19).
4. The production of **serotonin** from tryptophan and of **dopamine** from tyrosine involves decarboxylation of amino acids.

C. Products derived from tryptophan

Serotonin, melatonin, and the nicotinamide moiety of **NAD** and **NADP** are formed from tryptophan (see Fig. 8.10).

1. **Tryptophan** is hydroxylated in a **tetrahydrobiopterin**-requiring reaction, similar to the hydroxylation of phenylalanine. The product, 5-hydroxytryptophan, is decarboxylated to form **serotonin.**
2. **Serotonin** undergoes acetylation by acetyl-CoA and methylation by SAM to form **melatonin** in the pineal gland.
3. Tryptophan can be converted to the nicotinamide moiety of **NAD** and **NADP** (see Fig. 8.10), although the major precursor of nicotinamide is the vitamin niacin (nicotinic acid). Thus, to a limited extent, tryptophan can spare the dietary requirement for niacin.

D. Products derived from phenylalanine and tyrosine

Phenylalanine can be hydroxylated to form **tyrosine** in a reaction that requires **tetrahydrobiopterin**. Tyrosine can be hydroxylated to form **dopa** (Fig. 8.21).

1. The **thyroid hormones**, triiodothyronine (**T_3**) and thyroxine (**T_4**), are produced in the thyroid gland from tyrosine residues in **thyroglobulin** (see Chapter 9).
2. **Melanins**, which are pigments in skin and hair, are formed by the polymerization of oxidation products (quinones) of **dopa**.
   a. In this case, dopa is formed by the hydroxylation of tyrosine by an enzyme that uses copper rather than tetrahydrobiopterin.

CLINICAL CORRELATES

In **albinism**, tyrosinase is defective and tyrosine cannot be converted to the skin pigment melanin. The characteristics are very **light colored skin, hair**, and **eyes**.

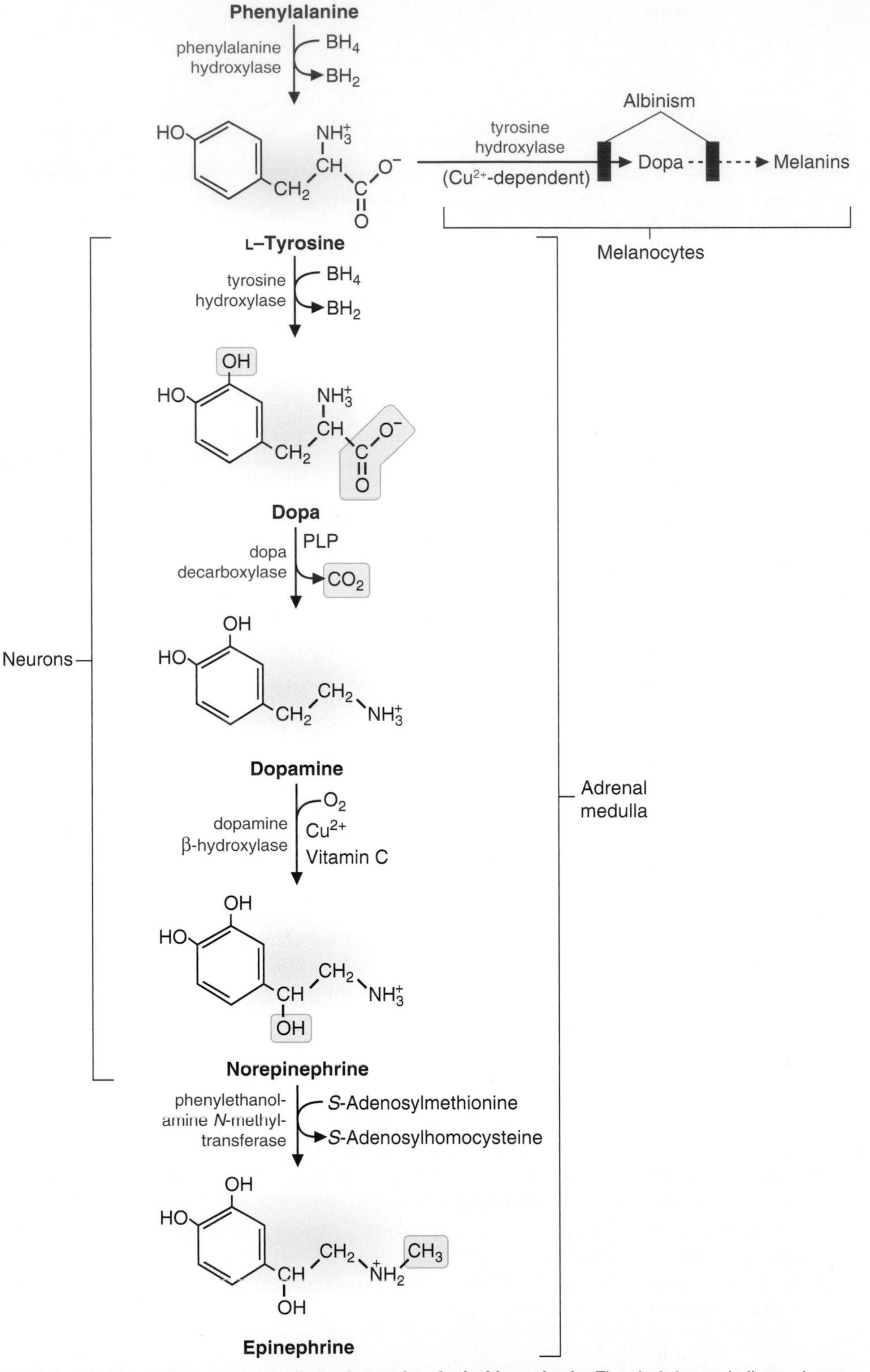

FIGURE 8.21. The pathways of catecholamine and melanin biosynthesis. The dark boxes indicate the enzymes, which, when defective, lead to albinism. BH_4, tetrahydrobiopterin; PLP, pyridoxal phosphate; dopa, dihydroxyphenylalanine.

1. The **catecholamines** (dopamine, norepinephrine, and epinephrine) are derived from tyrosine in a series of reactions (see Fig. 8.21).
   a. Synthesis of the catecholamines
      (1) Phenylalanine forms tyrosine, which forms dopa. In this case, both these hydroxylation reactions require **tetrahydrobiopterin**.
      (2) The decarboxylation of dopa forms the neurotransmitter **dopamine**.

CLINICAL CORRELATES In **Parkinson disease**, dopamine levels are decreased because of a deficiency in conversion of dopa to dopamine (see Fig. 8.21). The common characteristics are **tremors**, difficulty initiating voluntary movement, a **masked face** with a staring expression, and a **shuffling gait**.

      (3) The hydroxylation of dopamine by an enzyme that requires copper and vitamin C yields the neurotransmitter **norepinephrine**.
      (4) The methylation of norepinephrine in the adrenal medulla by SAM forms the hormone **epinephrine**.

CLINICAL CORRELATES **Pheochromocytomas** are tumors that produce and release episodically norepinephrine and/or epinephrine. The hypersecretion of these neuropeptides can lead to transient episodes of **hypertension**.

   b. Inactivation of the catecholamines (Fig. 8.22)
      (1) The catecholamines are inactivated by monoamine oxidase (**MAO**), which produces NH_4^+ and H_2O_2 and converts the catecholamine to an aldehyde, and by catecholamine *O*-methyltransferase (**COMT**), which methylates the 3-hydroxy group.
      (2) The major urinary excretory product of the deaminated, methylated catecholamines is **VMA** (vanillylmandelic acid, or 3-methoxy-4-hydroxymandelic acid).

CLINICAL CORRELATES **Nitric oxide**, a potent second messenger, is made from arginine by the enzyme nitric acid synthase, producing NO and ornithine. NO is a potent vasodilator, and individuals experiencing **angina** (chest pain) can take nitroglycerin tablets, which will be acted upon by mitochondrial aldehyde dehydrogenase to produce NO, thereby alleviating the symptoms.

E. Purine and pyrimidine metabolism

1. De novo purine and pyrimidine biosynthesis occurs in the liver and, to a limited extent, in the brain.
2. The nucleotides that are produced in the liver are converted to nucleosides and bases, which travel in red blood cells to other tissues where they are reconverted to nucleotides and further metabolized.
   a. **Purine synthesis** (Fig. 8.23)
      (1) The **purine base** is synthesized on the **ribose moiety.**
         (a) 5′-Phosphoribosyl 1′-pyrophosphate (**PRPP**), which provides the ribose moiety, reacts with **glutamine** to form phosphoribosylamine.
            1. This first step in purine biosynthesis produces N9 of the purine ring and is inhibited by AMP and GMP.
         (b) The entire **glycine** molecule is added to the growing purine precursor. Then C8 is added by **formyl-FH_4**, N3 by **glutamine**, C6 by CO_2, N1 by **aspartate**, and C2 by **formyl-FH_4** (see Fig. 8.23A).
         (c) **IMP**, which contains the base hypoxanthine, is generated.
            1. IMP is cleaved in the liver. Its free base, or nucleoside, travels to various tissues where it is reconverted to the nucleotide.

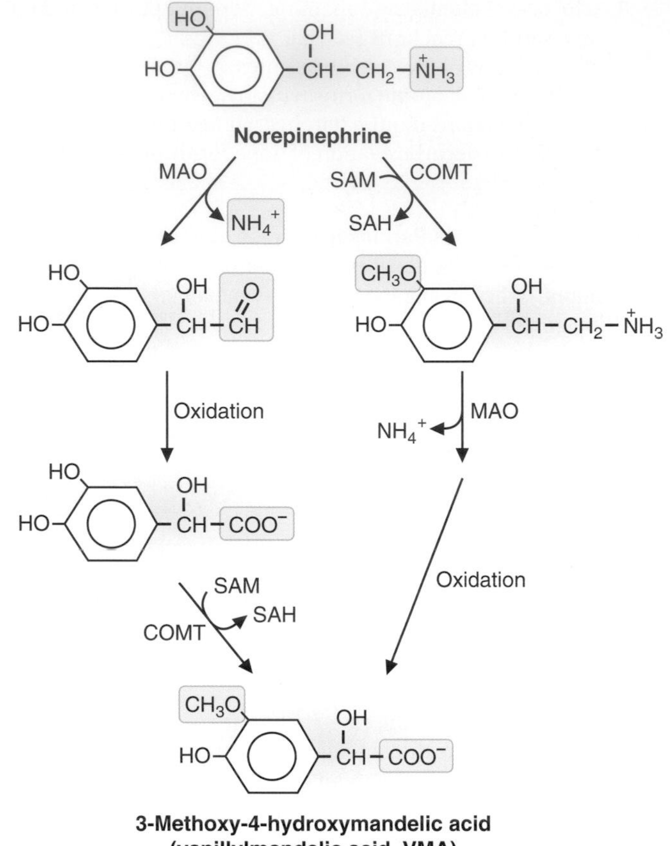

FIGURE 8.22. **The inactivation of catecholamines**. Methylation and oxidation may occur in any order. Their final compounds are secreted in the urine. COMT, catechol *O*-methyltransferase; MAO, monamine oxidase; SAH, *S*-adenosylhomocysteine; SAM, *S*-adenosylmethionine.

(2) **IMP** is the precursor of both **AMP** and **GMP.**
- **(a)** Each product, by feedback inhibition, regulates its own synthesis from the IMP branch point as well as inhibits the initial step in the pathway.
- **(b)** AMP and GMP can be phosphorylated to the triphosphate level.
- **(c)** The nucleotide triphosphates (ATP and GTP) can be used for energy-requiring processes or for **RNA synthesis**.

(3) **The reduction of the ribose moiety to deoxyribose** occurs at the diphosphate level and is catalyzed by **ribonucleotide reductase**, which requires the protein thioredoxin.
- **(a)** After the diphosphates are phosphorylated, dATP and dGTP can be used for **DNA synthesis**.

(4) **Purine bases** can be salvaged and converted between free bases, nucleotides, and nucleosides by a series of reactions, as indicated in Figure 8.24.

CLINICAL CORRELATES

A number of disorders result from mutations in purine salvage enzymes. **Severe combined immunodeficiency disease** (SCID) can result from a loss of activity of **adenosine deaminase** (another version of SCID, X-linked SCID, is due to the lack of a common subunit for various cytokine receptors). In adenosine deaminase deficiency, deoxyadenosine derivatives accumulate, leading to a lack of DNA synthesis in immune cell precursors. The thymus is virtually absent, and there is no T- and B-cell production. A **partial immune deficiency** results from a loss of activity of **purine nucleoside phosphorylase** activity. Only T-cell function is lost in this disorder, with a normal B-cell function. **Lesch–Nyhan syndrome** is caused by a defective hypoxanthine guanine phosphoribosyltransferase (**HGPRT**). Purine bases cannot be salvaged (i.e., reconverted to nucleotides). The purines are converted instead to uric acid, which rises in the blood and can lead to gout. **Mental retardation** and **self-mutilation** are characteristics of the disease.

A

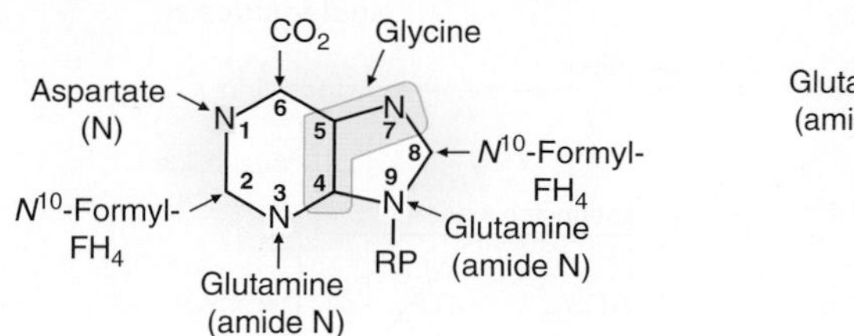

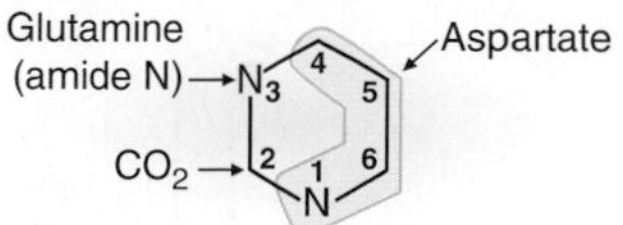

B

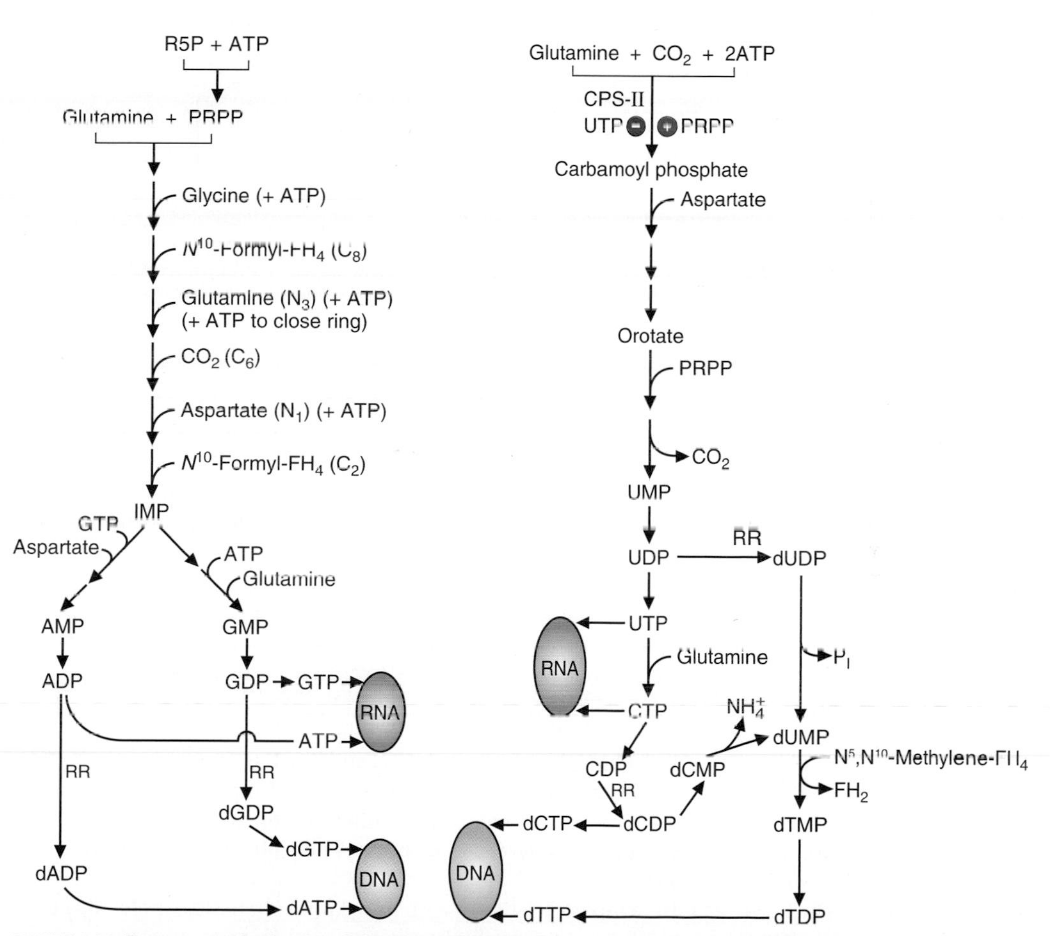

FIGURE 8.23. **De novo synthesis of purines and pyrimidines**. Ribonucleotide reductase (RR) catalyzes the reduction of the ribose moiety in ADP, GDP, and CDP to deoxyribose. The source of each of the atoms is indicated in part **(A)** of the figure. In hereditary orotic aciduria, the enzymes converting orotate to UMP are defective; FH_4, tetrahydrofolate; PRPP, 5′-phosphoribosyl 1′-pyrophosphate.

b. Purine degradation (Fig. 8.25)

(1) In the degradation of the purine nucleotides, **phosphate** and **ribose** are removed first; then the nitrogenous base is oxidized.

(a) The degradation of **guanine** produces **xanthine**.

(b) The degradation of **adenine** produces **hypoxanthine**, which is oxidized to **xanthine** by xanthine oxidase; this enzyme requires molybdenum.

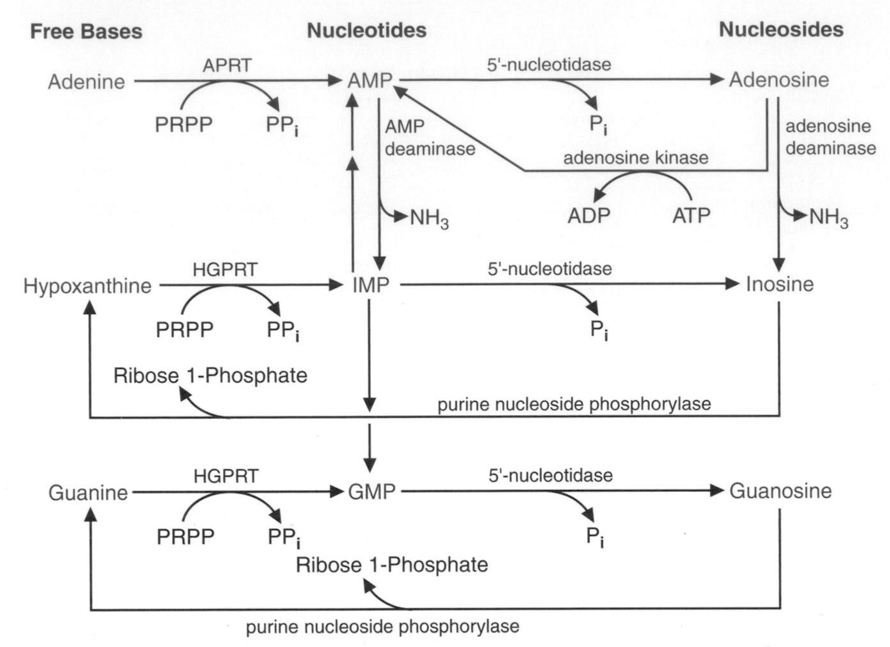

FIGURE 8.24. **The salvage of the purine bases**. The key enzymes are phosphoribosyltransferase, 5′-nucleotidase, adenosine deaminase, and purine nucleoside phosphorylase.

CLINICAL CORRELATES **Gout** is a group of diseases caused by an increased conversion of purine bases to uric acid or a decreased excretion of **uric acid** by the kidney. (Lead and metabolic acids–lactic acid, acetoacetic acid, and 3-hydroxybutyric acid–interfere with uric acid excretion.) An accumulation of uric acid, which is very insoluble, results in the precipitation of urate crystals in the joints. An **acute inflammatory arthritis** results. Chronic cases are treated with allopurinol, a base that forms a nucleotide that inhibits xanthine oxidase and prevents hypoxanthine and xanthine from being converted to uric acid (see Fig. 8.25).

(c) Xanthine is oxidized to **uric acid** by xanthine oxidase.

(d) Uric acid, which is not very water soluble, is **excreted** by the **kidneys**.

c. Pyrimidine synthesis (see Fig. 8.23)

(1) The **pyrimidine base** is synthesized prior to the addition of the ribose moiety.

(a) In the first reaction, **glutamine** reacts with CO_2 and **2 ATP** to form **carbamoyl phosphate.** This reaction is analogous to the first reaction of the urea cycle. However, for pyrimidine synthesis, glutamine provides the nitrogen and the reaction occurs in the cytosol, where it is catalyzed by **carbamoyl phosphate synthetase II**, which is inhibited by UTP.

(b) The entire **aspartate** molecule adds to carbamoyl phosphate. The molecule closes to yield a ring, which is oxidized, forming orotate.

(c) Orotate reacts **with PRPP**, producing orotidine 5′-phosphate, which is decarboxylated to form uridine monophosphate (UMP).

CLINICAL CORRELATES In **hereditary orotic aciduria**, orotic acid is excreted in the urine because the enzymes that convert it to UMP (orotate phosphoribosyl transferase and orotidine 5′-phosphate decarboxylase) are defective (see Fig. 8.23). Pyrimidines cannot be synthesized, and therefore, **growth retardation** occurs. Oral administration of **uridine** bypasses the metabolic block and provides the body with a source of pyrimidines.

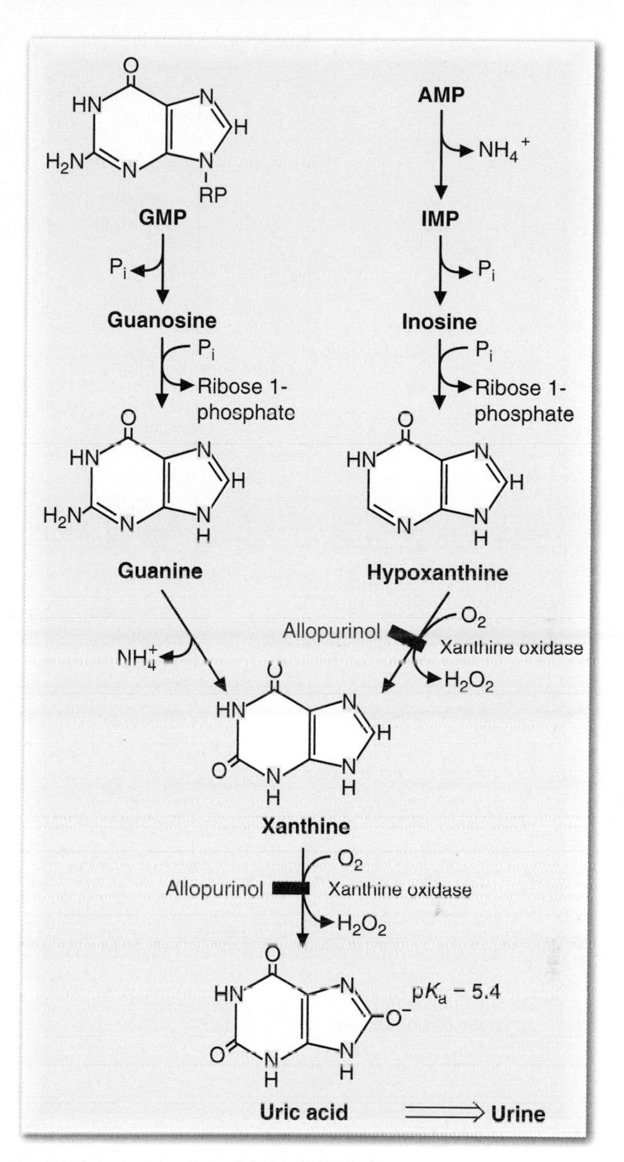

FIGURE 8.25. **Purine degradation.** Allopurinol (AP), which inhibits xanthine oxidase, is used to treat gout. Gout occurs when uric acid crystals precipitate in joints because of an increased concentration in the blood. R-1-P, ribose 1-phosphate.

(2) **UMP is phosphorylated to UTP**, which obtains an amino group from glutamine to form CTP. UTP and CTP are used in the synthesis of **RNA**.

(3) The ribose moiety of CDP is reduced to **deoxyribose**, forming dCDP. **Ribonucleotide reductase** is the enzyme.

(a) **dCDP** is deaminated and dephosphorylated to form **dUMP**.

(b) **dUMP** is converted to **dTMP** by methylene-FH_4.

(c) Phosphorylations produce dCTP and dTTP, which are the precursors of **DNA**.

d. **Pyrimidine degradation**

(1) In pyrimidine degradation, the carbons produce CO_2 and the nitrogens produce urea.

F. Heme metabolism

Heme, which consists of a **porphyrin ring** coordinated with **iron**, is found mainly in **hemoglobin**, but is also present in **myoglobin** and the **cytochromes**.

1. **Heme synthesis** (Fig. 8.26)

a. In the first step of heme synthesis, **glycine** and **succinyl-CoA** condense to form δ-aminolevulinic acid (δ-ALA). **PLP** is the cofactor for δ-ALA synthase. Glycine is decarboxylated in this reaction.

b. Two molecules of **δ-ALA** condense to form the pyrrole, porphobilinogen.

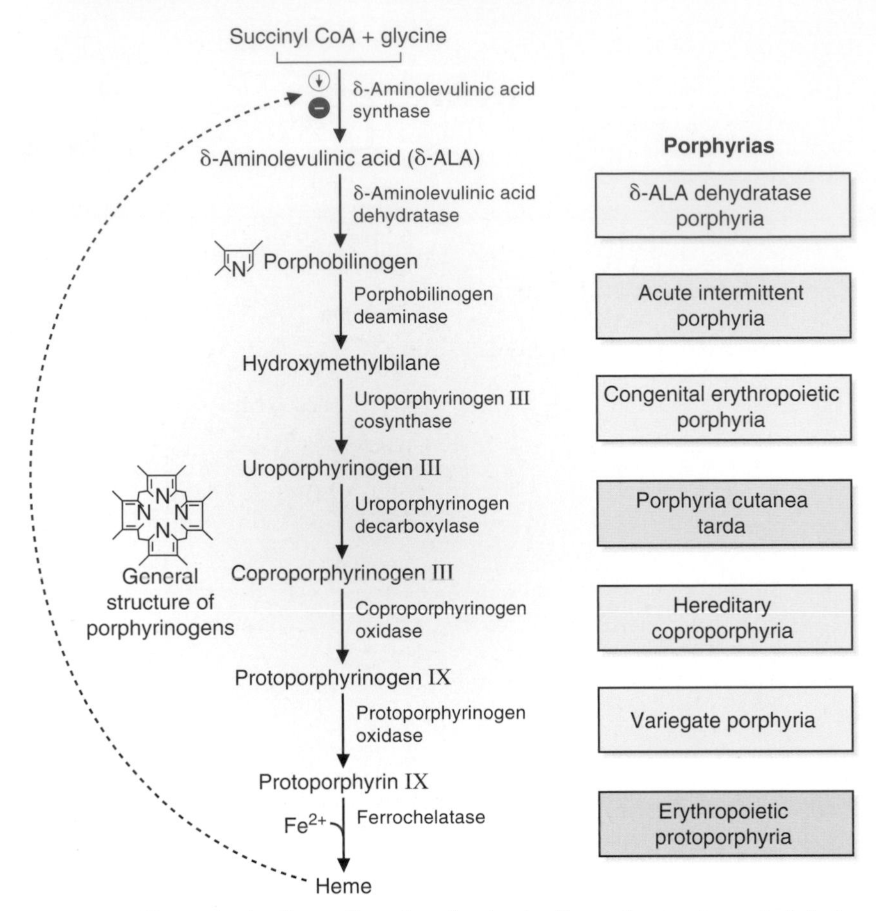

FIGURE 8.26. **The synthesis of heme**. To produce 1 molecule of heme, 8 molecules each of glycine and succinyl-CoA are required. Heme regulates its own production by repressing the synthesis of δ-ALA synthase, and by directly inhibiting the activity of this enzyme. Deficiencies of enzymes in the heme biosynthetic pathway result in a series of disorders known as porphyrias (listed on the right, beside the deficient enzyme).

CLINICAL CORRELATES In **lead poisoning**, δ-ALA and protoporphyrin IX accumulate because lead inhibits δ-ALA dehydratase and ferrochelatase. Heme production is decreased, and anemia results from a lack of hemoglobin.

c. Four **porphobilinogens** form the first in a series of porphyrins.

d. The **porphyrins** are altered by decarboxylation and oxidation, and protoporphyrin IX is formed.

CLINICAL CORRELATES **Porphyrias**, a group of rare inherited disorders, are caused by deficiencies of enzymes in the pathway of heme biosynthesis. **Neuropsychiatric symptoms** result from the accumulation of intermediates in the pathway, which have toxic effects on the nervous system. **Photosensitivity** results from the accumulation of porphyrinogens, which react with light to form porphyrins that produce oxygen radicals that damage the skin.

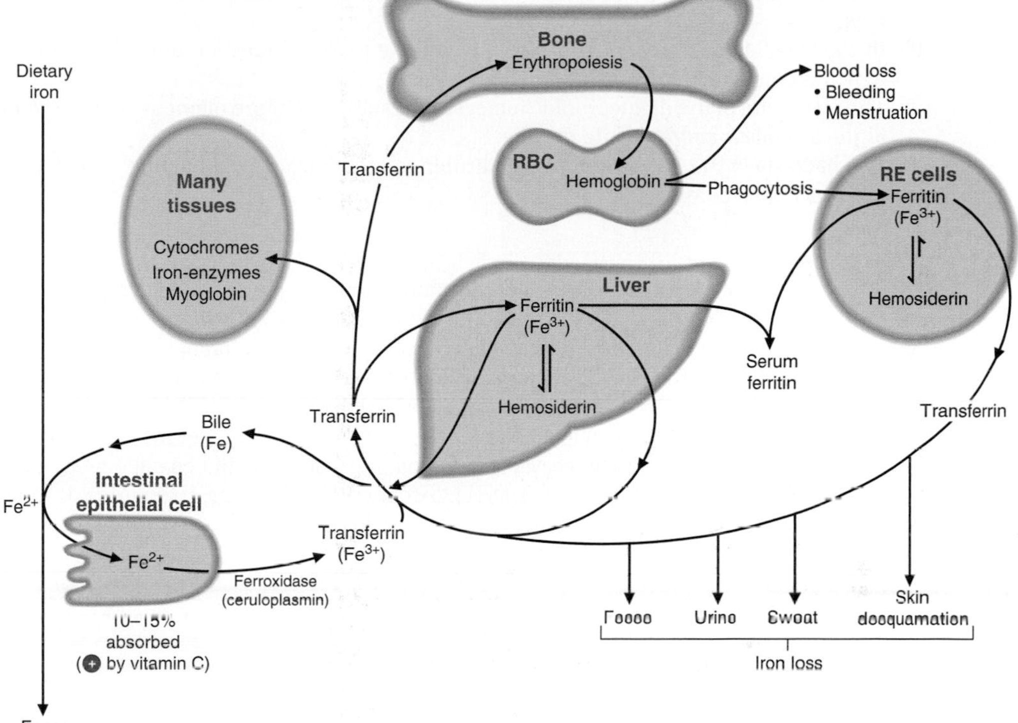

FIGURE 8.27. **Iron metabolism**. Iron is absorbed from the diet, transported in the blood bound to transferrin, stored as ferritin, and used for the synthesis of cytochromes, iron-containing enzymes, hemoglobin, and myoglobin. RE, reticuloendothelial.

e. **Protoporphyrin IX** binds **iron,** forming **heme.**
   (1) **Iron** is obtained from the diet, travels in the blood in the protein **transferrin**, and is stored as **ferritin** in tissues such as the liver and spleen (Fig. 8.27).
   (2) **Vitamin C** increases the uptake of iron from the intestinal tract.
   (3) **Ceruloplasmin**, a protein that contains copper, is involved in the oxidation of iron.
   (4) **Excess iron** is stored as **hemosiderin.**

CLINICAL CORRELATES **Dietary deficiencies of iron or of vitamin B_6** result in the production of small, pale red blood cells. A microcytic, hypochromic anemia results.

f. **Heme regulates** its own **production** by repressing the synthesis of δ-ALA synthase in the liver.
g. **Erythropoietin** induces heme synthesis in bone marrow.
h. **Heme stimulates** the synthesis of the protein **globin** by maintaining the translational initiation complex on the ribosome in its active state.

2. **Heme degradation**
   a. After **red blood cells**, which contain hemoglobin, reach their life span of about 120 days, they are **phagocytosed** by the cells of the reticuloendothelial system. Globin is released and converted to amino acids. Heme is degraded to **bilirubin**, which is excreted in the bile.
      (1) **Heme** is oxidized and cleaved to produce carbon monoxide and biliverdin, a green pigment.
      (2) **Iron** is released, oxidized, and returned by transferrin to the iron stores of the body.

(3) **Bilirubin**, produced by reduction of biliverdin, is carried by the protein albumin to the liver.
(4) In the liver, bilirubin reacts with **UDP-glucuronate** to form bilirubin monoglucuronide, which is converted to the diglucuronide.
(5) The formation of the diglucuronide increases the solubility of the pigment, and **bilirubin diglucuronide** is **secreted** into the **bile**.
(6) The **bacteria** in the intestine convert bilirubin to **urobilins** and **stercobilins**, which give feces its brown color.

CLINICAL CORRELATES

Excessive levels of bilirubin or bilirubin diglucuronide result in a yellowish discoloration of various tissues of the body (**jaundice**), which is particularly evident in the whites of the eyes. The causes of jaundice include overproduction of bilirubin due to **hemolytic anemia**, an inability of the liver to conjugate and excrete bilirubin due to **liver disease**, or an **obstruction of the bile duct** that prevents bilirubin diglucuronide from passing from the liver into the intestine. **Neonatal jaundice** is common, and is caused by an **immaturity of the system for conjugating and excreting bilirubin**. As the enzymes in the pathway develop, the condition resolves. Phototherapy, which converts bilirubin to other compounds that can be excreted in the urine, is used to prevent permanent damage to the nervous system (kernicterus).

Review Test

Questions 1 to 10 examine your basic knowledge of basic biochemistry and are not in the standard clinical vignette format.

Questions 11 to 35 are clinically relevant, USMLE-style questions.

Basic Knowledge Questions

1. A deficiency of which one of the following proteolytic enzymes would have the greatest effect on the digestion of proteins?

(A) Trypsin
(B) Chymotrypsin
(C) Carboxypeptidase A
(D) Pepsin
(E) Aminopeptidase

2. In liver disease, the enzymes AST and ALT leak into the blood from damaged liver cells. Both of these enzymes have which one of the following in common?

(A) They both transfer ammonia to α-keto acids to form amino acids.
(B) They both form intermediates of glycolysis from amino acids.
(C) They both require thiamine pyrophosphate as a cofactor.
(D) They both catalyze irreversible reactions.
(E) They both convert α-ketoglutarate to glutamate.

3. Which one of the following occurs in the urea cycle?

(A) Carbamoyl phosphate is derived directly from glutamine and CO_2.
(B) Ornithine reacts with aspartate to generate argininosuccinate.
(C) The α-amino group of arginine forms one of the nitrogens of urea.
(D) Ornithine directly reacts with carbamoyl phosphate to form citrulline.
(E) *N*-acetylglutamate is a positive allosteric effector of ornithine transcarbamoylase.

4. Starting with glucose, the synthesis of which one of the following would require the participation of isocitrate dehydrogenase?

(A) Serine
(B) Alanine
(C) Aspartate
(D) Glutamate
(E) Cysteine

5. The properties of glutamate are best represented by which one of the following?

	A precursor for histidine synthesis	Is produced from:	Donates a nitrogen to the purine ring	Delivers nitrogen to the liver	An organic compound that can fix ammonia
A	No	α-Ketoglutarate	No	No	Yes
B	No	Oxaloacetate	Yes	Yes	No
C	No	α-Ketoglutarate	No	No	No
D	Yes	Oxaloacetate	Yes	Yes	No
E	Yes	α-Ketoglutarate	No	No	No
F	Yes	Oxaloacetate	Yes	Yes	Yes

6. Which one of the following best represents the events that occur during the metabolism of the branched-chain amino acids?

	Requires vitamin B_1	Requires vitamin B_6	All produce a ketogenic product	Requires tetrahydrobiopterin	The primary tissue of oxidation is:
A	Yes	Yes	No	No	Liver
B	No	No	Yes	Yes	Liver
C	Yes	Yes	No	No	Muscle
D	No	Yes	Yes	No	Muscle
E	Yes	No	No	Yes	Liver
F	No	Yes	Yes	Yes	Muscle

7. Which one of the following best represents the kidney's utilization of amino acids?

	Produces urea from amino acid nitrogen	Produces creatine phosphate	Creates ammonia to buffer urine	Uses alanine to produce glutamate
A	No	No	Yes	Yes
B	Yes	No	No	Yes
C	No	No	Yes	No
D	Yes	Yes	No	No
E	No	Yes	Yes	Yes
F	Yes	Yes	No	No

8. Which one of the following best represents de novo pyrimidine synthesis in a liver cell?

	Built as a nucleotide	Requires FH_4	Requires an enzyme that fixes ammonia	Requires aspartic acid
A	No	No	Yes	Yes
B	Yes	Yes	Yes	No
C	No	No	No	Yes
D	Yes	Yes	No	Yes
E	No	No	No	No
F	Yes	Yes	Yes	No

9. Which one of the following best represents the metabolism of bilirubin?

	Contains iron	Excreted in the urine	In the formation of bilirubin, a carbon is lost as:	Solubility enhanced by:
A	Yes	Yes	CO	Glucuronic acid
B	Yes	No	CO_2	Glucose
C	Yes	Yes	CO_2	Glucuronic acid
D	No	No	CO	Glucuronic acid
E	No	Yes	CO_2	Glucose
F	No	No	CO	Glucose

10. In the biosynthetic pathways for the synthesis of heme, creatine, and guanine, which one of the following amino acids directly provides carbon atoms that appear in the final product?

(A) Serine
(B) Aspartate
(C) Cysteine
(D) Glutamate
(E) Glycine

Board-style Questions

11. An infant who appeared normal at birth began to develop lethargy, hypothermia, and apnea within 24 hours. An analysis of blood components indicated high levels of ammonia and citrulline, and low levels of urea. The most likely defective enzyme in this child is which one of the following?

(A) Carbamoyl phosphate synthetase I
(B) Ornithine transcarbamoylase
(C) Argininosuccinate synthetase
(D) Argininosuccinate lyase
(E) Arginase

12. A 5-day-old infant initially began feeding poorly, grew irritable, then very lethargic. The diaper, when changed, had a musky, sweet odor to it. At the emergency department of the local hospital, the child was suspected of having an inborn error of metabolism, and blood work was ordered. You would expect to see an elevation of which one of the following given the conditions of the child's case?

(A) Phenylalanine
(B) Phenylketones
(C) Dopa
(D) Isoleucine
(E) Acetone

13. A common polymorphism in the US population is a variant of N^5,N^{10}-methylene-FH_4 reductase, which has a reduced activity at 37°C as compared to 32°C. A person expressing this variant enzyme would have difficulty producing which one of the following at the nonpermissive temperature?

(A) Creatine phosphate from creatine
(B) Pyrimidines required for RNA synthesis
(C) The thymine nucleotide required for DNA synthesis
(D) Phosphatidylcholine from diacylglycerol and CDP-choline
(E) All deoxyribonucleotides

14. A young boy has been eating five to six raw eggs a day in an attempt to be like Rocky, the fictitious prize fighter. The boy, however, becomes tired between meals, and during one of these lethargic periods, he is taken to the emergency department. His blood glucose levels were 50 mg/dL. Which one of the following would be expected to be elevated in this boy's blood?

(A) Methylmalonic acid
(B) Malonic acid
(C) Propionic acid
(D) Succinate
(E) Acetic acid

15. A 65-year-old man visits his primary physician because of tingling in his hands and feet, and a sense that he is forgetting things more than usual. A CBC indicates a mild anemia. The patient states that his diet has not changed, other than eating more red meat than before. This patient can be best treated by which one of the following?

(A) Oral administration of vitamin B_{12}
(B) Oral administration of folic acid
(C) Oral administration of methionine
(D) Injections of B_{12}
(E) Injections of folic acid
(F) Injections of methionine

16. An individual who has been treated for Type 2 diabetes for the past 24 years had, as part of his annual physical, a 24-hour urine collection. Reduced levels of creatinine were found, which is most likely due to which one of the following?

(A) A decreased dietary intake of creatine
(B) A higher-than-normal muscle mass resulting from weight lifting
(C) A genetic defect in the enzyme that converts creatine phosphate to creatinine
(D) Kidney failure
(E) A lower-than-normal muscle mass due to a low-protein diet

17. A 3-year-old child from Russia, after immigrating to the United States, was found to have developmental delays and severe mental retardation. Elevated levels of phenylalanine and phenylpyruvate were found in the blood. The child was placed on a low-phenylalanine diet, but there was no improvement in the child. Given this information, the child would be expected to have difficulty in undergoing which one of the following conversions?

(A) Phenylalanine to phenylketones
(B) Tyrosine to dopamine
(C) Dopa to melanin
(D) Serotonin to melatonin
(E) Norepinephrine to epinephrine

18. A 42-year-old woman has been diagnosed with liver cancer, and is being treated with 5-FU. 5-FU is successful in destroying the tumor cells because it blocks the production of which one of the following?

(A) FH_4
(B) dTMP
(C) UMP
(D) Methylcobalamin
(E) CTP

Questions 19 and 20 are based on the following case:

A medical student has been exposed to a patient with tuberculosis and developed a positive tuberculin test (PPD), but exhibited a normal chest X-ray. He is placed on a 6-month course of prophylactic treatment, but subsequently develops peripheral neuropathies.

19. Which of the following vitamins would be considered a treatment for the neurotoxicity?

(A) B_1
(B) B_2
(C) B_3
(D) B_6
(E) B_{12}

20. Considering the potential vitamin deficiency in this patient, which class of molecules would be most affected by the lack of the vitamin?

(A) Carbohydrates
(B) Amino acids
(C) Fatty acids
(D) Cholesterol
(E) Phospholipids

Questions 21 and 22 are based on the following case.

A 20-year-old male is new to your practice, and you notice he has white hair, white skin, and nystagmus.

21. A defect in the metabolism of which one of the following compounds is responsible for this presentation?

(A) Branched-chain amino acids
(B) Histidine
(C) Tryptophan
(D) Tyrosine
(E) Methionine

22. The amino acid(s) pathway that contained a mutation as indicated in the previous question is also a precursor for which one of the following?

(A) Serotonin
(B) Norepinephrine
(C) Testosterone
(D) Aldosterone
(E) Histamine

Questions 23 and 24 are based on the following case:

A 45-year-old male is concerned about his risk of a heart attack since his brother just had a heart attack at the age of 46. His physician orders an HbA_1C, lipid panel, and homocysteine level.

23. A high homocysteine level could be associated with a deficiency of which one of the following vitamins?

(A) B_3
(B) B_{12}
(C) C
(D) E
(E) B_1

24. A reduction in the metabolism of which one of the following amino acids can lead to elevated homocysteine levels in the blood?

(A) Alanine
(B) Methionine
(C) Phenylalanine
(D) Glutamate
(E) Cysteine

25. A patient presents with fatigue, and a blood count reveals a macrocytic, hyperchromic anemia. Which one of the following may account for this type of anemia?

(A) Lead poisoning
(B) Folate deficiency
(C) Hereditary spherocytosis
(D) Sideroblastic anemia
(E) Iron deficiency

26. A 42-year-old male has fatigue, pale skin, and shortness of breath with exercise. Blood work shows a macrocytic, hyperchromic anemia with hypersegmented neutrophils and normal folate levels. The patient has been taking omeprazole for over 3 years to treat gastric reflux disease. One method to treat this patient is to do which one of the following?

(A) Give injections of vitamin B_6
(B) Give injections of intrinsic factor
(C) Give injections of vitamin B_{12}
(D) Give oral folic acid
(E) Give oral intrinsic factor

27. A 30-year-old female with severe psoriasis has been nonresponsive to topical steroid creams, coal tars, and UV light, but is controlled well on a small dose of medication she takes orally once a week. This systemic treatment leads to an accumulation of which one of the following?

(A) FH_4
(B) FH_2
(C) N^5-methyl-FH_4
(D) dTTP
(E) dATP

28. A 34-year-old female has a history of intermittent episodes of severe abdominal pain. She has had multiple abdominal surgeries and exploratory procedures with no abnormal findings. Her urine appears dark during an attack and gets even darker if exposed to sunlight. The attacks seem to peak after she takes erythromycin, due to her penicillin allergy. This patient most likely has difficulty in synthesizing which of the following?

(A) Heme
(B) Creatine phosphate
(C) Cysteine
(D) Thymine
(E) Methionine

29. A 30-year-old male has had multiple episodes of sudden, severe pain, redness, and swelling of metatarsophalangeal joint of his great toes. These problems seem to occur after the man has had a night out on the town with his friends, when they go barhopping, and the night usually ends with a cab ride home for the group. This problem would also be exacerbated if the man eats which one of the following during his night out?

(A) Hamburger
(B) Chicago hot dog
(C) Chopped liver
(D) Nachos and salsa
(E) Chicken wings

30. A 3-year-old child has mental retardation, poor muscle control, gout, chronic renal failure, facial grimacing, and lip and finger biting. This child has an inability to catalyze which one of the following reactions?

(A) Thymine + deoxyribose 1-phosphate yields deoxythymidine + inorganic phosphate
(B) Adenine + 5′-phosphoribosyl 1′-pyrophosphate yields AMP + pyrophosphate
(C) Uracil + ribose 1-phosphate yields uridine + phosphate
(D) Guanine + 5′phosphoribosyl 1′-pyrophosphate yields GMP + pyrophosphate
(E) IMP + NAD+ yields XMP + NADH

31. A newborn infant develops its first cold, and is fussy and cannot eat. After missing a few feedings, the child becomes quite lethargic, and the parents rush the child to the emergency department. Blood analysis indicates elevated levels of lactate and uric acid, and significantly decreased levels of glucose. After stabilizing the child with glucose infusions, a glucagon challenge is given to the infant, and blood glucose levels do not increase, but decrease slightly. The accumulation of which metabolite in the liver is most responsible for the elevated uric acid seen in the circulation?

(A) Glucose
(B) Lactate
(C) Ribose 5-phosphate
(D) Thymidine
(E) Acetoacetate

32. Parents of a newborn baby girl were concerned when they saw black spots in her diaper after the child had urinated. At their next meeting with the pediatrician, they were told that the disorder is one that can lead to arthritis in the spine and large joints, and the child may have heart problems and a propensity for kidney stones. The child has inherited an inborn error in the metabolism of which one of the following amino acids?

(A) Phenylalanine
(B) Tryptophan
(C) Proline
(D) Methionine
(E) Histidine

33. A 5-year-old boy has had episodic periods during which areas of his skin would develop a rash, which would spontaneously resolve in a week to 10 days. An astute pediatrician told the boy's parents to give him niacin the next time this occurred and a high-protein diet, and when they did, the rash resolved in a day or two. A likely disorder that this child has is which one of the following?

(A) Cystinuria
(B) Hartnup disease
(C) Myasthenia gravis
(D) Alkaptonuria
(E) Jaundice

34. A 9-month-old infant had been in and out of the hospital owing to frequent infections. Blood work demonstrated the virtual lack of B and T cells, and the almost complete absence of a thymic shadow on a chest X-ray. Measurement of metabolites in the blood would be expected to show elevated levels of which one of the following?

(A) Uric acid
(B) Orotic acid
(C) Deoxyadenosine
(D) NADPH
(E) dGTP

35. A 45-year-old man developed severe pain in his back, which, upon going to the emergency department, turned out to be due to kidney stones. A stone chemical analysis indicated a buildup of oxalic and glyoxalic acids. These compounds can accumulate due to a problem in the metabolism of which one of the following amino acids?

(A) Alanine
(B) Tryptophan
(C) Isoleucine
(D) Glycine
(E) Glutamine

Answers and Explanations

1. **The answer is A.** Trypsin cleaves and, thus, activates the pancreatic zymogens, converting chymotrypsinogen to the active form, chymotrypsin, and the procarboxypeptidases to the active carboxypeptidases. If trypsin were inactive, the other proteases could not be activated, as enteropeptidase is specific for trypsinogen. Pepsin is found in the stomach, whereas aminopeptidases are intestinal enzymes found on the brush border membrane, facing the lumen of the intestine.

2. **The answer is E.** These transaminases convert amino acids to their corresponding α-keto acids in reactions that are readily reversible. α-Ketoglutarate and glutamate serve as the other α-keto acid/amino acid pair. Pyruvate (the end product of glycolysis) is the α-keto acid corresponding to alanine, and oxaloacetate (an intermediate of the TCA cycle) is the partner of aspartate. PLP is the cofactor. Thus, AST will convert aspartate and α-ketoglutarate to oxaloacetate and glutamate, and ALT will convert alanine and α-ketoglutarate to pyruvate and glutamate.

3. **The answer is D.** Carbamoyl phosphate within the mitochondria is formed from NH_4^+, CO_2, and ATP. Carbamoyl phosphate synthetase II catalyzes carbamoyl phosphate synthesis from glutamine for pyrimidine synthesis in the cytoplasm. Carbamoyl phosphate reacts with ornithine to form citrulline, which reacts with aspartate to form argininosuccinate. Fumarate is released from argininosuccinate, and arginine is formed. Urea is produced from the guanidinium group on the side chain of arginine, not from the amino group on the α-carbon. Ornithine is regenerated. *N*-Acetylglutamate is an allosteric activator of carbamoyl phosphate synthetase I. Ornithine transcarbamoylase is not a regulated enzyme in mammals, and in bacteria *N*-acetylglutamate is not an allosteric effector of ornithine transcarbamoylase.

4. **The answer is D.** The formation of glutamate from glucose involves the TCA cycle intermediate α-ketoglutarate, which is formed from isocitrate in a reaction catalyzed by isocitrate dehydrogenase. α-Ketoglutarate is converted to glutamate either by glutamate dehydrogenase or a transaminase. The formation of serine, alanine, aspartate, and cysteine from glucose does not require the activity of isocitrate dehydrogenase. Serine is derived from 3-phosphoglycerate; alanine from pyruvate; aspartate from oxaloacetate; and cysteine from methionine (only the sulfur) and serine (the carbon atoms). The oxaloacetate needed for aspartate synthesis can be generated from pyruvate via the pyruvate carboxylase reaction.

5. **The answer is A.** Glutamate cannot produce histidine, as histidine is an essential amino acid in humans. Glutamate can fix ammonia to form glutamine in a reaction catalyzed by glutamine synthetase. Glutamate can be synthesized from α-ketoglutarate either through a transamination reaction or by glutamate dehydrogenase (which fixes ammonia into α-ketoglutarate). Glutamine donates nitrogens for purine ring synthesis, but glutamate does not. Glutamine is a nitrogen carrier in the blood, whereas glutamate is not.

6. **The answer is C.** Valine, isoleucine, and leucine (the branched-chain amino acids) are transaminated (which requires vitamin B_6) and then oxidized by an α-keto acid dehydrogenase (that requires lipoic acid as well as vitamin B_1 [thiamine], coenzyme A, FAD, and NAD^+). Four of the carbons of valine and isoleucine are converted to succinyl-CoA, which is a glucogenic product. Isoleucine also produces acetyl-CoA, a ketogenic product. Leucine is converted to HMG-CoA, which is cleaved to acetoacetate and acetyl-CoA, and is strictly ketogenic. Branched-chain amino acid metabolism occurs primarily in the muscle, as muscle contains the highest levels of the transaminase and dehydrogenase. Tetrahydrobiopterin is required for ring hydroxylations, which is not applicable to the metabolism of the branched-chain amino acids, which do not contain ring structures.

7. **The answer is C.** Glutaminase acts on glutamine to release ammonia, which enters the urine and serves as a buffer by forming NH_4^+. The kidney produces guanidinoacetate, which travels to the liver to produce creatine. The muscle and brain then utilize the creatine to produce creatine phosphate. The kidney takes up glutamine and releases serine and alanine into the blood. Most of the urea that is excreted by the kidney is produced in the liver; the kidney has a limited ability to produce urea.

8. **The answer is C.** Pyrimidines are first built as a free base (orotic acid) before being converted to a nucleotide (OMP), which is the opposite of that of purine synthesis, in which the purine is built upon ribose 5-phosphate to first produce the nucleotide IMP. FH_4 does not provide carbons for pyrimidine synthesis (the precursors are carbamoyl phosphate, derived from glutamine and carbon dioxide, and aspartic acid). Carbamoyl phosphate synthetase II, which produces carbamoyl phosphate in the cytoplasm for pyrimidine synthesis, does not fix ammonia into the product, but rather uses glutamine as the nitrogen donor. Aspartate is required to build the pyrimidine ring.

9. **The answer is D.** Bilirubin is produced by the oxidation of heme after its iron is released; CO is produced in this reaction. Bilirubin diglucuronide, which contains two glucuronic acid (not glucose) residues, is excreted into the bile (not the urine) by the liver. The bilirubin is modified in the intestinal tract and eventually excreted in the feces.

10. **The answer is E.** Glycine reacts with succinyl-CoA in the first step of heme synthesis and with arginine in the first step of creatine synthesis. The entire glycine molecule is incorporated into the growing purine ring. Serine is not utilized for the biosynthesis of either heme, creatine, or purines. Aspratate is used for purine ring synthesis only (one nitrogen of the purine ring is derived from aspartate). Neither cysteine nor glutamate is directly involved in the synthesis of heme, creatine, or purines.

11. **The answer is C.** The patient has a defect in argininosuccinate synthetase. Citrulline, the substrate for the reaction, accumulates and can be measured in the blood. A carbamoyl phosphate synthetase I deficiency would block carbamoyl phosphate formation, and citrulline would neither be synthesized nor accumulated. An ornithine transcarbamoylase deficiency would lead to orotic acid accumulation (carbamoyl phosphate made in the mitochondria would diffuse into the cytoplasm, thereby activating pyrimidine synthesis and overproducing ortoic acid). An argininosuccinate lyase deficiency would lead to elevated argininosuccinate, which is not observed, and an arginase deficiency would lead to elevated arginine, which was also not observed. Defects in argininosuccinate lyase and arginase also do not have as elevated ammonia levels as do defects in previous enzymes of the cycle, since two nitrogens have been disposed of in the synthesis of argininosuccinate.

12. **The answer is D.** The child has the symptoms of classic MSUD. In MSUD, the branched-chain amino acids (valine, leucine, and isoleucine) can be transaminated but not oxidatively decarboxylated because the α-keto acid dehydrogenase is defective. Therefore, these amino acids and their transamination products (the corresponding α-keto acids) will be elevated. Phenylalanine and phenylketones are elevated in PKU not in MSUD. The musky sweet odor of the urine strongly suggests MSUD. Dopa levels (dopa is derived from tyrosine) should be normal in MSUD, as there is no mutation in the phenylalanine degradation pathway. Acetone, derived from acetoacetate, is exhaled and is not found in the blood. Depending on the feeding state of the child, ketosis could develop, but the acetone would still be exhaled, while blood levels of acetoacetate and ß-hydroxybutyrate might be elevated.

13. **The answer is C.** The only pyrimidine that requires folate for its synthesis is thymine (dUMP → dTMP). Folate is required for the incorporation of carbons 2 and 8 into all purine molecules. The synthesis of creatine phosphate and of phosphatidylcholine do not require folate. Folate deficiencies during pregnancy can lead to neural tube defects (e.g., spina bifida) in the fetus. Deoxyribonucleotide synthesis requires ribonucleotide reductase, which uses thioredoxin, and does not require a folate derivative.

14. **The answer is C.** The conversion of propionyl-CoA to methylmalonyl-CoA requires biotin, and the conversion of methylmalonyl-CoA to succinyl-CoA requires vitamin B_{12}. The boy is

consuming large amounts of avidin, which binds to biotin and inhibits biotin-containing enzymes, such as propionyl-CoA carboxylase. The hypoglycemia results from the inhibition of pyruvate carboxylase, blocking a key enzyme necessary for gluconeogenesis.

15. **The answer is D.** The patient has a deficiency of B_{12}, caused by inadequate intrinsic factor production for the absorption of dietary vitamin B_{12}, which is required for the conversion of methylmalonyl-CoA to succinyl-CoA and of homocysteine to methionine. A vitamin B_{12} deficiency results in the excretion of methylmalonic acid in the urine and an increased dietary requirement for methionine. The methyl group transferred from vitamin B_{12} to homocysteine to form methionine comes from N5-methyl-FH_4, which accumulates in a vitamin B_{12} deficiency, causing a decrease in free folate levels and symptoms of folate deficiency, including increased levels of FIGLU and decreased purine biosynthesis. Both B_{12} and folate deficiencies will lead to anemia, but only a B_{12} deficiency will give rise to the mental-status changes and tingling in the hands and feet. As individuals age, a variety of conditions can give rise to reduced intrinsic factor production by the stomach. Since intrinsic factor is low, B_{12} needs to be delivered by injection, not by the oral route of administration.

16. **The answer is D.** The patient has diabetes-induced nephropathy. As the kidneys lose function, the ability of creatinine to be absorbed into the urine decreases, and its excretion would be low. Creatine is synthesized from glycine, arginine, and SAM. In muscle, creatine is converted to creatine phosphate, which is nonenzymatically cyclized to form creatinine. The amount of creatinine excreted by the kidneys each day depends on the body muscle mass. Weight lifting and increasing muscle mass would increase the levels of creatinine in the urine. A low-protein diet would not reduce the muscle mass, nor affect the creatinine excretion.

17. **The answer is B.** The child has nonclassical PKU, which is due to a defect in the biosynthesis of tetrahydrobiopterin. Tetrahydrobiopterin is required for ring hydroxylation reactions, such as the conversion of phenylalanine to tyrosine, tyrosine to dopa, and tryptophan to serotonin. However, tetrahydrobiopterin is not required for the conversion of dopa to melanin, serotonin to melatonin, and norepinephrine to epinephrine, which is a methylation reaction. A deficiency of tetrahydrobiopterin would cause phenylalanine to be converted to phenylketones rather than to tyrosine.

18. **The answer is B.** 5-FU inhibits the thymidylate synthase reaction, which produces dTMP from dUMP. It does not inhibit dihydrofolate reductase, which would block the production of FH_4. Methotrexate inhibits dihydrofolate reductase. UMP is made via the de novo pyrimidine biosynthetic pathway, which is not inhibited by 5-FU. Methylcobalamin is produced from N^5-methyl-FH_4 and vitamin B_{12}, and 5-FU does not block this activation of B_{12}. CTP is produced from UTP, using glutamine as the nitrogen donor, and that reaction is also not affected by the presence of 5-FU.

19. **The answer is D.** The treatment for a positive PPD test (tuberculosis) is isoniazid, which can interfere with vitamin B_6 (pyridoxine) function in cells. Pyridoxine is activated to PLP in cells (the active form of the vitamin), and isoniazid blocks this activation. A deficiency of B_6 can lead to peripheral neuropathy, as B_6 is required for the conversion of tryptophan to niacin. In many cases, vitamin B_6 is given along with isoniazid to prevent these side effects from occurring (by providing more substrate than the isoniazid can bind to). Isoniazid does not affect thiamine (B_1), riboflavin (B_2), niacin (B_3), or cobalamin (B_{12}) metabolism, although riboflavin is required to activate pyridoxine.

20. **The answer is B.** PLP (from vitamin B_6) is the major coenzyme of amino acid metabolism, as it participates in amino acid decarboxylation, amino acid racemization, ß-elimination reactions, ß-addition reactions, transaminations, and γ-elimination reactions. PLP is also a required cofactor for glycogen phosphorylase, but the majority of reactions that utilize this cofactor have amino acids as a substrate.

21. **The answer is D.** The patient has the signs of albinism, a lack of melanin. Melanin is produced from tyrosine (tyrosine is hydroxylated in melanocytes to form dopa, which then enters the pathway for melanin production). The defective enzyme in oculocutaneous albinism (the type exhibited by this patient, as opposed to ocular albinism, which only affects the eyes) is

tyrosinase. This is a melanocyte-specific genetic deficiency–neuronal cells also contain tyrosinase, which is needed to produce the catecholamines, and that isozyme is normal in patients with albinism. A defect in branched-chain amino acids would lead to MSUD; in phenylalanine, PKU; and in tryptophan, low serotonin levels. A defect in histidine metabolism may lead to reduced levels of histamine.

22. **The answer is B.** Tyrosine is a precursor of the catecholamines dopamine, epinephrine, and norepinephrine. The pathway for catecholamine production is normal in individuals with albinism, as it is the melanocyte isozyme of tyrosinase that is mutated, not the form in the cells that produce the catecholamines.

23. **The answer is B.** High homocysteine levels may be a risk factor for early atherosclerotic disease. The high levels appear to be associated with low B_6, B_{12}, and/or folate levels. Vitamin B_{12} is required for the methionine synthase reaction (homocysteine + N^5-methyl-FH_4 yields methionine and FH_4), vitamin B_6 is required for the cystathionine ß-synthase reaction (homocysteine + serine yields cystathionine), and folate is required for the methionine synthase reaction. Vitamins B_3 (niacin), C, E, and B_1 (thiamine) are not required for homocysteine metabolism.

24. **The answer is B.** Methionine, obtained from the diet, is activated by the reaction with ATP to form SAM, the universal methyl donor. After donating a methyl group, SAM is converted to *S*-adenosylhomocysteine. A hydrolase removes the adenosine, generating homocysteine. Homocysteine then has two fates–it can react with N^5-methyl-FH_4 (with B_{12}) to regenerate methionine, or it can react with serine (in the presence of B_6) to form cystathionine, which goes on to form cysteine. A defect in either the enzyme that forms methionine (methionine synthase) or the enzyme that forms cystathionine (cystathionine ß-synthase) will lead to elevated homocysteine levels.

25. **The answer is B.** A folate deficiency leads to a megaloblastic anemia since DNA synthesis is inhibited in the absence of folate. Large cells are generated in the bone marrow because the cell enlarges in preparation for division, but the DNA does not replicate and the cell does not divide. All of the other answer choices suggested create a microcytic, hypochromic anemia. Lead poisoning interferes with heme synthesis, as does an iron deficiency. Under these conditions, the red cells released are small in size, as the final size is, in part, dependent on the intracellular concentration of heme (so if heme levels are low, the cell will be small). Sideroblastic anemia also results from a disruption in heme synthesis, for a variety of causes. Hereditary spherocytosis is due to mutations in red cell membrane proteins, which lead to early removal of these cells from the spleen.

26. **The answer is C.** The patient is displaying megaloblastic anemia due to a deficiency of vitamin B_{12}. The use of omeprazole to reduce acid production in the stomach also reduces the ability of B_{12}, bound to ingested proteins, to be released by the proteins to be bound by intrinsic factor for effective absorption into the blood. Providing injections of B_{12} will bypass the need for separation of B_{12} from its binding proteins and will allow B_{12} to circulate throughout the body and reach its intracellular targets and proteins. Oral B_{12} would also work under these conditions. The patient is unlikely to have an intrinsic factor problem (due to his age), and intrinsic factor cannot be given orally or via injection (since it needs to work in the intestine). For lack of intrinsic factor, injections of B_{12} are also required. The patient has normal folate levels, so giving more folate will not help the anemia, and vitamin B_6 is not involved in these reactions.

27. **The answer is B.** One of the systemic drugs given for psoriasis is methotrexate, which inhibits dihydrofolate reductase. The rationale behind using this drug is to kill the skin cells that are giving rise to the condition, thereby alleviating the symptoms. Methotrexate inhibits dihydrofolate reductase, preventing FH_2 (which is produced in the thymidylate synthase reaction) from being converted to FH_4. This leads to a functional loss of folic acid, and cells treated with methotrexate will not be able to synthesize dTMP from dUMP, or perform de novo synthesis of the purine ring. Since dTMP cannot be made, dTTP will not accumulate, nor will dATP (in the absence of dTTP, ribonucleotide reductase cannot generate dGTP, which means that dATP

will also not be made). The lack of FH_4 production from FH_2 means that FH_4 levels will be low, along with N^5-methyl-FH_4 levels.

28. **The answer is A.** The patient has acute intermittent porphyria, which is a defect in one of the early steps leading to heme synthesis. The buildup of the intermediate that cannot continue along the pathway leads to the dark urine, and it turns darker when UV light interacts with the conjugated double bonds in the molecule. Erythromycin is metabolized through an induced P450 system, which requires increased heme synthesis. This leads to metabolite buildup to the level where the abdominal pain appears. The defect in heme synthesis does not affect creatine phosphate, cysteine, thymine, or methionine levels.

29. **The answer is C.** The man is experiencing gout attacks, due to a buildup of uric acid. Uric acid is the end product of purine degradation. The liver contains larger levels of nucleic acids (DNA, RNA) than do the other foods listed as answers, and the intestinal epithelial cells will convert the purines in the food to uric acid, and release it into the circulation. The alcohol the man has consumed leads to dehydration, which raises the uric acid concentration to the point where it will precipitate in the blood, leading to the painful episodes.

30. **The answer is D.** The child exhibits the symptoms of Lesch–Nyhan syndrome, which is a deficiency of HGPRT activity. HGPRT will convert the free bases hypoxanthine, or guanine, plus PRPP (5′-phosphoribosyl 1′-pyrophosphate) to the nucleotides IMP, or GMP, plus pyrophosphate. The reactions in answer choices A and C are part of the pyrimidine salvage pathways, using pyrimidine nucleoside phosphorylase, in which a nucleoside is formed from the free base and (deoxy)ribose 1-phosphate. The reaction in E is the IMP dehydrogenase step, the first step on the pathway to de novo GMP production. The reaction in the choice B is catalyzed by adenine phosphoribosyltransferase, and is analogous to the HGPRT reaction, other than adenine is the substrate, and not hypoxanthine or guanine.

31. **The answer is C.** The child has von Gierke disease, a lack of glucose-6-phosphatase activity. The lack of glucose-6-phosphatase leads to glucose 6-phosphate accumulation under conditions of glucose export from the liver, and the excess glucose 6-phosphate has three potential fates–synthesize more glycogen (due to allosteric activation of phosphorylated glycogen synthase), produce lactate (glycolysis is forced to be active by high substrate concentrations, and gives rise to lactate formation), and ribose 5-phosphate (through the oxidative reactions of the HMP shunt pathway). As ribose 5-phosphate accumulates, PRPP synthetase will convert the ribose 5-phosphate to PRPP, which then initiates the de novo purine biosynthetic pathway, and the purines are overproduced. With excess purines, they are then degraded to uric acid, which increases in concentration and leads to precipitation, and gout.

32. **The answer is A.** The girl has the disorder alkaptonuria, which is a defect in homogentisic acid oxidase, part of the phenylalanine/tyrosine degradative pathway. Homogentisic acid accumulates, and is removed from the body in the urine. Upon contact with the atmosphere, homogentisic acid is oxidized, and turns black. The constant presence of homogentisic acid in the circulation can lead to slow, but steady, deposits in the spine and joints, leading to arthritis in early adulthood. Calcification of the coronary arteries may also occur in these patients.

33. **The answer is B.** Hartnup disease is a transport defect, manifest in both the kidney and intestinal epithelial cells. The transporter is for large, neutral amino acids, and even though many amino acid transport systems have overlapping specificities, tryptophan uptake can be limiting with this disorder. Under such conditions, the body may not produce sufficient NAD for its needs (the nicotinamide ring can be produced from dietary niacin, as well as tryptophan), and a pellagra-like illness can develop. Giving the child niacin will allow the NAD to be regenerated, and a high-protein diet may increase tryptophan levels such that the transporter can transport tryptophan from the intestinal lumen into the bloodstream. Cystinuria is a different transport defect that will not allow cystine to be absorbed from the diet, or removed from the urine and returned to the blood in the kidney (which can give rise to kidney stones). Myasthenia gravis is due to autoantibodies directed against the acetylcholine receptor. Alkaptonuria is due to a

defect in homogentisic acid oxidase, and jaundice results from an inability to add glucuronic acid residues to bilirubin in the liver.

34. The answer is C. The child has inherited mutations in the genes for adenosine deaminase, and cannot convert adenosine to inosine (and deoxyadenosine to deoxyinosine). The deoxyadenosine is toxic and will accumulate in the blood cells, eventually forming dATP through salvage reactions. The dATP will, in part, inhibit ribonucleotide reductase, and the cells with the high dATP levels will not be able to proliferate when signaled to do so due to the lack of deoxyribonucleotide precursors. This is a form of SCID. Orotic acid builds up in hereditary orotic aciduria, but immune defects are not associated with that condition. Uric acid accumulation leads to gout without affecting the formation of the immune system. NADPH is required for the ribonucleotide reductase reaction, but its levels are not altered, nor is it secreted into the blood in an adenosine deaminase deficiency. dGTP levels do not increase with an adenosine deaminase deficiency.

35. The answer is D. Glycine metabolism involves a variety of pathways, one of which is a reversible transamination of glycine to form glycoxylic acid. The enzyme is glycine aminotransferase (also known as alanine-glyoxylate aminotransferase), and is defective in the disorder primary hyperoxaluria type 1. Glyoxalate can be produced from glycine by two different enzymes: the first is D-amino acid oxidase, and the second is the glycine aminotransferase. Glyoxylate is oxidized to oxalate, which forms calcium salts in the kidney and precipitates, forming kidney stones. A defect in any enzyme, which may lead to an accumulation of glyoxylate, will lead to kidney stone formation.

chapter 9

Molecular Endocrinology and An Overview of Tissue Metabolism

This chapter describes the basics behind endocrine disease processes, endocrine tumors, and the rationale for their treatment. It also reviews the basics of the regulation of tissue metabolism, which have been presented, although not as heavily emphasized, in the previous chapters of the book. The questions focus on the endocrinology, while the review comprehensive exam will focus on the material throughout the book.

OVERVIEW

- Communication between cells is essential for human survival, and is provided mainly by the nervous and endocrine systems.
- Endocrine glands produce hormones that travel through the blood to other tissues where they elicit a response (Table 9.1).
- The action of hormones at the molecular level involves receptors, as discussed previously in Chapter 4.
- Hormones often act as a chain of chemical messengers. For example, a hormone produced by the hypothalamus may stimulate the anterior pituitary to produce another hormone that subsequently causes an endocrine gland to produce yet another hormone that ultimately acts on its target cells.
- The various tissues of the body have different biochemical functions.
- These functions change with the physiologic state (e.g., fed, fasting, starving, exercise) and under different pathologic conditions.
- Biochemical measurements can be made to determine whether the body is in a normal or an abnormal state.

I. SYNTHESIS OF HORMONES

- An amino acid can be converted to a compound that serves as a hormone (e.g., epinephrine, thyroid hormone), or a series of amino acids may be joined by peptide bonds to produce a polypeptide hormone (e.g., insulin, prolactin).
- The steroid nucleus can be biochemically modified to produce a number of hormones (e.g., progesterone, cortisol).

table 9.1 Abbreviations for Hormones and Related Compounds

ACTH	Adrenocorticotropic hormone
ABP	Androgen binding protein
ADH	Antidiuretic hormone (also known as VP)
ANP	Atrionatriuretic peptide (or atriopeptin)
CRH	Corticotropin-releasing hormone
1,25-DHC	1,25-Dihydroxycholecalciferol
DHEA	Dehydroepiandrosterone
DHT	Dihydrotestosterone
E_2	Estradiol
FSH	Follicle-stimulating hormone
GH	Growth hormone
GRH	Growth hormone–releasing hormone (GHRH)
GnRH	Gonadotropin-releasing hormone
hCG	Human chorionic gonadotropin
IGF	Insulin-like growth factor
LH	Luteinizing hormone
LPH	Lipotropin
MSH	Melanocyte-stimulating hormone
POMC	Pro-opiomelanocortin
PRH	Prolactin-releasing hormone
PRIH	Prolactin release–inhibiting hormone (PIH)
PRL	Prolactin
PTH	Parathyroid hormone
T_3	Triiodothyronine
T_4	Thyroxine (tetraiodothyronine)
TRH	Thyrotropin-releasing hormone
TSH	Thyroid-stimulating hormone
VP	Vasopressin (also known as ADH)

A. Epinephrine (see Fig. 8.21)

1. **Tyrosine**, produced by the hydroxylation of the essential amino acid phenylalanine, is further hydroxylated to form **dihydroxyphenylalanine** (dopa), which is subsequently decarboxylated to form **dopamine**.
2. An additional hydroxylation reaction produces **norepinephrine**, which is methylated (mainly in the adrenal medulla) to produce **epinephrine**.

B. Thyroid Hormones (Fig. 9.1)

1. The follicular cells of the thyroid gland produce the protein **thyroglobulin**, which is secreted into the colloid (Fig. 9.2).

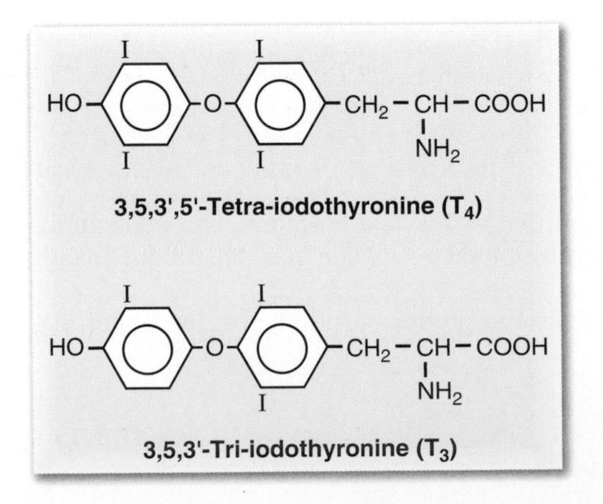

FIGURE 9.1. The thyroid hormones T_3 and T_4. T_4, thyroxine.

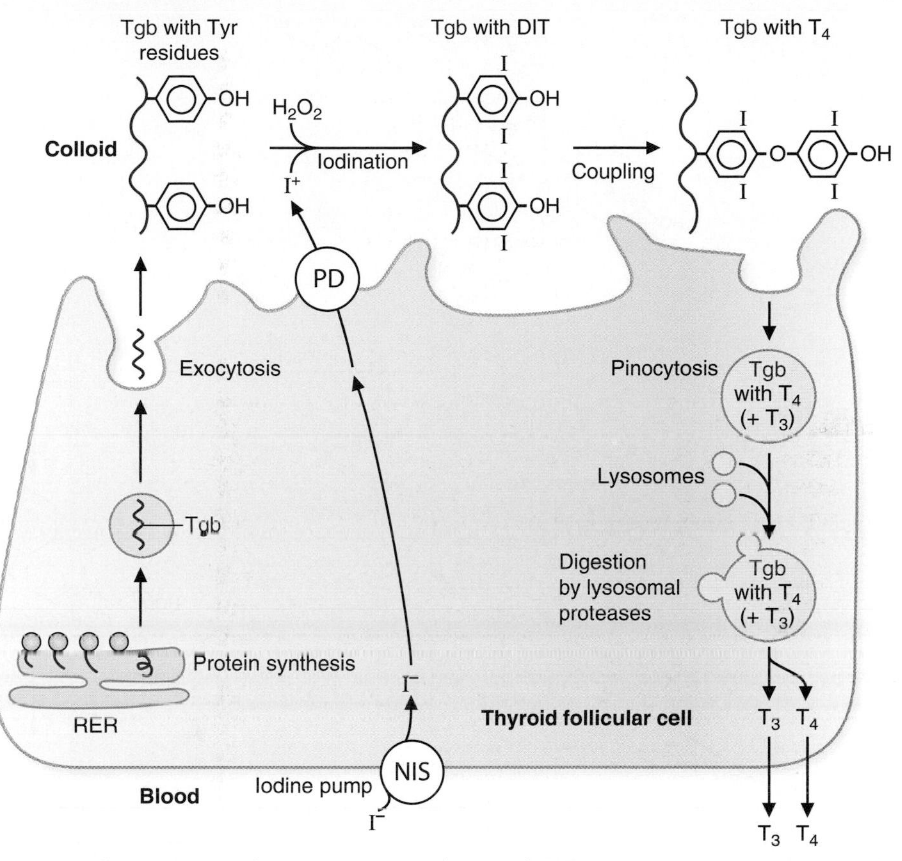

FIGURE 9.2. The synthesis of the thyroid hormones (T_3 and T_4). RER, rough endoplasmic reticulum; NIS, sodium–iodide symporter; PD, pendrin (a sodium-independent iodide transporter); Tgb, thyroglobulin.

2. **Iodine**, concentrated in the follicular cells by a pump in the cell membrane, is oxidized by a peroxidase. Iodination of **tyrosine residues** in thyroglobulin produces monoiodotyrosine (MIT) and diiodotyrosine (DIT), which undergo **coupling reactions** to produce 3,5,3′-triiodothyronine (T_3) and 3,5,3′,5′-tetraiodothyronine (T_4), which is also known as thyroxine.
3. Thyroid-stimulating hormone (**TSH**) stimulates the **pinocytosis** of thyroglobulin, and **lysosomal proteases** cleave peptide bonds, releasing free $\mathbf{T_3}$ and $\mathbf{T_4}$ from thyroglobulin. These hormones enter the blood.

C. Polypeptide hormones

1. Polypeptide hormones are **gene products**.
2. mRNA is transcribed from the gene and translated on ribosomes attached to the rough endoplasmic reticulum (RER). A polypeptide **precursor** (or preprohormone) larger than the active hormone is usually formed.
3. Removal of the **signal peptide** in the RER produces the **prohormone**.
4. Modification of the prohormone occurs in the Golgi complex, and the **mature hormone** is secreted from the cell by the process of **exocytosis**.

D. Steroid hormones

Steroid hormones are derived from **cholesterol** (Fig. 9.3), which forms **pregnenolone** by cleavage of its side chain.

1. **Progesterone** is produced by the oxidation of the A ring of **pregnenolone**.
2. **Testosterone** is produced from **progesterone** by the removal of the side chain of the D ring. Testosterone is also produced from **pregnenolone** via dehydroepiandrosterone (DHEA).

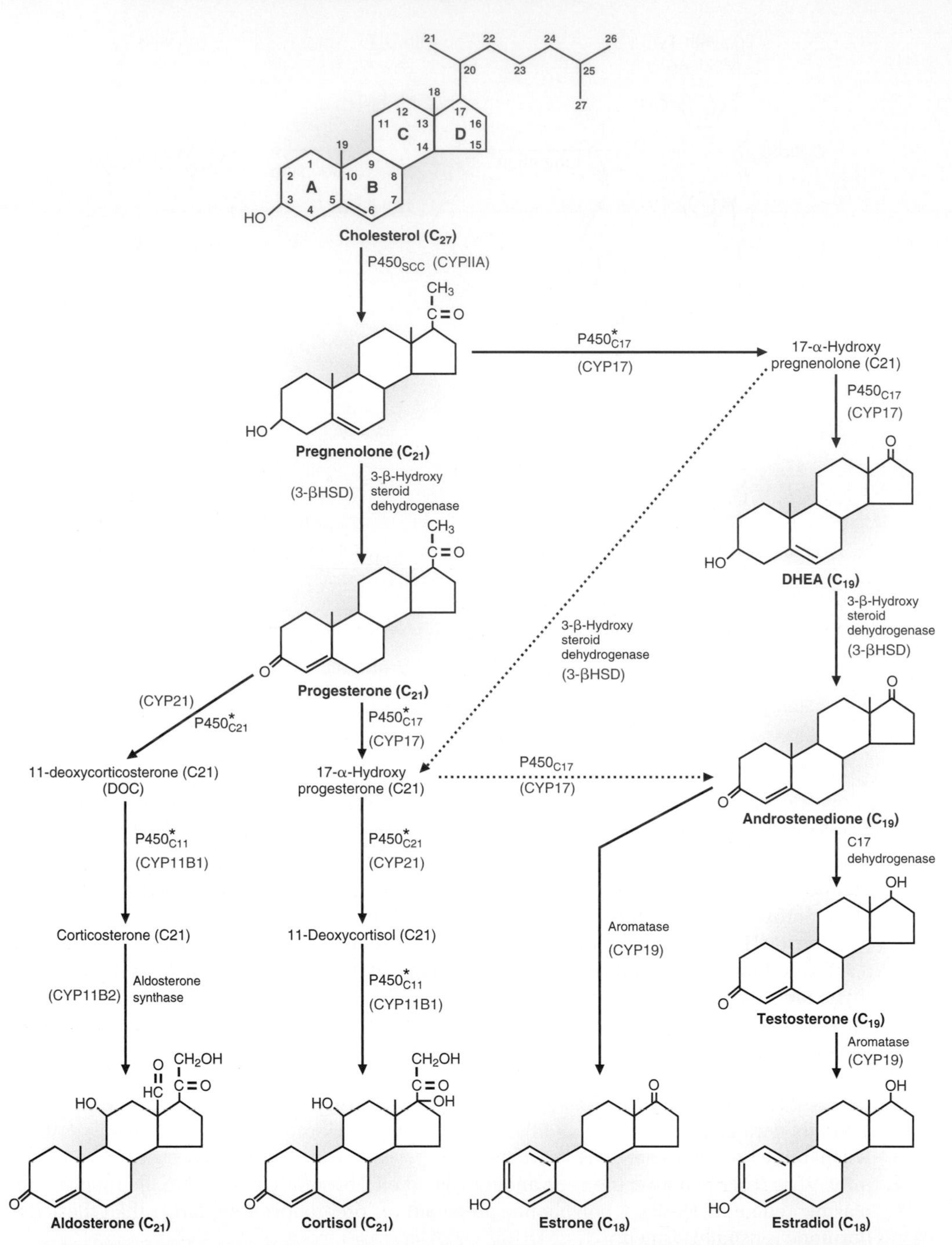

FIGURE 9.3. The synthesis of the steroid hormones. The rings of the precursor, cholesterol, are *lettered*. Dihydrotestosterone is produced from testosterone by reduction of the C–C double bond in ring A. The *dashed* lines indicate alternative pathways to the major pathways indicated. The *starred* enzymes are those that may be defective in congenital adrenal hyperplasia. DHEA, dehydroepiandrosterone.

3. **17β-Estradiol** (E_2) is produced from **testosterone** by the aromatization of the A ring.
4. **Cortisol** and **aldosterone**, the adrenal steroids, are produced from **progesterone**.
5. **1,25-Dihydroxycholecalciferol** (1,25-DHC or calcitriol), the active form of vitamin D_3, can be produced by two hydroxylations of **dietary vitamin D_3** (cholecalciferol). The first hydroxylation occurs at position 25 (in the liver), and the second occurs at position 1 (in the kidney). In addition,

7-dehydrocholesterol, a precursor of cholesterol produced from acetyl-CoA, can be converted by **ultraviolet light** in the **skin** to cholecalciferol and then hydroxylated to form 1,25-DHC.

II. GENERAL MECHANISMS OF HORMONE ACTION (ONLY A SUMMARY IS PROVIDED HERE AS THIS HAS ALREADY BEEN COVERED IN CHAPTER 4.)

- Hormones interact with **receptors** that are located either inside the cell or within the cell membrane.
- Cells are exposed to many hormones. Whether a given hormone will elicit a response in a particular cell depends on the complement of receptors that the cell contains.
- Insulin, a polypeptide hormone, binds to the insulin receptor on the cell membrane, causing the receptor to **phosphorylate** itself on **tyrosine residues** and then to phosphorylate intracellular proteins, initiating a series of events that result in cellular responses.
- In general, other polypeptide hormones and epinephrine act through **second messengers**. The hormone (the first messenger) reacts with the receptors in the cell membrane, altering the intracellular concentration of compounds known as second messengers (e.g., cyclic AMP [cAMP], cyclic GMP [cGMP], inositol trisphosphate [IP_3], diacylglycerol [DAG], and Ca^{2+}). These second messengers permit an external signal from the hormone to produce intracellular effects.
- The steroid and thyroid hormones, 1,25-DHC, and retinoic acid cross the cell membrane and bind to **intracellular receptors**, forming complexes that activate or inactivate genes (they act as **transcription factors**).

III. REGULATION OF HORMONE LEVELS

- In order to maintain homeostasis or to repeat physiologic processes such as the menstrual cycle, hormone levels must be regulated.

A. Regulation of hormone synthesis and secretion

1. The **release of hormones** is stimulated either by the changes in the environment or physiologic state or by a stimulatory hormone from another tissue that acts on the cells that release the hormone. For example:
   a. A **decrease** in **blood pressure** initiates a sequence of events that ultimately cause the adrenal gland to release **aldosterone**.
   b. In response to **stress**, the hypothalamus releases **corticotropin-releasing hormone (CRH)**, which stimulates the anterior pituitary to release **adrenocorticotropic hormone (ACTH)**. ACTH stimulates the adrenal gland to release **cortisol** (Fig. 9.4).
2. The physiologic effect of the hormone or the hormone itself causes a **decrease in the signal** that initially promoted the synthesis and release of the hormone. For example:
   a. **Aldosterone** causes an increased resorption from the kidney tubule of Na^+, and consequently of water, increasing blood pressure.
   b. **Cortisol** feeds back on the hypothalamus and the anterior pituitary, inhibiting the release of CRH and ACTH (see Fig. 9.4).

B. Hormone inactivation

1. After hormones exert their physiologic effects, they are inactivated and excreted or degraded.
2. Some hormones are converted to compounds that are no longer active and may be readily **excreted** from the body.

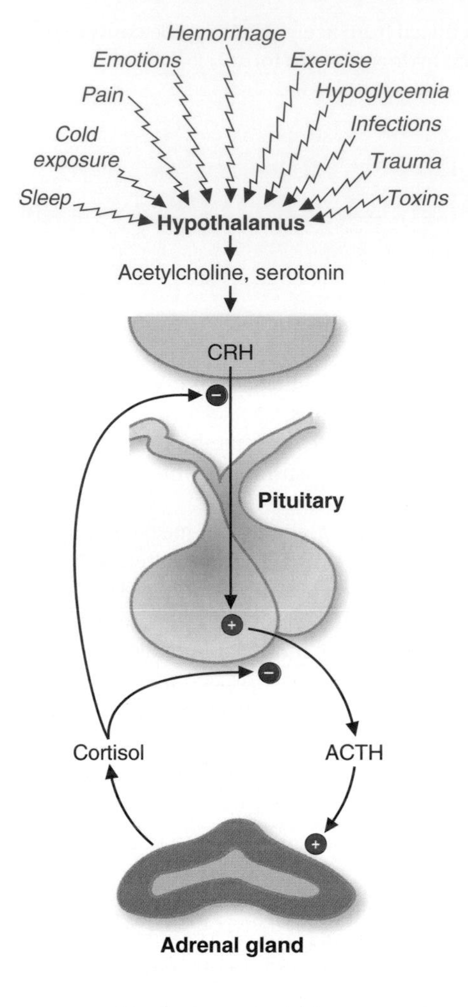

FIGURE 9.4. The regulation of cortisol secretion. Various factors act on the hypothalamus to stimulate the release of CRH. CRH stimulates the release of ACTH from the anterior pituitary, which stimulates the release of cortisol from the adrenal cortex. Cortisol inhibits the release of CRH and ACTH via negative feedback loops.

   a. Cortisol, a steroid hormone, is reduced and conjugated with glucuronide or sulfate and excreted in the urine and the feces.

3. Some hormones, particularly the polypeptides, are taken up by cells by the process of endocytosis and subsequently **degraded by lysosomal enzymes**.

   a. The **receptor**, which is internalized along with the hormone, can either be **degraded** by lysosomal proteases or it can be **recycled** to the cell membrane.

IV. ACTIONS OF SPECIFIC HORMONES

- Tissues that produce hormones include the hypothalamus, anterior and posterior pituitary, adrenal cortex and medulla, gonads, thyroid and parathyroid glands, heart, brain, cells of the gastrointestinal tract, and the pancreas.

A. Hypothalamic hormones (Fig. 9.5)

The hypothalamus produces **vasopressin (VP)** and **oxytocin (OT)**, and it produces **other hormones** (mainly peptides and polypeptides) that regulate the synthesis and release of hormones from the anterior pituitary.

B. Hormones of the posterior pituitary (see Fig. 9.5)

VP (also called antidiuretic hormone [ADH]) and **OT** are synthesized in the hypothalamus and travel through nerve axons to the posterior pituitary where they are stored, each complexed with a neurophysin. They are released into the blood in response to the appropriate stimulation.

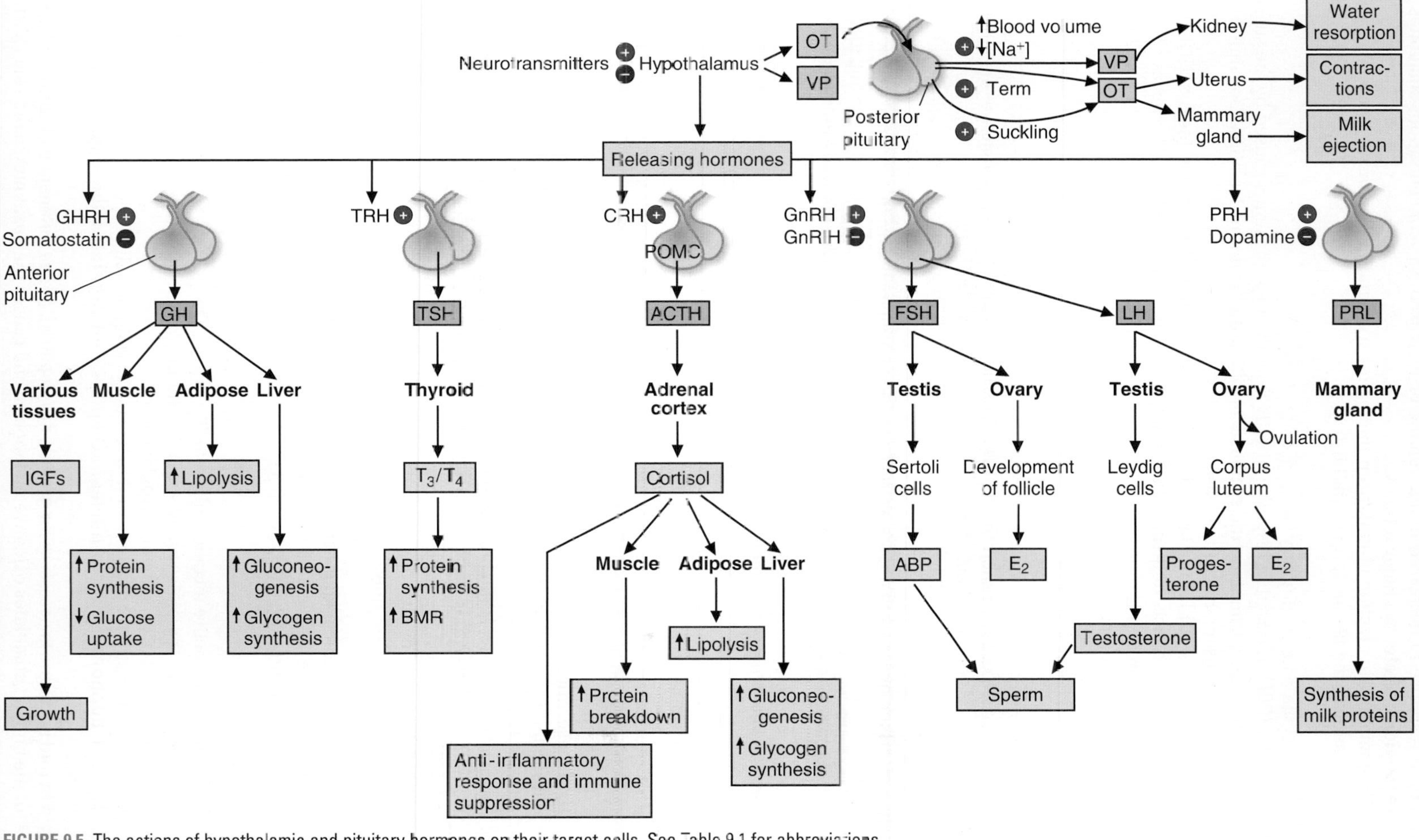

FIGURE 9.5. The actions of hypothalamic and pituitary hormones on their target cells. See Table 9.1 for abbreviations.

1. **VP**, in response to decreased blood volume or increased Na^+ concentration, **stimulates** the **resorption of water** by kidney tubules.
2. **OT promotes** the **ejection of milk** from the mammary gland in response to suckling and the **contraction of the uterus** during childbirth.

C. Hormones of the anterior pituitary (see Fig. 9.5)

1. **Prolactin (PRL)**, released in response to PRH from the hypothalamus caused by the suckling action of an infant, stimulates the **synthesis of milk proteins** during lactation. Dopamine from the hypothalamus inhibits PRL release.
2. **Growth hormone (GH)** stimulates the **release of insulin-like growth factors** (IGFs) and **antagonizes** the **effects of insulin on carbohydrate and fat metabolism**. The release of GH is stimulated by growth hormone–releasing hormone (GHRH) and inhibited by somatostatin from the hypothalamus.

CLINICAL CORRELATES **Excessive secretion of GH** occurs as a result of a **benign tumor** of the **anterior pituitary gland**. If the hypersecretion begins prior to closure of the growth centers in the long bones, excessive height (**gigantism**) occurs. If hypersecretion begins after the growth centers have closed, the bones grow in bulk and width, leading to a condition called **acromegaly**. Soft-tissue overgrowth occurs as well, leading to **organomegaly, thickness of the skin**, and **coarseness of the facial features**. Chronic GH excess may lead to **glucose intolerance** and **diabetes mellitus (DM)** because GH stimulates gluconeogenesis and slows down glucose uptake by muscle. If the pituitary tumor grows beyond the confines of the sella turcica, the tumor may encroach on the optic nerves, causing **visual difficulties**, or may lead to other cranial nerve dysfunction, progressive headaches, and eventually, symptoms of increased intracranial pressure.

3. **TSH**, produced in response to TRH from the hypothalamus, **stimulates** the release of $\mathbf{T_3}$ and $\mathbf{T_4}$ from the thyroid gland.
4. Luteinizing hormone (**LH**) and follicle-stimulating hormone (**FSH**) **stimulate** the **gonads** to release hormones that are involved in reproduction. The release of LH and FSH is stimulated by gonadotropin-releasing hormone (GnRH) and inhibited by GnRIH from the hypothalamus.
5. The protein product of the **pro-opiomelanocortin (POMC)** gene, produced in response to CRH from the hypothalamus, is cleaved to generate a number of polypeptides.
   a. **ACTH** stimulates the production of cortisol, and has a permissive effect on the production of aldosterone by the adrenal cortex.
   b. **Lipotropin (LPH)** may be cleaved to form melanocyte-stimulating hormone and endorphins.
   c. **Melanocyte-stimulating hormone (MSH)**, which is part of ACTH and LPH, stimulates the production of the pigment melanin by the melanocytes in the skin.
   d. **Endorphins** produce analgesic effects.

D. Thyroid hormone

1. $\mathbf{T_3}$ is much more active metabolically than T_4.
   a. Although the thyroid secretes some T_3, the majority is produced by **deiodination of** $\mathbf{T_4}$, a process that occurs in nonthyroidal tissue.
   b. During starvation, T_4 is converted to reverse T_3 (rT_3), which is not active.
2. **Thyroid hormone** binds to nuclear receptors and **regulates the expression of many genes**.
3. Thyroid hormone is necessary for **growth, development,** and **maintenance** of almost all tissues of the body. It **stimulates** oxidative metabolism and causes the **basal metabolic rate** (BMR) to increase.

CLINICAL CORRELATES Both too much thyroid hormone (**hyperthyroidism**) and too little thyroid hormone (**hypothyroidism**) can lead to disease. In patients with hypothyroidism, the stimulatory effect of thyroid hormone on the oxidation of fuels is diminished. As a consequence, the generation of ATP is reduced, causing a sense of **weakness, fatigue**, and **hypokinesis**. The **reduced BMR** is associated with diminished heat production, causing **cold intolerance** and **decreased sweating**. With less demand for the delivery of fuels and oxygen to peripheral tissues, the circulation is slowed down, causing a reduction in heart rate and, when far advanced, a reduction in blood pressure. In hypothyroidism, **TSH levels are elevated** and an enlarged thyroid (**goiter**) can occur. When the **thyroid gland secretes excessive quantities of thyroid hormone**, the rate of oxidation of fuels by muscle and other tissues is increased (i.e., the **BMR** is **increased**). With enhanced oxidative metabolism, heat production is increased, leading to a sense of **heat intolerance** and the need to dissipate heat through **increased sweating**. Thyroid hormone excess raises the tone of the sympathetic (adrenergic) nervous system, **raising** the **heart rate** and **systolic blood pressure**. In addition, **tremulousness**, a sense of **restlessness**, and **insomnia** often occur. Since stored fuels in muscle and fat tissue are being utilized at an excessive rate, **weight loss** occurs despite increased caloric intake. In hyperthyroidism, **TSH levels are low** and an enlarged thyroid (**goiter**) can occur.

E. Hormones that stimulate growth

1. **Insulin** and **GH** stimulate growth and promote protein synthesis.
2. However, **GH antagonizes** many of the metabolic actions of **insulin**, stimulating gluconeogenesis and promoting lipolysis. The result is that alternative fuels are made available so that muscle protein (i.e., growth) can be preserved.

CLINICAL CORRELATES Whenever **insulin levels** in the blood are chronically **elevated**, which occurs, for example, in patients with insulin-secreting pancreatic tumors or in patients inadvertently given excessive quantities of exogenous insulin, the transport of glucose from the blood into tissues such as skeletal muscle and fat cells is enhanced, leading to **hypoglycemia**. The clinical manifestations of a **reduction in blood glucose levels** are those related to the stimulation of the sympathetic nervous system by **hypoglycemia** (e.g., **sweating, palpitations** of the heart, **tremulousness**) and those due to inadequate delivery of glucose to the brain, also known as **neuroglycopenic sequelae** (e.g., irritability, slurring of speech, confusion, drowsiness, and eventually, coma).

F. Hormones that mediate the response to stress

Glucocorticoids (particularly cortisol) and **epinephrine** act in concert to supply fuels to the blood so that energy can be produced to combat stressful situations.

1. **Glucocorticoids** (Fig. 9.6)
   a. In response to ACTH, the adrenal cortex produces glucocorticoids. **Cortisol** is the major glucocorticoid in humans.
      (1) Glucocorticoids have **anti-inflammatory effects**.
         (a) They induce the synthesis of **lipocortin**, a protein that inhibits phospholipase A_2, the rate-limiting enzyme in prostaglandin, thromboxane, and leukotriene synthesis (see Fig. 7.20).
      (2) Glucocorticoids **suppress the immune response** by causing the lysis of lymphocytes.
      (3) Glucocorticoids **influence metabolism** by causing the movement of fuels from peripheral tissues to the liver, where gluconeogenesis and glycogen synthesis are stimulated (see Figs. 9.5 and 9.6).
         (a) **Amino acids** are released from muscle protein.
         (b) **Lipolysis** occurs in adipose tissue.
         (c) In addition to providing amino acids and glycerol as carbon sources, **glucocorticoids promote gluconeogenesis** by **inducing** the synthesis of the enzyme phosphoenolpyruvate carboxykinase (**PEPCK**).

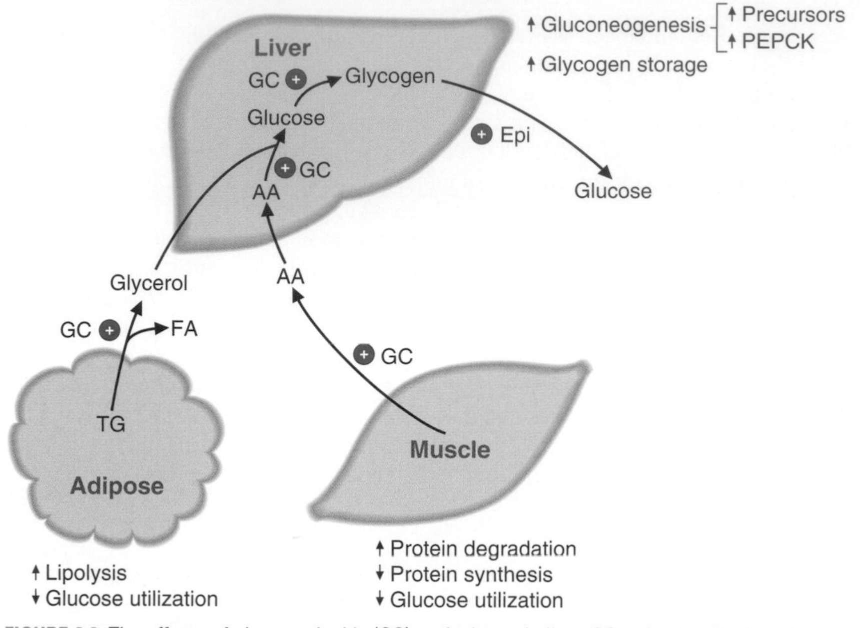

FIGURE 9.6. The effects of glucocorticoids (GC) on fuel metabolism. GCs stimulate lipolysis in adipose tissue and the release of amino acids from muscle protein. In liver, GCs stimulate gluconeogenesis and the synthesis of glycogen. Acute stress causes the release of epinephrine (Epi), which stimulates the breakdown of liver glycogen to produce blood glucose as fuel for "fight or flight." Epinephrine also stimulates glycogen breakdown in muscle to produce ATP for muscle contraction, and it stimulates lipolysis in adipose tissue and gluconeogenesis in the liver. PEPCK, phosphoenolpyruvate carboxykinase; TG, triglyceride; Epi, epinephrine; AA, amino acid.

(d) **Glucose**, produced by gluconeogenesis promoted by glucocorticoids, is **stored as glycogen** in the liver.

(e) Glucocorticoids prepare the body during stressful conditions so that fuel stores are ready for the "alarm" reaction mediated by epinephrine.

CLINICAL CORRELATES **Hypercortisolemia (glucocorticoid excess)** has an adverse effect on virtually every tissue of the body. Central nervous system effects range from **hyperirritability** to **depression**. The catabolic effect on protein-containing tissues leads to a reduction in the ground substance of bone and eventually to **osteoporosis**, loss of muscle protein causing **weakness**, and thinning and tearing of dermal and epidermal structures, which is manifest as reddish stripes, or **striae**, over the lower abdomen, the lateral thorax, and other areas where skin tension is increased. Similar catabolic effects on the elastin of vessel walls lead to vascular fragility with **easy bruising** and hemorrhaging of the skin. A suppressive effect on immunocompetence may increase the likelihood of infection. The diabetogenic actions of cortisol may lead to **glucose intolerance** or overt **DM**. A peculiar tendency for the disposition of fat in the face (**moon facies**), posterior neck (**buffalo hump**), thorax, and abdomen, while sparing the distal extremities, causes a distinct **"central obesity."** This constellation of clinical signs and symptoms resulting from chronic hypercortisolemia is referred to as **Cushing syndrome** if the condition is caused by the excessive production of **cortisol** by an **adrenal tumor** or by the intake of **exogenous glucocorticoids**. **Cushing disease** refers to hypercortisolemia caused by the excessive secretion of **ACTH** by a **pituitary tumor** (recall that ACTH stimulates cortisol release from the adrenal gland).

2. Epinephrine

a. Epinephrine increases blood glucose by **stimulating liver glycogenolysis** (see Fig. 9.6). It also **stimulates lipolysis** in adipose tissue and **glycogen degradation in muscle**. Overall, it makes fuels available for "fight or flight."

G. Hormones that regulate salt and water balance

In addition to **VP** (see Section IV B) and **ANP** (see Section II A 2 c), **aldosterone** is involved in regulating the salt and water balance.

1. Synthesis of aldosterone

a. **Renin** (produced by the juxtaglomerular cells of the kidney in response to decreased blood pressure, blood volume, or sodium ion concentration) **cleaves angiotensinogen to angiotensin I.**

b. **Angiotensin I** is **cleaved** to **angiotensin II** by angiotensin-converting enzyme (**ACE**), which is made in the lung.
   (1) Further cleavage to angiotensin III occurs.

c. **Angiotensin II** acts directly on vascular smooth muscle cells, causing **vasoconstriction,** which increases blood pressure.

d. **Angiotensin II and III** (and also decreased serum [Na^+] and increased serum [K^+]) **stimulate** the glomerulosa cells of the adrenal cortex to produce and secrete **aldosterone.**
   (1) ACTH has a permissive effect (i.e., it maintains cells so that they can respond to angiotensin II).

2. Action of aldosterone

a. Aldosterone causes the production of proteins in cells of the distal tubule and the collecting ducts of the kidney.
   (1) A **permease** is **produced** that allows Na^+ to enter the cells from the lumen.
   (2) **Citrate synthase** is **induced,** which increases the capacity of the tricarboxylic acid (TCA) cycle for the generation of ATP.
   (3) Energy is thus provided to drive the **Na^+–K^+ ATPase,** which may also be induced.

b. Overall, K^+ and H^+ are lost; Na^+ is retained; water is resorbed; and blood volume and pressure are increased.

CLINICAL CORRELATES A **deficiency of adrenocortical secretion of aldosterone** is usually accompanied by a reduction in the secretion of other adrenal steroid hormones as well. The loss of adrenocortical steroids is known as **Addison disease.** The mineralocorticoid deficiency leads to a net loss of sodium ions and water into the urine with a reciprocal retention of potassium ions (**hyperkalemia**) and hydrogen ions (**mild metabolic acidosis**). The subsequent contraction of the effective plasma volume may lead to a **reduction in blood pressure.** If volume loss is profound, perfusion of vital tissues such as the brain could lead to lightheadedness and possible loss of consciousness.

H. Hormones that control reproduction (see Fig. 9.5)

The hypothalamus produces **GnRH**, which causes the anterior pituitary to release **FSH** and **LH,** which act on both the **ovary** and the **testis.**

1. The action of FSH and LH on the ovary

a. **The menstrual cycle**
   (1) Initially, **FSH acts on** the **follicles** to promote the maturation of the ovum and to stimulate estradiol (E_2) production and secretion.
   (2) **Estradiol acts on** the uterine **endometrium,** causing it to thicken and vascularize in preparation for the implantation of a fertilized egg.
   (3) A **surge of LH** at the midpoint of the menstrual cycle **stimulates** the ripe **follicle to ovulate,** leaving the residual follicle, which forms the **corpus luteum** and **secretes** both **progesterone and estradiol.**
   (4) **Progesterone** causes the endometrium to continue to thicken and vascularize and increase its secretory capacity.

b. **Events in the absence of fertilization**
   (1) The **corpus luteum regresses** owing to declining LH levels. It produces diminishing amounts of progesterone and estradiol.
   (2) Because of the low steroid hormone levels, the cells die and the degenerating **endometrium is sloughed** into the uterine cavity and excreted (**menstruation**).

(3) The low levels of estradiol and progesterone cause feedback inhibition to be relieved, and the hypothalamus releases GnRH, initiating a new menstrual cycle.

c. Events following fertilization

(1) The **corpus luteum** is **maintained** initially by **human chorionic gonadotropin (hCG)**, produced by the cells of the developing embryo (trophoblast).

(2) Subsequently, the **placenta produces hCG and progesterone.**

(3) After the corpus luteum dies, the **placenta** continues to produce large amounts of **progesterone.**

(4) Near **term**, hCG and, subsequently, **progesterone levels fall.**

(5) Fetal cortisol may cause the decline in progesterone.

(6) **Prostaglandin $F_{2\alpha}$ ($PGF_{2\alpha}$)** and **OT** (released from both maternal and fetal pituitaries) stimulate uterine contractions, and the infant is delivered.

2. The action of FSH and LH on the testis

a. LH stimulates Leydig cells to produce and secrete testosterone.

b. FSH acts on **Sertoli cells** of the seminiferous tubule to promote the synthesis of androgen binding protein (**ABP**).

c. ABP binds testosterone and transports it to the site of spermatogenesis, where **testosterone is reduced** to the more potent androgen, dihydrotestosterone (**DHT**).

d. Testosterone plays a role in **spermatogenesis** in the adult male.

(1) Testosterone is responsible for **masculinization** during early development.

(2) At puberty, testosterone promotes the **sexual maturation** of the male.

CLINICAL CORRELATES **Congenital adrenal hyperplasia** (CAH) is a group of diseases caused by a genetically determined deficiency in a variety of enzymes required for cortisol synthesis. The most common deficiency is that of 21-α-hydroxylase (CYP21), the activity of which is necessary to convert progesterone to 11-deoxycorticosterone and 17-α-hydroxy progesterone to 11-deoxycortisol. Thus, this deficiency reduces both aldosterone and cortisol production without affecting androgen production. If the enzyme deficiency is severe, the precursors for aldosterone and cortisol production are shunted to androgen synthesis, producing an overabundance of androgens, which leads to prenatal masculinization in females and postnatal virilization of males. Another enzyme deficiency in this group of diseases is that of 11-β-hydroxylase (CYP11B1), which results in the accumulation of 11-deoxycorticosterone. An excess of this mineralocorticoid leads to hypertension (through binding of 11-deoxycorticosterone to the aldosterone receptor). In this form of CAH, 11-deoxycortisol also accumulates, but its biologic activity is minimal, and no specific clinical signs and symptoms result. The androgen pathway is unaffected, and the increased ACTH levels may increase the levels of adrenal androgens in the blood. A third possible enzyme deficiency is that of 17-α-hydroxylase (CYP17). A defect in 17-α-hydroxylase leads to aldosterone excess and hypertension; however, because adrenal androgen synthesis requires this enzyme, no virilization occurs in these patients.

I. Hormones that promote lactation (see Fig. 9.5)

Many hormones are necessary for the development of the mammary glands during adolescence.

1. Preparation of the mammary gland for lactation

a. During pregnancy, **PRL, glucocorticoids, and insulin** are the major hormones responsible for the differentiation of mammary alveolar cells into secretory cells capable of producing milk.

b. PRL stimulates the **synthesis of milk proteins**, particularly casein and α-lactalbumin.

(1) **α-Lactalbumin**, the major protein in human milk, serves as a **nutrient.**

(2) α-Lactalbumin binds to galactosyltransferase, decreasing its K_m for glucose and, thus, **stimulating the synthesis** of the milk sugar **lactose.**

c. Progesterone inhibits milk protein production and secretion during pregnancy.

d. At term, when progesterone levels fall, the inhibition of milk protein synthesis is relieved.

2. Regulation of milk secretion during lactation

a. PRL causes milk proteins to be produced and secreted into the alveolar lumen.

b. OT causes contraction of the myoepithelial cells surrounding the alveolar cells and the lumen, and **milk is ejected** through the nipple.

c. The **secretion** of both PRL and OT by the pituitary is stimulated by the **suckling** action of the infant and by other factors.

CLINICAL CORRELATES The most common secretory neoplasm of the anterior pituitary gland is a **prolactin-secreting adenoma** (prolactinoma). An early symptom of prolactin excess is a **milky discharge from the breasts** (galactorrhea). Because hyperprolactinemia suppresses the secretion of gonadotropic hormones (e.g., LH, FSH), **menstrual irregularity, amenorrhea,** and **infertility** can occur in women. In men, gonadotropic hormone dysfunction can lead to reduced libido, sexual impotence, and infertility.

J. Hormones involved in growth and differentiation

1. **Retinoids are produced** in the body from dietary **vitamin A**. The major dietary source, β-carotene, is cleaved to two molecules of retinal.
2. **Retinal** (an aldehyde) and **retinol** (an alcohol) are interconverted by oxidation and reduction reactions. **Retinoic acid** is produced by the oxidation of retinal and cannot be reduced.
3. **Retinol,** the **transport** form, is stored as retinyl esters.
4. **Retinal** is a functional component of the reactions of the **visual cycle.**
5. **Retinoic acid** is involved in the **growth, differentiation,** and **maintenance** of **epithelial tissue.** The functions of **retinoic acid** result from its ability to **activate genes** (i.e., it acts like a steroid hormone).

K. Hormones that regulate Ca^{2+} metabolism

Calcium has many important functions. It is involved in blood coagulation, activation of muscle phosphorylase, and secretory processes. It combines with phosphate to form the hydroxyapatite of bone. Parathyroid hormone (**PTH**), **1,25-DHC**, and **calcitonin** are the major regulators of Ca^{2+} metabolism.

1. **PTH**, produced in response to low calcium levels, acts to **increase Ca^{2+}** levels in the extracellular fluid.
   a. PTH promotes Ca^{2+} and phosphate mobilization from **bone.**
   b. PTH acts on **renal tubules** to resorb Ca^{2+} and excrete phosphate.
   c. PTH stimulates the **hydroxylation of 25-hydroxycholecalciferol** to form 1,25-DHC, the active hormone.
2. **1,25-DHC** stimulates the synthesis of a protein involved in **Ca^{2+} absorption** by **intestinal** epithelial cells. 1,25-DHC acts synergistically with PTH in **bone resorption** and promotes resorption of Ca^{2+} by the **renal tubular cells.**
3. **Calcitonin lowers Ca^{2+}** levels by inhibiting its release from bone and stimulating its excretion in the urine.

CLINICAL CORRELATES When the secretion of PTH is excessive **(hyperparathyroidism)**, the physiologic effects of this peptide on the gut, the skeleton, and the kidney tubules are enhanced. The percentage of dietary calcium absorbed into the circulation is increased; calcium ions are released from bone and enter the blood more rapidly; and the renal tubules reabsorb more calcium than usual from the luminal urine, all leading to **hypercalcemia**. Chronic hypercalcemia is associated with vague generalized musculoskeletal pain, fatigue, and eventually, slowed mentation. The osteolytic effect of excessive PTH action may lead to demineralization of the skeleton (**osteoporosis**) and fracture. Chronic renal filtration of blood rich in calcium leads to the saturation of the tubular fluid with calcium salts; as a consequence, **renal calculi** (kidney stones) can occur.

L. Hormones that regulate the utilization of nutrients

1. **Gut hormones** (for a full listing, see Table 9.2)
   a. **Gastrin** from the gastric antrum and the duodenum stimulates gastric acid and pepsin secretion.

table 9.2 Gastrointestinal-derived Hormones That Affect Fuel Metabolism Directly

Hormone	Primary Cell/Tissue of Origin	Actions	Secretory Stimuli (and inhibitors)
Amylin	Pancreatic β cell, endocrine cells of stomach and small intestine	1. Inhibits arginine-stimulated and postprandial glucagon secretion 2. Inhibits insulin secretion	Cosecreted with insulin in response to oral nutrients
Calcitonin gene-related peptide (CGRP)	Enteric neurons and enteroendocrine cells of the rectum	Inhibits insulin secretion	Oral glucose intake and gastric acid secretion
Galanin	Nervous system, pituitary, neurons of gut, pancreas, thyroid, and adrenal gland	Inhibits the secretion of insulin, somatostatin, enteroglucagon, pancreatic polypeptide, and others	Intestinal distension
Gastric inhibitory polypeptide/glucose-dependent insulinotropic polypeptide (GIP)	Neuroendocrine K cells of duodenum and proximal jejunum	1. Increases insulin release via an "incretin" effect 2. Regulates glucose and lipid metabolism	Oral nutrient ingestion, especially long-chain fatty acids
Gastrin-releasing peptide (GRP)	Enteric nervous system and pancreas	Stimulates the release of cholecystokinin; GIP, gastrin, glucagon, GLP-1, GLP-2, and somatostatin	
Ghrelin	Central nervous system, stomach, small intestine, and colon	Stimulates GH release	Fasting
Glucagon	Pancreatic α cell, central nervous system	Primary counterregulatory hormone that restores glucose levels in hypoglycemic state (increases glycogenolysis and gluconeogenesis as well as protein–lipid flux in liver and muscle)	Neural and humoral factors released in response to hypoglycemia
Glucagon-like peptide-1 (GLP-1)	Enteroendocrine L cells in ileum, colon, and central nervous system	1. Enhances glucose disposal after meals by inhibiting glucagon secretion and stimulating insulin secretion 2. Acts through second messengers in β cells to increase the sensitivity of these cells to glucose (an incretin)	1. Oral nutrient ingestion 2. Vagus nerve 3. GRP and GIP 4. Somatostatin inhibits secretion
Glucagon-like peptide-2 (GLP-2)	Same as for GLP-1	Stimulates intestinal hexose transport	Same as GLP-1
Neuropeptide Y	Central and peripheral nervous system, pancreatic islet cells	Inhibits glucose-stimulated insulin secretion	Oral nutrient ingestion and activation of sympathetic nervous system
Neurotensin (NT)	Small intestine N cells (especially ileum), enteric nervous system, adrenal gland, pancreas	In brain, modulates dopamine neurotransmission and anterior pituitary secretions	1. Luminal lipid nutrients 2. GRP 3. Somatostatin inhibits secretion
Pituitary adenylate cyclase–activating peptide (PACAP)	Brain, lung, and enteric nervous system	Stimulates insulin and catecholamine release	Activation of central nervous system
Somatostatin	Central nervous system, pancreatic δ cells, and enteroendocrine δ cells	1. Inhibits secretion of insulin, glucagon and PP (islets), and gastrin, secretin, GLP-1, and GLP-2 (in gut) 2. Reduces carbohydrate absorption from gut lumen	1. Luminal nutrients 2. GLP-1 3. GIP 4. PACAP 5. VIP 6. β-Adrenergic stimulation
Vasoactive intestinal peptide (VIP)	Widely expressed in the central and peripheral nervous systems	May regulate the release of insulin and pancreatic glucagon	1. Mechanical stimulation of gut 2. Activation of central and peripheral nervous systems

table 9.3 Actions of GLP-1 and GIP Relevant to Glucose Control

	GLP-1	GIP
Pancreas		
Stimulates glucose-dependant insulin release	+	+
Increase insulin biosynthesis	+	+
Inhibits glucagon secretion	+	–
Stimulates somatostatin secretion	+	–
Induces β cell proliferation	+	+
Inhibits β cell apoptosis	+	+
Gastrointestinal tract		
Inhibits gastric emptying	+	–
Inhibits gastric acid secretion	+	+
Central nervous system		
Inhibits food and water intake	+	–
Promotes satiety and weight loss	+	–
Cardiovascular system		
Improves cardiovascular function after ischemia	+	–
Adipose tissue		
Insulin-like lipogenic actions	–	+
Lipid storage	–	+

b. **Cholecystokinin (CCK)** from the duodenum and jejunum stimulates the contraction of the gallbladder and the secretion of pancreatic enzymes.
c. **Secretin** from the duodenum and jejunum stimulates the secretion of bicarbonate by the pancreas.
d. **Ghrelin** from the central nervous system, stomach, small intestine, and colon stimulates GH release under fasting conditions.
e. **Gastric inhibitory polypeptide (GIP)** from the small bowel enhances insulin release and inhibits the secretion of gastric acid.
f. **Glucagon-like peptide 1 (GLP-1)** from the neuroendocrine cells in the ileum, colon, and central nervous system enhances glucose disposal after meals by inhibiting glucagon secretion and stimulating insulin secretion. The actions of GIP and GLP-1 relevant to glucose control are summarized in Table 9.3 and Figure 9.7.
g. **Vasoactive intestinal polypeptide (VIP)** from the pancreas relaxes smooth muscles and stimulates bicarbonate secretion by the pancreas.
h. **Somatostatin** from the central nervous system, pancreatic δ cells, and enteroendocrine δ cells inhibits the secretion of insulin and glucagon from the pancreas, inhibits the release of gastrin, secretin, and GLP-1 from the gut, and reduces carbohydrate absorption from the gut lumen.

2. **Insulin and glucagon**
   a. The two major hormones that **regulate fuel metabolism**, insulin and glucagon, are produced by the pancreas. Their actions (discussed extensively in Chapters 6, 7, and 8) are summarized in Table 9.4.

V. BIOCHEMICAL FUNCTIONS OF TISSUES

A. The use of biochemical measurements to diagnose pathologic conditions

Measurements can be made on body fluids and tissues and used to draw conclusions about clinical problems.

1. **Measurements can be made on:**
   a. **Compounds** that **enter** the body (e.g., carbohydrates, fats, proteins).
   b. **Samples of blood, urine, feces,** and various **secretions.**
   c. **Cells** obtained from **blood, scrapings,** or **biopsies** of tissues.

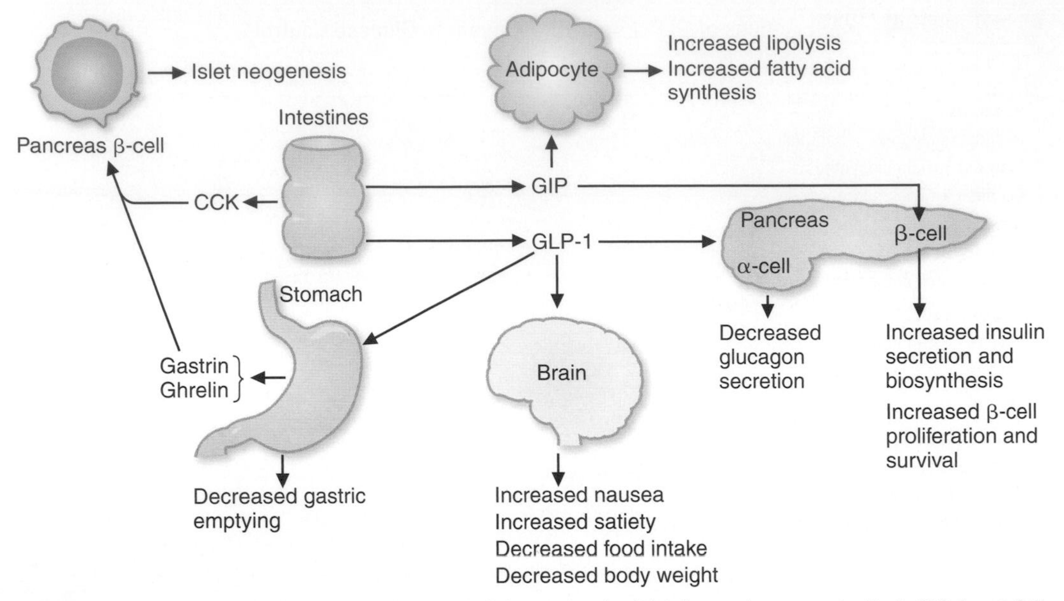

FIGURE 9.7. The actions of selected peptides on vital tissues involved in glucose homeostasis. Both GLP-1 and GIP increase insulin secretion and β cell survival. GLP-1 has additional actions related to glucose metabolism. In contrast, gastrin and CCK do not acutely regulate plasma glucose levels, but appear to increase β cell proliferation.

2. Examples of types of measurements

a. The **amounts** of various **chemical compounds** can be measured (e.g., blood glucose, blood urea nitrogen, urinary creatinine).

b. Enzyme assays can be performed (e.g., creatine kinase, aspartate or alanine transaminase, alkaline phosphatase).

c. DNA, **RNA**, or **proteins** can be identified (e.g., by hybridization with DNA probes or by reaction with antibodies), or they can be sequenced.

d. Substances can be given to a patient to determine whether the body **processes** them normally (e.g., a glucose tolerance test, radioactive iodine uptake by the thyroid gland).

3. Conclusions from measurements

a. From the results of biochemical measurements, conclusions can be drawn about the patient's condition, and, in conjunction with a medical history and physical examination,

table 9.4 Actions of Insulin and Glucagon

Insulin	Glucagon
Elevated in the fed state	Elevated during fasting
Promotes the storage of fuels: glycogen and triacylglycerol	Increases the availability of fuels (glucose and fatty acids) in the blood
Stimulates:	
Glycogen synthesis in liver and muscle	Glycogen degradation in liver, but not in muscle
Triacylglycerol synthesis in liver and conversion to very low density lipoprotein (VLDL)	Gluconeogenesis
Triacylglycerol storage in adipose tissue	Lipolysis (breakdown of triacylglycerols) in adipose tissue
Glucose transport into muscle and adipose cells	
Protein synthesis and growth	

the information can be used to formulate a **diagnosis** or to determine what **additional tests** need to be done.

b. In order to draw useful conclusions, a physician must understand the **biochemical relationships among tissues** and how they are altered by the changes in the physiologic state (e.g., fed, fasting, starving, exercise) and the effects of pathologic changes.

(1) The intertissue relationships are summarized in Chapter 1 and presented in more detail throughout the book.

(2) Some of the effects of pathologic changes are described in the Clinical Correlations that appear throughout each chapter.

(3) Many of the questions you have already worked in the book have depended on these values to help focus you on the appropriate answer.

B. Biochemical functions of tissues

1. To review the functions and sort them by tissue, we will take a tour of the body, starting in the mouth and ending in the kidney. This will act as an overall review for the biochemistry that has been discussed in the previous chapters.

2. We will consider the major biochemical functions of various tissues and organs and the consequences of abnormalities in these functions caused, for example, by physical or chemical damage, infection, dietary deficiencies or excesses, and mutations in genes.

3. We will illustrate, in addition, how biochemical measurements can pinpoint a disorder and allow one to determine which enzyme is functioning improperly.

a. Mouth

(1) Salivary glands produce **α-amylase**, which cleaves α-1,4 bonds between glucose residues in dietary starch. (Pancreatic α-amylase catalyzes the same reaction.)

b. Stomach

(1) Chief cells produce the proteolytic enzyme **pepsin**, as its inactive precursor pepsinogen. Pepsin digests proteins.

(2) Parietal cells produce hydrochloric acid (**HCl**) and **intrinsic factor**.

(a) HCl causes pepsinogen (the precursor of pepsin) to cleave itself (autocatalysis), producing **pepsin**.

(b) Intrinsic factor binds dietary **vitamin B_{12}** and aids in its absorption.

(c) Vitamin B_{12} is the cofactor for the conversion of **homocysteine to methionine** and **methylmalonyl-CoA to succinyl-CoA**.

CLINICAL CORRELATES A **deficiency of intrinsic factor** causes a condition known as **pernicious anemia**, which results in a deficiency of vitamin B_{12} in the body because of inadequate absorption. A **deficiency of vitamin B_{12}** is characterized by a **megaloblastic anemia**, which is caused by a secondary folate deficiency and by **neurologic problems** caused by the demyelination of nerves.

c. Gallbladder

(1) Bile salts, synthesized in the liver from cholesterol, pass through the gallbladder into the intestine, where they aid in lipid digestion.

(2) Bilirubin diglucuronide, produced in the liver from bilirubin (the excretory product of heme degradation), passes through the gallbladder into the intestine.

CLINICAL CORRELATES A decrease in bile salt availability can be caused by **problems with bile salt production** in the liver or **blockage of secretion** from the gallbladder. The result is an increase of fat in the stool (**steatorrhea**) because of decreased fat digestion and absorption (which can lead to **caloric deficits** and **fat-soluble vitamin** and **essential fatty acid deficiencies**). **Obstruction of the flow of bile** into the small intestine is one cause of **jaundice** (the accumulation of bilirubin in tissues, producing a yellow color, which is especially obvious in the whites of the eyes).

d. **Pancreas**

(1) The pancreas **produces bicarbonate (HCO_3^-)**, which neutralizes stomach acid as it enters the intestinal lumen. The subsequent increase in pH in the lumen allows more extensive **ionization of bile salts** (so they serve as better detergents) and increases the **activity of digestive enzymes**.

(2) The pancreas produces **digestive enzymes** (e.g., trypsin, chymotrypsin, the carboxypeptidases, elastase, α-amylase, lipase).

(3) The B (or ß) cells of the endocrine pancreas produce **insulin** (the hormone that stimulates the storage of fuels in the fed state), and the A (or α) cells produce **glucagon** (the hormone that stimulates the release of stored fuels during fasting).

CLINICAL CORRELATES **Pancreatic disorders** that cause a decrease in bicarbonate release lead to decreased activity of digestive enzymes and increased excretion of bile salts in the feces (because they are less ionized, they are more readily excreted). Therefore, **steatorrhea** occurs (increased excretion of dietary fat in the feces). **Obstruction of the flow of pancreatic secretions** into the intestinal lumen (caused, e.g., by a cancer of the head of the pancreas or by cystic fibrosis) leads to **decreased digestion** of nutrients (and, therefore, to deficiencies of calories, vitamins, and other essential nutrients in the body). **Decreased production of insulin**, which is usually caused by an autoimmune destruction of pancreatic β cells, results in **Type 1 DM**. Type 1 diabetes is characterized by **hyperglycemia**, the result of decreased uptake of glucose by cells and increased output of glucose by the liver (due to low insulin and high glucagon levels in the blood). **Decreased release of insulin** from the pancreas or **decreased sensitivity** of tissues to insulin (insulin resistance) results in **Type 2 DM**. This condition also is characterized by **hyperglycemia. Excessive** levels of **insulin**, caused by an insulin-secreting tumor of the pancreas (an insulinoma) or by injection of too much insulin, results in **hypoglycemia**.

e. **Intestine**

(1) The **enzymes** from the **exocrine pancreas digest food** in the intestinal lumen.

(2) The **digestive enzymes** are bound to the brush borders of **intestinal epithelial cells** (aminopeptidases, di- and tripeptidases, lactase, sucrase, maltases, and isomaltases).

(3) The **absorption** of digestive products occurs through the intestinal epithelial cells.

(4) The intestinal epithelial cells produce **chylomicrons** from the digestive products of dietary fat (fatty acids and 2-monoacylglycerols) and secrete the chylomicrons into the lymph.

(5) Most **bile salts** are **resorbed** in the ileum and recycled by the liver. Only **5% are excreted** in the feces. This excretion of bile salts, along with cholesterol secreted by the liver into the gut via the gallbladder, is the major means by which the body disposes of the cholesterol ring structure (sterol nucleus).

CLINICAL CORRELATES There are a variety of disorders that are related to the intestine. Excessive **proliferation of bacteria** within the intestinal lumen causes increased **deconjugation** and **dehydroxylation** of **bile salts**, leading to increased bile salt excretion and steatorrhea. **Lactase deficiency** is very common in the population. In lactase deficiency, ingestion of milk products results in **diarrhea** (caused by the osmotic effect of excess lactose in the lumen of the gut) and bloating (caused by excessive **gas** production that results from the metabolism of lactose by intestinal bacteria).

Decreased absorption of **neutral amino acids** (due to a deficient transport protein) leads to **Hartnup disease. Celiac disease** (in children) and **nontropical sprue** in adults are caused by the absorption of protein fragments that cause **allergic reactions. Cholera** is caused by the production of an enterotoxin that ADP-ribosylates G-proteins. Adenylate cyclase remains active. Increased cAMP causes increased phosphorylation of transport proteins and, thus, inhibits active transport of Na^+ into the intestinal cells. As a result, water accumulates in the gut, causing **severe diarrhea**.

table 9.5 Flowchart of Changes in Liver Metabolism

When blood sugar increases:	When blood sugar decreases:
Insulin is released, which leads to the **dephosphorylation** of:	Glucagon is released, which leads to the **phosphorylation** of:
• PFK-2 (kinase activity now active)	• PFK-2 (phosphatase activity now active)
• Pyruvate kinase (now active)	• Pyruvate kinase (now inactive)
• Glycogen synthase (now active)	• Glycogen synthase (now inactive)
• Phosphorylase kinase (now inactive)	• Phosphorylase kinase (now active)
• Glycogen phosphorylase (now inactive)	• Glycogen phosphorylase (now active)
• Pyruvate dehydrogenase (now active)	• Pyruvate dehydrogenase (now inactive)
• Acetyl-CoA carboxylase (now active)	• Acetyl-CoA carboxylase (now inactive)
Which leads to **active**	Which leads to **active**
• Glycolysis	• Glycogenolysis
• Fatty acid synthesis	• Fatty acid oxidation
• Glycogen synthesis	• Gluconeogenesis

f. Liver (the enzymic regulation of liver function is summarized in Tables 9.5 and 9.6

(1) Functions of the liver include:

(a) Storage of **glycogen** produced from dietary carbohydrate.

(b) Synthesis of very low density lipoprotein (**VLDL**), mainly from dietary carbohydrate.

(c) Production of high-density lipoprotein (**HDL**), which transfers C_{II} and E apoproteins to chylomicrons and VLDL, converts cholesterol to cholesterol esters (via the lecithin-cholesterol acyltransferase or LCAT reaction), and reduces blood cholesterol levels by participating in the process by which cholesterol and cholesterol esters are transported from tissues to the liver (reverse cholesterol transport).

(d) Maintenance of blood glucose levels during fasting via glycogenolysis and gluconeogenesis.

(e) Production of **urea** from nitrogen derived, in part, from amino acids as they are being converted to glucose (via gluconeogenesis) during fasting.

(f) Production of **ketone bodies** from fatty acids derived from lipolysis of adipose triacylglycerols during fasting.

(g) Synthesis of **cholesterol** (which is also made in other tissues).

(h) Conversion of cholesterol to **bile salts.**

(i) Production of many **blood proteins** (e.g., albumin, blood-clotting proteins).

(j) Production of purines and **pyrimidines**, which are transported to other tissues via red blood cells.

(k) Degradation of purines (to uric acid) and **pyrimidines** (to CO_2, H_2O, and urea).

(l) Oxidation of drugs and other **toxic compounds** via the cytochrome P450 system.

(m) Conjugation of bilirubin and excretion of bilirubin diglucuronide into the bile.

(n) Oxidation of alcohol via alcohol and acetaldehyde dehydrogenases and the microsomal ethanol oxidizing system (MEOS).

(o) Synthesis of **creatine** (from guanidinoacetate), which is used to produce creatine phosphate, mainly in muscle and brain.

(p) Conversion of **dietary fructose** to glycolytic intermediates.

(2) If liver cell function is compromised (e.g., in viral hepatitis or alcoholic cirrhosis):

(a) NH_4^+, which is **toxic** (particularly to the central nervous system), **increases** in the blood.

(b) The blood urea nitrogen (**BUN**) level **decreases** because the liver has a decreased capacity to produce urea.

(c) Blood glucose decreases because of decreased glycogenolysis and gluconeogenesis.

(d) Blood **cholesterol** levels **decrease** due to an inability to produce and secrete VLDL.

(e) The production of **bile salts decreases.**

(f) Bilirubin levels **increase** in the body (causing jaundice) owing to reduced conjugation with glucuronic acid.

table 9.6 Regulation of Liver Enzymes Involved in Glycogen, Blood Glucose, and Triacylglycerol Synthesis and Degradation

Liver Enzymes Regulated by Activation/Inhibition

Enzyme	Activated by	State in Which Active
Phosphofructokinase-1	Fructose 2,6-bisP, AMP	Fed
Pyruvate carboxylase	Acetyl-CoA	Fed and fasting
Acetyl-CoA carboxylase	Citrate	Fed
Carnitine:palmitoyltransferase I	Loss of inhibitor (malonyl-CoA)	Fasting

Liver Enzymes Regulated by Phosphorylation/Dephosphorylation

Enzyme	Active Form	State in Which Active
Glycogen synthase	Dephosphorylated	Fed
Phosphorylase kinase	Phosphorylated	Fasting
Glycogen phosphorylase	Phosphorylated	Fasting
Phosphofructokinase-2/fructose 2,6-bisphosphatase (acts as a kinase, increasing fructose 2,6-bisP levels)	Dephosphorylated	Fed
Phosphofructokinase-2/fructose 2,6-bisphosphatase (acts as a phosphatase, decreasing fructose 2,6-bisP levels)	Phosphorylated	Fasting
Pyruvate kinase	Dephosphorylated	Fed
Pyruvate dehydrogenase	Dephosphorylated	Fed
Acetyl-CoA carboxylase	Dephosphorylated	Fed

Liver Enzymes Regulated by Induction/Repression

Enzyme	State in Which Induced	Process Affected
Glucokinase	Fed	Glucose → TG
Citrate lyase	Fed	Glucose → TG
Acetyl-CoA carboxylase	Fed	Glucose → TG
Fatty acid synthase	Fed	Glucose → TG
Malic enzyme	Fed	Production of NADPH
Glucose-6-phosphate dehydrogenase	Fed	Production of NADPH
Glucose-6-phosphatase	Fasted	Production of blood glucose
Fructose 1,6-bisphosphatase	Fasted	Production of blood glucose
Phosphoenolpyruvate carboxykinase	Fasted	Production of blood glucose

AMP, adenosine monophosphate; fructose 2,6-bisP, fructose 2,6-bisphosphate; TG, triacylglycerol.

(g) Lysis of damaged liver cells allows **enzymes** to **leak** into the blood.

1. Lactate dehydrogenase (**LDH**) increases.
2. Alanine aminotransferase (**ALT**) increases.
3. Aspartate aminotransferase (**AST**) increases.
4. **Alkaline phosphatase** increases.

(h) **Chronic** liver problems result in a **decreased protein synthesis.**

1. **Serum proteins** (e.g., albumin) decrease.
2. **VLDL** production decreases because of decreased apoprotein B-100, and triacylglycerols accumulate in the liver. A fatty liver results.

CLINICAL CORRELATES There are a large number of specific diseases that affect the liver, and a number (not all) of them are summarized here. **Glycogen storage diseases** include **Von Gierke,** due to a glucose-6-phosphatase deficiency. Blood glucose cannot be produced by glycogenolysis or gluconeogenesis, resulting in severe hypoglycemia during fasting. **Hers disease** is due to a liver phosphorylase deficiency. Glycogen cannot be converted to blood glucose, and blood glucose is maintained only by gluconeogenesis. A third glycogen storage disease is **Pompe**, which is due to a lysosomal α-glucosidase deficiency. Glycogen accumulates in membrane-enclosed vesicles (residual bodies) and interferes with liver function. The metabolism of alcohol occurs within the liver, and **alcoholism** is a major disorder. The oxidation of ethanol **produces NADH** within the liver (see Fig. 7.24), and ingesting alcohol with food intake over an extended period of time results in a high ratio of NADH to NAD^+, which can cause increased conversion of pyruvate to lactate, producing a **lactic acidosis**; an inhibition of gluconeogenesis, leading to **hypoglycemia**; increased levels of glycerol-3-phosphate, which combines with fatty acids from adipose triacylglycerols to form VLDL, which can lead to a **hyperlipidemia**. Eventually, with chronic alcoholism, **protein synthesis decreases** in the liver, leading to reduced VLDL secretion, and the formation of a **fatty liver** (accumulation of triacylglycerol in the liver). **DM** also affects the liver, as low insulin levels (Type 1) or insensitivity to insulin (Type 2) results in increased glycogenolysis and gluconeogenesis, which contribute to the **elevated blood glucose levels**. Increased ketone body production can lead to **diabetic ketoacidosis** (DKA), particularly in Type 1 DM. Ketone body synthesis increases because of increased release of fatty acids from adipose triacylglycerols, and an accumulation in the blood since the brain is still using glucose as an energy source. The liver is also the site of **fructose intolerance**, which is due to a **deficiency of aldolase B**. This results in elevated fructose 1-phosphate and low inorganic phosphate (P_i) levels after fructose ingestion. (Glycolysis is not affected even though the same aldolase is involved.) Glycogenolysis and gluconeogenesis are inhibited, and **hypoglycemia** results.

g. Brain

(1) **Glucose** is the major fuel for the brain.

(2) The brain can use **ketone bodies**, but only **after 3 to 5 days of fasting when blood ketone body levels are elevated**.

(3) The brain needs energy to **think** (i.e., memory involves RNA synthesis), conduct **nerve impulses**, synthesize **neurotransmitters**, etc.

CLINICAL CORRELATES Interruptions in the brain's metabolism can have deleterious effects. Abrupt decreases in blood glucose levels can result in **coma** from **lack of ATP**. Highly elevated blood glucose levels can cause a **hyperosmolar coma**. The synthesis of some neurotransmitters (e.g., GABA, serotonin, dopamine) involves decarboxylation of amino acids and requires pyridoxal phosphate (from vitamin B_6). **Deficiencies of vitamin B_6** can lead to **convulsions**.

h. Red blood cells

(1) Red blood cells **lack mitochondria**, so they have no TCA cycle, β-oxidation of fatty acids, electron transport chain, and other pathways that occur in mitochondria.

(2) **Glucose** is the **major fuel** for red blood cells.

(a) Glucose is converted to pyruvate and lactate.

(3) Red blood cells carry bases and nucleosides from the liver to other tissues.

(4) The major function of red blood cells is to **carry O_2** from the lungs to the tissues and to aid in the **return** of **CO_2** from the tissues to the lungs.

CLINICAL CORRELATES Even though the red blood cells are metabolically simple cells, and have few functions, those functions are essential, and many diseases have been associated with the red blood cells. The first of these are the h**emoglobinopathies**. **Sickle cell anemia** occurs because valine replaces glutamate at position 6 in the ß-globin chain. This hydrophobic amino acid permits interactions to occur between globin chains of different hemoglobin

(continued)

molecules. Cells become misshapen ("sickled") and can block the blood flow through capillaries. Painful vaso-occlusive crises can occur. The misshapen cells are phagocytosed by the spleen, and anemia results. **Thalassemias** are caused by deficiencies of α- or ß-globin chains (an imbalance in the synthesis of the chains), which result from many different types of mutations in genes. Decreased hemoglobin levels result in anemia. **Iron deficiency anemia** can result from red blood cell precursors not producing normal amounts of heme. Cells are small (microcytic) and pale (hypochromic). **Vitamin B_6 deficiency** also gives rise to an anemia. Vitamin B_6 is required for the first reaction (glycine + succinyl-CoA $\rightarrow$ δ-ALA) in heme biosynthesis in red blood cell precursors. Decreased heme production leads to decreased hemoglobin levels in mature red blood cells and a microcytic, hypochromic anemia. **Megaloblastic anemia** results from either a folate or B_{12} deficiency. **Folate** is required for the production of thymine (by the conversion of dUMP to dTMP) and for the production of the purine bases (carbon 2 and carbon 8 come from the tetrahydrofolate 1-carbon pool). In a **folate deficiency**, the precursors of red blood cells cannot divide because of a lack of thymine and of purines for DNA synthesis. Megaloblasts form, and a **megaloblastic anemia** results. In a **vitamin B_{12} deficiency**, folate accumulates as methyltetrahydrofolate because the methyl group cannot be transferred to B_{12} (the methyl trap theory). Thus, there is decreased thymine for DNA synthesis and decreased purines for DNA and RNA synthesis. Decreased cell growth and division occur, and red blood cell precursors form **megaloblasts**. In a **B_{12} deficiency** (but **not in a folate deficiency**), decreased conversion of methylmalonyl-CoA to succinyl-CoA occurs, resulting in **neurologic problems**. **A deficiency of folate or vitamin B_{12}** results in **decreased** conversion of **homocysteine to methionine**. Unless dietary methionine is increased, low levels of methionine result in **decreased** synthesis of *S*-adenosylmethionine (**SAM**), and, thus, decreased conversion of norepinephrine to epinephrine, phosphatidyl ethanolamine to phosphatidylcholine, guanidinoacetate to creatine, etc. **Deficiency of pyruvate kinase**, a key glycolytic enzyme, leads to decreased ATP production from glycolysis. Red blood cells (which depend on glycolysis for energy) cannot maintain their membranes. They crenate (shrink) and are phagocytosed, and an anemia results. However, some patients can be asymptomatic due to the accumulation of 2,3-bisphosphoglycerate within the erythrocyte. **Glucose-6-phosphate dehydrogenase deficiency**, under appropriate conditions, can lead to a **hemolytic anemia**. Glucose-6-phosphate dehydrogenase, the first enzyme of the pentose phosphate pathway, produces **NADPH**. In **glucose-6-phosphate dehydrogenase deficiency**, NADPH production is decreased. Therefore, **glutathionine** is not reduced at a normal rate. Problems arise when the demand for NADPH is high (e.g., when drugs that require NADPH for their metabolism are used, such as primaquine for malaria). As hemoglobin (Fe^{2+}) is slowly oxidized to methemoglobin (Fe^{3+}), superoxide (O_2^-) is produced. Superoxide dismutase converts superoxide to hydrogen peroxide (H_2O_2) and O_2. Glutathionine peroxidase converts H_2O_2 to H_2O, producing oxidized glutathionine (GS-SG), which must be converted by NADPH to reduced glutathione (GSH). If reduction does not occur rapidly enough, oxidative damage causes red blood cells to lyse and a **hemolytic anemia** results. **Hemolytic anemias** cause **bilirubin** and **lactate dehydrogenase** to be released into the blood. If bilirubin is produced faster than the liver can conjugate and excrete it, **jaundice** can occur. **Stercobilin**, a pigment produced from bilirubin by intestinal bacteria, increases in the feces, causing the stool to turn dark brown.

i. Adipose tissue

(1) The **major fuel** of adipose tissue is **glucose**.

(2) **Insulin** stimulates the **transport of glucose** into adipose cells.

(3) The function of adipose tissue is to **store triacylglycerol** in the fed state and release it (via **lipolysis**) during fasting.

(a) **In the fed state**, insulin stimulates the synthesis and secretion of lipoprotein lipase (LPL), which degrades the triacylglycerols of chylomicrons and VLDL in the capillaries. Fatty acids from these lipoproteins enter adipose cells and are converted to triacylglycerols and stored. Glucose provides the glycerol moiety. (Glycerol is not used, because adipose cells lack glycerol kinase.)

(b) **During fasting**, a hormone-sensitive lipase (phosphorylated and activated via a cAMP-mediated mechanism) initiates lipolysis in adipose cells.

CLINICAL CORRELATES **In DM**, low insulin levels (Type 1) or insulin resistance (Type 2) result in decreased degradation of the triacylglycerols of chylomicrons and VLDL (because of decreased LPL). **Hypertriglyceridemia** results. In addition, lipolysis occurs, causing excessive amounts of fatty acids and glycerol to reach the liver, where they produce VLDL, thus adding to the hypertriglyceridemia. In addition, the **uptake of glucose** by adipose tissue is low because of decreased insulin (Type 1) or insulin resistance (Type 2). This decreased uptake of glucose, coupled with decreased uptake of glucose by the muscle, contributes to the elevated blood glucose levels found in the diabetic condition.

j. **Muscle**
   (1) Muscle uses **all fuels** that are available (glycogen stores, and fatty acids, glucose, ketone bodies, lactate, and amino acids from the blood) to obtain energy for contraction.
   (2) During fasting, muscle protein is degraded to provide amino acids (particularly alanine) for **gluconeogenesis**.
   (3) **Creatine phosphate** transports high-energy phosphate from the mitochondria to actinomyosin fibers and provides ATP for muscle contraction.
   (4) **Creatinine** is produced nonenzymatically from creatine phosphate, and a **constant amount** (dependent on the body muscle mass) is released into the blood each day and excreted by the kidneys.
   (5) **Muscle glycogen phosphorylase** differs from liver phosphorylase but catalyzes the same reaction (glycogen + $P_i \leftrightarrow$ glucose-1-phosphate).
   (6) **Insulin** stimulates the **transport of glucose** into muscle cells.

CLINICAL CORRELATES **Muscle damage** results in the release of **creatine kinase** (CK) into the blood, the enzyme that catalyzes the reversible reaction–creatine + ATP ↔ creatine phosphate. As indicated below, the release of the heart isozyme of CK is confirmation that a heart attack has taken place. **McArdle disease** is a glycogen storage disease in which muscle phosphorylase is deficient. Muscle glycogen cannot be degraded at a normal rate. Glycogen stores increase, and lactate is not produced during exercise. Vigorous exercise quickly produces fatigue; however, mild exercise, using fatty acids and glucose from the blood, can be tolerated. In **Type 1 and Type 2 DM**, the uptake of glucose by muscle is low, and is one of the factors that contribute to hyperglycemia (along with reduced glucose into adipose tissue).

k. **Heart**
   (1) The heart is a specialized muscle that uses **all fuels** from the blood.
   (2) The muscle–brain **(MB) isozyme** of CK is found in heart muscle. Its release can be used to monitor a heart attack.

CLINICAL CORRELATES **Heart disease** can result from a variety of insults. **Atherosclerotic plaques** can occlude blood vessels, blocking the flow of nutrients and O_2. Muscle tissue beyond the block suffers from a lack of energy and can die. When the amount of functional cardiac muscle tissue that remains is insufficient to pump blood through the body at a normal rate, **heart failure** occurs. The damaged cells release the **MB isozyme** of CK into the blood. **High blood cholesterol** levels are associated with increased risk of a heart attack (or a stroke, which is caused by a similar process in the brain). Cholesterol is carried in the blood lipoproteins, and is elevated in a group of conditions known as the **hyperlipidemias**.

Type I: chylomicrons are elevated; thus, triacylglycerols are high.
Type II: LDL receptors are defective; cholesterol is high.

(continued)

Type III: a spectrum of partial degradation products of VLDL appear in the blood (a broad β band).
Type IV: VLDL is elevated; thus, triacylglycerols are elevated. Type IV is often associated with DM.
Type V: chylomicrons and VLDL are elevated.

The treatments for high cholesterol include **a diet low in fat** (particularly saturated fat), **decreased dietary carbohydrate** (to lower VLDL), and **HMG-CoA reductase inhibitors** (statins) to decrease blood cholesterol levels (by decreasing cholesterol synthesis). Decreased cellular levels of cholesterol increase the synthesis of LDL receptors. Thus, more LDL (composed of almost 50% cholesterol and cholesterol esters) is taken up from the blood. **Bile salt sequestrants** are also used to reduce elevated cholesterol levels by increasing the excretion of bile salts, a major means by which the cholesterol ring structure is removed from the body. **Niacin** can also be taken in pharmacological doses to decrease lipolysis in adipose tissue and VLDL synthesis in the liver.

I. Kidney

(1) The kidney **excretes substances** from the body via the urine, including **urea** (produced by the urea cycle in the liver), **uric acid** (from purine degradation), **creatinine** (from creatine phosphate), **NH_4^+** (from glutamine via glutaminase), **H_2SO_4** (produced from the sulfur of cysteine and methionine), and **phosphoric acid.**

(2) Daily **creatinine** excretion is **constant** and depends on the body muscle mass. It is used as a measure of kidney function (the creatinine-clearance rate).

(3) **Glutaminase** action increases during **acidosis** and produces NH_3, which enters the urine and reacts with H^+ to form NH_4^+. NH_4^+ buffers the urine and removes acid (H^+) from the body.

(4) **Uric acid** excretion is inhibited by lead (Pb) and metabolic acids (ketone bodies and lactic acid). High blood uric acid can result in **gout.** Gout can be caused either by increased production or by decreased excretion of uric acid. Deficiency of the base salvage enzyme hypoxanthine guanine phosphoribosyltransferase (HGPRT) in **Lesch–Nyhan syndrome** results in increased production of uric acid.

(5) **Kidney dysfunction** can lead to increased BUN, creatinine, and uric acid in the blood, and decreased levels of these compounds in the urine.

(6) During ketoacidosis, **ketone bodies** are excreted by the kidney, and during lactic acidosis, **lactic acid** is excreted.

(7) Elevated blood glucose levels (over 180 mg/dL) in DM results in the **excretion of glucose** in the urine.

CLINICAL CORRELATES

Enzyme deficiency diseases that affect various tissues of the body often result in the **excretion** of substances in the **urine**. Examples of these types of diseases include p**henylketonuria (PKU)**, a defect in either phenylalanine hydroxylase or tetrahydrobiopterin production. Decreased conversion of phenylalanine to tyrosine results in the appearance of phenylalanine and its degradation products (e.g., phenylketones) in the urine. **Maple syrup urine disease (MSUD)** is due to an α-keto acid dehydrogenase deficiency. The branched-chain amino acids (valine, isoleucine, and leucine) are excreted. **Alcaptonuria** is due to homogentisic acid oxidase deficiency. The oxidized products of homogentisic acid give urine a dark color. The condition is oftentimes associated with arthritis. **Cystinuria** is due to a deficiency of the intestinal and kidney transport protein for cystine. Cystine is not resorbed and accumulates in the urine, forming kidney stones. **Homocystinuria** will result from deficiencies of either methionine synthase (homocysteine to methionine), cystathionine synthase (homocysteine + serine → cystathionine), or of the cofactors for these enzymes (folate and vitamin B_{12} for methionine synthase, and vitamin B_6 for cystathionine synthase). Homocysteine accumulates, is oxidized to homocystine, and excreted. Homocystinemia (high levels of homocysteine in the blood) is associated with atherosclerotic vascular disease. **Cystathionuria** is due to a deficiency of cystathionase. Cystathionine is excreted. High doses of vitamin B_6, the precursor of the cofactor (pyridoxal phosphate) for the enzyme, are sometimes beneficial. **Benign fructosuria** is due to a deficiency of fructokinase. Fructose is not phosphorylated, and thus it accumulates and is excreted. There are no deleterious consequences. **Galactosemia** can be due to deficiencies of either galactokinase or galactose-1-phosphate uridyltransferase. In a galactokinase deficiency, galactose accumulates, causes cataracts, and spills into the urine. In the uridyltransferase deficiency (classic galactosemia), hypoglycemia, jaundice, and other problems result, in addition to cataracts and the appearance of galactose in the urine. **MCAD deficiency** is due to a deficiency of the medium-chain fatty acyl-CoA dehydrogenase. Fatty acids cannot be completely oxidized. Consequently, more glucose must be oxidized for energy, resulting in hypoglycemia. Increased ω-oxidation of fatty acids produces dicarboxylic acids that are excreted in the urine.

Review Test

Questions 1 to 10 examine your basic knowledge of basic biochemistry and are not in the standard clinical vignette format.

Questions 11 to 35 are clinically relevant, USMLE-style questions.

Basic Knowledge Questions

Questions 1 through 6 are matching questions. For each numbered question, select the one-lettered option that is most closely associated with it. An answer may be used once, more than once, or not at all.

1. Has its release inhibited by thyroxine.
2. Binds to the receptors on Leydig cells.
3. Stimulates the production of insulin-like growth factor (IGF).
4. Stimulates the synthesis of milk proteins.
5. Stimulates the production of progesterone by the corpus luteum.
6. Stimulates the production of estradiol by the immature ovarian follicle.

(A) Luteinizing hormone (LH)
(B) Prolactin (PRL)
(C) Thyroid-stimulating hormone (TSH)
(D) Growth hormone (GH)
(E) Follicle-stimulating hormone (FSH)

Questions 7 through 10 are matching questions. For each description of a patient given below, choose the most likely cause from the lettered list. An answer may be used once, more than once, or not at all.

7. A large, protruding jaw, large hands and feet, normal height, and an elevated blood glucose level.
8. Thin limbs, central obesity, fat cheeks, a ruddy complexion, and an elevated blood glucose level.
9. Hypertension and heart disease.
10. Galactorrhea, amenorrhea, and blurred vision.

(A) Elevated blood levels of aldosterone and renin resulting from an atherosclerotic plaque in a renal artery.
(B) Hyperprolactinemia due to a pituitary tumor.
(C) Acromegaly due to a growth hormone (GH)-producing tumor that developed in adulthood.
(D) Cushing syndrome due to an adrenal tumor.

Board-style Questions

11. A 35-year-old female presents with central obesity, a "buffalo hump," Type 2 diabetes, abdominal striae, and "moon facies." The hormone being overproduced in this syndrome is produced from which one of the following compounds?

(A) Progesterone
(B) Testosterone
(C) Estradiol
(D) Aldosterone
(E) Chenocholic acid

12. Patients with both Graves disease and Cushing syndrome are overproducing hormones that have which one of the following in common?

(A) Reacting with receptors in the cell membrane
(B) Utilizing second messengers
(C) Binding to intracellular receptors
(D) Binding to RNA to produce physiologically active proteins
(E) Inducing rRNA to ablate a particular genes expression

13. A 32-year-old female presents with 3 months of irregular menstrual periods and a milky discharge from both breasts. Under the microscope, the discharge has many fat globules, but no red blood cells. She is on no medications, has had a bilateral tubal ligation, and her pregnancy test is negative. Which ONE of the following would inhibit the release of the hormone that is being overproduced in this patient?

(A) GnRH
(B) Dopamine
(C) TRH
(D) Somatostatin
(E) Estradiol

14. A 20-year-old male presents with weight loss, heat intolerance, bilateral exophthalmos, a lid lag, sweating, and tachycardia. These symptoms are due to an increased production and secretion of a hormone that is derived from which one of the following?

(A) Cholesterol
(B) Dopamine
(C) Tryptophan
(D) Tyrosine
(E) Glutamate

15. A 60-year-old female has been diagnosed with temporal arteritis and is being treated with oral prednisone. Which one of the following mechanisms is the basis for this treatment?

(A) Causing the lysis of lymphocytes
(B) Inducing the synthesis of PEPCK
(C) Suppressing the immune response
(D) Inducing the synthesis of lipocortin
(E) Inhibiting oxidative phosphorylation in immune cells

16. A 68-year-old patient developed atrophic gastritis, and 2 years later developed a macrocytic, hyperchromic anemia. His anemia has most likely occurred due to which one of the following reasons?

(A) The atrophic gastritis leads to vitamin B_{12} malabsorption.
(B) The atrophic gastritis raises the pH in the duodenum, leading to folate malabsorption.
(C) The terminal ileum is also involved, so iron is malabsorbed.
(D) The atrophic gastritis leads to increased red blood cell absorption by the spleen.
(E) Heme synthesis is inhibited under these conditions.

Questions 17 through 19 are based on the following case presentation:

A patient complains of nervousness, palpitations, sweating, and weight loss without loss of appetite and has a goiter. Suspecting a defect in thyroid function, the physician orders a total serum T_4. The test is performed by radioimmunoassay. The standard curve for the assay, which measures T_4 in 0.1 mL of serum, is shown below. The normal levels of T_4 are 4 to 10 μg/dL. In an assay of 0.1 mL of the patient's serum, 15% of the radioactive T_4 was bound to the antibody.

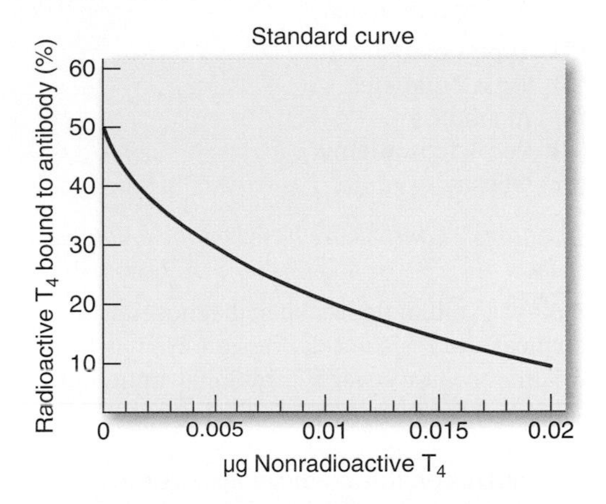

17. According to the radioimmunoassay, the approximate blood levels of T_4 are which one of the following?

(A) 0.015 μg/dL
(B) 0.15 μg/dL
(C) 15 μg/dL
(D) 20 μg/dL
(E) 30 μg/dL

18. The patient is best classified as which one of the following?

(A) Hypothyroid
(B) Normal
(C) Hyperthyroid

19. The TSH levels in the patient's blood were also measured by radioimmunoassay. If pituitary function is normal, the patient's TSH levels will most likely be which one of the following?

(A) Higher than normal
(B) Normal
(C) Lower than normal

20. A 23-year-old female overdosed on acetaminophen in a suicide attempt. She was brought to the emergency room (ER) and given a chemical as an emergency treatment. This treatment is designed to increase the synthesis of which one of the following?

(A) NAPQI (*N*-acetyl-p-benzoquinone imine)
(B) Glutathione
(C) Glycine
(D) Alanine
(E) NAPDH

21. Which one of the following diseases or conditions predisposes a patient to acetaminophen toxicity?

(A) Type 1 diabetes
(B) Type 2 diabetes
(C) Alcoholism
(D) Pernicious anemia
(E) Obesity

Questions 22 and 23 are based on the following case:

A 32-year-old male has been diagnosed with an atypical and resistant depression. He has failed treatment with several traditional antidepressants and has been placed on selegiline.

22. Patients on this medication can have a hypertensive crisis if they consume foods containing which one of the following?

(A) Tryptophan
(B) Tyramine
(C) Alanine
(D) Glycine
(E) Threonine

23. Which one of the following foods should the above-mentioned patient avoid?

(A) Milk
(B) Fresh fish
(C) Ice cream
(D) Aged cheeses
(E) Romaine lettuce

24. A 43-year-old male high-level executive presents to his physician due to a persistant dizziness when standing quickly. He also has been exhibiting chronic fatigue, some muscle weakness, and an unusual craving for salty foods. The physician notices that the pateint has a constant "bronze tan," and blood obtained during the visit demonstrated normal cholesterol levels but also hypoglycemia, hyperkalemia, and hyponatremia. The underlying cause for these problems is which one of the following?

(A) Adrenal insufficiency
(B) Pituitary insufficiency
(C) Lack of insulin
(D) Lack of glucagon
(E) Lack of GH

25. A 30-year-old female presents with a need to progressively buy larger and wider shoes. She also cannot wear any of her rings anymore because they are too small. A physical exam shows a prominent brow, protruding lower jaw, and spaces between all of her teeth. This woman may have a tumor in which one of the following organs/tissues?

(A) Hypothalamus
(B) Bone marrow
(C) Adrenal glands
(D) Pancreas
(E) Adipose tissue

26. A 45-year-old female has had three vertebral fractures and received a diagnosis of "early" osteoporosis. She has also had a number of kidney stones. Routine labs revealed hypercalcemia. Increased secretion of a hormone from which gland/organ may lead to this condition?

(A) Thyroid
(B) Adrenal
(C) Hypothalamus
(D) Pancreas
(E) Parathyroid

27. A patient with central obesity, thin limbs, and purple striae on the abdomen complained of muscle weakness, depression, and blurred vision. The patient's blood glucose level was 280 mg/dL (the reference range is 80 to 100 mg/dL). No ketone bodies were present in the urine. Plasma cortisol levels were 56 μg/mL (the reference range is 3 to 31 μg/mL), and plasma ACTH levels were 106 pg/mL (the reference range is 0 to 100 pg/mL). A low dose (1 mg) of dexamethasone (a synthetic glucocorticoid) was administered in the evening. This dose failed to suppress the plasma cortisol level by the next morning. After a high-dose (8 mg) dexamethasone suppression test, the plasma cortisol level was 21 μg/mL. On the basis of this information, if the patient's problem is due to a single cause, the most likely diagnosis is which one of the following?

(A) Type 2 DM
(B) Type 1 DM
(C) A secretory tumor of the anterior pituitary
(D) A secretory tumor of the posterior pituitary
(E) A secretory tumor of the adrenal cortex

28. A 24-year-old accountant complains of a white discharge from his breasts. He is most likely experiencing which one of the following?

(A) A tumor of the posterior pituitary that could be surgically removed.
(B) Excessive production of OT in the hypothalamus.

(C) Deficient testosterone receptors in the mammary glands.
(D) A prolactinoma that would decrease its secretory activity in response to bromocriptine (an analog of dopamine).
(E) A glucagonoma.

29. A female patient presents to her primary care doctor with thinning scalp hair, excessive facial hair, and high ACTH and low cortisol levels in her blood. The physician should seriously consider which one of the following problems?

(A) A tumor of the anterior pituitary that produces abnormally large amounts of POMC.
(B) A tumor of the adrenal medulla that secretes abnormally large amounts of its normal hormone product.
(C) A genetic deficiency of an enzyme in the pathway for cortisol synthesis.
(D) A mutation in the gene for CRH that decreases the production of this polypeptide.
(E) An insulinoma.

30. Following an automobile accident, a 14-year-old girl developed diabetes insipidus (excessive loss of water in the urine), amenorrhea (lack of menses), and cold intolerance, and her growth rate slowed markedly. Her symptoms best correspond to which one of the following?

(A) They are all related to damage to her kidneys and ovaries caused by the accident.
(B) They are all related to damage to her pituitary gland.
(C) The symptoms could be relieved by treatment with oral doses of anterior and posterior pituitary hormones.
(D) The symptoms would require long-term treatment with estrogen, thyroxine, and glucocorticoid tablets and also restriction of water intake.
(E) Damage to the pancreas caused by airbag deployment.

31. A 69-year-old man has been feeling lethargic, and at his yearly physical exam had his testosterone levels determined. The blood work came back with a value of 168 ng/dL (the normal range is 250 to 800 ng/dL, depending on age), so the patient agreed with his physician's suggestion of biweekly testosterone injections. Which one of the following best represents some properties of testosterone in this patient?

	Precursor to a more active androgen	**Receptor location:**	**Estradiol is a precursor**	**Represses glucocorticoid synthesis**	**Stimulates GnRH release**
A.	No	Cytoplasm	Yes	Yes	No
B.	No	Nucleus	No	Yes	Yes
C.	No	Cytoplasm	Yes	Yes	No
D.	Yes	Nucleus	No	No	Yes
E.	Yes	Cytoplasm	Yes	No	Yes
F.	Yes	Nucleus	No	No	No

32. A patient of Mediterranean ancestry was given primaquine to protect against malaria when going on an overseas trip. The patient rapidly developed a hemolytic anemia due to a mostly silent mutation in which one of the following pathways or enzymes?

(A) Malic enzyme
(B) Glycolysis
(C) Hexose monophosphate shunt
(D) Gluconeogenesis
(E) Fatty acid oxidation

33. A vitamin B_{12} deficiency may affect heme synthesis by reducing the concentration of which of the following? Choose the one best answer.

(A) Acetyl-CoA
(B) Succinyl-CoA
(C) Glycine
(D) Alanine
(E) Succinate

34. An individual has a mutation such that GnRH release is greatly reduced. This mutation would have its greatest effects on the release of which one of the following?

(A) GH
(B) T_3 and T_4
(C) PRL
(D) IGF
(E) LH and FSH

35. A teenager in Australia has developed a goiter, due to a dietary deficiency. Which one of the following would be expected in the case of this dietary deficiency? Choose the one best answer.

	Salt balance in the blood is altered	**TSH secretion is:**	**TRH secretion is:**	**Heat production is:**
A.	No	Increased	Increased	Increased
B.	Yes	Increased	Decreased	Increased
C.	No	Increased	Increased	Decreased
D.	Yes	Decreased	Decreased	Decreased
E.	No	Decreased	Increased	Decreased
F.	Yes	Decreased	Decreased	Increased

Answers and Explanations

1. **The answer is C.** Thyroxine (the end product) inhibits the release of TSH by the anterior pituitary. Thyroxine has no effect on the release of LH, PRL, GH, or FSH.
2. **The answer is A.** LH binds to the receptors on Leydig cells and stimulates the release of testosterone. PRL, TSH, GH, and FSH do not bind to Leydig cells.
3. **The answer is D.** GH stimulates the release of IGF by the liver and other tissues. The other hormones listed do not stimulate IGF release from any tissue.
4. **The answer is B.** PRL is the only hormone listed that stimulates the synthesis of milk proteins. PRL will induce α-lactalbumin, which will stimulate lactose production.
5. **The answer is A.** LH stimulates the corpus luteum to produce progesterone.
6. **The answer is E.** FSH stimulates the maturation of the ovarian follicle, which produces estradiol.
7. **The answer is C.** In acromegaly, the height is normal, but the jaw, hands, feet, and soft tissues grow. DM frequently occurs because of the stimulation of gluconeogenesis by GH and the inability of glucose to suppress the release of GH by the tumor.
8. **The answer is D.** Excess production of cortisol causes muscle breakdown (hence thin limbs), fat deposition in the face and abdomen, thin skin (due to protein breakdown), and an increase in red blood cells resulting in a red complexion. Blood glucose levels are elevated because cortisol stimulates gluconeogenesis. These symptoms do not appear due to elevated renin, aldosterone, PRL, or GH levels.
9. **The answer is A.** The plaque in the renal artery causes the release of renin, which elevates aldosterone, thus raising blood pressure. Atherosclerotic plaques are also probably the cause of the heart disease. PRL (milk production), GH, or cortisol release will not lead to hypertension and heart disease.
10. **The answer is B.** A tumor of the anterior pituitary that produces PRL could cause milk production (galactorrhea) and inhibit LH and FSH production, causing amenorrhea. If the tumor grew over the optic nerves, blurred vision would result. Aldosterone/renin, GH, or cortisol overproduction would not lead to these symptoms.
11. **The answer is A.** The patient has Cushing syndrome, and is overproducing the glucocorticoid cortisol. Cortisol, testosterone, and aldosterone are all produced from progesterone. Estradiol is produced from testosterone. Chenocholic acid is one of the bile salts, and is derived from cholesterol.
12. **The answer is C.** Graves disease is due to the hypersecretion of thyroid hormone, whereas Cushing syndrome is an overproduction of cortisol. Steroid hormones and thyroid hormones cross the cell membrane and bind to intracellular receptors. The hormone–receptor complex binds to DNA, not RNA. Polypeptide hormones and epinephrine react with the receptors in the cell membrane triggering second messengers to transmit the signal that the receptor is occupied. Ribosomal RNA (rRNA) does not ablate gene expression; the induction of micro RNA has that ability.
13. **The answer is B.** The patient has a prolactin-releasing microadenoma of the anterior pituitary, causing altered menstrual periods and galactorrhea. Dopamine from the hypothalamus inhibits PRL release. GnRH inhibits LH and FSH (which could affect her menstrual periods, but should not give galactorrhea). TRH stimulates TSH, which stimulates the production of T_3 and

T_4. Somatostatin inhibits the release of GH and can be used in the treatment of insulinomas. Estradiol does not affect the secretion of PRL as dopamine does.

14. **The answer is D.** The patient has Graves disease hyperthyroidism, an overproduction of thyroid hormone, which is derived from tyrosine. Hyperthyroidism increases the rate of oxidation of fuels by muscle and other tissues, increasing heat production, and causes a sense of heat intolerance and increased sweating. The heart rate and blood pressure are also increased, as is weight loss in spite of a healthy appetite. Dopamine (a catecholamine) is also derived from tyrosine and can be hydroxylated to norepinephrine, which can then be methylated to epinephrine, but the catecholamines cannot be transformed to thyroxine. Tryptophan can be metabolized into serotonin and melatonin. Cholesterol is the basis of the steroid hormones progesterone, testosterone, estradiol, cortisol, and aldosterone. Glutamate gives rise to GABA via a decarboxylation reaction.

15. **The answer is D.** Prednisone is a synthetic glucocorticoid, and it is being given as an anti-inflammatory agent in this case. Glucocorticoids exert their anti-inflammatory effects by stimulating lipocortin, a protein that inhibits phopholipase A_2, the rate-limiting enzyme in prostaglandin, thromboxane, and leukotriene synthesis. Phospholipase A_2 removes the initiating fatty acid (usually arachidonic acid) from a phospholipid, leaving a lysophospholipid behind. Glucocorticoids also suppress the immune response by causing the lysis of lymphocytes. This would be useful in treating something like Lupus, but is not classified as an anti-inflammatory effect. Glucocorticoids promote gluconeogenesis by inducing the synthesis of PEPCK (which is not an anti-inflammatory effect). Glucocorticoids do not affect oxidative phosphorylation, although they can alter the regulation of glycolysis in certain cell types.

16. **The answer is A.** This patient has trouble absorbing vitamin B_{12} for two reasons. The first is that the stomach no longer produces adequate levels of intrinsic factor. Atrophic gastritis leads to the loss of the glandular cells of the stomach, and replacement of those cells with intestinal and fibrous tissues. This will lead to a loss of function of the stomach cells, and less intrinsic factor will be produced and secreted. Additionally, B_{12} bound to proteins in the diet will have difficulty in being removed from their carrier proteins due to the lack of stomach acid. After the body's stores of B_{12} are exhausted (about 2 years), the anemia will develop due to the inability of reticulocytes to grow and divide. The loss of acid would fail to stimulate pancreatic bicarbonate, so the pH would stay the same in the duodenum and would not increase, such that folate absorption continues normally. The terminal ileum is not involved in atrophic gastritis, but can be involved in Crohn disease. Iron deficiency would give a microcytic, hypochromic anemia, not the observed anemia. Heme synthesis is not impaired under these conditions.

17. **The answer is C.** If 15% of the radioactive T_4 is bound to antibody, the amount of T_4 in 0.1 mL of the patient's serum is 0.015 μg/0.1 mL or 15 μg/dL. (1 dL = 100 mL).

18. **The answer is C.** The patient's T_4 level is above the normal range, which indicates that the patient is hyperthyroid.

19. **The answer is C.** Thyroid hormone suppresses TSH secretion by the anterior pituitary. If thyroid hormone levels are elevated, TSH levels will be lower than normal. This patient probably has Graves disease, in which thyroid-stimulating antibodies promote T_3 and T_4 production by the thyroid gland. These thyroid hormones inhibit the release of TSH from the anterior pituitary.

20. **The answer is B.** Acetaminophen is detoxified through a sulfotransferase (which sulfates acetaminophen for excretion) route or a glucuronyltransferase route (which adds glucuronate to acetaminophen for excretion), as well as a P450 enzyme that produces toxic NAPQI. NAPQI is detoxified by conjugation with glutathione, which then allows the conjugate to be excreted in the urine. High doses of NAPQI overwhelm the normal cellular levels of glutathione. The chemical given is *N*-acetylcysteine, which will boost glutathione levels to conjugate with NAPQI in order to detoxify the compound. Increasing the level of NAPQI would be counterproductive, and increasing levels of amino acids or NADPH will not help to reduce NAPQI levels to subtoxic amounts.

21. **The answer is C.** The enzyme that produces the toxic NAPQI, CYP2E1, is induced by alcohol. Diabetes (either type 1 or type 2) will not lead to an induction of CYP2E1, nor will pernicious anemia (intrinsic factor deficiency) or obesity (in the absence of alcoholism). Chronic alcoholics have a reduced tolerance for acetaminophen due to the induction of the MEOS, which results in increased levels of CYPE21.

22. **The answer is B.** Monoamine oxidase (MAO) is a key enzyme in the inactivation of catacholamines and serotonin. Selegiline is an MAOI (monoamine oxidase inhibitor). MAOIs inhibit the catabolism of some dietary amines. If tyramine is consumed, a hypertensive crisis can occur, as tyramine will induce the release of norepinephrine into the circulation. If tryptophan is consumed, a serotonin syndrome can occur. Alanine, threonine, and glycine are not catabolized by MAO, and taking the drug would have no effect on their metabolism. Tyramine is found in food that is aged, such as certain cheeses and wine.

23. **The answer is D.** Tyramine is produced by the decarboxylation of tyrosine during fermentation. Fresh and properly refrigerated protein-rich foods are not affected. Aged wines and chocolate are also high in tyramine.

24. **The answer is A.** The patient is exhibiting the symptoms of Addison disease, initiated by an adrenal insufficiency. Due to the problem in the adrenal glands, cortisol and aldosterone cannot be produced and released. The lack of cortisol leads to hypoglycemia and fatigue. The lack of aldosterone leads to the alteration in salt balance in the blood and the craving for salt. The hyperpigmentation occurs due to excessive ACTH production, which gives rise to the MSH. The ACTH levels are high due to reduced repression of ACTH synthesis, which is due to the lack of cortisol release from the adrenal cortex. A reduction of insulin levels would lead to diabetes, not the constellation of symptoms observed in this patient. A lack of GH (in the adult) would lead to increased body fat, high cholesterol, and reduced bone density. A lack of glucagon would lead to hypoglycemia, but not the other symptoms observed in this patient.

25. **The answer is A.** The patient is displaying the symptoms of acromegaly, caused by excessive or episodic secretion of GH in the adult (if the growth plates had not closed, then gigantism would result). GH is made in the pituitary and secreted from there in response to GHRH, which is produced in the hypothalamus. A tumor of the hypothalamus, which secretes GHRH, would lead to excessive GH secretion, and the symptoms observed. Tumors of the bone marrow (leukemia), adrenal glands (a variety of endocrine disorders), pancreas, or adipose tissue would not lead to elevated GH levels.

26. **The answer is E.** The patient has hyperparathyroidism, an oversecretion of PTH from the parathyroid gland. PTH leads to increased calcium absorption from the diet, and the release of calcium from the bone, thereby weakening the bones. The increased calcium in the kidney can lead to the formation of insoluble salts, leading to kidney stone formation. Because there is elevated calcium in the circulation, hypercalcemia is observed. These symptoms do not arise if hormones from the thyroid, adrenal gland, hypothalamus, or pancreas are released in large levels.

27. **The answer is C.** Because ACTH and cortisol were initially elevated, the most likely cause is a tumor of the anterior pituitary that is overproducing ACTH, causing cortisol to be overproduced by the adrenal gland. This conclusion is supported by the fact that the administration of a very high dose of a glucocorticoid (dexamethasone) caused the plasma cortisol level to decrease. (Glucorticoids inhibit the release of ACTH.) The patient's hyperglycemia was caused by the elevated cortisol. Diabetes (either type 1 or 2) would not lead to purple striae on the abdomen, muscle weakness, or depression (retinopathy may develop, leading to blindness in either type of diabetes).

28. **The answer is D.** The accountant has galactorrhea (inappropriate production of milk) caused by a prolactinoma (a tumor of the anterior pituitary that secretes PRL). Dopamine, the major regulator of PRL secretion, inhibits PRL production and release by the anterior pituitary. Bromocriptine is a drug that acts like dopamine to inhibit PRL release. While OT stimulates the ejection of milk from the mammary gland, PRL is necessary for milk to be produced in the gland. Neither glucagon nor testosterone will affect milk production. PRL is produced in the anterior pituitary, not the posterior pituitary.

29. **The answer is C.** Excess production of ACTH by the anterior pituitary is caused by a lack of suppression by cortisol, because cortisol levels are low. A tumor that produces POMC (the precursor of ACTH) would result in high ACTH and high cortisol levels. CRH levels are probably high (because of lack of suppression by cortisol), resulting in the high ACTH levels. The problem must be in the adrenal cortex (not the medulla), which is failing to produce adequate amounts of cortisol in response to ACTH. Because of a deficient enzyme in the cortisol pathway, cortisol precursors are being used to produce excess androgens, which are causing the patient's secondary male sexual characteristics.

30. **The answer is B.** An injury to her pituitary gland that resulted in decreased hormone production could explain all her symptoms. A low vasopressin (ADH) level results in diabetes insipidus. Decreased levels of LH and FSH cause amenorrhea. A decreased TSH leads to low thyroid hormone levels, which result in a decreased BMR (and a reduced production of body heat). Lack of GH slows down growth. Anterior and posterior pituitary hormones are small peptides that would be digested by proteolytic enzymes in the gut if taken orally. Thyroxine and glucocorticoids would alleviate some of her problems, but estrogen alone would not restore menses, and water intake would have to be increased, not restricted.

31. **The answer is F.** Testosterone is reduced to DHT, the more active androgen. Testosterone is a steroid hormone, thus it binds to intracellular receptors (nuclear) and the hormone–receptor complex acts as a transcription factor. Testosterone is a precursor of estradiol, not the other way around. Testosterone inhibits the synthesis of GnRH, and does not repress glucocorticoid synthesis.

32. **The answer is C.** The patient most likely has glucose-6-phosphate dehydrogenase deficiency, one of the enzymes that produces NADPH. When glucose-6-phosphate dehydrogenase is defective, both it and the subsequent enzyme in the hexose monophosphate shunt pathway (6-phosphogluconate dehydrogenase) do not produce NADPH. The red blood cells have insufficient NADPH; under these conditions, to keep glutathione in its reduced, protective form, and in the presence of a strong oxidizing agent (primaquine), red cell membrane damage occurs and the cells lyse, producing the hemolytic anemia. Mutations in glycolysis, gluconeogenesis, and fatty acid oxidation do not fit this pattern of damage in the presence of a strong oxidizing agent. Mutations in glycolysis (such as pyruvate kinase) can lead to anemia, but that occurs in the absence of oxidizing agents as well. The red blood cells do not oxidize fatty acids, nor do they carry out gluconeogenesis.

33. **The answer is B.** Vitamin B_{12} participates in two reactions in the body–conversion of homocysteine to methionine and conversion of methylmalonyl-CoA to succinyl-CoA. Methylmalonyl-CoA is produced via various amino acid degradation pathways, and from odd-carbon chain fatty acid oxidation. In the absence of B_{12}, succinyl-CoA would only be produced as an intermediate of the TCA cycle, and if it were removed from the cycle for heme synthesis, energy production may suffer. Glycine is usually obtained from the diet, although in a B_{12} deficiency a functional folate deficiency may also develop, leading to an inhibition of serine hydroxymethyltransferase, the enzyme that converts serine to glycine, and requires free tetrahydrofolate. Succinyl-CoA and glycine are the precursors for heme synthesis. A B_{12} deficiency would not inhibit the production of acetyl-CoA, succinate, or alanine.

34. **The answer is E.** GnRH stimulates the release of two pituitary hormones, LH and FSH. GHRH stimulates the release of GH; TSH, the release of T_3 and T_4; and PRH, the release of PRL. IGF stimulates cell growth, and its production is stimulated by the release of GH. Somatostatin reduces GH secretion, which would also lead to reduced IGF secretion.

35. **The answer is C.** The patient has a dietary iodine deficiency. When iodine is deficient in the diet, the thyroid does not make normal amounts of thyroid hormone. Consequently, there is less feedback inhibition of TRH and TSH production and release; thus, there is an increase in release of both of these hormones. Low levels of thyroid hormone result in decreased heat production, not an increase in heat production. Aldosterone affects salt levels in the blood, not thyroid hormones.

chapter 10

Human Genetics—An Introduction

The major clinical uses of this chapter are to understand inheritance patterns, how to give advice to couples on the odds of having a child with a particular disease, and to understand gamete formation and aspects of gamete formation that can lead to miscarriages.

OVERVIEW

- Human cells are diploid, with half the chromosomes coming from the mother and the other half from the father.
- A karyotype displays all the chromosomes of a cell.
- Mendelian inheritance, based on independent assortment of chromosomes during meiosis, can be classified as dominant expression, recessive expression, and X-linked expression (which can be either dominant or recessive).
- Mitochondrial inheritance is always from the mother.
- Alleles reside at specific loci on chromosomes.
- Mutations are alterations in the allele sequence that can lead to disease.
- Nondisjunction events in meiosis (or mitosis) can lead to aneuploidy, which often leads to disease.
- Gene dosage is important, as over- or underexpression of genes, due to trisomy, or monosomy, or copy number variation, can lead to disease.
- Abnormalities of chromosome structure include inversions, duplications, insertions, formation of isochromosomes, deletions, and translocations.
- The Hardy–Weinberg equation allows one to make estimates of allele frequencies, and heterozygote frequencies, within a given population.
- Multifactorial diseases involve significant interactions between the genes and the environmental factors.
- Triplet repeat expansions in and around genes can lead to disease if the expansion gets too large. Anticipation refers to an increase in the number of triplet repeats in successive generations, which correlates with the severity of the disease in successive generations.
- Imprinting refers to altering the expression of an allele without altering the nucleotide sequence of the allele. Imprinting is sex-specific. Males and females imprint alleles differently. The imprint remains throughout the life of the cell and its progeny. Imprinting is reset when germ cells are produced.
- Tumor suppressors, if analyzed through pedigrees, display an autosomal dominant pattern of inheritance, yet the molecular mechanism is recessive. The loss of a functional allele is known as loss of heterozygosity, and occurs through a variety of mechanisms.

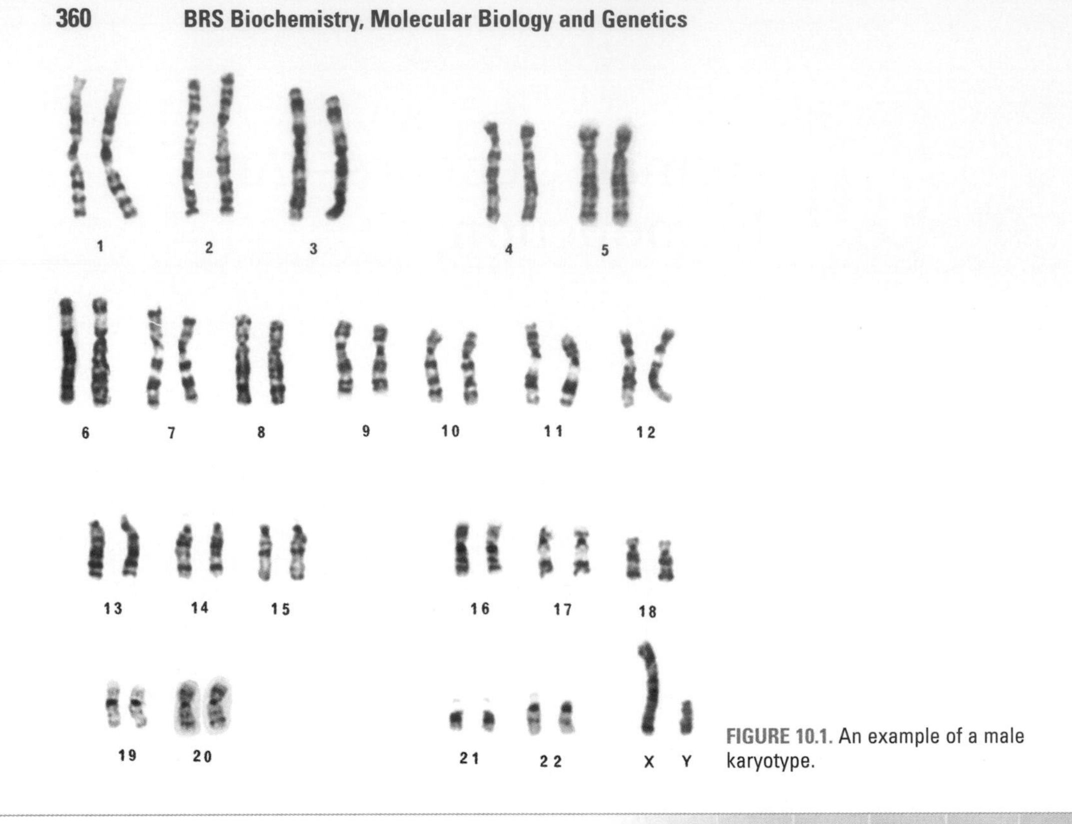

FIGURE 10.1. An example of a male karyotype.

I. MENDELIAN INHERITANCE PATTERNS

1. Humans are **diploid** organisms, meaning that each somatic cell contains two copies of each chromosome, one from each parent.
2. Human somatic cells contain 46 chromosomes (two copies of each of the autosomes, chromosomes 1 to 22, and two sex chromosomes, either XX [female] or XY [male]).
3. A **karyotype** displays all of the chromosomes in a cell, obtained from a metaphase spread (Fig. 10.1).
4. Germ cells (eggs and sperm) are produced by meiosis and contain a haploid number of chromosomes (23) that consists of one copy each of the autosomes, and one sex chromosome.
5. An important principle of Mendelian genetics is that of **independent assortment** of the chromosomes during meiosis: each chromosome in a pair is randomly sorted into a daughter cell. There is no linkage of chromosomes when they segregate during meiosis.
6. The principle of independent assortment allows the calculation of probabilities concerning the transmission of a mutant allele through an extended family.
7. The inheritance patterns of disease can be traced through pedigrees through multiple generations. The common symbols used in pedigrees are shown in Figure 10.2.

II. GENES

1. Genes reside at specific locations, known as **loci** (plural) or **locus** (singular), on a particular chromosome.
2. The form of a gene at a given locus is an **allele**; thus each locus has two alleles (one per chromosome).
   a. **Homozygous** refers to the two alleles being identical.
   b. **Heterozygous** refers to the two alleles having a different nucleotide sequence, which may be caused by mutations.

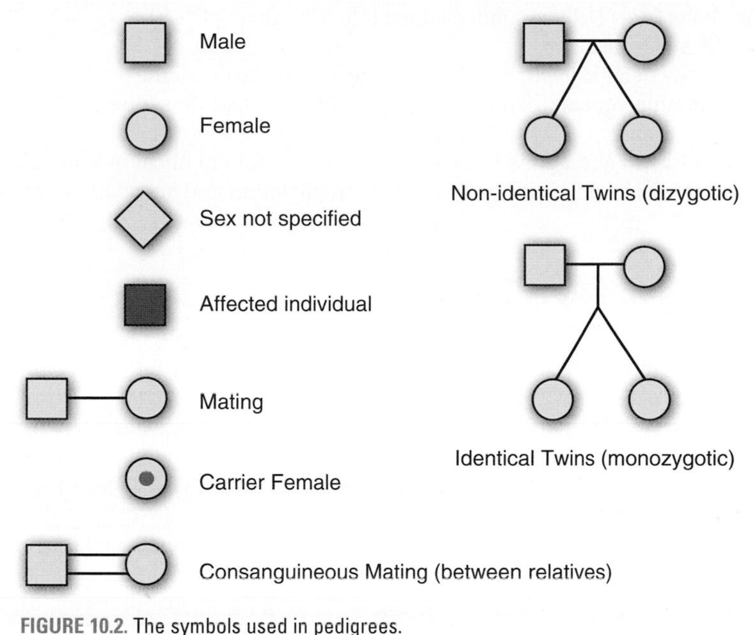

FIGURE 10.2. The symbols used in pedigrees.

3. **Phenotype**–the observable traits of the individual, produced by the interaction of the genes and the environment.
4. **Genotype**–the genetic composition of the individual.
5. **Heritability**–the capability to inherit a trait from a parent depends on two variables, a genetic component and an environmental component.
   a. A genetic component of 100% indicates no influence of the environment on the inheritance of the phenotype of the individual.
   b. A genetic component of 10% would indicate that the major determinant of the phenotype is environmental rather than genetic.
6. **Dominant** trait–one that manifests itself even when an allele is in a heterozygous state.
7. **Codominant** traits–both alleles in a heterozygous pair are expressed (an example are the blood group antigens).
8. **Recessive** trait–one that manifests only when the gene is in a homozygous state.
9. **Sex-linked** trait–when the mutant allele is located on the X chromosome, the disease will generally be expressed in males (since males have one X chromosome). Females with one mutant allele on the X chromosome are carriers of the disorder, and generally do not express the disease because the "normal" allele is present.
10. **Penetrance** refers to an individual expressing a phenotype when inheriting a particular mutated allele. A 100% penetrance means that everyone who inherits that mutant allele will express the disease.
11. **Variable expressivity** refers to the severity of the expressed phenotype caused by a mutated allele. Even if penetrance is 100%, the same mutation may exhibit different phenotypes (variable expressivity) in different members of the same family who inherit the allele.

III. MUTATIONS

1. Alterations in the DNA sequence (a mutation) of an allele can give rise to a nonfunctional or nonregulated gene product (see Chapter 3).
2. Mutations can be classified as the following:
   a. **Point mutations** (a change in one base of the DNA).
   b. **Deletions** (a loss of bases in the DNA).

c. **Insertions** (new DNA sequence added into the existing DNA).
d. **Loss of a chromosome.**
e. **Extra copy** of an allele (**trisomy** for an autosomal chromosome leads to disease).
f. **Trinucleotide repeats**–expansion of specific trinucleotide sequences in a gene can give rise to a disease.
g. **Epigenetic**–no alterations in the base sequence, but chemical modifications to the DNA and/or histones do occur (gain or loss of methylation and acetylation patterns, e.g.).

IV. INHERITANCE PATTERNS

1. **Autosomal dominant** inheritance
   a. A sample pedigree of autosomal dominant inheritance is shown in Figure 10.3.
      (1) An affected individual has an affected parent (unless the affected is expressing a new mutation).
      (2) The affected individuals are heterozygotes, as homozygosity for these traits is statistically very unlikely.
      (3) An affected parent has a 50% chance of passing the affected allele to their offspring.
      (4) The transmission of the trait is sex-independent, and both sexes can express the disease.
   b. The Punnet square analysis will aid in calculating the probabilities of passing the altered allele to the children (Fig. 10.4).
   c. For autosomal dominant inheritance patterns, 50% of the children will be affected, whereas 50% are not (thus, there is a one-in-two chance of inheriting the mutated allele).

CLINICAL CORRELATES Examples of autosomal dominant disorders include **achondroplasia (dwarfism,** due to a mutation in an FGF receptor), **Huntington disease, type 2** (due to a triplet repeat expansion in the *HTT* gene), **hypercholesterolemia** (due to a mutation in the LDL receptor), **Marfan syndrome** (due to a mutation in the fibrous protein fibrillin), **polycystic kidney disease** (due to mutations in a number of membrane proteins), **tuberous sclerosis** (although many examples of tuberous sclerosis are due to new mutations in the *TSC1* and *TSC2* genes), and **neurofibromatosis, type 1 (NF-1)** (due to mutations in the *NF1* gene, which is a GTPase-activating protein; many examples of NF-1 are due to new mutations in the *NF1* gene).

2. **Autosomal recessive** inheritance
   a. Figure 10.5 indicates a pedigree of autosomal recessive inheritance.
      (1) Both sexes are affected equally.
      (2) The transmission of the trait must occur from both parents.

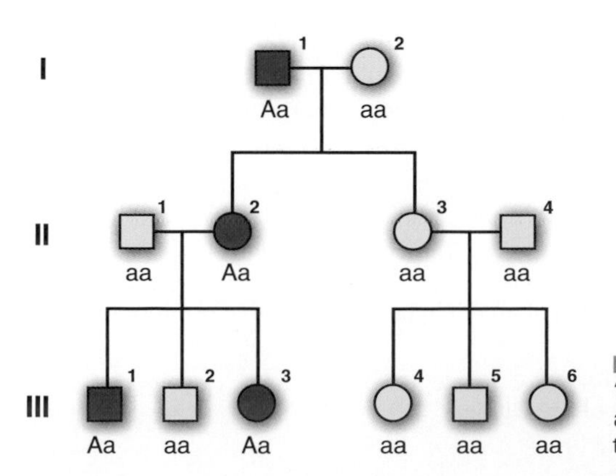

FIGURE 10.3. An autosomal dominant pedigree. The large "A" is the mutated allele, and the small "a" is the normal allele. The darkened boxes indicate the individuals with the disease.

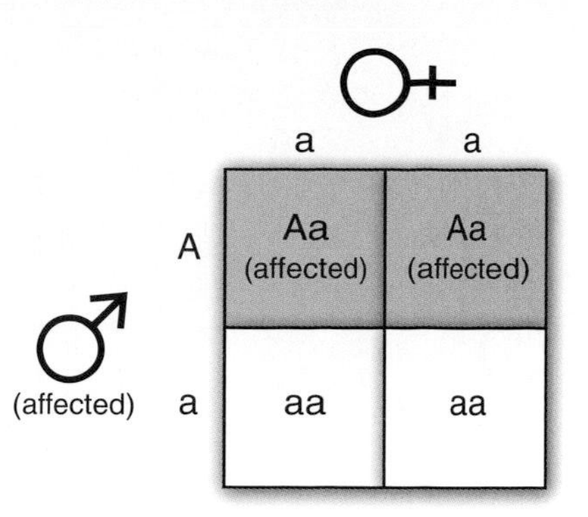

FIGURE 10.4. A Punnet square analysis of an autosomal dominant disorder. The disease gene is indicated by the large "A." Note that 50% of the offspring will inherit the disease.

(3) Generations may be skipped in the expression of the disease if all offspring are heterozygotes.

(4) Figure 10.6 displays a Punnet square analysis of autosomal recessive inheritance. Note that one in four children will be affected with the disease, and two in four children will be a carrier of the disease.

CLINICAL CORRELATES Examples of autosomal recessive disorders include **albinism** (which affects 1 in 20,000 live births, and is due to a loss of melanocyte tyrosinase), **cystic fibrosis** (which affects 1 in 2,500 live births of individuals of northern European ancestry, and is due to a mutation in the cystic fibrosis transmembrane conductance regulator (CFTR) protein), **phenylketonuria** (which affects 1 in 14,000 live births, and is primarily due to a mutation in phenylalanine hydroxylase), **hemochromatosis** (iron storage disease, affecting about 1 in 600 live births, due to a variety of mutations involved in iron absorption and transport), and **sickle cell disease** (which affects 1 in 400 live births in the African-American and many African populations, and is an E6V amino acid change in the β chain of globin).

3. X-linked inheritance

a. Males are **hemizygous** for genes on the X chromosome.

b. X-linked recessive disorders

(1) In such a pedigree, there is no male-to-male transmission (Fig. 10.7).

(2) Females are usually asymptomatic (but see exception below, Lyon hypothesis).

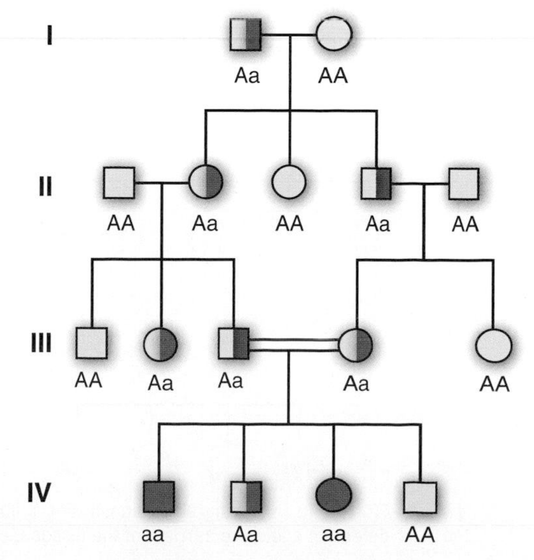

FIGURE 10.5. An autosomal recessive inheritance pedigree. In this case, the small "a" reflects the disease allele; a person with the genotype aa will express the disease, whereas the genotype Aa reflects a carrier of the disease.

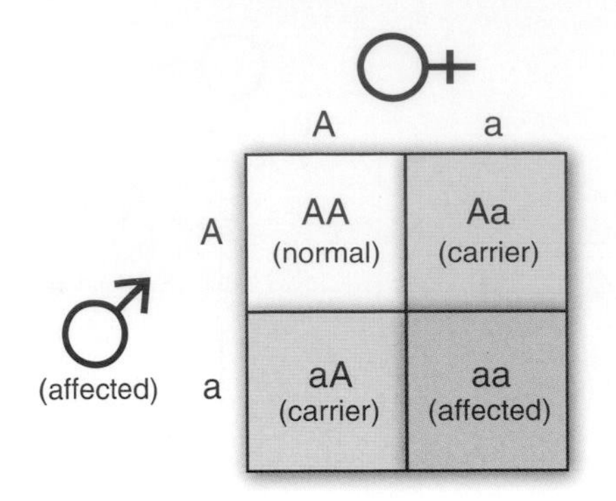

FIGURE 10.6. A Punnett square analysis of autosomal recessive inheritance. Note that one in four of the offspring will inherit both the mutated alleles, and express the disease.

(3) Sons and daughters both have a 50% chance of inheriting the mutated allele from their mothers, but sons will express the disease and the daughters will not.

(4) The Punnet square analysis of X-linked recessive disorders is shown in Figure 10.8.

CLINICAL CORRELATES Examples of X-linked recessive disorders include **hemophilia A** (1 in 12,500 live births, due to a mutation in the factor VIII gene, necessary for blood clotting), **Duchenne muscular dystrophy** (1 in 100,000 live births, due primarily to large deletions in the *DMD* gene), **red-green color blindness** (1 in 20 males of northern European ancestry, due to mutations in cone photoreceptors), certain forms of **Alport syndrome** (due to mutations in a specific collagen gene), and **ornithine transcarbamoylase deficiency** (1 in 100,000 live births, the most common inborn error of the urea cycle).

c. Gene dosage and the **Lyon hypothesis**

(1) The expression of genes on an additional chromosome (trisomy), or loss of expression of genes on a chromosome (monosomy, due to a chromosomal deletion) appears to be detrimental for human development (see Section VI for a further discussion). However, alterations in sex chromosome numbers are tolerated.

(2) The X chromosome is about five times larger than the Y chromosome, and if females expressed all of the genes on both X chromosomes, they would express more genes than males.

(3) To compensate, and keep the gene dosage equal between the sexes, **X-inactivation** occurs.

(a) One X chromosome in each cell is inactivated, condensed, and is known as a **Barr body** (Fig. 10.9). Inactivation is random as to the origin (maternal or paternal) of the chromosome.

(b) Only a small piece of the Barr body is transcriptionally active.

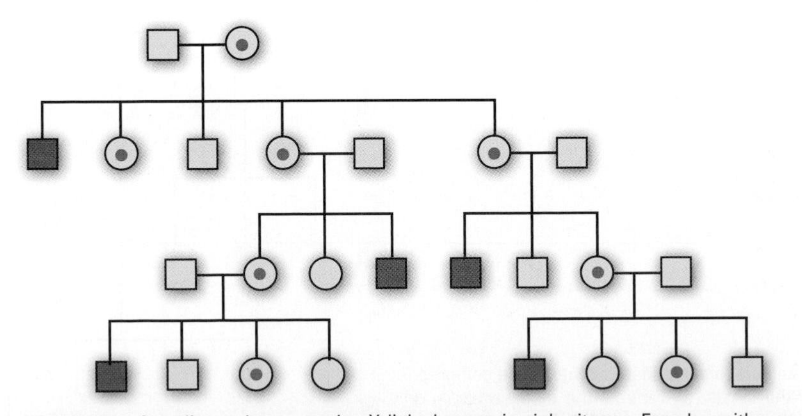

FIGURE 10.7. A pedigree demonstrating X-linked recessive inheritance. Females with one defective allele are carriers of the disease.

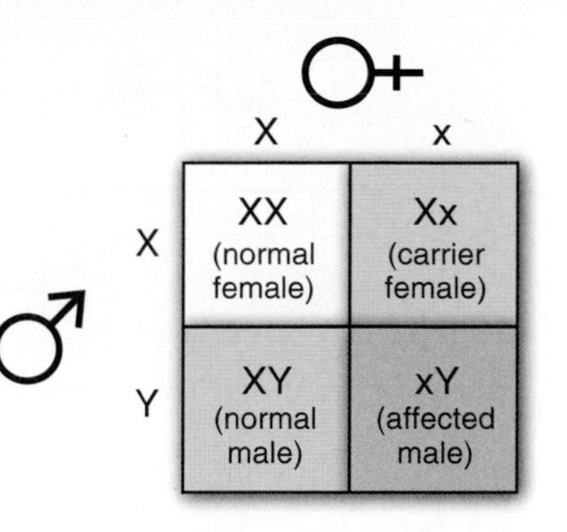

FIGURE 10.8. A Punnett square analysis of X-linked recessive allele transmission.

- **(c)** If a woman has inherited three X chromosomes (karyotype is XXX), two X chromosomes will be inactivated. There is no phenotype associated with multiple copies of the X chromosome in women.
- **(d)** The X-inactivation was hypothesized by Mary Lyon, and is known as the **Lyon hypothesis**.

(4) If unequal X-inactivation occurs (more cells inactivate the paternal X chromosome as opposed to the maternal X chromosome), a female may express symptoms of an X-linked recessive disorder.

d. X-linked dominant disorders (Fig. 10.10)

(1) Women carrying an X-linked dominant disorder will express symptoms of the disease.

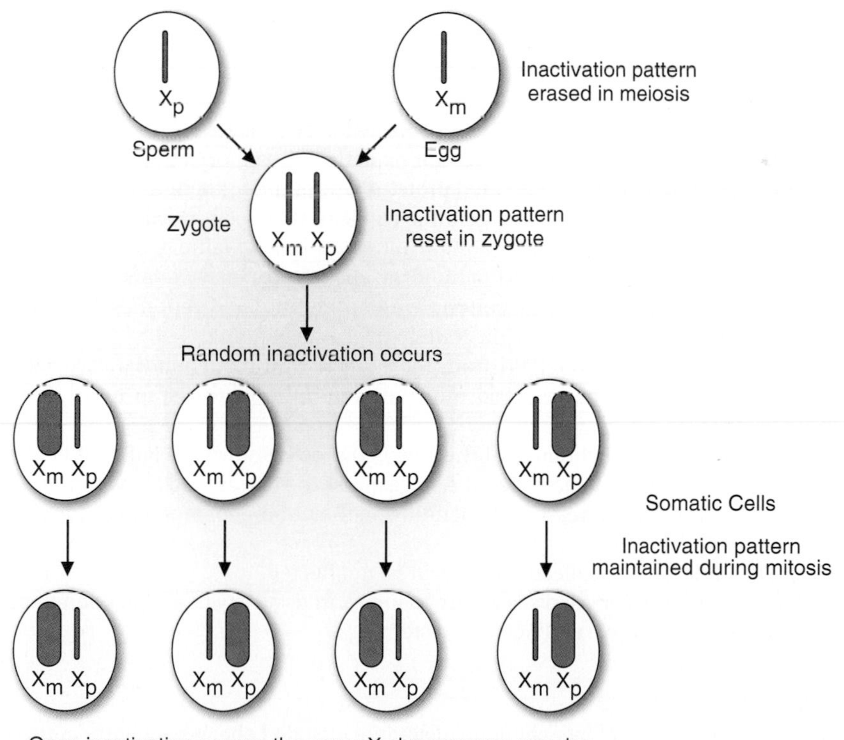

FIGURE 10.9. The Lyon hypothesis. The dark bar represents the inactivated chromosomes. The Lyon hypothesis explains how equal numbers of active genes are maintained in males and females. Inactivation at the 16-cell stage is random. Once an X chromosome is inactivated in a cell, all subsequent daughter cells have the same pattern of X-inactivation. In the zygote, both maternal (X_m) and paternal (X_p) X chromosomes are active.

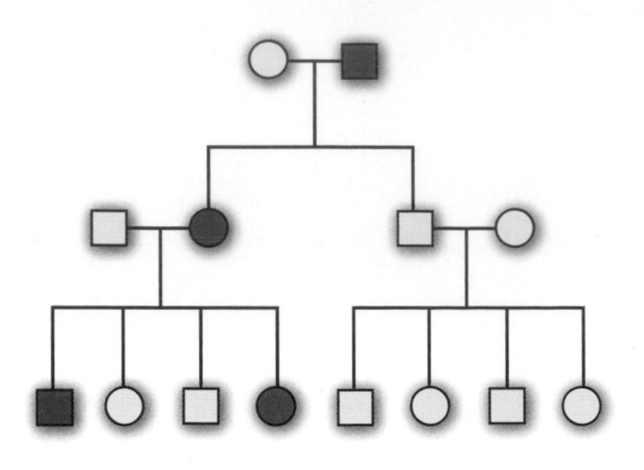

FIGURE 10.10. An example of X-linked dominant inheritance. The male in generation I has an X-linked dominant disease, and passes it to his daughter, but not his son. The daughter then passes the defective allele to one of her sons, and to one daughter.

(2) Affected males will give the disease to their daughters 100% of the time, but not to their sons.

(3) Females may express less severe symptoms than males due to the presence of a corresponding nonmutated allele.

CLINICAL CORRELATES Examples of X-linked dominant disorders include **hypophosphatemic rickets** (due to mutations in the *PHEX* gene) and **incontinentia pigmenti type 1** (the latter disease is fatal in males, but not in females. The mutation is in the *IKBKG* gene, which regulates a family of transcription factors.).

4. Mitochondrial inheritance

a. Mitochondria have a genome of 16,569 bp, which encodes a number of proteins and tRNA molecules. The proteins are needed for oxidative phosphorylation, and the tRNA to synthesize the mitochondrial-encoded proteins within the mitochondria.

b. Mutations in the mitochondrial genome can lead to defects in **oxidative phosphorylation** and reduction of energy production by mitochondria containing a mutated genome.

c. A cell has multiple copies of mitochondria, and **heteroplasmy** refers to the fact that some mitochondria contain normal genomes, and other mutated genomes. **Homoplasmy** is the term used if all the mitochondria contain the same genome.

d. All mitochondria are inherited from the mother (the mitochondria associated with the sperm do not enter the egg), so mitochondrial inheritance is an example of maternal inheritance.

e. An example of a mitochondrial-inherited pedigree is shown in Figure 10.11.

(1) All children of an affected female will express the disease (100% penetrance), but there will be variable expressivity depending on the number of mutated mitochondria inherited by each child.

(2) The children of an affected male will be unaffected.

(3) The symptoms of the diseases are manifest in tissues with a high energy requirement, such as the muscle and nervous system.

CLINICAL CORRELATES Examples of mitochondrial disorders include **Leber's hereditary optic neuropathy (LHON)**, due to a mutation in a protein-encoding gene, **myoclonic epilepsy with ragged red fibers (MERRF)**, which is due to a mutation in a mitochondrial tRNA gene, **mitochondrial encephalomyopathy, lactic acidosis, and stroke-like episodes (MELAS)**, also due to a mutation in a different mitochondrial tRNA gene, and **Kearns–Sayre disease**, characterized by muscle weakness, cerebellar damage, and heart failure, due to a deletion of part of the mitochondrial genome.

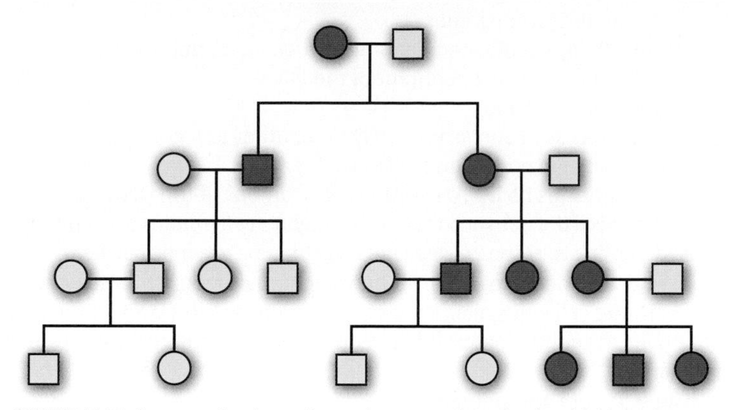

FIGURE 10.11. An example of a pedigree demonstrating mitochondrial inheritance.

V. A SUMMARY OF INHERITANCE PATTERNS IS GIVEN IN TABLE 10.1.

table 10.1 Summary of Inheritance Patterns (Pedigree Analysis)

	Frequency of the Child Inheriting the Trait	Sex of Affected Individuals	Pedigree Patterns in Families	Other
Autosomal dominant	50%	Both	Vertical, no skipping of generations	Father-to-son transmission occurs
Autosomal recessive	25%	Both	Horizontal, skipping of generations	Consanguinity
Sex-linked: X-linked recessive	50% for sons of a carrier female, and a 50% chance of a daughter being a carrier, 0% for sons of an affected male, although 100% of daughters are carriers	Males	Horizontal, do see skipping of generations	Absence of father-to-son transmission, affected females are rare
Sex-linked: X-linked dominant	50% sons affected, and 50% daughters affected if mother affected; 0% sons and 100% daughters affected if father affected.	Both	Vertical, disease phenotype seen in generation after generation	Absence of father-to-son transmission, expression usually less severe in heterozygous females than in affected males
Mitochondrial disorders	100% if mother has disease, 0% if father has disease	Both	Vertical, transmitted maternally	Absence of father-to-child transmission

VI. CYTOGENETICS

1. Chromosome abnormalities (alterations that are large enough to be seen under the microscope) are responsible for a significant number of diseases, occurring with a frequency of 1/150 live births.
   a. The leading cause of mental retardation.
   b. The leading cause of miscarriage (spontaneous abortion).
      (1) Chromosomal abnormalities are seen in 50% of first-trimester pregnancy loss.
      (2) Chromosomal abnormalities are observed in 20% of second-trimester pregnancy loss.
2. **Mitosis** and cell division
   a. Human cells contain 46 chromosomes, 22 pairs of autosomes (numbered 1 to 22), and two sex chromosomes (XX is female, XY is male).
   b. The haploid number of chromosomes is 23, also termed *n*.

c. Diploid is $2n$, or 46 chromosomes.
d. In mitosis, the DNA is replicated, leading to a $4n$ state, and then cell division creates two daughter cells containing a $2n$ content of the DNA.

3. **Meiosis** and gamete formation
   a. Meiosis is the process of converting $2n$ DNA content to n content.
   b. This occurs through two meiotic divisions.
   c. The cell first replicates its DNA, creating a $4n$ content of the DNA.
   d. In meiosis I, the cell divides such that each daughter cell obtains a $2n$ content of the DNA, but as opposed to mitosis, at the $2n$ stage the cell contains exact duplicates of each chromosome except for regions where crossing over between homologous chromosomes has occurred.
   e. In meiosis II, the cell splits again, creating n content of the DNA in the germ cells.
   f. Crossing over of genetic material between homologous chromosomes occurs during the meiosis I stage.
4. Chromosome structure and nomenclature
   a. Chromosomes contain **centromeres**, which is where homologous chromosomes are attached during cell division.
      (1) **Metacentric** chromosomes have the centromere in the middle of the chromosome.
      (2) **Submetacentric** chromosomes have the centromeres between the middle of the chromosome and the tip of the chromosome.
      (3) **Acrocentric** chromosomes have the centromeres at the tips of the chromosomes.
   b. **Telomeres** are the ends of the chromosomes.
   c. The short arm of a chromosome is known as "**p**," and the long arm as "**q**."
   d. Appropriate staining of chromosomes can further subdivide the chromosome into regions; for example, 14q32 refers to the second band in the third region of the long arm of chromosome 14.
   e. The karyotype nomenclature is summarized in Table 10.2.
5. Abnormalities of chromosome number
   a. **Euploid** refers to cells with a multiple of 23 chromosomes (23 is haploid, 46 is diploid, 69 is triploid, and 92 is tetraploid).
   b. Triploidy and tetraploidy are incompatible with human life.
   c. **Aneuploid** refers to conditions when the total chromosome number is not a multiple of 23.
      (1) Aneuploidy involves primarily **monosomies** (one copy of one chromosome) or **trisomies** (three copies of one chromosome).
      (2) Autosomal monosomies are always lethal.

table 10.2 Chromosome and Karyotype Nomenclature

Designation	Meaning
1–22	Autosome numbers
X, Y	Sex chromosomes
p	Short arm of chromosome, petit
q	Long arm of chromosome
Del	Deletion of chromosomal material
Der	Derivative, a structurally rearranged chromosome
Dup	Duplication of a part of a chromosome
Ins	Insertion of DNA into a chromosome
Inv	Inversion of the DNA within a chromosome
/	A designation to indicate mosaicism; two different types of cells within one individual. The first karyotype goes before the /; the second karyotype after the /
t	Translocation; the regions that are translocated are described after the t symbol
Ter	Terminal (also seen as pter, or qter, when referring to the terminal end of a specific chromosome arm)
r	Ring chromosome (the two ends are joined to form a ring structure)
+ or −	Placed before the chromosome number, these symbols indicate either additions (+) or loss (−) of a whole chromosome; placed after the chromosome number, these symbols indicate gain or loss of a chromosome part, e.g., 5p− indicates a loss of part of the short arm of chromosome 5; however del (5p) is the preferred nomenclature for such a loss

Reprinted with permission from ISCN. An International System for Human Cytogenetic Nomenclature. Mitelman F, ed. Basel, Switzerland: S. Karger; 1995.

(3) Autosomal trisomies are often lethal with only a few exceptions (Table 10.3). Sex chromosome aneuploidies are better tolerated.

d. Aneuploidy can arise via **nondisjunction**.

(1) Nondisjunction is unequal chromosome sorting during meiosis I or II, in which the "wrong" number of chromosomes is distributed to the daughter cells (one too many, or one too few).

(2) Figure 10.12 depicts how a nondisjunction event can occur during meiosis.

(3) If the appropriate chromosomal markers are available, one can determine whether a nondisjunction event occurred in meiosis I or II.

6. Abnormalities of chromosome structure

a. Inversions: two breaks in a single chromosome and an inversion of genetic material between the breaks.

table 10.3 Major Chromosome Aneuploidy Syndromes Compatible with Live Births

Syndrome	Chromosomal Abnormality	Major Features
Patau syndrome	Trisomy 13	Cleft lip and palate, severe central nervous system anomalies, polydactyly; occurs in 1 out of 10,000 live births, 90% die within 12 mo
Edward syndrome	Trisomy 18	Low birth weight, central nervous system anomalies, heart defects; occurs in 1 out of 6,000 live births, 90% die within 12 mo
Down syndrome	Trisomy 21	Hypotonia, characteristic facial features, developmental delay, mental retardation, heart abnormalities, increased risk of leukemia; occurs in 1 out of 800 live births
Turner syndrome	Monosomy X	Short stature, amenorrhea, lack of secondary sex characteristics; occurs in 1 out of 5,000 live female births
Klinefelter syndrome	XXY	Small testes, infertility, tall stature, learning problems; occurs in 1 out of 1,000 live male births
Triple-X	XXX	Learning disabilities, no major physical anomalies; occurs in 1 out of 1,000 live female births
XYY	XYY	Learning and behavioral problems in some individuals, occurs in 1 out of 1,000 live male births

Adapted with permission from Korf BR. *Human Genetics: A Problem-Based Approach.* 2nd ed. Boston, MA: Blackwell Science; 2000.

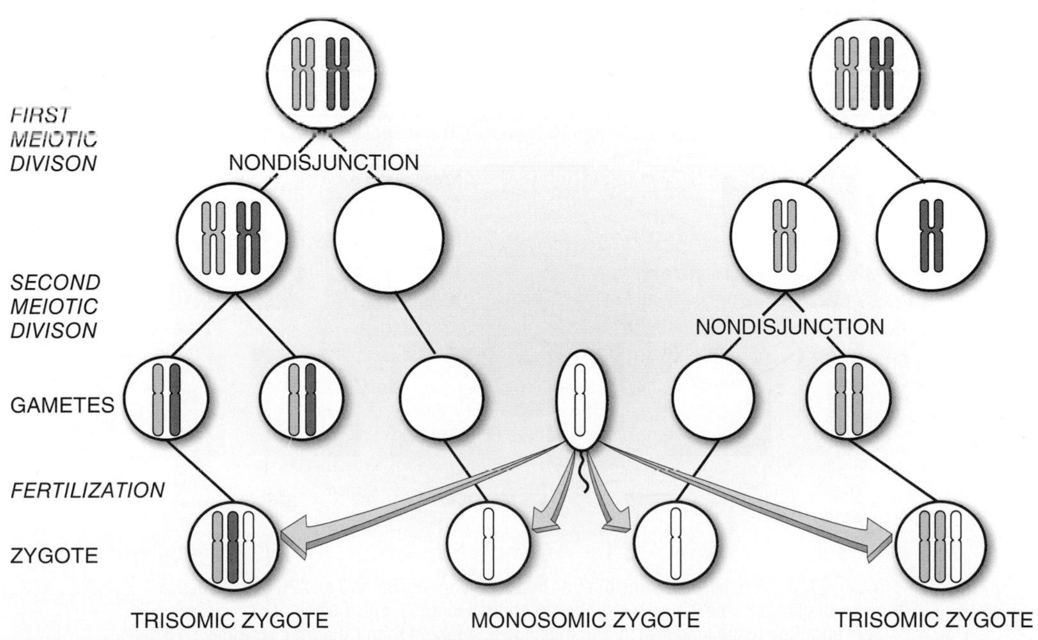

FIGURE 10.12. Examples of nondisjunction in meiosis I and II. Adapted from Gelehrter TD, Collins FS. *Principles of Medical Genetics.* Baltimore, MD: Williams & Wilkins; 1990: 165.

b. Duplications: part of a chromosome is duplicated and inserted into the same chromosome.

c. Insertions: a section of one chromosome is inserted into another chromosome, and there does not have to be a loss of genetic material.

d. Isochromosomes: an abnormal centromere division can lead to chromosomes in which entire arms are swapped between homologous chromosomes, such that an isochromosome can contain two p arms or two q arms derived from the two copies of a particular chromosome.

e. Translocations

(1) Reciprocal translocations (Fig. 10.13) occur when there are breaks in two different chromosomes and the material between the breaks is mutually exchanged.

(a) No genetic material is lost in a reciprocal translocation (a balanced translocation).

(b) The individuals with balanced translocations are phenotypically normal, but gamete formation can be compromised, leading to multiple miscarriages due to fetal monosomies and/or trisomies.

CLINICAL CORRELATES Chromosomal translocations can lead to disease, particularly if a gene is regulated inappropriately, or if a fusion protein is created on one of the translocated chromosomes. **Chronic myelogenous leukemia** (CML), due to a translocation between chromosomes 9 and 22, results from the creation of the fusion protein bcr-abl, which is an unregulated tyrosine kinase. **Burkitt lymphoma**, a translocation between chromosomes 8 and 14, leads to dysregulated myc expression, and uncontrolled cell proliferation.

(2) Robertsonian translocations (Fig. 10.14)

(a) Robertsonian translocations occur between acrocentric chromosomes (13, 14, 15, 21, 22) in which the short arms (satellites) are lost and the long arms are fused between the two chromosomes.

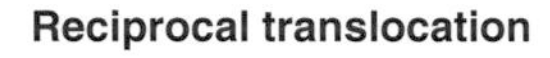

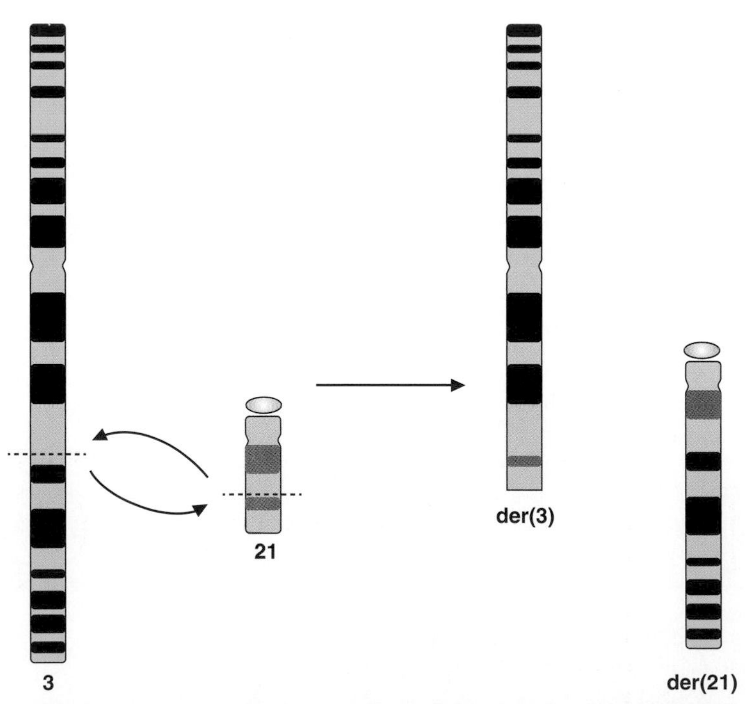

FIGURE 10.13. A reciprocal translocation between chromosomes 3 and 21. In this case, a piece of chromosome 3 is now attached to chromosome 21, and a piece of chromosome 21 is attached to the long arm of chromosome 3. Adapted from Gelehrter TD, Collins FS. *Principles of Medical Genetics.* Baltimore MD: Williams & Wilkins; 1990: 167.

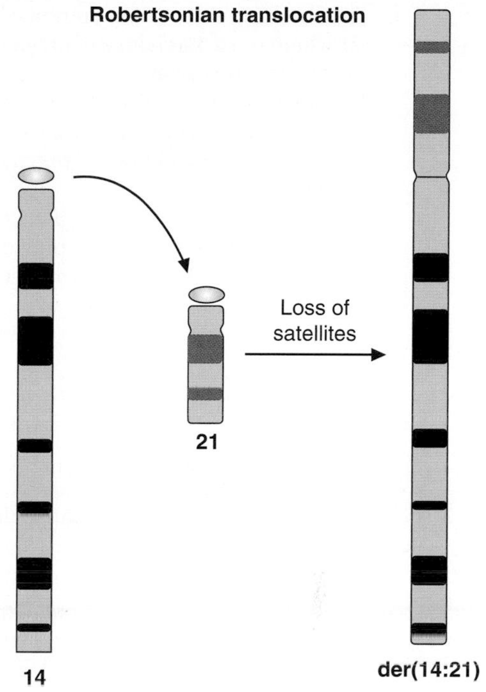

Robertsonian translocation

Loss of satellites

21

14

der(14:21)

FIGURE 10.14. An example of a Robertsonian translocation between chromosomes 14 and 21. Adapted from Gelehrter TD, Collins FS. *Principles of Medical Genetics.* Baltimore MD: Williams & Wilkins; 1990: 168.

table 10.4 Partial Listing of Chromosome Microdeletion Syndromes

Syndrome	Features	Deletion
Langer–Giedion syndrome	Sparse hair, bulbous nose, cone-shaped epiphyses, cartilaginous exostoses, mental retardation	8q24.1
WAGR	Wilms tumor, aniridia, genital and renal anomalies	11p13
Retinoblastoma	Retinoblastoma, mental retardation, dysmorphic facies	13q14.1
Prader–Willi	Hypotonia, eating disorder, obesity, mental retardation	15q11 (paternal)
Angelman	Hypotonia, seizures, inappropriate laughter, uncoordination	15q11 (maternal)
Miller–Dieker	Lissencephaly, dysmorphic facies	17p13.3
Smith–Magenis	Characteristic facies, mental retardation	17p11.2
Alagille	Biliary dysplasia, pulmonary stenosis, vertebral anomalies, dysmorphic facies	20p11
DiGeorge	Congential heart defect, hypoplasia of parathyroid and thymus, facial anomalies	22q11

Adapted from Jorde C, White M. *Medical Genetics.* 1995 [Table 6-3]; Korf BR. *Human Genetics: A Problem-Based Approach.* 2nd ed. Boston, MA: Blackwell Science; 2000. [Table 5-3.]

(b) The loss of genes from the short arms does not cause a problem as the genes lost are primarily those for rRNA, which are present in multiple copies throughout the genome.

(c) The individuals carrying a Robertsonian translocation have 45 chromosomes, and are phenotypically normal.

(d) As with reciprocal translocations, gamete formation can be compromised, leading to multiple miscarriages due to fetal monosomies or trisomies.

f. Microdeletion syndromes: syndromes with a consistent but complex phenotype associated with a small (less than five megabases) chromosomal deletion.

(1) FISH (fluorescent in situ hybridization) is often necessary to detect such microdeletions.

(2) A partial list of microdeletion syndromes is presented in Table 10.4.

7. Prenatal cytogenetics

a. The indications for use include advanced maternal age (>35 years), family history of a child with a chromosomal disorder, and an abnormal prenatal study.

b. Amniocentesis (gold standard)

(1) Can be performed at 15 to 16 weeks of gestation.

(2) Obtain fetal cells from amniotic fluid, culture them, and karyotype them.
(3) There is a 0.5% chance of pregnancy loss.

c. **Chorionic villus sampling**
(1) Can be performed at 10 to 12 weeks of gestation.
(2) Obtain placental cells, purify fetal cells, and karyotype them.
(3) There is a 0.5% chance of pregnancy loss.

d. **Cordocentesis**
(1) Umbilical cord blood sampling after 18 weeks of gestation.
(2) Used if previous methods provide ambiguous results.
(3) There is 1% to 2% chance of pregnancy loss.

VII. POPULATION GENETICS

1. The **Hardy–Weinberg** equilibrium is based on five assumptions for a large population (over 1,000 individuals).
   a. Random mating between individuals
   b. Fairly large populations
   c. Negligible mutation rate between the wild-type allele and the disease allele
   d. Negligible migration in and out of the population
   e. No selection, and all genotypes viable and fertile
2. Consider two alleles, A and a, at a particular locus; aa represents a disease phenotype for an autosomal recessive disorder.
   a. Let p = frequency of allele A in the population.
   b. Let q = frequency of allele a in the population.
   c. $p + q = 1$.
   d. If one squares $p + q = 1$, the Hardy–Weinberg equation is realized, $p^2 + 2pq + q^2 = 1$.
      (1) p^2 represents the frequency of AA homozygotes in the population (wild type).
      (2) q^2 represents the frequency of aa homozygotes in the population (those with the autosomal recessive disease).
      (3) $2pq$ represents the frequency of carriers (heterozygotes) in the population.
3. Problem-solving exercise
   a. The frequency of individuals with cystic fibrosis (CF), in individuals of northern European ancestry, is 1 in 2,500. What is the carrier frequency for CF in that population?
   b. The term $q^2 = 1/2{,}500$, so q (the disease allele frequency) $= 1/50$.
   c. Since $p + q = 1$, $p = 49/50$, which can be approximated to 1.
   d. The carrier frequency (heterozygote frequency) is $2pq$, or $2 \times 1/50 \times 1$, or 1 in 25 individuals within this population.

VIII. MULTIFACTORIAL DISEASES (COMPLEX TRAITS)

1. Multifactorial diseases result from the interaction of multiple genes (the genotype) with the environment; they are complex to study due to the gene interactions required to promote the observed phenotype (it is more than just one mutated allele).

Examples of complex trait disorders include **cardiovascular disease, cancer, asthma, emphysema, hypertension, Type 2 diabetes, and birth defects.**

2. Risk versus relative risk

a. Risk is the probability that a child will be born with a particular trait.

b. Relative risk is the ratio of the risk of having the trait in the person being studied to the risk of having the trait in a random person in the population. The higher the relative risk, the greater is the chance that a couple will have a child with a particular trait.

c. Environmental issues are important, even in the womb.

(1) For identical twins (same genotype), if one is born with a cleft palate, the chance of the other twin having a cleft palate is not 100%, but is 40%.

(2) The relative risk of a monozygotic twin having a cleft palate when the other does is 400; however, the rate of cleft palate is 0.1%, so this translates to a 40% probability that both twins will have a cleft palate.

d. In a simplistic way, one can view birth defects, or other complex traits, as requiring a certain number of mutated alleles to accumulate for the proband to express the defect.

(1) Thus, if a couple has a child with a **cleft palate**, for example, their relative risk for a second child with a cleft palate is increased relative to the general population since it is known that the couple contains, between them, sufficient mutated alleles to bring about this phenotype.

(2) Another form of birth defect, **pyloric stenosis**, is more frequently seen in males than females. Thus, if a mother has had pyloric stenosis as a child, then the relative risk of her passing on the disorder to her son is greater than if the father had pyloric stenosis as a child.

IX. TRIPLET REPEAT EXPANSIONS

1. A number of diseases result from a triplet repeat expansion within a transcribed region of a gene.
2. The number of repeats inherited correlates with the severity of the disease, leading to variable expressivity of the disease.
3. The diseases are not necessarily 100% **penetrant**; the individuals in a pedigree may inherit a repeat, but do not express the disease, and thus they can transmit the triplet repeat to their children, who would express the disease.
4. Within a pedigree, the transmission of the disorder follows an autosomal dominant mode of inheritance.
5. Southern blots can determine the size of the expansions through the identification of appropriate restriction length polymorphisms, or by PCR using primers that surround the repeated regions.
6. It has been observed for diseases in which triplet repeat expansions are important that the length of the expansion increases with every generation, and the severity of the disease is worse in subsequent generations (earlier onset, and more severe symptoms). This phenomenon is known as **anticipation**. The triplet expansion tends to cause problems during DNA replication, and additional triplets can be added to this region during gamete formation.

CLINICAL CORRELATES Diseases due to triplet repeat expansions include **myotonic dystrophy (DM), fragile X syndrome, Huntington chorea, spinocerebella ataxia, and spino and bulbar muscular dystrophy**. DM is due to a CTG repeat in the 3′-untranslated region of the gene, which encodes a protein kinase. The triplet expansion may inhibit transcription of the gene, or it may inhibit translation of the mRNA produced from the gene. Fifty or more copies of the repeat will lead to a mild form of DM. Greater than 100 copies will lead to classical DM, and greater than 1,000 copies will lead to congenital DM.

X. IMPRINTING

1. Imprinting refers to a modification of a gene's ability to be expressed via a means other than changing the base sequence of the DNA.
   a. Imprintable alleles are transmitted in Mendelian fashion.
   b. The expression of the gene will be determined by the sex of the parent.
      (1) **Maternal imprinting** refers to the alleles that females modify during gamete formation, which may never be expressed in their children.
      (2) **Paternal imprinting** refers to the alleles that males modify during gamete formation, which may never be expressed in their children.
      (3) For the purposes of this text, we will assume that the imprinting event has blocked expression of the allele, although that is not universally true for all imprinted genes.
2. If a fetus is created with two sets of **paternal** chromosomes (two sperm fertilizing an egg that lacks maternal DNA), the result is an **aborted hydatidiform mole (androgenetic).**
3. If a fetus is created with two sets of **maternal chromosomes** (cell division problem during female meiosis), the result is an **aborted ovarian teratoma (gynogenetic).**
4. Figure 10.15 displays examples of maternal and paternal imprinting in pedigrees (recall that maternal imprinting is being interpreted to mean that alleles are inactivated by the mother, whereas paternal imprinting is the inactivation of alleles by the father).
   a. Maternal imprinting: when eggs are made, certain alleles are marked (one example of marking genes is methylation of adenine bases, or cytosine bases in CpG islands in the genes promoter), which leads to those genes not being expressed when the egg is fertilized, and in all progeny cells.
   b. The corresponding allele from the father is not inactivated during maternal imprinting, so for those genes inactivated maternally, the fetus is hemizygous for those alleles.

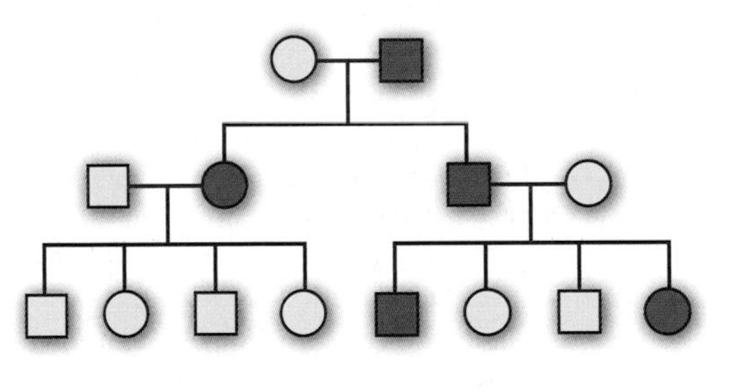

Maternal Imprinting pattern, in which the paternal gene has a mutation.

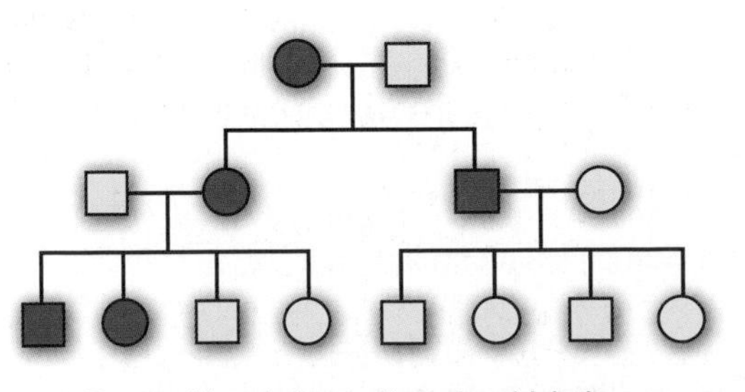

Paternal Imprinting pattern, in which the maternal gene has a mutation.

FIGURE 10.15. Idealized imprinting pedigrees, with the **(top)** pedigree an example of maternal imprinting, and the **(bottom)** pedigree an example of paternal imprinting.

c. If the father contained an inactivating mutation in a maternally imprinted gene, then the fetus would lack expression of that gene (the maternal gene is imprinted, the paternal gene is mutated), and a disease may result.
d. The probability of passing such a mutated allele to children would be 50% (the father has one normal gene, and one mutated allele at the imprinted locus).

5. Disease examples of imprinting at the same locus are **Prader–Willi disease and Angelman syndrome.**
   a. Prader–Willi syndrome occurs in about 1 out of 10,000 live births.
   b. The karyotype displays a small deletion in the long arm of chromosome 15 (15q), always of paternal origin.
   c. Angelman syndrome displays the same deletion, but always of maternal origin.
   d. The symptoms of the two diseases are quite distinct (Fig. 10.16).
   e. These findings indicate that there are both maternally and paternally imprinted genes in this region of chromosome 15, and loss of a corresponding allele (through the deletion) can lead to a disease, which differs depending on which genes are missing and/or inactivated.

CLINICAL CORRELATES Other examples of imprinted diseases include certain types of cancer (**Wilms tumor** with maternally derived deletions on chromosome 11), **retinoblastoma** (the first mutation most often occurs on the paternal chromosome), and **Huntington chorea** (about 5% to 10% of paternally derived triplet repeat expansions lead to symptoms that are not observed in maternally derived triplet repeat expansions).

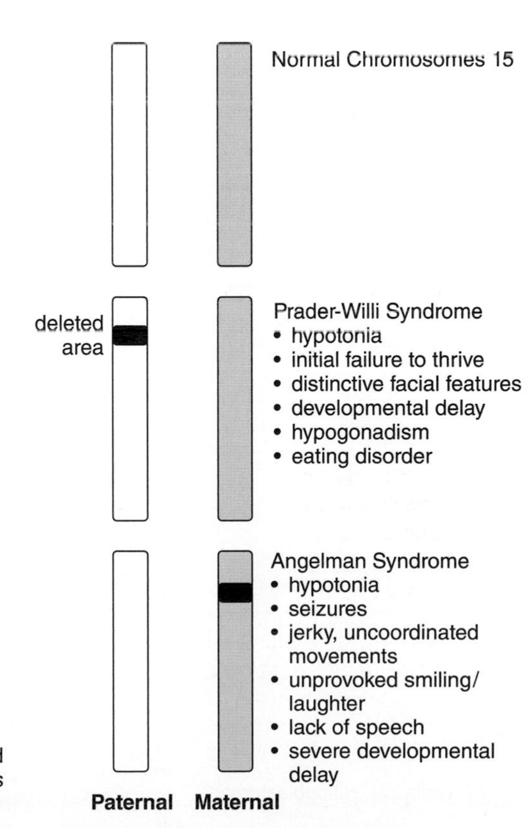

FIGURE 10.16. A schematic illustration of Prader–Willi and Angelman syndromes, indicating the different symptoms depending on which chromosome contains the deletion.

XI. THE GENETICS OF TUMOR SUPPRESSORS

1. As discussed in Chapter 4, tumor suppressors are recessive oncogenes, yet the pedigree of such diseases exhibits a dominant type of transmission (Fig. 10.17).
   a. The example shown is **retinoblastoma (Rb)**, which can be hereditary or sporadic. The sporadic variety accounts for 60% of the disease.
   b. The genetic locus is 13q14.
   c. Note the autosomal dominant pattern of expression in the hereditary form of Rb, and that bilateral Rb predominates (osteosarcomas can also result); in contrast, the sporadic form is cancer in only one eye, and tumors are not found in other locations.
2. In hereditary Rb (and other tumors due to tumor suppressors), one of the Rb alleles is already mutated in all cells, and, over time, there is a 100% chance that the corresponding normal allele will lose activity, leading to tumor formation.
3. In sporadic Rb, both an initial mutation and the second event have to occur from scratch, which makes these tumors less frequent, and not seen bilaterally (Fig. 10.18).
4. The **Knudson** two-hit model explained these results. A tumor ensues when two events have occurred in the same cell lineage that knocks out both copies of a gene. Two rare events must occur to produce a sporadic tumor. A person who is born with the first mutation, however, needs to acquire only one additional mutation to develop a tumor.
5. The loss of the corresponding normal allele is known as **loss of heterozygosity**. This can occur in a variety of ways.
   a. **Nondisjunction** such that a cell loses the normal Rb allele (an example of mitotic nondisjunction).
   b. Nondisjunction followed by **reduplication** of the existing chromosome (now there are two identical copies of the chromosome in the cell, both with the Rb mutation).
   c. **Mitotic recombination** such that the Rb mutation is present in both the alleles within one cell.
   d. **Gene conversion**, an event that can occur during mitosis (although more commonly seen in meiosis) in which the genetic information from one chromosome is transferred to the

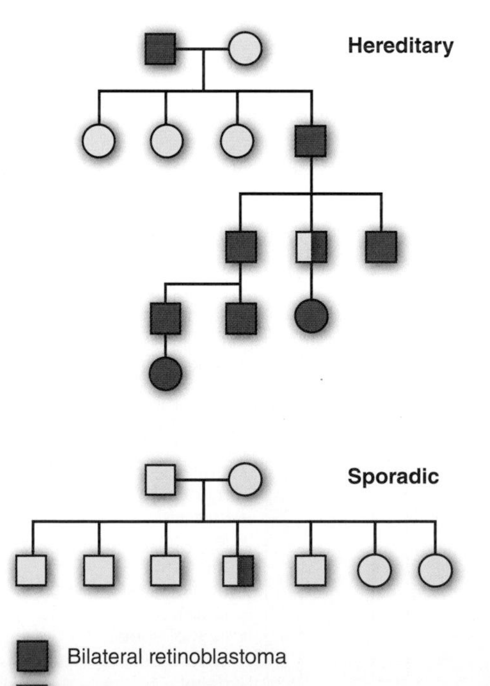

FIGURE 10.17. A pedigree analysis indicating both hereditary **(top)** and sporadic Rb.

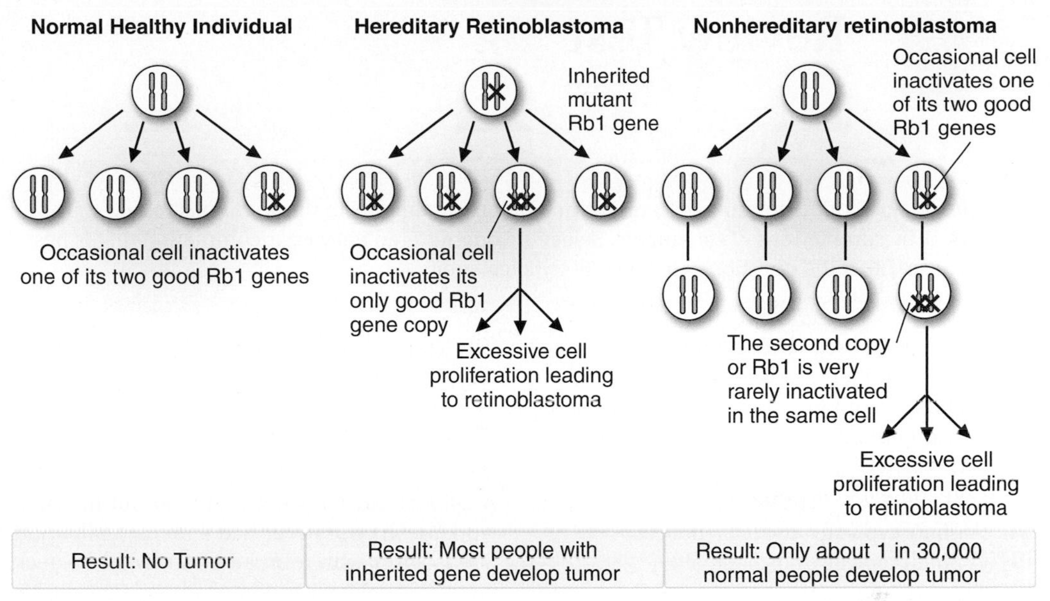

FIGURE 10.18. An illustration of the manner in which Rb mutations can lead to disease. Not the differences between hereditary and nonhereditary Rb transmission.

other chromosome; in this case, copying the mutated Rb allele and replacing the wild type Rb allele on the corresponding chromosome.

  e. A **deletion** of the functional Rb allele from the corresponding chromosome.
  f. A **point mutation** in the functional Rb allele such that it becomes nonfunctional.

6. While Rb was the disease of choice, these principles apply to all tumor suppressor genes (such as p53, *APC*, *BRCA1*, *BRCA2*, *NF1* [neurofibromas] and *NF2* [gives rise to acoustic neuromas], *WT1* [Wilms tumor], and p16 [can give rise to melanomas]).

Review Test

Directions: Each of the numbered items or incomplete statements in this section is followed by answers or by completions of statements. Select the one-lettered answer or completion that is best in each case. All of the questions are USMLE-style questions.

1. A 40-year-old pregnant woman is very concerned about chromosomal abnormalities in her fetus. She undergoes amniocentesis, and the fetus is demonstrated to have the karyotype 47 XXX. Which of the following is the most likely outcome of this pregnancy?

(A) A phenotypically normal female.
(B) A phenotypically normal male.
(C) Since trisomies are almost always lethal, there will be a spontaneous abortion.
(D) Since trisomies are almost always lethal, this will end in a stillbirth.
(E) Since trisomies are almost always lethal, the baby will die shortly after birth.

2. A 27-year-old male underwent a workup for what appeared to be a feminizing disorder since he was tall with sparse facial hair, small testes, and gynecomastia. He also had poor coordination and language and reading difficulties. During the workup, it was discovered that he had an abnormal karyotype. Which of the following best describes the reason for his abnormal karyotype?

(A) Trinucleotide repeat
(B) Nondisjunction
(C) Point mutation
(D) Translocation
(E) Deletion

3. A 16-year-old female has a constant sense of hunger, obesity, almond-shaped eyes, strabismus, mental retardation, and retarded puberty development. Which of the following best describes the reason for this chromosomal abnormality?

(A) Trinucleotide repeat
(B) Nondisjunction
(C) Point mutation
(D) Translocation
(E) Deletion

Questions 4 and 5 are based on the following case:

A 20-year-old female presents for an infertility workup. She has never had a menstrual period. She is short with a broad chest, webbed neck, and low-set ears. It is demonstrated that she has an abnormal karyotype.

4. Which one of the following best describes the cause of this genetic abnormality?

(A) Trisomy
(B) Monosomy
(C) Trinucleotide repeat
(D) Translocation
(E) Point mutation

5. The cause of the woman's abnormal karyotype is which one of the following?

(A) Maternal nondisjunction
(B) Paternal nondisjunction
(C) Both maternal and paternal nondisjunction
(D) Either maternal or paternal nondisjunction
(E) A reciprocal translocation
(F) A Robertsonian translocation

Questions 6 and 7 are based on the following case:

A 22-year-old male has a long face, large forehead, ears and jaw, large testes, autism, speech and language delays, and hand flapping. His only sister is normal except for extreme shyness.

6. Which one of the following best describes the basis of this chromosomal abnormality?

(A) Autosomal dominant
(B) Codominant

(C) Autosomal recessive
(D) Trinucleotide repeat
(E) Mitochondrial inheritance

7. The phenotypic pattern of this man's disorder is best described as which one of the following?

(A) Full expressivity
(B) Variable expressivity
(C) Gene dosage effect
(D) Homoplasmy
(E) Heteroplasmy

8. A 19-year-old male, at a routine physical exam for sports activities (long-distance running) at his college, is noticed to have elevated fasting blood glucose levels (about 7.5 mM). Measurements of C-peptide and insulin levels were close to normal under fasting conditions. After eating, blood glucose levels are only slightly elevated above the normal fasting levels before stabilizing at the fasting levels. The student indicates that he is not drinking or urinating excessively, but that he remembers that his mother had gestational diabetes when pregnant with him. This alteration in glucose homeostasis is best typified by which one of the following types of inheritance?

(A) Autosomal dominant
(B) Autosomal recessive
(C) Sex-linked
(D) Mitochondrial
(E) Multifactorial

9. A 35-year-old male woodcutter and politician from Illinois is very tall and thin with arachnidactyly, pectus excavadum, a high arched palate, lens dislocation, and aortic insufficiency. His wife is of normal height and weight, and does not exhibit any of the same symptoms as the politician. What is the probability that one of their children will exhibit features similar to the father's?

(A) 100%
(B) 75%
(C) 50%
(D) 25%
(E) 0%

10. A 32-year-old female has had multiple deep venous thrombosis and pulmonary emboli, especially during her pregnancies. She is on chronic warfarin therapy. Her father died of a pulmonary embolus after a retinal detachment operation. Neither of her three siblings nor her two children have had any clotting problems. Which one of the following inheritance patterns/descriptions best typifies this genetic problem?

(A) Autosomal dominant
(B) Autosomal incomplete dominant
(C) Codominant
(D) Autosomal recessive
(E) Sex-linked

Questions 11 and 12 are based on the following case:

A 22-year-old male's lifelong dream was to be a fighter pilot for the US Air Force. He passed his vision test without a problem, but failed the color portion of the standard Snellen chart.

11. What is the probability that the subject's brother would have the same problem?

(A) 100%
(B) 75%
(C) 67%
(D) 50%
(E) 33%
(F) 25%
(G) 0%

12. If the above patient marries a normal female and has one son and one daughter, which one of the following inheritance patterns will occur?

(A) The son will be color-blind, but the daughter will be normal.
(B) The daughter will be color-blind, but the son will be normal.
(C) Both children will be color-blind.
(D) The son will be a carrier, but the daughter will be normal.
(E) The daughter will be a carrier, but the son will be normal.

Questions 13 and 14 are based on the following case:

A 20-year-old African-American male presents to the ER with severe abdominal, low back, and rib pain. He has had similar episodes all his life. Blood work reveals severe microcytic, hypochromic anemia. A peripheral smear shows abnormally shaped red blood cells and a hemoglobin electrophoresis confirms his abnormality.

13. Which of the following best describes this genetic abnormality?

(A) Trinucleotide repeat
(B) Point mutation
(C) Trisomy
(D) Deletion
(E) Translocation

14. If the above patient has children with a partner with normal hemoglobin genes, which one of the following patterns will occur?

(A) All sons will have the same disease as the father, while all daughters will be carriers.
(B) All children will have the same disease as the father.
(C) All children will be carriers for the disease (contain the mutated gene, but do not express the disease phenotype).
(D) All children will have normal hemoglobin, like their mother.
(E) All daughters will have the same disease as their father, while all sons will be carriers.

15. A 35-year-old nonsmoking male has been diagnosed with emphysema. His father died of emphysema at age 30, but he smoked. His father also had cirrhosis and recurrent pancreatitis but did not drink alcohol. Which one of the following inheritance patterns typifies this disease process?

(A) Autosomal dominant
(B) Incomplete dominance
(C) Codominant
(D) Autosomal recessive
(E) Sex-linked

Questions 16 to 18 refer to the following case:

A 22-year-old male reports to his eye doctor that he lost the center portion of the vision in his right eye over a week's time and 2 weeks later, the same thing happened in his left eye. He has two older brothers who had the same thing happen to them at age 21 and 23, and one older sister who had the same eye problems starting at age 24.

16. Which one of the following typifies this type of genetic abnormality?

(A) Autosomal dominant
(B) Autosomal recessive
(C) Sex-linked
(D) Mitochondrial
(E) Trinucelotide repeat

17. If the patient has children, which one of the following patterns will occur?

(A) All of his children will have the disease.
(B) None of his children will have the disease.
(C) Only his sons will have the disease.
(D) Only his daughters will have the disease.
(E) All of his children will be carriers of the disease.

18. If the patient's sister has children, which of the following patterns will occur?

(A) All of her children will have the disease.
(B) None of her children will have the disease.
(C) Only her sons will have the disease.
(D) Only her daughters will have the disease.
(E) All of her children will be carriers of the disease.

Questions 19 to 21 are based on the following case:

A 1-year-old boy has been hospitalized twice for lung infections, was often wheezing and out of breath, and was lagging in both height and weight in his growth charts, despite a very good appetite. A subsequent test found that his sweat contained elevated levels of chloride ions.

19. The inheritance pattern for this disorder is which one of the following?

(A) Autosomal dominant
(B) Autosomal recessive
(C) X-linked recessive
(D) X-linked dominant
(E) Mitochondrial
(F) Multifactorial

20. If the frequency of affected individuals for this disease is 1 in 2,500 in a specific population, what is the frequency of carriers within that population?

(A) 1 in 10
(B) 1 in 25
(C) 1 in 50
(D) 1 in 100
(E) 1 in 1,250

21. The 1-year-old boy has a sister who is phenotypically normal. What is the probability that the sister is a carrier of the disease?

(A) 100%
(B) 75%
(C) 67%
(D) 50%
(E) 33%
(F) 25%
(G) 0%

Questions 22 and 23 are based on the following case:

A couple is having trouble bringing a pregnancy to term. They have experienced three miscarriages in the past 2 years and want to understand what the problem may be.

22. An initial test that should be run on the couple is which one of the following?

(A) Karyotype analysis of the female
(B) Karyotype analysis of the male
(C) Karyotype analysis of both the potential parents
(D) FISH analysis of the mother using chromosome-specific probes
(E) FISH analysis of the father using chromosome-specific probes
(F) FISH analysis of both potential parents using chromosome-specific probes

23. The miscarriages were most likely caused by which one of the following?

(A) Multiple X chromosomes
(B) Autosomal monosomies
(C) Autosomal trisomies
(D) Either autosomal monosomies or trisomies
(E) Trisomy 21
(F) Lack of the Y chromosome

24. The family in the pedigree shown has one family member with an autosomal recessive disease. Assuming that this is a rare disorder, what is the probability that individual III-1 is a carrier of the mutated allele?

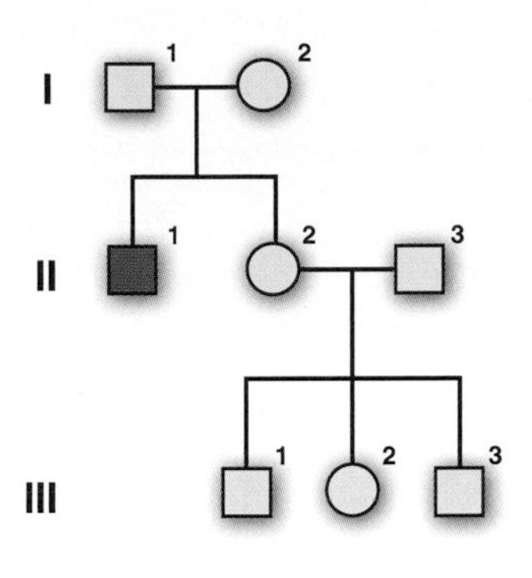

(A) 100%
(B) 75%
(C) 67%
(D) 50%
(E) 33%
(F) 25%
(G) 0%

25. Which one of the following observations would rule out a sex-linked trait in an extended family pedigree?

(A) Males expressing the disease
(B) Females expressing the disease
(C) Female-to-male transmission
(D) Male-to-male transmission
(E) The disease-skipping generations in the pedigree

26. Consider the pedigree shown in which the affected individuals are expressing a rare autosomal recessive disease. What is the probability that the unborn child of the marriage will express the disease?

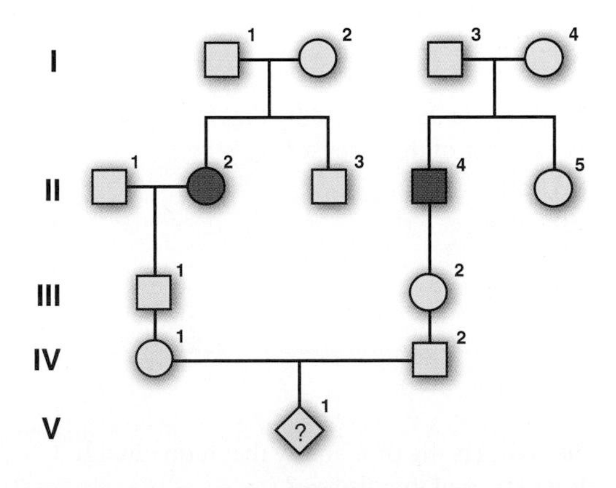

(A) 1/2
(B) 1/4
(C) 1/8
(D) 1/16
(E) 1/32
(F) 1/64
(G) 1/128

Questions 27 to 30 are based on the following pedigree:

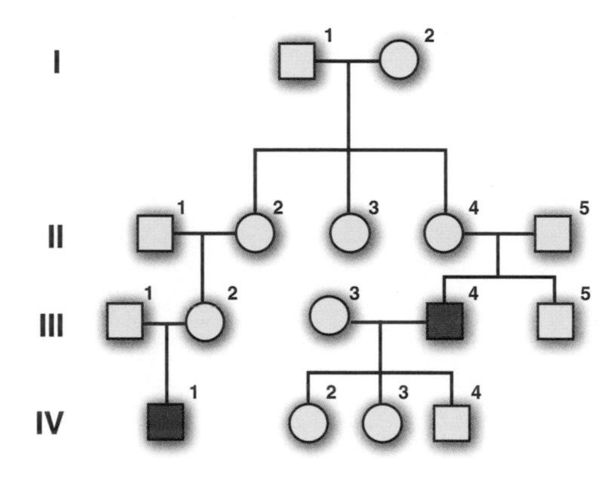

27. What is the most likely mode of transmission of this disease?

(A) Autosomal recessive
(B) Autosomal dominant
(C) Mitochondrial
(D) X-linked recessive
(E) X-linked dominant

28. What is the probability that individual III-5 is a carrier of the disease?

(A) 0%
(B) 25%
(C) 33%
(D) 50%
(E) 67%
(F) 75%
(G) 100%

29. What is the probability that individual IV-3 is a carrier of the disease?

(A) 0%
(B) 25%
(C) 33%
(D) 50%
(E) 67%
(F) 75%
(G) 100%

30. What is the probability that individual II-3 is a carrier of the disease?

(A) 0%
(B) 25%
(C) 33%
(D) 50%
(E) 67%
(F) 75%
(G) 100%

31. A couple in their mid-30s has had three children, two boys and one girl. One boy and one girl have severe hearing loss, but the other boy has only a mild hearing loss. The father has normal hearing, but the mother does wear a hearing aid on her right ear. The mother's brother is also hard of hearing, but he has three children who have no hearing loss. A likely mode of inheritance for this disorder is which one of the following?

(A) Autosomal recessive
(B) Autosomal dominant
(C) X-linked dominant
(D) X-linked recessive
(E) Mitochondrial
(F) Triplet repeat expansion

32. An X-linked recessive disorder is found in a particular family. Using the glucose-6-phosphate dehydrogenase allele as a marker, which contains two polymorphic forms, A and B, all family members of the pedigree were genotyped for the presence of either the A, or B, or both alleles. Considering the pedigree shown, what is the probability that individual IV-1 will express this disease?

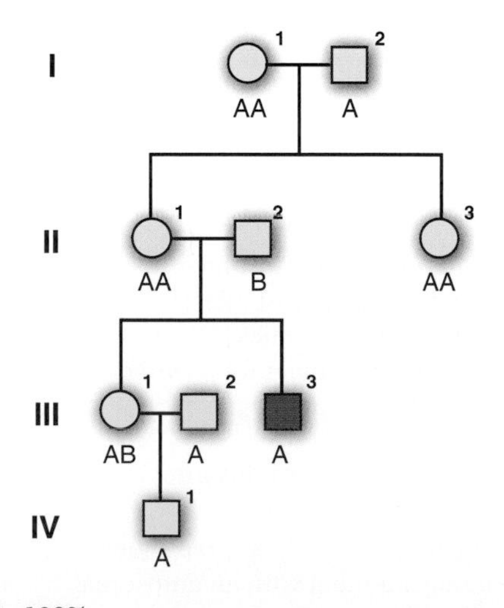

(A) 100%
(B) 75%

(C) 50%
(D) 25%
(E) 0%

33. The gene frequency for an X-linked recessive disease is 1 in 1,000 in the general population. What is the frequency of affected males in this population?

(A) 1 in 10
(B) 1 in 100
(C) 1 in 500
(D) 1 in 1,000
(E) 1 in 2,000

34. The disease frequency for sickle cell anemia in the African-American population is 1 in 400. What is the carrier frequency in this population?

(A) 1 in 5
(B) 1 in 10
(C) 1 in 20
(D) 1 in 40
(E) 1 in 50
(F) 1 in 100

35. A phenotypically normal woman underwent a karyotype analysis for difficulties in conceiving. She was found to contain three Barr bodies, but no translocations or large deletions. Her karyotype would be best represented by which one of the following?

(A) 48 XXXXY
(B) 46 XX
(C) 48 XXXX
(D) 48 XXXY
(E) 48 XXYY

Answers and Explanations

1. **The answer is A.** The baby will be born a phenotypically normal female as two of the X chromosomes will be inactivated per cell, and become Barr bodies. In this manner, there are no gene dosage effects occurring during the development of the fetus. There is no Y chromosome, so this baby will not be male. Trisomies for autosomal chromosomes are lethal except for trisomy 21, trisomy 18, and trisomy 13. However, polysomies of the X chromosome can be asymptomatic.

2. **The answer is B.** This patient has Klinefelter syndrome or 47 XXY. This is caused by a nondisjunction of the X chromosomes in either the egg, which is then combined with a Y chromosome during fertilization, or the sperm, in which case an XY sperm combines with an egg carrying one X chromosome. Even though one of the X chromosomes will become a Barr body, men with the karyotype 47 XXY have reduced testosterone levels as compared to men who are 46 XY. While symptoms are not seen in all individuals, the ones described in this question are the classic symptoms for Klinefelter syndrome. Trinucelotide repeats, point mutations, translocations, or deletions will not give rise to the 47 XXY genotype. That can only occur through a nondisjunction event in either the formation of the mother's eggs or father's sperm. The nondisjunction event that leads to 47 XXY appears to occur evenly between the mother and father.

3. **The answer is E.** This patient has the classic phenotype of Prader–Willi syndrome that is caused by the inheritance of a partial deletion (microdeletion) of chromosome 15 from the father. The corresponding maternal alleles in the deleted regions have been imprinted, and some inactivated, leading to a loss of gene expression and the phenotype observed in the patient.

4. **The answer is B.** This patient has Turner syndrome or 45 XO. She is missing a second sex (X) chromosome. She would not have a Barr body and would be infertile. She is phenotypically female.

5. **The answer is D.** Monosomy X (Turner syndrome) can be due to either maternal nondisjunction (an egg is created lacking an X chromosome, whereas another egg has two X chromosomes) or paternal nondisjunction (a sperm lacks the X chromosome, whereas another sperm contains both the X and Y chromosomes). Turner syndrome is not caused by translocation events, either reciprocal or Robertsonian (recall that Robertsonian translocations only occur amongst acrocentric chromosomes, and the X chromosome if not acrocentric).

6. **The answer is D.** The patient has fragile X syndrome, the most common inherited mental retardation disorder in males. There is a trinucleotide repeat of the *FMR1* gene (in the 5'-untranslated region) on the X chromosome. The more repeats, the more severe the signs and symptoms. Males are more severely affected than females (only one X chromosome), and many women with fragile X appear asymptomatic except for excessive shyness (awkward social interactions). Fragile X syndrome was named for chromosome breaks in this region of the chromosome when cells are cultured in a folate-deficient medium.

7. **The answer is B.** Fragile X syndrome, the disorder displayed by the patient, exhibits variable expressivity depending upon the number of repeats inherited. Some males may be asymptomatic, but a grandson of a male who is asymptomatic may have full expression of the disease (remember, males cannot transmit an X chromosome to their sons, only to their daughters). Though the abnormality may be present in individuals, they may show no features of the disease. With full expression, everyone with genetic abnormalities would show the full spectrum of the disease process (100% penetrance and expressivity). Homoplasmy and heteroplasmy are terms used in mitochondrial inheritance. Heteroplasmy refers to the fact that some

mitochondria contain normal genomes and other mutated genomes. Homoplasmy refers to all mitochondria containing the same genome. Gene dosage refers to the expression of too many genes such as in trisomies, which is not the case in triplet repeat disorders.

8. **The answer is A.** The patient is displaying the signs of maturity onset diabetes of the young (MODY), which can be due to a mutation in the pancreatic glucokinase gene, such that its K_m is increased. The increase in the K_m for glucokinase would lead to glucose only being metabolized at higher-than-normal levels. Once glucose is metabolized in the β cells of the pancreas, and ATP levels increase, then insulin can be released. The glucokinase mutation causes insulin release to occur at higher-than-normal glucose levels. The mother also expresses the mutant glucokinase gene. During pregnancy, the effect of placental hormones tends to inhibit insulin's action, and in a mother with MODY, in which insulin is not being released appropriately owing to the glucokinase mutation, blood glucose levels rise significantly during the pregnancy, leading to gestational diabetes. MODY, in terms of the glucokinase mutation, is transmitted in an autosomal dominant manner.

9. **The answer is C.** The politician has the phenotype of an individual with Marfan syndrome, which is inherited in an autosomal dominant pattern. The mutation is in the protein fibrillin, which is a glycoprotein found in elastic fibers in connective tissue. Since this is an autosomal dominant disorder, the man has a one-in-two chance (50%) of passing the mutation on to one of his children.

10. **The answer is B.** The patient has Factor V Leiden, which is the most common hereditary hypercoagulable disorder in the United States. Individuals who inherit one copy of this mutation (heterozygote) are at an increased risk for clotting, but that risk is less than someone who is homozygous for the mutation. Thus, the mutated allele is not completely dominant (since the homozygous state increases the risks of clots), and is termed incomplete dominance. Not everyone who inherits just one mutated allele will develop a clotting problem. Codominant would imply two mutations of the gene (either or both of which would produce the disease process). The mutated gene for Factor V Leiden is not on the X or Y chromosome, so it is not sex-linked. Since a person inheriting just one mutated allele can express disease symptoms, the inheritance process is not autosomal recessive.

11. **The answer is D.** Red and green are the standard colors on a Snellen eye chart, and are used to screen for the most common "color blindness" in boys, red-green. This is an X-linked recessive trait. Females can be carriers with no color-perception problems, but can have red-green color blindness if they are homozygous for the mutation. Half of a carrier female's sons will be normal and half will be red-green color-blind. None of an affected male's sons will have the problem, but all of his daughters will be carriers. The mother of the person in question is a carrier for red green color blindness, and she would have a 50% chance of passing the mutated allele on one of her X chromosomes to the patient's brother.

12. **The answer is E.** Red-green color blindness is an X-linked recessive disorder. The male who has the disease will transmit his X chromosome (with the mutation) to his daughter, who will then be a carrier of the disease (since the X chromosome inherited from the mother contains a normal allele). The father transmits his Y chromosome to his son, who inherits his X chromosome from his mother, so the son does not inherit the mutation and will be normal in terms of red-green color discrimination.

13. **The answer is B.** Sickle cell anemia is due to a point mutation on the β-globin chain of the hemoglobin where glutamic acid is replaced by valine in the sixth position (E6V). Under deoxygenated conditions, the valine in this position can form hydrophobic interactions with another deoxygenated hemoglobin molecule, leading to the polymerization of the hemoglobin within the cell. The long hemoglobin polymer alters the shape of the red cell, and leads to a loss of red cell elasticity, hemolysis, and "sludging." The altered, or sickle, shape of the cells prevents them from entering capillaries, leading to vaso-occlusive crises and the typical symptoms observed by individuals in a sickle cell crisis. This is an autosomal recessive disease, and the probability that two carriers will have a child affected with sickle cell disease is 25%.

14. **The answer is C.** Since the father has two mutated hemoglobin genes (both of his β-globin genes contain the E6V mutation) and his partner has two normal β-globin genes, all offspring will have one normal and one sickle gene and therefore will be carriers (the children will have sickle trait). Since this is an autosomal recessive trait and not a sex-linked trait, the sex of the offspring is irrelevant.

15. **The answer is C.** Alpha-1 antitrypsin (A1AT) deficiency is a codominant process. Codominance means that multiple versions of the gene may be active or expressed, and the genetic trait is due to the effects of both the expressed alleles. The *SERPINA1* gene on chromosome 14 codes for A1AT, which is a protein that protects tissues (especially the lungs) from neutrophil elastase. A1AT is synthesized in the liver and secreted into the circulation. Under normal conditions, neutrophils in the lung engulf and destroy particulate matter in the air we breathe. At times, the protease elastase escapes from the neutrophils, and is inactivated by A1AT. In the absence of functional A1AT, the elastase destroys the lung cells, and chronic obstructive pulmonary disease (COPD) will occur very early in life. If the patient smokes, the condition is greatly exacerbated. A common mutation in A1AT leads to misfolding and accumulation of the misfolded form of the protein in the liver. The accumulation of this inactive, misfolded protein can lead to cirrhosis of the liver, and eventual liver failure.

16. **The answer is D.** LHON is due to a mutation in the mitochondrial genome, so it is classified as mitochondrial inheritance. All mitochondria is inherited from the mother since the mitochondria associated with sperm does not enter the egg. All of the offspring of an affected mother will have the disease (100% penetrance), although the expressivity is quite variable depending on the degree of heteroplasmy exhibited by each child. Males cannot transmit mitochondrial diseases.

17. **The answer is B.** The disease is a mitochondrial disorder, LHON. Mitochondrial diseases are maternally inherited, as all of the mitochondria in a developing embryo are derived from the egg, and none from the sperm. Since the patient is a male, none of his mutant mitochondria will enter the egg, and none of his children will express the disease, nor will they be carriers of the disease.

18. **The answer is A.** As the patient has a mitochondrial disorder (which he inherited from his mother), the patient's sister also has the same mitochondrial disorder. As a female, the woman will pass on mutant mitochondria to all of her children, who will express the disease.

19. **The answer is B.** The boy has the classic symptoms of CF, an autosomal recessive disorder. The prevalence of CF in the northern European population is 1 in 2,500, with a carrier frequency of 1 in 25. The mutated protein is the CFTR, which regulates chloride transport across membranes. The drying of the pancreatic duct leads to a reduction of secretions from the pancreas reaching the intestine, which leads to the digestive problems exhibited by patients with CF.

20. **The answer is B.** Using the Hardy–Weinberg equilibrium, q^2 (the disease frequency) is equal to 1 in 2,500, so q equals 1 in 50. The heterozygote (carrier) frequency is $2pq$ (q is 1 in 50, and p is very close to 1), or 1 in 25. The disease the child is exhibiting is CF, and the population frequencies are for those individuals of northern European heritage.

21. **The answer is C.** A sibling of an affected individual, in whom the disease is due to autosomal recessive transmission, has a two-in-three chance of being a carrier. The allele distribution from the parents leads to four possibilities: having the disease, being homozygous for the wild-type allele, and two possible ways of being a carrier (either inherit the variant allele from the mother or father). Since the sister does not express the disease, she has two chances to be a carrier, and one chance to be homozygous normal, or a two-in-three chance of carrying the mutated allele.

22. **The answer is C.** One of the potential parents most likely has a translocation, which is causing the formation of abnormal gametes, in terms of chromosome number, during meiosis. FISH analysis may miss translocations and would not be able to detect this problem. A chromosomal translocation in either parent (either reciprocal or Robertsonian) can lead to the problems in conception the couple is experiencing.

23. **The answer is D.** The translocations would cause problems during meiosis, leading to gametes either lacking a chromosome (or a portion of a chromosome) or gained an additional chromosome (or portion thereof). This would lead to autosomal monosomies or trisomies after fertilization, the majority of which are incompatible with life and lead to early termination of the pregnancy. Multiple X chromosomes are tolerated, and would not lead to pregnancy termination. Trisomy 21 leads to Down syndrome, which is compatible with life and does not lead to miscarriage. Lack of the Y chromosome would lead to females, which does not lead to pregnancy termination.

24. **The answer is E.** Individual II-3 does not contribute to this calculation, as he is from outside the family, and the disease is rare in the population, so his risk of being a carrier is very low. The probability that individual II-2 is a carrier is 2/3, as she is an unaffected sibling of an affected individual. There is a 50% chance that II-2 will transmit the mutated allele to her son (III-1), such that the probability that III-1 will inherit this allele is 2/3 × 1/2, or 1/3 (33%).

25. **The answer is D.** Sex-linked traits are passed via the X chromosome. Males transmit their Y chromosome to their sons, and their X chromosomes to their daughters. Thus, an affected male cannot transmit a mutated X allele to his son, so the presence of male-to-male transmission in a pedigree categorically eliminates X-linked transmission as the genetic pattern of inheritance.

26. **The answer is F.** The probability that individual II-2 transmitted the mutated allele to III-1 is 50%, and the probability that III-1 transmitted the mutated allele to IV-1 is also 50%. The probability that IV-1 has transmitted the mutated allele to V-1 is 50%, so the overall probability of the disease allele in II-2 being transmitted to V-1 is 1/8. The same is true for the chances of II-4 transmitting to III-2, to IV-2, and to V-1. For V-1 to express the disease, all of these events have to occur, such that there is a 1/64 chance that V-1 will express the disease (1/8 × 1/8).

27. **The answer is D.** This is an example of an X-linked recessive disease. Only males express the disease, and they obtained the mutated allele from their mothers, who do not express the disease (so it cannot be an X-linked dominant disorder). It is also observed that certain generations can be skipped, but that the gene is still passed through the pedigree via the female members of the family.

28. **The answer is A.** As this is an X-linked recessive disorder, any male who has the mutated allele will express the disease, and would not be a carrier for the disease. Since individual III-5 does not express the disease, he does not carry the mutated allele, and cannot pass the mutated allele on to his daughters.

29. **The answer is G.** Individual IV-3's father has the disease, and he has passed on his X chromosome (which carries the mutated allele) to his daughters. The daughter does not express the disease because she also carries a normal allele that was inherited from her mother. All daughters of fathers with X-linked disorders will be carriers of the disease.

30. **The answer is D.** Since individuals IV-1 and III-4 are expressing the disease, their mothers (III-2 and II-4) must be carriers of the disease. III-2 also must have acquired the disease allele from her mother, II-2. II-3 is the sister of both II-2 and II-4, and would have had a 50% chance of inheriting the mutated allele from her mother, I-2.

31. **The answer is E.** This is an example of mitochondrial inheritance in that all transmission is from the female (all children of an affected female display the trait, with variable expressivity), and an energy-intensive organ is most severely affected. The family history indicates that male transmission of the disorder does not occur (the mother's brother), and that variable expressivity is evident due to the degree of heteroplasmy inherited by each child.

32. **The answer is C.** The glucose-6-phosphate dehydrogenase alleles allow one to trace X chromosomes throughout the pedigree, and to determine the probabilities that someone has inherited the X allele that leads to the disease. Analyzing individual III-3, who has the disease, one can determine that an "A" polymorphic form of glucose-6-phosphate dehydrogenase travels with the disease locus. III-1 inherited his "A" allele from his mother (II-1), who contains

two "A" alleles, one on each X chromosome. Since we do not know which X chromosome in II-1 contains the mutated allele, II-1 has a 50% chance of passing on the X chromosome with the mutated allele to her daughter, III-1 (the "B" allele in III-1 came from her father). When III-1 and III-2 have their child, IV-1, the "A" allele in IV-1 had to have come from the mother, as the father passed the Y chromosome to IV-1, and not his X chromosome. This is the same chromosome that has a 50% chance of carrying the mutation, so there is a 50% chance that IV-1 will express the disease.

33. The answer is D. If the disease frequency in the population is 1 in 1,000, and the disorder is X-linked, that means that out of 1,000 men, one would have the disease, as each man contains one X chromosome. Since women contain two X chromosomes, the carrier frequency is 1 in 500, as 500 women would contain 1,000 X chromosomes, one of which would contain the mutated allele.

34. The answer is B. If the disease frequency for this autosomal recessive disorder (sickle cell anemia) is 1 in 400, then $q^2 = 1/400$, and $q = 1/20$. The heterozygote frequency, according to the Hardy–Weinberg equilibria, is $2pq$, or $2 \times 1 \times 1/20$, or 1 in 10. For the purposes of this calculation, we are considering p, which is really 19/20, to be functionally equivalent to 1.

35. The answer is C. If the woman is expressing three Barr bodies, then she has four X chromosomes per cell, three of which have been inactivated. This would give her a total of 48 chromosomes, and four of those would be X chromosomes, for a karyotype of 48 XXXX. Any karyotype with a Y chromosome would be a male.

Comprehensive Examination

Directions: Each of the numbered items in this section is followed by potential answers of the statement. Select the one-lettered answer that is best in each case.

1. A patient who needed to lose weight began eating at fast-food restaurants. He did not change his exercise level. However, the composition of his diet was altered in that his carbohydrate intake decreased by 50 g/day and his fat intake increased by 50 g/day. Otherwise, his diet remained the same. Which one of the following best reflects what would occur to this patient after following this diet for 3 months?

	Weight loss?	Calorie change in diet?
A	No	No change
B	Yes	Additional 250 kcal/day
C	No	Reduced 250 kcal/day
D	Yes	No change
E	No	Additional 250 kcal/day
F	Yes	Reduced 250 kcal/day

Questions 2 and 3 are based on the following case:

A patient who is obese and has hypertension requires a weight-reduction diet. She weighs 176 lb and has a sedentary lifestyle.

2. What is the approximate number of calories the patient burns each day at this weight?

(A) 1,920
(B) 2,500
(C) 3,350
(D) 4,220
(E) 5,490

3. The patient has been advised to reduce her caloric intake to 10% less than what is suggested for someone her weight and activity level. This change can be accomplished by changing which one of the following in her diet?

(A) Reducing protein intake by 45 g/day
(B) Reducing carbohydrate intake by 45 g/day
(C) Reducing fat intake by 45 g/day
(D) Reducing carbohydrate intake by 17.5 g/day, and fat intake by 20 g/day
(E) Reducing fat intake by 17.5 g/day, and carbohydrate intake by 20 g/day

4. A patient is on a very low calorie liquid diet and must take supplements to ensure that he has the essential vitamins and minerals to maintain his health. Under the appropriate conditions, which one of the following compounds can be synthesized in humans, and would not need to be supplemented to the extent that the others are?

(A) Riboflavin
(B) Linoleic acid
(C) Leucine
(D) Thiamine
(E) Niacin

5. A person who accidentally ingested a mold toxin that completely inhibited phosphoenolpyruvate carboxykinase could still form substantial amounts of blood glucose from which one of the following?

(A) Muscle glycogen stores
(B) Lactate produced by red blood cells
(C) Ingested fructose
(D) Ingested galactose
(E) Ingested fructose or galactose

6. A solution contains 2×10^{-3} mol/L of a weak acid ($pK = 3.5$) and 2×10^{-3} mol/L of its conjugate base. Its pH is best approximated by which one of the following?

(A) 4.1
(B) 3.9
(C) 3.5
(D) 3.1
(E) 2.7

7. The cytochrome P_{450} system of liver in normal individuals has a capacity (V_m) to oxidize approximately 10 nmol of drug X per minute per gram of liver. When the concentration of drug X in the liver is 2 μM, oxidation products are formed at the rate of 4 nmol/minute/g of liver. What is the K_m of cytochrome P_{450} for this drug?

(A) 4 nM
(B) 5 nM
(C) 10 nM
(D) 2 μM
(E) 3 μM

8. Enzyme Y was purified from a tissue sample obtained from a patient. The kinetic properties of this enzyme and those of the same enzyme isolated from a normal individual are shown in the graph. Which one of the following statements best describes these results?

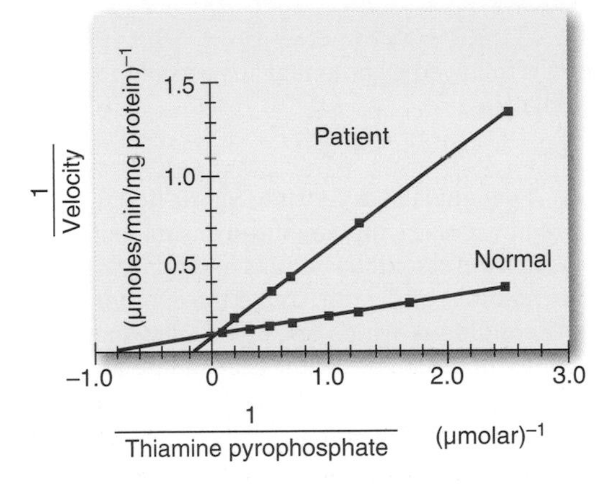

(A) The enzymes do not have the same V_{max}.
(B) A lower concentration of thiamine pyrophosphate (TPP) is required to saturate the patient's enzyme.
(C) The patient's enzyme has a K_m for TPP that is less than that for the normal enzyme.
(D) Administration of thiamine to the patient should result in a greater proportion of the enzyme in the active enzyme–TPP complex.
(E) Administering TPP reduces the V_{max} of the enzyme without altering the K_m.

9. A lack of pantothenic acid would most directly affect the reaction catalyzed by which one of the following enzymes?

(A) Citrate synthase
(B) Isocitrate dehydrogenase
(C) Succinate dehydrogenase
(D) Fumarase
(E) Malate dehydrogenase

10. A college wrestler has been fasting for 5 days in order to make weight for an upcoming tournament. Under these conditions, glucose is the major fuel for which one of the following tissues? Choose the one best answer.

(A) Muscle
(B) Brain
(C) Liver
(D) Red blood cells
(E) Kidney

11. Adult hemoglobin A (HbA) is best described by which one of the following?

	Total number of iron atoms per molecule of HbA	Types of subunits	Protons stabilize:	2,3-bisphosphoglycerate stabilizes:	Has a higher affinity for binding to HbA
A	4	2α, 2β	Deoxy form	Deoxy form	CO
B	4	2α, 2γ	Deoxy form	Oxygenated form	O_2
C	4	2α, 2β	Deoxy form	Deoxy form	CO_2
D	2	2α, 2γ	Oxygenated form	Oxygenated form	CO
E	2	2α, 2β	Oxygenated form	Deoxy form	O_2
F	2	2α, 2γ	Oxygenated form	Oxygenated form	CO_2

Questions 12 and 13 are based on the following case:

A homeless man was seen at a clinic because of bleeding gums and loosening teeth. His dietary history revealed that he has been consuming only chocolate milk and fast-food hamburgers for the past 6 months.

12. The patient is exhibiting these symptoms due to which one of the following?

(A) Reduced synthesis of fibrillin
(B) Reduced synthesis of collagen
(C) Reduced hydrogen-bond formation in collagen
(D) Increased hydrogen-bond formation in collagen
(E) Reduced disulfide-bond formation in collagen
(F) Increased disulfide-bond formation in collagen

13. The patient, due to his diet, had become deficient in which one of the following vitamins, which would lead to the symptoms observed?

(A) Vitamin A
(B) Vitamin C
(C) Vitamin B_1
(D) Vitamin B_2
(E) Vitamin B_6

14. Which one of the following best represents fuels and the use of fuels within the body after an overnight fast?

	The brain utilizes:	The red blood cells utilize:	The muscle utilizes:	Ketone bodies are synthesized by:
A	Glucose	Glucose	Fatty acids	Muscle
B	Fatty acids	Fatty acids	Ketone bodies	Brain
C	Ketone bodies	Ketone bodies	Glucose	Liver
D	Glucose	Glucose	Fatty acids	Liver
E	Fatty acids	Fatty acids	Ketone bodies	Brain
F	Ketone bodies	Ketone bodies	Glucose	Muscle

15. Which one of the following best represents aspects of eukaryotic protein synthesis?

	Initiating amino acid	Coupled transcription–translation	Ribosome size	Occurs in this location
A	Met	No	80S	Nucleus
B	*N*-formyl-met	No	70S	Nucleus
C	Met	No	80S	Rough ER
D	*N*-formyl-met	Yes	70S	Rough ER
E	Met	Yes	80S	Rough ER
F	*N*-formyl-met	Yes	70S	Nucleus

Questions 16 and 17 are based on the following finding:

The sequence for a portion of a gene responsible for a lysosomal storage disease (Tay–Sachs) has been determined. The normal gene sequence and the mutant gene sequence are given below. (There is a dot above every fifth base and a number above every 10th base.)

```
            •    10   •    20   •
Normal  CGTATATCCTATGGCCCTGACCCAG
Mutant  CGTATATCTATCCTATGGCCCTGAC
```

The sequence shown is in frame, beginning with the base at the 5′-end of the sequence shown.

16. The sequence of the encoded amino acids in this region is which one of the following?

(A) R-I-S-Y-G-P-D
(B) A-I-S-T-G-P-P
(C) D-P-G-Y-S-I-R
(D) There is no sequence as the first coding sequence is a termination signal.
(E) There is no sequence, as the base T is not in RNA.

17. In this instance, Tay–Sachs disease comes about because of which one of the following?

(A) A conservative amino acid substitution in the mutant gene.
(B) A nonconservative amino acid substitution in the mutant gene.
(C) A truncated protein produced by the mutant gene.
(D) A greatly larger-than-normal protein produced by the mutant gene.
(E) A deletion leads to an almost normal-sized protein being made, but in the incorrect reading frame.
(F) An insertion leads to an almost normal-sized protein, but in the wrong reading frame.

18. A scientist has created a eukaryotic cell line that was unable to synthesize mRNA at 42°C, but normal synthesis was observed at 35°C. Assuming this is due to a single base change in the DNA, which one of the following is a possible target of this mutation?

(A) RNA polymerase is mutated so it cannot bind directly to the promoter region of DNA at the nonpermissive temperature.
(B) The capping enzyme does not function at the elevated temperature.
(C) The splicesome can no longer recognize the AGGU sequence at a splice junction at the higher temperature.

(D) Ribonucleotide reductase is inactive at the higher temperature.
(E) TFIID contains a mutation in one of its proteins such that it does not bind to DNA.

Questions 19 and 20 refer to the following experiment:

A variant temperature-sensitive cell line was created in which it was observed that when the cells are placed at the nonpermissive temperature they grow for two to three generations, then slowly stop growing and die. An analysis of the DNA of cells grown at the higher temperature indicated a significant level of uracil in the DNA, which was not present in cells grown at the lower (permissive) temperature.

19. What base change is occurring in the DNA at the higher temperature that will eventually lead to mutations, and cell death?

(A) An A:T to a T:A
(B) A C:G to a T:A
(C) A T:A to a C:G
(D) A G:C to a C:G
(E) An A:U to a C:G

20. An enzyme most likely defective at the nonpermissive temperature is which one of the following?

(A) Ligase
(B) DNA polymerase
(C) Reverse transcriptase
(D) An endonuclease
(E) A glycosylase

21. An individual with an autosomal recessive disease has had a section of their DNA sequenced to determine the nature of the mutation. Two restriction enzymes were used, Kpn1, with a sequence recognition sequence of GGTACC, and EcoR1, with a recognition sequence of GAATTC. Previous work had demonstrated that the mutant gene showed a restriction length polymorphism that allowed a diagnostic test to be developed. Which one of the following best reflects the data shown in the attached gels?

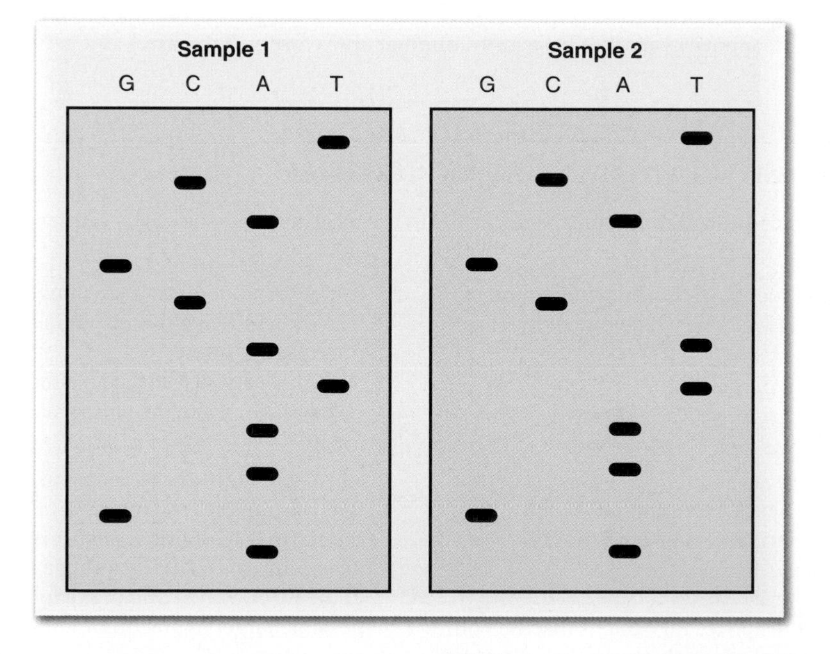

	Restriction enzyme used to create the fragments for sequencing	Restriction enzyme that creates the polymorphism	The DNA from this sample is the mutant sample
A	Kpn1	EcoR1	2
B	EcoR1	EcoR1	2
C	Kpn1	EcoR1	1
D	EcoR1	Kpn1	1
E	Kpn1	Kpn1	1

22. A researcher has developed a prokaryotic cell line that would not divide at 42°C, but does at 35°C. An analysis of the cellular DNA at the nonpermissive temperature found a large number of short, single-stranded nucleic acids that consisted of both RNA and DNA. Much longer single-stranded DNA was also found. The mutated enzyme is most likely which one of the following?

(A) DNA ligase
(B) DNA polymerase I
(C) DNA polymerase II
(D) DNA primase
(E) DNA gyrase

23. A student, worried about exams, has not eaten for 48 hours; however, her blood glucose levels are still at normal fasting levels. At this point in time, the blood glucose is being produced by which one of the following pathways?

(A) Liver glycolysis
(B) Muscle glycolysis
(C) Liver gluconeogenesis
(D) Muscle gluconeogenesis
(E) Liver glycogenolysis
(F) Muscle glycogenolysis
(G) Liver pentose phosphate pathway
(H) Red blood cell pentose phosphate pathway

24. An individual accidentally ingests an overdose of supplements containing chlorogenic acid, a chemical that inhibits glucose-6-phosphatase. After an overnight fast, this individual, compared with a healthy person, would exhibit which one of the following?

(A) An increased rate of gluconeogenesis
(B) An increased rate of glycogenolysis
(C) An increased level of liver glycogen
(D) An increased level of blood glucose
(E) A decreased level of serum fatty acids

25. To determine the genetic father of a child (C), a laboratory test was performed that depended on a restriction fragment length polymorphism (RFLP) in a region upstream from a known gene. This region contains a variable number of tandem repeats (VNTR). Consequently, *HpaI* produces restriction fragments that differ in size. The restriction fragment map is shown below.

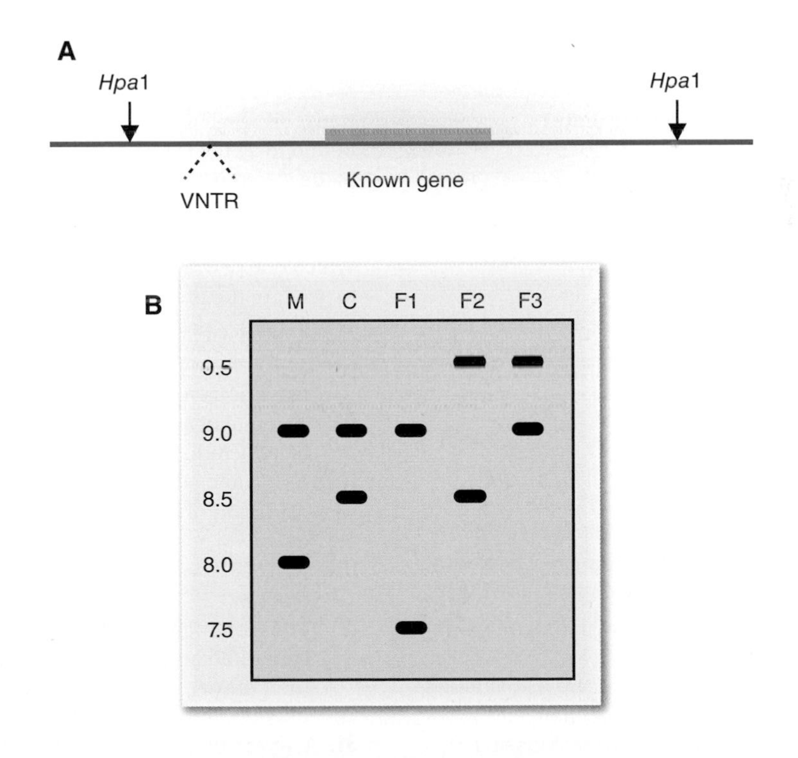

DNA, extracted from a child (C), the child's mother (M), and three men (F1, F2, and F3) was digested with HpaI and subjected to electrophoresis. A probe that hybridized to the known gene was used so that only bands containing this gene were visualized. The gel is shown. The numbers on the left refer to the size in kilobase (kb) pairs of the *HpaI* restriction fragments that bound the probe. On the basis of this data, the child's father is most likely to be which one of the following?

(A) F1
(B) F2
(C) F3
(D) Any one of these men; the test cannot distinguish between them
(E) None of these men

26. A scientist has developed a triglyceride analog that acts as a potent suicide inhibitor of pancreatic lipase. Animal studies using this inhibitor would be expected to show which one of the following results in animals fed a normal diet?

	Steatorrhea	Eventual prostaglandin deficiency	Eventual chylomicron levels	Develops symptoms of cystic fibrosis
A	Yes	Yes	Low	Yes
B	No	No	Low	Yes
C	Yes	Yes	Low	No
D	No	No	Elevated	No
E	Yes	Yes	Elevated	Yes
F	No	No	Elevated	No

Questions 27 and 28 are based on the following equations in which the standard free energy change ($\Delta G^{o\prime}$) for each of the reactions in the conversion of malate to isocitrate is shown.

Malate + $NAD^+ \rightarrow$ oxaloacetate + NADH + H $\Delta G^{o\prime} = +7.1$ kcal/mol

Oxaloacetate + acetyl-CoA $\rightarrow$ citrate + CoASH $\Delta G^{o\prime} = -7.7$ kcal/mol

Citrate $\rightarrow$ isocitrate $\Delta G^{o\prime} = +1.5$ kcal/mol

27. What is the $\Delta G^{o\prime}$ (in kcal/mol) for the net conversion of malate to isocitrate?

(A) −13.3
(B) −2.1
(C) −0.9
(D) + 0.9
(E) + 1.5

28. The equilibrium constant for the net conversion of malate to isocitrate is which one of the following?

(A) 0
(B) >1
(C) <1

29. As a runner begins a 1,500-m race, increased ATP hydrolysis during muscle contraction can lead to which one of the following?

(A) A decrease in the rate of palmitate oxidation to acetyl-CoA
(B) A decrease in the rate of NADH oxidation by the electron transport chain
(C) Activation of phosphofructokinase-1 (PFK-1)
(D) An increased proton gradient across the inner mitochondrial membrane
(E) A decrease in the rate of lactate formation

30. A scientist was studying the electron transport chain and developed a compound that bound to all the cytochromes in the proteins of the electron transport chain. Once bound, the compound blocked electron acceptance by the iron in the heme group of the cytochrome. Which one of the following would be most likely to occur in the presence of an oxidizable substrate?

(A) The amount of heat generated from NADH oxidation would increase.
(B) The rate of succinate oxidation would not be affected.
(C) ATP could still be generated from the oxidation of NADH by transfer of electrons to O_2.
(D) Cell death would result from a lack of ATP.
(E) The electrochemical gradient could still be generated across the inner mitochondrial membrane.

31. A 3-year-old boy was found to have reduced red blood cell (RBC) numbers, yet exhibited very few signs of anemia. An analysis of labeled red blood cells indicated a greatly reduced ATP yield as compared to someone without the anemia. In this child, which one of the following would be expected to increase in RBC?

(A) The life span of the red blood cells
(B) The rate of fatty acid oxidation
(C) ATP production
(D) The NADH/NAD^+ ratio
(E) The activity of hexokinase

Questions 32 and 33 are based on the following case:

A 10-year-old boy displays exercise intolerance, complaining of leg and arm pain after only a short period of exercise. The pain is worse if anaerobic activities are being attempted; for the more aerobic activities, the boy can complete the task, but at a much slower rate than others in the class. Hemolytic anemia was not observed in this child at any time. A muscle biopsy displayed elevated levels of glycogen in the muscle. A stress test in which the boy squeezed a rubber ball over and over again demonstrated a reduced production of lactate as compared to someone who did not display exercise intolerance.

32. The most likely enzyme deficiency in this boy is which one of the following?

(A) Liver phosphorylase
(B) Muscle phosphorylase
(C) Liver glucose-6-phosphatase
(D) Muscle glucose-6-phosphatase
(E) Liver PFK-1
(F) Muscle PFK-1

33. Neither of the boy's parents is exercise-intolerant. What is the probability that the boy's parents will have another child with the same phenotype as this child?

(A) 0%
(B) 25%
(C) 33%
(D) 50%
(E) 100%

Questions 34 and 35 are based on the following case:

A woman develops severe abdominal cramps and flatulence whenever she eats dairy products, so she has decided to eliminate all such products from her diet. She has been this way all her life, and had been on a special diet while growing up, but got tired of it when she entered college.

34. What enzyme deficiency is most likely in this woman?

(A) Fructokinase
(B) Aldolase
(C) Lactase
(D) Glucoamylase
(E) Pancreatic amylase

35. Which one of the following is an accurate statement concerning sugar metabolism in this woman?

(A) She cannot produce mucopolysaccharides that contain galactose.
(B) She cannot produce lactose during lactation.
(C) She is likely to suffer from a calcium deficiency.
(D) She is likely to have high levels of galactose-1-phosphate.
(E) She is likely to have elevated levels of fructose 1-phosphate.

Questions 36 and 37 are based on the following case:

A person with Type 1 diabetes went on a trip and ran out of insulin. Instead of going through the trouble of transferring her prescription, she decided to go without insulin for 5 days, at which point she would be back home. However, after 4 days she felt lethargic, nauseous, and had difficulty standing. She was rushed to the emergency department for further evaluation.

36. Blood tests of the woman would indicate elevated levels of which of the following? Choose the one best answer.

(A) Glucose
(B) Ketone bodies
(C) Insulin
(D) Glucose and ketone bodies
(E) Ketone bodies and insulin
(F) Glucose and insulin

37. After appropriate treatment, which one of the following liver enzymes would be reduced in activity as compared to before treatment?

(A) Fructose 1,6-bisphosphatase
(B) Pyruvate kinase
(C) Pyruvate dehydrogenase
(D) PFK-1
(E) PFK-2 (kinase activity)

38. Which one of the following best represents some properties of proteoglycans?

	Contains repeating disaccharide units	Degradation occurs in lysosomes	Carbohydrate is covalently linked to proteins	Sulfation occurs prior to sugar addition to the proteoglycan
A	Yes	Yes	Yes	Yes
B	Yes	No	Yes	Yes
C	Yes	Yes	Yes	No
D	No	No	No	No
E	No	Yes	No	Yes
F	No	No	No	No

Questions 39 and 40 are based on the following case:

An 18-month-old boy is exhibiting failure to thrive despite a healthy appetite, and foul-smelling and frequent stools. A sweat test for the presence of chloride ions was positive.

39. Digestion of which of the following substances would most likely be impaired in this child? Choose the one best answer.

	Lactose	Sucrose	Starch	Triglycerides	Proteins
A	Yes	Yes	Yes	No	No
B	No	Yes	No	No	Yes
C	Yes	Yes	Yes	No	No
D	No	No	Yes	Yes	Yes
E	Yes	No	No	Yes	No
F	No	No	No	Yes	Yes

40. The boy has a 3-year-old sister who displays none of the symptoms as does the boy. What is the probability that she is a carrier for this disease?

(A) 25%
(B) 33%
(C) 50%
(D) 67%
(E) 100%

41. A woman suffered through intermittent pain in her stomach area, particularly after eating. Imaging tests revealed a large gallstone blocking the upper part of the bile duct. Such a blockage would cause an increase in which one of the following?

(A) The formation of chylomicrons
(B) The recycling of bile salts
(C) The excretion of bile salts
(D) The excretion of fat in the feces
(E) The formation of VLDL

42. A 20-year-old woman with diabetes mellitus (DM) was admitted to the hospital in a semiconscious state with fever, nausea, and vomiting. Her breath smelled of acetone. Which of the following best represents the various aspects of her condition?

	Blood glucose levels	Initial treatment of the condition	Origin of the acetone	Blood pH
A	>100 mg/dL	Glucose infusion	Acetyl-CoA	<7.4
B	<100 mg/dL	Glucagon injection	Hydroxybutyrate	>7.4
C	>100 mg/dL	Insulin injection	Acetoacetate	<7.4
D	<100 mg/dL	Glucose infusion	Acetyl-CoA	>7.4
E	>100 mg/dL	Glucagon injection	Hydroxybutyrate	<7.4
F	<100 mg/dL	Insulin injection	Acetoacetate	>7.4

43. Which one of the following best represents de novo fatty acid biosynthesis starting with cytosolic citrate?

	NADH is required	Major product is:	Occurs at a glucagon:insulin ratio best described as:	Required cofactors
A	No	Palmitic acid	Low	Biotin
B	No	Stearic acid	High	Coenzyme A
C	No	Palmitic acid	Low	Biotin and coenzyme A
D	Yes	Stearic acid	High	Biotin
E	Yes	Palmitic acid	Low	Coenzyme A
F	Yes	Stearic acid	high	Biotin and coenzyme A

44. The synthesis of triacylglycerol from glucose in the liver is best represented by which one of the following?

	Transports acetyl-CoA across the mitochondrial membrane	Reducing equivalents can be derived from:	2-monoacylglycerol is an intermediate	The compound below is a necessary intermediate in the synthesis of triglycerides
A	Citrate	Malic enzyme	Yes	Phosphatidylcholine
B	Malate	TCA cycle	No	Phosphatidylserine
C	Oxaloacetate	Pyruvate dehydrogenase	Yes	Phosphatidic acid
D	Citrate	Malic enzyme	No	Phosphatidic acid
E	Malate	Pyruvate dehydrogenase	No	Phosphatidylserine
F	Oxaloacetate	TCA cycle	Yes	Phosphatidylcholine

45. A scientist has developed an adipocyte cell line that, at 42°C, cannot degrade triglycerides to glycerol and free fatty acids. At 25°C, the triglyceride degradation is normal. Which one of the following best reflects enzymes in which a temperature-sensitive mutation may lead to this phenotype?

	Protein kinase A	Adenylate kinase	Insulin receptor	Glucagon receptor	Glycerol kinase
A	Yes	No	No	Yes	Yes
B	No	No	Yes	No	Yes
C	Yes	No	No	Yes	No
D	No	Yes	Yes	No	No
E	Yes	Yes	No	Yes	Yes
F	No	Yes	Yes	No	No

Questions 46 through 48 are based on the following case:

A 40-year-old man suffered a heart attack, and his cholesterol levels were determined to be 322 mg/dL. The patient's father had died of a heart attack at the age of 47, and his 38-year-old brother had experienced a heart attack last year.

46. These family members are most likely heterozygotes for a mutation in which one of the following?

(A) HMG-CoA reductase
(B) HMG-CoA lyase
(C) Apolipoprotein B48
(D) Apolipoprotein B100
(E) LDL receptor
(F) The microsomal triglyceride transfer protein

47. The patient was placed on a statin, and after 4 months on the drug, his circulating cholesterol levels were 180 mg/dL. Statins can produce this effect due to which one of the following?

(A) They cause cellular levels of squalene to increase.
(B) They cause cellular levels of HMG-CoA to decrease.

(C) They cause LDL receptor numbers to increase on the cell surface.
(D) They cause blood triacylglycerol levels to increase.
(E) They cause cellular acyl:cholesterol acyltransferase (ACAT) activity to increase.

48. What would be the probability that a sibling of the patient would experience a myocardial infarction before the age of 20?

(A) 0%
(B) 25%
(C) 33%
(D) 50%
(E) 67%
(F) 100%

49. Which one of the following best represents some properties of the bile salts?

	Can contain glycine or serine	Most effective as detergents at a pH <3.0	They are amphipathic molecules	They are resorbed in the ileum, and synthesized in the liver	They are derived from cholesterol
A	No	No	Yes	Yes	Yes
B	No	Yes	Yes	No	Yes
C	No	No	No	Yes	Yes
D	Yes	Yes	No	No	No
E	Yes	No	No	Yes	No
F	Yes	Yes	Yes	No	Yes

50. A 2-day-old newborn was very fussy and lethargic, and refused to eat. The child was taken to the emergency department, and blood work indicated hyperammonemia and hypercitrullinemia. Which one of the following enzymes is most likely defective in this child?

(A) Argininosuccinate synthetase
(B) Carbamoyl phosphate synthetase I
(C) Formiminotransferase
(D) Ornithine transcarbamoylase
(E) Arginase

51. A 3-day-old male infant became increasingly fussy, lethargic, and difficult to feed. The parents rushed the child to the emergency department, where blood work demonstrated hyperammonemia, and urinalysis indicated large quantities of orotic acid. Neither parent ever expressed any of these abnormal blood and urine results. The mode of transmission of this disease is which one of the following?

(A) Autosomal recessive
(B) Autosomal dominant
(C) Mitochondrial
(D) Multifactorial
(E) X-linked recessive
(F) X-linked dominant

52. An inactivating mutation in which one of the following enzymes would disrupt the glucose–alanine cycle during fasting?

(A) Liver citrate synthase
(B) Muscle citrate synthase
(C) Liver pyruvate dehydrogenase
(D) Muscle pyruvate dehydrogenase
(E) Liver pyruvate kinase
(F) Muscle pyruvate kinase

53. An 8-year-old child is taken for a physical exam due to pressure from the boy's teachers that something is "not right," and that the child is underperforming in class. The boy is remarkable for being tall and thin, with long limbs, and a small curvature to his spine (scoliosis). He displays pectus carinatum, and a dislocation of the lens in one eye. He has also had a number of recent blood clots that required treatment. After examination, and the running of some tests, the boy was placed on pharmacological doses of vitamin B_6, and this treatment appeared to alleviate some of the symptoms. The enzyme that is most likely to be defective in this child is which one of the following?

(A) Cystathionine β-synthase
(B) *S*-adenosylhomocysteine hydrolase
(C) Methionine synthase
(D) N^5,N^{10}-methylenetetrahydrofolate reductase
(E) SAM synthetase

54. A 12-year-old Russian girl immigrated to the United States with her parents. She was slow in her mental milestones, displayed arm and leg tremors, along with periods of hyperactivity. A blood test indicated elevated levels of phenylpyruvate. A potential defective enzyme in this child is which one of the following?

(A) Tyrosinase
(B) Homogentisic acid 1,2-dioxygenase
(C) Branched-chain α-keto acid dehydrogenase
(D) Monoamine oxidase
(E) Dihydrobiopterin reductase

55. A deficiency in vitamin B_6 would be expected to reduce the synthesis of which of the following neurotransmitters? Choose the one best answer.

	γ-Aminobutyric acid (GABA)	Serotonin	Epinephrine	Dopamine	Histamine
A	No	Yes	Yes	No	Yes
B	No	No	No	Yes	No
C	No	Yes	No	No	Yes
D	Yes	Yes	Yes	Yes	Yes
E	Yes	No	Yes	No	No
F	Yes	Yes	Yes	Yes	No

56. In comparing the de novo synthesis of IMP and UMP, which one of the following best represents commonalities in the pathways?

	Both require PRPP	Both require folate derivatives	Both require glutamine	Both require glycine	Both require aspartic acid
A	Yes	No	Yes	No	Yes
B	Yes	No	No	No	Yes
C	Yes	Yes	Yes	No	No
D	No	Yes	No	Yes	Yes
E	No	No	Yes	Yes	No
F	No	Yes	No	Yes	Yes

57. A college professor, celebrating his 60th birthday, had too much foie gras and wine at his celebratory dinner. The next morning, he awakened with both a hangover and a severe pain in his right great toe, a condition he had experienced a number of times previously. Being somewhat absentminded, the professor then remembered that he had forgotten to take his maintenance medication for this condition for over 2 weeks. His maintenance medication most likely blocks which one of the following reactions?

(A) IMP to GMP
(B) Adenosine to inosine
(C) Hypoxanthine to xanthine
(D) dUMP to dTMP
(E) Cytosine to uracil

Questions 58 to 60 are based on the following case:

A man of Mediterranean descent has just consumed a large bowl of fava beans, as it was a delicacy that he had never tried before. A few hours later, the man complained of being tired and very lethargic. He was taken to the emergency department of a local hospital, and it was determined that his red blood cell count was low, and on microscopic analysis it appeared as if the red cells had burst, which contributed to the low red blood cell count.

58. A likely enzyme defect in this patient is which one of the following?

(A) Glucokinase
(B) Glucose-6-phosphate dehydrogenase
(C) Pyruvate dehydrogenase
(D) Pyruvate kinase
(E) Spectrin
(F) α-Ketoglutarate dehydrogenase

59. A close examination of the patient's eyes allows detection of a yellow tinge. This is most likely due to the accumulation of which one of the following?

(A) Hemoglobin
(B) Glycosylated hemoglobin
(C) Bilirubin
(D) Bilirubin diglucuronide
(E) Stercobilin

60. The probability of this patient passing his defective allele to his son is which one of the following?

(A) 0%
(B) 25%
(C) 50%
(D) 67%
(E) 100%

61. An individual has been on a fad diet for 6 weeks, and has begun to develop a number of skin rashes, diarrhea, and forgetfulness. These symptoms could have been less severe if the diet contained a high content of which one of the following?

(A) Tyrosine
(B) Tryptophan
(C) Thiamine
(D) Thymine
(E) Riboflavin

62. Pyridoxal phosphate is a key cofactor in metabolism. Which one of the following best represents reactions that require this cofactor?

	Glycogen$_n$ converted to glucose-1-phosphate and glycogen$_{n-1}$	Pyruvate plus aspartate producing alanine and oxaloacetate	Homocysteine plus N^5-methyl-THF produces methionine and THF	Homocysteine plus serine produces cystathionine	Histidine produces histamine
A	Yes	Yes	No	Yes	Yes
B	Yes	Yes	Yes	No	Yes
C	Yes	Yes	No	Yes	No
D	Yes	No	Yes	No	No
E	No	No	No	Yes	Yes
F	No	No	Yes	No	No

63. A 52-year-old patient with a round face, acne, and a large hump on the back of his neck complains that he is too weak to mow his lawn. His fasting blood glucose level is 170 mg/dL (the reference range is 80 to 100 mg/dL); plasma cortisol level is 62 μg/mL (the reference range is 3 to 31 μg/mL); and plasma ACTH level is 0 pg/mL (the reference range is 0 to 100 pg/mL). If the patient's condition is due to a single cause, the most likely diagnosis is which one of the following?

(A) Type 1 DM
(B) Type 2 DM
(C) A secretory tumor of the anterior pituitary
(D) A secretory tumor of the posterior pituitary
(E) A secretory tumor of the adrenal cortex

64. A 25-year-old female patient underwent a complex surgery that resulted in damage to the pituitary gland, such that hypopituitarism resulted. Part of the treatment for this patient, postsurgery, should be which one of the following?

(A) TSH and ACTH should be given orally.
(B) Water intake should be restricted to compensate for low vasopressin levels.
(C) Thyroxine tablets should be prescribed and taken regularly by the patient.
(D) Cortisol should be administered daily except during periods of increased stress.
(E) Estrogen and progesterone are the only hormones needed if the patient wishes to remain fertile.

65. A woman whose thyroid gland was surgically removed was treated daily with 0.10 mg of thyroxine (tablet form). After 3 months of treatment, her serum TSH levels were constant at 6 MIU/mL (the reference range is 0.3 to 5 MIU/mL). She complained of fatigue, weight gain, and hoarseness. Her dose of thyroid hormone should be adjusted in which manner?

(A) Increased
(B) Decreased
(C) Remain the same

Questions 66 to 69 are matching questions. Indicate whether the blood levels of the compounds below would be higher, lower, or the same in a person with Type 1 (insulin-dependent) DM who fails to take insulin for 2 days compared with a normal person who has just finished dinner.

66. Glucose
67. Glucagon
68. Urea
69. Ketone bodies

(A) Higher
(B) Lower
(C) The same

Questions 70 to 72 are matching questions. Match each antibiotic below with the appropriate step in translation that it inhibits in prokaryotes.

70. Tetracycline
71. Streptomycin
72. Erythromycin

(A) Initiation
(B) Binding of aminoacyl-tRNA to the "A" site on the ribosome
(C) Peptide-bond formation
(D) Translocation

73. A scientist was studying a fibroblast cell line obtained from a child who had died at 2 years of age because of lysosomal dysfunction. An analysis of the cell line demonstrated that many lysosomal enzymes were secreted from the cell, as opposed to being located in the lysosomes. Addition of lyosomal enzymes from normal cells to the growth medium of the patient's cells led to the enzymes being incorporated by the cells and finding their way into the lysosomes. The cellular compartment in which the enzyme defect exists in the patient is which one of the following?

(A) The cytoplasm
(B) Mitochondria
(C) Golgi apparatus
(D) Peroxisome
(E) Endoplasmic reticulum
(F) Lysosome

74. A woman who suffers from frequent and severe migraine headaches has had five children, all of whom have experienced, beginning between the ages of 8 and 12, stroke-like episodes compounded with exercise intolerance and lactic acidosis. The father of the children does not suffer from migraines, nor is he exercise-intolerant. A target of a mutation that can explain these findings is most likely which one of the following?

(A) Mitochondrial tRNA
(B) Cytoplasmic tRNA
(C) Cytochrome c
(D) Pyruvate dehydrogenase
(E) Mitochondrial porin

75. Simian sarcoma virus produces a PDGF homolog that is necessary for cellular transformation. The basic mechanism of action of this homolog is best described by which one of the following?

(A) Autocrine
(B) Endocrine
(C) Paracrine

Questions 76 to 79 are matching questions. The blood levels of glucose, galactose, and fructose were measured in normal persons and in persons with various enzyme deficiencies soon after they drank a milk shake made with milk and sugar. Match each comparison below with the appropriate sugar(s). An answer may be used once, more than once, or not at all.

76. Lower in the blood of a person with a lactase deficiency than in a normal person
77. Higher in the blood of a person with a galactose-1-phosphate uridyltransferase deficiency than in a normal person
78. Higher in the blood of a person with a fructokinase deficiency than in a normal person
79. Higher in the blood of a person with an aldolase B deficiency than in a normal person

(A) Glucose
(B) Galactose
(C) Fructose
(D) Glucose and galactose
(E) Glucose, galactose, and fructose

Questions 80 to 82 are based on the following case:

A chronic alcoholic was found unconscious near a local bar, and was taken to the emergency department at the local hospital. Lab work demonstrated a blood glucose level of 44 mg/dL (the normal fasting level is 80 to 100 mg/dL). Thiamine levels were also determined to be significantly below normal. The physical exam indicated a yellow sclera, and a slight yellowing of the skin.

80. In order to determine thiamine levels, the lab would be able to assay which one of the following enzymes, both in the absence and presence of exogenous thiamine?

(A) Isocitrate dehydrogenase
(B) Glycogen phosphorylase
(C) Adenosine deaminase
(D) Aldolase
(E) Transketolase

81. As the patient had not eaten for a week, other than drinking alcohol, the disruption in the glucose–alanine cycle is occurring at which one of the following steps?

(A) Pyruvate to alanine in the muscle
(B) Alanine to pyruvate in the liver
(C) Pyruvate to oxaloacetate in the liver
(D) PEP to 2-phosphoglycerate in the liver
(E) Glucose 6-phosphate to glucose in the liver

82. The yellow sclera and skin is an indication that which one of the following reactions is occurring at a slower-than-normal rate?

(A) Hemoglobin to Hemoglobin A1C
(B) Hemoglobin to glucuronidated hemoglobin
(C) Hemoglobin to bilirubin
(D) Bilirubin to glycosylated bilirubin
(E) Bilirubin to glucuronidated bilirubin

Questions 83 to 88 are matching questions. A deficiency of each substance below can result in an anemia. For each substance, choose the type of anemia that would occur if the substance were deficient. An answer may be used once, more than once, or not at all.

83. Iron
84. Intrinsic factor
85. Pyridoxine
86. B_{12}
87. Folate
88. NADPH

(A) Megaloblastic anemia
(B) Hypochromic, microcytic anemia
(C) Hemolytic anemia
(D) Sickle cell anemia

Questions 89 to 93 are matching questions. For each untreated condition below, select the blood or urine value that best distinguishes that condition from the others. All values are measured after an overnight fast and are compared with those of a normal individual. An answer may be used once, more than once, or not at all.

89. Type 1 (insulin-dependent) DM
90. Myocardial infarction
91. Hepatitis
92. Renal failure
93. Alcoholism

(A) Increased MB fraction of serum creatine kinase (CK)
(B) Increased blood ketone bodies
(C) Decreased creatinine in the urine
(D) Decreased blood lactate
(E) Decreased blood urea nitrogen (BUN)

Questions 94 and 95 are based on the following case:

A 42-year-old man was feeling tired and lethargic, and blood work indicated an iron deficiency. A colonoscopy indicated a significant right-sided mass in his proximal colon. His family history indicated that his grandfather and mother both had colon cancer in their late 40s. There were very few polyps viewed during the colonoscopy.

94. This patient most likely has an initiating mutation in which one of the following genes?

(A) *BRCA1* or *BRCA2*
(B) Retinoblastoma
(C) The *APC* gene
(D) DNA mismatch repair genes
(E) The ras gene

95. The analysis of the pedigree for the patient in the previous question would reveal which type of inheritance pattern?

(A) Autosomal dominant
(B) Autosomal recessive
(C) X-linked dominant
(D) X-linked recessive
(E) Mitochondrial

96. A gain-of-function mutation within a gene, which can lead to increased cellular proliferation and potential tumor formation, is best classified as which one of the following? Choose the one best answer.

(A) Tumor suppressor
(B) G-protein
(C) Transcription factor
(D) Oncogene
(E) Integral membrane protein

Questions 97 to 100 should be answered from the lettered list below. Each pathologic condition below can be associated with one of the blood hormone levels listed as potential answers. For each condition, choose the most appropriate hormone. An answer may be used once, more than once, or not at all.

97. Hypothyroidism caused by a viral infection of the thyroid gland
98. Elevated levels of somatostatin
99. Low levels of dopamine produced by the hypothalamus
100. Hypotension

(A) High prolactin (PRL)
(B) High thyroid-stimulating hormone (TSH)
(C) High cortisol
(D) Low growth hormone (GH)
(E) Low aldosterone

Questions 101 to 105 are matching questions. For each condition below, choose the major fuel that is being used. An answer may be used once, more than once, or not at all.

101. By the brain after 1 day of fasting
102. By red blood cells following an overnight fast
103. By the liver after 2 days of fasting
104. By the brain after 1 week of fasting
105. By the chest muscles used during weight lifting

(A) Ketone bodies
(B) Blood glucose
(C) Fatty acids
(D) Glycogen
(E) Alanine

Questions 106 to 109 are matching questions. Match each description below with the appropriate lipid–protein complex. An answer may be used once, more than once, or not at all.

106. The major donor of cholesterol to peripheral tissues
107. The first lipoprotein to increase in concentration in the blood after ingestion of 400 g of jelly beans (carbohydrate)
108. The site of the LCAT reaction
109. Composed mainly of triacylglycerols synthesized in intestinal epithelial cells

(A) Chylomicrons
(B) VLDL
(C) LDL
(D) HDL
(E) Fatty acid–albumin complexes

110. A 6-month-old boy has had frequent infections, some of which required hospitalization. Blood work demonstrated an almost complete lack of B and T cells. Deoxyadenosine levels were nondetectable. A family history indicated that the boy's mother had a brother with similar symptoms, and who had died at 2 years of age. The most likely molecular defect in this child is which one of the following?

(A) An inactivating mutation in purine nucleoside phosphorylase
(B) An inactivating mutation in adenosine deaminase
(C) An inactivating mutation in Bcl-2
(D) An inactivating mutation in the retinoblastoma gene product
(E) An inactivating mutation in cytokine receptors

Questions 111 and 112 are based on the following case:

A 6-year-old boy developed a tumor in his right eye, and 3 years later, one in his left eye. FISH analysis for a suspected mutated gene indicated two signals in normal cells from the boy, but only one signal from tumor cells derived from the boy.

111. The tumor most likely arose from which one of the following?

(A) Inhibition of DNA repair
(B) Stimulation of DNA repair
(C) Ras activation
(D) Enhanced tyrosine kinase activity
(E) Dysregulation of gene transcription

112. What is the significance of the FISH results in the child?

(A) An indication that the mutated gene is an oncogene.
(B) An indication that a fused chromosome has occurred.
(C) An indication that loss of heterozygosity has occurred.
(D) An indication that gene duplication has occurred.
(E) An indication that polyploidy is evident.

Questions 113 to 117 are matching questions. Each condition listed below (the numbered question) can be caused by a problem with the metabolism of a particular compound. Match the condition with the appropriate compound. An answer (lettered choices) may be used once, more than once, or not at all.

113. Ehlers–Danlos syndrome
114. Parkinson's disease
115. Tay–Sachs disease
116. McArdle's disease
117. Maple syrup urine disease

(A) Glycogen
(B) Collagen
(C) Dopamine
(D) Valine
(E) A sphingolipid

Questions 118 to 121 are matching questions. A dietary deficiency of a vitamin can cause each of the conditions below. Match each condition with the appropriate vitamin. An answer may be used once, more than once, or not at all.

118. Pellagra
119. Scurvy
120. Beriberi
121. Rickets

(A) Vitamin C
(B) Niacin
(C) Vitamin D
(D) Biotin
(E) Thiamine

122. A scientist was studying a fibroblastic cell line derived from a patient's tumor and found that the MAP kinase activity (erk) was constitutively active. Activating mutations in which of the following would lead to such a phenotype?

	raf	mek	SMAD	STAT	Receptor tyrosine kinase
A	Yes	Yes	No	No	Yes
B	Yes	No	Yes	No	Yes
C	Yes	Yes	No	Yes	Yes
D	No	No	Yes	Yes	No
E	No	Yes	No	No	No
F	No	No	Yes	Yes	No

Questions 123 to 132 are based on the longitudinal history given below. As later questions will, by the nature of this series of questions, give the answers to the previous questions, answer these questions in order before consulting the answers.

123. A 6-year-old boy has become lethargic, was not eating well, but drinking copious amounts of water. He also urinated frequently, sometimes wetting the bed while sleeping. One morning he was difficult to rouse, and his parents took him to the emergency department, where a stat glucose showed 650 mg/dL, and a strip test for ketones was positive. The emergency-room physicians immediately placed the boy on an iv drip. For appropriate treatment, the iv drip should contain which one of the following at initial treatment?

(A) Glucose
(B) Glucagon
(C) Insulin
(D) Fatty acids
(E) Vitamins

124. The elevated glucose level exhibited by the boy is due, primarily, to which one of the following?

(A) Overeating of sweets
(B) Reduced number of glucose transporters in the brain
(C) Competition for glucose entry into the liver by fructose
(D) Reduced use of glucose by the muscle
(E) Reduced glucose filtration by the kidneys, resulting in more glucose in circulation

125. The major reason for the boy's lethargy was which one of the following?

(A) An alkalosis
(B) An acidosis
(C) Hypoglycemia
(D) Hyperlipidemia
(E) Absence of fatty acids in the blood

126. The alteration in blood pH occurred due to which one of the following?

(A) Lactate release by the red blood cells
(B) Lactate release by the muscle
(C) Brain use of glucose as its fuel supply
(D) Muscle use of glucose as its fuel supply
(E) Bicarbonate release by the pancreas
(F) Bicarbonate release by the gall bladder

127. Upon release from the hospital, the boy was put on an insulin regime in order to stabilize his blood glucose levels. On a day when the boy's blood glucose levels are well-regulated, measurement of his C-peptide levels, as compared to someone who does not have Type 1 diabetes, would be which one of the following?

(A) Greater
(B) The same
(C) Reduced

128. The boy is taking two different types of insulin, one long-acting, but slow to initiate action, and the other fast-acting, but not as long-lasting. The major difference between these two insulin forms is best described by which one of the following?

(A) Ability to complex with zinc
(B) Ability to dissociate from the zinc
(C) Protease susceptibility at the injection site
(D) Difference in the injection site
(E) One is oral, the other is injected subcutaneously

129. At the boy's 6-month visit to the endocrinologist, blood work indicated that a fasting blood glucose level was 131 mg/dL, and the HbA1c was 7.2. These results indicate which one of the following?

(A) Excellent glycemic control
(B) Poor glycemic control for 3 months
(C) Abnormal hemoglobin synthesis
(D) An anemia is about to develop
(E) An elevated insulin:glucagon ratio

130. When the boy was 16, he was at a sleepover with some friends, and they were busy playing video games and eating pizza. The patient had taken a shot of fast-acting insulin, but was so engrossed in his video game that he did not eat any pizza. The next morning, he was difficult to awaken, and his friends called 911 in a panic. The reason for this occurring is which one of the following?

(A) Severe hypoglycemia
(B) Severe hyperglycemia
(C) Dehydration
(D) Insulin-induced anemia
(E) Ketoacidosis

131. When the paramedics arrived to treat the boy, he was injected immediately with which one of the following?

(A) Insulin
(B) Glucagon
(C) Cortisol
(D) Epinephrine
(E) Norepinephrine

132. When the patient reached his mid-20s, his physician noticed that when doing the micro-filament test on the soles of the patient's feet there was a reduced response. This has resulted from which one of the following?

(A) Reduced glucose levels in the blood
(B) Nonenzymatic glycosylation of neurons
(C) Elevated glucagon levels
(D) Nonenzymatic glycosylation of hemoglobin
(E) Elevated LDL levels in the blood

Questions 133 to 144 are based on the following patient scenario:

A 55-year-old Native American female had difficulty conceiving when she was young and was diagnosed with polycystic ovarian syndrome. When she did get pregnant, she was diagnosed with gestational diabetes, but successfully delivered a healthy 10-lb baby boy. She had difficulty controlling her weight over the next several years, but her blood glucose checks were always in the "normal" range and her only diagnosis prior to age 40 was sleep apnea. She has a strong family history of diabetes. At age 40, at a health fair, she had a finger-stick random blood glucose of 240. She made an appointment with her primary care doctor who ordered a fasting blood glucose (150) and a hemoglobin A1c (7.4). Other lab values were an HDL cholesterol of 35, total cholesterol of 210, triglycerides of 350, slightly elevated ALT, AST, and uric acid with normal BUN, creatinine, bilirubin, electrolytes, and alkaline phosphatase. Her blood pressure on repeated checks was consistently 150/90. For the next 15 years, she was tried on multiple different medications with only partial success. Currently, her height is 5′10″ (1.8 m), weight 220 lb (100 kg), BP 138/80, HbA1c 8.2, and creatinine 2.0. Her lipid values have not changed. Over the past 2 years, she has been hospitalized three times—for an MI, a community-acquired pneumonia (CAP), and gouty arthritis.

133. Which one of the following is this patient's most complete diagnosis?

(A) Type 1 DM
(B) Type 2 DM
(C) Hypertension
(D) Insulin resistance syndrome
(E) Fatty liver
(F) Sleep apnea

134. Which one of the following would be the most appropriate treatment for her complete diagnosis?

(A) Metformin
(B) Insulin
(C) Weight loss
(D) Statin
(E) Diuretic

135. Which one of the following is closest to her calculated BMI?

(A) 25
(B) 31
(C) 35
(D) 41
(E) 45

136. The woman's BMI places her in which of the following categories?

(A) Underweight
(B) Normal weight
(C) Overweight
(D) Obese
(E) Morbidly obese

137. Her physician again advises diet and weight loss. She does no physical exercise or activity. Which of the following would be the maximum amount of calories per day she could ingest and still lose 1 lb in a week?

(A) 2,400
(B) 2,620
(C) 3,120
(D) 3,930
(E) 3,620

138. What is the patient's LDL value?

(A) 52.5 mg/dL
(B) 105 mg/dL
(C) 125 mg/dL
(D) 150 mg/dL
(E) 210 mg/dL

139. The medication she would most likely be prescribed for her lipid abnormalities to reduce the risk of another heart attack has which one of the following mechanisms of action?

(A) Blocks the absorption of ingested cholesterol
(B) Leads to the upregulation of LDL receptors
(C) Blocks the reabsorption of bile salts
(D) Inhibits the production of VLDL
(E) Significantly reduces serum triglycerides

140. The most commonly recommended outpatient treatment for her CAP has which one of the following mechanisms of action?

(A) Binding of the drug to the 60S ribosomal subunit

(B) Binding of the drug to the 50S ribosomal subunit
(C) Binding of the drug to the 40S ribosomal subunit
(D) Binding of the drug to the 30S ribosomal subunit
(E) Inhibiting folate synthesis

141. Prior to her being admitted to the hospital for her worsening CAP, she had been treated with a sulfa antibiotic orally. Sulfa antibiotics have which of the following mechanisms of action?

(A) Binding of the drug to the 60S ribosomal subunit
(B) Binding of the drug to the 50S ribosomal subunit
(C) Binding of the drug to the 40S ribosomal subunit
(D) Binding of the drug to the 30S ribosomal subunit
(E) Inhibiting folate synthesis

142. Ingestion of food that has a high level of which one of the following components has the potential for worsening one of her diagnoses?

(A) Purines
(B) Histidine
(C) Leucine
(D) Methionine
(E) Pyrimidines

143. The medication most likely used to try to prevent the problem in the previous question works by blocking the metabolism of which one of the following?

(A) Arginine
(B) Ornithine
(C) Xanthine
(D) Carbamoyl phosphate
(E) Argininosuccinate

144. Which of the following is the inheritance pattern for this patient's diagnosis?

(A) Autosomal dominant
(B) Autosomal recessive
(C) Sex-linked
(D) Mitochondrial
(E) Multifactorial

145. A third-year medical student was celebrating the end of a particularly stressful rotation and went on a "binge" of drinking alcohol, without eating any food. Upon awakening the next morning, he felt tremulous and very hungry. Realizing the symptoms of hypoglycemia, he performed a finger-stick glucose and found that his blood glucose level was very low. This most likely occurred because of which one of the following?

(A) Elevated NADH levels in the liver
(B) Elevated NADPH levels in the liver
(C) Elevated NAD^+ levels in the liver
(D) Elevated $NADP^+$ levels in the liver
(E) Acetaldehyde inhibition of gluconeogenesis

146. The number of distinct human chromosomes is which one of the following?

(A) 22
(B) 23
(C) 24
(D) 25
(E) 26

147. A population in Hardy–Weinberg equilibrium has certain individuals expressing a rare autosomal recessive disease. The frequency of affected individuals in the population is 1 in 90,000. What is the frequency of carriers in this population?

(A) 1 in 100
(B) 1 in 150
(C) 1 in 200
(D) 1 in 250
(E) 1 in 300

148. Consider the pedigree shown. The most likely mode of inheritance is which one of the following?

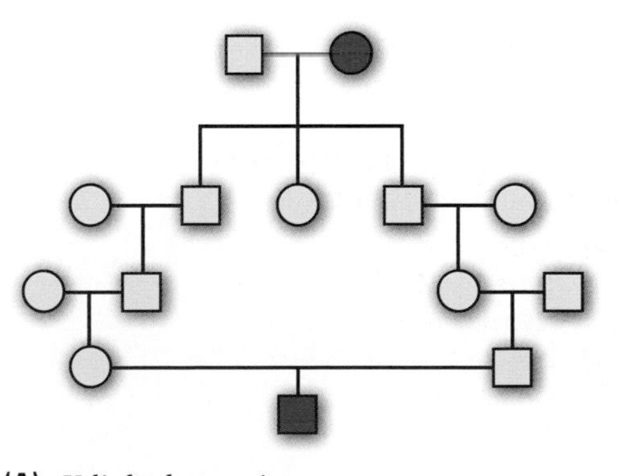

(A) X-linked recessive
(B) X-linked dominant
(C) Mitochondrial
(D) Autosomal dominant
(E) Autosomal recessive

149. A woman is a carrier for an X-linked disease. The disease frequency in the population is 1 in 2,000. What fraction of women in this population are carriers for this particular disease?

(A) 1 in 500
(B) 1 in 1,000
(C) 1 in 1,500
(D) 1 in 2,000
(E) 1 in 4,000

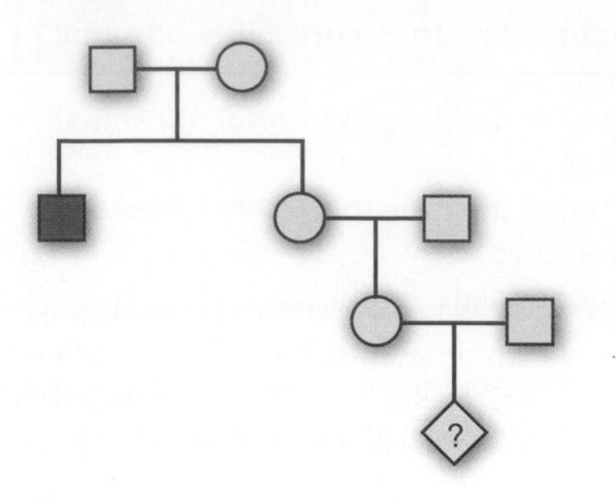

150. The pedigree shown is for a very rare autosomal recessive disease. What is the probability that individual IV-1 will be a carrier of the disease?

(A) 1 in 2
(B) 1 in 4
(C) 1 in 6
(D) 1 in 8
(E) 1 in 10

Answers and Explanations

1. **The answer is E.** Fat contains 9 kcal/g, whereas carbohydrate contains 4 kcal/g. Therefore, when the patient substituted 50 g of fat for 50 g of carbohydrate, he took in 250 more calories per day and, over a 3-month period, gained weight. At an additional 250 kcal/day, the patient would require about 14 days to gain 1 lb (3,500 kcal/lb), or about 6 lb in the 3-month period.

2. **The answer is B.** The patient's weight (176 lb $\times$ 0.454 lb/kg) is 80 kg. Since the basal metabolic rate (BMR) is approximately 24 kcal/kg/day, her BMR (24 kcal/kg/day $\times$ 80 kg) is 1,920 kcal/day. She requires 30% more calories for her activity (sedentary), or 1,920 $\times$ 1.3 = 2,500 kcal/day.

3. **The answer is D.** The patient requires 2,500 g/day to maintain her weight. A 10% reduction per day is a loss of 250 kcal/day from the diet. Proteins and carbohydrates contain 4 kcal/g, and fat contains 9 kcal/g. If the patient reduces her carbohydrate intake by 17.5 g/day, that translates to 70 kcal less per day. If she also reduces her fat intake by 20 g/day, that is an additional 180 kcal less per day. The sum of 70 and 180 is 250 kcal less per day. If these numbers were reversed (a loss of 17.5 g of fat, and 20 g of carbohydrate), then the loss per day would be 237.5 kcal, slightly less than the 10% required.

4. **The answer is F.** Although niacin is a vitamin, it can be synthesized to a limited extent from tryptophan. None of the other vitamins indicated can be synthesized in humans to any extent.

5. **The answer is E.** Both fructose and galactose can be converted to blood glucose in the absence of phosphoenolpyruvate carboxykinase, but this enzyme is required for the conversion of lactate to blood glucose. Muscle glycogen is not converted to blood glucose. PEPCK converts oxaloacetate to PEP, and is used to bypass the irreversible pyruvate kinase step of glycolysis. Neither fructose nor galactose metabolism needs to generate pyruvate to produce glucose, so their gluconeogenic pathway does not require PEPCK. Lactate, however, is converted to pyruvate in its first step of gluconeogenesis, so PEPCK is required to convert lactate to glucose.

6. **The answer is C.** The pH and pK are related as follows: pH = pK + log ([A−]/[HA]). Thus, when the concentrations of a weak acid and its conjugate base are equal, the pH equals the pK. The pK is defined as the pH at which [A−] = [HA].

7. **The answer is E.** According to the Michaelis–Menten equation, the velocity (v) is related to the concentration of substrate, [X], as follows: $v = V_m [X]/(K_m + [X])$. If one uses all concentrations (in nmoles), one gets $4 = (10 \times 2{,}000)/(K_m + 2{,}000)$. This leads to $4 K_m + 8{,}000 = 20{,}000$. $4 K_m = 20{,}000 - 8{,}000 = 12{,}000$. $K_m = 12{,}000/4 = 3{,}000$ nM, which is 3.0 μM.

8. **The answer is D.** The Y intercept ($1/V_m$) is the same for the normal enzyme and the patient's enzyme, and thus the enzymes have the same V_m. However, the X intercepts ($-1/K_m$) differ. The normal enzyme has a K_m of 1 μM, whereas the patient's enzyme has a K_m of 5 μM. Therefore, more TPP is required to saturate the patient's enzyme, and raising the body levels of thiamine should cause more of the patient's enzyme to be in the active complex. The mutation may be in the binding site for TPP, reducing its affinity for the enzyme.

9. **The answer is A.** Acetyl-CoA, which contains pantothenic acid as part of its coenzyme A moiety, is a substrate for citrate synthase. If coenzyme A levels were limiting due to the reduction in pantothenic acid levels, acetyl-CoA could not be synthesized, and the citrate synthase reaction would slow down. The other TCA cycle enzyme that requires pantothenic acid (as a part of coenzyme A) is α-ketoglutarate dehydrogenase, which converts α-ketoglutarate to succinyl-CoA, carbon dioxide, and NADH.

10. **The answer is D.** Red blood cells use glucose (via glycolysis) as their only energy source because they do not have an active tricarboxylic acid (TCA) cycle; they lack mitochondria.

The other tissues have mitochondria and can use other fuels. Even though the brain can reduce its glucose needs by 40% under starvation conditions, the majority of the brain's energy needs still come from glucose. Red blood cells cannot reduce their dependence on glucose under any conditions, as the mature red blood cells lack mitochondria and always generate their energy from glycolysis.

11. **The answer is A.** Hemoglobin contains two α chains and two β chains, which combine to form the quaternary structure. Each mole of a subunit binds 1 mole of heme, which binds 1 mole of O_2. Therefore, 1 mole of HbA binds 4 moles of O_2, contains 4 moles of heme, and 4 moles of iron. Both protons (creating a reduced pH) and 2,3-bisphosphoglycerate stabilize the deoxygenated form of HbA. Carbon monoxide (CO) has a higher affinity (about 250 times) for the iron in heme than does oxygen, even when the distal histidine is present to force CO to bind to the iron at an angle.

12. **The answer is C.** The homeless man has developed scurvy owing to a lack of vitamin C in his diet. Vitamin C is a required cofactor for the hydroxylation of proline and lysine within the collagen molecule. The lack of hydroxyproline reduces the stability of the collagen because of reduced hydrogen-bonding capabilities within the collagen triple helix. The lack of vitamin C does not affect disulfide-bond formation, which is required to initiate triple-helix formation within the cell. Fibrillin is not altered by the lack of vitamin C; it is the protein mutated in Marfan's syndrome.

13. **The answer is B.** Scurvy is due to a lack of vitamin C, which is obtained from citrus fruits, which have been lacking in the diet. The patient may also become deficient in the other vitamins listed, but the lack of those vitamins will not lead to the symptoms characteristic of scurvy.

14. **The answer is D.** After an overnight fast, fatty acids are released from adipose tissue, oxidized by muscle (but not the brain), and converted to ketone bodies in the liver. Glucose is still the choice of fuel by the red blood cells (since they lack mitochondria) and the brain (the brain will not begin switching to some ketone body utilization until about 72 hours after the onset of a fast).

15. **The answer is C.** The initiating amino acid in eukaryotic protein synthesis is methionine; *N*-formylmethionine is used by prokaryotes as the initiating amino acid. Coupled transcription–translation only occurs in prokaryotes, as the nuclear membrane in eukaryotes requires the export of the mRNA from the nucleus into the cytoplasm to begin translation. Eukaryotic ribosomes are 80S (a combination of the 40S small ribosomal subunit and the 60S large ribosomal subunit), whereas prokaryotic ribosomes are 70S (30S and 50S). Eukaryotic protein synthesis occurs in both the cytoplasm and on the rough endoplasmic reticulum (for targeted and secreted proteins).

16. **The answer is A.** The sequence shown is the coding strand of DNA, so it needs to be converted to RNA (replace the Ts with Us, but otherwise the sequence is identical to that of the DNA) in order to read the triplet code. The first codon would be CGU, which encodes arginine (R). There is no stop codon in frame with this sequence. The amino acids in the answers are represented by the single-letter code.

17. **The answer is C.** The mutant gene has a four-base insertion (TATC) starting at position 9. Consequently, a frameshift occurs, and the mutant gene encodes a protein with a different amino acid sequence beyond this point. The insertion causes the sequence TGA at position 22 to come into frame. This corresponds with UGA, a termination codon in the mRNA. Therefore, the mutant protein will be shorter than the normal protein. The mutation does not lead to a conservative or nonconservative substitution of a single amino acid, or the production of a normal-sized, or extended, protein.

18. **The answer is E.** Since there is no RNA synthesis at the nonpermissive temperature, there most likely is a problem with RNA polymerase initiating transcription. RNA polymerase does not bind directly to DNA at the promoter region, rather it will bind to TFIID (which contains the TATA binding protein), which is bound to the DNA. Thus, the best answer is a mutation

in one of the components of TFIID, which prevents appropriate binding of TFIID to the DNA, or to RNA polymerase. Capping and splicing occur on RNA molecules that have already been synthesized, and the lack of either of those activities will not block transcription (although the lack of activity may affect the stability of the RNA produced). Ribonucleotide reductase makes dNTPs, which are not required for RNA synthesis.

19. **The answer is B.** The uracil in the DNA is derived from the spontaneous deamination of the base cytosine, which creates an internal U:G base pair. Normally, the U is removed and replaced by a C, but the mutation in this cell line does not allow this mismatch to be corrected. Therefore, when the DNA is replicated, one of the daughter cells will contain a U:A base pair at this position, which will become a T:A base pair after repair or another round of replication. The changing of a C:G base pair to a T:A base pair will lead to mutations, and eventually cell death.

20. **The answer is E.** The uracil base needs to be removed by base excision repair. In base excision repair, the damaged base is first removed by a glycosylase (uracil-DNA glycohydrolase). Then an endonuclease cleaves the apyrimidinic (AP) site. Nucleotides are removed and replaced by a DNA polymerase, and DNA ligase seals the repaired region to the rest of the DNA strand. Reverse transcriptase uses an RNA template to synthesize DNA, and is not involved in base excision repair.

21. **The answer is C.** The gene sequence is read 5′ to 3′ from the bottom to the top of the gel. The normal and mutant sequences are the same except for a point mutation that converted an A to a T. Thus, gene #1 contains the sequence 5′-GAATTC-3′, which is cleaved by EcoR1, but gene #2 does not. EcoR1, therefore, is the enzyme that allows the polymorphism to be seen, and gene #1 is the mutant gene. Kpn1 would not cleave within the sequences shown on the gels. Kpn1 was used to cleave these two sequences from the larger fragments of DNA. One can see, in both samples, at the 5′-end, the same AGAA sequence, which is part of the Kpn1 cleavage site.

22. **The answer is B.** Okazaki fragments are synthesized on the lagging strand in the 5′ to 3′ direction, moving away from the replication fork and copying the parental DNA strand in the 3′ to 5′ direction. This is due to the activity of DNA polymerase III. The synthesis starts with an RNA primer (created by DNA primase) to which DNA precursors are attached by polymerase III. The RNA is subsequently removed by the 5′-3′ exonuclease activity of DNA polymerase I, and replaced with DNA, and the fragments are joined by DNA ligase. A mutation in DNA polymerase I would lead to the observations, as DNA ligase will not link a ribonucleotide to a deoxyribonucleotide.

23. **The answer is C.** Blood glucose is maintained after about 2 hours of fasting by liver glycogenolysis, which is subsequently supplemented by liver gluconeogenesis. However, after about 1 day of fasting, liver glycogen is depleted, so thereafter, gluconeogenesis is solely responsible for maintaining blood glucose levels. The muscle lacks glucose-6-phosphatase, so does not export free glucose to maintain blood glucose levels, either from glycogenolysis or gluconeogenesis. Glycolysis metabolizes glucose, and does not produce glucose. The pentose phosphate pathway (hexose monophosphate shunt pathway) does not produce glucose for export.

24. **The answer is C.** After an overnight fast, glycogenolysis and gluconeogenesis act to maintain blood glucose levels in a normal person. Both pathways produce glucose 6-phosphate and require glucose-6-phosphatase to produce free glucose. If the phosphatase is inhibited, blood glucose levels will be lower and liver glycogen stores higher than normal. (When glucose-6-phosphatase is genetically deficient, a similar set of circumstances occurs, and the individual has a glycogen storage disease—von Gierke's disease.) Due to the decrease in blood glucose levels, the amount of fatty acids in the circulation will increase because of the activation of hormone-sensitive lipase by protein kinase A.

25. **The answer is B.** Because there are two copies (alleles) of this gene in the genome, two fragments containing this gene are produced from each person. These two restriction fragments have a different number of tandem repeats; one fragment is inherited from the mother and the other from the father. The child received the 9-kb fragment from the mother. An 8.5-kb fragment could only have come from F2; therefore, he is most likely to be the father.

26. **The answer is C.** Pancreatic lipase catalyzes the breakdown of dietary triacylglycerols into free fatty acids and 2-monoacylglycerols, an essential step in the digestion of dietary lipids. Since prostaglandins are produced from linoleate, an essential fatty acid found in the triacylglycerols of dietary plants (or plant oils), a deficiency of pancreatic lipase would eventually cause a prostaglandin deficiency. Since triglycerides cannot be digested, they would exit in the feces, creating steatorrhea. Chylomicron levels would be low, as dietary triglycerides would not be digested, and their associated fatty acids and 2-monoacylglycerol would not be entering the intestinal epithelial cells to produce chylomicrons. Patients with cystic fibrosis can exhibit steatorrhea, but this is due to a blockage of the pancreatic duct with dried mucous, preventing the digestive enzymes from reaching the lumen of the intestine. An inhibitor directed against pancreatic lipase would not affect CFTR, the protein which is mutated in cystic fibrosis.

27. **The answer is D.** For a series of coupled reactions, the individual $\Delta G^{o\prime}$ values may be added to give the value of $\Delta G^{o\prime}$ for the overall reaction. This yields +0.9 kcal/mol overall for the conversion of malate to isocitrate.

28. **The answer is C.** $\Delta G^{o\prime} = -2.303\,RT \log K_{eq}$. For the conversion of isocitrate to malate, $\Delta G^{o\prime} = +0.9$ and $\log K_{eq}$ is negative. Therefore, K_{eq} is <1. An alternative way to analyze this question is to realize that the $\Delta G^{o\prime}$ is a positive value, which means that the overall reaction is unfavorable thermodynamically. For an unfavorable reaction, the ratio of the concentration of the product to the concentration of the reactant at equilibrium would be <1 (a value of 1 would mean the [product] = [substrate], and the overall $\Delta G^{o\prime}$ would be 0).

29. **The answer is C.** A decrease in the concentration of ATP stimulates the processes that generate ATP. The proton gradient across the inner mitochondrial membrane decreases; NADH oxidation by the electron transport chain increases; and fuel utilization increases. Palmitate is oxidized, and glycolysis increases because of the activation of PFK-1 by AMP. As ATP decreases, AMP rises due to the adenylate kinase reaction, in which 2 ADP can react to produce ATP and AMP (or vice versa). Lactate production will increase as the race continues, in which bursts of speed will require anaerobic glycolysis to generate energy, which produces lactate.

30. **The answer is D.** If the cytochromes are inhibited, ATP production, the electrochemical potential, heat production from NADH oxidation, and succinate oxidation all decrease, as electron flow is not permitted. The rapid drop in ATP levels will lead to cell death.

31. **The answer is D.** The child has an inherited pyruvate kinase deficiency. This step converts phosphoenolpyruvate (PEP) to pyruvate, producing ATP. A pyruvate kinase deficiency would slow down glycolysis, and less ATP would be produced. The NADH/NAD^+ ratio would rise because less pyruvate would be available for conversion to lactate (the reaction in which NADH is converted back to NAD^+). The intermediates of glycolysis before the blocked step would accumulate, and glucose 6-phosphate would inhibit hexokinase. Because of a lack of ATP, the life span of the cells would decrease. Fatty acids cannot serve as a source of energy because red blood cells lack mitochondria, which is the site of fatty acid oxidation.

32. **The answer is B.** The child has McArdle's disease due to a lack of muscle phosphorylase activity. In this disorder, muscle glycogen cannot be oxidized during exercise and glycogen accumulates within the muscle. As such, lactate levels would be low, and the person could not tolerate intense exercise of brief duration and would rely on fuels from the blood (glucose, fatty acids, and ketone bodies) for energy. The person could engage in mild exercise of long duration, using these blood fuels. The liver would not be affected because it contains a different phosphorylase isozyme. While a muscle PFK-1 mutation can also lead to these symptoms (Tarui's disease), a PFK-1 deficiency also leads to hemolytic anemia, due to the poor ATP yield from glycolysis in the red blood cells due to the lack of PFK-1 activity in those cells. Hemolytic anemia was not observed in the patient described in this question. Glucose-6-phosphatase is only present in the liver (not in the muscle), and a mutation in glucose-6-phosphatase will affect fasting blood glucose levels, but not exercise tolerance.

33. **The answer is B.** The boy has McArdle's disease, a deficiency of muscle glycogen phosphorylase, which is an autosomal recessive disorder. For the boy to have the disease, each one of his

parents must be a carrier for the disease (neither parent has the disease as they do not express the exercise-intolerance characteristic of the disease). There is a 50% chance that each parent will pass the mutated allele along to their next child, and for both parents to do this at the same time, the probability is 50% × 50%, or 25% (a one-in-four chance that the child will be homozygous for the mutated allele).

34. **The answer is C.** The woman has a lactase deficiency, and she cannot convert lactose (the disaccharide in milk and dairy products) into glucose and galactose. The lactose travels through the intestines and is metabolized by bacteria in the large intestine, which produces the symptoms the woman exhibits. A deficiency of amylase would lead to an inability to digest starch, not just dairy foods. Glucoamylase also only cleaves glucose-glucose bonds, not lactose. Deficiencies in fructokinase or aldolase B would lead to deficiencies in metabolizing fructose, so eating foods containing sucrose (such as fruits) would be the offending agent, not dairy products.

35. **The answer is C.** Due to the patient's inability to metabolize dairy products, and her avoidance of them, she is likely to be calcium-deficient because of her diet. Lactase is a digestive enzyme that cleaves lactose to galactose and glucose. However, galactose is not required in the diet. It can be produced from glucose and would be metabolized normally in this woman. Glucose 6-phosphate is converted to glucose-1-phosphate, then used to form the nucleotide sugar UDP-glucose, which is epimerized to form UDP-galactose. The UDP-galactose then combines with free glucose to form lactose. Due to the woman's inability to obtain galactose from the diet, she would have low levels of galactose-1-phosphate. Elevated fructose 1-phosphate levels is due to a mutation in aldolase B, which is a distinct enzyme from lactase. The woman would be expected to metabolize fructose normally.

36. **The answer is D.** The patient, by not taking any insulin for 4 days, yet eating a normal diet, developed diabetic ketoacidosis. Both glucose and ketone bodies would be elevated in the blood. Insulin levels are very low, due to the nonproduction of insulin in the pancreas, and lack of insulin injections. The lack of insulin prevented muscle and fat uptake of circulating glucose, and instructed the liver to produce ketone bodies. Since the brain had adequate levels of glucose to use, the ketone bodies were not utilized, and they accumulated in the circulation, causing the blood pH to drop, and an acidosis to develop.

37. **The answer is A.** The patient will be treated with insulin to reverse the effects of diabetic ketoacidosis. Insulin stimulates the activation of pyruvate kinase (through dephosphorylation), pyruvate dehydrogenase, and phosphofructokinase-2 (PFK-2, the kinase activity, also by dephosphorylation). PFK-2 then catalyzes the formation of fructose 2,6-bisphosphate, which is an activator of PFK-1 and an inhibitor of fructose 1,6-bisphosphatase, a gluconeogenic enzyme. Thus, of the enzymes listed, only fructose 1,6-bisphosphatase will have a reduction in activity when insulin is administered to the patient.

38. **The answer is C.** The glycosaminoglycans (mucopolysaccharides) of proteoglycans contain long chains of repeating disaccharide units that are covalently linked to a protein. Sulfation occurs after the monosaccharides are incorporated into the glycosaminoglycan chain. Proteoglycans are degraded by lysosomal enzymes. Deficiencies of these enzymes result in diseases known as mucopolysaccharidoses.

39. **The answer is D.** Lactose and sucrose are digested by disaccharidases on the brush border of intestinal epithelial cells. Starch is digested by salivary and pancreatic α-amylase. Therefore, its digestion would be affected by a lack of pancreatic juice, as the pancreatic amylase would be missing from the lumen of the intestine. Fat is only digested in the intestinal lumen utilizing pancreatic lipase and pancreatic colipase. In the absence of lipase and colipase in the intestinal lumen, the triglyceride cannot be digested, fatty acids and 2-monoacylglycerol would not be produced for absorption, and the triglycerides would leave the body in the feces. A common finding in cystic fibrosis is steatorrhea. Proteolytic enzymes are also released by the pancreas (such as trypsinogen, chymotrypsinogen, and proelastase), and thus protein digestion in the intestine will also be impaired if the pancreatic duct is blocked. Pepsin digestion of ingested proteins would still occur in the stomach, but the peptides produced by pepsin would have difficulty in being hydrolyzed in the intestinal lumen.

40. The answer is D. The boy's sister is an unaffected sibling of an individual who expresses an autosomal recessive disease (cystic fibrosis). As such, there is a two-in-three chance (67%) that the sister carries one mutated CF allele. One of the four possibilities is that the child will inherit both mutated alleles from the parents. Since the sister does not express the symptoms of CF, this did not happen. This leaves three possible genotypes for the sister, two of which include inheriting a mutated allele from either the mother or the father. The third genotype is inheriting the wild-type gene from both parents.

41. The answer is D. In this situation, bile salts could not enter the digestive tract. Therefore, recycling and excretion of bile salts, digestion of fats, and formation of chylomicrons would all decrease. As a consequence, fat in the feces would increase (steatorrhea). VLDL synthesis, which is dependent on dietary carbohydrate intake, would not be affected under these conditions, as the triglyceride in VLDL is derived primarily from endogenous triglycerides, and not those obtained in the diet.

42. The answer is C. The woman is experiencing a bout with diabetic ketoacidosis. The acetone on her breath is produced from decarboxylation of acetoacetate, one of the ketone bodies. Her blood glucose levels would be high because her insulin levels are too low to stimulate glucose transport into muscle and adipose tissue and to stimulate glycogen and triacylglycerol synthesis in the liver. An insulin injection would reduce her blood glucose levels and decrease the release of fatty acids from adipose triacylglycerols. Consequently, ketone body production would decrease. Glucagon administration would make her condition worse by increasing her blood glucose and ketone body levels. As the ketone bodies are acids, her blood pH would be <7.4, and could potentially be as low as 6.8, at which point the acidosis becomes life-threatening.

43. The answer is C. Fatty acid synthesis is maximal in the fed state when insulin is elevated and the glucagon:insulin ratio is low. Glucose is converted to citrate, which moves from the mitochondrion to the cytosol, where it is cleaved to oxaloacetate and acetyl-CoA. The regulatory enzyme acetyl-CoA carboxylase requires biotin and converts acetyl-CoA to malonyl-CoA, which provides the 2-carbon units for the elongation of the fatty acyl chain by the fatty acid synthase complex. Thus, biotin (for the carboxylation reaction) and Coenzyme A (to produce acetyl-CoA) are the required cofactors for fatty acid synthesis. NADPH provides reducing equivalents. The major product of the fatty acid synthase complex is palmitic acid.

44. The answer is D. Citrate transports acetyl units from the mitochondrion to the cytosol. NADPH is provided by the pentose phosphate pathway and the malic enzyme. Phosphatidic acid is an intermediate in the synthesis of triglyceride. 2-Monoacylglycerol is produced only in intestinal cells. The glycerol portion of the triglyceride can be derived from glycerol (liver, as it is the only tissue that expresses glycerol kinase) or dihydroxyacetone phosphate (other tissues).

45. The answer is C. The hormone-sensitive lipase of adipose tissue is responsible for initiating the degradation of triglyceride. The enzyme is activated by glucagon via a cAMP-mediated process (the activation of protein kinase A). Thus, mutations in PKA or the glucagon receptor may interfere with the eventual phosphorylation and activation of the hormone-sensitive lipase. Adenylate kinase is not involved in this cascade (although a mutation in adenylate cyclase would be involved, as that is the enzyme that produces cAMP). The insulin receptor promotes triglyceride synthesis and storage in the adipocyte, so a mutation in the insulin receptor would not promote triglyceride degradation. Glycerol kinase is only expressed in the liver and is not present in adipose cells.

46. The answer is E. On the basis of the symptoms the family is exhibiting, familial hypercholesterolemia, due to a mutation in the LDL receptor, is the most likely diagnosis. Individuals who are homozygous for this mutation would have even higher levels of circulating cholesterol, and would suffer heart attacks at a much younger age. A mutation in HMG-CoA reductase (the target of the statin drugs) would reduce cholesterol synthesis within cells, and lead to an upregulation of LDL receptors, which would actually reduce circulating cholesterol levels, not lead to elevated circulating cholesterol levels. HMG-CoA lyase is needed for ketone body synthesis, but is not involved in cholesterol metabolism. A mutation in apolipoprotein B48 would affect chylomicron metabolism, but would not alter VLDL or LDL metabolism. A mutation in the microsomal triglyceride transport protein gives rise to abetalipoproteinemia, a lack of chylomicron and VLDL synthesis. LDL levels are not elevated under these conditions.

47. **The answer is C.** Statins inhibit HMG-CoA reductase, the rate-determining step of cholesterol biosynthesis. The inhibition of HMG-CoA reductase leads to an increase in cellular levels of HMG-CoA and a decrease in squalene (an intermediate beyond this step) and cholesterol. The decreased cholesterol levels in cells cause ACAT activity to decrease (the formation of cholesterol esters) and the synthesis of LDL receptors to increase. An increased number of receptors cause more LDL to be taken up from the blood. Consequently, blood cholesterol levels decrease, but blood triacylglycerol levels do not decrease much, since LDL does not contain much triacylglycerol.

48. **The answer is B.** If the sibling inherits a mutated LDL-receptor allele from both parents, the cholesterol levels are so high that he or she will suffer their first myocardial infarction before the age of 20. Statins will not reduce cholesterol levels in these patients, as there are no functional LDL receptors to be upregulated. Since the disorder is autosomal recessive, there is a one-in-four chance (25%) that the sibling would inherit the mutated allele from both parents.

49. **The answer is A.** Bile salts are synthesized from cholesterol, but only in the liver. Glycine or taurine (not serine) is conjugated to the carboxyl group on the side chain in order to reduce the pK_a of the carboxylic acid group. As such, the molecule consists of both a hydrophobic portion and an aliphatic portion, and is considered amphipathic. Ionization occurs with a pK of about 4 for the glycoconjugates and 2 for the tauroconjugates. Above the pK, they carry a negative charge and are most effective as detergents. Bile salts are secreted in the bile, participate in lipid digestion, are resorbed in the ileum, and are recycled by the liver.

50. **The answer is A.** If argininosuccinate synthetase activity is low, less citrulline will be converted to argininosuccinate than normal. Citrulline levels will rise, and citrulline will be excreted in the urine. A deficiency of CPSI or OTC would lead to decreased citrulline levels. A deficiency of formiminotransferase would lead to increased FIGLU in the urine. A deficiency of arginase would lead to elevated arginine, which would be excreted. The child has an inborn error in argininosuccinate synthetase, which reduces its activity.

51. **The answer is E.** The child has ornithine transcarbamoylase deficiency, the most common inherited urea-cycle defect. Orotic acid accumulates due to the stimulation of cytoplasmic pyrimidine synthesis, which requires carbamoyl phosphate. The mitochondria accumulate carbamoyl phosphate (since it cannot condense with ornithine to form citrulline), and the carbamoyl phosphate leaks into the cytoplasm, bypasses the committed step of pyrimidine synthesis, and begins to overproduce orotic acid, an intermediate on the pathway to UMP formation. The OTC gene is located on the X chromosome, and is considered an X-linked recessive disorder. Women may express mild manifestations of the disease, but that is due primarily to the random X-inactivation effect early in embryogenesis.

52. **The answer is F.** During fasting, the muscle generates pyruvate (which requires functional pyruvate kinase activity), which is transaminated with an amino acid to form alanine, which travels to the liver to deliver the nitrogen for disposal in the urea cycle, and the carbons for gluconeogenesis. The glucose produced in the liver can then be exported and used by the muscle or other tissues. Gluconeogenesis in the liver, starting with pyruvate, does not require pyruvate kinase activity (pyruvate carboxylase and PEP carboxykinase are the enzymes used to convert pyruvate to PEP). Citrate synthase and pyruvate dehydrogenase are not required in either muscle or liver as part of the glucose–alanine cycle.

53. **The answer is A.** The boy has the recessive disorder homocystinuria, in which homocystine accumulates (due to homocysteine accumulation). Homocysteine is produced when *S*-adenosylhomocysteine (SAH) releases adenosine. Homocysteine is used in the synthesis of cystathionine, requiring the enzyme cystathionine β-synthase. Homocysteine accepts a methyl group (from N^5-methyltetrahydrofolate via vitamin B_{12}) to form methionine. Classical homocystinuria is due to a defect in cystathionine β-synthase, which requires pyridoxal phosphate as a cofactor. Most often, the mutation in the enzyme leads to a reduction in affinity

of the cofactor for the enzyme. Increasing cellular levels of pyridoxal phosphate, by taking pharmacological doses of vitamin B_6, can overcome the reduction in enzyme activity and reduce homocysteine, and homocystine, levels. None of the other enzymes listed as potential answers would regain activity in the presence of large amounts of vitamin B_6. An inactivating mutation in *S*-adenoyslhomocysteine hydrolase would reduce homocysteine levels. A defect in methionine synthase (which requires vitamin B_{12}) would raise homocysteine levels, but could not be reversed by B_6 treatment. A defect in N^5,N^{10}-methylenetetrahydrofolate reductase would also raise homocysteine levels, but that enzyme requires NADH, and would not be rescued by elevated B_6 levels. A defect in SAM synthetase would lead to elevated methionine levels, and increasing the concentration of B_6 would not overcome such a mutation.

54. **The answer is E.** The child is displaying the symptoms of hyperphenylalanemia, or phenylketonuria (PKU). The child was never tested for the disorder as an infant, and has developed symptoms (mental retardation) indicative of the high phenylalanine levels in the blood. Due to the high phenylalanine levels, some transamination of phenylalanine occurs, generating phenylpyruvate. PKU can be caused by mutations in phenylalanine hydroxylase (the classical form, in which phenylalanine is converted to tyrosine), which requires the cofactor tetrahydrobiopterin (THB), which donates electrons during the reaction to form dihydrobiopterin (DHB). The DHB is then converted back to THB by dihydrobiopterin reductase. A mutation in dihydrobiopterin reductase will reduce the functional levels of THB significantly, such that the phenylalanine hydroxylase reaction cannot take place, giving rise to nonclassical PKU, which is the disease the patient is exhibiting. A mutation in tyrosinase will lead to albinism (if the isozyme is in the melanocytes), and would not lead to elevated phenylalanine levels. A mutation in homogentisic acid 1,2-dioxygenase would lead to alkaptonuria, an accumulation of homogentisic acid. Maple syrup urine disease occurs because of a defect in the branched-chain α-keto acid dehydrogenase, and will not lead to elevated phenylalanine levels. Monoamine oxidase is utilized to inactivate various neurotransmitters, and its absence will not lead to elevated phenylalanine levels.

55. **The answer is D.** In the synthesis of each of the compounds listed, a decarboxylation of an amino acid occurs. Amino acid decarboxylation reactions, as well as transaminations, require pyridoxal phosphate. γ-Aminobutyric acid (GABA) is synthesized via a decarboxylation of glutamic acid. Serotonin requires a decarboxylation of a tryptophan derivative during its synthesis. Epinephrine and dopamine require a decarboxylation of a tyrosine derivative. Histamine is produced from the decarboxylation of histidine.

56. **The answer is A.** In pyrimidine biosynthesis, carbamoyl phosphate, produced from glutamine, CO_2, and ATP, reacts with aspartate to form a base that, after oxidation, reacts with phosphoribosyl pyrophosphate (PRPP) to form a nucleotide. This nucleotide is decarboxylated to form UMP. In purine biosynthesis, the base is produced on ribose-5-phosphate, which is generated from PRPP. Glycine is incorporated into the precursor, and tetrahydrofolate derivatives donate carbons 2 and 8 as IMP is formed. Glutamine is a nitrogen donor for both purine and pyrimidine biosynthesis. Aspartate is required for both UMP and IMP synthesis. Aspartate is also required in the conversion of IMP to AMP.

57. **The answer is C.** The professor is suffering from gout, due to an accumulation of uric acid in the blood, and precipitation of the uric acid in his big toe. Xanthine oxidase is involved in the conversion of the purine bases to uric acid. It catalyzes the oxidation of hypoxanthine to xanthine and of xanthine to uric acid. Allopurinol is used as a maintenance medication in the treatment of gout, and acts as a suicide substrate that blocks xanthine oxidase activity. This results in lower concentrations of uric acid (so the uric acid in the blood will not precipitate), and higher concentrations of the more water-soluble compounds hypoxanthine (derived from adenine) and xanthine (derived from guanine).

58. **The answer is B.** The patient lacks glucose-6-phosphate dehydrogenase activity (G6PDH), and develops a hemolytic anemia upon eating or ingesting compounds that contain, or are, strong oxidizing agents. Fava beans contain vicine and covicine, glycosides that can be converted to strong oxidizing agents by various cell types. The poisons will oxidize the reduced glutathione in the red blood cell membranes, but the oxidized glutathione cannot be converted back to the

protective, reduced form because of the lack of NADPH in the red blood cell. The enzyme that produces NADPH in the red cell is G6PDH, and when NADPH levels are low, glutathione reductase is missing its necessary cofactor, NADPH. While both a pyruvate kinase deficiency (low energy production) and a mutation in spectrin (hereditary spherocytosis) will lead to hemolytic anemia, those anemias are not triggered by oxidizing agents—they are always present. The red blood cell does not contain the enzymes glucokinase (the red cell form is hexokinase), pyruvate dehydrogenase, and α-ketoglutarate dehydrogenase (the latter enzymes are mitochondrial enzymes, and the red blood cell lacks mitochondria).

59. **The answer is C.** When red blood cells are destroyed (e.g., by lysis or by phagocytosis), hemoglobin is degraded. Bilirubin is produced from heme at an increased rate. The liver converts bilirubin to the diglucuronide at a rapid rate and excretes it into the bile. In the intestine, bacteria convert bilirubin to stercobilins, which give stool its brown color. Due to the large amount of hemoglobin released, the liver cannot keep up with metabolizing the hemoglobin to bilirubin, and then adding glucuronic acid residues to the bilirubin to increase its solubility. The nonconjugated bilirubin, which is fat-soluble, will diffuse into the tissues and will provide a yellow tint to the skin and the eyes.

60. **The answer is A.** The allele for glucose-6-phosphate dehydrogenase is on the X chromosome (the disorder is an X-linked recessive disorder). As such, the man will not pass his X chromosome on to his son, as he passes the Y chromosome to his sons. All of his daughters will be carriers for this disease.

61. **The answer is B.** The individual has developed pellagra due to a lack of dietary niacin. Although dietary niacin is the major source of the nicotinamide ring of NAD, it may also be produced from excess tryptophan. Tyrosine, thiamine, thymine, and riboflavin cannot contribute to the synthesis of the nicotinamide ring of NAD.

62. **The answer is A.** Pyridoxal phosphate is required for the activity of glycogen phosphorylase (the phosphorolysis of glycogen to glucose-1-phosphate), transaminases (such as transferring the nitrogen from aspartate to pyruvate to produce oxaloacetate and alanine), β-elimination and addition reactions (exemplified by the condensation of serine and homocysteine to form cystathionine), and amino acid decarboxylations (histidine to histamine). Pyridoxal phosphate is not required for the conversion of homocysteine to methionine, as that reaction requires N^5-methyltetrahydrofolate and vitamin B_{12}.

63. **The answer is E.** The patient most like has a secretory tumor of the adrenal cortex. The key to answering this question is the nondetectable ACTH levels, but elevated cortisol levels. Under normal conditions, ACTH release stimulates cortisol release from the adrenal cortex. In the absence of ACTH, a secretory tumor of the adrenal cortex would release cortisol, which signals the liver to export glucose, which resulted in the elevation of blood glucose levels seen in the patient.

64. **The answer is C.** Thyroxine can be taken orally to compensate for low TSH levels. TSH and ACTH are polypeptide hormones; therefore, they would be digested by pancreatic proteases in the gut. Low vasopressin causes water to be lost in the urine (diabetes insipidus); thus, water intake should be increased. Cortisol is required particularly during stress. GnRH, FSH, and LH are required for the production of a mature egg. Estrogen and progesterone, taken alone, will suppress ovulation.

65. **The answer is A.** Thyroid hormone feeds back on the anterior pituitary and inhibits the release of TSH. This patient's TSH levels are elevated, and thus her thyroid hormone levels are too low, and her dose should be increased.

66. **The answer is A.** A person with Type 1 DM who is not taking insulin behaves metabolically like a person who is fasting, except that glucose levels are elevated. The low insulin levels cause decreased transport of glucose into muscle and adipose cells, decreased conversion of glucose to glycogen and triacylglycerols in liver, and increased production of glucose by the liver via glycogenolysis and glyconeogenesis. Thus, the blood glucose levels of the person with untreated Type 1 diabetes will be higher than the normal person who has just finished dinner.

67. **The answer is A.** As insulin decreases, glucagon rises and stimulates glycogenolysis and gluconeogenesis. In a person who has eaten, insulin levels increase while glucagon levels decrease. In the diabetic patient, glucagon levels are higher because of the complete absence of insulin.

68. **The answer is A.** When insulin is low and glucagon is high, the carbon skeletons of amino acids derived from muscle protein are converted to glucose in the liver by gluconeogenesis. The amino acid nitrogen is converted to urea. This is occurring in the diabetic patient, but not in the normal individual who has just eaten a meal.

69. **The answer is A.** When the insulin:glucagon ratio is low, fatty acids are released from adipose tissue and converted to ketone bodies by the liver. Ketone body levels rise in the diabetic patient due to the high blood glucose levels, and the brain continues to use its preferred fuel as an energy source (glucose) rather than use the ketone bodies that are also available. The liver keeps producing ketone bodies, but the brain does not utilize them for energy. The well-fed individual will not be producing ketone bodies under conditions in which insulin levels are elevated (such as after eating a meal).

70. **The answer is B.** Tetracycline prevents aminoacyl-tRNA from binding to the "A" site on the ribosome. Streptomycin prevents the formation of the initiation complex, and erythromycin blocks translocation. Chloramphenicol will block peptide-bond formation in prokaryotic cells.

71. **The answer is A.** Streptomycin prevents the formation of the initiation complex, whereas erythromycin blocks translocation. Tetracycline prevents aminoacyl-tRNA binding to the "A" site of the ribosome. Chloramphenicol would block peptide-bond formation.

72. **The answer is D.** Erythromycin prevents translocation. Tetracycline prevents aminoacyl-tRNA binding to the "A" site on the ribosome. Streptomycin prevents the formation of the initiation complex, and chloramphenicol will block peptide-bond formation in prokaryotes.

73. **The answer is C.** The patient had I-cell disease, an inability to target lysosomal enzymes to the lysosomes, leading to the accumulation of nondigested material in the lysosomes, the forming of inclusion bodies, and cell death. The disease is due to the lack of *N*-acetylglucosamine phosphotransferase activity. This enzyme uses UDP-*N*-acetylglucosamine as a substrate and adds a phospho-*N*-acetylglucosamine to a mannose residue on lysosomal enzymes in the Golgi apparatus. A second enzyme then removes the *N*-acetylglucosamine, leaving behind mannose-6-phosphate. Mannose-6-phosphate binds to the mannose-6-phosphate receptor, which then targets the proteins to the lysosomes. The plasma membrane contains mannose-6-phosphate receptors, so when mature lysosomal enzymes are added to the outside of cells they will bind to these receptors and are targeted to the lysosomes. The glycosyl transferase activity lacking in I-cell disease is only found in the Golgi apparatus.

74. **The answer is A.** The children have all inherited defective mitochondria from their mother and are experiencing MELAS (mitochondrial encephalomyopathy, lactic acidosis, and stroke-like episodes). The degree to which the children express these symptoms is dependent on the segregation of normal and mutant mitochondria during embryogenesis (heteroplasmy). Eighty percent of MELAS mutations are due to alteration in the mitochondrial $tRNA^{leu}$, leading to improper translation of proteins from the mitochondrial genome. One key to mitochondrial inheritance is that all children from an affected mother will be variably affected (variable expressivity, even though the disorder is 100% penetrant), whereas no children of an affected father would exhibit symptoms of the disease.

75. **The answer is A.** Simian sarcoma virus transforms cells through autocrine stimulation. The virus codes for the PDGF-like growth factor, which binds to cellular PDGF receptors and stimulates the cell to grow. This is due to internal growth factor—receptor interactions. The growth factor does not need to leave the cell to stimulate nearby cells (a paracrine mechanism), nor does it need to travel through the blood to target tissues (an endocrine mechanism).

76. **The answer is D.** Galactose and glucose would be lower in a person with a lactase deficiency, because lactose in the milk would not be cleaved to produce galactose and glucose. The sucrose in the diet will produce fructose and glucose in both the person with lactase deficiency

and the normal individual; however, the lack of glucose production from lactose will result in a lower blood glucose concentration after the meal for the lactose-intolerant person as compared to the normal person.

77. **The answer is B.** Galactose would be higher, because in classic galactosemia (uridyltransferase deficiency), galactose can be phosphorylated but it cannot be metabolized further. Galactose-1-phosphate and its precursor, galactose, increase. This disorder would not affect glucose or fructose levels in the blood.

78. **The answer is C.** Fructose derived from the table sugar (sucrose) would not be converted to fructose 1-phosphate, thus it accumulates in the blood and is excreted in the urine, producing a benign fructosuria. Glucose and galactose levels in the blood are not affected under these conditions.

79. **The answer is C.** An aldolase B deficiency (fructose intolerance) results in a decreased ability to cleave fructose 1-phosphate. This compound increases in liver cells and its precursor, fructose, increases in the blood. Glucose and galactose would not be affected under these conditions.

80. **The answer is E.** Of the enzymes listed, only transketolase requires thiamine for activity. Transketolase is easily obtained from red blood cells, and upon measuring its activity in the presence and absence of exogenous thiamine, one can easily determine whether the patient is thiamine-deficient. If the addition of thiamine increases transketolase activity, the patient was deficient in thiamine.

81. **The answer is C.** Alcohol metabolism generates NADH, and skews the NADH/NAD^+ ratio to a very high number. Under these conditions, reactions that utilize the NADH/NAD^+ cofactors favor the direction of the reaction that regenerates NAD^+ such that alcohol metabolism can continue. If pyruvate is generated in the liver (from alanine derived from the muscle), then the lactate dehydrogenase reaction will go in the direction of lactate formation, which utilizes NADH and generates NAD^+. This reduces the amount of pyruvate that can be used for gluconeogenesis, and helps to lead to the hypoglycemia observed. The other key reactions for gluconeogenesis are also impaired, as oxaloacetate (derived from aspartate or asparagine) is directed toward malate production (to generate NAD^+) and glycerol-3-phosphate (produced from glycerol obtained from triacylglycerol degradation) is not converted to dihydroxyacetone phosphate, as that reaction requires NAD^+, which is in short supply.

82. **The answer is E.** Due to the patient's chronic drinking, his liver has been damaged, and the liver's ability to detoxify bilirubin has been impaired. Bilirubin is the degradation product of hemoglobin, which is constantly released from red blood cells that have been removed from circulation by the spleen. In order to increase the solubility of bilirubin, the liver conjugates it to two glucuronic acid residues. A damaged liver has a reduced capacity to glucuronidate hemoglobin, leading to free bilirubin in the circulation. The bilirubin, which has a yellow color, is fat-soluble and can diffuse to the skin and sclera, giving a yellow appearance. Hemoglobin A1C is glycosylated hemoglobin (nonenzymatic addition of glucose) and reflects the glycemic control of the patient. Hemoglobin does not get glucuronidated.

83. **The answer is B.** An iron-deficiency anemia is characterized by small, pale red blood cells. The lack of iron reduces the synthesis of heme, so the red cells cannot carry as much oxygen (which gives them the pale color). The cells are small in order to maximize the concentration of hemoglobin present in the cells. A megaloblastic anemia is due to deficiencies in either vitamin B_{12} or folic acid (a lack of intrinsic factor will lead to a B_{12} deficiency named pernicious anemia). These cells are large because the vitamin deficiency interferes with DNA synthesis, and the cells double in size without being able to replicate their DNA. Once the anemia begins, the large blast cells are released by the marrow in an attempt to control the anemia. Hemolytic anemia occurs when the red cell membrane fragments, which can occur with pyruvate kinase deficiencies or a lack of glucose-6-phosphate dehydrogenase activity (which results in reduced NADPH levels). Sickle cell anemia is caused by a point mutation in the β-globin gene, substituting a valine for a glutamic acid.

84. The answer is A. Intrinsic factor is required for the absorption of dietary vitamin B_{12}. Lack of B_{12} (or folate) results in a megaloblastic anemia. In a B_{12} deficiency, irreversible neurologic problems (due to demyelination) also occur. When decreased intrinsic factor causes a B_{12} deficiency, the condition is called pernicious anemia. An iron-deficiency anemia is characterized by small, pale red blood cells. The lack of iron reduces the synthesis of heme, so the red cells cannot carry as much oxygen (which gives them the pale color). The cells are small in order to maximize the concentration of hemoglobin present in the cells. Hemolytic anemia occurs when the red cell membrane fragments, which can occur with pyruvate kinase deficiencies or a lack of glucose-6-phosphate dehydrogenase activity (which results in reduced NADPH levels). Sickle cell anemia is caused by a point mutation in the β-globin gene, substituting a valine for a glutamic acid.

85. The answer is B. Pyridoxine is required for the formation of pyridoxal phosphate, the cofactor for the first reaction in heme formation. The lack of heme means that the red cells cannot carry as much oxygen (which gives them the pale color). The cells are small in order to maximize the concentration of hemoglobin present in the cells. A megaloblastic anemia is due to deficiencies in either vitamin B_{12} or folic acid (a lack of intrinsic factor will lead to a B_{12} deficiency named pernicious anemia). These cells are large because the vitamin deficiency interferes with DNA synthesis, and the cells double in size without being able to replicate their DNA. Once the anemia begins, the large blast cells are released by the marrow in an attempt to control the anemia. Hemolytic anemia occurs when the red cell membrane fragments, which can occur with pyruvate kinase deficiencies or a lack of glucose-6-phosphate dehydrogenase activity (which results in reduced NADPH levels). Sickle cell anemia is caused by a point mutation in the β-globin gene, substituting a valine for a glutamic acid.

86. The answer is A. A vitamin B_{12} deficiency results in a megaloblastic anemia plus demyelination of nerves, due to reduced levels of SAM in the nervous system. These cells are large because the vitamin deficiency interferes with DNA synthesis, and the cells double in size without being able to replicate their DNA. Once the anemia begins, the large blast cells are released by the marrow in an attempt to control the anemia. A hypochromic, microcytic anemia can result from the lack of iron, or lack of pyridoxal phosphate. Both conditions lead to a reduction in the synthesis of heme, so the red cells cannot carry as much oxygen (which gives them the pale color). The cells are small in order to maximize the concentration of hemoglobin present in the cells. Hemolytic anemia occurs when the red cell membrane fragments, which can occur with pyruvate kinase deficiencies or a lack of glucose-6-phosphate dehydrogenase activity (which results in reduced NADPH levels). Sickle cell anemia is caused by a point mutation in the β-globin gene, substituting a valine for a glutamic acid.

87. The answer is A. Folate deficiency results in a megaloblastic anemia because of decreased production of purines and the pyrimidine thymine. Thus, lack of folate causes decreased DNA synthesis. In contrast with a vitamin B_{12} deficiency, neurologic problems do not occur in a folate deficiency. A hypochromic, microcytic anemia can result from the lack of iron, or lack of pyridoxal phosphate. Both conditions lead to a reduction in the synthesis of heme, so the red cells cannot carry as much oxygen (which gives them the pale color). The cells are small in order to maximize the concentration of hemoglobin present in the cells. Hemolytic anemia occurs when the red cell membrane fragments, which can occur with pyruvate kinase deficiencies or a lack of glucose-6-phosphate dehydrogenase activity (which results in reduced NADPH levels). Sickle cell anemia is caused by a point mutation in the β-globin gene, substituting a valine for a glutamic acid.

88. The answer is C. Deficiency of glucose-6-phosphate dehydrogenase results in decreased production of NADPH by the pentose phosphate pathway during oxidative stress. The components of cell membranes are oxidized, and a hemolytic anemia results, as the protective glutathione (the reduced form) cannot be regenerated due to the lack of NADPH. A hypochromic, microcytic anemia can result from the lack of iron, or lack of pyridoxal phosphate. Both conditions lead to a reduction in the synthesis of heme, so the red cells cannot carry as much oxygen (which gives them the pale color). The cells are small in order to maximize the concentration of hemoglobin present in the cells. Hemolytic anemia occurs when the red cell membrane fragments, which can occur with pyruvate kinase deficiencies or a lack of glucose-6-phosphate dehydrogenase activity (which results in reduced NADPH levels). Sickle cell anemia is caused by a point mutation in the β-globin gene, substituting a valine for a glutamic acid.

89. The answer is B. In the absence of insulin, a person with Type 1 DM will behave metabolically like a person undergoing prolonged starvation, except that blood glucose levels will be elevated. Lipolysis in adipose tissue will produce fatty acids, which will be converted to ketone bodies in the liver if the diabetes is left untreated. Since the blood glucose levels are also elevated, the brain continues to use glucose as a fuel, and the ketone bodies accumulate in the blood. The MB fraction of creatine kinase (CK) is the heart muscle–specific isozyme, and when the heart is damaged, intracellular enzymes can leak into the circulation. Elevated MB fractions would be indicative of a myocardial infarction. Hepatitis and chronic alcoholism damage the liver, which would interfere with the urea cycle, leading to decreased BUN, and increased ammonia in the circulation. Renal failure would display reduced creatinine in the urine, and elevated creatinine in the blood.

90. The answer is A. CK is found in large amounts in muscle cells. When a tissue is damaged, cellular enzymes leak into the blood. Heart muscle has more of the MB isozyme of CK than skeletal muscle (which contains mainly the MM isozyme) and the brain (which consists mostly of the BB isozyme). In the absence of insulin, a person with Type 1 DM will behave metabolically like a person undergoing prolonged starvation, except that blood glucose levels will be elevated. Lipolysis in adipose tissue will produce fatty acids, which will be converted to ketone bodies in the liver if the diabetes is left untreated. Since the blood glucose levels are also elevated, the brain continues to use glucose as a fuel, and the ketone bodies accumulate in the blood. Hepatitis and chronic alcoholism damage the liver, which would interfere with the urea cycle, leading to decreased BUN, and increased ammonia in the circulation. Renal failure would display reduced creatinine in the urine, and elevated creatinine in the blood.

91. The answer is E. Urea is made in the liver. If the liver is infected, less urea will be produced and the BUN will decrease. Consequently, ammonia levels will increase. In the absence of insulin, a person with Type 1 DM will behave metabolically like a person undergoing prolonged starvation, except that blood glucose levels will be elevated. Lipolysis in adipose tissue will produce fatty acids, which will be converted to ketone bodies in the liver if the diabetes is left untreated. Since the blood glucose levels are also elevated, the brain continues to use glucose as a fuel, and the ketone bodies accumulate in the blood. The MB fraction of CK is the heart muscle–specific isozyme, and when the heart is damaged, intracellular enzymes can leak into the circulation. Elevated MB fractions would be indicative of a myocardial infarction. Chronic alcoholism damages the liver, which would interfere with the urea cycle, leading to decreased BUN, and increased ammonia in the circulation. Renal failure would display reduced creatinine in the urine, and elevated creatinine in the blood.

92. The answer is C. The kidney excretes nitrogenous waste products, including urea, ammonia, creatinine, and uric acid. If the kidneys fail, these waste products will not be excreted into the urine. Creatinine is excreted in proportion to muscle mass, so a decrease in creatinine excretion, without a concomitant loss of muscle mass, is a measure of kidney failure. In the absence of insulin, a person with Type 1 DM will behave metabolically like a person undergoing prolonged starvation, except that blood glucose levels will be elevated. Lipolysis in adipose tissue will produce fatty acids, which will be converted to ketone bodies in the liver if the diabetes is left untreated. Since the blood glucose levels are also elevated, the brain continues to use glucose as a fuel, and the ketone bodies accumulate in the blood. The MB fraction of CK is the heart muscle–specific isozyme, and when the heart is damaged, intracellular enzymes can leak into the circulation. Elevated MB fractions would be indicative of a myocardial infarction. Hepatitis and chronic alcoholism damage the liver, which would interfere with the urea cycle, leading to decreased BUN, and increased ammonia in the circulation.

93. The answer is E. A long-term exposure of the liver to alcohol can cause cirrhosis. Because the amount of functional liver tissue decreases, urea production decreases. In the absence of insulin, a person with Type 1 DM will behave metabolically like a person undergoing prolonged starvation, except that blood glucose levels will be elevated. Lipolysis in adipose tissue will produce fatty acids, which will be converted to ketone bodies in the liver if the diabetes is left untreated. Since the blood glucose levels are also elevated, the brain continues to use glucose as a fuel, and the ketone bodies accumulate in the blood. The MB fraction of CK is the heart muscle–specific isozyme, and when the heart is damaged, intracellular enzymes can leak into the circulation. Elevated MB fractions would be indicative of a myocardial infarction. Hepatitis damages the liver, which would interfere with the urea cycle, leading to decreased BUN, and increased ammonia in the circulation. Renal failure would display reduced creatinine in the urine, and elevated creatinine in the blood.

94. The answer is D. The patient has hereditary nonpolyposis colorectal cancer (HNPCC), which is primarily right-sided and due to mutations in any of seven different genes involved in mismatch repair (which is distinct from the generalized DNA repair system, mutations in which lead to xeroderma pigmentosum). These genes are all tumor suppressor genes. A mutation in the *APC* gene would lead to adenomatous polyposis coli, another form of colon cancer, which consists of many polyps, and is not restricted to the right side of the colon. Mutations in *BRCA1* and *BRCA2*, both of which are involved in DNA repair, are linked to breast cancer, not to colon cancer. Mutations in the Rb gene do not lead to colon cancer. While the ras gene may become activated in HNPCC, it is not the initiating mutation that leads to the disease–that is due to the loss of a component of the DNA mismatch repair system.

95. The answer is A. The genes that lead to HNPCC are tumor suppressor genes, but upon analysis in a pedigree the disease appears to be autosomal dominant. This is due to the finding that if an individual inherits a mutation in one of the genes involved in mismatch repair, there is close to a 100% probability that a loss of heterozygosity will occur, such that the person contracts the disease. This conundrum was first identified by Knudson, who formed the hypothesis to explain how recessive oncogenes can appear dominant in a pedigree. The loss of heterozygosity can occur in multiple ways, but the end result is the loss of activity of the functional allele, and a loss of mismatch repair activity.

96. The answer is D. A gain-of-function mutation in a gene, which can lead to uncontrolled cellular proliferation, is an oncogene. A loss-of-function mutation in a gene, which leads to cancer, is a tumor suppressor. Mutations in transcription factors can lead to an oncogene, but choice D is a better answer than C. Mutations in G-proteins have been linked to various cancers, but there are specific instances. The answer oncogene would cover these specific mutations as well.

97. The answer is B. Low levels of thyroid hormone result in increased production of TSH by the anterior pituitary, in an attempt to raise thyroid hormone levels.

98. The answer is D. Somatostatin, produced by the hypothalamus, inhibits growth hormone production and secretion by the anterior pituitary. Somatostatin will also reduce the secretion of TSH, insulin, and glucagon.

99. The answer is A. Dopamine is an inhibitor of prolactin production and secretion. Thus, if dopamine is low, more prolactin will be produced and secreted by the anterior pituitary.

100. The answer is E. Low levels of aldosterone (which occurs in Addison's disease) results in low blood pressure (hypotension).

101. The answer is B. Blood glucose is the major fuel for the brain, except when, after 3 to 5 days of fasting, ketone bodies in the blood reach a concentration at which the brain can begin to utilize them. The brain does not take up, to an appreciable extent, fatty acids, and fatty acid oxidation does not provide much energy for the brain. The brain's glycogen stores are very small, and alanine is not used as an energy source under any conditions by the nervous system.

102. The answer is B. Red blood cells lack mitochondria and thus are dependent on glucose for energy under all conditions. Neither ketone bodies nor fatty acids can be metabolized by the red blood cells (due to the lack of mitochondria), and red blood cells lack glycogen, and will not use alanine as an energy source (again, due to the lack of mitochondria).

103. The answer is C. Fatty acids are the major fuel for the liver during fasting. The liver is exporting glucose at this time, so glycogen is not being used as an energy source as the glucose residues in glycogen are being exported to the blood. The liver lacks the enzyme needed to oxidize ketone bodies for energy, and the liver, using the glucose–alanine cycle, will convert the carbons of alanine back to glucose, and place the nitrogen from alanine into urea.

104. The answer is A. Ketone bodies are used by the brain after 3 to 5 days of fasting. This is done to spare protein degradation to provide carbons for gluconeogenesis. Approximately 60% of the brain's energy needs will still be dependent on glucose, but the other 40% can be realized from ketone body oxidation. The brain will not oxidize fatty acids, as they have trouble crossing the blood–brain barrier, and the brain has very low levels of glycogen.

105. The answer is D. Muscles, particularly those with a preponderance of white fibers, use their glycogen stores when exercising. Weight training is an anaerobic activity, and the muscles will degrade their own glycogen stores to provide glucose for energy. Blood glucose will not become available until the AMP levels increase significantly in the muscle.

106. The answer is C. LDL is the cholesterol-rich particle in the blood. Cells take up LDL by LDL binding to the LDL receptor, and the LDL–LDL receptor complex enters the cell via endocytosis. Once inside the cell, the cholesterol is released from cholesterol esters by a lysosomal enzyme, and enters the cytoplasm of the cell for further processing. Chylomicrons are produced from dietary lipids and cholesterol in the intestinal epithelial cells, whereas VLDL contains newly synthesized triglyceride and cholesterol produced by the liver. HDL is involved in reverse cholesterol transport, which requires the LCAT reaction. Fatty acids, since they are so hydrophobic, cannot travel freely in the blood, and need to be bound to albumin for transport through the circulatory system.

107. The answer is B. VLDL is synthesized by the liver from dietary sugar (present in jelly beans). LDL is the cholesterol-rich particle in the blood. Cells take up LDL by LDL binding to the LDL receptor, and the LDL–LDL receptor complex enters the cell via endocytosis. Once inside the cell, the cholesterol is released from cholesterol esters by a lysosomal enzyme, and enters the cytoplasm of the cell for further processing. Chylomicrons are produced from dietary lipids and cholesterol in the intestinal epithelial cells. HDL is involved in reverse cholesterol transport, which requires the LCAT reaction. Fatty acids, since they are so hydrophobic, cannot travel freely in the blood, and need to be bound to albumin for transport through the circulatory system.

108. The answer is D. HDL removes cholesterol from cell membranes. The lecithin-cholesterol acyltransferase (LCAT) reaction, which converts the cholesterol to cholesterol esters, occurs on HDL. LDL is the cholesterol-rich particle in the blood. Cells take up LDL by LDL binding to the LDL receptor, and the LDL–LDL receptor complex enters the cell via endocytosis. Once inside the cell, the cholesterol is released from cholesterol esters by a lysosomal enzyme, and enters the cytoplasm of the cell for further processing. Chylomicrons are produced from dietary lipids and cholesterol in the intestinal epithelial cells, whereas VLDL contains newly synthesized triglyceride and cholesterol produced by the liver. Fatty acids, since they are so hydrophobic, cannot travel freely in the blood, and need to be bound to albumin for transport through the circulatory system.

109. The answer is A. Chylomicrons contain triacylglycerols synthesized from dietary lipid in intestinal epithelial cells. LDL is the cholesterol-rich particle in the blood. Cells take up LDL by LDL binding to the LDL receptor, and the LDL–LDL receptor complex enters the cell via endocytosis. Once inside the cell, the cholesterol is released from cholesterol esters by a lysosomal enzyme, and enters the cytoplasm of the cell for further processing. VLDL contains newly synthesized triglyceride and cholesterol produced by the liver. HDL is involved in reverse cholesterol transport, which requires the LCAT reaction. Fatty acids, since they are so hydrophobic, cannot travel freely in the blood, and need to be bound to albumin for transport through the circulatory system.

110. The answer is E. The child has X-linked SCID (severe combined immunodeficiency disease). The mutation in this form of SCID is the γ chain of cytokine receptors (this is a common subunit in many cytokine receptors, so all receptors that require this subunit are inactive). The mother is a carrier of this mutation, as her brother died from this disease. Mutations in purine-nucleoside phosphorylase lead to a partial immune defect, whereas mutations in adenosine deaminase lead to SCID. However, the mechanism whereby SCID comes about in ADA mutations is different than in X-linked SCID. ADA mutations will lead to an accumulation of deoxyadenosine, which proves to be toxic to immune cells. Deoxyadenosine levels are not elevated in this case. An inactivating mutation in Bcl-2 can lead to premature cell death due to apoptosis, but has not been linked to SCID. Loss of the Rb gene product will lead to retinoblastoma (tumors in the eyes), but not an immune defect.

111. The answer is E. The mutated protein that led to retinoblastoma in the child is the Rb protein, which regulates the activity of the E2F family of transcription factors. If Rb activity is lost, E2F is active in the "wrong" areas of the cell cycle, and uncontrolled cellular proliferation can ensue. Mutations in DNA repair enzymes can lead to cancer, but retinoblastoma will only result if a mutation occurs in the Rb gene. Ras activation and enhanced tyrosine kinase activity are other mechanisms of tumor formation, but not the underlying cause of retinoblastoma.

112. The answer is C. The retinoblastoma gene is a tumor suppressor. A loss of activity of the Rb gene product is necessary for the tumor to develop. If one Rb gene is mutated (and for the child to have bilateral retinoblastoma, it is highly likely that he inherited one mutated allele from one of his parents), the functional Rb gene needs to be inactivated before a tumor develops. For this to occur, a loss of heterozygosity has to occur. In the patient's case, this occurred by the loss of the functional gene from the tumor cells (since there was only one signal in FISH analysis). If two Rb genes were present in the cell, then two signals should be seen, which would occur with fused chromosomes, gene duplication, and polyploidy.

113. The answer is B. A defect in the synthesis or processing of collagen will lead to a variety of diseases, of which Ehlers–Danlos syndrome is one (osteogenesis imperfecta is another). Parkinson's disease is due to low levels of dopamine in the nervous system. In the initial stages of Parkinson's disease, giving dihydroxyphenylalanine (DOPA) can reduce the severity of the symptoms, as DOPA can be decarboxylated to form dopamine. DOPA can easily enter the brain, whereas dopamine cannot. Tay–Sachs disease results from an inability to degrade GM2, a ganglioside (it is also a sphingolipid). The enzyme missing is β-hexosaminidase. McArdle's disease is due to a defective muscle glycogen phosphorylase, such that the muscle cannot generate glucose from glycogen, leading to exercise intolerance. Maple syrup urine disease is due to the lack of branched-chain α-keto acid dehydrogenase activity, a necessary step in the metabolism of the branched-chain amino acids (leucine, isoleucine, and valine).

114. The answer is C. Parkinson's disease is due to low levels of dopamine in the nervous system. In the initial stages of Parkinson's disease, giving DOPA can reduce the severity of the symptoms, as DOPA can be decarboxylated to form dopamine. DOPA can easily enter the brain, whereas dopamine cannot. A defect in the synthesis or processing of collagen will lead to a variety of diseases, of which Ehlers–Danlos syndrome is one (osteogenesis imperfecta is another). Tay–Sachs disease results from an inability to degrade GM2, a ganglioside (it is also a sphingolipid). The enzyme missing is β-hexosaminidase. McArdle's disease is due to a defective muscle glycogen phosphorylase, such that the muscle cannot generate glucose from glycogen, leading to exercise intolerance. Maple syrup urine disease is due to the lack of branched-chain α-keto acid dehydrogenase activity, a necessary step in the metabolism of the branched-chain amino acids (leucine, isoleucine, and valine).

115. The answer is E. Because a hexosaminidase is deficient in Tay–Sachs disease, partially degraded sphingolipids (gangliosides) accumulate in lysosomes. A defect in the synthesis or processing of collagen will lead to a variety of diseases, of which Ehlers–Danlos syndrome is one (osteogenesis imperfecta is another). Parkinson's disease is due to low levels of dopamine in the nervous system. In the initial stages of Parkinson's disease, giving DOPA can reduce the severity of the symptoms, as DOPA can be decarboxylated to form dopamine. DOPA can easily enter the brain, whereas dopamine cannot. McArdle's disease is due to a defective muscle glycogen phosphorylase, such that the muscle cannot generate glucose from glycogen, leading to exercise intolerance. Maple syrup urine disease is due to the lack of branched-chain α-keto acid dehydrogenase activity, a necessary step in the metabolism of the branched-chain amino acids (leucine, isoleucine, and valine).

116. The answer is A. McArdle's disease is due to a defective muscle glycogen phosphorylase, such that the muscle cannot generate glucose from glycogen, leading to exercise intolerance. Glycogen accumulates in the muscle because of this enzyme deficiency (McArdle's disease is one of the glycogen storage disease). A defect in the synthesis or processing of collagen will lead to a variety of diseases, of which Ehlers–Danlos syndrome is one (osteogenesis imperfecta is another). Parkinson's disease is due to low levels of dopamine in the nervous system. In the initial stages of Parkinson's disease, giving DOPA can reduce the severity of the symptoms, as DOPA can be decarboxylated to form dopamine. DOPA can easily enter the brain, whereas dopamine cannot. Tay–Sachs disease results from an inability to degrade GM2, a ganglioside (it is also a sphingolipid). The enzyme missing is β-hexosaminidase. Maple syrup urine disease is due to the lack of branched-chain α-keto acid dehydrogenase activity, a necessary step in the metabolism of the branched-chain amino acids (leucine, isoleucine, and valine).

117. The answer is D. In maple syrup urine disease, the α-keto acid dehydrogenase involved in metabolism of the branched-chain amino acids (valine, isoleucine, and leucine) is defective. McArdle's disease is due to a defective muscle glycogen phosphorylase, such that the muscle cannot generate glucose from glycogen, leading to exercise intolerance. Glycogen accumulates in the muscle because of this enzyme deficiency (McArdle's disease is one of the glycogen storage disease). A defect in the synthesis or processing of collagen will lead to a variety of diseases, of which Ehlers–Danlos syndrome is one (osteogenesis imperfecta is another). Parkinson's disease is due to low levels of dopamine in the nervous system. In the initial stages of Parkinson's disease, giving DOPA can reduce the severity of the symptoms, as DOPA can be decarboxylated to form dopamine. DOPA can easily enter the brain, whereas dopamine cannot. Tay–Sachs disease results from an inability to degrade GM2, a ganglioside (it is also a sphingolipid). The enzyme missing is β-hexosaminidase.

118. The answer is B. Pellagra is due to a dietary deficiency of niacin, beriberi is due to a lack of thiamine (vitamin B_1), scurvy due to a lack of vitamin C, and rickets from a lack of vitamin D.

119. The answer is A. Scurvy is caused by lack of vitamin C. Pellagra is due to a dietary deficiency of niacin, beriberi is due to a lack of thiamine (vitamin B_1), and rickets from a lack of vitamin D.

120. The answer is E. Lack of thiamine (vitamin B_1) in the diet causes beriberi. Scurvy is caused by lack of vitamin C. Pellagra is due to a dietary deficiency of niacin, and rickets from a lack of vitamin D.

121. The answer is C. A dietary deficiency of vitamin D causes rickets. Scurvy is caused by lack of vitamin C. Pellagra is due to a dietary deficiency of niacin, and beriberi is due to a lack of thiamine (vitamin B_1).

122. The answer is A. The MAP kinase can be activated through a cascade initiated by the activation of a receptor tyrosine kinase activity (such as the EGF and PDGF receptors). Once the receptor tyrosine kinase activity is turned on, ras is activated through the activation of SOS (a guanine nucleotide exchange factor) by phosphorylation. The activation of ras leads to the activation of a MAP kinase kinase kinase, an example of which is the raf kinase. Raf then activates the MAP kinase kinase (mek is an example of a MAP kinase kinase), and mek activates the MAP kinase. SMAD proteins are necessary for TGF-β signaling, which works through a serine–threonine kinase activity. The SMAD proteins will act as transcription factors. STAT proteins are associated with cytokine receptors that have the tyrosine kinase JAK bound to them. STAT, when phosphorylated, acts directly as a transcription factor.

123. The answer is C. The boy has developed the signs and symptoms of untreated Type 1 diabetes. His pancreas is no longer producing insulin because of an autoimmune destruction of the β cells, so in response to elevated blood glucose levels insulin cannot be released. This leads to hyperglycemia, to the point that the kidney cannot reabsorb all of the glucose, and glucose is lost in the urine. The glucose in the urine causes an osmotic problem, and excess water is lost in the urine, leading to dehydration. This causes the frequent urination and thirst of the individual. The dehydration is leading to the lethargy and difficulty to awaken. Normal blood glucose levels should be 80 to 100 mg/dL in the fasting state. He has also developed a diabetic ketoacidosis, as the liver is producing ketone bodies in response to glucagon, but the brain is not utilizing them due to the high blood glucose levels. By giving the boy insulin, his blood glucose levels will stabilize as the muscle and fat cells begin to take glucose out of the circulation. Once blood glucose levels are below 300 mg/dL, glucose can be added to the iv bag. Rehydration is also occurring with normal saline as a part of the iv treatment.

124. The answer is D. Insulin is required to stimulate glucose entry into muscle and fat cells. In the absence of insulin, a number of metabolic derangements occur. First, the liver believes the body is in fasting mode (since only glucagon is present), and the liver begins to export glucose from glycogenolysis and gluconeogenesis. Insulin is required for muscle and fat cells (but not brain cells) to take up glucose from the blood (through the increase in GLUT4 transporters in the membranes of those tissues). In the absence of insulin, and the inability of muscle and fat to use glucose in the circulation, glucose levels become elevated in the blood. The kidney actually allows glucose to enter the urine because of its high concentration, rather than keeping all of the glucose in the circulation.

125. The answer is B. The boy is hyperglycemic, so the brain is getting plenty of fuel for its metabolic needs. Hyperglycemia can lead to lethargy as it will lead to severe dehydration, but this is not an answer choice. Fatty acids are actually increased in the blood due to the activation of hormone-sensitive lipase in the adipose cells. A hyperlipidemia would not lead to lethargy symptoms. The acidosis is what leads to the lethargy, as the acid–base balance of the blood and tissues is altered under these conditions. The acidosis results from excessive ketone bodies in circulation, since the brain is ignoring them in favor of the high blood glucose levels present.

126. The answer is C. The pH has dropped because of the high levels of ketone bodies in the blood, which occurs because the brain is using glucose as an energy source (since glucose levels are so high), and the brain is not using the ketone bodies. Lactate release by brain and muscle does not contribute to the acidosis, as the liver will utilize the lactate to generate glucose in the liver (since gluconeogenesis is stimulated due to the presence of glucagon, and the absence of insulin). Bicarbonate release could lead to an alkalosis, but not an acidosis. Under type-1-diabetic conditions, in the absence of insulin, the muscle cannot use glucose as an energy source, and it will use fatty acids for its energy needs (and ketone bodies to a limited extent).

127. The answer is C. Endogenous insulin is synthesized as preproinsulin. As part of the maturation of preproinsulin to insulin, a number of proteolytic cleavages occur, releasing the C-peptide, and the A- and B-peptides are linked by disulfide bonds, and form a mature insulin. Injected insulin, due to the way the recombinant insulin is produced, lacks the C-peptide. Thus, individuals with Type 1 diabetes usually have nondetectable C-peptide levels, whereas individuals without diabetes will have detectable C-peptide levels, in proportion to the amount of insulin they produce.

128. The answer is B. The fast-acting insulin will dissociate from the zinc at a faster rate than the long-lasting insulin. Both the insulin forms are complexed with zinc as hexamers, and when injected, the rate at which the hexamer dissociates to the dimer, then the monomer, limits how rapidly the insulin enters the bloodstream. The fast-acting insulin forms the dimer at a faster rate, due to an alteration in the primary structure of the recombinant insulin. This dissociation then allows the insulin to enter the blood at a faster rate. Insulin is injected subcutaneously, and there are limited proteases in that region. Insulin is not taken orally, as it will be degraded within the stomach and intestine. Both long-acting and rapid-acting insulin are injected at the same sites.

129. The answer is B. The fasting level of blood glucose is too high, as it should be between 80 and 100 mg/dL. The HbA1c represents the percentage of hemoglobin that has been nonenzymatically glycosylated. Normal values are below 6%; a value above 6% indicates hyperglycemia for extended periods of time. The average life span of a red blood cell is 3 months, so an elevated

HbIAc indicates that for the past 3 months blood glucose levels were higher than normal. Hemoglobin can also be glycosylated through high postprandial blood glucose levels (about 140 mg/dL). Glycosylated hemoglobin may not function as well as nonglycosylated hemoglobin, but the difference between a nondiabetic person and a diabetic person with poor glycemic control is small (1% to 2%), so an anemia will not develop, and glycosylated hemoglobin has no effect on hemoglobin synthesis. These effects are occurring because the boy has taken inadequate amounts of insulin, and the insulin-to-glucagon ratio is lower than it needs to be.

130. The answer is A. The boy is entering a hypoglycemic coma, caused by taking insulin without eating. The insulin will stimulate glucose transport into the muscle and fat cells, thereby reducing blood glucose levels. The insulin also, paradoxically, blocks the liver from undergoing glycogenolysis and gluconeogenesis, so the liver cannot maintain blood glucose levels. This leads to a dramatic and drastic lowering of fasting blood glucose levels. Under these conditions, ketone bodies are not being produced (the injection of insulin blocks hormone-sensitive lipase from degrading triglyceride to provide fatty acids as an energy source), blood glucose levels are lowered, and dehydration does not occur (there is no osmotic diuresis occurring). Insulin does not induce an anemic condition.

131. The answer is B. Since the boy is in an insulin induced hypoglycemic coma, an insulin counterregulatory hormone should be administered, and the best choice is glucagon. Glucagon will stimulate the liver to export glucose, which will raise blood glucose levels so the brain will get adequate energy sources. While cortisol and epinephrine are also considered insulin counterregulatory hormones, the time frame of cortisol action is too slow to be effective under these conditions (recall that cortisol is a steroid hormone, and works through the induction of new gene synthesis, a slow process). Giving the boy epinephrine or norepinephrine, even though they are insulin counterregulatory hormones, will exacerbate his symptoms, as it is epinephrine release in response to the hypoglycemia that leads to his sweating, tremors, and tachycardia. Giving even more epinephrine or norepinephrine will put the boy at a high risk for a heart attack.

132. The answer is B. After many number of years with poor glycemic control, the nonenzymatic glycosylation of protein in neurons eventually becomes harmful to the function of the neurons, leading to diabetic neuropathy in a variety of tissues. This occurs frequently in the extremities, where long neurons are found. The formation of HbA1c is an estimate of how poor glycemic control is. Reduced blood glucose levels would reduce the levels of nonenzymatic glycosylation in the neurons. LDL levels in the blood have not been associated with neuropathy in diabetic individuals or normal individuals.

133. The answer is D. This patient has the classic constellation of diagnoses that comprise the Insulin Resistance Syndrome (IRS). Some patients will display only some of these components while others will display the entire syndrome. The entire syndrome of the IRS consists of obesity (especially central obesity), Type 2 DM, hypertension, low HDL, high TG, fatty liver, sleep apnea, high urate (with or without gout), elevated plasminogen activator inhibitor (PAI-1), and (in women) polycystic ovarian syndrome. Her major pathophysiology is insulin resistance at the cellular level caused by obesity. She would actually display hyperinsulinemia to attempt to overcome the insulin resistance, so she does not have Type 1 DM (lack of insulin). She does have Type 2 DM, hypertension, fatty liver, and sleep apnea, but since all of these are only components of her entire IRS, these are not the most complete diagnosis. Type 2 DM is only one component of IRS, and this patient displays all the classic risk factors for Type 2 DM—age >40, Native American heritage (highest prevalence of any cultural group), obesity, family history, history of large-for-gestational-age baby, gestational diabetes, PCOS, "prediabetes," low HDL, high triglycerides, and hypertension.

134. The answer is C. The best treatment for IRS is to reduce weight to the "normal" range. With weight loss, all components of IRS can revert to normal even without medication. The drugs listed under answer choices A, B, D, and E may be appropriate treatments for individual parts of the IRS, but not for the entire syndrome. Both a sulfonylurea (glipizide) and insulin can raise insulin levels, but if the target tissues are nonresponsive (due, perhaps, to persistent downregulation of the insulin receptors), the addition of those drugs will be nonproductive. Statins may lower blood LDL levels, but would not affect the elevated blood glucose levels. A diuretic would not address the high lipids and blood glucose levels the woman is displaying.

135. The answer is B. BMI (body mass index) is calculated as weight in kilograms over height in meters squared. The calculation in this patient would be 100 kg/(1.8 m)^2 or 100/3.24, which would equal 30.9.

136. The answer is D. A BMI $<$18.5 is considered underweight, 18.5 to 24.9 as normal weight, 25 to 29.9 as overweight, and $>$30 as obese. Some older classifications used $>$40 to signify morbidly obese. The calculated BMI and obesity classification can be important for insurance purposes especially if obesity surgery is considered.

137. The answer is B. In order to help advise a patient in diet and weight loss, it is important to calculate how many calories they need each day to maintain their weight and how many calories need to be reduced in order to lose weight. This patient's daily energy expenditure would be her resting metabolic rate or basal metabolic rate (1 kcal/kg/hour) plus physical activity (for her very sedentary lifestyle, 30% of resting metabolism rate). The contribution of diet-induced thermogenesis (10% or less of intake) is difficult to calculate and will be ignored in the calculation. The calculation would be 1 kcal/kg/hour $\times$ 100 kg $\times$ 24 hour/day. This equals 2,400 kcal/day. Her physical activity would be 30% of 2,400 kcal/day, or 720 kcal. RMR plus physical activity would be 2,400 + 720, or 3,120 kcal. Thus, a daily consumption of 3,120 kcal will allow her to maintain her present weight. In order to lose 1 lb, she would need to consume $<$3,500 kcal (1 lb = 3,500 kcal). To lose 1 lb in a week, she would need to reduce 500 kcal/day (500 kcal/day $\times$ 7 days/week). So, she could consume a maximum of 3,120 − 500 kcal, or 2,620 kcal/day in order to lose 1 lb in a week. Choice A would be her RMR; C would be her RMR plus physical activity, which is sedentary; D would be RMR plus physical activity, if she were very active; and E would be daily expenditure plus 500 kcal/day, which would cause her to gain 1 lb in a week.

138. The answer is B. Her calculated LDL would be derived from the formula Total cholesterol (TC) = LDL cholesterol + HDL cholesterol + the triglyceride reading divided by 5 (the approximate value of VLDL cholesterol). In the patient's case, this works out to be 210 = LDL + 35 + 350/5. Her LDL would equal 105 mg/dL, which is greater than recommended, and places the woman at risk for a second heart attack.

139. The answer is B. Because she has Type 2 DM, it is recommended that her LDL be reduced below 100 mg/dL, and since she has had an MI, her LDL should be reduced below 70 mg/dL. The most effective medication to produce this degree of LDL reduction would be a "statin," which is an HMG-CoA reductase inhibitor. By inhibiting HMG-CoA reductase, the cell becomes cholesterol-starved, and upregulates its LDL receptors, allowing LDL to be removed from circulation at a faster rate. Statins have been shown to reduce a second occurrence of a heart attack. Answer A describes the mechanism of ezetimibe and answer C describes bile acid–binding resins, neither of which would reduce LDL to the level needed in this patient. In addition, neither method suggested in answer choices A and C has been shown to reduce the frequency of second heart attacks. Answers D and E would reduce triglycerides, which is the mechanism of action of the fibrate class of drugs. While the patient's triglycerides are high, reduction to normal values has not been shown to reduce the risk of heart attacks.

140. The answer is B. The most commonly recommended antibiotic for outpatient treatment of CAP is an erythromycin class of antibiotic (macrolide). Macrolides work by binding to the 50s subunit of ribosomes in bacteria and blocks the translocation step in protein synthesis. Humans do not have a 50s or 30s ribosomal subunit, but instead contain 60s and 40s subunits, so this class of antibiotic affects bacteria but not human cell protein synthesis. The penicillin class of antibiotics inhibits bacterial cell wall synthesis, but this was not one of the options. The aminoglycoside class of antibiotics binds to the 30S ribosomal subunit, and sulfonamides block folate synthesis in bacteria (and have no effect in humans, since folate is an essential vitamin for humans).

141. The answer is E. Bacteria must synthesize folic acid from paraaminobenzoic acid (PABA) since folate will not diffuse across bacterial cell walls as it can in humans. Sulfa drugs compete with PABA for incorporation into folic acid in bacteria, but not in humans. The sulfa drugs do not bind to bacterial ribosomes (the 30S and 50S subunits), nor do they bind to human ribosomal subunits (the 40S and 60S subunits).

142. The answer is A. She has a history of elevated uric acid and at least one episode of true gout. Persons with gout should not consume purines since they degrade to poorly soluble uric acid, which can crystallize in the blood, leading to gout. Foods with a high purine content include organ meats, such as liver (any tissue with large levels of RNA and DNA). Pyrimidines degrade to water-soluble products and do not cause or worsen gout. Histidine, leucine, and methionine are all essential amino acids and are needed in the diet. Their metabolism also does not contribute to the formation of uric acid.

143. The answer is C. In order to prevent gouty attacks or treat elevated uric acid, xanthine oxidase inhibitors (such as allopurinol) are used. Xanthine oxidase catalyzes two reactions prior to the synthesis of uric acid. The first is the conversion of hypoxanthine to xanthine (part of the adenine degradative pathway), and the other is xanthine to uric acid (part of the guanine degradative pathway, as well as adenine). In the presence of such an inhibitor of xanthine oxidase, both hypoxanthine and xanthine accumulate, but they are more water-soluble than uric acid, and precipitation of those compounds does not occur in the blood. Because xanthine oxidase inhibitors are not uricosuric (a drug that increases renal clearance of uric acid), such drugs can be used in patients with reduced renal function as in this patient with chronic renal failure. Arginine, ornithine, carbamoyl phosphate, and argininosuccinate are all compounds found in the urea cycle, and are not involved in uric acid synthesis. Blocking the metabolism of these compounds would lead to hyperammonemia, but would have no effect on uric acid formation.

144. The answer is E. This patient has IRS, a multifactorial disease process. While this "runs in families," there is no one mutated gene or combination of genes that have been discovered as the cause of this very common problem. There is no evidence to support that IRS is due to an autosomal dominant or recessive disorder, a mitochondrial mutation, or a sex-linked gene. However, it does appear as if multiple genes are involved in generating this syndrome, and that different combinations of mutated genes may lead to different phenotypes of the disorder being expressed.

145. The answer is A. Due to alcohol metabolism, NADH/NAD^+ ratios are elevated. Elevated NADH levels inhibit fatty acid oxidation and the TCA cycle. Lactic acidosis, hyperuricemia, and hypoglycemia occur because of the high NADH levels in the liver. Elevated NAD^+ would have the opposite effect. NADPH and $NADP^+$ are involved in the pentose phosphate pathway producing 5-carbon sugars. Acetaldehyde does not inhibit gluconeogenesis—the high NADH favors lactate formation over pyruvate conversion, favors malate over oxaloacetate formation, and favors glycerol-3-phosphate over dihydroxyacetone formation. This reduces the ability of the major three gluconeogenic precursors form moving along gluconeogenesis and producing glucose under these fasting conditions.

146. The answer is C. A haploid human cell has 23 chromosomes, whereas a diploid human cell contains 46 chromosomes. There are 22 autosomes (numbered 1 through 22, according to size) and two sex chromosomes (X and Y). Given the 22 autosomes, and two sex chromosomes, there are 24 distinct human chromosomes. Within a diploid cell, there are two copies of each autosome, and two sex chromosomes, either XX or XY, for the total of 46 chromosomes.

147. The answer is B. For a population in Hardy—Weinberg equilibrium, the formula $p^2 + 2pq + q^2 = 1$ applies, where p^2 refers to the frequency of wild-type homozygotes, $2pq$ to the frequency of carriers, and q^2 to the frequency of mutant homozygotes. "*P*" is the gene frequency for the wild-type allele, and "*q*" is the gene frequency for the mutated allele. In this problem, $q^2 = 1$ in 90,000, such that the mutant gene frequency equals 1 in 300. To determine the carrier frequency, we know that $p + q = 1$, so $p = 299/300$, which will be rounded to 1. The carrier frequency (heterozygote frequency) is $2pq$, or $2 \times 1 \times 1/300$, or 1 in 150 individuals within this population.

148. The answer is E. The pedigree most resembles that of an autosomal recessive disease. The key features that indicate this are the disease not being present in every generation (in this case, it is manifest in generations 1 and 4), either sex can express the disease (a female in generation 1, a male in generation 4), and consanguinity, keeping the mutant gene within the extended family. If this were autosomal dominant, one would expect to see individuals expressing the disease in every generation, and if it were mitochondrial, the three children of the affected female in generation 1 should be expressing the disease.

149. The answer is B. For an X-linked disease, it is important to realize that the disease frequency is the same as the gene frequency. Males only have one X chromosome, so if the disease frequency is 1 in 2,000, then 1 in 2,000 X chromosomes will carry the mutation. As females contain two X chromosomes, 1,000 women will represent 2,000 X chromosomes, so the carrier frequency is 1 in 1,000. Using Hardy–Weinberg equilibrium, $q = 1$ in 2,000, p is very close to 1 (1,999/2,000), so $2pq$ is 1 in 1,000.

150. The answer is C. Given, from the pedigree, that individual II-1 has the rare autosomal recessive disorder, then his parents each have to be a carrier for the disease. The probability that the affected individual's sister is a carrier is 2/3, as she does not express the disease (there are four possible genotypes for the sister, AA, homozygous for the wild-type genes, Aa and aA, a carrier for the disease, and aa, an affected individual. Since the last case is not true, of the remaining genotypes, two of the three are carrier status). The probability that the sister will pass the mutated allele to her daughter is one-half, so her daughter has a 1/3 chance of being a carrier ($2/3 \times ½$). The daughter has a 50% chance of passing the mutated allele to the child in question, so the probability that the child in question will be a carrier is $1/3 \times ½$, or 1/6. Since this is a very rare disease, the probability of obtaining the mutated gene from outside the family is considered zero.

Index

Page numbers followed by t indicate table; those in italics indicate figure.